Fundamentals of Ground Radar

Fundamentals of Ground Radar
For Air Traffic Control Engineers and Technicians

Ronald Bouwman
FAA (Retired)

SCITECH PUBLISHING, INC.

Scitech Publishing, Inc.
Raleigh, NC
scitechpub.com

SciTech Publishing, Inc.,
911 Paverstone Drive, Suite B
Raleigh, NC 27615
(919) 847-2434, fax (919) 847-2568
scitechpublishing.com

Editor: Dudley R. Kay
Production Director: Susan Manning
Production Coordinator: Robert Lawless
Cover Design: Doug Bond
Typesetting: Aptara

ISBN: 9781891121753

Library of Congress Cataloging-in-Publication Data
Bouwman, Ronald.
Fundamentals of ground radar for air traffic control engineers and technicians / Ronald Bouwman.
p. cm.
Includes bibliographical references and index.
ISBN 978-1-891121-75-3 (hardcover : alk. paper)
1. Radar air traffic control systems. 2. Radar in aeronautics. 3. Radar—History. 4. Radar—Equipment and supplies. I. Title.
TL696.R25B68 2009
629.136'6—dc22

2009002923

Dedication

Abraham Roelfs Bouwman
and
Simon Derk Bouwman

This book is dedicated to my grandfather, Abraham Roelfs Bouwman, and to my father, Simon Derk Bouwman.

My grandfather was called "Me-fixee," by those who sought his services as a locksmith, gunsmith, machinist, and tool-and-die maker. He acquired the nickname from his usual way of answering a customer after he had inspected some piece of machinery in need of repair. In 1908, he brought my father, uncle, and grandmother to America in search of opportunities not available to him in his Netherlands homeland. He had been trained as a mechanical engineer, but an industrial accident rendered him unemployable in his country. He had a cleft lip, soft palate, and glass eye. "Grampa" taught me that creativity and invention are the keys to success. Never did he allow his handicaps, even worsened by his "Groninger" accent, stop him. He reveled in creating seemingly foolish ideas that proved to be the best answers. Only inspired by the memory of his ingenuity and drive would I have attempted this book.

My father was just six years old when his family arrived in Bloomfield, Nebraska from the Netherlands. Partly out of a perceived necessity, he learned to speak "Oxford English" with the impeccable finesse of a statesman, and he became a country schoolteacher. Even for those times, his high character was uncommon, and it would be unbelievable to today's cynics. By his example, I know that the praiseworthy accounts of Washington and Lincoln are far less preposterous than lesser men can even comprehend. Dad lived an undeservedly hard life. He passed away at the young age of fifty-one, when I was just sixteen. I then swore an oath to myself that I would honor his name for the rest of my life.

Ronald Bouwman

Table of Contents

List of Illustrations

NOTE

All illustrations in this book are artistic depictions by the author, unless specifically annotated to the contrary. As such, they may contain minor, but inconsequential, technical flaws. They are sufficiently accurate and adequate for the intended purpose of conveying general theoretical information.

Preface

This text and reference has been written for use by technicians and engineers of a wide variety of ages, educational backgrounds, and experience in work with electronics equipment. Radar is a wide and diverse field, and air traffic control ground radar is but a small yet important part of that field. It is the only subject of this book. The content has evolved from the author's fifty years of learning, teaching, writing about, and working on ground radar systems. It is an assemblage of previously documented knowledge, carefully reconstructed into a logical order and progression, and, it is hoped, restated in a manner especially understandable to the technician. It has been my purpose, after all, to write and illustrate in order to teach, not to impress. The way to make work enjoyable is to make it interesting, and the way to be competent is to learn all that one can learn. Thus I have tried to capitalize on my accumulated knowledge to make this book interesting, comprehensive, and set within the context of the great pioneers who led the way. So with this book I hope to have attained my primary objective: to create a single training foundation that will stand on its own as a reference source, without reckless revision, abridgement, and "dumbing down" to mere tasks and procedures.

Rationale for the book

As a long-time technician, technical writer, artist, and technical instructor, I have witnessed a growing need for a standard reference for FAA and military air traffic control radar technicians. Even though radar theory has grown to enormous heights since the first air traffic control radars were deployed under the emergency of World War II, technical knowledge among technicans has been in decline. This is the result of increasing sophistication, particularly circuit miniaturization and integration and also computer-terminal interfaces with radar systems. The "down side" has been an evolution of a technical approach away from knowledge-based problem solution to the simpler remedies of lowest-replaceable units (LRU). Under the current approach, radars are restored more rapidly, but costs and waste are increasing. The result has been fewer technicians, maintaining a growing number of systems, knowing less about their equipment. While employed by the government, my efforts to encourage the early pride in knowledge of radar pioneers were only marginally successful. My solution was to create a permanent indestructible record of all I had learned and experienced for those who would follow me. Accordingly, I worked long hours to self-publish *Radar Systems for Technicians* in 1994, two years after my retirement from the FAA, and I began selling it in 1996. By 2006, USAF Colonel Kerry Bowers, a radar training instructor, had purchased a copy and recommended it to Mr. Dudley Kay, President of SciTech Publishing, a leader in radar book publishing and distribution, particularly airborne radar. I had already been considering a second edition to include more latter-day technology, and I began this work immediately after conferring with Mr. Kay, and his enthusiasm for a ground radar book, in the fall of 2006.

Organization

To begin using this book, you may find it helpful to know a little about its design and intent. Although it is organized for a progressive study from front to back, each chapter has been designed to stand alone for your future reference. That intent causes some repetition but relieves you of the need to leave the chapter to find background information.

The book begins with an interesting introduction of the history of radar, oriented, of course, toward air traffic control systems. Chapter Two deals with professional ethics. Chapter Three introduces the use and manipulation of deciBels, fundamentally essential to all radar engineers and technicians. The fourth chapter is about the basic physical scientific principles of radar. Chapter Five divides air traffic control "primary" radar into two major types, the original "magnetron" system, and the performance-superior, but more costly, "synthesis" system. Chapter Six is a block-diagram-level discussion of a hypothetical air traffic control surveillance radar, acquainting the reader with hardware and terminology. Chapter Seven provides substantial information on "secondary" radar principles, evolution, and hardware. Chapter Eight is based on waveguide and cavity theory obtained from USAF Manual 52-8, published in 1951, supplemented with commentary on modern-day hardware. Chapter Nine addresses synchronizers, somewhat limited to general principles because of the great variety of system design requirements. Chapter Ten, on transmitters, contains information on dc line-charging, magnetrons, power klystrons, and more. Chapter Eleven introduces receiver design and hardware, with particular attention to the effects of noise and bandwidth. Chapter Twelve delves into basic MTI concepts, its ties to Doppler theory, and the development of velocity response to moving targets. Chapter Thirteen is based on an ASR-8 digital MTI system to illustrate an application of the theories introduced in Chapters Eleven and Twelve. Chapter Fourteen explains the ASR-9 MTD system, the late 1980's Doppler-filter, and greatest improvements to earlier MTI systems. Chapter Fifteen describes radar indicator (display) methods from the earliest WWII displays to the current digital scan converters. Each chapter of the book concludes with a set of questions and answers immediately following to allow the reader to check his understanding.

Glossary and Appendices

An annotated glossary of many radar terms and acronyms is located in the back of the book. A brief explanation is provided for each term, and references to pages in the book are offered for those items addressed in the text. In some cases, illustrations are provided, and in other cases, there is some treatment of subject matter not contained in the text. Should you encounter something unfamiliar, you may find additional information in the glossary, and it may direct you to other places in the text, making a time-consuming search unnecessary.

You may find other information in the appendices useful. There is a listing of several radar equations, of Greek symbols, and of conversions and constants. Most don't use trigonometry frequently enough to remember all the identity relationships, so there is a quick reference to assist you in recalling them. Should you not remember or recognize the application of a trigonometric equation, you may wish to consult a trigonometry textbook. This book is intentionally general in nature and could be no more without becoming something approaching the size of an encyclopedia. After WWII, such an "encyclopedia of radar" called the *MIT Radiation Laboratory Series* was written as a multi-book series. By this time, those twenty-some volumes based on vacuum-tube theories are mostly valuable as interesting archives, even though much of the physics and mathematics are still valid. Should you wish greater detail and depth, look for publications addressing specific areas of radar theory. Many fine radar publications are offered on the SciTech Publishing website. No two radar systems are identical, and learning all the detailed theory of operation of a single system requires the manufacturer's instruction book. There are many fundamental ideas and design concepts contained here that will help you to rapidly comprehend the manufacturer's design and reasoning,

Summary of Features

- To assure that this book will be useful, long after the technician has completed training, the book is a single ground radar reference source, containing definitions, equations, conversions, and useful data
- Built-in aids for self-learning: chapter questions with answers, extensive glossary, data appendices
- The book may very well contain the most thorough, yet simply understandable, information available anywhere on MTI, MTD, and Air Traffic Control Radar Beacon Systems
- The book contains a unique thorough treatment of the development of the radar maximum range equation and the effects upon maximum range by variations within the equation
- To encourage the development of professional pride, responsibility, and quest for knowledge, the book includes rarely included chapters on professional ethics and history

This book is intended only as general information. Neither I nor SciTech assume any liability for reduced equipment performance resulting from its use. Radar engineers or technicians must always follow the manufacturer's instruction books and/or official government directives to align, repair, and evaluate radar systems. Great care has been taken to achieve accuracy in this book. However, should any errors or oversights be discovered by readers, students, or instructors, the author assumes responsibility for them, and invites your feedback. This is important in helping us to better future printings or editions.

Acknowledgements

I first thank Vicki Bouwman, my wife of forty-six years, for her tolerance of these home projects, both this book and the first edition. I call her "Mary Poppins" because she is "practically perfect in every way." Secondly, I thank all who purchased and otherwise supported the first edition of this book. A number of colleagues were instrumental in shaping my technical knowledge and the resulting first edition of the book, and I am saddened that they are no longer with us to see this second edition: Alfred Eno was a career technical representative for ITT Gilfillan. Al was a member of the first WWII GCA crew, US Army Air Corps, Airways and Air Communication Service (AACS). Al provided personal, first-hand accounts for historical information. Robert Holland, Sparks, Oklahoma, provided critical review and support of the first edition. Pete Larson, Tuttle, Oklahoma, provided valuable software recommendations. Even though available applications have improved and multiplied manifold since the first edition was published in 1996, I still use upgraded versions of an obscure graphics program for my illustrations. I am grateful to all of those who loaned me books, materials, diagrams, and other documents. Kerry Bowers, formerly a radar and EW instructor with the USAF and now president of KdB Training Inc., considered my self-published first edition quite valuable to his students and brought it to the attention of SciTech Publishing. Learning that SciTech is a leader in publishing and distributing books on radar, I have entrusted the Second Edition to their care. Dudley Kay, President and Founder of SciTech, has been my valued and supportive editor, as well as one who can take and give puns in equal measure to my own delight in a good play on words. Susan Manning and her assistant Robert Lawless, have helped steer the book through the difficult production phases of page design and layout. Doug Bond created the stunning cover. Taken all together, it was a pleasurable and rewarding experience, not something every author can say!

Ronald Derk Bouwman
Oklahoma City, OK
radarbook@aol.com

CHAPTER 1

Radar's Rich History and Development

The quality of a professional that distinguishes him from a tradesman is his formal knowledge, which includes an understanding of the background and origins of his business. A chemist should know about Robert Boyle, Antoine Lavoisier, John Dalton, Amadeo Avogadro, Henry Louis Le-Chatelier, and many, many, more. A medical doctor should know about Hippocrates, Anton van Leeuwenhoek, Louis Pasteur, Ignaz Philipp Semmelweis, Joseph Lister, and more. If for no other reason, those who made contributions to the science must never be forgotten, for to forget them is to dishonor them, and to dishonor the roots of the science is to discredit the profession. Science is a characteristic of a highly developed civilization and intellect; its growth and maintenance relies upon perpetual regard for the past. In most cases, the founders of science were geniuses who made incredible discoveries with only primitive knowledge; we owe them eternal gratitude, for there are very few of us, if any at all, who could achieve what they have achieved. It is the responsibility of every member of a profession to keep his or her memory and respect alive.

Although most think of radar as a precipitate of World War II, its beginnings are traceable to the late nineteenth and early twentieth centuries, a brief period of unparalleled discovery, ingenuity, and invention in the history of man. And although the US government was the major entity responsible for the development of modern microwave radar as we know it today, this credit is due to the imposition of urgency of World War II, and to the gift of the multicavity magnetron by the British government in 1940.

The Basic Physical Science

The original discovery of electricity as a force is attributed to Benjamin Franklin (1706–1790) [3] between 1727 and 1746. The physical science concepts behind radar began perhaps as early as 1831, when the English scientist Michael Faraday (1791–1867) [1] first discovered the relationships between electromagnetism and electric current. The British physicist James Clerk Maxwell (1831–1879) [2], using Faraday's experimental findings, determined and quantified the mathematical relationships between current and magnetic fields. Maxwell further concluded that electric and magnetic lines of force would cause changes in the space surrounding the operating circuit, providing the first suggestion that electrical signals could be radiated into space.

Thomas Alva Edison (1847–1931) [4] deserves much credit in the invention of early electrical and electronic devices; however even though modern technology could not have progressed to its present state without his inventions, he is more regarded by the scientific world as a very successful trial-and-error experimenter.

Propagation of Electrical Waves

In 1842, the German physicist Christian Johann Doppler (1803–1853) [5] first described the apparent shift of a propagated frequency caused by motion; the Doppler effect is a major element of the science of physics, and of radar theory today. Doppler's work deserves recognition in the invention of radar, since the comparison of transmitted and received frequencies was the foundation of the earliest ***bistatic*** echo detectors, which simply relied upon the interference between transmitted and reflected signals. Further, Doppler shift is the phenomenon upon which ***moving-target-indicator (MTI)*** systems and weather-velocity detection radars are based. Without MTI, detection of moving targets is nearly impossible in any locations where ground clutter echoes are received. Weather-velocity radar detectors make it possible to detect violent weather to issue warnings or take evasive action.

In 1879, David E. Hughes of London successfully generated and received radio waves [8]. The German physicist Heinrich Rudolph Hertz (1857–1894) [6] further expanded on Maxwell's work, proving that electrical and light waves behaved identically, and documenting his discovery in an 1890 paper entitled ***Uber die Beziehungen zwischen Licht und Electrizitat.*** Hertz had also determined that electrical waves could be reflected, as could light waves, providing probably the first prediction that radar was possible. In the 1960s, the words describing oscillations, "cycles per second," were universally changed to ***Hertz (Hz)*** in his honor.

Inventions in Early Radio Communications

An immigrant to the United States from Serbia, Nikola Tesla (1856–1943) [7] developed alternating-current theory and outlined, in 1892, the basic manner in which radio transmission and reception could be accomplished. Tesla's genius was too early for the time, and he was unable to immediately put his ideas to practical applications worthy of patent, mostly because working components were not yet available, rather than because his ideas were faulty. Research proceeded throughout the world. In 1894, Sir Oliver Lodge of England [8] had developed a radiotelegraph system, and in 1895, Alexander Popov of Kronshtadt, Germany, had independently devised another system [8].

In 1896, the Italian electrical engineer Guglielmo Marconi (1874–1937) [9] left Italy after the government had refused recognition of his radiotelegraph invention; he went to England to obtain his first patent on wireless telegraphy. In 1898, he achieved radiotelegraph communication across the English channel; in 1901, his radio waves crossed the Atlantic from Cornwall, England, to St John's, Newfoundland. In 1915, Tesla failed in legal actions against Marconi's patents, and was not vindicated until 1943, when the US Supreme Court made a decision recognizing that Tesla's work preceded Marconi's, and invalidating Marconi's patents [7]. One can only speculate as to the influence upon the court by World War II; Marconi was a member of Italy's Fascist party and held high office in Mussolini's government [9].

Component inventions contributed enormously to the development of radio. The vacuum tube was invented by John Ambrose Fleming [8]; Lee DeForest (1873–1961) [10] added a "grid" called the "Audion element" to the vacuum tube. The grid ultimately permitted low voltages to control the larger current through the tube, and the concept of amplified current control became the foundation of both vacuum tube and transistor technology. DeForest further invented radio transmitters and antennas, leading to regular transatlantic communication by 1907. By 1910, an audio-modulated DeForest transmitter first broadcast the voice of Enrico Caruso from the New York Metropolitan Opera; by 1915, DeForest transmitted human voice from Arlington, VA, to Paris, France.

Radiolocation

The growth of radiotelegraphy and radio broadcast was paralleled by work in radiolocation, a predecessor to radar. The word "radar" was actually coined by the United States; some sources attribute it to the US Navy; it may also have come from the Massachusetts Institute of Technology's (MIT) Radiation Laboratories. Before the universal acceptance of the word "radar," it had been called "radio direction finder" and "radio position finder." A significant radiolocation event occurred in World War I, on May 31, 1916, when the British Navy attacked German warships off the western coast of Denmark, after detecting a 1.5° change in radio signals from the German fleet [8]. The conflict is known as the great naval battle of Jutland and was a turning point in the war; the German fleet remained in port afterward, even though England had suffered greater losses [11].

All early experimentation with reflected radio signals dealt with the interference, or beating, of continuous wave received and transmitted signals; systems based on this principle are now called "bistatic" [12]. In 1903, a German engineer named Christian Hulsemier patented a device which emitted a spark and received an echo; he intended that it be used by ships to detect obstacles in its path [8, 18]. It could not measure range and is not recognized as radar.

Pulsed Radar

During World War I, in 1917, Tesla had talked of the possibility of transmitting a high-frequency burst of energy to be reflected and received, and then displayed on a fluorescent screen; this was probably the first visualization of range-measurement pulsed radar as we know it today. Tesla had again been ahead of his time. It does not appear that any great interest in radio reflection was created until Marconi, in June of 1922, lectured in New York about the "reflection and deflection" of short radio waves [8].

By September of 1922, two American civilian scientists, Dr. A. Hoyt Taylor and his assistant Leo Young, working for the Department of The Navy, discovered that short radio waves transmitted across the Potomac were interrupted by a ship [13]. The first successful application of pulsed echoing was accomplished in 1925 by Gregory Breit of the Carnegie Institution of Washington and Merle Tuve of Johns Hopkins University [8]; they used a 1-ms transmitter [8] made by the Naval Research Lab to measure the height of the ionosphere [18]. The development of radar followed, paralleled in Germany, Great Britain, and the United States.

The Cathode Ray Tube

Apart from transmitters and receivers, probably the most important invention to make pulsed radar possible was the cathode ray tube (crt), necessary to display high-speed radar information; it is also the major component of a television display. Research reveals that so many were involved in several stages of development of this device that it would be unfair to assign credit to all in a brief history such as this. As early as 1875, an English scientist, Sir William Crookes [8], had developed a vacuum tube with electrodes at both ends and gained current flow through the tube. By 1895, the German physicist Wilhelm Konrad Roentgen (1845–1923) [14] found that rays from the tube were found to cause photographic exposure, and called them "X-rays" (it may be significant to note that Roentgen died of cancer). Roentgen was, of course, the founder of X-ray technology. The crt, using phosphor instead of film as a beam target, was significantly developed at Bell Laboratories by 1922 [8], and a home television receiver was first offered by Allen Dumont in 1939; experiments on color television began as early as 1940. The fields of television and radar have been related in many respects since those earliest days because of the crt, high-speed video reproduction, and phase detection.

Radar Developments from 1930 through 1940

As early as the summer of 1933, German naval scientists recorded an echo from a small surface vessel at a range of 8 miles [8]. In December of 1934, the US Naval Research staff had obtained a photograph of a pulsed radar echo from an aircraft at a 2 or 3 miles range [8]. In 1935, the German scientist Dr. Wilhelm Runge obtained an echo from a Junkers Ju.52 [8], a large trimotor aircraft used as a heavy bomber and cargo transport. The Ju.52 is illustrated in Figure 1-1.

In late 1934, the British Air Ministry created a Committee for the Scientific Survey of Air Defense [8, 18], directed by Sir Robert Watson-Watt [8, 12, 18]. One proposal recommended that Britain line its coast with low-frequency antennas in a "home chain" of antennas [8,18]. An operational system had been developed for use by the Royal Air Force by March of 1938 [8, 18]. This historic 22–28 MHz ***chain home*** pulsed system could detect aircraft as distant as 90 miles [12], well into parts of Germany. By September of 1938, the system was in round-the-clock service, and remained so throughout the war [12]. The system exhibited poor range and azimuth resolution, but was sufficient to bring about severe losses to the Nazis in the early years of the war, and is credited with discouraging a German invasion of Great Britain [8, 12, 13].

By 1936, the US Army had developed a 75-W, 100-MHz pulsed radar, operating at a pulse repetition rate of 20 kHz. In May of 1937, the US Army Air Corps requested the development of a "long-range detector and tracker" [18]. It evolved into the SCR-270 early-warning radar, which was used to detect the Japanese surprise attack on Pearl Harbor [12, 13, 18]. In 1938, testing of the SCR-268 began. It was to be used for searchlight and anti-aircraft gun control; this was probably the first ***fire-control*** radar. The SCR-268 was designed and built by the US Army Signal Corps Laboratories [18].

In 1939, the Radio Corporation of America was granted a contract to build six shipboard systems [12]. The first permanent shipboard radar, the CXAM, operating at 195 MHz [12], was installed on the battleship "New York" [12, 18] and tested in 1940. An original purchase of six systems was made [18]. Nineteen of these shipboard systems were operational by 1941 [12].

In October of 1940, a British Technical Team went to the United States for the purpose of information exchange and gave all British radar research and development information to the US government with an appeal for assistance [12, 13].

In November of 1940, the MIT Radiation Laboratories was created by the US Department

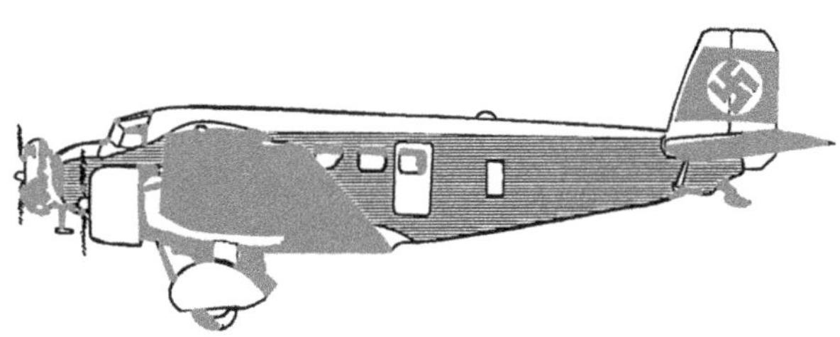

FIGURE 1-1

Junkers JU.52—the first aircraft detected by German radar.

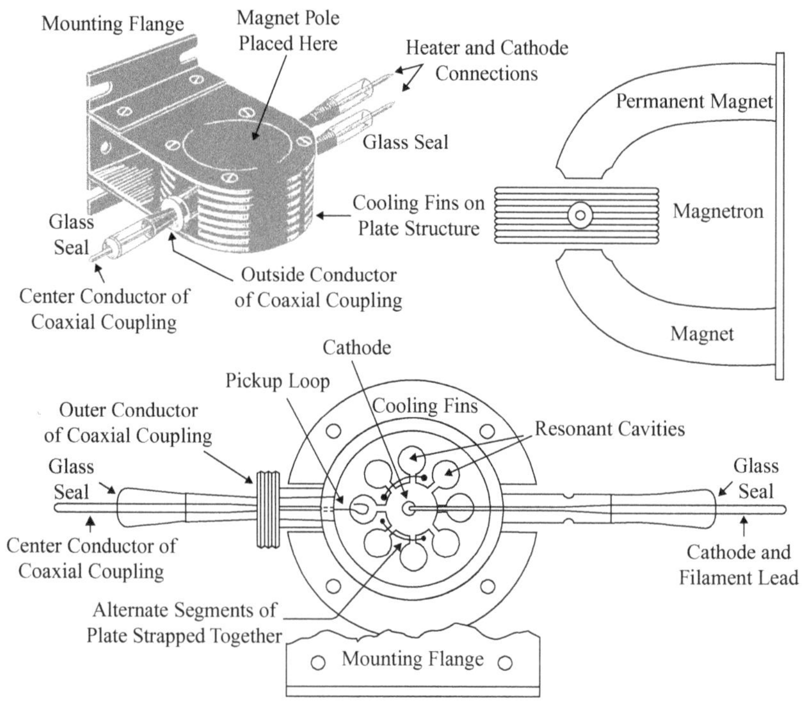

FIGURE 1–2

The magnetron. Based on USAF Manual 52-3.

of War [12, 18]. Dr. Lee A DuBridge of the University of Rochester was appointed director, and it remained the nation's central authority on radar for over 5 years [18]. The organization started with 40 people [12, 13, 18].

Before the Japanese surprise attack on Pearl Harbor on December 7, 1941, the US Army had purchased 100 SCR-270 early warning radars [17]. Six of these were on the island of Oahu before the attack [12].

Microwave Radar and "The Greatest Shipment"

The greatest limitations to the early radars of the 1930s were related to their low operating frequencies, long wavelengths, and large antennas. Azimuth bearing resolution depended on the antenna radiation pattern, which, in turn, depended upon antenna size and transmitter wavelength. Radar development in Great Britain was concentrated on airborne radar, where size requirements made short wavelengths absolutely essential. Throughout the world, all attempts at transmitting short wavelengths at high power had failed until November of 1939, when three English physicists, Sir John Randall, Henry Albert (Harry) Boot [12, 13], and James Sayers, working on the Birmingham University team [8] in Great Britain, developed probably the second most significant invention in all the history of radar, the ***multicavity magnetron***, which made possible the use of high frequencies with 10 cm wavelengths (λ), at power levels of 1 kW [12]. A magnetron is illustrated in Figure 1-2. In 1940, Great Britain was in great peril, and was seeking help from the United States [13]. In October of 1940, a magnetron was taken to the US government aboard a ship, in a black container, by a member of the British radar team [13]. Many say that singular action had more to do with the Allied victory than any other, including the development of the atomic bomb, and that the discreet transfer of the magnetron was the most important shipment of the war [13].

FIGURE 1–3

SCR-270 detected attack.

The Pearl Harbor Surprise Attack

Even though the MIT Radiation Laboratories had successfully created an operational microwave radar within 3 months after the delivery of the British magnetron, interest in radar had somewhat ebbed [13] until the surprise attack on US installations at Pearl Harbor and Hickam field on Sunday morning, December 7, 1941. See Figures 1-3 and 1-4. The approaching Japanese aircraft were detected by a low-frequency Westinghouse SCR-270 early warning radar system at Opana Station above Kahuku Point [17] on the island of Oahu; the Opana radar station was one of six on the coast of Oahu, and the US Army had purchased over 100 of these systems before the

Pearl Harbor attack [17]. No less than 19 radar detections [17] were recorded from 7:02 am to 7:39 am, and the approaching attack force was twice that reported by Private Joseph Lockard of the US Army [13] to an officer at Fort Shafter [17].

Some accounts indicate that, when first detected, the approaching Japanese aircraft were over 200 miles away, and only 183 of the 353 attack planes had been launched. However, the power output and pulse repetition frequency of the SCR-270 (100 kW and 621 Hz) [18] make it unlikely they could have been more than 130 miles out. Some of the Japanese aircraft probably passed directly over the radar site. The attack began at 7:55 am, 16 min after the final radar detection and 53 min after the first.

FIGURE 1–4

Pearl Harbor, December 7, 1941.

The radar detection reports were disregarded, partly because a group of 11 B-17s from Hamilton Field, CA, were due into Hickam Field at the time of the attack [13, 17]. In defense of the officer who concluded the approaching aircraft were the B-17s. They did indeed follow a course very similar to the Japanese attack route, arriving at Hickam field at 8:20 am to find themselves, disarmed and defenseless, in the middle of a battle, and between the first and second attack waves [18]; two were shot by Japanese aircraft, one on the Hickam runway. One B-17 crew member at the waist gun opening took a photograph of a Japanese dive bomber flying above and forward of his aircraft.

Although the Japanese force was considerably larger than the B-17 flight, the SCR-270 was a primitive radar with a range resolution of 0.8 to 2.0 nmi and an azimuth resolution of 28° [18]; the number of aircraft could not have been determined. One B-17 landed on the Kahuku golf course [17], just below the Opana radar station. The Pearl Harbor disaster, and the recognition that radar detection of the surprise attack had provided nearly an hour to prepare a defense against it, brought radar to the attention of the world forever.

The Reflex Klystron

Scientists in the United States had been working on another type of microwave oscillator tube in the late 1930s. Efforts were devoted toward a tube which operated on a principle of electron repulsion and bunching within a cavity. This ***reflex klystron*** could not produce adequate power for radar transmitters, but was a continuous-wave device that proved essential as a receiver local oscillator for many years. Further development of klystrons led in postwar years to the ***klystron drift tube***, a high-power amplifier tube exhibiting many performance advantages over magnetrons. Klystron drift tubes have been used in radar systems for years, but are gradually being replaced by solid-state technology.

The Explosion of Radar Science in World War II

World War II created a phenomenal growth in radar science and manufacture. The MIT Radiation Laboratories grew from 40 to 4,000 employees, and finally occupied 15 acres of floor space [13, 18]. By the end of the war, nearly 3 billion dollars of radar equipment had been purchased by the Army and Navy [13, 18]. For many years afterward, several volumes of engineering books created by the organization after the war, entitled ***MIT Radiation Laboratory Series*** [18], served as a national encyclopedia for the field of radar. Although the science was founded on 1930s' developments by the British, the feverish effort at development went far past what Great Britain had envisioned.

When the United States entered the war after Pearl Harbor, crash programs had already begun at the MIT Radiation Laboratories to develop radar systems, monitored by Secretary of War Henry Stimpson's representative, Edward L Boles, an electrical engineering professor at MIT [13]. One major objective was to develop automatic tracking radars to accurately point anti-aircraft guns at enemy airplanes and "buzzbombs." These were to become known as ***fire-control*** radars; the first radars of this type simply provided information to aim searchlights at aircraft, so that manually operated guns could be fired at them. One of these, the AN/TPL-1, is illustrated in Figure 1-5, and a prewar gun-laying radar is shown in Figure 1-6. Development led to the SCR-584 radar system, containing an electromechanical connection through an analog computer to anti-aircraft guns. Another objective was to develop ***airborne*** radar, which permitted aircraft to detect and destroy German U-boats on surfacing.

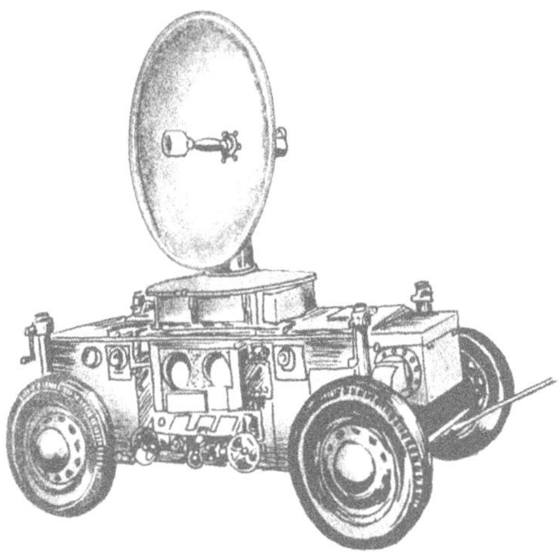

FIGURE 1-5

A/N TPL-1: searchlight-directing radar.

Shipboard radars were also needed for the Navy, for defense warning, searching for the enemy, fire control, and directing returning aircraft to carriers.

Still another objective was to improve ***early warning*** radars to provide for detection of approaching enemy aircraft, V-1 buzzbombs, and V-2 rockets. Yet another objective was to develop a "landing radar" to permit operations against the Axis in all types of weather. Both the landing radar and early warning radar have evolved to become the two major components of today's ***air traffic control*** radar system.

Frequencies used for these early systems were originally based on the 10-cm λ British magnetron, but a 3-cm λ magnetron was necessary and quickly developed for precision airborne radar ground imaging and for the landing radar. The 3-cm λ also allowed for small components and antennas in aircraft. Early in the war, development of the P-61A radar fighter shown in Figure 1-7 was started as a means to counter Nazi fighters as they attacked bombers returning to Great Britain. For purposes of wartime secrecy, the radars were assigned letter designations to describe the bands; those band designations are still in use today. The 10-cm λ was named "S" band for wartime security purposes; it is widely used today for airport surveillance radar (ASR) systems and weather radars. The 3-cm λ was similarly named "X" band.

FIGURE 1-6

SCR-268: early "gun laying" radar.

Germany also used radar in World War II, but their systems were the large-antenna, low-frequency, long-range type used to detect approaching enemy aircraft. In 1943, the Nazis captured a microwave radar from a downed British aircraft and reproduced the system. Unfortunately for Hitler and his Nazis, it was too late in the war [13]. Russia, Japan, France, and Italy also had primitive radar systems operating at the frequencies below microwave [12].

Electronic Countermeasures and Aircraft Identification

German radar and German interference with Allied radar led the United States to the development of electronic countermeasures (ECM). A major ECM technique during the war required the use of hundreds of tons of tin foil strips, called "chaff," to cause large blocks of radar clutter which would obscure radar detection of the Allied aircraft [13]. Nazi aircraft were also able to capitalize on Allied radar by following Allied aircraft back to England; this led to the deployment of an equipment called ***IFF***, for "identification friend or foe." IFF equipment incorporated a two-pulse ***interrogation*** from the ground, and a multipulse ***reply***, at a different frequency, from the aircraft. A greatly expanded version of the IFF equipment has become the ***Air Traffic Control Radar Beacon System (ATCRBS)***, now used throughout the world for civil aircraft identification and altitude reporting.

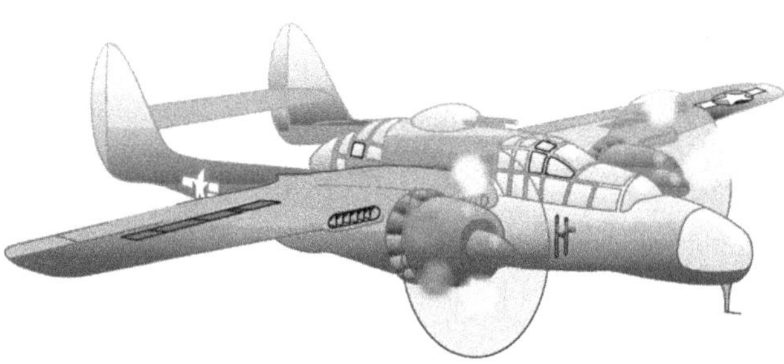

FIGURE 1-7

WWII radar night fighter P-61A.

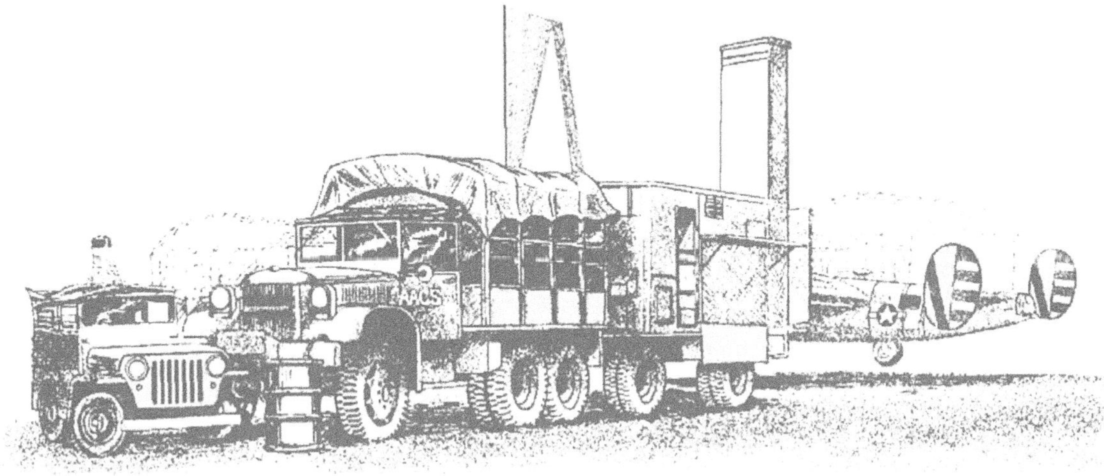

FIGURE 1–8

AN/MPN-1 ground-controlled-approach (GCA) system.

The Landing Radar

To accomplish the many radar projects, MIT Radiation Laboratories worked with over 70 manufacturers capable of converting the scientific creations into working hardware [13]. One of the Radiation Laboratories' major programs was instituted in 1942 to create a "landing radar." One small company, previously devoted to building radios and electrical tools, the Gilfillan Corporation of Los Angeles, CA, was subsequently assigned this landing radar project [15]. Gilfillan built the first ***Ground-Controlled-Approach (GCA)*** radar systems, delivered the first system in December of 1943 [15], and remained the leader in these types of systems for many years after the war. These systems actually contained two types of radars, a ***search*** or ***surveillance*** (both terms often used) system to permit observation of all traffic in the vicinity and a ***precision*** system used to "zero in" on the ***final approach*** of aircraft to the runway. The radar was assigned the official US Army Air Corps designation, AN/MPN-1; it is illustrated in Figure 1-8.

The MPN-1 was an incredible engineering achievement for the time, and a major milestone in radar history. Consider the overall state of technology in the United States in the 1940s. Across the country, farms were still without electricity. Homes that did have electricity usually had a single living room, a-m radio, if any at all. Model A Fords were still the only means of transportation for thousands. New Chevrolets relied on scoop-and-splash (no oil pump) engine lubrication. The most common cargo aircraft was the DC-3/C-47. Yet, this MPN-1 contained both 10-cm λ and 3-cm λ magnetrons and klystrons. Even beyond that, the precision approach GCA antennas were a motor-driven, variable, phased array, sometimes called the ***squeezing waveguide***, and waveguide itself was new. Antennas identical to the original squeezing waveguide are still in use today, and they were the foundation for today's exotic, phased-array tracking radars.

The MPN-1 was the first air traffic control radar system, and the first radar to contain a form of phased-array beam steering. It permitted Allied aircraft to operate at night and in low-visibility weather, thereby increasing the frequency of attacks and the subsequent pressure on both Nazi Germany and Japan; it played a great part in winning the war. The significance of GCA is apparent in an account by a Pacific Theatre GCA technician telling of a B-29 making a GCA landing under conditions of a 100-foot ceiling, visibility under a mile, crosswinds, turbulence, and an obstacle near the approach course; a safe landing would have otherwise been impossible [15].

GCA operations were based on the use of two separate radars, called "search" and "precision" (see Figure 1-9). The precision was used to monitor the aircraft's final approach to the runway, and both azimuth and elevation antennas were used to observe course and altitude. Because the precision system was dedicated to accuracy, its radar observation was of only a small wedge of space, and the search system was necessary to initially detect the aircraft, providing the controllers a means to guide the aircraft into the precision antenna patterns.

Even though the MPN-1 was such an incredible achievement for its time, it was an infant in the development of air traffic control radar, and it was severely limited in usefulness. The search radar had a range of only 30 miles, and the precision, 10 miles. Ground clutter echoes on the search system obscured aircraft echoes at close ranges, where it was most important to see them; controllers used a "timed approach" technique to

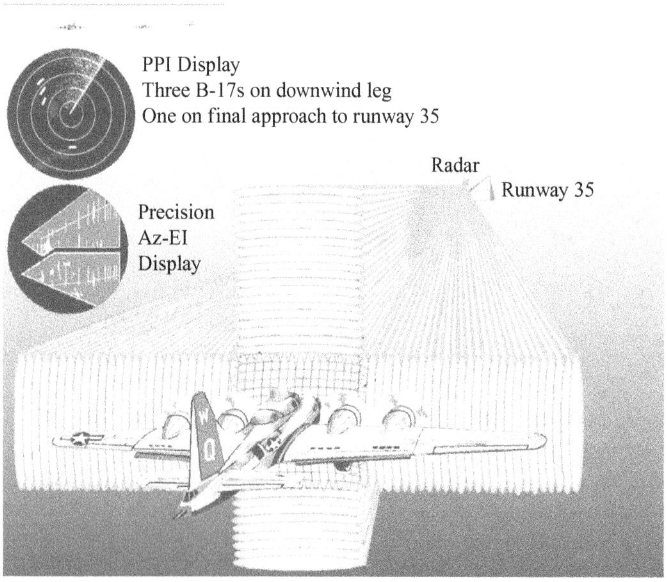

FIGURE 1–9

Azimuth and elevation precision approach beam scans.

calculate when and where aircraft would emerge from the clutter, or used traffic patterns which avoided clutter areas. When wind changes made it necessary to use different runways, the entire radar had to be turned around or moved by the "prime mover," the truck which towed it, and which contained the engine generator. There were no spare radar channels, and upon a failure, the radar would be out of service until it could be repaired.

The US Army Air Corps and its successor, the US Air Force, made such extensive use of the AN/MPN-1 that every US air base in the world had one by the early 1950s. However, because of the clutter problem, the range limitations, poor surveillance azimuth resolution, and a list of improvement needs to be expected of the first experience with a new system, it was naturally destined to be replaced with an improved system. Work had been done by the Radiation Laboratories and Gilfillan to reduce clutter returns by detecting an absence of Doppler shift in clutter echoes; the new type of system was called ***moving target indicator (MTI)***. See Figure 1-10. A new GCA system employing MTI, the Gilfillan AN/CPN-4, was used in the 1948–1949 Berlin airlift [16] and then deployed on a wide-scale basis in the early 1950s. However, the ***MIT Radiation Laboratory Series*** contains substantial information on MTI, evidence that significant work on it had been done during the war. A secret wartime project called "Rosebud" at the Gilfillan Los Angeles plant was probably dedicated to the development of MTI.

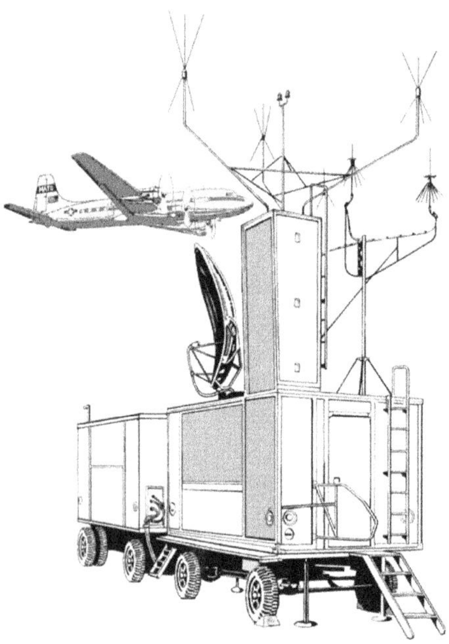

FIGURE 1–10

The A/N CPN-4 used in the Berlin airlift.

MTI is probably the greatest single milestone development in the history of air traffic control radar (see Figure 1-11). MTI removes ground clutter from radar presentations, yet displays aircraft targets flying over that ground clutter. The AN/CPN-4 system became the basis for the ***Civil Aeronautics Administration's (CAA)*** radar program. The search system became the first ***airport surveillance radar (ASR)*** ASR-1, and the precision radar became the CAA's first ***precision approach radar (PAR)***, the PAR-1. ASR radars now comprise the greatest share of FAA radars, but PAR systems are now rare, partly because more economical ***instrument landing systems (ILS)*** are widely used. There is some hazard involved with the total reliance on a single system, particularly one which relies on operational equipment in the aircraft.

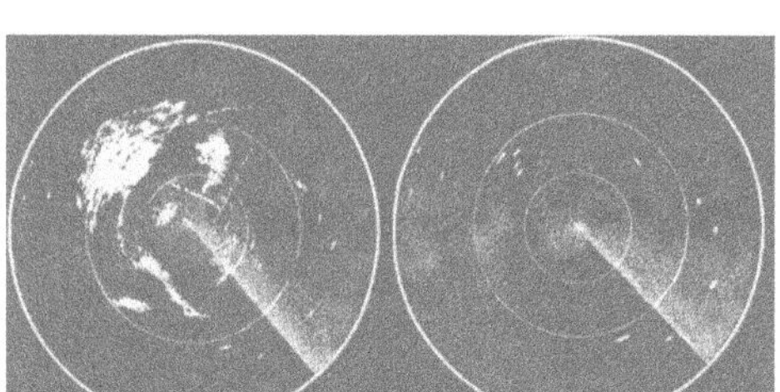

FIGURE 1–11

Normal vs. MTI displays.

The US military and other governments still use PARs, but often use a system called the ***quadradar***, as illustrated in Figure 1-12. The quadradar was first built in the 1950s to offer an economical alternative to air traffic control radar needs for low-volume airports. Quadradars do not use the phased-array precision antennas, have no MTI, and do not operate simultaneously in search and precision. Military designations of the quadradar were AN/FPN-33, -36, -40, and FSQ-84. Other Gilfillan radars of similar design were used for the carrier-controlled approach and were designated AN/SPN-XX. Still others were built for prompt tactical deployments. Hundreds of these radars have been sold throughout the world; among major purchasers have been the US Army, Navy, Marine Corps, Canada, and West Germany.

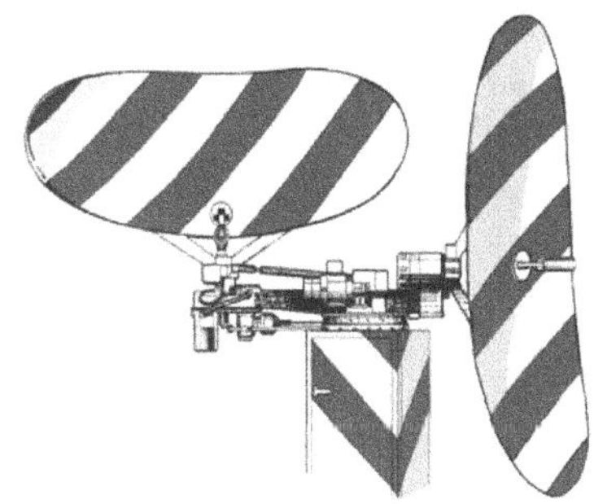

FIGURE 1–12

Gilfillan quadradar.

The Berlin Airlift

GCA radar may well have been the single factor to prevent war from breaking out over the Berlin blockade through 1948 and 1949. Josef Stalin ordered the total blockade of West Berlin, and citizens of the city lost all means to obtain such basic necessities as food and coal. As a nonaggressive solution, the United States launched the largest air traffic control operation in history, the Berlin airlift [16]. At West Berlin's Templehof Air Force Base, two MPN-1 systems were installed at each end of the runways. One radar served as a backup for another, and runway direction could be reversed without relocating the system. From June of 1948 until May of 1949, 277,264 flights [15] to Templehof delivered 2 million tons of supplies [16] from other airports in West Germany, Western Europe, and the Great Britain. Aircraft landed at a rate of one every 2 to 3 min; on one emergency occasion, 26 C-54 aircraft landed in 26 min. Over a 10-month period, 54,000 GCA landings were made [15].

An AN/CPN-4 was deployed at Templehof in late 1948. The CPN-4 contained an MTI system, which afforded the United States a new radar advantage. Aircraft approaching or departing West Berlin were restricted to narrow corridors over buildings, and terrain which caused ground clutter. Soviet fighters harassed the freight aircraft [16], and air traffic controllers could not monitor conflicts while the aircraft were obscured by clutter. However, the approaching fighters could be detected by the CPN-4 in spite of the clutter, and defensive actions thwarted the Soviet harassment.

GCA and Civilian Air Traffic Control

The US Air Force, and its predecessor, the Army Air Corps, developed air traffic control radar for wartime use. It naturally followed that the technology would be put to use for civil aviation by the CAA, then a part of the Department of Commerce. A GCA system was used by the CAA at the Indianapolis Airport in 1946 [15]. In 1948, President Harry Truman's airplane made a GCA landing under conditions of less than 3/8-mile visibility, attracting the attention of the aviation world [15]. In the early 1950s, the federal government initiated a development program called "Project Friendship," to supply both the Air Force and CAA with air traffic control radar. Many cities throughout the country had both air bases and commercial airports, and a single search radar on the air base served for traffic control. The facilities used for this purpose were called ***RAPCON***, an acronym for ***radar approach control.***

The Project Friendship RAPCONs employed an AN/CPN-18 surveillance radar, which was very similar to the CPN-4 search and ASR-1, but manufactured by Bendix. The precision radar was an AN/FPN-16, a non-transportable system used only by the military. The FPN-16 was mounted on a large turntable to allow the entire radar set, including the building, to be electrically rotated for different runway alignments. The AN/FPN-16 was nearly identical to the CPN-4 precision, and was also manufactured by Gilfillan. Search and precision display equipment was installed in a common RAPCON building on the air base. In a joint-use RAPCON, CAA/FAA controllers performed all ASR control, and USAF controllers performed the PAR approaches to the military base. The first operational RAPCON was at the Offutt Air Force Base Strategic Air Command Headquarters near Omaha, NE, and many others followed. This author began work as a GCA technician at the McConnell Air Force Base RAPCON in Wichita, KS, in 1956.

Obviously, every city does not have an air base, and all civilian traffic could not be controlled from joint-use RAPCONs. The CAA installed ASR-1s, -2s and -3s at municipal airports across the country. In the 1960s, the FAA built new control facilities and grew away from the RAPCONs. Although the CAA/FAA did use a limited number of PAR radars, the ground-controlled approach was unpopular in the civilian world, and it is now practically extinct in civil aviation. Its replacement was the ***instrument landing system (ILS)***, and now global positioning satellite (GPS) approaches are being made. This has been a controversial subject, as some are opposed to excess reliance on equipment aboard the aircraft, since acts of war could destroy the satellites and create havoc.

The early ASRs and CPN-18s were replaced with ASR-4, -5, and -6 improved dual-channel systems, and ***Automated Radar Terminal System (ARTS)*** computers were installed to handle increasing volumes of traffic and information. Some of the ASR-3s at low-volume airports remained in operation until they were replaced with ASR-7s or ASR-8s in the 1970s. Radar facilities became increasingly complicated with ARTS computers, video-mapping equipment, television scan conversion equipment, air traffic control radar beacon systems (ATCRBS), multiple radio communications channels, telephone switching equipment, and more.

Other Air Traffic Control Radars

Although the MPN-1 was the first air traffic control radar, other systems followed. In the early 1960s, the US Navy deployed AN/SPN-35 ***carrier-controlled-approach (CCA)*** systems for aircraft carriers; these radars were mounted on "stable tables," which were gyro-controlled level platforms to prevent the radar from rocking with the ship's motion. Many of the first CCA radars were very similar to the Quadradar. Today, CCA radars have become sophisticated automatic tracking systems. CCA and GCA radar has now advanced with the state of the art, and phased-array automatic tracking systems are now in operation. It is noteworthy that an "automatic GCA" system had been developed early in the 1950s.

Ground-Controlled Intercept (GCI) Radar Systems

Long-range, early-warning radar systems are among the oldest; the Pearl Harbor attack was even detected by such a radar. The basic principle of operation was very similar to the GCA surveillance system, with a continuously rotating antenna, but several performance characteristics differed to serve the purpose. The antenna rotation rate was slower to permit the beam to dwell on distant targets, the transmitter required a greater power output, and lower L-band frequencies have usually been used for several reasons. These military systems were also called ***Aircraft Control and Warning (AC&W).*** Large-scale deployment of these systems within the United States began in the early 1950s. In addition to the long-range surveillance radar, AC&W facilities were also characterized by ***height-finder*** radars, necessary to determine the altitude of intruders to direct fighter intercept operations.

Air Route Surveillance Radar (ARSR)

The ASR radar is used by the FAA for airport arrival and departure control, but the FAA has an additional need for radar for another purpose. Once an aircraft has departed from an airport traffic area to begin a cross-country flight, it becomes an ***enroute*** aircraft, and its flight progress is monitored by an ***Air Route Traffic Control Center (ARTCC).*** Enroute aircraft traverse great distances, and radar surveillance requires the use of greater range, higher power systems. ARSR systems employ a 200-mile range, in contrast to the 60-mile range of ASR systems. The ARSR-3 antenna is illustrated in Figure 1-13.

The Joint-Use ARSR Program.

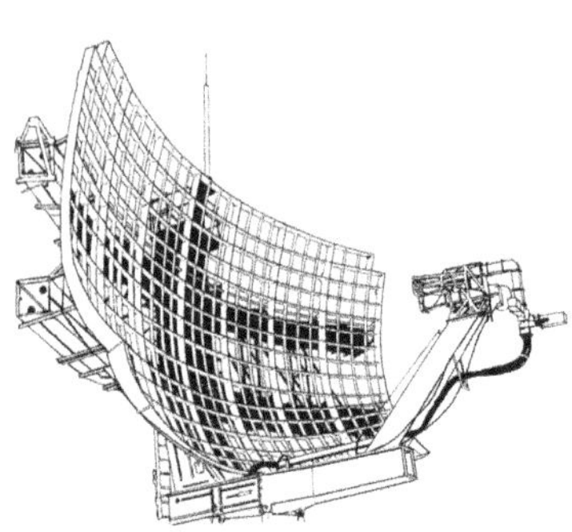

FIGURE 1–13

FAA ARSR-3.

The requirements for ARSRs and GCI radars were so similar that separate systems for both air defense and air traffic control were clearly unnecessary. The FAA and USAF then entered into joint-use agreements in which the FAA would maintain military long-range radar systems and supply radar data to both Air Force air defense centers and FAA ARTCCs. Initially, most of these military ARSRs were the AN/FPS-20; to this day, it has been one of the finest, most stable and reliable of all systems. The FPS-20 was among the very first ***synthesis*** radar systems, a type of radar that did not use a magnetron transmitter, but instead ***synthesized*** the transmitter frequency by combining oscillator outputs and then amplifying the frequency to high power with a klystron "drift" tube. Most new radar systems are now of the synthesis type, as the MTI performance of these systems is superior to that of magnetron systems.

FIGURE 1–14

AN/FPS-35.

In the 1960s, several very sophisticated radar systems were incorporated into the joint-use ARSR program to thwart any possible Soviet air attack on the continental United States (see Figure 1-14). Across the country, the air defense radar network comprised radars operating in several different frequency bands to make ***electronic countermeasures (ECM)*** against the entire system difficult, if not impossible. Because of the use of the different bands, the program was called the ***Frequency Diversity (FD)*** radar program. Each of the FD systems also contained many special ***electronic counter-countermeasures (ECCM)*** features to defeat the effects of jamming at its own frequency, and some of the techniques appear in many modern radar systems today. Among these were ***pulse compression, video integration, frequency agility,*** and many others. Systems included the AN/FPS-24 (VHF), FPS-35 (UHF), FPS-7 (L band), FPS-27 (S band), and upgraded versions of the L band FPS-20, designated FPS-64, -65, -66, and -67.

The immensity of the FD radar systems deserves some description. The FPS-35, for instance, was contained in a 60′ × 60′ five-story concrete building. The UHF frequency band necessitated an 86-ton antenna approximately 150′ wide, driven by six 100 hp, 440-V electric motors. The antenna was too large for a radome; in high winds, it was shut down when the total motor current reached 900 A per leg on the three phases. One channel of the radar was installed on the fourth floor, the other on the third floor. The first floor contained all the cooling pumps and mechanical equipment; the fifth floor contained the large UHF waveguide assemblies, and the antenna drive and lubrication equipment. Maintenance of the system required 24-h attendance; each of three shifts comprised a supervisor, three radar technicians, and an electro-mechanical technician.

The Anti-Ballistic Missile (ABM) treaty prohibited further use of the sophisticated FD systems, and the ARSR program returned to simpler systems in the 1970s. Many of the FPS-64, -65, -66, and -67 systems were retained for use as ARSRs with the ECCM features deactivated, and the ARSR network exclusively used the L band. However, the sophistication achieved in these systems made an impact on radar and radar technicians forever.

The FAA also utilized a number of purely civilian ARSRs, called the ARSR-1 and ARSR-2. These systems were deployed on a large scale following the collision of two airliners over the Grand Canyon in the late 1950s. The ARSR-1 and -2 were magnetron transmitter systems that employed another microwave amplifier tube, called the ***amplitron,*** to achieve the necessary high transmitter power. The amplitron is also known as a ***crossed-field amplifier (CFA).***

In the late 1970s, an ARSR-3 system was deployed; the antenna is illustrated in Figure 1-11. It was a major state-of-the-art upgrade, containing microprocessor circuitry, and using a new klystron drift tube. The ARSR-4, a 1990s' radar, is a ***rho-theta-phi*** (three-dimensional; range, azimuth, and altitude) phased-array feedhorn, ***CHIRPed (compressed high resolution pulse)*** system, for FAA/military joint use. The ARSR-4, deployed mostly around the US borders, does have a height-finding capability to detect and challenge intruding aircraft.

Early in radar development, defense radars had also contained a "secondary" aircraft-recognition radar system called ***identification, friend or foe (IFF)***; it was also ***called selective identification feature (SIF).*** An expanded version of these systems, compatible with both military and civilian systems, was to later become civil aviation's ATCRBS. The systems have also been expanded into data-link systems. Military data-link systems are called "Air Traffic Control Radar Beacon System, Identification Friend or Foe, Mark XII System," given the acronym ***AIMS***. Civilian data-link systems introduced in the 1980s are called ***Mode S***, and military challenges to potential intruders are accomplished with ***Mode 4 Interrogations.***

Airport Surface Detection Equipment (ASDE)

Airport runways and taxiways have always been dangerous places for both aircraft and ground vehicles, and the increased traffic and landing speeds over the years have worsened the hazard. Serious collisions and many deaths have been caused by "runway incursions," the term often used to describe a vehicle or aircraft entering an active runway as an aircraft is landing. In the earliest days of aviation, high-intensity "light guns" in the air traffic control towers were used to signal aircraft pilots that they had been cleared to land. Colored light guns were later used to signal vehicles on the ground, green for "proceed," red for "don't proceed," and white for "clear the area immediately." Many vehicles regularly on the airfield were not equipped with mobile radios. For clearance to cross a runway, it was common practice to flash headlights at the tower to gain attention, or to begin driving in circles, if that did not work. I was once erroneously cleared to drive an Air Force truck onto a runway in front of a landing B-47; had it become anything other than a very close call, I would not be writing this today.

Other incidents such as my own prompted the installation of mobile radios in all airfield vehicles, but even that was far from adequate. In bad weather, a vehicle driver could become disoriented, and fail to realize that he should contact the tower, and controllers in the tower could not see the vehicle. Some true disasters occurred when landing airliners fell upon other aircraft on the runway.

ASDE radar was envisioned in the 1940s and first deployed in the 1950s. For echo definition, a very short pulse and high operating frequency (originally K_a band) became necessary, creating many engineering design challenges. The short-range and high pulse repetition rate offered a large number of hits per scan, and the antenna could be rotated at a high rate (the ASDE-3, 60 rpm) for rapid update. The antenna requires a very high gain and narrow beam width (0.25° for the ASDE-3) to achieve azimuth definition, and that requirement had much to do with the choice of high operating frequencies. A traveling-wave tube was employed in the ASDE-3 transmitter. The earliest ASDEs operated in the K_a band. For the ASDE-3, the frequency was reduced to the K_u band, and then for the ASDE-X, to the X band. Precipitation has always plagued the ASDE systems, and that was the motivation for frequency reduction. Among further efforts to reduce the derogating effects of precipitation, the ASDE-3 antenna reflector was shaped to limit elevation coverage to just 200′ above the ground level. Because of the high antenna rotation rate, the ASDE-3 antenna is enclosed in a radome that rotates with the antenna. The radome is aerodynamically designed to provide lift when rotating, so as to reduce stress, wear, noise, drag, and drive-motor current.

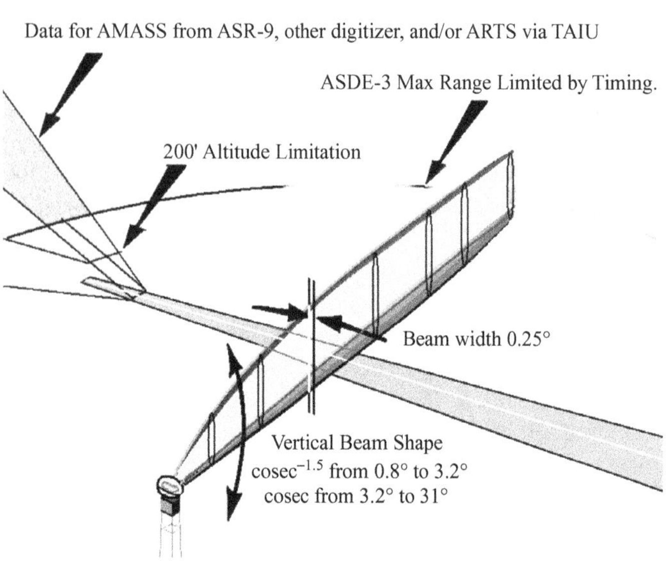

FIGURE 1–15

ASDE, AMASS, and TAIU.

Because of increasingly high landing speeds, it became necessary to see approaching aircraft outside the ASDE range and above its antenna pattern (see Figure 1-15). An interface system was created to merge ASR-9 digital track data, ARTS (Automated Radar Terminal System) flight plan data, and ASDE radar data. A terminal interface unit (TAIU) usually located at an ARTS facility provides digital data for approaching aircraft to an airport movement area surveillance system (AMASS). When aircraft courses and velocities indicate a developing hazard, the automated system provides both aural and visual alarms in the air traffic control tower.

References

1. J. R. Partington, "Michael Faraday" Contribution to *Collier's Encyclopedia*, Crowell-Collier Publishing Co., U.K. and Canada, 1964.
2. R. L. Smith-Rose, "James Clerk Maxwell" Contribution to *Collier's Encyclopedia*, Crowell-Collier Publishing Co., U.K. and Canada, 1964.
3. T. R. Hay, "Benjamin Franklin" Contribution to *Collier's Encyclopedia*, Crowell-Collier Publishing Co., U.K. and Canada, 1964.
4. H. Gordon Garbedian, "Thomas Alva Edison" Contribution to *Collier's Encyclopedia*, Crowell-Collier Publishing Co., U.K. and Canada, 1964.
5. D. Park, "Doppler Effect" Contribution to *Collier's Encyclopedia*, Crowell-Collier Publishing Co., U.K. and Canada, 1964.
6. R. Todd, "Heinrich Rudolph Hertz" Contribution to *Collier's Encyclopedia*, Crowell- Collier Publishing Co., U.K. and Canada, 1964.
7. L. I. Anderson, "Nikola Tesla" Contribution to *Collier's Encyclopedia*, Crowell-Collier Publishing Co., U.K. and Canada, 1964.
8. F. J. Kerr, "Radar" Contribution to *Collier's Encyclopedia*, Crowell-Collier Publishing Co., UK. and Canada, 1964.
9. C. W. Davis, "Guglielmo Marconi" Contribution to *Collier's Encyclopedia*, Crowell-Collier Publishing Co., U.K. and Canada, 1964.
10. C. W. Davis, "Lee DeForest" Contribution to *Collier's Encyclopedia*, Crowell-Collier Publishing Co., U.K. and Canada, 1964.
11. M. Kraus, "Battle of Jutland" Contribution to *Collier's Encyclopedia*, Crowell-Collier Publishing Co., U.K. and Canada, 1964.
12. M. Skolnik, *Introduction to Radar*, McGraw-Hill, New York, 1990.
13. L. Garmon, *Echoes of War,* television production for Public Broadcasting System.
14. I. Behrens, "Wilhelm Konrad Roentgen" Contribution to *Collier's Encyclopedia*, Crowell-Collier Publishing Co., U.K. and Canada, 1964.
15. *ITT Gilfillan Scanner*, Vol. 28, No. 3, Spring 1992.
16. C. C. Held and K. Schroeder, "Berlin" Contribution to *Collier's Encyclopedia* Crowell-Collier Publishing Co., U.K. and Canada, 1964.
17. L. Kimmet and M. Regis, *The Attack on Pearl Harbor*, Navigator Publishing, Seattle, WA, 1991.
18. L. N. Ridenour, *Radar System Engineering: Volume I, MIT Radiation Laboratory Series,* McGraw-Hill, New York, 1947.

Review Questions

1. Those listed below all had some part in the development of radio and/or radar. Name the things each accomplished:

   Benjamin Franklin
   Christian Johann Doppler
   Michael Faraday
   James Clerk Maxwell
   Sir William Crookes
   Wilhelm Roentgen
   Heinrich Rudolph Hertz
   Nikola Tesla
   Christian Hulsemier
   Guglielmo Marconi
   Thomas Edison
   John Ambrose Fleming
   Lee DeForest
   Dr. A. Hoyt Taylor
   Leo Young
   Randel and Boot
   Joseph Lockhard

2. The first scientist to propose that electromagnetic energy would be propagated in the same manner as light was ______________.
3. Microwave technology began with the invention of the ______________ in (country) ______________ by ______________.
4. The first scientist known to have expressed the idea of pulsed radar was ______________.
5. The first invention leading to the cathode ray tube was the ______________ by ______________.
6. Japanese aircraft attacking Pearl Harbor were detected ______________ miles north of Oahu by an ______________ radar system.
7. The first air traffic control radar was the ______________ built in (time period) ______________.
8. Define the following terms or acronyms:

   AC&W
   AMASS
   ARSR
   ASR
   ARTS
   ASDE
   ATCRBS
   Fire Control Radar
   Squeezing Waveguide
   GCA
   GCI
   MTI
   Microwave
   PAR
   Quadradar
   TAIU

Answers to Review Questions

1. Those listed below all had some part in the development of radio and/or radar. Name the things each accomplished:

    Benjamin Franklin ***discovered electricity***.

    Christian Johann Doppler ***proved and quantified frequency shift by motion.***

    Michael Faraday ***performed early experiments with electrical charges.***

    James Clerk Maxwell ***developed equations to quantify Faraday's findings.***

    Sir William Crookes ***made current flow through space in a tube.***

    Wilhelm Roentgen ***used Crookes' findings to develop the X-ray.***

    Heinrich Rudolph Hertz ***showed that electromagnetic energy could be propagated as light.***

    Nikola Tesla ***pioneered alternating-current theory, and devised several ideas before their time.***

    Christian Hulsemier ***invented a spark device to receive echoes.***

    Guglielmo Marconi ***used others' findings to create a transatlantic telegraph.***

    Thomas Edison, ***an experimenter who developed many devices.***

    John Ambrose Fleming ***invented the vacuum tube.***

    Lee DeForest ***invented amplitude-modulated radio transmitters.***

    Dr. A. Hoyt Taylor ***with assistant Leo Young first began work to create radar in the United States.***

    Leo Young ***was an assistant to Dr. Taylor.***

    Randel and Boot ***created the magnetron.***

    Joseph Lockhard ***reported the radar detection of Japanese aircraft approaching Pearl Harbor.***

2. The first scientist to propose that electromagnetic energy would be propagated in the same manner as light was ***Heinrich Rudolph Hertz.***
3. Microwave technology began with the invention of the ***magnetron in England by Randel and Boot.***
4. The first scientist known to have expressed the idea of pulsed radar was ***Nikola Tesla.***
5. The first invention leading to the cathode ray tube was the ***X-ray by Wilhelm Roentgen.***
6. Japanese aircraft attacking Pearl Harbor were detected over ***100 miles*** north of Oahu by ***an SCR-270*** radar system.
7. The first air traffic control radar was the ***AN/MPN-1, built in World War II.***
8. Define the following terms or acronyms:

    AC&W: ***Air Control and Warning, 200-mile radar systems for detection and interception.***

    AMASS: ***Airport Movement Area Surveillance System.***

    ARSR: ***Air Route Surveillance Radar, 200-mile "en route" air traffic control radar.***

    ASR: ***Airport Surveillance Radar, 60-mile "terminal" radar for airport arrival and departure control.***

    ARTS: ***Automated Radar Terminal System (or Service). Computer handling of radar data for airport arrival and departure control.***

    ASDE: Airport Surface Detection Equipment. ***A very short range radar for surveillance of ground traffic on the airport.***

    ATCRBS: ***Air Traffic Control Radar Beacon System (or Service).***

    Fire Control Radar: ***Radar used to automatically track targets and fire guns or missiles.***

    Squeezing Waveguide: ***A variable-width waveguide used for beam-steering in GCA systems.***

GCA: ***Ground-Controlled Approach.***

GCI: ***Ground-Controlled Intercept.***

MTI: ***Moving Target Indicator.***

Microwave: ***Short-wavelength technology requiring waveguides and other special devices.***

PAR: ***Precision Approach (landing) Radar***.

Quadradar: ***A four-mode radar for use at low-density airports.***

TAIU: ***Terminal Automation Interface Unit for AMASS to allow ASR and ARTS data to enhance ASDE safety margin.***

CHAPTER 2

The Professional Radar Engineer/Technician

"Professional"

The word "professional" is regularly misused today, and its use in the title of this chapter may appear to the outsider to be yet one more abuse. However, the work of a radar technician truly falls within proper definitions of the term. Some work is called a "trade," and some is called a "profession." A "trade" is defined as skilled work, but is differentiated from a profession by dictionary definitions. Definitions for "profession" include such concepts as "involving mental rather than manual work." Generally, "trade" would describe such occupations as plumbers, electricians, welders, mechanics, and machinists where skill in performing tasks is more necessary than extensive knowledge. "Profession" would describe such occupations as doctors, nurses, dentists, chemists, physicists, engineers, attorneys, etc. A ***Webster's Dictionary*** definition of the word "technician" is "one versed in the technicalities of some subject, specifically, an artist, writer, musician, etc. who has great technical skill or knowledge." The "knowledge" part of the general definition of "technician" places it in the "professions" category.

A "profession" may require certain manual skills, but it is characterized principally by formal, academic, or scientific knowledge, and by a requirement to arrive at conclusions through application of that knowledge. The radar technician's job most certainly falls within that definition. He must be able to correctly theorize the source of a problem, even when he cannot see it. He must be able to document his actions in a manner adequate to allow readers to unquestionably understand him; this requires the correct and accurate use of accepted technical and scientific terms and definitions. He must also be able to speak of technical and scientific matters with the same precise, accurate, and correct use of terms and definitions, as problem solving and correction may often involve telephone communication with other technicians in locations perhaps hundreds of miles away, or even more.

Any electronics technician needs a good knowledge of basic electricity and electronics, but there is a world of difference between, on one hand, the bench technician who, for instance, may repair circuit cards or some other type of electronic assembly and, on the other hand, the radar systems' technician who must first be able to isolate problems to a single system, and then proceed with isolation within that single system. The single system may be the radar set itself, ancillary equipment such as a secondary radar, a digitizer, a data communications link, a computer, or display equipment; each of these systems is complex in itself, and each requires a considerable amount of study to understand precisely what it does, and how it does it.

Even though circuit troubleshooting to a single resistor or capacitor is being outdated by integrated circuits, computers, and multi-layered or expendable printed circuit cards, the radar technician's job today requires every bit as much, if not even more, knowledge than in the past. In the 1940s and 1950s, various radar systems were all similar in many respects. Today, many of the original shortcomings have been corrected with new techniques and expansions upon the original comparatively simple ideas; the technician must know both the old, simple ways, and the reasons behind the newer ways to have a good understanding of his work. There are those who would diminish the profession to a trade, who say that knowledge is unnecessary. This is equivalent to a medical doctor who proposes that surgeons need only know where to cut.

Radar Technicians in World War II

The radar technician's occupation began with World War II, when radar was developed under the urgency to create Allied defenses against the Axis. Because of its secret new technology, and the scientific knowledge required to maintain and operate it, the army and navy selected some of the most well-educated, knowledgeable, and promising recruits to serve on radar crews. Because of the great threat to the nation, the country's entire effort was devoted to the war, and all able young men entered the military. Those entering the field of radar were, therefore, among the country's very brightest.

After World War II, the economy cooled, headed back toward the Great Depression that had preceded the war. Jobs were scarce, and those fortunate few who had learned radar and other electronics systems during the war found opportunities not available to others. Television was appearing on the scene, there were corporations manufacturing radar systems for national defense, airlines were in need of electronics experts, and the Civil Aeronautics Administration was instituting an air traffic control radar program utilizing many electronics systems. Outside these types of jobs, and the professions of law and medicine, the nation was still at a state in which there was little at which one could do well. The most common occupations were farming, automobile and farm equipment repair, teaching, and trades requiring long apprenticeships; dreary work for those who had been exposed to the excitement of growing with a new science.

The Korean War

On June 25, 1950, North Korean forces crossed the 38th Parallel to invade South Korea, and the Korean War began. Many of those who had served in World War II were called back into service, and the draft was resumed. In many cases, this second war so interrupted their lives that they would return home to drop earlier career dreams in favor of earning a satisfactory living in radar, which by that time had been their major interest in life for several years. And once again, the brightest of the draftees entered the field. When the war was over, all were faced with the choice between learning a new skill and taking advantage of lucrative new opportunities in electronics. Needless to say, many of these young radar technicians stuck to their newly learned skill in a science which was obviously in its infancy.

The Post-Korean Period

After the Korean War, the United States, for several years, maintained a universal military service law; the law required all able young men to serve a total of 8 years of military service, some of it on active duty. Those from families of little means, which at that time was the majority of the population, often enlisted for 4-year tours in the navy or air force, in the hopes of gaining useful knowledge or skills. Many were taught radar by instructors and supervisors who had served in World War II and Korea, and many of those had been the best of the field.

Radar today has many applications, but the two largest US users are undoubtedly the US military and the Federal Aviation Administration. This book will be principally directed toward those types of radar systems used by the Federal Aviation Administration. These radar systems are limited in type, simple in comparison to some military systems, and most likely to be of interest to students using this book because of employment opportunities.

The Division of Professions

The interesting history of air traffic control radar will be discussed in another chapter. When radar traffic control began with the World War II deployment of the AN/MPN-1 ***ground control approach (GCA)*** system, traffic volume was low in comparison to modern levels, and the main concern was the operation and upkeep of the radar itself. The original radar crews comprised soldiers who both operated and maintained the radar. As the tasks of both air traffic controller and radar maintenance became more demanding, the CAA required, in 1953, that the duties be separated into two occupations. GCA radar specialists were then given the choice to become either full-time air traffic controllers or technicians. Since that time, both professions have grown immensely in knowledge and skills requirements, and now bear little of the original similarity to each other.

The Scientific Approach

The scientific approach, once called "inductive reasoning," is nearly as old as science itself; it is attributable to a medical doctor, Francis Bacon (1561–1626). In some respects, the scientific approach to problem-solving by a radar technician might be compared to the chemist's analysis of a solution or compound. There is no way for the chemist to immediately visually recognize the chemical components, and he subjects the questioned substance to a series of tests which may include processes such as reagent application, distillation, centrifuge, and many others. The tests can only be performed by the educated chemist who understands the invisible molecular reactions he is creating.

Similarly, since most electronics operations are not visible to the naked eye, the radar technician must be able to understand the system or circuitry operations without first being able to see them. When his knowledge of normal operation leads him to believe that something is not right, he must begin analyzing the system to confirm his suspicions. Such an analysis will require inspection of the system, possible only with test equipment, which may, itself, require considerable knowledge and operating skill. Even the oscilloscope requires special techniques in radar applications.

For just one example of the tests or adjustments that a radar technician may deem necessary, consider the implications of an air-traffic-controller complaint that aircraft targets are occasionally disappearing, remembering also that this is only one of many possible complaints. The following sources are all possible, and even many more could be listed:

transmitter power
low-noise amplifier
parametric amplifier
preselector filter
local oscillator frequency
receiver mixer crystal
i-f amplifier bandpass
i-f amplifier adjustment
rotary joint
antenna tilt
antenna radiation pattern
antenna polarizer
MTI system subclutter visibility
MTI velocity response shape selection
MTI blind phase
MTI blind speed
MTI tangential effect
canceller gain
integrator feedback
digitizer threshold
line driver/compensator gain
indicator gain or intensity

Obviously, the cause cannot be found by trial and error; it may be a component failure, it may be a simple adjustment, or it may even be a normal occurrence. To attempt repair of complex equipment, a less competent technician might use a "shotgun" approach, replacing multiple components in the hope that he may luckily happen to get the correct one; this approach is common in the automobile repair trade, and often costs customers/victims a great deal of money. This "shotgun method" could be an endless task in a radar system. There is a good possibility that the technician will replace one maladjusted unit with another, never realizing that he had stumbled over the cause; there is another possibility that he may introduce problems while looking for the cause, and there is still another possibility that there was nothing really wrong in the first place.

Only a methodical, scientific, thoughtful, knowledgeable analysis of the system can reveal the source of a performance degradation, if, in fact, there is even one. Principally because of propagation phenomena, radar is imperfect, and investigation of performance complaints often involves only scientifically proving, beyond a shadow of a doubt, that the system is operating normally. This scientific proof may involve, for example, some of the following, and many more:

video-level measurements with an oscilloscope
system-sensitivity measurements with a signal generator
noise-figure measurements with a noise-figure test set
transmitter-spectrum measurements with a spectrum analyzer
transmitter average power measurements with a power meter
transmitter pulse width measurements with a spectrum analyzer
transmitter pulse width measurements with a video detector
transmitter peak power measurements by mathematical calculation
voltage standing wave ratio (VSWR) meas urements
local oscillator tuning check with an echo box
i-f bandpass tests with a swept signal generator
subclutter visibility measurements with an echo box and signal generator
canceller velocity response shape testing with a swept audio oscillator

All of the test equipment in the preceding list requires both a theoretical knowledge of its operation and an operator's skill.

Webster's Dictionary Definition of "Scientific"

Webster's dictionary contains at least five different definitions of the word "scientific," all of which bear many similarities to each other. Perhaps the most appropriate for this application is one which states, "according to, based on, or using the rules or principles of science; systematic and exact; . . . ." An excellent example of orderly scientific proof is the study of plane geometry, where the mathematical relationships of angles, circles, etc. are all conclusively proven through a systematic progression of proofs built upon each other; each proof is beyond question because of the ones that preceded it. Scientific thinking may somewhat be paralleled by the thought processes required by the game of chess; one careless move can lose the game for one who makes it. Precision, care, and thoughtfulness are all qualities of a scientific approach. Some of the rules which the technician must follow in any scientific investigation or undertaking are described in following sections.

Observing the Electronic Event in Question

This may require some creativity. It may simply involve oscilloscope viewing, or it may require that you use special techniques ranging from oscilloscope synchronization tricks to fabricated test devices. Special problems may require special methods, and there is always room for ingenuity.

Using Standard Language

There are hundreds of unique, special, or peculiar terms to be defined for use in radar applications, and accurate professional communication depends upon common understandings. Radar is peculiar in its use of slang originating in World War II; these slang terms will be defined in this book. However, the use of local slang or personalized definitions is not responsible and can only bring about confusion. Many terms are defined in this book; more are defined in other publications. The methods of abbreviated expression have also been determined in both technical writing specifications and IEEE standards. In general, abbreviations of common terms in written documents are done in lowercase letters, except the case where the first letter of an individual's name is used. For example, the deciBel is expressed as "dB" because the "B" is from "Alexander Graham Bell." Also, it is a usual practice to capitalize all the letters when naming a control, test point, indicator, or mnemonic; this practice makes those names immediately apparent in text or procedures.

Guesswork

Any conclusion without hard evidence, such as a measurement or series of measurements, is simply unprofessional and irresponsible. Beyond that, it will likely lead you away from your objective, rather than toward it; the usual expression is "up a blind alley."

Making Up Your Own Stories

To do so will quickly discredit you. This is not to say that you cannot make your own discoveries, but you must do so with a methodical method of reliable evidence and conclusive proof founded on known facts and supported by recognized authority. This will often involve research of technical books and documents, or manufacturer's instruction books. Supporting your conclusions by reference to accepted authority adds legitimacy to those conclusions and prevents others from becoming unnecessarily suspicious. In radar professions, a major engineering reference is Merrill Skolnik's ***Radar Handbook.*** Skolnik's book contains many more references to other information. To find information on air traffic control radar systems being manufactured throughout the world, consult ***Jane's Airport & ATC Equipment.***

Ego Versus Truth

Scientific endeavor is no place for ego, competition, or politics. The object of any investigation must always be only to find the truth, and never to discredit another, for such pursuits can hide the truth. The source of an idea should never be considered; the idea should be judged only on its scientific merit. Truth and proof are absolutes;

there can only be debate when the proof is unfounded or unstable, and the proof can only be questioned when there is no evidence, mathematical basis, or authority behind it.

Mathematics and Radar

The operation of a radar system is founded upon the exact science of physics. The subjects of mathematics, physics, and radar are inseparable. It is short-sighted, unprofessional, and simply wrong to assume that the technician must only know how to replace failed parts, or that mathematics is used only in design work. At the least, the radar technician should be well versed in algebra, plane geometry, and trigonometry. A few trigonometric identities are commonly used in several applications within radar systems, and logarithms are used in both everyday measurements and design techniques.

The operation of a radar system is based on many mathematical relationships, and an understanding of these relationships is essential to performance analysis. Everything that occurs in a radar system is applied mathematics and can be indisputably proven with formulas. For example, the receiver bandwidth and system noise are determined by the transmitter burst width and spectral width. A change in either the burst width or bandwidth can degrade target detection to a measurable degree, and both can be adjusted.

Some radar technicians will progress to become writers, instructors, equipment experts, and engineers. On entering the technician profession, neither the student nor his superiors can predict whether the student will advance to one of these positions; the most bumbling beginner may become highly competent, and the brightest one may not progress for laziness. If all students are not initially well founded in mathematical and scientific responsibility, those few who become "experts" will begin to write and teach incorrectly, and those who learn from them will compound the misunderstandings. It is relatively easy for one with an inadequate foundation in scientific responsibility to begin creating his own erroneous theories. The end result will be a universal incompetence and a dishonor of the profession.

The need for mathematical basis for radar theory must, in contrast, be tempered with some practical reason. An excessively heavy reliance on the mathematical aspects of equipment theory with inadequate text to explain circuit operations can quickly discourage understanding by the technician. Engineering textbooks and documents often exhibit an appearance that the author must have been more interested in impressing others with his knowledge and prowess than with providing an understandable explanation of how and why the equipment really works. These authors would undoubtedly defend themselves by charging that the reader had inadequate background. (In defense of those competent, well-intentioned writers who do offer reasonable explanation, it might also be suggested that nuclear physics cannot ordinarily be taught to a ditch digger with an eighth-grade education.) It will be the intent of this book to provide a reasonable balance, using equations from recognized authority, but offering clear verbal explanations, so as to make the subject understandable to the widest possible audience.

Because most radars are pulsed systems, both average and peak power calculations must be made regularly. Gains and losses throughout the system will be described in deciBels, which are based on logarithms. The radar technician must be proficient in the use of logarithms and deciBel calculations, and should never rely on conversion charts or graphs, since he can never be certain that these "crutches" will be available, and because regular use of them will destroy his calculation proficiency. Any unnecessary reliance on charts is unprofessional.

Talents Required of the Radar Technician

The radar technician is in a position to make good use of a wide variety of talents. Of course, the major thing the radar technician must be able to do is to knowledgeably connect and operate electronic test equipment, and to repair electronic circuitry on circuit cards and chassis. However, the radar system comprises mechanical systems as well; for examples, there are oil and ethylene-glycol cooling systems for transmitters and mechanical drive assemblies for antennas. The technician may find it necessary to use such tools as micrometers, depth gauges, torque wrenches, taps and dies, presses, chain hoists, oil pumps, and others. Some radar applications may also require the technician to operate large or special-purpose vehicles.

Additionally, it will be necessary for the radar technician to document and describe his work, and a certain degree of literacy and competence as a technical writer will be required. Because of the use of mechanical drawings, block diagrams, and schematics, skills as an artist or draftsman are also useful. And, just as in the business world, documentation and communications are all becoming dependent upon the use of computers; some familiarity with general computer terminology, hardware, and common software is essential. In fact, newer radar systems are beginning to utilize built-in computers to accomplish radar data processing, or to access the system for performance evaluation or adjustments. Still further, the radar technician must be a teacher, for once he becomes experienced, he will find it necessary to share his experience with less experienced technicians; if he avoids this, he is derogating the competence of his staff and sabotaging his employer's mission.

The field of radar provides an endless fascination for those who enjoy working with new ideas and expanding their knowledge; however, it is very unpleasant work for those who would seek to devote minimal mental effort to their occupation. The career radar technician must accept the notion that he will be a perpetual student. If he remains in the same position at the same location, there will always be new equipment, modifications, and expansions. If he changes jobs or transfers to another location, he will very likely need to learn the operation of different equipment. Each time he encounters new equipment, he will learn new techniques that he can apply elsewhere, or his previous knowledge may help him to expand upon existing knowledge of the new equipment. As he becomes more experienced, the technician must increasingly discipline himself to maintain his scientific curiosity, never closing his mind to new ideas, or to new approaches to old ideas; once he believes he knows all, he ceases to learn and deepen his knowledge.

The human mind has an infinite capacity, and there is no such thing as too much knowledge, or as unnecessary knowledge; anything one knows about any subject can be used to discover something else. No harm is done by knowledge that may never be applied, but much can be lost by knowledge never acquired; it is impossible to make intelligent advance judgments about what knowledge may or may not later prove useful. The more the technician knows, the more competent he will be; to enjoy his work, his frame of mind must be such that he is endlessly, tirelessly fascinated with learning.

Professional Responsibility

The privileges one enjoys by being a member of a profession, such as respect and income, depend largely on the public image of that profession. It then follows that each member of the profession must conduct himself in such a manner as to encourage respect, maintaining high standards of scientific responsibility, ethics, decency, and principled treatment toward colleagues, dress, grooming, speech; all those characteristics generally attributed to confident, principled, well-educated, intellectual individuals with high self-esteem.

Review Questions

1. Explain the difference between "trade" and "profession."
2. When did the radar technician profession begin?
3. Why is theoretical knowledge necessary to the technician?
4. Explain the connection between personal behavior and income.
5. What is the "shotgun approach"?
6. Explain "scientific responsibility."
7. Explain "scientific approach."
8. At what point may you conclude that you fully understand a technical subject?
9. Name some requisites of a radar technician.
10. Why does the technician need to know mathematics?
11. Why should a technician avoid the use of conversion charts and tables?

Answers to Review Questions

1. Explain the difference between "trade" and "profession."
   A "trade" generally describes manual work and skill; a "profession" requires formal knowledge and analytical skills.
2. When did the radar technician profession begin?
   Mostly with World War II.
3. Why is theoretical knowledge necessary to the technician?
   One cannot fix or adjust something he cannot see or understand.
4. Explain the connection between personal behavior and income.
   Image ultimately determines what employers are willing to pay. Every member of a profession owes it to his colleagues to maintain a good image.
5. What is the "shotgun approach?"
   The replacement of great quantities of parts based on the hope that one of those will be the one that is needed.
6. Explain "scientific responsibility."
   Taking care that one's judgments are based on sound evidence, authority, or mathematical proof.
7. Explain "scientific approach."
   A methodical analysis in which conclusions are based on a series of step-by-step proofs, each founded upon the preceding one.
8. At what point may you conclude that you fully understand a technical subject?
   Never. There is always more to learn.
9. Name some requisites of a radar technician.
   Literacy, mathematical ability, mechanical aptitude, and knowledge of physics.
10. Why does a technician need to know mathematics?
    It is required for everyday use.
11. Why should a technician avoid the use of conversion charts and tables?
    They become "crutches," and the technician loses his competence to perform tasks without them.

CHAPTER 3

Logarithms, DeciBels, and Power

This is a practical book for use by technicians or engineers who need to know everyday radar equipment theory of operation and maintenance methods. It is not intended to be a source for a mathematics course, and it is written with the assumption that the reader has foundation knowledge in algebra, geometry, trigonometry, physics, alternating-current mathematics, direct-current mathematics, and fundamental electronic theory. Nevertheless, everyday work in radar does require the frequent use of logarithms and deciBels. Still further, the remainder of this book will frequently refer to quantities in deciBels. The use of deciBels must therefore be reviewed before proceeding with any further discussion.

The DeciBel

It is a measurement tool originally devised by Alexander Graham Bell. Since the "Bel" portion of the word is derived from Bell's name, the "B" is always capitalized, even when the word is abbreviated to its common form, "dB." The "deci" portion of the word is related to powers of 10, as the dB is 10 times the value of a base-10 logarithm. In later days, binary circuitry has precipitated the use of base-2 logarithms, but these are rarely, if ever, used in connection with test equipment measurements.

Logarithms

Radar equipment measurements require frequent calculations of gain and attenuation. In earlier times, electronic calculators were not available, and these gain or attenuation measurements required time-consuming multiplication and division. These operations were greatly simplified by the use of logarithms. Adding two logarithms achieves a result which is the log of a multiplication product. In subtraction of logarithms, the subtrahend is the log of an antilog divisor, and the difference is the logarithm of an antilog quotient. Dividing a logarithm by any number provides a simple method to find a root. For instance, to halve a logarithm provides a means to find the square root of its antilog, or to divide a logarithm by 3 provides a means to find a cube root of its antilog.

Finding logarithms or their antilogs previously required the use of a book of tables, but the electronic calculator now makes the task quick, easy, and more accurate. Scientific calculators have a "log" function which yields the base-10 logarithm of a number, or a "10 ×" function to find the antilog of a base-10 logarithm.

DeciBels to Express Gain or Attenuation

A value expressed in deciBels is 10 times the value of a base-10 logarithm (see Table 3-1). Its purpose is in use as an indicator of gain or attenuation to power. Microwave attenuators, directional couplers, and other devices are marked in a dB value to indicate gain or loss. A value in dB, then, is an expression of the input–output ratio of a device. If an attenuator reduces a power level to one-half of its input value, then it has a 2:1 ratio, and the output is 0.5 times the input. The logarithm of 0.5 is –0.301, which is –3.01 dB. If the device were an amplifier with a gain of 2, the log of 2 would be +0.301 and the gain would be expressed as +3.01 dB. In short, 3 dB may be used to halve or double power (expressed in dBm or dBW). The 0.301 log value of 0.5 or 2 is an exponent of 10, and could also be expressed as $1 \times 10^{0.301}$. Note that the change of logarithm sign represents a reciprocal; the absolute value of 0.301 is the log of either "2" or "1/2."

TABLE 3–1 From Exponents to DeciBels

Multiplier expressed as an exponent (power of 10)	Multiplier 1 raised to power of 10 (exponent)	Multiplier in dB = 10 log of exponent (dB)
1×10^{0}	1	0
$1 \times 10^{0.301}$	2	3
1×10^{1}	10	10
1×10^{3}	1,000	30
1×10^{6}	1,000,000	60
$1 \times 10^{-0.301}$	0.5	−3.01
$1 \times 10^{-0.602}$	0.25	−6.02
1×10^{-2}	−100	−100

Since logarithms and deciBels express the exponent to which 10 has been raised or lowered and since the multiplication of numbers with exponents requires that the exponents be added, deciBels may be simply added, as in the following examples:

1. A 10-dB attenuator and a 3-dB attenuator are connected in series. Together they offer 13 dB of attenuation, or a −13-dB gain. Since a dB is 10 times the log of the ratio, the actual loss is the antilog of −1.3, or 0.05 of the original.
2. An amplifier consists of seven stages, each having a gain of 3 dB. The total gain of the amplifier is 21 dB. Since a dB is 10 times the log of the ratio, the actual gain is the antilog of 2.1, which is 125.89.
3. A radar system contains a ***waveguide directional coupler*** to measure transmitter power. The coupler is marked "47 dB." The transmitter power in the waveguide is 47 dB greater than the available power at the port connector on the directional coupler.
4. A radar waveguide directional coupler is used to inject a test signal into the system. The coupler is marked "42.2 dB." A test signal from a signal generator is injected into the directional coupler; the test signal in the waveguide will be 42.2 dB less than that at the directional coupler port connector.

Expression of Power in dBW or dBm

As dB is used to express an exponent of a number for use in attenuation or amplification, it may also be used to express power, providing a starting point, or reference, is known. For instance, 3 dB greater than 1 W would be 2 W, 6 dB greater than 1 W would be 4 W, and 10 dB greater than 1 W would be 10 W. Those values may be expressed in dBW; the "W" indicates that 1 W is the reference. For example, 0 dBW represents 1 W, 3 dBW represents 2 W, and 10 dBW represents 10 W.

Many measurements will be in dBm, where the reference is 1 mW. In this case, 0 dBm = 1 mW = 1/1,000 W = −30 dBW.

Note that there are 1,000 times the number of mW as W, so a dBm quantity will always be 1×10^{3} = 30 dB greater than a dBW quantity. Some common power levels are:

- 1 mW = 0 dBm = −30 dBW
- 1 μW = −30 dBm = −60 dBW
- 1 pW = 1×10^{-12} W = −90 dBm = −120 dBW
- 0.1 pW = −100 dBm = −130 dBW
- 1 W = 30 dBm = 0 dBW
- 10 W = 40 dBm = 10 dBW
- 1,000 W = 60 dBm = 30 dBW
- 1,000,000 W = 1 MW = 90 dBm = 60 dBW

TABLE 3–2 Gain, Attenuation, and Power

Multiplier expressed as an exponent (power of 10)	Multiplier 1 raised to power of 10 (exponent)	Multiplier in dB = 10 log of exponent (dB)	Reference = 1 W (dBW)	Reference = 1 mW (dBm)	1 W in dBm + multiplier in dB (dBm)
1×10^{0}	1	0	0	0	30
$1 \times 10^{0.301}$	2	3	3.01	3.01	33.01
1×10^{1}	10	10	10	10	40
1×10^{3}	1,000	30	30	30	60
1×10^{6}	1,000,000	60	60	60	90
$1 \times 10^{-0.301}$	0.5	−3.01	−3.01	−3.01	26.99
$1 \times 10^{-0.602}$	0.25	−6.02	−6.02	−6.02	23.98
1×10^{-2}	−100	−20	−20	−20	10

Using dBm or dBW in Combination with dB

Recall that dB expresses gain or loss, and that dBm or dBW expresses a power level related to a reference value (see Table 3-2). If a power level is raised or lowered by gain or attenuation, it is multiplied or divided by that gain or attenuation. For instance, 1 W applied to an amplifier with a gain of 2 will be amplified to 2 W. In terms of deciBels, a power level in dBm or dBW may be raised or lowered by adding or subtracting the gain or attenuation in dB. Using the same example as previously, a 1-W input is a 0-dBW input, a gain of 2 is 3 dB, and the resulting 3-dBW output is 2 W. Following are some application examples:

1. A power meter is connected to a 40-dB directional coupler. The power meter indicates +4 dBm. There is 44 dBm in the waveguide, and the antilog of one-tenth of 44 dBm is 25,118.8 mW, or 25.1 W. Another approach would be to convert the 44 dBm to 14 dBW, and then find the antilog of 1.4, which is 25.1 W.
2. A power meter is connected, through a 20-dB attenuator, to a 50-dB directional coupler in the radar system waveguide. The power meter indicates +1.5 dBm. There is 1.5 dBm + 20 dB + 50 dB = 71.5 dBm = 41.5 dBW = 14.125 kW in the waveguide.
3. A power meter reads 0, and is connected to a 40-dB directional coupler. It was believed that there was about 30 kW in the waveguide. If the 30 kW were present, there would have been +74 dBm in the waveguide, and +34 dBm would have been applied to the power meter. Most power meter inputs are limited to a maximum of +10 dBm or less. The reason the meter read 0 is that the thermistor mount was instantly destroyed when it was connected. *Never connect test equipment to a power source until you have verified the anticipated maximum power and the test equipment allowable input.* Use attenuators to reduce the meter input to a safe level, placing the smaller attenuators nearest the power source. Some thermistor mounts are made more sensitive than others for the purpose of measuring very low power levels, and are even more delicate.
4. A signal generator is producing a −55-dBm signal, connected through a cable with 2.4 dB of attenuation, to a 50-dB directional coupler. A power level of −107.4 dBm is injected into the waveguide.

Mental Approximations

Often, you will only need a rough idea of the power level. One case would be a situation in which you were making point-to-point checks, in search of a significant loss in signal. Another would be the one where you wished to estimate a maximum possible output before connecting test equipment. Remember that a whole exponent multiplied by 10, such as 10 dBW, 20 dBW, 60 dBW, etc., represents 1×10^{1} W, 1×10^{2} W, 1 MW, etc. Also remember that 3 dB may be used to halve or double a power level. Proceed mentally, and then in 3-dB steps from the nearest whole-number exponent of 10 to the approximate power estimate, as in the following examples:

60 dBW = 1 MW	52 dBW = 156.25 kW
If 60 dBW = 1 MW, then 63 dBW = 2 MW	60 dBW = 1 MW, so
If 63 dBW = 2 MW, then 66 dBW = 4 MW	57 dBW = 500 kW
If 66 dBW = 4 MW, then 69 dBW = 8 MW	54 dBW = 250 kW
70 dBW = 10 MW, so	51 dBW = 125 kW
67 dBW = 5 MW	50 dBW = 100 kW, so
64 dBW = 2.5 MW	53 dBW = 200 kW
61 dBW = 1.25 MW	56 dBW = 400 kW
58 dBW = 625 kW	59 dBW = 800 kW
55 dBW = 312.5 kW	

In dBm, the power measurements are simply 30 dB greater than that in dBW. As more 3-dB steps are used to arrive at a figure, a small increasing error is introduced because a gain or loss by 2 is actually 3.01 dB rather than 3 dB.

Actual Attenuation of Devices

The attenuation of microwave devices will often be frequency dependent. Always look for information to indicate the actual attenuation for the operating frequency of the radar system. There may simply be a few numbers from which you may find it necessary to interpolate, or there may be a graph to provide more precision.

Any coaxial cable used in microwave measurements will exhibit significant attenuation, and must be calibrated. The usual method is to measure the output of a known source and then connect the cable to that source and measure the output at the end of the cable. The additional loss, of course, is the cable attenuation. Test cable attenuations as high as 6 dB are not unusual.

Attenuation in power measurement provides an additional advantage in increasing accuracy for two reasons. Using more sensitive scales on the meter makes the meter more precise, as small variations are more obvious. For example, a change from –20 dBm to –21 dBm is much smaller than a change from 0 dBm to –1 dBm. Further, the additional attenuation minimizes reflections caused by mismatches between the test equipment and source, as the reflections are attenuated on each trip between the devices.

Peak and Average Power Measurement.

Radar power measurements are made at waveguide directional couplers, permanently installed in the radar system waveguide hardware. There will normally be two of these: one called the ***incident power coupler*** and other called the ***reflected power coupler***. The incident coupler is used for most measurements; the reflected power coupler is provided for measuring reflections from the antenna rotary joint or other hardware. The reflected power is used in VSWR calculations.

A pulsed radar system's transmitter produces a short burst of energy and then remains idle for a long "listening" time. In this book, the duration of the burst is designated t_p, and the period from the beginning of one burst to the beginning of the next is designated T_r. Then, there exists a ratio between t_p and T_r, called either the ***duty cycle*** or ***duty ratio***. The duty ratio is defined as t_p/T_r. The duty ratio may also be described in terms of a percentage equal to $100(t_p/T_r)\%$. For example, if a radar transmitter burst were 3 μs wide and the period were 1,100 μs, the duty ratio would be 2.73×10^{-3} or 0.273%. The duty ratio may also be expressed in dB, simply because it is a ratio, and the example would become $10 \log(2.73 \times 10^{-3}) = -25.6$ dB.

A power meter thermistor is a heat-sensing device and only responds to average power. Pulsed radar power measurements will therefore be power averages, but it is necessary to know the ***peak power in Watts (P_t)***, which is the emitted power only during the duration of the pulse. To convert the average power to peak, you may multiply the ***average power in Watts (P_{ave})*** by the reciprocal of the ***duty cycle* (DC)**, or you may subtract the ***dB value of the duty cycle* (DC_{dB})** from the average power in dBW or dBm. For example:

$$P_{ave} = P_t \, DC$$

$$P_t = \frac{P_{ave}}{DC} = P_{ave} \frac{1}{DC}$$

$$P_{ave_{dB}} = P_{t_{dB}} + DC_{dB}$$

$$P_{t_{dB}} = P_{ave_{dB}} - DC_{dB}$$

where

P_t = peak pulse power in W or mW

$P_{t_{dB}}$ = peak pulse power in dBW or dBm

P_{ave} = average pulse power in W or mW

$P_{ave_{dB}}$ = average pulse power in dBW or dBm

DC = duty cycle, a fraction

DC_{dB} = duty cycle expressed as a negative dB multiplier.

The ***pulse width*** (t_p) may be measured by different methods, one of which is simply to observe the transmitter burst envelope on an oscilloscope. This is accomplished by installing a crystal detector, with adequate attenuation, on the directional coupler, and then cabling the output to the oscilloscope vertical input. The oscilloscope input must be terminated with either 50 or 75 Ω, depending upon the detector. The pulse width is measured between the 70% amplitude points, as those points represent half-power, –3 dB. Pulse width can also be determined with an echo box or spectrum analyzer; this will be addressed at another point in the book.

Accuracy in power measurement depends on accuracy in the duty ratio calculation, which, in turn, depends upon the accurate measurement of T_r. Simple measurement between transmitter triggers with the oscilloscope graticule and main time base is inadequate. An oscilloscope with a calibrated delay sweep, or with timebase cursors, offers more accuracy. A counter offers still greater accuracy; the reciprocal of the measured repetition frequency is T_r.

Half-Power and 70% Amplitude

When voltage is measured on an oscilloscope, it is the result of a drop across a resistance, and both the resistance and power are involved. The 70% voltage points represent the half-power points because 0.707 is the reciprocal of the square root of 2. The relationship is as follows:

$$P = \frac{E^2}{R} \Rightarrow \frac{P}{2} = \frac{1}{2} \times \frac{E^2}{R} \Rightarrow \frac{P}{2} = \frac{E^2}{2R} \Rightarrow 2E^2 = 2PR \Rightarrow E^2 = 2\left(\frac{PR}{2}\right)$$

$$\Rightarrow E = \sqrt{2}\sqrt{\frac{PR}{2}} = 1.414\sqrt{\frac{PR}{2}} \Rightarrow \frac{E}{1.414} = \sqrt{\frac{PR}{2}} \Rightarrow 0.707E = \sqrt{\frac{PR}{2}}.$$

Decibels in Voltage (dBV)

Because of the exponent of 2 in the power equation, voltage relationships may also be expressed in dB, but dBV = 20 log of the voltage ratio, as shown below:

$$\text{dBV} = 20 \log E$$

because

$$E = \sqrt{PR} \Rightarrow \log E = \frac{\log(PR)}{2} \Rightarrow 2 \log E = \log(PR) \Rightarrow 20 \log E = 10 \log(PR).$$

Review Questions

1. What is a base-10 logarithm?
2. What is a deciBel?
3. A radar technician is preparing to connect a power meter to the incident power directional coupler port. A graph on the directional coupler indicates that the attenuation at the operating frequency is "55.3 dBm." The maximum safe input to the power meter thermistor is 0 dBm. The average power in the waveguide may be as high as 600 W. How much additional attenuation is needed to protect the meter?
4. A technician makes two power measurements at the same point, one with the power meter on the 0-dBm scale, and one with the power meter on the –20-dBm scale, with an additional 20 dB of attenuation connected to the thermistor mount. There is a 0.5-dB difference in the two readings. Which is the most likely accurate, and what might be the source of the error?
5. What is the value in watts, of a power level of 88.3 dBm?
6. Without resorting to a calculator or tables, determine the approximate power in watts for each of the following:

   70 dBm _______
   71 dBm _______
   72 dBm _______
   73 dBm _______
   74 dBm _______
   75 dBm _______

7. A technician is using a signal generator to measure the receiver minimum discernible signal. The directional coupler attenuation for the operating frequency is –47.4 dB, the test cable loss is –5.5 dB, and the signal generator output is –54 dBm. What is the test signal level injected into the waveguide?
8. A technician is measuring the transmitter power output at an incident power directional coupler. The coupler attenuation at the operating frequency is 28 dB. The power meter is on the +5-dBm scale and indicates –2 dBm. The transmitted pulse width at the –3-dB points is 2 μs and the interval is 3,000 μs. What is the peak power output in watts?
9. Express the ratio of 10 mV to 10 V in dB.
10. Why might the 0.707 voltage points on a waveform represent half-power?

Answers to Review Questions

1. What is a base-10 logarithm?

   An exponent, 1×10^n, where n is the logarithm.

2. What is a deciBel?

   Ten times the logarithm of a ratio.

3. A radar technician is preparing to connect a power meter to the incident power directional coupler port. A graph on the directional coupler indicates that the attenuation at the operating frequency is, "55.3 dBm." The maximum safe input to the power meter thermistor is 0 dBm. The average power in the waveguide may be as high as 600 W. How much additional attenuation is needed to protect the meter?

   $P_{avddBW} = 10 \log(600) = 27.8 \text{ dBW} = 57.8 \text{ dBm}$

   $57.8 \text{ dBm} - 55.3 \text{ dB} = 2.5 \text{ dBm to meter}$

   A 3-dB attenuator provides inadequate safety margin and a 6-dB attenuator should be safe; a 10-dB attenuator would be better.

4. A technician makes two power measurements at the same point, one with the power meter on the 0-dBm scale, and one with the power meter on the –20-dBm scale, with an additional 20 dB of attenuation connected to the thermistor mount. There is a 0.5-dB difference in the two readings. Which is the most likely accurate, and what might be the source of the error?

   The lower scale measurement is more likely accurate. The source of the error might be an impedance mismatch.

5. What is the value in watts, of a power level of 88.3 dBm?

   $88.3 \text{ dBm} = 58.3 \text{ dBW}$

   $\text{antilog } 5.83 = 676.082 \text{ kW}$

6. Without resorting to a calculator or tables, determine the approximate power in watts for each of the following:

   $$70 \text{ dBm} = 40 \text{ dBW} = 1 \times 10^4 = 10 \text{ kW}$$
   $$71 \text{ dBm}: \text{ three 3-dB steps from 80 dBm} = 12.5 \text{ kW}$$
   $$72 \text{ dBm}: \text{ six 3-dB steps from 90 dBm} = 15.625 \text{ kW}$$
   $$73 \text{ dBm} = \text{twice the value at 70 dBm} = 20 \text{ kW}$$
   $$74 \text{ dBm} = \text{twice the value at 71 dBm} = 25 \text{ kW}$$
   $$75 \text{ dBm}: \text{ five 3-dB steps from 90 dBm} = 31.25 \text{ kW}$$

7. A technician is using a signal generator to measure the receiver minimum discernible signal. The directional coupler attenuation for the operating frequency is –47.4 dB, the test cable loss is –5.5 dB, and the signal generator output is –54 dBm. What is the test signal level injected into the waveguide?

   $-54 \text{ dBm} - 5.5 \text{ dB} - 47.4 \text{ dB} = -106.9 \text{ dBm} = -136.9 \text{ dBW} = 2.04 \times 10^{-14} \text{ W} = 0.024 \text{ pW}$

8. A technician is measuring the transmitter power output at an incident power directional coupler. The coupler attenuation at the operating frequency is 28 dB. The power meter is on the +5-dBm scale, and indicates –2 dBm. The transmitted pulse width at the –3-dB points is 2 μs and the interval is 3,000 μs. What is the peak power output in watts?

   (a) Since the power meter only measures averages, the duty cycle must be computed to facilitate conversion to peak power:

   $$DC_{dB} = 10 \log\left(\frac{t_p}{T_r}\right) = 10 \log\left(\frac{2}{3{,}000}\right) = -31.76 \text{dB}$$

(b) ***Since the power meter is on the +5-dBm scale, and indicates –2 dBm on the meter face, a power of +3 dBm is indicated.***

(c) ***Total the power, attenuation, and duty cycle reciprocal:***

+3 dBm + 28 dB + 31.76 dB = 62.76 dBm
62.76 dBm = 32.76 dBW = 1,888 W

9. Express the ratio of 10 mV to 10 V in dB.

$$20 \log \frac{10 \times 10^{-3}}{1 \times 10^{1}} = -60\,\text{dBV}$$

10. Why do the 0.707 voltage points on a waveform represent half-power? ***0.707 is the reciprocal of the square root of 2, the "2" being the "2" in "1/2," and the square root being a result of the exponent in*** E^2.

CHAPTER 4

The Science Behind Radar

Introduction

This chapter is a very basic introduction to the principle of pulsed radar. Obviously, getting a beginning concept of the main principle is one objective. Another is to introduce many terms commonly used in the radar language. Those terms which are the most essential to remember are printed in bold italics to assist you.

The basic principle of pulsed radar is not complex (see Figure 4-1). A transmitter emits a burst of rf energy, and objects in its path will reflect the pulse back to the transmitting antenna as an ***echo***. A steering device, called a ***duplexer***, allows the transmitter burst to go to the antenna without damaging the receiver, and then isolates the transmitter during a "listening" time. The listening time is usually called ***live time***. During live time, the echoes from any objects, called ***targets***, are routed into the receiver.

Echo Time

The time required for an echo is measurable and representative of target ***range***, the distance from the radar antenna to the target. The transmitted burst travels at the speed of light, and the distance of the target may be calculated by the basic physics formula, distance = rate × time ($d = rt$). Since the transmitter burst is echoed, the two-way travel requires that the speed of light be halved in using it as "r" in the $d = rt$ equation. Thus, $d = rt$ becomes $d = (c/2)t$.

Radar ranges are usually expressed in either meters or nautical miles depending upon the radar application. Where navigation is the application, the distances are normally measured in nautical miles (nmi), since the nmi is one-sixtieth of 1° of the circumference of the earth and is used extensively throughout the world. It takes 12.3552×10^{-6} s (12.3552 µs) for a transmitted burst to travel 1 nmi to a target, to be reflected, and then to return to the receiver. Although semantically incorrect because it is time rather than distance, 12.36 µs is widely called the ***radar mile***. Similarly, other radar time-for-echo-for-distance values are expressed below:

1 US statute mile (5,280 feet) = 10.736 µs
1 international nmi (6,076.11549 US feet) = 12.3552 µs
1 US foot = 2.03 ns
1 US inch = 169.45 ps
1 cm = 66.7 ps
1 m = 6.67 ns
1 km = 6.67 µs

Synchronization

The ***synchronizer*** provides the means to ***temporally*** (having to do with time) link the transmitter, receiver, and display circuitry. It generates timing ***triggers*** and ***gates*** for use throughout the radar system, and these signals occur repeatedly at a regular rate. A "trigger" is an abrupt signal (for instance, 0.5 µs) with a sharp risetime, used to "trip" other operations. A "gate" is generally longer than a trigger, is used to enable another signal, and is so named because its application resembles the gate in a fence, allowing some other signal to pass when it is "open."

One of the synchronizer triggers goes to the ***pulse modulator***, which develops a high-voltage pulse (for example, 10 kV to 100 kV or more) for the transmitter final power amplifier tube. The time when the

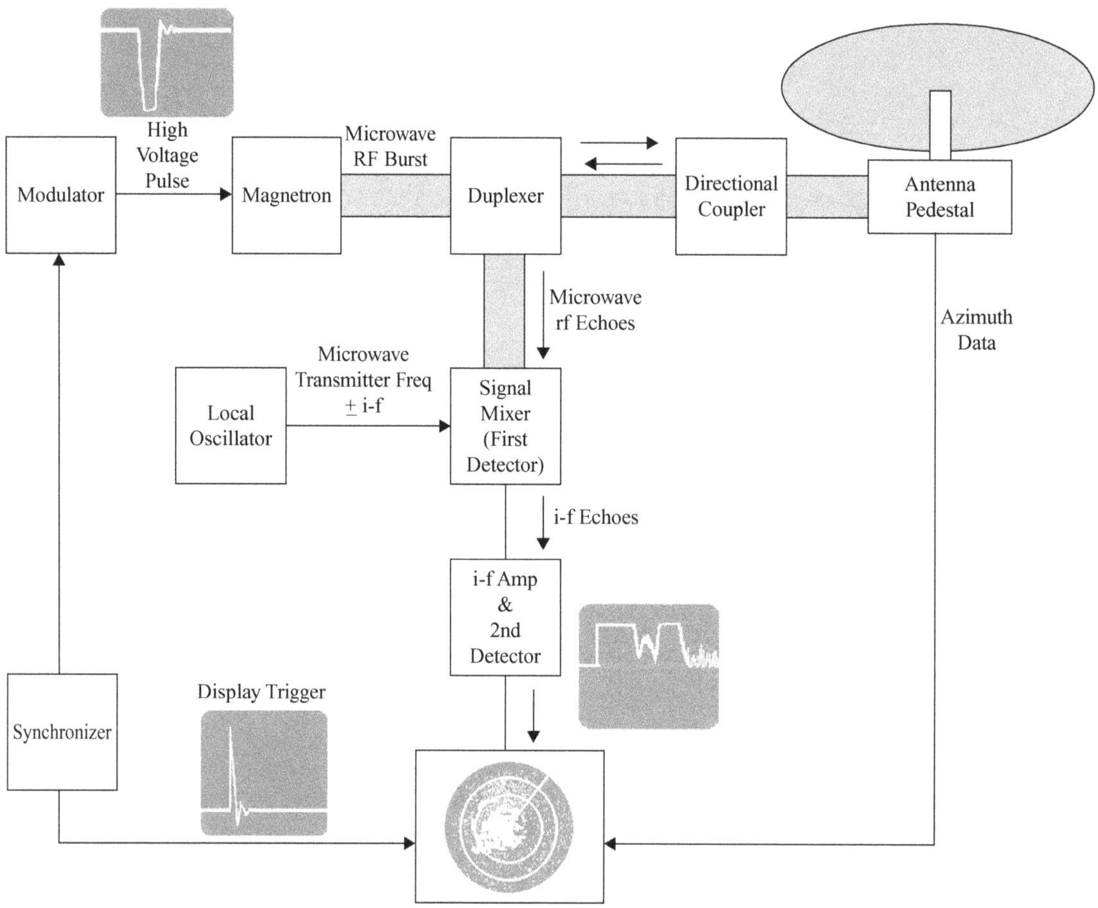

FIGURE 4–1

Basic components of a radar system.

transmitter is pulsed is usually called ***time zero*** (T_0). The pulse modulator builds a high-voltage charge during the interval between pulses and then rapidly discharges at T_0.

Displaying the Received Echoes

The radar illustrated in Figure 4-1 contains only one receiver; this was the type of receiver originally used in prewar and World War II (WWII) radars. Because it is basic and fundamental, it is called the ***normal*** receiver, to distinguish it from a ***moving-target-indicator (mti)*** receiver or a ***logarithmic-gain (log)*** receiver. The normal receiver uses simple amplitude detection, and radar echoes on an oscilloscope would appear as shown in Figure 4-2. Radar receivers are most often a ***superheterodyne*** type, in which the rf echoes are compared to a ***local oscillator*** to produce an ***intermediate frequency (i-f)*** for additional amplification and detection. Some ***multiple-conversion*** radar systems may contain more than one intermediate frequency.

In Figure 4-2, the oscilloscope has been synchronized with a trigger that occurs before T_0, so that a little ***deadtime*** appears before the video begins.

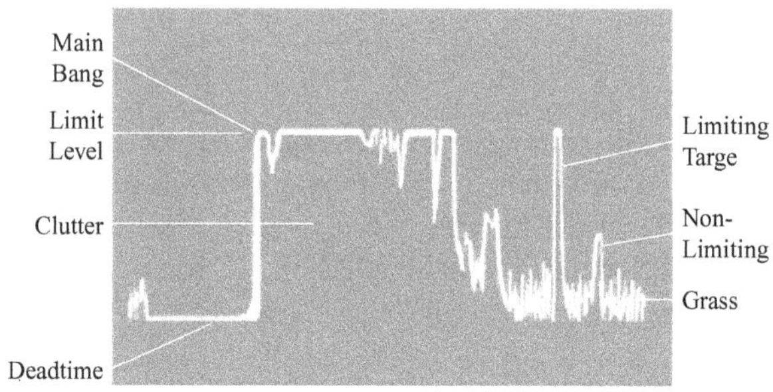

FIGURE 4–2

"Normal" radar video on an oscilloscope.

Deadtime

It is the time between (1) the end of the live time in which the receiver is "listening" and (2) time T_0.

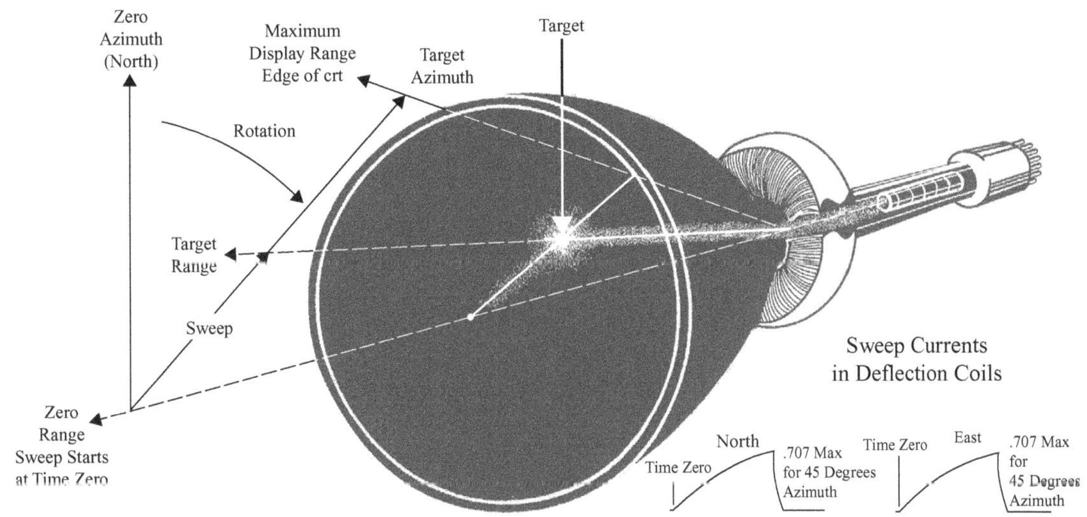

FIGURE 4–3

Displaying radar video on a crt (planned position indicator method).

Because Figure 4-2 is an oscilloscope presentation, the echoes cause vertical deflection to the sweep. Where no echoes are received, the oscilloscope video is caused by receiver noise. Because it resembles fine-bladed grass waving in a gentle wind, it is called ***grass***. The very first part of the video is a result of an echo from the antenna reflector, leakage from the duplexer, and other minor reflections of the rf energy, and is called the ***main bang***. This is one of the oldest slang terms in the radar world, and probably began with early transmitters that fired when a spark gap discharged a high voltage.

The ***display*** equipment may take different forms (see Figures 4-3 and 4-4). In an ***analog real-time planned position indicator***, analog video is displayed with minimal delay at the same rate the echoes are received. It is a range-direction, ***rho-theta*** display. The real-time ppi has been the most common form of display until recent years, but it is gradually being replaced by television technology, although the final appearance may remain similar. This older type of display is a good starting place for study, since all temporal relationships to transmitting time remain intact. A trace is generated on a cathode ray tube by a "flying spot," beginning at ***video time zero*** and ending at an established maximum time at the edge of the tube. Echo video intensifies the trace as echoes are received, leaving intensified areas. The position of the spots on the trace then indicates the ***range*** of the targets. The trace starts at the center of the crt, sweeps to the edge, and rotates with the antenna. A single trace is called a ***sweep*** and a revolution is called a ***scan***.

To obtain the directional information, called ***azimuth***, a means of pointing the sweep in the same direction as the antenna is necessary. The sweep deflection is therefore somehow linked to the antenna position. Two of the most common ways of doing this are with ***azimuth pulse generators (apg)*** or ***synchro-data systems***. Synchro systems are now rare. For a radar antenna which continuously rotates in the same direction, the apg is the preferred type, but synchro systems may be required if the antenna may be reversed or stopped as a normal operating function. Other azimuth-data coupling systems may be necessary where electronically positioned, ***agile-beam*** antennas are used. Figure 4-4 illustrates the appearance of a normal video on an analog real-time ppi display.

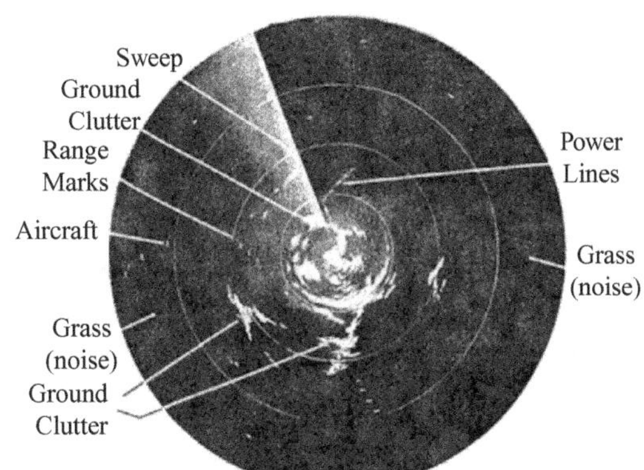

FIGURE 4–4

"Normal" video on a ppi.

Summary of the Basic Radar Principle

The basic radar system block diagram is shown in Figure 4-5. The synchronizer runs continuously, regularly producing all the triggers and gates for the system. Each time it produces a trigger to create the high-voltage modulator

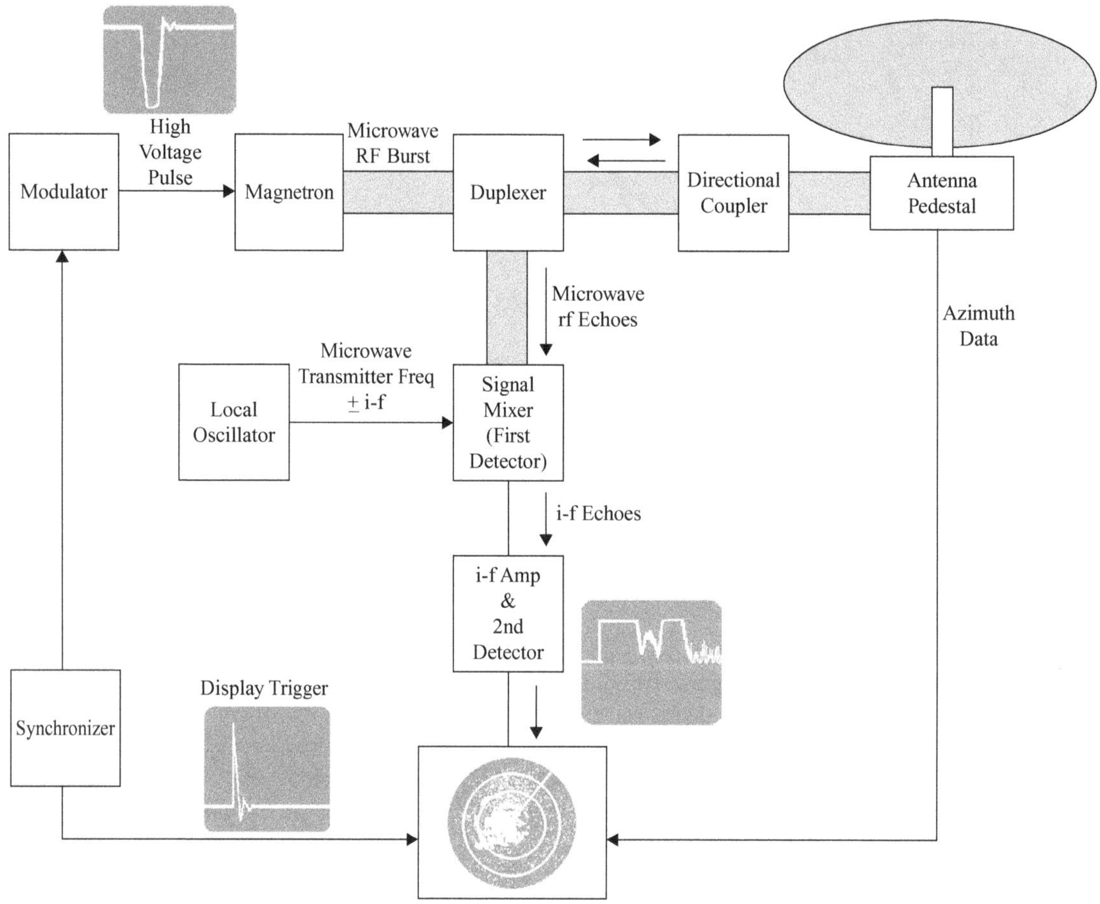

FIGURE 4–5

Simple block diagram.

pulse, another trigger starts a sweep on a display. As echoes are received, they are mixed with a local oscillator to produce a heterodyned i-f, further amplified and detected, and then displayed on the ppi at positions determined by the antenna azimuth and target range.

Repetition Rate and Period

Many factors are involved in the choice of the repetition rate of the synchronizer timing signals, but the predominant requirement is that the period between transmitter triggers be greater than the desired radar range. For instance, an FAA ASR radar uses a 60-nmi range, so a live time of (12.3552 μs) (60 nmi) = 741.3 μs is needed. Some amount of deadtime is also necessary, and the total period between transmitter triggers must be at least as long as the necessary live time plus the minimum necessary deadtime. A period of 1,000 μs would not be unusual for the 60-nmi ASR radar.

There are many terms used to define the repetition rate and period. The student may encounter any of those, depending upon employer, government service, equipment manufacturer, or textbook; it is to his advantage to be able to recognize all of them; they are listed in Table 4-1.

Radar Detection Range

The remainder of this chapter will be devoted to the many factors that determine the realistic expectations of radar echoes. The effects of each will be quantitatively analyzed so that the reader may recognize which factors are within his control and the performance degradations that may occur. Among those factors are antenna gain and aperture, receiver bandwidth, transmitter spectrum, transmitter peak power, receiver sensitivity, and target size.

TABLE 4–1 Variation in Terms

Rate			*Period*		
Term		**Source**	**Term**		**Source**
Pulse repetition frequency	PRF	USAAC/USAF, FAA, manufacturers	Pulse repetition time	PRT	USAAC/USAF, FAA, manufacturers
Pulse repetition rate	PRR	US Navy, manufacturers	Pulse repetition interval	PRI	US Navy, manufacturers
Repetition frequency	f_r	Academia, FAA, textbooks	Interpulse period	IPP	Westinghouse, FAA, USAF
Pulse frequency	f_p	Academia, textbooks	Time of repetition	T or T_r	Academia, textbooks

The Antenna Radiation Pattern (Beam)

To provide for target definition in azimuth or elevation, the rf burst must be radiated in narrow dimensions. Air traffic control radar designs usually strive to achieve narrow azimuth dimensions and tall elevation dimensions. Other systems are designed for narrow dimensions in both planes; such a radiation pattern is sometimes called a ***pencil beam***.

The beam dimensions are measured at the point where the transmitted power concentration is half that in the center, most dense, area. Power measurements in radar are generally expressed in ***deciBels (dB)***, and half-power is 3 dB less than maximum, so the beam dimensions are called ***the 3-dB points***, as illustrated in Figure 4-6.

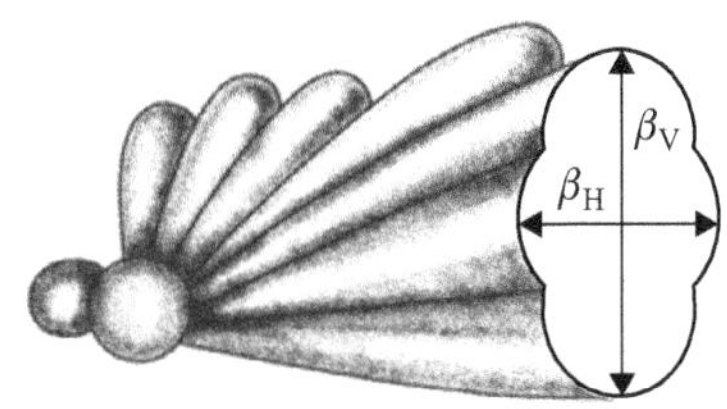

FIGURE 4–6

Beam dimensions.

No matter what type of antenna is used, the shaping of the antenna radiation pattern is accomplished by precisely phasing a large number of radiating sources so the individual radiation patterns from all the sources combine to make a concentrated, directional, wave front. This may be accomplished by an array of dipoles or waveguide slots, or with a horn-fed reflector, precisely shaped to create a constant-phase wavefront plane in space (see Figures 4-7 and 4-8).

As frequencies increase, wavelengths decrease, antenna element size decreases, and smaller elements may be more closely spaced to achieve phasing. In WWII, the development of microwave technology was urgent because small antennas were needed for use in aircraft; pre-microwave, low-frequency radars required very large antennas.

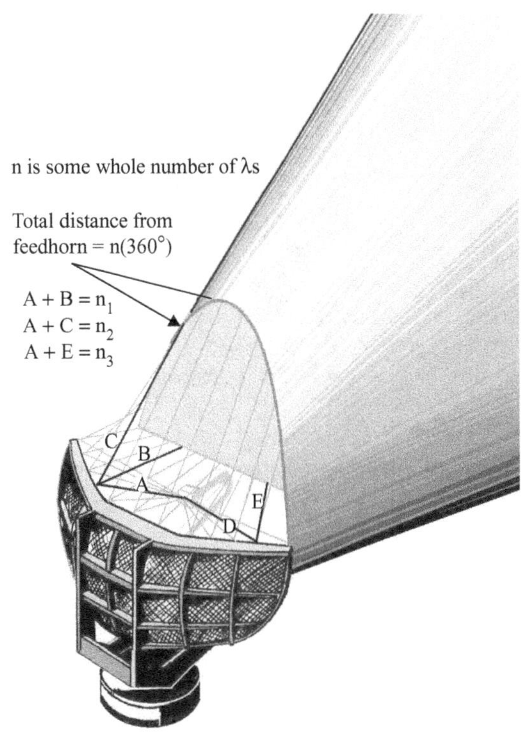

FIGURE 4–7

Forming the radiation pattern.

The Isotopic Source Reference

As the antenna radiating surface becomes larger, the outgoing, transmit, one-way radiation pattern becomes more densely focused, so the size becomes a factor in the ***antenna*** *directivity (D)* (see Figure 4-9). The directivity is the ratio of the concentrated power P_d to the hypothetical power P_i that would be present had the energy been radiated from a single-point ***isotropic source*** in a perfectly spherical pattern. Since this isotropic reference is used, radiation pattern formulas will often contain the expression $4\pi R^2$, the surface area

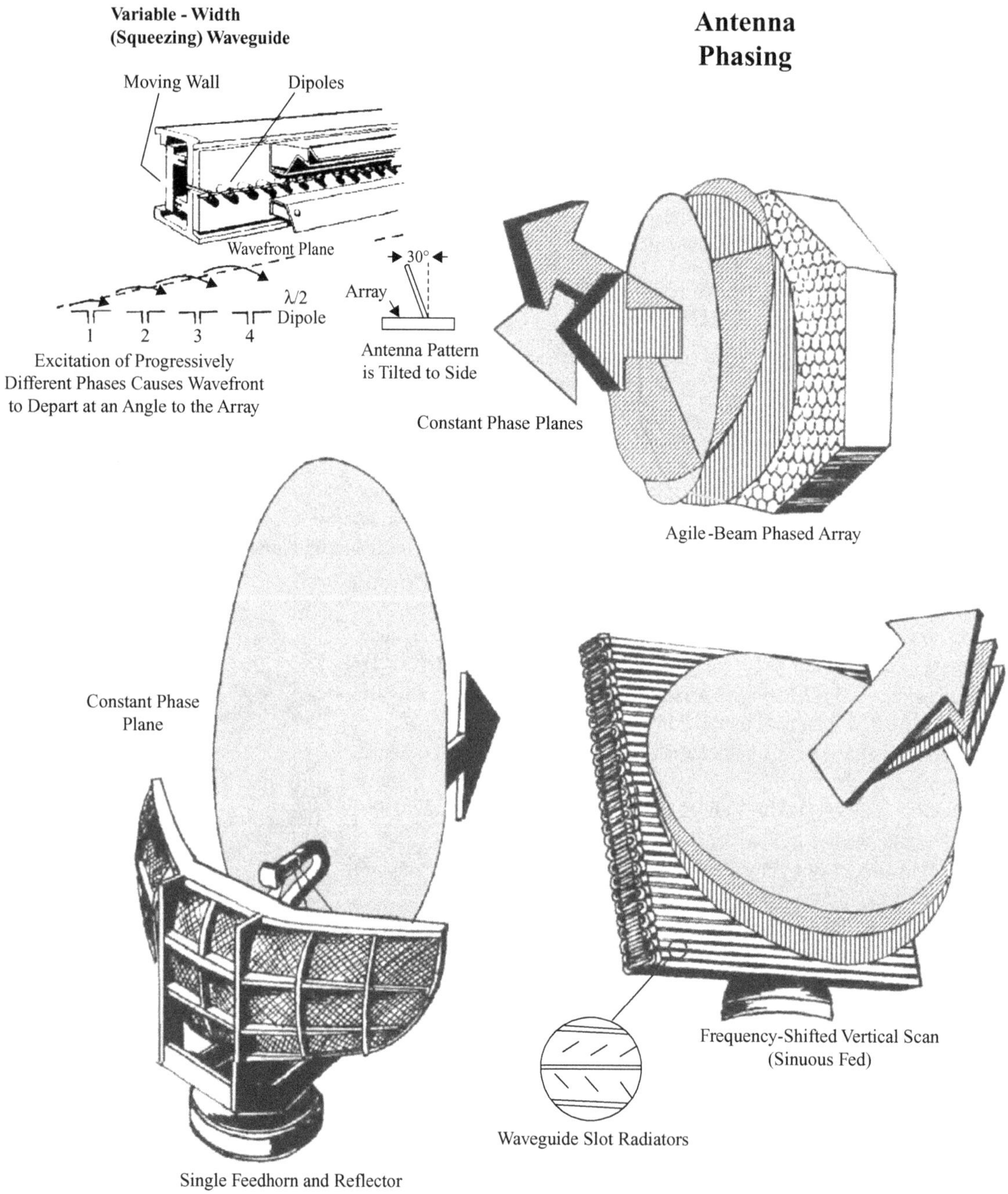

FIGURE 4–8

Variety of means in forming the constant-phase plane.

of a sphere. As the sphere expands with range, the transmitted energy, at a given point, and from an isotropic source, would proportionally decrease. However, the power decrease with range for a directional antenna will be less.

It should be clear that there is a relationship between directivity and the beam dimensions. The area between the 3-dB points can be closely approximated by $\beta_h\beta_v R^2$, where β_h and β_v are the horizontal and vertical beam dimensions in radians, and r is the range. The power concentration in the $\beta_h\beta_v R^2$ area will be increased to the total power concentration that would be present over the entire sphere of an isotropic pattern. Thus, there is a ratio between the two areas, and it may be expressed as follows.

Isotropic source versus directional power:

$$D_{a}=\frac{4\pi R^2}{\beta_h \beta_v R^2}=\frac{4\pi}{\beta_h \beta_v}. \tag{4-1}$$

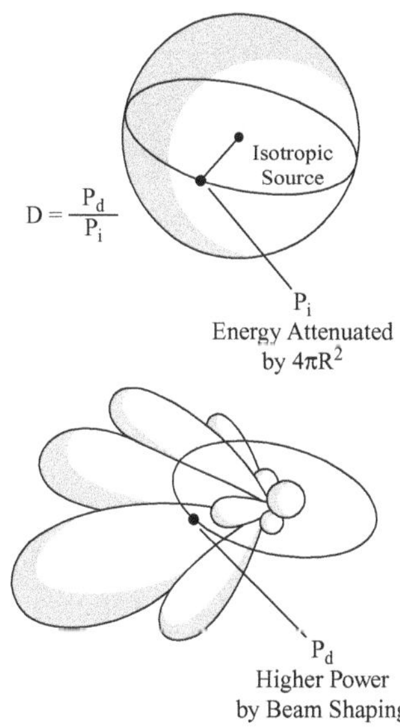

FIGURE 4–9

Isotropic versus directional power.

The Antenna Gain (G_t)

It is very similar to directivity, but it is based on power ***inputs*** to the directional and isotropic antennas, the difference being that the efficiency, k, of the directional antenna enters the equation. Gain, therefore, is expressed as

$$G_t = kD. \tag{4-2}$$

Aperture

In radar, the two-way radiation pattern is of greatest interest. The target becomes the source of return radiation, and the size of the antenna determines how much of the echo wavefront may be recovered. The area of the receiving surface becomes the ***effective aperture*** A_e of the antenna. The gain of a receiving antenna is the same as the gain of a transmitting antenna, and the antenna gain may also be expressed in terms of A_e as follows.

Antenna gain and effective aperture:

$$G_t = \frac{4\pi A_e k}{\lambda^2}. \tag{4-3}$$

Attenuation of the Echo

The transmitted burst of rf energy is subjected to severe attenuation by its expansion and dissipation as it is propagated. When a small, fractional portion of this expanded, dissipated, outgoing energy is reflected by a target, the target reflects the energy in many directions other than directly back toward the radar antenna. The resulting echoes are thus minuscule in respect to the transmitted burst. If the antenna radiation pattern were a perfect sphere, the antenna would be called an "isotropic source," and the energy would be reduced by the radius of the expanding sphere at the target. The radar range is the radius of the sphere, and spherical geometry puts the amount of signal dissipation at $4\pi R^2$. If the target were an isotropic source, the echoed burst would be further reduced by $4\pi R^2$ for an overall two-way attenuation of $(4\pi R^2)^2$, which is $16\pi^2 R^4$.

Since the gain G_t of the antenna will increase the burst power density at the target, the target cross-sectional area A_o will determine the amount of energy reflected to the radar antenna, and the antenna aperture A_e will determine the amount of recoverable received echo, these values must be taken into account to determine the predictable strength of an echo. The power density of the echo therefore becomes as shown in Equation (4-4).

Received power at a specific range:

$$P_r = \frac{P_t G_t A_o A_e}{16\pi^2 R^2}. \tag{4-4}$$

Receiver Sensitivity ($P_{r\,(min)}$)

A very sensitive receiver is required to recover these echoes. The greatest obstacle to receiver sensitivity is ***noise***, and the ***minimum discernible signal*** is the lowest peak pulse signal power that can be seen in the noise by the eye, or electronically retrieved from the noise.

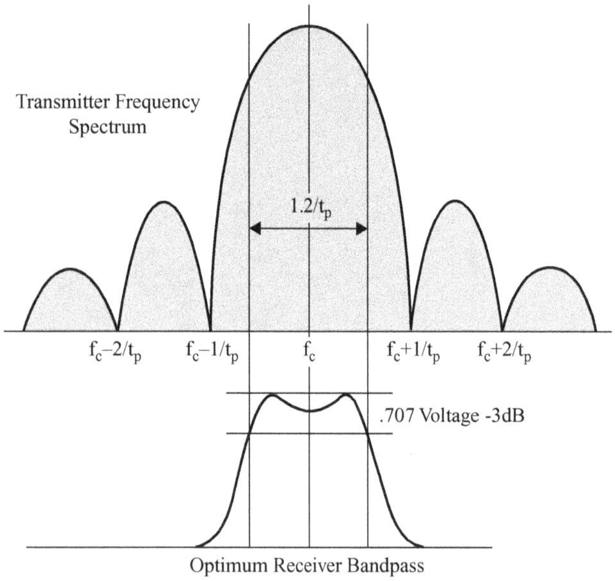

FIGURE 4–10

Transmitter spectrum versus receiver bandwidth.

Noise is supplied by two major sources, the atmosphere and the receiver itself. In the receiver, the predominate noise source is the local oscillator. Noise is present at all frequencies, so the received noise is directly proportional to the receiver bandwidth (Δf). Narrow Δf values are therefore desirable in respect to the noise problem, but another mitigating consideration must be made. The radar transmitter does not simply emit a burst of energy at a single frequency. Because it is pulsed, it is 100% amplitude-modulated. The modulating pulse contains a ***Fourier series*** of harmonics, and each of those harmonics becomes two sidebands to the carrier, which is the center frequency of the transmitter. The transmitter then emits a wide ***spectrum*** of frequencies, shown in Figure 4-10.

To recover and reproduce the transmitted-burst echo, the receiver must have a sufficient Δf to amplify a useful portion of the spectrum. As illustrated in Figure 4-11, once the Δf has exceeded $1.8/t_p$, no additional signal power is gained.

Since noise power is directly proportional to Δf, an excessive Δf value will increase noise power without improving upon signal power, and the signal-to-noise ratio will decrease. The greatest signal-to-noise ratio occurs where the Δf is $1.2/t_p$, called ***optimum bandwidth*** (Δf_{opt}). For a number of reasons, the receiver design may vary in either direction from Δf_{opt}; the normal (linear amplitude detection, not mti) receiver will usually most closely approach it.

The noise received from the atmosphere is unavoidable, and for purposes of standard calculations, it is assumed to be the noise present at a standard temperature of 290 K (62.6 F, 17°C). If the receiver were perfect, it would introduce no noise, and that ***available noise power*** P_a would be the only noise to compete with the signal. In such a perfect receiver, P_a would then be the only limitation to receiver sensitivity. Of course, the receiver is not perfect, and the minimum discernible signal, $P_{r(min)}$ will be greater than P_a. The P_a at standard temperature is expressed as in Equation (4-5).

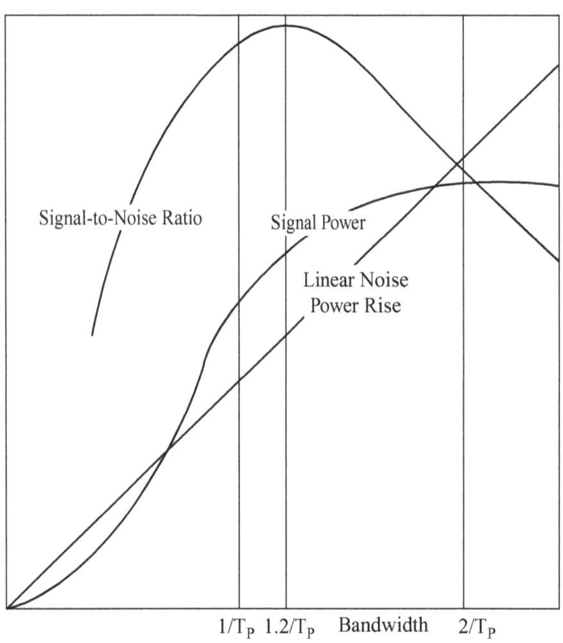

FIGURE 4–11

Noise and signal power versus receiver bandwidth.

Available noise power in a given passband:

$$P_a = kT_k \Delta f, \tag{4-5}$$

where
k = 1.38×10^{-23} J/K
T_k = standard temperature in kelvin = 290 K
Δf = receiver bandwidth in Hz.

Since kT_k may be substituted by fixed values, a single constant may be developed as shown in Equation (4-6):

$$kT_k = (1.38 \times 10^{-23})(290) = 4.002 \times 10^{-21} \text{ W} = -173.98 \text{ dBm}. \tag{4-6}$$

(Note: A watt is 1 J for 1 s, so k is in Watts.)

Receiver bandwidth is a multiplier of noise, and may thus be expressed in dB, as 10 log Δf.

The entire equation, $P_a = kT_k\Delta f$ (in dBm), may therefore be expressed as shown in Equation (4-7):

$$P_a = (4.002 \times 10^{-21}\ \text{W})(\Delta f)$$
$$P_{a_{dBm}} = -174\ \text{dBm} + 10 \log(\Delta f). \quad (4\text{-}7)$$

The minimum discernible signal $P_{r(min)}$ will be greater than P_a because of noise added by the receiver. The signal-to-noise receiver input ratio divided by the signal-to-noise receiver output ratio is called the noise figure, F. Since the input ratio must always be greater than the output ratio, the noise figure must be greater than unity. As a ratio, it may be expressed in dB, and when added to $P_{a(dBm)}$, Equation (4-8) for $P_{r(min)(dBm)}$ results.

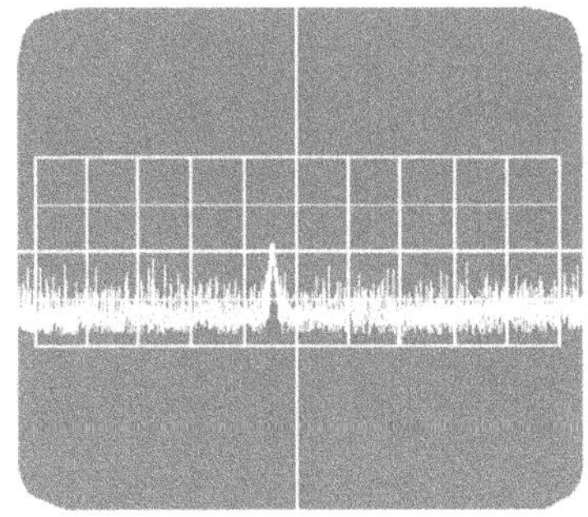

FIGURE 4–12

Test signal just above minimum discernible.

Incorporating the receiver noise figure:

$$P_a = (4.002 \times 10^{-21}\ \text{W})(\Delta f)$$
$$P_{r_{min}} = -174\ \text{dBm} + 10 \log(\Delta f) + F_{dB}. \quad (4\text{-}8)$$

The second part of Equation (4-8) is invalid if the Δf versus t_p combination produces less than an optimum bandwidth of $1.2/t_p$. Below Δf values of $1.2/t_p$, signal harmonic reconstruction will produce a test pulse of less than reliable measurement amplitude. In actual practice, the $P_{r(min)}$ will be measured with a signal generator test burst injected at the directional coupler. Before reducing it to a point where it is barely visible, its appearance on an oscilloscope connected to the receiver output will resemble that illustrated in Figure 4-12. As shown, it is at ***tangential mds***, where the 0.707 points of the test pulse are 3 dB above the 0.707 average noise. Tangential mds is somewhat more realistic, partly because it is the lowest point at which a pulse may be thresholded and separated for use in a digitizer. For tangential mds, use –171 dBm in the $P_{r(min)}$ equation.

Radar Maximum Range

At the maximum range at which a signal may be detected, the echo power is equal to $P_{r(min)}$, and the range may be described as R_{max}. Figure 4-13 is the ***range equation*** derived from preceding equations.

In making calculations of performance expectation from the range equation, A_o, A_e, and R_{max} must be in like quantities (see Figure 4-14). If R_{max} is to be in nautical miles, first calculate the value in feet, and then divide by 6,076.11549. In that case, A_o and A_e must be in square feet. If R_{max} is to be in kilometers, as in many radars other than air traffic control, then the metric system is similarly applied.

$$P_{r_{min}} = \frac{P_t G_t A_o A_e}{16\pi^2 R^4_{max}}$$

$$R^4_{max} = \frac{1}{16\pi^2} \; \frac{P_t G_t A_o A_e}{P_{r_{min}}}$$

$$\sqrt[4]{R^4_{max}} = \sqrt[4]{\frac{1}{16\pi^2}} \; \sqrt[4]{\frac{P_t G_t A_o A_e}{P_{r_{min}}}}$$

$$R_{max} = \sqrt[4]{\frac{1}{16\pi^2}} \; \sqrt[4]{\frac{P_t G_t A_o A_e}{P_{r_{min}}}}$$

$$R_{max} = 0.2821 \sqrt[4]{\frac{P_t G_t A_o A_e}{P_{r_{min}}}}$$

note on solving for value:

$$\sqrt[4]{\frac{1}{16\pi^2}}$$

$$\pi^2 = 9.8696$$
$$16\pi^2 = 157.9136$$
$$\frac{1}{16\pi^2} = 0.0063325768$$
$$\log 0.0063325768 = -2.19842$$
$$\frac{-2.19842}{4} = -0.549604883$$
$$\text{antilog} -0.549604883 = 0.2821$$

FIGURE 4–13

The radar maximum range equation.

Parameter	Typical Value	Standardized	Logarithm
P_t	5 MegaWatts	67 dBW	6.7
G_t	34 dB	34 dB	3.4
A_o	$1m^2$	10.7639 sq ft	1.03196544
A_e	480 sq ft	480 sq ft	2.68124
$P_{r_{min}}$	−105 dBm	−135 dBW	−13.5

$$R_{max} = 0.2821 \sqrt[4]{\frac{P_t G_t A_o A_e}{P_{r_{min}}}}$$

Log 0.2821 = −0.549596914

$$R_{max} = \text{antilog}\left[-0.54956914 + \left(\frac{6.7 + 3.4 + 1.032 + 2.681 - (-13.5)}{4}\right)\right]$$

fourth root

= 1899681.791 feet

= 312.647 nautical miles

FIGURE 4–14

Application and manipulation of the range equation.

Figure 4-14 illustrates calculations with logarithms. Values expressed in dBm must be converted to dBW (subtract 30 dB) and then divided by 10. G_t is normally in dB, and is divided by 10 to express the logarithm. The standard 1-m target A_o becomes 10.7639 square feet, and the log of 10.7639 is 1.03196544. $P_{r(min)}$ is −105 dBm, which converts to −135 dBW, and the logarithm for use in the equation is −13.5. Since this logarithm represents a divisor, it becomes a negative number (−(−13.5)), the sign is reversed, and it is added to the sum of the logarithms of P_t, G_t, A_o, and A_e, originally in the numerator of the antilog equation. Dividing the sum of the logs of P_t, G_t, A_o, A_e, and $-P_{r(min)}$ by four yields the log of the fourth root. The log of the 0.2821 constant is −0.549569 and is added to the log of the entire quantity within the fourth root radical before taking the antilog.

Using the calculations in Figure 4-14 to establish a set of initial values, the quantity under the radical is varied to illustrate the effects on R_{max} from 90% to 10% (see Figure 4-15). Also shown are values of 200% and 6.3%. Increasing the value under the radical by a multiplier of two, as might be achieved by a 3-dB increase in transmitter power, yields only a 19% increase in R_{max}. A reduction of 6.3% to the total value under the radical will halve R_{max}. The log of .063 is −1.2, and the deciBel value is −12 dB. The −12 dB multiplier is used in developing and calculating ***sensitivity time control (stc)*** curves. Stc is a means of "leveling" echo strength to prevent strong targets from saturating the receiver at near ranges. Additional information on stc is presented at more advanced points in this book.

12 dB per Octave Curves

See Figure 4-16, illustrating a curve to represent the decay of expected echoes with range to R_{max}. Such a 12-dB curve, used to attenuate the receiver input, terminates at a recovery point less than R_{max}, somewhat leveling average echo power out to that recovery point. Figure 4-17 illustrates a 12-dB per octave stc desensitization curve. Figures 4-18 and 4-19 illustrate that the electrical stc recovery point can be moved inwared to provide less attenuation at distant ranges. Attenuation greater than 40 dB is counterproductive, and the stc circuits generally contain a 40 dB limiter. Beacon receivers use 6-dB per octave curves because of the one-way ***downlink*** travel time from the aircraft transponder.

Parameter	Typical Value	Standardized	Logarithm
P_t	5 MegaWatts	67 dBW	6.7
G_t	34 dB	34 dB	3.4
A_o	1m^2	10.7639 sq ft	1.03196544
A_e	480 sq ft	480 sq ft	2.68124
$P_{r_{min}}$	−105 dBm	−135 dBW	−13.5

$$R_{max} = 0.2821 \sqrt[4]{\frac{P_t G_t A_o A_e}{P_{r_{min}}}}$$

$$R_{max} = \text{antilog}\left[-0.54956914 + \left(\frac{6.7 + 3.4 + 1.032 + 2.681 - (-13.5)}{4}\right)\right]$$

Log 0.2821

fourth root

Numerator total = 27.313

= 1899681.791 feet
= 312.647 nautical miles

	Value of Change	Logarithm	log of total Value Under Radical	log divided by 4 (fourth root)	+ log (0.282) − 0.549569	antilog = R_{max} in feet	R_{max} in nautical miles	change in R_{max}
	200%	0.302	27.615	6.90375	6.354181	2260377.627	372.0	+18.98%
Reference	100%	0	27.313	6.82825	6.278681	1899682.404	312.65	Reference
	90%	−0.046	27.267	6.81675	6.267181	1850039.495	304.48	−2.6%
	80%	−0.097	27.175	6.79375	6.244181	1754611.615	288.77	−7.6%
	70%	−0.15	27.163	6.79075	6.241181	1742532.954	286.78	−8.27%
	60%	−0.22	27.093	6.77325	6.223681	1673713.041	275.46	−11.9%
	50%	−0.30	27.013	6.75325	6.203681	1598383.545	263.06	−15.86%
	40%	−0.399	26.914	6.7285	6.178931	1509840.254	248.48	−20.52%
	30%	−0.523	26.790	6.6975	6.147931	1405824.151	231.37	−26%
	20%	−0.699	26.614	6.6535	6.103931	1270372.255	209.08	−33%
	10%	−1	26.313	6.57825	6.028681	1068269.92	175.81	−43.76%
	6.3%	−1.2	26.113	6.52825	5.978681	952096.569	156.69	−50.1%

FIGURE 4–15

Variations in R_{max} from a reference.

Other Methods for Ground Clutter Reduction

It is undesirable to attenuate receiver inputs in areas where that is unnecessary, simply because weak echoes can be lost (see Figure 4-20). Conversely, excessive clutter strength can so exceed the strength of aircraft echoes over clutter that the aircraft echo cannot present a useful output of the mti phase detector. Many systems provide a number of options to lessen unnecessarily severe attenuation.

- Azimuth-gated stc: a choice of different curves.
- Range-azimuth receiver gain gates: attenuation in selected areas.
- Passive receiver high beam gate: a receive-only antenna feedhorn with an elevated pattern. Reduces ground-clutter echo strength, may offer no loss or improvement to aircraft returns.

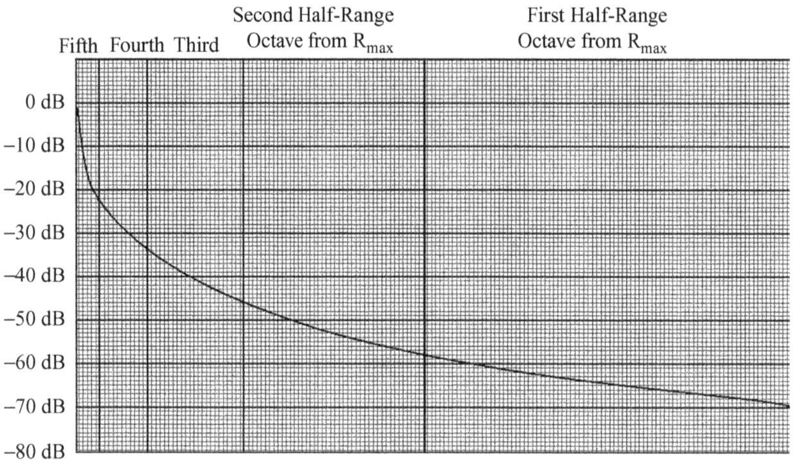

FIGURE 4–16

12-dB per octave echo attenuation curve to R_{max}.

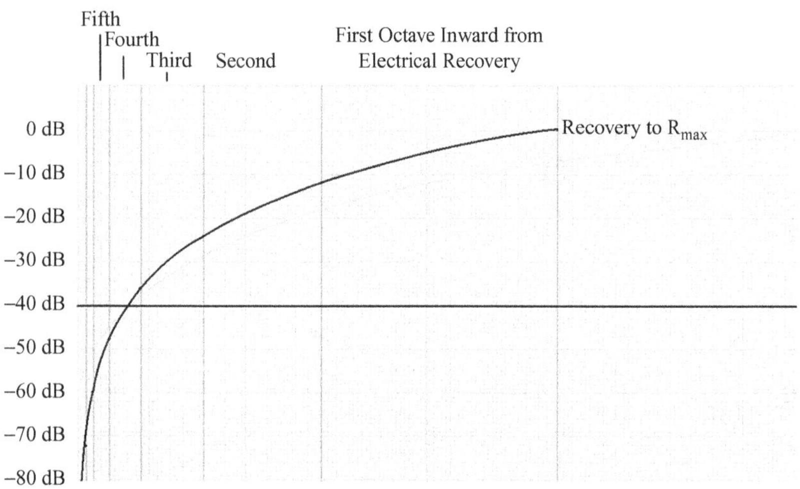

FIGURE 4–17

12-dB per octave stc curve with long recovery point.

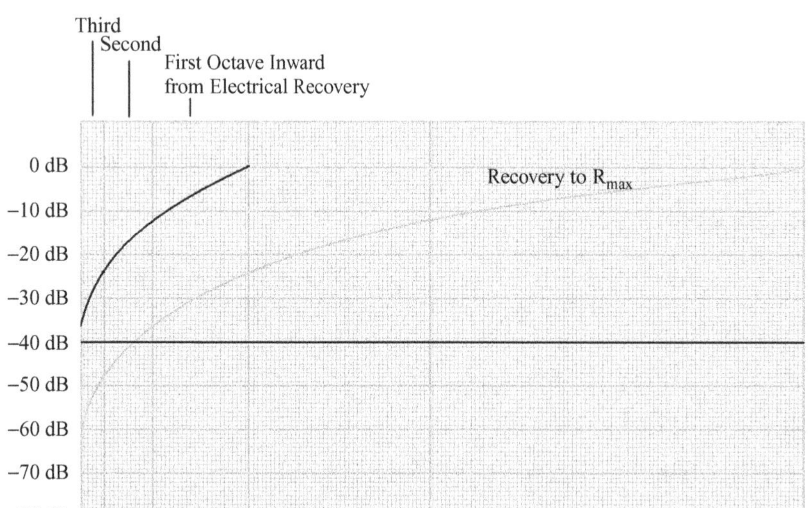

FIGURE 4–18

Steep curve, rapid recovery.

Mitigations to the Range Equation

In actual conditions, performance outcome predictions may be tempered by many variables. Among these are irregularities in the reflecting surface and its angle or angles to the radar antenna. A 1-meter square target is often used as a standard A_o, but aircraft targets may be considerably larger. The antenna radiation pattern may be altered by terrain, altering the G_t value and $P_{r(min)}$. $P_{r(min)}$ may also be significantly altered by atmospheric conditions. Where mti is employed, the echo from a target over clutter may be substantially attenuated by ***subclutter visibility***. Antennas may be polarized horizontally, vertically, or both, and the choice may affect echo strength from targets of different shapes. Latter-day receivers may have a very narrow Δf to reduce noise; the target strength can be "recovered" by a process called ***"coherent integration"*** in a ***moving target detector (mtd)*** system. Westinghouse, manufacturer of the ASR-9, claimed an 8 dB improvement to $P_{r(min)}$ by coherent integration. Special receivers incorporating narrow Δf, severe limiting, and pulse width discrimination have also been used. More information on most of these subjects will appear in following chapters.

CHIRP

The range equation was developed through research at the MIT Radiation Laboratories in World War II, and intended for expectations from pulsed transmitters of a single frequency (see Figure 4-21). In the late 1950s development of the FD program, long-duration, frequency-modulated, high-power transmitter bursts were used to obtain strong echoes; those echoes were then "compressed" to short-duration pulses in frequency-sensitive delay lines in the receiver. Purely for rounding and to avoid any accurate revelations, but to yet convey the principle, envision a 30 μs, 7 MW, transmitter burst compressed to 2 μs. Many latter-day radars employ ***compressed high-resolution pulse (CHIRP)*** transmitter–receiver systems; the FAA ASR-11 and ARSR-4 are in that category. These systems apply frequency modulation to the transmitter burst and ***pulse compression*** to the received echo. Pulse compression circuitry delays portions of the pulse by a variety of values to cause the echo amplitude to be greater, somewhat simulating the consequence of greater peak transmitter power. CHIRP systems are increasing in usage, because they now permit solid-state, low-voltage transmitters to replace the aging high-power systems.

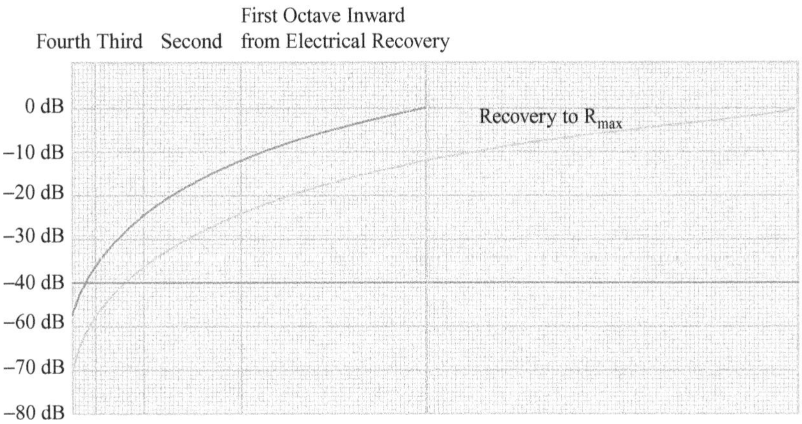

FIGURE 4–19

Electrical Recovery at one-half R_{max}.

Diplexed Transmitters

Although CHIRP systems have been in use since the early 1960s, they were not widely used until the advantage of lower power, solid-state transmitters presented opportunities for increased reliability and lower costs (see Figures 4-21, 4-22, and 4-23). Preceding modern CHIRP techniques in ATC radars was ***diplexing***, a method of routing the bursts and echoes from two transmitters of different frequencies through tuned waveguide filters into a common waveguide to and from the antenna. An initial assumption might resemble a doubling of peak transmitter power and maximum range, but the very best any R_{max} improvement could be is 19%. Inspection of the range equation will reveal that increasing peak power by a factor of 2 yields an increase of 1.19, the fourth root of 2.

Well beyond any increase in R_{max}, a major objective in diplexing two transmitters was to overcome the effects of lobes and nulls in the radiation pattern, caused by the effects of terrain or anomalous propagation (see Figures 4-22 and 4-23). When anomalous propagation, an atmospheric condition of thermal layers, causes lobing, it

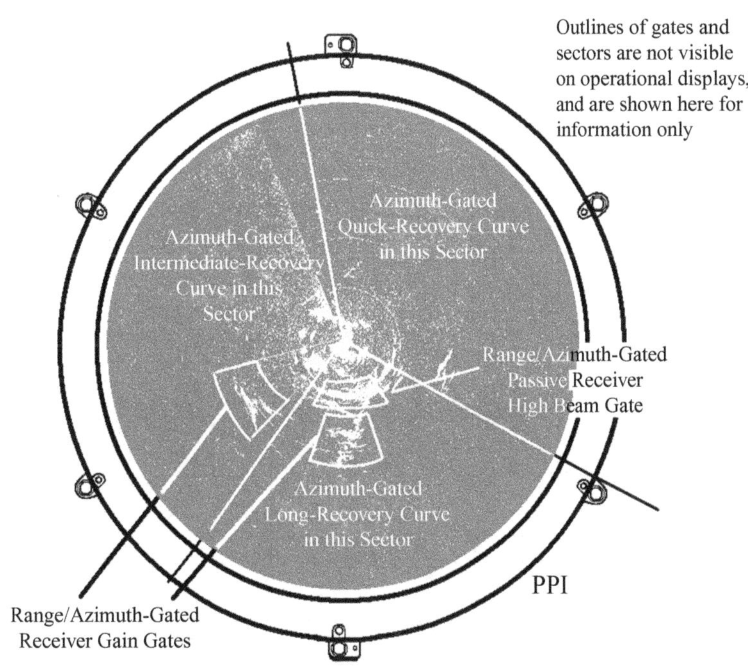

FIGURE 4–20

Techniques for reducing clutter returns.

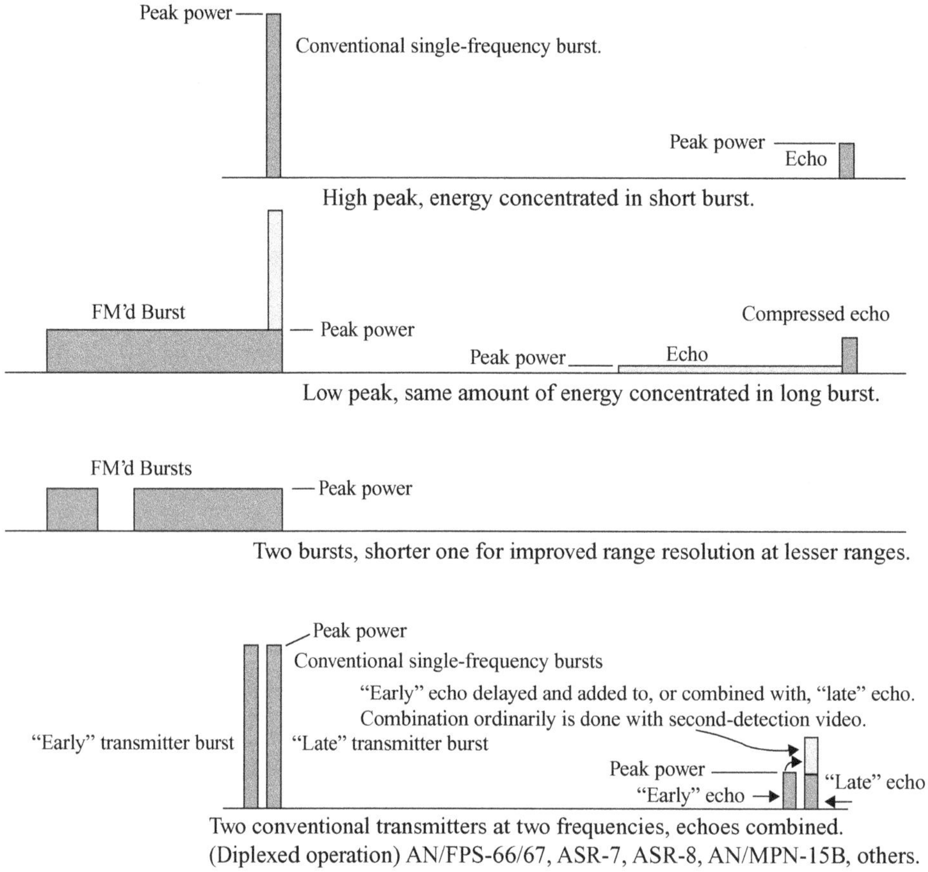

FIGURE 4–21

Pulse compression and diplex principles.

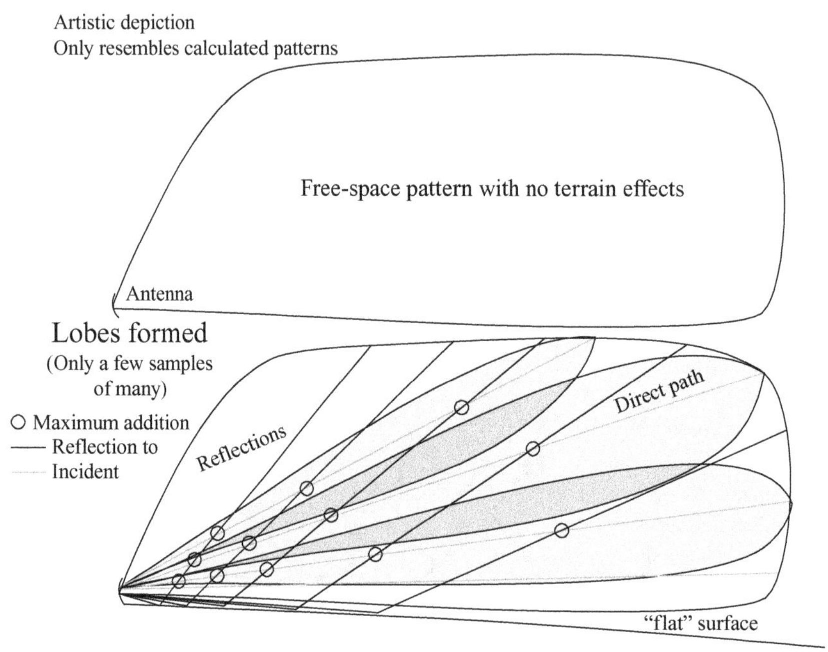

FIGURE 4–22

Lobes and nulls formed by terrain.

is called ***ducting.*** Reflections of radiated energy from the ground will recombine with the incident pattern, and the phase at points of recombination may attenuate or intensify the incident radiation to create new lobes and nulls. This condition is most severe over smooth, flat terrain, or large expanses of thermal differences. Of course, the phase combinations are frequency dependant, so the use of diplexed transmitters may help the lobes of one to fill the nulls of another. Since lobing may exist, the combined echo signal strength may well seldom be sufficient for the 19% increase in maximum range.

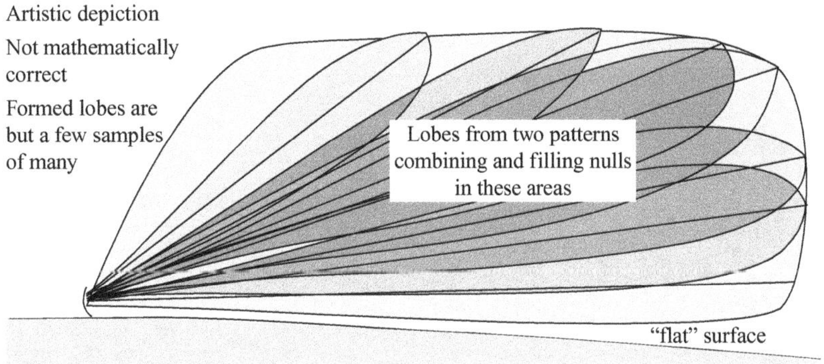

FIGURE 4-23

Combining radiation patterns from two diplexed transmitters.

R_{max} and stc in Secondary Radars

Secondary radars, also called "beacon, IFF, ATCRBS, and others," are "answer-back" systems, and are affected by the one-way attenuation of the transmitted signals in either the radar-to-aircraft or aircraft-to-radar paths (also called, "uplink," and "downlink," respectively). The maximum range is a square-root function instead of the fourth-root for primary radar, and a corresponding stc curve would be at 6 dB per octave. Further, the radar maximum range equation is not applicable, because A_o is not a factor, because G_t or A_e will be different values for the two links, and because different values of $P_{r(min)}$ and P_t will apply to the two links. As a precipitate, the stc recovery points are set to a greater range. R_{max} is a lesser consideration in beacon systems in comparison to primary radars, because the two unidirectional paths provide ample signal strength, and the downlink half of the system is the aircraft transponder with its own variables. In fact, excessive power in both uplink interrogators at the radar transmitter site and downlink transponders in the aircraft has been a major source of multiple system problems since the earliest days, and transmitter power is now substantially less than it was in the earlier decades of the system. Of course, periodic measurements of transmitter power, receiver sensitivity measurements, and other parameters remain a legitimate requirement on the owners of equipment on both ends of the system, to insure that the components are operational and within established standards and tolerances.

R_{max} Inversely Proportional to Transmitter Frequency

Note that, in comparing an L-band and S-band antenna of the same gain, the larger effective aperture A_e of the lower frequency L-band antenna will yield a somewhat greater maximum range, providing yet one more reason to use lower frequencies for greater range radars. Even further, consider the wider pulse and narrower spectrum of the long-range system in development of the transmitted radiation pattern. Still further, the narrowed spectrum offers lower values of Δf_{opt} to reduce receiver noise.

More Forms of CHIRP

Early in the 1960s, CHIRP was also employed in phased-array antennas as a means to provide altitude determination in defense radars, commonly called 3D radar (see Figure 4-24). By transmitting a series of nine discrete frequencies in one burst, nine "pencil beams" at different elevation angles provided real time altitude data. These systems necessitated such complicated synchronization that computers became essential. As the beam elevation changed, the frequency change also altered the horizontal angle, causing an azimuth "skew," and necessitating an azimuth correction. As the ship pitched or rolled, the altitude computations required automated correction. It is of interest that the computers used by the Navy for 3D radar evolved into a different application for air traffic control, eventually becoming the airport radar terminal (or tracking) system ARTSIIIA.

Early CHIRP systems such as the Sperry AN/FPS-35 (late 1950s) were designed primarily for the benefits of pulse compression, and for ECCM purposes. The transmitted pulse was frequency modulated either upward

FIGURE 4–24

Discrete frequencies and phased-array "beam fan."

or downward by a linear sawtooth (now called "ramp") and the feedhorn-reflector style antenna was similar to many others. The up-or-down transmitted CHIRP was made downward by electronic selection of an "above" or "below" second local oscillator frequency. This type of system was called ***linear CHIRP*** because of the modulation ramp. Another type of CHIRP was by ***Barker code***, a series of phase changes to a single transmitted frequency. The Barker-coded modulation provided a type of source identity to echoes, allowing pulse compression while rejecting interference.

Only a few years after the FPS-35 production, stepped-frequency-pulse CHIRP and phased-array antennas were developed to provide a means of altitude detection in rotating surveillance antennas. Echoes from different beams could be separated by virtue of their frequencies, and then the received signal strengths were compared to obtain an altitude determination. Perhaps the best known of the systems is the venerable US Navy's AN/SPS-48, first designed and manufactured in the early 1960s by ITT Gilfillan, the WWII pioneer in phased arrays. Today, the FAA utilizes CHIRP for altitude detection in the ARSR-4.

The need for frequency-sensitive phased arrays in the FAA is primarily in its joint-use cooperation with the military. Most ARSR-4 systems are located near the US coastlines, where intruders without transponders must be detected and challenged. Altitude determination by ATCRBS means is considerably more accurate for air traffic control purposes, but 3D radar is essential for defense and acquisition.

Review Questions

1. The radar synchronizer produces __________ and _________ for use throughout the system.
2. A simple, amplitude-detection receiver is called a __________ receiver.
3. Most radar receivers are of the _______________ type.
4. A radar receiver system employing more than one intermediate frequency is called ___________ ___.
5. A single trace across a radar ppi is called a ___________.
6. A 360° rotation of the radar trace is called a ___________.
7. Of the radar system interval, that portion of time used to receive and process echoes is called _____________.
8. A radar echo is produced by a target at ranges listed in the following table. Calculate the time from time zero at which the target would appear on a display.
   375 nmi___________________
   200 US statute miles___________________
   93 inches___________________
   400 feet___________________
   15 cm___________________
   90 km___________________
9. The outgoing transmitted radar energy is attenuated by a maximum amount of $4\pi R^2$. What is the origin of this expression?
10. A radar system can transmit a peak power of 2 MW, has a 1 μs pulse width, an antenna gain of 30 dB, an effective aperture of 400 square feet, and a receiver noise figure of 3.5 dB. What is the maximum range of a 1 m^2 target?
11. A radar receiver has a bandwidth of 4 MHz, and a noise figure of 4 dB. What is the minimum discernible signal?
12. What are some reasons for:
    CHIRP
    Diplexing
    Frequency-modulated transmitter burst.
13. What is the connection between CHIRP and phased arrays?
14. Why are lower frequencies used in greater range radars?

Answers to Review Questions

1. The radar synchronizer produces ***triggers and gates*** for use throughout the system.
2. A simple, amplitude-detection receiver is called a ***normal*** receiver.
3. Most radar receivers are of the ***superheterodyne*** type.
4. A radar receiver system employing more than one intermediate frequency is called ***multiple conversion***.
5. A single trace across a radar ppi is called a ***sweep***.
6. A 360° rotation of the radar trace is called a ***scan***.
7. Of the radar system interval, that portion of time used to receive and process echoes is called ***live time***.
8. A radar echo is produced by a target at ranges listed in the following table. Calculate the time from time zero at which the target would appear on a display.

   375 nmi $\boldsymbol{375 \times 12.3552\ \mu s = 4.633\ ms}$
   200 US statute miles $\boldsymbol{200 \times 10.736\ \mu s = 2.147\ ms}$
   93 inches $\boldsymbol{93 \times 169.45\ ps = 15.732\ ns}$
   400 feet $\boldsymbol{400 \times 2.03\ ns = 812\ ns}$
   15 cm $\boldsymbol{15 \times 66.7\ ps = 1\ ns}$
   90 km $\boldsymbol{(90)(1000)(6.67\ ns) = 600\ \mu s}$

9. The outgoing transmitted radar energy is attenuated by a maximum amount of $4\pi R^2$. What is the origin of this expression?

   The expression describes the surface area of a sphere. The reference in computing power concentration is the attenuation that would have occurred had the radiation been in a perfectly expanding sphere from a single-point isotropic source.
10. A radar system can transmit a peak power of 2 MW, has a 1 μs pulse width, an antenna gain of 30 dB, an effective aperture of 400 square feet, a receiver noise figure of 3.5 dB, and a receiver bandwidth at optimum. What is the maximum range of a 1 m^2 target?

    $$\begin{aligned}
    &\textit{1 foot} && = 0.3048 \text{ meters}\\
    &\textit{1 square foot} && = .09290304 \text{ square meters}\\
    &\textit{400 square feet} && = 37.16 \text{ square meters}\\
    &P_{r(min)\,(dBm)} && = -174\text{ dBm} + 10\log(\Delta f) + F_{dB}\\
    & && = -174\text{ dBm} + 10\log(1.2/1 \times 10_{-6}) + 3.5\\
    & && = -174\text{ dBm} + 60.8\text{ dB} + 3.5\text{ dB} =\\
    & && = -109.7\text{ dBm}\\
    & && = -139.7\text{ dBW}\\
    & && = 1.07 \times 10_{-14}\text{ W}\\
    & && = 457.8\text{ km}
    \end{aligned}$$

    There are 3.28 feet per meter, so 457.8 km = 1,501,968.5 feet
    There are 6,076.11549 feet per nmi, so 1,501,968.5 feet = 247.2 nmi

11. A radar receiver has a bandwidth of 4 MHz, and a noise figure of 4 dB. What is the minimum discernible signal?

    Mds = –174 dBm + 10 log 4 MHz + 4 = –174 dBm + 66 dB + 4 dB = –104 dBm
12. What are some reasons for:

    CHIRP

    The wide transmitter burst and narrow received pulse permits the use of low peak-power solid-state transmitters.

Diplexing

Transmitting on two well-separated frequencies can reduce terrain and atmospheric ducting effects.

Frequency-modulated transmitter burst.

If purely linear, the objective is to compress the receiver pulse to a higher amplitude. If coded in specific steps, it may be for altitude detection.

13. What is the connection between CHIRP and phased arrays?

    For a phased array antenna, the coded transmitter burst can cause a fan of beams for altitude determination.

14. Why are lower frequencies used in greater range radars? ***The necessarily higher power requires larger transmitter components, and lower frequencies offer that. The larger antenna reflector provides a greater A_e. The wider pulse and narrower spectrum offer a lessened Δf_{opt}.***

CHAPTER 5

Magnetron versus Synthesis

Radar has become an enormous field, and it would not be practical to explore all types of systems in great detail. Such a project would require many authors, and is not the intent of this book. Should you wish to find more about a specific type of system not adequately addressed in this book, a good beginning reference is Merrill Skolnik's ***Radar Handbook***. In addition to providing general information, it may also provide you with further reference information to direct you to other sources.

This book is directed toward basic practical applications of technical information for the technician. Even so, this requires that the technician have some fundamental understandings of system types for performance analysis, measurements, and corrective actions. Since the information is oriented to the novice technician, it will mostly use, as examples, simpler single-frequency (as opposed to pulse-compression) transmitters and fixed-beam (as opposed to agile-beam, phased-array, or frequency-scanned) antenna systems.

Early in World War II, two British physicists, Sir John Randall (1905–1984), and Henry Albert Howard (Harry) Boot (1917–1983), invented the cavity magnetron, the only device at that time capable of providing microwave energy at power levels sufficient for radar echoing at any significant distance. The magnetron is still in use today and has advantages over other transmitters in low production costs, ease of tuning, and ease of replacement. It is doubtful that it will cease to be a part of radar systems for many years to come. Coincidentally, it has also become the main component of microwave ovens.

Although the easiest to repair, and most economical, the magnetron is inferior to others in several aspects. The magnetron has a lesser power output, and other types of latter-day transmitting tubes can produce considerably greater power. In earlier long-range magnetron systems, a microwave ***crossed-field amplifier (CFA)***, also called an ***amplitron***, tube was used to boost the magnetron power. The magnetron also may produce poor or unbalanced frequency spectrums, and there is usually little that can be done to correct this, other than replacing the tube. Since it is a high-power oscillator, its output is independent of other frequencies generated in the radar set, and some method of keeping the local oscillator tuned to provide the appropriate difference intermediate frequency is necessary. In mti systems, ***phase coherence*** is necessary to intelligibly link the phases of the transmitter burst, rf echoes, and i-f echoes; the magnetron transmitter requires special circuitry to achieve that.

In contrast to the magnetron system is a ***synthesis*** class of systems, which actually includes nearly all other pulsed radars. Synthesis systems were produced after practical high-power microwave final amplifiers had been developed. In a synthesis system, the transmitter tube is not an oscillator, but amplifies an input ***drive*** burst. The usual output tube for these systems is a high-power ***klystron electron tube***, but other devices have also been used. In a single-conversion synthesis system, the local oscillator output is combined with a second oscillator output to ***synthesize*** a third frequency for use in the transmitter. The transmitter drive chain often contains doublers or triplers to arrive at the microwave drive to the final power amplifier. Another oscillator frequency is equal to the i-f, and because of its original application in mti systems, is often called the ***coherent oscillator (coho)***. A synthesis system may also incorporate more than one stage of conversion, and may utilize several different intermediate frequencies. The basic single-conversion synthesis system is simplest and most common, and it will be given the most attention in this book.

Synthesis System Block Diagram

The synthesis of the transmitter frequency begins in a central rf oscillator, most often called the ***stalo*** (see Figure 5-1). "Stalo" is an old abbreviation for ***stabilized local oscillator***. When mti systems were first built, great effort was necessarily devoted to achieving greater stability to the microwave local oscillator for good ***coherence,*** a continuing intelligible relationship between the transmitted burst, echoes, and local oscillator, in a magnetron

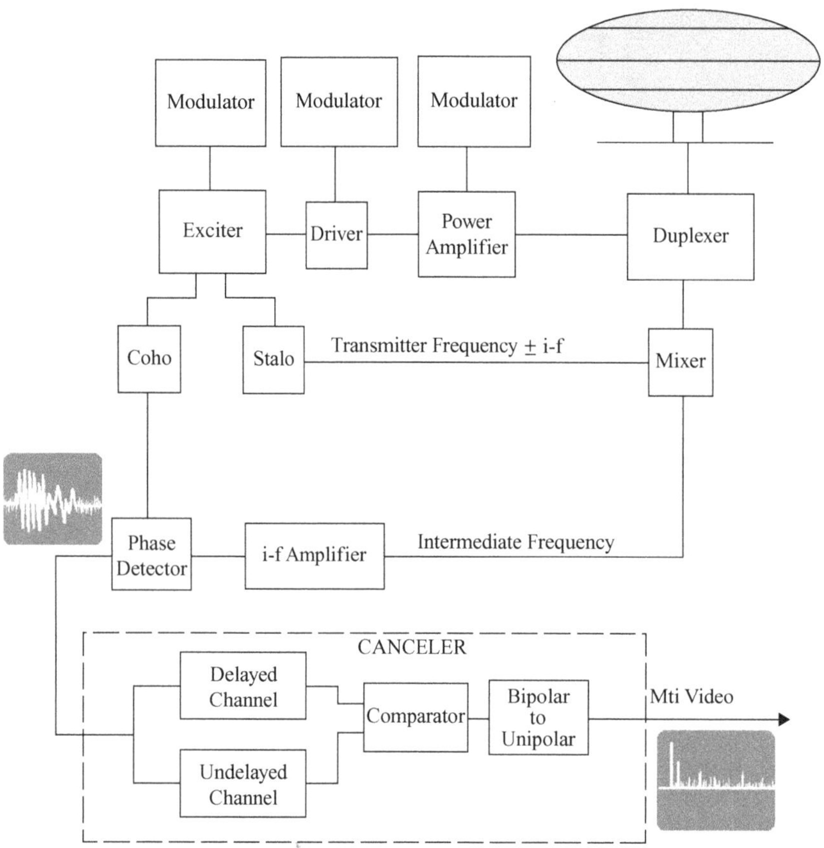

FIGURE 5–1

Simplified mti canceler-type synthesis system.

system. In synthesis systems, coherence is natural and inherent; however, use of the magnetron-system terms, "coho" and "stalo," have been continued even though a magnetron phase-locking technique is unnecessary.

Figure 5-1 is a single-conversion, single-frequency, synthesis mti system. In the ***exciter*** or ***buffer-mixer*** (nomenclature is manufacturer dependent), the stalo rf will be combined with the ***coho*** signal to produce either a sum or difference transmitter frequency. In older systems, the coho frequency was generally at 30 MHz. In newer systems, it will often be at 31.07 MHz, as system timing is often obtained by frequency division of the coho, and 31.07 MHz may be divided by 24 to produce 1.2946-MHz clocks. The period of 1.2946-MHz clocks is 772.45 ns, which is 1/16 nmi. In digital processors, binary division of the radar mile results in power-of-two fractions, such as 1/2, 1/4, 1/8, 1/16, 1/32, etc; so a 1/16 nmi division is compatible with such processors.

Stalo–Coho–Transmitter Phase Relationships

Both the stalo and coho are uninterrupted, continuous-wave oscillators, and therefore bear a predictable phase relationship between each other at any given time. For instance, consider a stalo operating at 2,725 MHz, and a coho operating at 31.07 MHz. In 1,123 μs, the stalo, coho, and transmitted frequency phases would predictably change by a degree calculated as follows:

(2,725.5 MHz)(1,123 × 10^{-6})= 3,060,736.5 cycles. A half cycle is 180°, so the stalo phase will have changed by 180° after 1,123 μs. For instance, if it were at 103.45° at one point, it would be at 283.45° after 1,123 μs.

(31.07 MHz)(1,123 × 10^{-6}) = 34,891.61 cycles
(0.61 cycle)(360° per cycle) = 219.6°
The coho phase will be 219.6° different after 1,123 μs. If it were at 10° at the beginning of the 1,123-μs period, it would be at 229.6° at the end.

If the stalo and coho were added to achieve the transmitter frequency, the resultant frequency would be 2,725.5 MHz + 31.07 MHz = 2,756.57 MHz. After 1,123 μs, the phase of the transmitter frequency would have changed by (2,756.57 MHz)(1,123 × 10^{-6}) = 3,095,066.11 cycles, or 39.6°. Recall that the stalo changed by 180°, and the coho changed by 219.6°; the sum of these changes is 399.6°, or 39.6°. Using the previous examples, if the sum of the two phases happened to be 103.45° + 10° = 113.45° at the beginning of the 1,123-μs period, it would be 283.45° + 229.6° = 513.05° = 153.05° at the end. The starting combined phase of 113.45° + the change of 39.6° = 153.05°.

The predictable amount of phase change of the coho, stalo, and the transmitter output over a given period of time, and the stability and continuous operation of the two oscillators make the synthesis system ***inherently coherent***. This coherence will become the basis for the detection of Doppler shift and the cancellation of ground clutter.

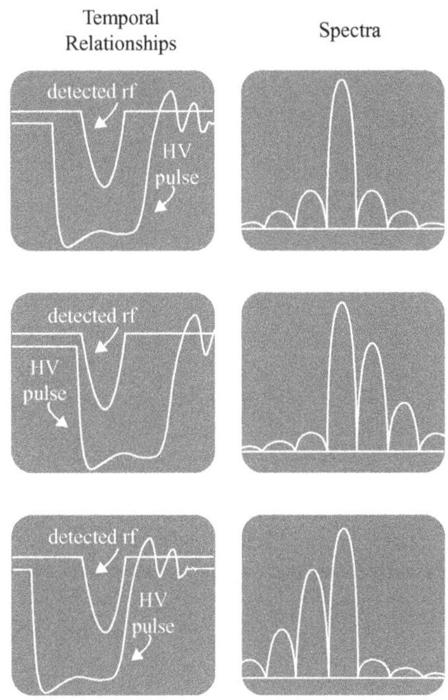

FIGURE 5–2

High voltage versus drive burst.

The Transmitter Burst

In a synthesis system, the transmitter final power amplifier is energized by a high-voltage pulse somewhat wider than the drive burst (see Figure 5-2). The high-voltage pulse timing is such that it envelops the input drive burst, causing it to be amplified. Timing of the high-voltage pulse in relationship to the drive burst must be adjustable, as the high-voltage pulse is subject to variable component delays. In some radars, a manual adjustment is provided; in others, the adjustment may be automatically regulated. As a general rule, the high-voltage pulse, not the drive burst, is adjusted, because temporal changes to the drive burst could alter all the relationships to radar time zero. The most accurate way to finally adjust the timing relationship is to monitor the transmitter frequency spectrum for balanced side lobes.

The shape of the drive burst and tuning of the final power amplifier tube determine the frequency spectrum of the power output. The ***rf driver*** stage permits adjustment of the drive-burst shaping and spectrum. Excessively abrupt rise or fall times cause unnecessarily broad transmitter frequency spectrums, and the drive burst is carefully shaped to preclude this.

Phase Detection

In an mti system, the output of the signal mixer is applied to a low-noise preamplifier, or directly to an i-f amplifier containing a preamplifier. To avoid cable losses of the lowest level targets, the preamplifier will be physically located as closely as practical to the receiver signal mixer. Before detection, the amplified output is severely limited, and then applied to a ***phase detector,*** which is the second detector in this type of receiver. The phase detector compares the i-f echoes to the coho signal. The phase detector exhibits a cosinusoidal response; the output is a positive or negative voltage to represent the cosine of the difference between the two input signals. For instance, if a limited i-f echo signal were compared to a coho signal at a 45° difference, and the phase detector maximum possible outputs were +2 V and –2 V, the phase detector output would be a +1.414-V pulse, as illustrated in Figure 5-3.

Since the bipolar data amplitude and polarity is dependent upon the phase relationship between the i-f echo and coho, it should be clear that the phase of the i-f echo is of great importance. That *i-f echo phase is determined by the phase of the stalo at the time the rf echo arrives at the mixer, and by the phase of the rf echo, which in turn is determined by the phase of the transmitted burst.* All these interrelationships among the stalo, coho, transmitter, rf echo, i-f echo, and bipolar data are further aspects of ***coherence.***

The mti Canceler

Coherence causes the phase detector output for the echoes from any given target to represent the ***Doppler shift*** of the target, as illustrated in Figure 5-4.

If a target is in motion, the successive echoes from one interval to the next will cause varying phase comparisons, and the bipolar data for that specific target will vary at the rate of the Doppler shift. If the target is not in motion, there will be no Doppler shift, and the phase-detector data for the target will remain at a constant amplitude. The canceler contains a delayed and undelayed data channel, with the outputs of both applied to a comparator. The delay in the delayed channel is precisely the same as the transmitter pulse interval, and the comparator inputs are always the undelayed data and the delayed data from the same range and previous interval.

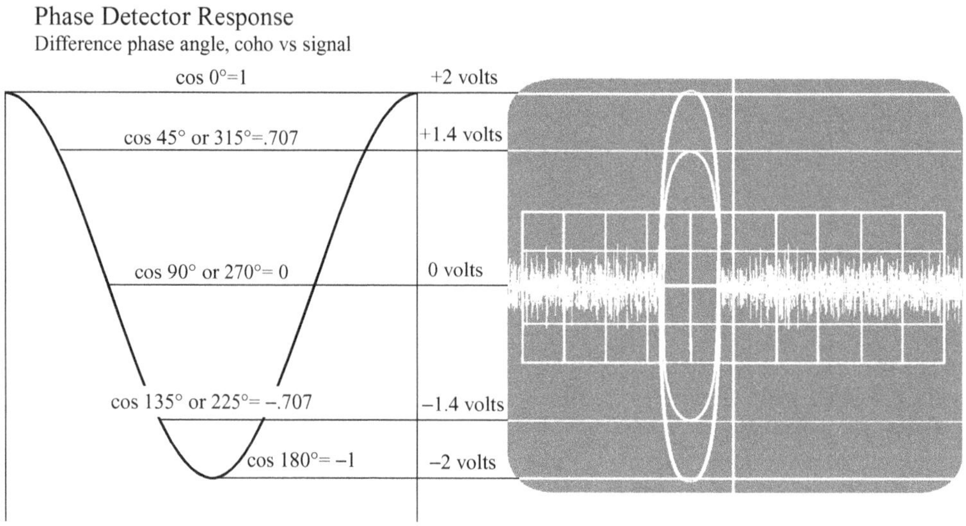

FIGURE 5–3

Phase detector response.

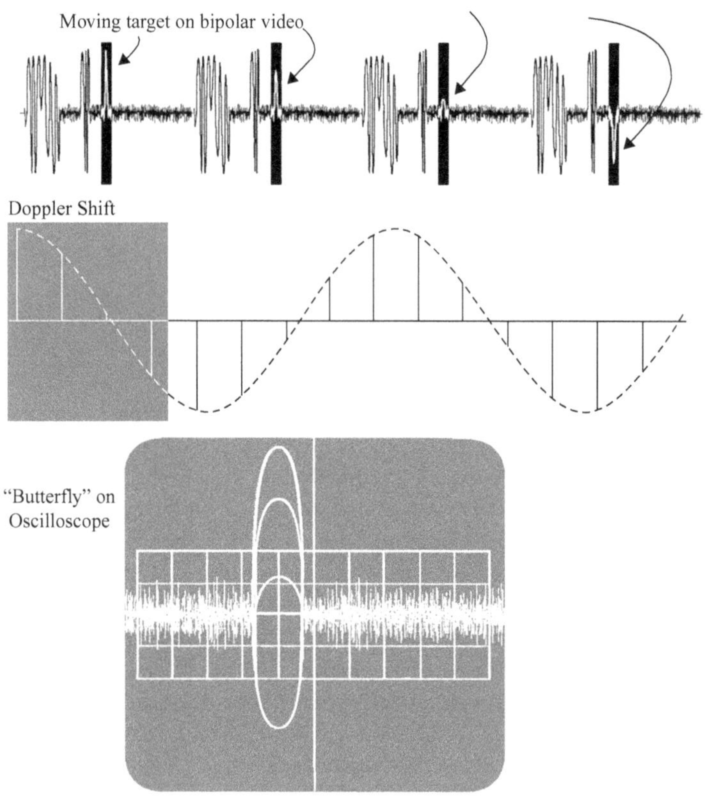

FIGURE 5–4

Doppler shift.

Fixed targets will "cancel" because the delayed and undelayed are identical, and moving targets will provide a difference output because the data is changing from one pulse to the next. After the comparison, the remainder bipolar data is converted to unipolar by inverting either the negative-going or positive-going portions (but not both) of the data. A unipolar video may be displayed on a radar indicator.

The Magnetron System

The magnetron was the first microwave transmitter tube, and produces oscillations within itself by a ring of microwave cavities in a grounded anode, circled around a center cathode (see Figure 5-5). A high-voltage negative pulse, for example, –22,000 V, causes current flow to excite the cavities, which ring at a frequency determined by the cavity volume. The output is taken by a coupling loop in one of the cavities.

Coherence

Because there is no natural relationship between the coho, stalo, and magnetron, there is no inherent coherence in such a system (see Figure 5-6). In simple systems containing only normal amplitude-detection receivers, this does not pose a problem, and a very economical radar design is possible. In mti systems, the coherence is essential, and additional circuitry must be provided to achieve it. This is achieved by mixing a sample of the transmitter burst with the stalo to produce an intermediate-frequency burst, and then using that burst to shift the coho phase. The i-f transmitter burst representation is called the ***coho lock pulse.*** A waveguide junction in the transmit-energy

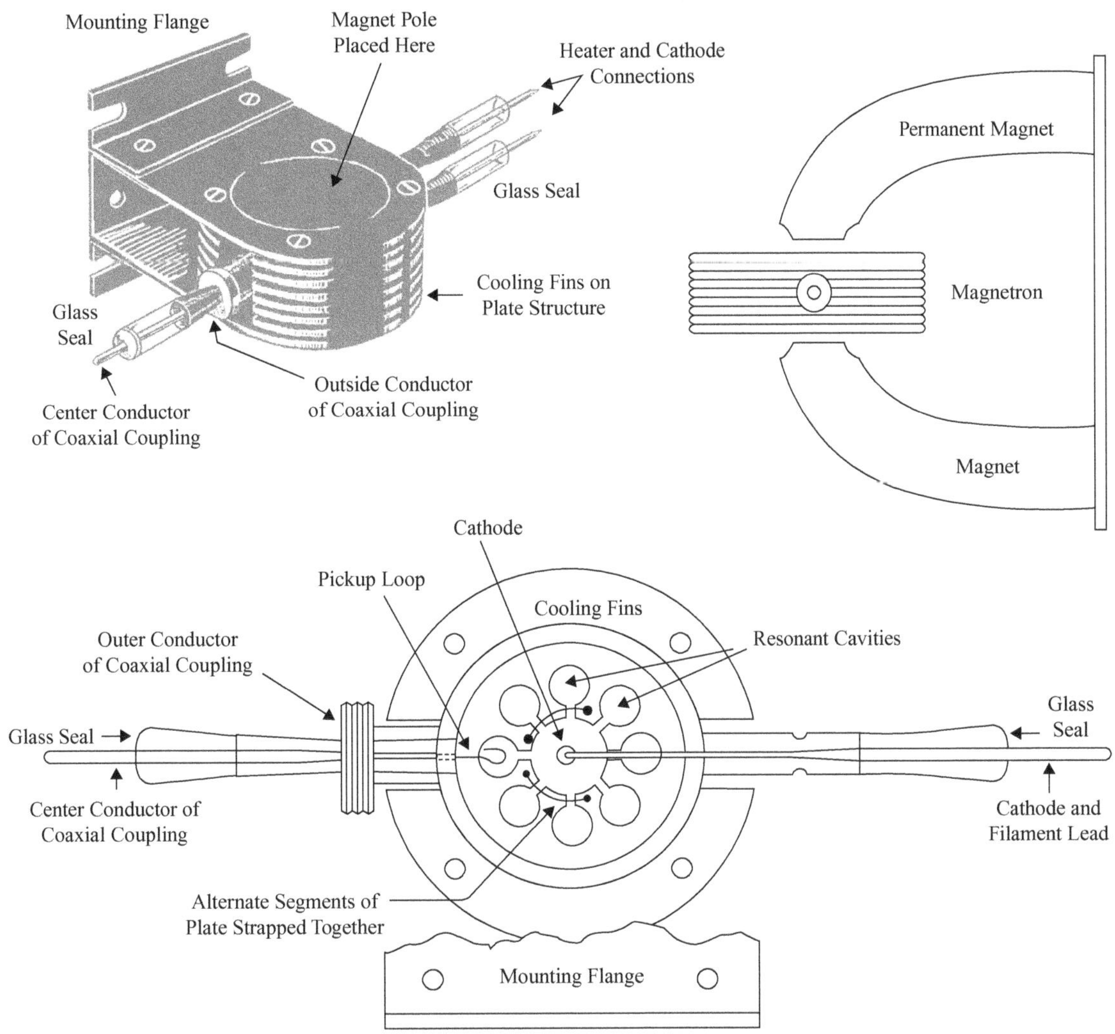

FIGURE 5–5

The multicavity magnetron. Based on USAF Manual 52-8.

path provides a route, through an attenuation of perhaps 100 dB or greater, to a ***coho mixer,*** which is similar, if not identical, to the signal mixer in the receiver path. Received echo signals are too severely attenuated to produce any appreciable output from the coho mixer.

Automatic Frequency Control (afc)

Since the magnetron oscillates at a frequency unrelated to the stalo output, there must be a provision to tune one of the two oscillators to provide the difference intermediate frequency to which the i-f amplifier is tuned, usually 30 MHz. The original mti systems were contained in ground-control-approach trailers positioned next to airport runways; the technician was also the air traffic controller, and had immediate access to the entire radar set. The technician/controller tuned the stalo as necessary.

As systems became more complex, air traffic control operations no longer were accomplished from the radar trailers and were located at remote ***instrument flight rule (IFR)*** rooms. Still further, the technician and controller occupations were separated, and controllers were no longer competent to perform maintenance or make critical adjustments to the system. A new system design then necessitated afc circuitry to automatically maintain the correct i-f.

In an mti radar system with afc, the coho lock pulse is used as a sample representation of the tuning; when it differs from the i-f frequency by a predetermined amount, the tuned oscillator, either the magnetron or stalo, is automatically and mechanically adjusted by a drive motor. It is most desirable to tune the stalo, rather than the magnetron, since transmitting frequencies are allocated by government agencies. Systems originally designed

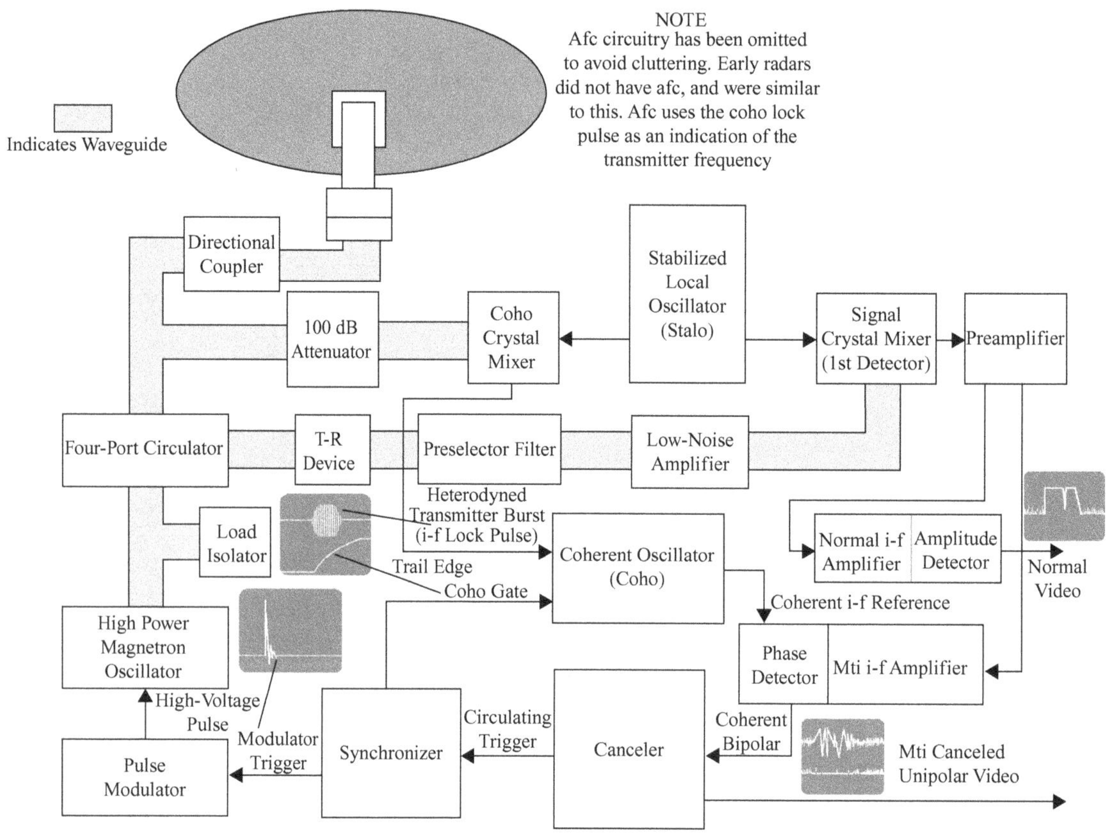

FIGURE 5–6

Magnetron System Block Diagram.

with afc will generally contain a stalo-tuned afc system; systems modified to add afc, or modified from synthesis to magnetron, may contain a magnetron-tuned afc system, as this could be easily implemented without stalo redesign.

There are different types of afc circuitry, but the most common is a "swept-bandpass" or "reactance-tube modulator" circuit. The center frequency of a tuned circuit is varied about the i-f center via a 60-Hz sine wave. The input to the tuned circuit is the coho lock pulse. If the coho lock pulse is at the i-f center, the output of the tuned circuit is a 120-Hz envelope. If the lock pulse is above or below, a 60-Hz component will appear at the output, the phase indicating the direction of error. The 60-Hz error becomes a drive for the stalo or magnetron tuning motor. Because tuning interrupts coherence, the error voltage does not continually drive the stalo or magnetron, and a threshold circuit requires that a minimum, often 100 kHz, error occur before tuning is effected. Because magnetron systems have been designed with a dual purpose of coherence and afc, many later systems will now refer to the i-f transmitter burst representation as the ***afc lock pulse, afc/coho lock pulse, or coho lock pulse.***

Radar System Frequencies, Applications, and Ranges

There are many considerations to be made in the choice of a frequency band in which to operate a radar set. In general, lower frequency bands are used for greater ranges. The following sections will detail all the interrelated reasoning behind this choice.

Component Size versus the Applied Voltage

Because waveguide and cavity sizes are determined by wavelength, component sizes are larger for lower frequencies. This permits the use of the higher voltages required for the higher power transmitters. Conversely, the small components used for the very high frequencies would be destroyed by the voltages required for high-power, long-range systems.

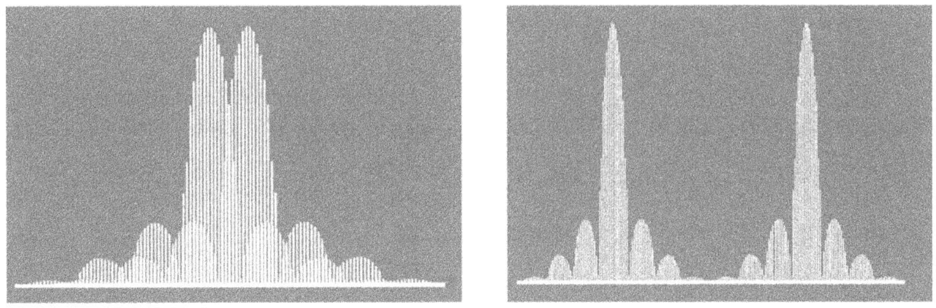

FIGURE 5–7

Interfering and Non-interfering spectra.

Receiver Bandwidth and Range Resolution

The lowest and best minimum discernible signal occurs when the receiver bandwidth is $1.2/t_p$, called ***optimum bandwidth.*** The pulse width determines the minimum range separation, called ***range resolution,*** at which two targets may be distinguished. The range resolution is the range in terms of the ***time-for-echo-for-distance*** T_{ed} value. For instance, if the transmitted pulse width is 1 μs, the range resolution is 1 μs divided by 2.033 ns per foot, which is 491.9 feet, slightly greater than 5/64 nmi.

Longer ranges, because of radar application, require less range resolution, and the radar pulse width may be wider, decreasing the value of $1.2/t_p$. This permits a more economical, narrower receiver bandpass, and also presents wide, bright targets when an analog display is set to maximum range.

Doppler Shift

The Doppler shift f_D to an echo is dependent upon the resultant radial target velocity V_r directly toward or away from the radar antenna, the transmitter frequency f_{xmtr}, and the velocity of light c. The Doppler shift bears the same relationship to the transmitter frequency as the radial velocity bears to the velocity of light, as

$$\frac{f_D}{f_{xmtr}} = \frac{V_r}{\frac{c}{2}} \tag{5-1}$$

where

f_D = Doppler shift

f_{xmtr} = transmitter frequency in hertz

$c/2$ = 1/2 velocity of light, 291.375 × 106 nmi/h.

V_t = radial velocity in nmi/h.

Spectrum and Interference

As more and more government, military, and commercial uses of radar develop, the frequency spectrums, particularly around large cities, become more crowded (see Figure 5-7). The frequency span, called spectral width, of the main spectral lobe is determined by the distance between the null on both its sides, or $2/t_p$. As range and power are lessened, and the pulse width is made narrower, the spectral width increases, but is less likely to interfere with another radar because of the lower power. Further, there are fewer users in the higher bands and interference is less likely for that reason, as well.

Radar Type Designations

Air Traffic Control and other radar types generally fall within two naming methods, FAA and military, as follows.

FAA:

- ARSR: Air route surveillance radar, 200-mile range
- ASDE: Airport surface detection equipment, 3-mile range
- ASR: Airport surveillance radar, 60-mile range
- ATCBI: Air traffic control beacon interrogator

- PAR: Precision approach radar, 20-mile range
- TDWR: Terminal Doppler weather radar

Military: Begins with A/N (Army/Navy) (those radars designated "SCR" preceded the nomenclature system).

Designation is three characters plus system number assignment, as follows.

The first character is the radar siting type:

- A: Airborne
- C: Cargo transportable
- F: Fixed
- G: Ground stationary
- M: Mobile
- S: Shipboard

The second character, "P," indicates "pulsed radar."

The third character shows the purpose

- N: Navigational
- Q: Fire control, mortar locating, bomb-sighting, etc
- S: Surveillance
- Y: Special purpose

New Band Designation

Although Table 5-1 is still in use, some entities are now using different methods, in which the alphabetical characters are arranged in alphabetical order with ascending frequency.

TABLE 5–1 IEEE Frequency Bands and Applications

Designator	Frequency	Usage
Radar operating frequencies		
HF	22–28 MHz	Great Britain's Chain Home
VHF	30–300 MHz	Formerly used for AN/FPS-24 ARSR. Early US Army SCR-268, SCR-270, USN CXAM, others
UHF	300–1,000 MHz	Formerly used for AN/FPS-35 ARSR
L	1,000–2,000 MHz	Beacon Interrogators 1,030 MHz. Beacon Transponders 1,090 MHz. ARSR; 1,215–1,450 MHz. Includes AN/FPS-20, 66, 67, etc.
S	2,000–4,000 MHz	ASR 2700–2900. Formerly AN/FPS-27 ARSR
C	4,000–8,000 MHz	Terminal Doppler Weather Radar (5.6–5.65 GHz). USAF height finders, others
X	8,000–12,000 MHz	PAR 9,000–9,600 MHz. Microwave Data Links
		FPN-16/CPN-4, others 9,010–9,080 MHz
K_u	12–18 GHz (under H_2O vapor resonance)	ASDE-3 14–17 GHz. Microwave Data Links
K	18–27 GHz	ASDE-1,2 23.8–24.27 GHz
K_a	27–40 GHz (above H_2O vapor resonance)	

Review Questions

1. There are two main types of radar frequency transmitting systems. They are: ________________ ________ and ________________________.
2. Coherence is ________________________.
3. Coherence in a system employing a magnetron is obtained by ____________________________.
4. Coherence in a system other than a magnetron is ____________________.
5. An automatic frequency control is used to control the transmitter frequency. This probably indicates that ________________________________.
6. A phase detector exhibits a ___________________ response to the phase difference between two input signals.
7. The output of a phase detector in an mti system is called __________________________.
8. An mti canceler compares ________________ to _________________.
9. A radar operating in the K_a band probably has a (high or low?) range resolution, a (very long or very short?) range, a (.040 μs or 6 μs?) transmitter burst, a (30 MHz, 250 kHz?) Δf, and transmits a (5 MW, 100 mW) peak burst.
10. Doppler ambiguity occurs when ____.
11. Compared to an mti radar designed for the same range, pulsed Doppler radars employ (high or low) f_p values and (high or low) f_{xmtr} values.
12. A pulsed Doppler radar is one in which ____________________ is the main objective.

Answers to Review Questions

1. There are two main types of radar frequency transmitting systems. They are: ***magnetron*** and ***synthesis.***
2. Coherence is ***an intelligible relationship among the transmitter, local oscillator, and intermediate frequencies.***
3. Coherence in a system employing a magnetron is obtained by ***shifting the coho phase to agree with the transmitter phase, each Tr.***
4. Coherence in a system other than a magnetron is ***inherent.***
5. An automatic frequency control is used to control the transmitter frequency. This indicates ***that the afc is probably a modification.***
6. A phase detector exhibits a ***cosinusoidal*** response to the phase difference between two input signals.
7. The output of a phase detector in an mti system is called ***bipolar data.***
8. An mti canceler compares ***undelayed bipolar data*** to ***delayed bipolar data.***
9. A radar operating in the K_a band probably has a ***high range resolution, a very short range, a .040-μs transmitter burst, a 30-MHz*** Δf***, and transmits a 100-mW peak transmitter pulse.***
10. Doppler ambiguity occurs when ***the Doppler shift exceeds*** $f_p/2$.
11. Compared to an mti radar designed for the same range, pulsed Doppler radars employ ***high*** f_p values and ***high*** f_{xmtr} values.
12. A pulsed Doppler radar is one in which ***velocity measurement*** is the main objective.

CHAPTER 6

Circuitry and Hardware

A Generic System (see Figure 6-1)

The intent of this chapter is to introduce the reader to the general configuration of FAA radar installations. The FAA radar network contains a wide variety of systems: ***Airport Surveillance Radar (ASR), Air Route Surveillance Radar (ARSR), Airport Surface Detection Equipment (ASDE), Precision Approach Radar (PAR)*** (use has been discontinued in favor of ***instrument landing systems (ILS)***), and ***Terminal Doppler Weather Radar (TDWR).*** Until deployment of the ASR-9, ASR-11, and ARSR-4, the ASR and ARSR, up to the video outputs, were very similar. This chapter will begin with a hypothetical, generic, synthesis systems as illustrated in Figure 6-1. This system will roughly resemble the ASR-8, ARSR-3, FPS-20, FPS-64, 65, 66, and 67. The system could also be converted to a magnetron system with the addition of coho phase-locking circuitry and automatic frequency control, and would then resemble the ASR-4, 5, 6, and 7, and the ARSR-1 and -2. Additionally, the USAF uses a modified version of the ASR-8 containing a magnetron and the necessary additions; in this system, the magnetron tuning is driven by the afc.

Other Types

Because of their ***moving-target-detector (MTD)*** systems, the ASR-9, ASR-11, and ARSR-4 differ radically from other radars. Further, the ASR-11 and ARSR-4 employ ***compressed high-resolution pulse (CHIRP)*** and altitude-detection capability. Earlier radars containing ***moving-target-indicator (MTI)*** systems exhibited several undesirable characteristics that could not be corrected until computer-age speed and hardware became available. Another entire chapter of this book is devoted to the ASR-9 MTD system.

The ASDE is a very short range system for monitoring airport runway and taxiway traffic; the newest version employs a 40-ns pulse, no mti, and a 60-rpm antenna. The ASDE is so dramatically different from all the ASRs and ARSRs that this chapter is not significantly representative of that system. The TDWR is a true pulsed Doppler weather radar with many special features and bears little resemblance to any other FAA radar. Information on the Doppler effect contained in this book may prove helpful to the reader interested in the TDWR.

Data Processing

In the early 1970s, a ***common digitizer (CD)*** system was installed in ARSR facilities to provide for the transmission of common digital modem data to joint FAA and USAF Air Defense Command users. These systems eliminated the need for the microwave links used to transfer analog radar information from the ARSRs to the ***Air Route Traffic Control Centers (ARTCCs).*** Since the modem data was transferred by audio-range telephone lines, and since the modulation of the microwave link channels by analog video caused X-band spectrum modulation, the digital systems were called ***narrow band*** and the analog systems were called ***broad band***.

Radar data is now transmitted almost exclusively in a narrow-band form. Newer systems, such as the ARSR-3 and -4, and the ASR-9, now have self-contained digitizers. Older systems at airports have been equipped with new digitizers. Most digitizers employ an ***azimuth-sliding-window detector,*** and the data messages sent to the user facility contain target center-of-azimuth and range information. This chapter contains a brief generic description of the azimuth-sliding-window process. The same process is also used in ***Automated Radar Terminal Systems (ARTS)*** computers, which have evolved from the US Navy computers first used in the 1960's AN/SPS-48 radar program. The ASR-9 digitizer uses the conventional azimuth-sliding-window detector for ***secondary radar (beacon)*** data, but, because of the MTD system, uses a special ***centroiding*** process to

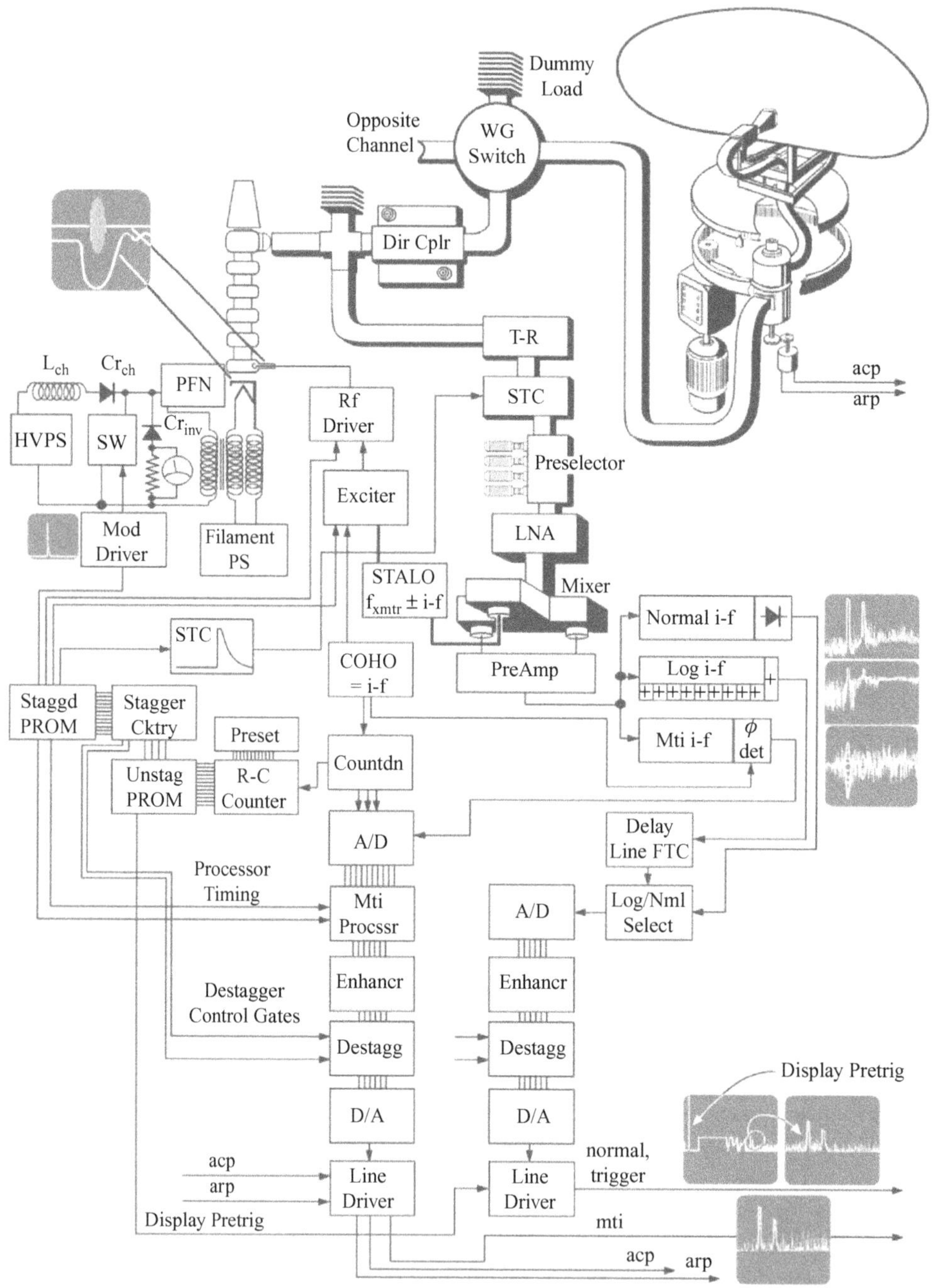

FIGURE 6–1

A generic FAA radar.

determine primary radar target azimuth. New ***Air Traffic Control Radar Beacon Interrogators (ATCRBI)*** have progressed to the ***monopulse secondary surveillance radar (MSSR)*** for improved accuracy and speed in centroiding. Chapter 7 of this book is devoted to beacon systems.

The Coherent Oscillator (Coho)

In latter-day synthesis systems, this unit is the beginning of the entire system operation. Originally, it was given its name in magnetron systems, because the phaselocking by the coho lock pulse made the system coherent. The synthesis system is inherently coherent, and the oscillator runs without interruption. The name endured, nevertheless. The coho is a crystal-controlled oscillator, running at 30 MHz in older systems, or at 31.07 MHz in newer systems. There may be several isolated outputs, supplying the phase detector(s), timing circuitry, and test-target generators used in MTI troubleshooting.

Countdown Circuitry

The countdown circuitry is ordinarily located in the synchronizer. The coho is divided to produce range-cell clocks. If the coho frequency were 31.07 MHz, division by 24 would provide 1.29 MHz clocks, with a period of 772 ns, which is 1/16 nmi. Signals from the countdown circuitry may also be used in the ***analog-to-digital (A/D)*** converter, where several different events will occur within a range cell, such as sampling gates and clocks.

The Range Cell Counter

The range cell counter, also a part of the synchronizer, is a binary counter which starts at a preset value, determined by the dual-in-line (DIP) switches. The DIP switches determine the f_p of the radar system.

The Timing Programmed Read-Only Memory (PROM)

This circuit, also a part of the synchronizer, is used to provide all the major triggers and gates in the radar system. As the counter addresses PROM, different outputs occur.

The Stagger Circuitry and PROM (see Figure 6-2)

MTI radars exhibit an undesirable characteristic called ***blind speed***; any moving target which produces a Doppler shift equal to the f_p or any multiple will be canceled as if it is a fixed target. By using different T_r's, the ***Tr-to-Tr change in phase*** ($\Delta\phi$) of the target changes from one pulse to the next, even though the Doppler does not, and blind speed targets become visible. The stagger has no effect on fixed targets, since they have neither Doppler nor $\Delta\phi$. The stagger circuitry contains another counter, which addresses another PROM. The staggered counter starts at different times each T_r. Generally, all signals for the transmitter and receiver will be staggered, and all timing having to do with destaggered and display data will be unstaggered. The ASR-7 is an exception; it provides staggered video and triggers to the indicator facility.

The Modulator Trigger

One output from the staggered PROM goes to an additional ***modulator driver,*** to provide a higher voltage trigger for the switch in the transmitter modulator. Such circuit may also be used to provide final manual or automatic temporal adjustment of the transmitter firing time. The modulator driver may also be called ***premodulator*** or (slang) ***baby modulator.***

Danger!

Lethal high voltages with great current capabilities are present in radar transmitters, even when the equipment has been turned off. Make liberal use of a shorting stick and keep one hand behind your back whenever possible! Except for safe panel adjustments and isolated testpoints provided by the manufacturer, never work on a live system and always discharge all components with a wellgrounded shorting bar before proceeding with any work. Do not take this lightly; people have been killed by radar transmitter modulators!

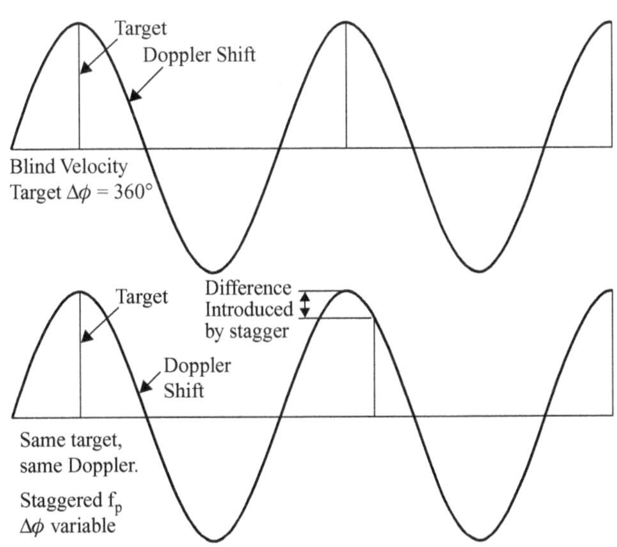

FIGURE 6–2

Staggered f_p introduces new Δf.

The Transmitter Modulator (see Figure 6-3)

This illustration combines photos, a block diagram, and partial schematics of an FPS-20 series modulator system, in an attempt to encourage appreciation of the size and capabilities of the transmitter system. Note the "triaxial" high-voltage pulse cable from the pulse-forming network in the modulator cabinet to the pulse transformer in the RF amplifier. The cable sometimes arcs and requires repair. It may well contain a lethal charge after failure.

The latest developments in radar transmitters do not use power amplifiers, but instead combine the outputs of a multitude of small transmitters onto a single line. These transmitters bear little resemblance to the ones described in this generic system; the HVPS may be as little as 48 V, and the transmitters are solid-state devices. The ARSR-4 and ASR-11 use this type of transmitter, and work is now under way to develop the technique for upgradation of older radars, as well.

The Pulse Transformer

The pulse transformer has a turns ratio that accomplishes two design objectives. One, of course, is to increase the high-voltage pulse to a sufficient amplitude for the desired PA power output. Another is to match the impedance of the PA to that of the pfn. The impedance reflected across a transformer changes with the square of the turns ratio. If the turns ratio were 1:3.5, and the impedance of the cathode were 600 Ω, then the impedance at the primary would be 49 Ω ($600/3.5^2$). Transmitter designs usually provide a slight impedance mismatch to create an ***inverse current.*** An ***inverse current diode,*** in parallel with the switch, conducts during the inverse current to ensure the appropriate beginning value of the next charge cycle. The inverse current is monitored with a front-panel meter. Either a high or low inverse current is an indication of fault.

The klystron filament voltage is applied through ***bifilar windings*** of the pulse transformer. These windings permit the filament current, but the fields created by the high-voltage pulse are in opposition, and the pulse creates no filament current. The pulse transformer is usually immersed in oil, in a container frequently referred to as the ***pulse tank.*** The oil may be circulated through an external cooling radiator.

The Filament Power Supply

The filament voltage and current are critical to the operation of the klystron, and the filament power supply must be closely regulated. The filaments have alternating current, but likely to have something other than a sine wave because of the type of regulation. In the ASR-9, the average filament current is the result of a rectangular wave, and the transition is timed so that it does not occur during the transmitter pulse; this was done to preclude the possibility of introducing any variations in the transmitted frequency.

Depending upon the power output of the radar transmitter, and usually only in ARSRs, the transmitter-modulator construction may include an ethylene-glycol/water cooling system, incorporating coolant pumps, radiators, blowers, etc. In some systems, the ethylene-glycol mixture may be circulated through a jacket of the power amplifier tube.

This circuit area includes the ***high-voltage power supply (HVPS), switch, pulseforming network (pfn),*** the ***pulse transformer*** connected to the ***klystron drift tube power amplifier (PA),*** the ***filament power supply,*** the klystron drift tube, and the ***rf driver.*** The HVPS output is applied, through a ***charging choke*** and a ***charging diode*** to the PFN. The PFN and charging choke comprise a resonant circuit, tuned so that a half cycle occurs in a little less time than the minimum T_r of the radar system. The PFN is predominately capacitive in relation to the charging choke. The switch, once always a thyratron, is only enabled during the transmitter pulse time.

The pfn will first charge to the value of the HVPS output (see Figure 6-4). When that has been accomplished, however, the charging choke has developed a magnetic field about it. Once the pfn charge reaches the HVPS voltage, the current through the charging choke begins to drop, but the magnetic field collapses, sustaining the current into the pfn until its charge reaches twice the value of the HVPS output. When the switch is enabled, the pfn, an artificial transmission line, is discharged over a period of time determined by its electrical length. The current through the pulse transformer primary creates the high-voltage pulse. While the switch is on, because the impedances of the pfn and pulse transformer are nearly equal, the two devices become a voltage divider, with approximately half the pfn charge dropped across the pulse transformer primary. Therefore, the peak of the primary pulse is nearly equal to the HVPS voltage, minus losses due only to efficiencies of

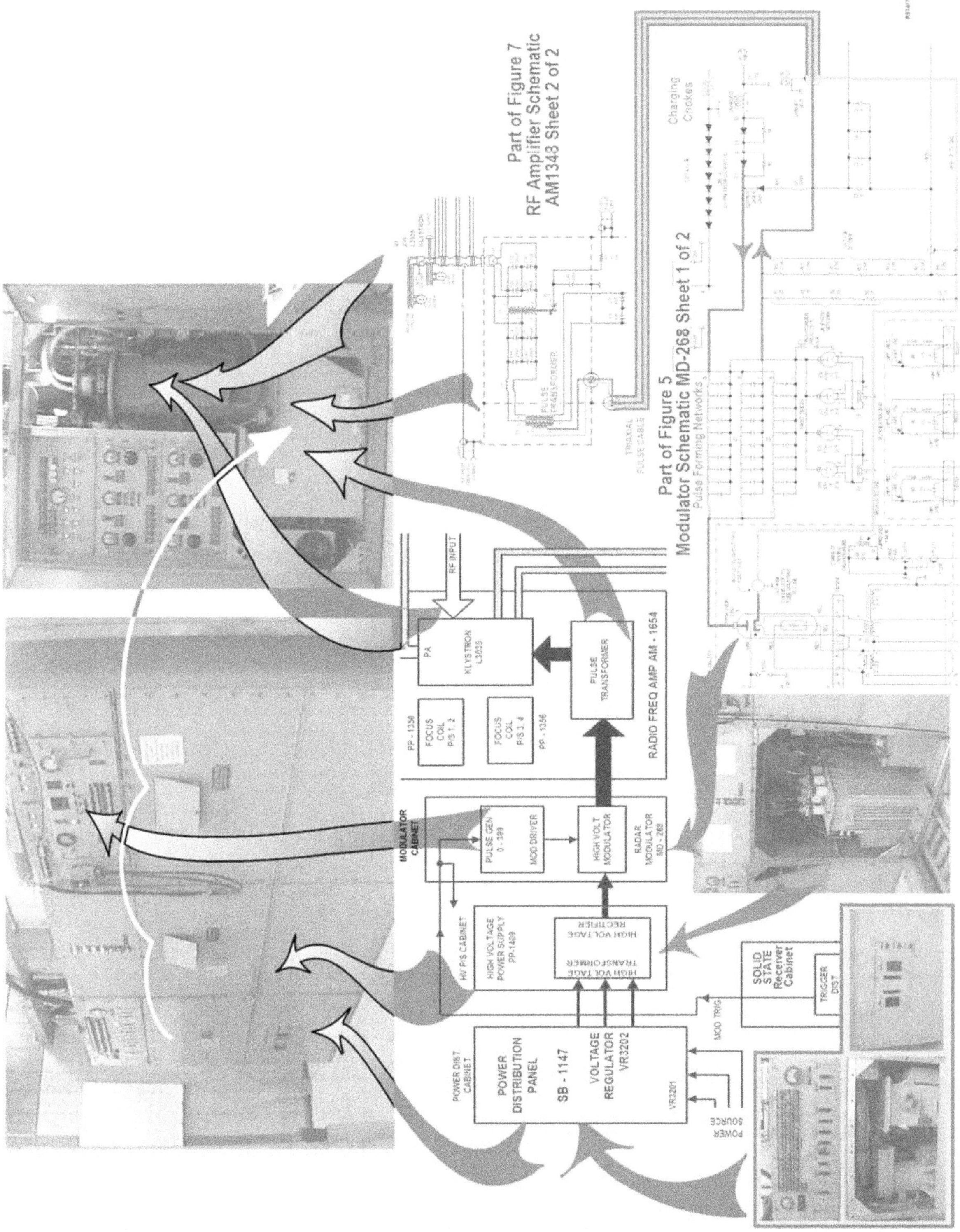

FIGURE 6–3

Partial illustration. FPS-20 series transmitter HV and modulator cabinets.

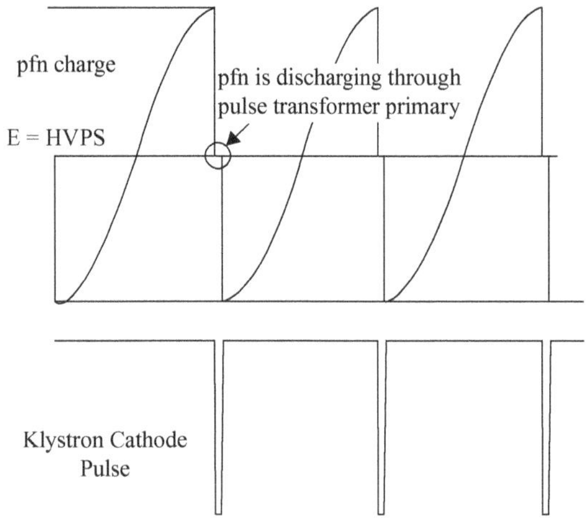

FIGURE 6-4

PFN charging.

the charging system. The pfn discharge time is determined by the inductive and capacitive values in the pfn, and the pulse-transformer primary serves as a nearly matched load. The pfn determines the width of the high voltage pulse; the discharge occurs in two waves: the first, from the pulse-transformer load toward the termination, and the second, from the termination toward the pulse-transformer load.

The Charging Diode

The charging waveform shown in Figure 6-4 is for optimum charging; the transmitter firing occurs as the pfn charge reaches peak. If T_r is made longer, the resonant circuit will attempt to ring in the opposite direction. In staggered operation, the T_r's differ, and the charging diode "disconnects" the charging choke to prevent the reverse cycle of the oscillation. The staggered charging waveform is illustrated in Figure 6-5.

The High-Voltage Power Supply

This unit is the major power source for the radar transmitter. It is generally supplied with three-phase power and contains six rectifiers. The HVPS has always been a major source of failure because of the arcing and breakdown from the high voltages; early systems produced as much as 35 kV. Transmitter-modulator design has been in a perpetual state of change since the early 1970s, and such enormous power-supply voltages will someday become a thing of the past. Prior to 1970, nearly all transmitter-modulator designs were almost identical, and the technician needed only to know one to know them all. The need for higher power, more reliability, and personnel safety led to new designs. For instance, the ASR-8 provides a peak high-voltage pulse of up to 72 kV, but the HVPS only produces a maximum of about 300 V. The power is derived from the collective currents in parallel circuits, and with saturable-transformer technology.

The klystron filament voltage is applied through ***bifilar windings*** of the pulse transformer. These windings permit filament current, but the fields created by the high-voltage pulse are in opposition, and the pulse creates no filament current. The pulse transformer is usually immersed in oil, in a container frequently referred to as the ***pulse tank.*** The oil may be circulated through an external cooling radiator.

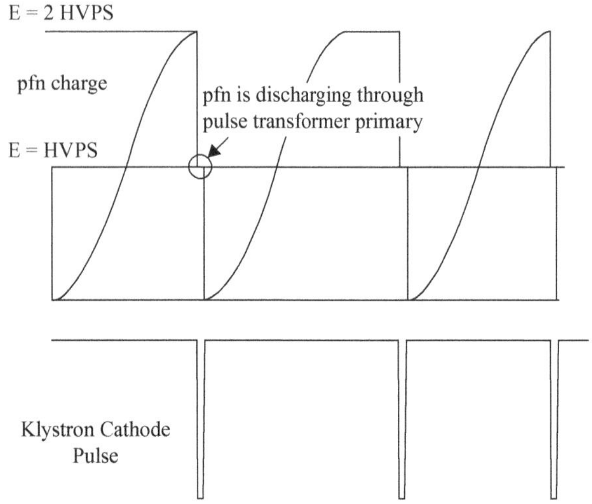

FIGURE 6-5

Staggered charging.

The Switch

The very first radars in the 1930s and early 1940s used a ***spark-gap modulator.*** An electric motor spun a circular element that passed in close proximity to another, once each revolution. Each time the elements reached proximity, the high voltage was discharged by arcing. The modulator pulse was then used as a radar time-zero reference trigger. Obviously, this left much to be desired. The ***hydrogen thyratron*** tube replaced the spark gap and was used as a switch for many years, and is still in limited use today. The greatest disadvantages to the thyratron are the limited life and critical adjustment.

In the thyratron tube, a hydrogen generator, called the ***capsule,*** produces hydrogen as voltage is applied. The amount of produced hydrogen is determined by the ***capsule voltage adjustment,*** critical to proper operation ***and life*** of the thyratron. The thyratron will be made to conduct when a trigger is applied to the grid. This ionizes the hydrogen, and the tube becomes a virtual short circuit to the pfn charge. The tube will continue to conduct until current ceases to flow at the end of the pfn discharge through the pulse transformer. One reason for the deliberate mismatch between the pfn and pulse transformer is to ensure that the thyratron will be completely de-ionized by the reverse current.

The trigger on the thyratron grid must be a high-level voltage, for instance, 700 or 1,000 V. Obviously, the trigger from the PROM shown in the generic system is inadequate, and there must be a method of amplifying it. To accomplish this, the older systems will have a ***premodulator, or modulator driver,*** often assigned the slang term ***baby modulator.*** The driver works in a similar manner as the main modulator, except with lower applied voltages. The load for the driver is a pulse transformer, the secondary of which supplies the thyratron grid with its trigger.

To escape the failures of short-lived electron tubes, particularly the thyratron, manufacturers have employed other devices. ***Reverse blocking diode thyristors (RBDTs)*** are used in the ARSR-3 and ASR-9. These PNPN devices block current in the reverse direction, as would a diode. However, when the reverse voltage exceeds a given value, the RBDT conducts in the reverse direction, until current drops to a minimum value. The RBDTs are employed in "stacks" to reduce the power, heat, and failure likelihood, of each. To ensure proper timing of the discharge, SCR stacks may operate in conjunction with the RBDTs, so as to place a rapid, timed, voltage change on the RBDT anodes.

The ASR-8 Magnetic Modulator

The ASR-8 uses a ***magnetic modulator,*** which employs saturable chokes and transformers. In this modulator, 12 ***pulse module*** circuit cards, each containing a charging capacitor and SCR switch, operate in parallel. The time constants of the capacitors (roughly equivalent to a pfn) are such that they reach the maximum charge in about one-third of a T_r. Because the outputs of all these modules are paralleled, a significant current capability is created. When the SCRs are triggered, the outputs of the pulse modules are routed through a ***holdoff inductor*** and ***saturable transformer*** to a bank of high-voltage capacitors. When the high-voltage capacitors are fully charged, the saturable transformer prohibits them from discharging in the opposite direction, and the discharge path is through the pulse transformer. The pulse modules are contained in an air-cooled rack called the ***air section,*** and the other devices are in an ***oil section,*** also sometimes called the ***pulse tank.*** In this modulator, the reactances and the consequent times required for the transfer of charges cause a considerable delay of roughly 14 μs between the time the pulse modules are triggered, and the time the actual high-voltage pulse is applied to the thyratron. The delay varies with temperature, and an automatic regulator provides temporal control of the SCR trigger.

The Power Klystron Tube

These tubes are sometimes called "drift tubes," but drift tubes are internal to the power klystron and connect cavities to each other. They have been introduced in ASR radars, beginning with the ASR-8 in the mid-1970s. However, they had been in use in the AN/FPS-20 since the 1950s, and had proven to be very dependable and long lived in that system. A large, copper collector shown at the top is grounded. The high negative pulse on the cathode creates a ***beam current,*** and electrons flow from the cathode to the collector. The beam current is modulated by ***rf drive,*** causing electrons in the beam to "bunch" into groups, or clouds. As each cloud passes a cavity on its way toward the collector, the cavity "rings," somewhat as a bottle whistles, when one blows across the opening. Of course, this ringing is at microwave frequencies. The ringing cavities create electric and magnetic (e and h) fields which cause additional bunching. By the time the clouds reach the collector, strong e and h fields, oscillating at microwave frequencies, exist in the tube, and these fields cause propagation through the output glass window into the waveguide.

The beam must be focused by a strong magnetic field, and one or more large coils, called either ***focus coils,*** or ***focus solenoids,*** surround the klystron. These coils are so large and heavy that they must be lifted off the tube with a hoist, even for ASR transmitters. A high current is passed through the focus coil(s) and the current must be precisely set to a value specified by the tube manufacturer. The AN/FPS-20 allowed for adjustments to these coils to optimize the klystron output. So, another major component of the transmitter modulator is the

Focus Coil Power Supply or ***Solenoid Power Supply.*** These power supplies will operate at a high current, but with a low voltage, such as 70 V dc or less. On newer systems, regulation is achieved by SCRs on the three-phase input to the supply.

The cavities in the power klystron are all tunable from the exterior. When delivered from the manufacturer, the tube comes with a data sheet listing the initial settings for several frequencies in the band. The numbers correspond to dial settings on an external tuning mechanism. When first installed, the klystron is energized at a minimum high-voltage setting, and initial refinements are made to the preliminary settings. While tuning, the technician observes (1) the power output with a meter, (2) the detected pulse shape with a detector on the incident connector of the directional coupler, and (3) the frequency spectrum, using a spectrum analyzer.

The klystron cavities are not tuned to the same frequencies, and the rf burst is shaped by the spectrum created by the cavities. The detected transmitted rf pulse is then considerably squared in comparison to the rf drive pulse.

Klystron and modulator stability is absolutely essential to satisfactory mti performance. In latter-day synthesis radar systems, the mti processor is extremely reliable, and ***the most likely cause of mti clutter residue is the transmitter modulator.*** Variations in the modulator pulse amplitude or unstable klystron tuning are the most frequent causes. The ASR-8 has a "de-Q'ing" modulator circuit to maintain constant pulse amplitude, and it is very easy to render the de-Q'ing circuit inoperative by a minor maladjustment.

The Exciter and rf Driver

These are the names assigned to the units in the ASR-8 by the manufacturers; they may be assigned other names. The roughly comparable names in the AN/FPS-20, and its descendants, AN/FPS-64, -65, -66, -67, and others, would be the ***buffer mixer*** and ***intermediate power amplifier (IPA).*** In the exciter, the coho is added to, or subtracted from, the stalo, to synthesize the transmitter frequency. The transmitter frequency is then pulse modulated, meaning gated and amplified, for application to the rf driver. The modulation by the gate in the exciter is by a gate substantially greater than the transmitter burst width. In the ASR-8, the transmitter burst width is 0.6 μs, but the exciter gate is 5 μs.

The rf driver further modulates and amplifies the transmitter frequency, but great attention is devoted to the shape of the modulating pulse. Because the rf driver output is the frequency-generating input for the klystron, its shape has a great effect on the radiated spectrum and the shape of the transmitted burst. The technician may find that there are several adjustments, and a manufacturer's procedure and specifications, for the rf driver pulse shape.

The Duplexer (see Figure 6-6)

A specific block entitled "duplexer" is not shown in Figure 6-1. The duplexer comprises the four-port circulator and T-R. To steer energy from the transmitter to the antenna, and from the antenna to the receiver, older radar systems used a duplexer with an ***anti-transmit-receive (ATR)*** tube, which energized with the transmitter burst to permit outward propagation, and a ***transmit-receive (T-R)*** tube, which provides a short circuit in the receive path during transmit time, so as to protect the delicate receiver components from the high-power transmitter burst. The ATR tube is no longer used. Both the T-R and ATR were gas-filled tubes, and the T-R tube required a keep-alive voltage of around –700 V so that it would be quickly energized by the first rising power in the lead edge of the transmitter burst. The T-R tube has been replaced with a newer T-R device, which works in much the same manner, but does not require the high negative keep-alive voltage.

The Four-Port Circulator

This is a steering device that essentially replaces the ATR tube. It directs the transmitter energy to the antenna and the received energy to the T-R device. Its main purpose is to isolate the transmitter from received echoes by offering a low impedance to outward wave propagation, but a high impedance to waves propagated from the antenna feedhorn.

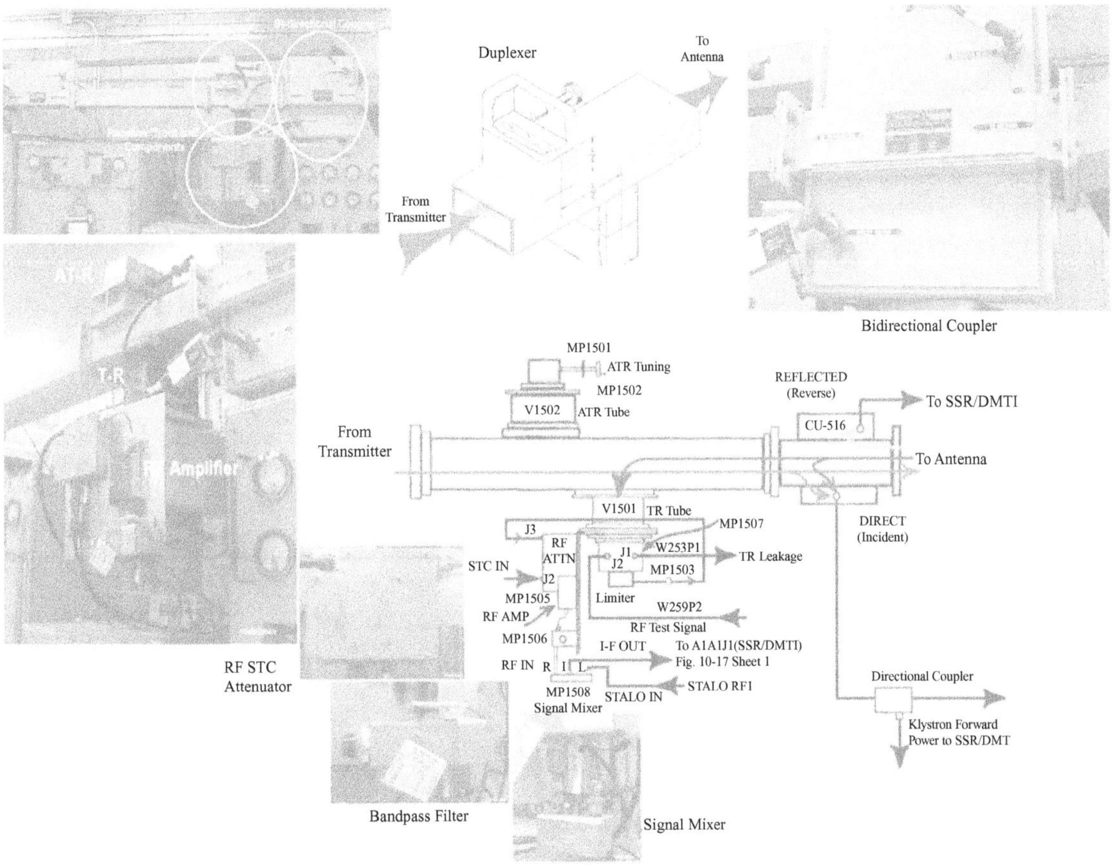

FIGURE 6–6

FPS-20 series duplexer, bidirectional coupler, and associated hardware.

Directional Coupler

The directional coupler in Figure 6-1 is more appropriately called a ***bidirectional coupler,*** since it contains two sections, one labeled INCIDENT, and the other REFLECTED (see also Figure 6-6 and Figures 8-50 and 8-55). Additionally, on each coupler, attached by small rivets or epoxy, there will be a metal tag displaying a chart, to indicate the specific attenuation to all frequencies in the band. The INCIDENT coupler is probably the single most utilized test connector in the radar system. It is used for power measurements, test signal injection, spectrum analysis, echo-box connection, and for connection of a video detector, so that the technician can view the shape of the transmitter pulse. The REFLECTED connector is used to compare reflected power to incident in determining VSWR. A very potential source of radar performance derogation is the rotary joint; excessive reflected power fluctuations with antenna rotation, called ***WOW,*** indicate the need for rotary joint repair.

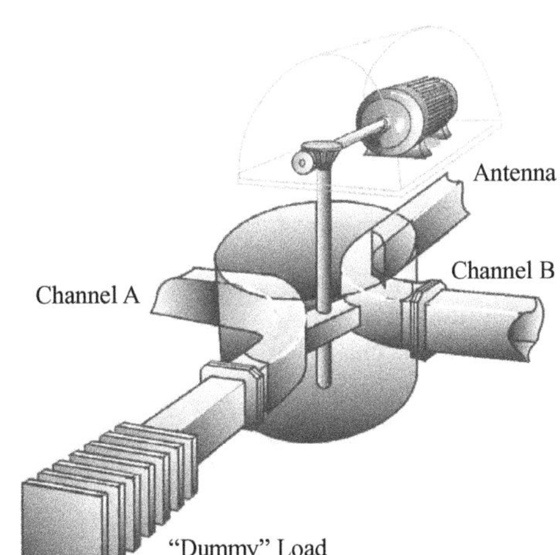

FIGURE 6–7

General waveguide switch principle, simplified.

The Waveguide Switch (see Figure 6-7)

FAA radar systems contain two complete radar sets, called ***channels.*** This is primarily to provide a backup system in case of failure, and secondarily, to provide for preventive maintenance, alignment, and adjustment of the off-line channel. The waveguide switch may route the transmitter output to a ***dummy load,*** or to the antenna. The simplified generic switch illustrated in Figure 6-7 can be rotated by a motor drive. Rotating the internal waveguide bends by 90° will switch one channel from antenna to dummy load, and the other from dummy load to the antenna. Not shown in the simplified diagram are speed reduction gearing and switches to stop the motor when the internal bends have been rotated to the commanded location. Other switch contacts provide readbacks to indicate that the bends are properly positioned. Those readbacks may indicate the selected channel, and their absence may indicate that the waveguide is in transition. While the switch is in transition, triggers to the transmitter may be interrupted by the readback absence to prevent arcing and other related system damage.

In FAA radars, one antenna online channel is electrically designated "master." The master channel provides synchronization and coho frequency signals to the opposite channel, to prevent crosschannel interference.

Many systems require several waveguide switches. One case is the diplexed ASR-8, where both transmitters may be placed on the antenna, or into the dummy load, or a combination of the two.

Each radar channel has three possible modes of operation, ***ON LINE,*** connected to the antenna for use, ***STANDBY,*** in dummy load, but available for immediate use by the using facility, and ***MAINTENANCE,*** where the standby channel has been ***electrically released*** for work by technicians. These may have other designations, depending upon the manufacturer, original purchaser (FAA or military), and time of purchase. Some of these are:

ON LINE	OFF LINE	UNAVAILABLE
OPERATE	STANDBY	MAINTENANCE
RADIATE	LOAD	MAINTENANCE

Three Modes of a Radar Channel

For some tests, the system must be in dummy load to prevent video from obscuring injected test signals. It should be noted that the dummy load is a resistive device, and heat is created when the transmitter output is routed to it. Heat generates noise in the waveguide system, and noise derogates receiver sensitivity. An mds measurement will be worse when taken on a standby channel with the transmitter radiating into the load, because of this additional noise.

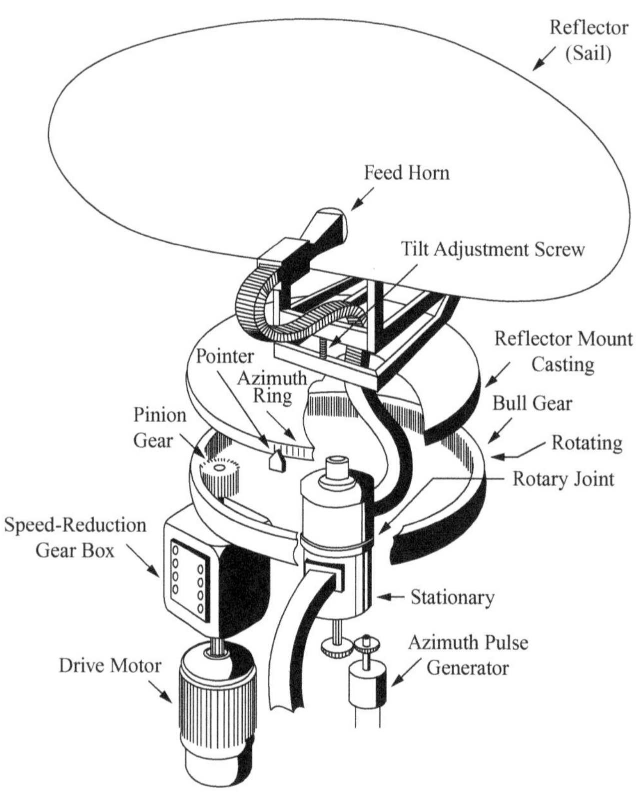

FIGURE 6–8

Major assemblies of a surveillance antenna.

The Antenna (see Figure 6-8)

The antenna is rotated by a three-phase ac motor and speed-reduction gear box. ARSR antennas rotate at 5 rpm. ASR antennas may rotate at several speeds from 12 to 15 rpm, the most common being 12.5 rpm. This has been reduced from the earlier choice of a standard 15 rpm, done both to increase the hits-per-scan for better detection and to reduce the ***pulse-to-pulse phase change*** **(**$\Delta\phi$**)** of clutter due to antenna rotation; such an effect causes ***mti clutter residue,*** which can make it difficult to distinguish moving targets on ppi displays.

On the output shaft of the speed reducer is a ***pinion***, a mechanical device defined by Webster's dictionary as "a small cogwheel, the teeth of which fit into a larger gear wheel or those of a rack." The pinion drives a ***bull gear,*** another mechanical engineering term, used to describe a large, main, gear that supplies the main rotation for mechanical equipment. The bull gear is mounted on bearings and is usually immersed in oil. Affixed to the bull gear is the entire rotating part of the antenna, including a ***spider casting*** to support the entire assembly. On the spider casting are the necessary lugs for affixing the reflector hardware, which includes pivots and a ***tilt adjustment. Leveling bubble indicators*** on either the rotating or fixed portion of the pedestal (depending upon the manufacturer) are provided; a very slight leveling error can create drastic consequences in radar coverage of a given area.

Affixed to the outer perimeter of the spider casting is an ***azimuth ring,*** and a pointer is affixed to the fixed portion of the pedestal. The arrangement may be opposite, with the pointer on the rotating portion. The azimuth ring is engraved in fractions of degrees, and the ring and pointer are aligned upon installation so that 0° is indicated at surveyed true north. To be in agreement with aircraft compasses, FAA radar display equipment must use magnetic north, but magnetic north is variable. An ***azimuth offset*** somewhere in the azimuth data circuitry will correct for the difference.

The Azimuth Pulse Generator (apg) (see Figures 6-8 and 6-9)

At the bottom of the ***rotary joint,*** the assembly which transfers rf energy from the fixed waveguide to the rotating assembly is an azimuth pulse generator (apg). There are different types of apg's; the one illustrated in Figure 6-9 was common into the late 1990s. The apg contains two LEDs, two photosensitive transistors, a silver-coated glass disk with 4,096 etched marks near the circumference, and a single etched mark to be used as the north azimuth reference. One LED transistor set is mounted to project light through the 4,096 marks, and the other is to project light through the north reference mark. The light is amplified, and the pulses created as the antenna rotates are standardized to a fixed width on the circuit card. The 4,096 pulses are called ***azimuth change pulses (acp's),*** and the north reference pulse is called the ***azimuth reference pulse (arp).*** The arp is sometimes called the ***north mark,*** but that term may be a source of confusion as a "north mark" has been used to define a strobe to indicate north on radar indicators which have received data via a microwave link. The north mark was used to alert air traffic controllers of a failure in azimuth data; the sweeps would continue to turn on the remote indicators, but the north mark would cease to point to north.

Aligning the apg

Until developments over the past decade or more, alignment of the apg was difficult. Newer digital systems now provide for relatively simple corrective offset to the azimuth data via a GUI. In the past, it was necessary to physically adjust the apg assembly to occur precisely at true north. The arp was difficult to see on an oscilloscope, because only one pulse is produced as the antenna passes through north, and only a single dim trace could appear on the oscilloscope. Inadequate even in darkness, daytime sunlight lessened the oscilloscope usefulness even more. One method is to first synchronize the oscilloscope sweep with an acp video input to set brightness and then slow the sweep speed so that the trace can remain on the tube long enough for the arp to be visible. At Dallas-Fort Worth International Airport, a creative radar technician designed a ***pulse catcher,*** a hand-held circuit box containing a flip-flop multivibrator and LED. When the arp occurs, the flipflop is triggered, the LED lights, and remains lit until manually reset. When the antenna is slowly turned by hand, the azimuth indicated by a pointer and the azimuth ring would be noted. Newer oscilloscopes now allow an event to be "captured" and remain on the screen.

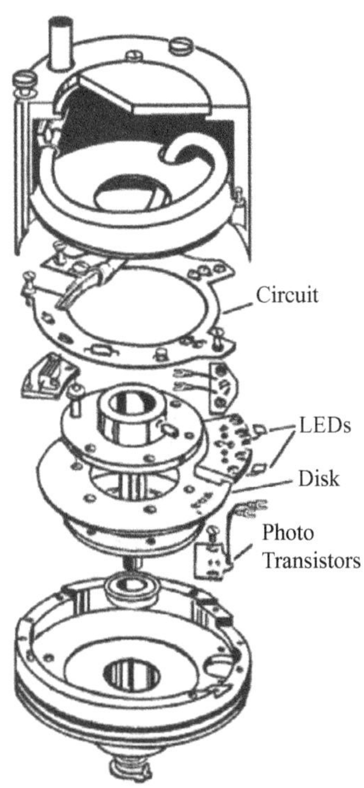

FIGURE 6–9

Azimuth pulse generator.

Newer 16,384-ACP Generators

Some ARSR facilities have used a magnetic pick-up device to create 4,096 acp's and an arp. A cog, or tooth, on a wheel interrupted a magnetic field to create the pulses. For the ASDE, or for latter-day monopulse beacon systems,

4,096 acp's are inadequate, and new 16,384-pulse apg's have been manufactured. The new device operates on a "Hall effect." The device is excited by an ac current from an oscillator. The oscillations produce magnetic fields, which are opposed or strengthened by a rotating element, producing two modulated ac waveforms with a sin–cos relationship. Circuits on a card use the sin–cos waveforms to produce the acp's. A "zero crossover" detector produces an output each time the sine or cosine voltage waves transition from positive to negative, or vice versa. This causes the generation of four acp's in the time previously needed for a single acp, multiplying 4,096 to 16,384.

The Rotary Joint

The rf principles of rotary joints are covered in more depth in Chapter 8. The waveguide from the duplexer is connected via a waveguide switch to the ***rotary joint,*** which provides conduction of the rf energy into the rotating part of the antenna. Voltages needed for functions on the rotating portion of the antenna are conducted through the rotary joint by hardened sliding contacts, usually called ***slip rings.***

Antenna Polarization and Control

From the top, rotating part of the joint, both rf energy and control voltages are applied to the ***polarizer assembly.*** Normally in the ***linear*** polarization mode, which allows the assembly to act as a standard feedhorn, the polarizer assembly may be placed in a configuration called ***circular*** polarization, used to attenuate precipitation echoes. In circular polarization, the transmitted energy is split into two perpendicular planes in a square waveguide, one delayed 90° from the other. In the square waveguide, the instantaneous e and h fields for a specific transmitted phase angle will have been multiplied by the cosine of the phase angle for one plane, and by the sine of the phase angle, for the other.

The Basic Principle of Circular Polarization

Since $\sin^2\theta + \cos^2\theta$ is equal to unity, the wave power is divided into the two planes. The average power of each plane is half or 3 dB less than the input. The feedhorn matches the impedance of the square waveguide to the impedance of free space, and the power is radiated in horizontal and vertical polarization, as illustrated in Figure 6-10. The polarizer is periodically tested for equal power in the horizontal and vertical planes. With the polarizer in the circular mode, a test fixture consisting of a feedhorn and radial dial is mounted on the reflector. The waveguide connecting to the polarizer assembly is disconnected, and a coaxial adapter is attached to the polarizer to permit connection of a power meter. A signal generator is connected to the test fixture, and the test signal is measured with the test fixture in two positions, 90° apart; the power should be nearly equal, within the manufacturer's tolerance. The procedure may have variances, depending upon the manufacturer and radar type; the primary object is to ascertain near equal propagation in both the vertical and horizontal planes.

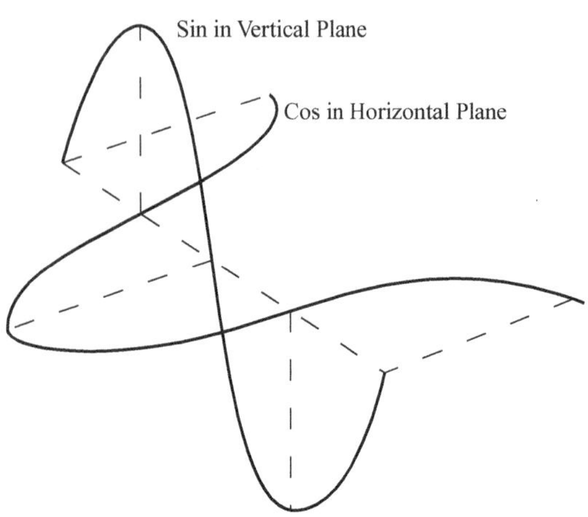

FIGURE 6–10

Power sharing between vertical and horizontal waves.

No matter how small, any object in space will reflect some portion of a radar burst, often even in directions other than back to the radiation source. Many objects are too small for their echo to be useful, but an accumulation of a multitude of small echoes at the same range can be significant, as with hordes of insects, or hundreds of millions of raindrops. In the case of raindrops, because their shapes approach circularity, the echoes for each droplet, and the accumulation of millions of such echoes, will reflect strong and nearly equal amounts of energy in both the horizontal and vertical polarizations. When these echoes return into the square waveguide, and are subjected to another 90° phase shift for a total of 180° for the round trip, the fields for the two echoes are in opposition in the polarizer assembly, and are severely attenuated. Noncircular targets are less affected because there will be a stronger echo on one polarization than on the other.

Active and Passive Feedhorns

The rotary joint illustrated in Figures 6-1 and 6-8 is in the simplest possible form, with one waveguide in and out. (See Figure 6-11.) In most FAA radars, the joint will contain several rf sections. Most systems now use two feedhorns, one called ***active*** and one called ***passive*** Each has a waveguide path which must pass through the rotary joint. Further, the beacon 1,030 MHz mode pair and 1,090 MHz transponder replies require a rotary joint section, as does a beacon omnidirectional antenna. The joint must also contain the sliprings to provide the electrical connections for polarizer control and readback.

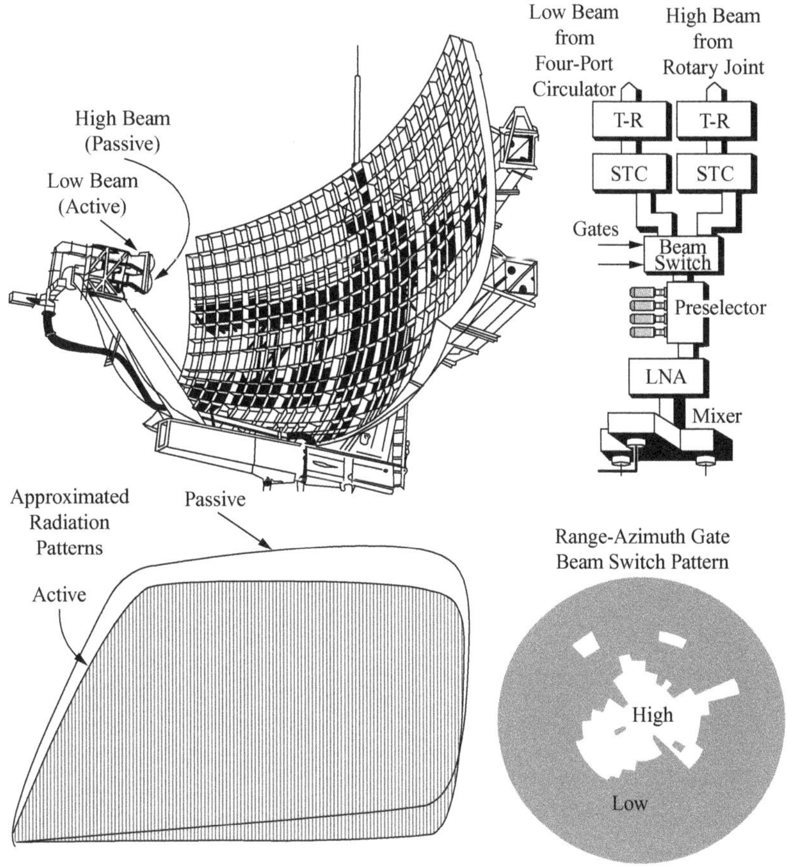

FIGURE 6–11

Active–passive receive systems.

The active feedhorn operates as previously discussed; it is called "active" because the waveguide system feeding it is the one ultimately connected to the transmitter. The "passive" waveguide system is a receive-only system, for use at near ranges. The passive feedhorn, also called the ***highbeam*** feedhorn, is physically mounted below the active feedhorn, also called ***low-beam,*** so that the feeds themselves may be as close together as physically possible. The passive horn is directed at a steeper upward angle than the active, so as to provide an elevated radiation pattern of approximately 3.5°. In the receiver, range-azimuth gating will select one of the two inputs; the passive input is used predominately at near ranges, and the active path at greater ranges.

The T-R Device

This gas-filled device is energized during transmit time to protect the receiver. It is important for the technician to note that the T-R exhibits a "recovery time" which should be periodically measured. After the transmitter burst, the gas in the device does not instantaneously deionize, and there is a significant attenuation to received echoes. The procedure to measure this is called the ***T-R recovery-time measurement,*** as illustrated in Figure 6-12. The technician connects a synchronized, pulsed, microwave signal generator to the INCIDENT POWER connector on the directional coupler of the standby radar channel that has been placed in dummy load, and connects an oscilloscope to the normal receiver output. The test signal delay is set to place the signal just beyond the area attenuated by the recovery time, and the signal level is set for an unlimited signal. The oscilloscope vertical positioning is used to set the top of the signal at a reference; then the signal generator output level is increased by 3 dB. The signal generator delay is then decreased until the signal decreases to the reference line.

The T-R tube is often the cause of poor mds measurements. An unsatisfactory T-R device might be revealed by troubleshooting with a portable noise figure test set, if available. T-R recovery time measurements are also

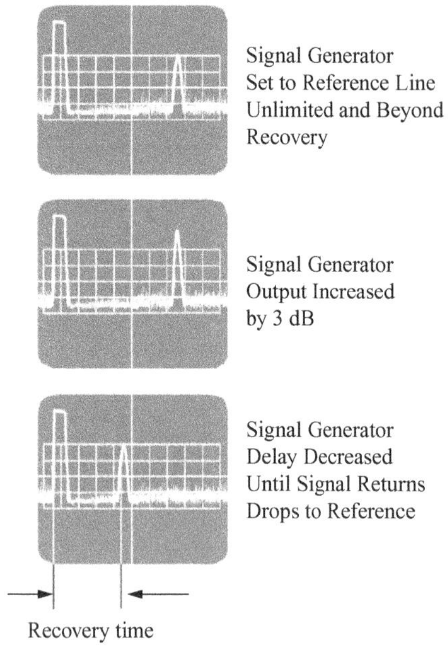

FIGURE 6–12

Recovery time measurement.

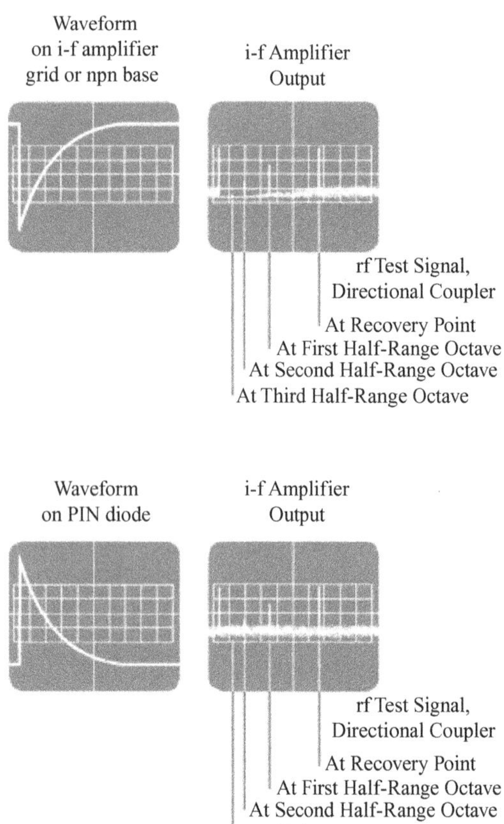

FIGURE 6–13

STC waveforms and measurements.

necessary on the passive receiver channel; the transmitted burst from the reflector creates a high rf power level in the passive waveguide system. This passive system measurement may require a special procedure, because the measurement would be meaningless when the system is in dummy load, but clutter will obscure the test pulse when the system is on line. It may be necessary to stop the antenna while it is radiating into a clear area. Further, because the beam switch pattern will most likely be set for the passive system at near ranges, the passive T-R recovery time may well be more significant to system performance than the active.

The stc Module

Sensitivity time control (stc) is among the oldest "fixes" in radar. In the earliest systems, it was discovered that the strongest radar echoes at close ranges would so saturate the receiver that it could not recover immediately; this was called ***receiver blocking.*** After the development of mti, stc became an important tool in reducing the loss of moving-target echoes in strong ground-clutter returns.

Scientists at MIT devised a method to desensitize the receiver at close ranges and then gradually recover to maximum with increasing time from the ***main bang,*** the slang term used to describe transmitter firing time. Thus, the sensitivity is increased with time, as the name implies.

The earliest stc was applied as a negative bias to the grids of amplifiers in the ***i-f strips,*** the chain of intermediate-frequency amplifiers in the receiver, preceding the ***second detector,*** which may be either a conventional amplitude detector for normal video, or a phase detector, in the mti receiver (see Figure 6-13).

The accepted method of testing stc operation is by injecting an rf test signal at the directional coupler, locating the precise range point at which the stc recovers to maximum sensitivity, then decreasing the range of the test signal in half-range ***octaves*** (dictionary definitions of this word do not support this usage). The unlimited test signal is set to a reference line at the recovery point; then the amount of attenuation at each descending half-range point is verified by increasing the signal generator level to bring the signal back to the reference line. Radar stc waveforms are usually based on a 12-dB-per-octave curve; beacon systems are based on a 6-dB-per-octave curve, because only one-way travel time to the transponder reply is involved, and the recovery point must be at a greater range.

In digital processors, it is necessary to maintain a constant noise rate throughout the listening interval; this noise rate is called the ***false alarm rate.*** The stc waveforms applied to the i-f strips caused a severe loss of noise at near ranges, and was incompatible with processors. Additionally, the greatest effect upon signal-to-noise ratios can be achieved at the most forward point, nearest the antenna, in the receiver ***front end,*** called the ***common receiver,*** in some radars manufactured by Texas Instruments. In the late 1960s and early 1970s, a special diode for use as an rf attenuator became available; it contains positive, intrinsic, and negative materials, leading to the name ***PIN diode.*** The PIN diode is contained in a manufactured waveguide assembly, called an ***stc attenuator.*** The first versions of these required a positive-going analog stc waveform input, and provided a specific attenuation for a specific voltage, expressed in dB per volt. Later versions contain a factory-calibrated D/A converter and operate with a parallel binary input.

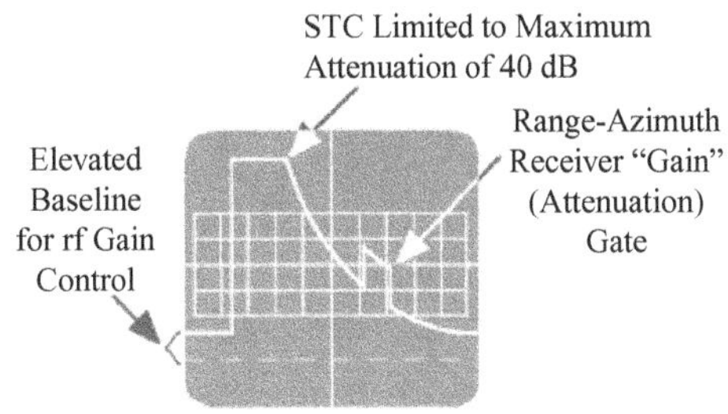

FIGURE 6–14

An ASR-8 STC waveform.

The PIN diode stc waveform in this illustration contains extra features, such as a range-azimuth gate, to cause additional attenuation in a programmed area (see Figure 6-14). Such a gate might be used to reduce the effect of strong clutter on mti; this can make aircraft flying over that clutter difficult to detect. In mti systems, even though clutter is canceled, moving targets must compete with the clutter for detection, and the moving targets become harder to detect as the clutter becomes stronger. The effect is known as ***subclutter visibility*** and is measurable.

The stc PIN diode may also be used as an ***rf gain control.*** By placing the waveform on a dc baseline, the dc value serves as a constant attenuation. This feature becomes very useful in the presence of ***anomalous propagation*** or light interference. When the feature is made available to users, it is absolutely essential that operators be made aware of the risks of decreased sensitivity.

Radar scientists have determined that there is no benefit, and potential detriment, to attenuating receiver inputs by more than 40 dB, an attenuation of 10,000. The very strongest clutter is in the neighborhood of 60 dB above system noise. For this reason, stc waveforms are limited to 40 dB of attenuation, causing a "flat spot" at the near ranges when the recovery point is set to more distant ranges.

The Beam Switch (see Figure 6-11)

This is another PIN diode device, but the PIN diodes are used as steering switches, rather than attenuators; they either enable or disable the active or passive waveguide paths, so that one is "open," and the other "closed." The beam switch may also be called the ***pattern switch.*** The switch is operated by a ***range-azimuth-gating (RAG)*** program.

The Preselector Filter

This device is located downstream of the beam switch, in order that only one is necessary. On the ASR-8, it is immediately before the crystal mixer, and on the ASR-9, it is ahead of the ***Low Noise Amplifier (lna).*** One intended purpose is to restrict the front-end bandpass to reduce noise. Another is to reject other frequencies which might result in a superheterodyned i-f near or equal to that for which the system is designed. In ASR radars, the preselector is tuned for a 10 MHz bandwidth at the 3 dB points. The preselector is tuned according to factory-attached charts or graphs. Touchup to the tuning while monitoring noise figure, or a very low test signal, may provide a slight improvement to performance; there may be slight interpolation errors in reading the chart. It is important that the technician remember the need to retune the preselector if the operating frequency is changed.

The Low-Noise Amplifier (lna)

These ***gallium-arsenide-FET (GAsFET)*** devices are off-the-shelf technology taken from the satellite-receiver industry. The first radar systems applied the waveguide echo from the T-R device to the signal mixer; the lower frequency systems employed a preamplifier ahead of the mixer. Since the first active stages in the receiver have the most significant effect on noise figure and mds, low-noise amplification in front of the signal mixer can make the minimum discernible signal considerably lower.

Before lnas became available, an exotic waveguide-circuit device, called a ***parametric amplifier,*** served the high-gain, low-noise requirement through the 1960s, 1970s and 1980s. The parametric amplifier operation was centered around two major components, a ***varactor diode*** and a high-frequency ***pump oscillator,*** which might operate as much as 11 times the radar transmitter frequency. The junction capacitance of the varactor could be varied at microwave frequencies; each time the capacitance was increased, the potential was increased,

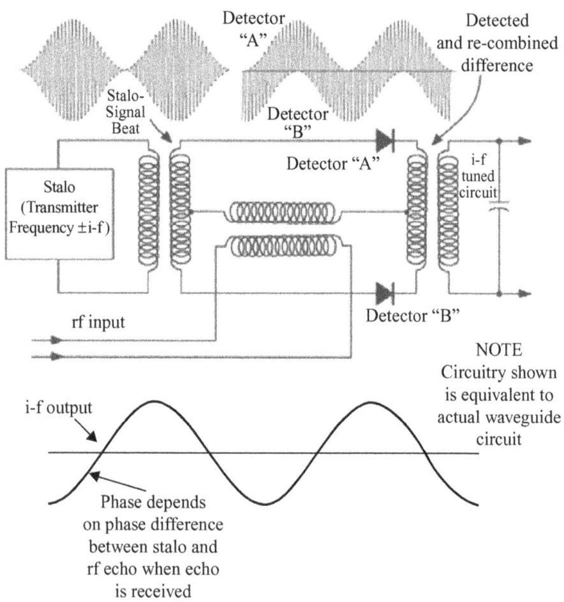

FIGURE 6–15

Equivalent circuit of balanced mixer.

"pumping" the received signal. Because the amplification was accomplished without the usual current-and-resistance circuitry, it generated very little noise while amplifying the signal. Amazing and ingenious as it was, the parametric amplifier was a maintenance problem, perhaps even because it was beyond most of those who worked on it. Nevertheless, it is rapidly becoming one more technology step to be remembered only in history books.

The Crystal Mixer

The mixer illustrated in Figure 6-1 is a ***double-balanced mixer*** and utilized a special waveguide device called the ***magic tee.*** There are now several types of mixers in radars. The most recent ones are contained in circuit packages in an rf amplifier module; nearly all of these are some type of double-balanced mixer. The magic-tee mixer was used for this generic system because it makes the design intent and function obvious. There are two crystal detectors, each supplied with a signal-stalo beat frequency.

The stalo injection to the magic tee creates a phase reversal in the two paths to the detectors, while the rf signal injection does not (see Figure 6-15). Since the stalo is an active oscillator, using active components, it is also a noise generator. Because the stalo input is of opposite phase to the two detectors, any noise spikes from the stalo will be of equal and opposite polarity as they are applied to the detectors.

When a signal is applied to the circuit, the signal, being the transmitter frequency, which is either greater or less than the stalo frequency, will be in phase with stalo in some instances, and out of phase in others. When in phase, the stalo and signal are additive, and when of opposite phase, a zero beat is achieved, and phases between those produce a similarly resultant output. The rate at which the beat frequency passes in and out of phase is the difference between the two frequencies, and will become the intermediate frequency.

The beat frequencies are rectified in the detectors and then recombined in the output transformer; any noise spikes will have been of equal and opposite polarity, and therefore canceled. The output circuitry passes only the intermediate-frequency component, and the i-f output results.

In any radar system utilizing phase detection, it is important to note that the starting phase of the i-f echo depends upon the phase of the stalo at the time the echo arrives at the mixer. This is true of all superheterodyne systems, but is insignificant in amplitude-detection systems. The signal mixer is often called the first detector; in phase detection systems, the signal mixer is the first phase detector.

The Preamplifier

The preamplifier is located as closely as possible to the signal mixer. In some systems, the recombination circuit for the double-balanced mixer is contained in the preamp, as is illustrated in Figure 6-1. The preamp is a low-noise amplifier, carefully engineered to provide a high gain to the low-level mixer output. Tube-type radars used a ***Wollman low-noise amplifier,*** a grounded-grid triode device now essentially extinct. In the ASR-9, both the mixer and preamp are contained in an ***rf amplifier module.*** As illustrated in Figure 6-1, and common practice in many systems, the preamp also serves as a distribution amplifier for the normal, log, and mti i-f amplifiers.

The Stalo

This microwave oscillator serves as the local oscillator and runs at a frequency different from that of the transmitter by the value of the intermediate frequency. This unit has changed so much over the years that latter-day stalos bear no resemblance to the ones first used in mti systems. The first stalos were based on microwave triode oscillators and resonant cavities. The earliest mti systems were all the magnetron type, and the stalo had to be frequently retuned. The AN/CPN-18, the surveillance radar used in joint-use RAPCONs in the 1950s and 1960s, contained a stalo system with a cavity, and two 6AU6 electron tubes for afc; it was so susceptible to mechanical shock and vibration that FAA technicians mounted the unit on a foam pad in front of the chassis. The CPN-18s were replaced with ASR-4, -5, and -6 systems employing afc-driven electromechanical tuning; the afc was a "swept-bandpass" type. The ASR-7 used an L-band oscillator, multiplied to S-band, with a discriminator-operated afc; this is now being replaced with a digitally based synthesizer.

Early in the development of 3-cm (X-band) radars in WWII, a ***reflex klystron*** was used as a stalo; development of this tube paralleled the magnetron development (see Figure 6-16). The reflex klystron principle was chiefly based on an "electron bunching" in a self-contained cavity in the tube. Electrons emitted from the cathode were accelerated to a high velocity by an accelerator grid, and into a cavity. A repeller plate with a high negative voltage would then reverse the electron trajectory to cause the bunching. The natural ringing of the cavity further enhanced the bunching to cause the oscillations. These klystrons are still widely used in microwave links.

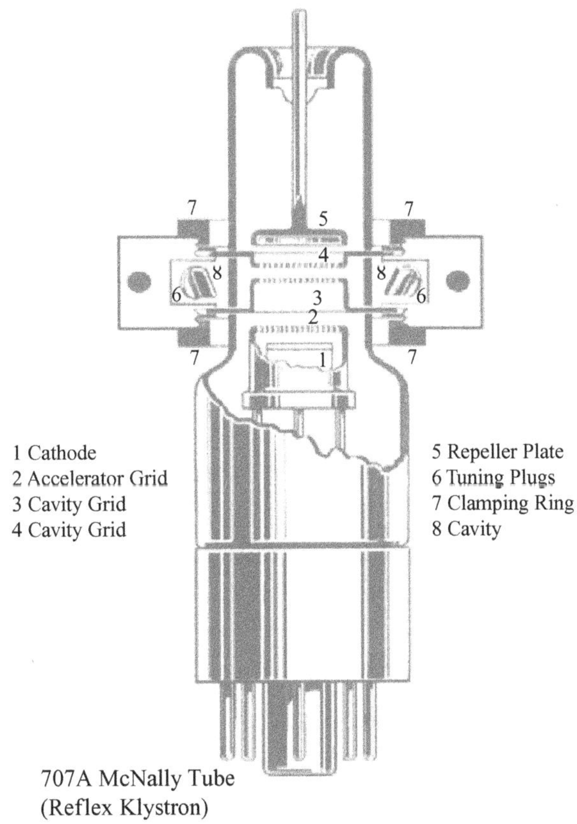

FIGURE 6–16

The reflex klystron. From USAF Manual 52-8.

The stalo in synthesis systems is both simpler and more stable, since it is the frequency-determining device for the radar channel and requires no afc. The early synthesis systems used triode oscillators similar to those in the magnetron systems, but later versions evolved to crystal-controlled oscillators, multiplied upward in frequency. In the ASR-8, the crystal-oscillator output was applied to a rectifier, creating a signal rich in harmonics. The microwave hardware circuitry is then tuned to the 28th or 29th harmonics and amplified to provide the stalo outputs to the signal mixer and exciter.

Most earlier radar receivers had an ***XTAL CURRENT*** meter wired into the preamplifier to indicate current in the signal detector diode(s). Nearly all the crystal current is caused by the stalo, and the crystal current is a reliable indication that the stalo is operating; typical values are 0.5 to 0.7 mA. Additionally, most of the receiver noise is created by the stalo, and the absence of grass on the receiver video is an indication that either the stalo or signal mixer may have failed. On first approaching and entering a radar facility, it should be an immediate, habitual, and unconscious act for the technician to verify that (1) the antenna is rotating, (2) the modulator is audible, and (3) there is a receiver crystal current. Immediately thereafter, he should energize the maintenance ppi to ensure that clutter-free mti video is present.

The i-f Amplifiers

The preamplifier outputs are distributed to three i-f amplifiers, and the preamplifier is the last unit in what is commonly called the ***front end*** of the receiver system. Texas Instruments, the manufacturer of the ASR-4 through ASR-8, named the front end the ***common receiver,*** since it was common to the three i-f amplifiers. A receiver system includes the front end and i-f amplifier, so there are three receiver systems in the generic radar of Figure 6-1.

The Normal i-f Amplifier (see Figure 6-17)

This amplifier is called "normal" because it is the original form of the radar receiver; it is a conventional i-f amplifier with amplitude detection. The amplifier stages may be tuned in doubles, triples, or quads, so as to achieve the desired bandpass. Of the three receivers, the bandpass of the normal receiver usually most closely

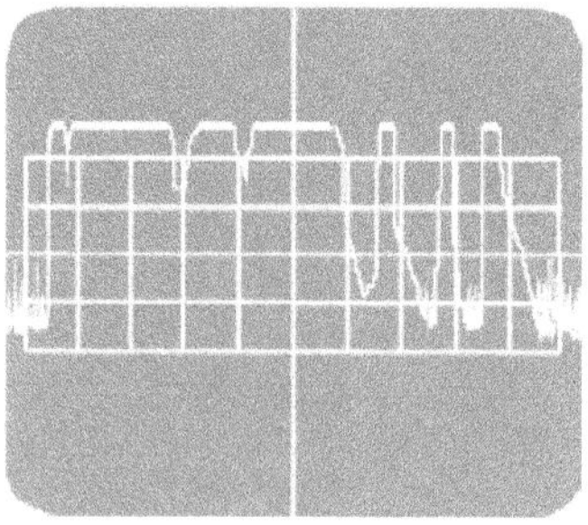

FIGURE 6–17

Normal video.

approaches $1.2/t_p$, and the mds for the normal receiver will be greater than that of the mti.

The normal receiver provides the best opportunity for target detection; there is a tendency for users to view the mti video only, and the practice should be discouraged. Not only is the normal receiver more sensitive, the mti processing can cause poor target visibility for many other reasons. Many systems contain range-azimuth gating circuitry to provide a mixed video called "mti," but which is actually processed mti only where programmed over clutter areas, and normal video in all other areas. There have been a number of systems employing ***clutter-gated mti,*** where the normal video is quantized and then used as a switch to enable mti only where clutter existed. A modern version of clutter-gated mti is used in the ASR-9 weather detection circuitry; a range-azimuth ***clear-day-map (cdm)*** memory is loaded with clutter-gate information during a special setup procedure, the information in the cdm is then used to switch between data that is equivalent to normal and mti.

See Figures 6-17 and 6-18. In appearance, normal video is characterized by the appearance of fine (as opposed to coarse) grass and a high percentage of limited targets. Depending upon the system and user, the video will have in the neighborhood of a 4:1 or 6:1 signal-to-noise ratio. An example of a 4:1 ratio would be grass of 0.5 V at the 70% average of the peaks, and video limited at 2 V.

Fast Time Constant (ftc)

Ftc circuits have been used since the 1940s as a means to attenuate large blocks of ground or weather clutter. The earliest circuits were simple differentiating networks in the normal video circuits, but the technology evolved to greater sophistication to achieve target survivability in the clutter. Target limiting in the linear normal i-f amplifier destroyed target data in clutter, leading to the development of an amplifier which exhibited decreasing gain with increasing input. More information on ftc and ftc types is to be found in other places in this publication.

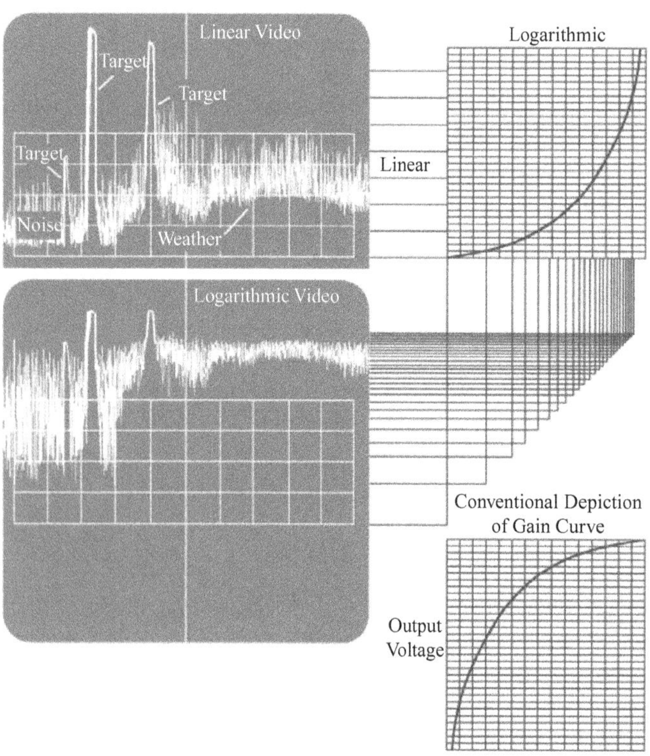

FIGURE 6–18

Logarithmic versus linear amplification.

The Logarithmic i-f Amplifier (see Figure 6-18)

The use of logarithmic information in radar circuitry has been increasing for, at least, the last three decades. Originally used as an electronic countermeasure, the log receiver now has several uses, some in weather detection and some as part of a weather-cancellation circuit. The log receiver illustrated in the generic system of Figure 6-1 is used for the latter purpose; its output is routed to a ***delay-line ftc*** circuit. Log video is also used for inputs to radar digitizers, where ***constant-false-alarm-rate (CFAR)*** thresholding is used to reduce false target message formation.

A major characteristic of the log i-f is that the signal may be detected at each stage, rather than only at the end of the strip, as in the normal i-f amplifier. Each detected output is placed onto a common summation line. As the signal grows stronger, the last stage will reach saturation, and the gain of the entire amplifier for still-increasing signals will be lessened by the gain of the saturated last stage. As the signal is increased further, the second-to-last stage becomes saturated, and the gain to increasing signals is lessened by the gain of the last stage, times the gain of the stage preceding it. As the signal is increased still further, stages continue to saturate, and the gain continues to decline. Because of this declining gain with increasing signal, the signal output is logarithmic. To one familiar with normal video, logarithmic video would appear to be unlimited and would appear to have a very poor signal-to-noise ratio. This appearance is due to the high gain to noise and the low gain to strong clutter. Figure 6-18 illustrates a comparison between normal and log videos. Note the pronounced difference in the appearance of weather.

Because the gain curve makes the signal logarithmic, the amplifier output becomes representative of the ***signal power in deciBels,*** and it follows that the purpose of the logarithmic amplifier is to convert the radar data to a deciBel scaling. When logarithms are added, their antilogs are multiplied, and when a logarithm is subtracted from another, the antilog of the subtrahend is a divisor of the number. With radar data, this becomes a means to amplify or attenuate a signal.

The log i-f amplifier requires special alignment procedures other than the usual swept-frequency generator method. Because the detected swept-frequency output is distorted by the logarithmic gain, the displayed bandpass may not be a good representation of the true frequency response. Of great concern in the log i-f performance is the ***dynamic amplitude response,*** a test of the input versus output signal amplitudes to verify the logarithmic gain. Manufacturers provide specific instructions for the manner in which this is to be done.

The ASR-9 has only one, linear, wide-dynamic-range i-f amplifier with a very narrow bandpass. This causes a very low grass level. Because technicians at the radar facility must be able to see a ppi display of the receiver video, an output of the i-f amplifier is applied to a logarithmic-gain circuit to decrease the signal-to-noise ratio to one which will better display the grass on the monitor PPI.

The mti i-f

The mti system depends on ***phase detection,*** and the mti i-f amplifier design contains two features to prepare the received signals for phase detection. It has a very wide bandpass, sometimes as much as $3.5/t_p$, and it is ***hard limited,*** so that all signals except the very weakest reach the limit level. Because the objective is phase detection and because the phase detector output voltage is representative of the phase difference between the i-f echo and the coho, the i-f inputs to the phase detector must be of a standardized amplitude for accurate detector outputs. Those weak, unlimiting signals from clutter at the edges of the beam cause amplitude variations at the canceler input, and are ***the main cause of mti antenna scanning clutter residue.*** Adjustment to the I-F GAIN control can place more targets into limit, but such adjustment also raises the grass level and makes targets over clutter more difficult to distinguish. The MTI I-F GAIN is one of two final adjustments to the MTI system; the controls are called the ***payoff controls.***

Older radar systems used a single phase detector with a ***triangular response*** at the output of the mti i-f amplifier, as illustrated in Figure 6-19. A triangular response is necessary wherever a single phase detector is used.

As illustrated in Figure 6-20, quadrature phase detectors were implemented to overcome a condition called ***phase ambiguity,*** also known as the ***blind phase effect.*** Although the condition and its correction had been recognized for years, the cost and bulk of circuitry

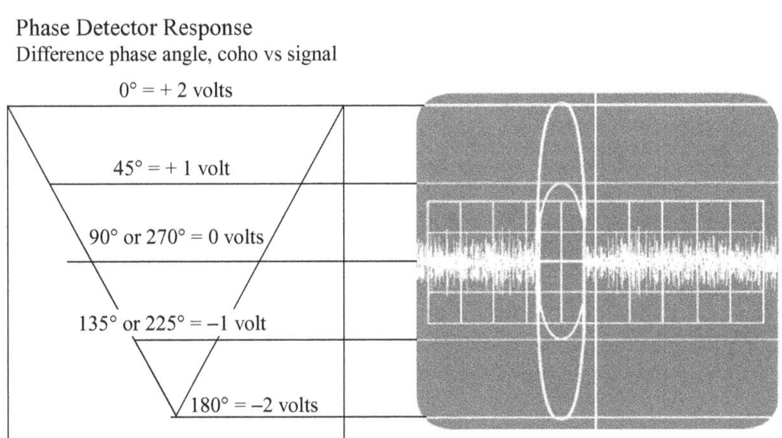

FIGURE 6–19

Triangular phase detector response.

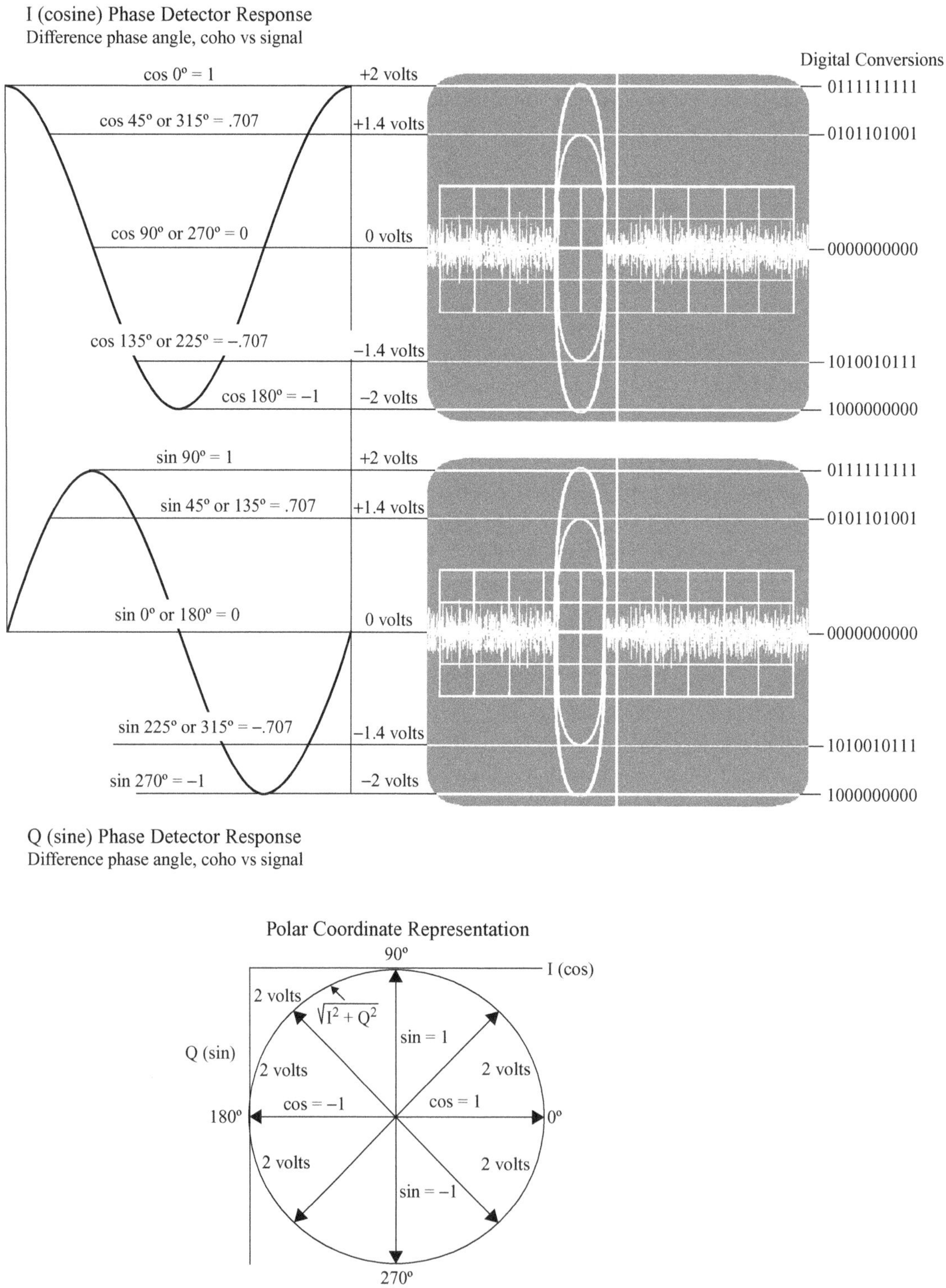

FIGURE 6–20

Quadrature phase detection.

made the fix impractical until integrated-circuit mti cancelers became feasible. The i-f output is applied to two phase detectors, instead of one. The coho applied to one of the phase detectors is shifted by 90°, and each detector offers a sinusoidally shaped response. The normal response of a detector is cosinusoidal, so the 90° shift causes the other phase detector to have a sinusoidal response. The sine or cosine are those of the angular difference between the signal and coho.

Quadrature Phase Detection

Quadrature phase detection and its applications can become an extensive subject. One of the most familiar applications is in the color television receiver, where the phase of a 3.58-MHz subcarrier determines the displayed color, the amplitude determines the hue (depth) of the color, and the baseline upon which it rides determines the luminance (brightness). To obtain the amplitude information, a single phase detector will not suffice, since the output amplitude is representative of phase angle. However, when quadrature detectors are used, the trigonometric identity $\sin^2\theta + \cos^2\theta$ = unity provides the means to achieve absolute magnitude information. As illustrated in the lower part of Figure 6-20, if the peak amplitude of the phase-detector outputs were 2 V (unity), the sum of the squares of the sin and cos of any angle times 2 V would also be equal to 2 V. Obviously, the same will apply for any amplitude.

Quadrature phase modulation and detection is also a major technique in a high-speed modem data transmission technique called ***phase-amplitude modulation (PAM).*** Voice-quality telephone lines are capable of only 2,400 bits/s, or 416 μs/bit. If the modem subcarrier has two possible amplitudes and eight possible phase angles during the 416 μs, then 16 possible combinations may be represented in the 416 μs. Of course, 4 bits can be represented by these 16 combinations, so 9,600 bits/s can be transmitted over 2,400 bits/s lines.

In color television, and in PAM, the cosine phase detector was called ***in-phase (I)*** and the sine phase detector was called ***quadrature (Q).*** The same designations are also assigned the two phase detectors in quadrature mti systems. In the literature regarding the earliest radar systems, the I detector was sometimes called the sine detector, but the in-depth analysis of the phase detector operation will make it apparent that this may have been erroneous.

If a single radar phase detector produces a target, the signal-to-noise ratio for that target is dependent upon the phase difference between the i-f echo and coho. However, because $\sin^2\theta + \cos^2\theta$ = unity, the combined output from the two phase detectors is representative of the magnitude of the target. In the case of the hard-limited mti amplifier, nearly all targets should be i-f limited, and the two phase-detector outputs will represent the i-f limit level.

MTD systems also use quadrature phase detection, but the i-f amplifier is not limited, and the output of the two phase detectors is deliberately utilized as a representation of signal strength.

In quadrature mti, the two phase-detector outputs will be used by two independent canceler systems called I and Q cancelers. These are not to be confused with cascaded cancelers 1 and 2, contained in each (I and Q) canceler. Chapter 13 of this book is devoted to mti processing.

The mti Analog-to-Digital (A/D) Converter

In digital mti systems, the phase-detector output is converted to a binary number, once each range cell; the optimum size of the range cell is $0.75t_p$. A larger range cell can result in missed targets; a smaller one places unnecessarily high requirements on amplifier bandpasses and data memory capability.

In synthesis systems, the time of a range cell is related to the coho frequency; in magnetron systems, another oscillator must be provided, since the coho must be interrupted each T_r. The ASR-8 uses a 30-MHz coho divided by 14, to produce range cells of 0.467 μs. Later systems use a 31.07-MHz coho divided by 24, to provide 0.7725 μs range cells, which are 1/16 nmi in radar range. This is useful in data processing, as 1/16 is 2^{-4}, and fractions such as 1/8, 1/4, and 1/2 are readily available.

Not shown in the generic system of Figure 6-1 is a "conditioner" between the phase detector and A/D converter. This device is used for the insertion of test signals and to raise the bipolar video baseline so that all the phase-detector data is either positive or negative; this baseline shift is only done to satisfy an operational requirement of the D/A converter. The baseline is digitally removed after the conversion.

After the A/D conversion and baseline restoration, which may be in the canceler, the binary number in each range cell will be polarized, as illustrated in Figure 6-17. The baseline will be all 0s, a maximum positive will be represented by a 0 and all 1s, and a maximum negative will be represented by a 1 and all 0s.

The A/D converter will, in most cases, comprise two ***quantizers,*** an ambiguous term used to describe two A/D converters: a ***primary quantizer*** used to develop the msbs and a ***secondary quantizer*** to utilize a primary quantizer error voltage to develop the lsbs. The term quantizer is ambiguous because it has been used for years to describe amplitude-standardization circuitry, in which video applied to a thresholded amplifier causes either a maximum or minimum output.

The mti Processor

This block may consist of many processing circuits, and Chapter 12 and 13 of this book are devoted to it. It may contain cascaded cancelers, delay shift registers, velocity-shaping feedback circuitry, a quadrature combiner, digital log conversion, ftc, antilog circuitry, and others. The basic purpose is to pass those targets exhibiting a significant Doppler shift, and to discriminate against those targets which do not. The Doppler shift is apparent as a pulse-to-pulse change in amplitude and/or polarity of bipolar targets. In digital systems, the cancellation of fixed targets is accomplished in a digital adder circuit, where the undelayed data is added to the complement of the delayed data, and so done with a forced carry to accomplish two's complement. In short, delayed is subtracted from undelayed. After cancellation, the bipolar mti data is converted to unipolar.

The Enhancer

The intent of the first of this type of device is somewhat debatable; it began as a ***video integrator,*** developed almost simultaneously for use as an electronic countermeasures device, and an improvement to air traffic control radar displays. The enhancer will contain a one-T_r delay device, such as a shift register. A pulse-to-pulse comparison will be made of the video, as in a canceler, but the delayed video will be approximately 95% of the undelayed, and will be ***added to the undelayed.*** The delayed video is also the sum of the delayed and undelayed, so a feedback loop is created; the 95% delayed amplitude is necessary to prevent oscillation. In the analog integrators, the feedback was adjustable, and the adjustment was very critical.

Figure 6-21 is an abbreviated drawing of the enhancer first used in the ASR-8; as depicted, there is no provision for enhancer-off operation. Figure 6-22 illustrates analog equivalents of the input, limited, and output videos, and of the differences in ppi presentations between enhancer-on and -off operations.

The limited data is applied to the feedback adder, where it is summed with data that has been delayed by one T_r and scaled by 15/16, which is 93.75%. The scaling is accomplished by right-shifting the delayed data by four places to achieve 1/16, complementing it and then adding it to the unaltered data with a forced carry. The end result is that 1/16 of the data is subtracted from the unaltered data for 15/16 feedback. The illustrated circuit does not contain adequate bit resolution, and the intent is to show the technique for right-shifting four places by appropriate wiring of the input to the adder. The actual circuitry used all six complemented bits and added zeros to the lsb inputs of the uncomplemented adder input for column alignment.

The output of the feedback adder is routed to both the delay shift register and the output circuitry. In the output circuitry, a bias may reduce the noise from the "bottom up" by clipping the baseline. A limiter also limits the peak output amplitude.

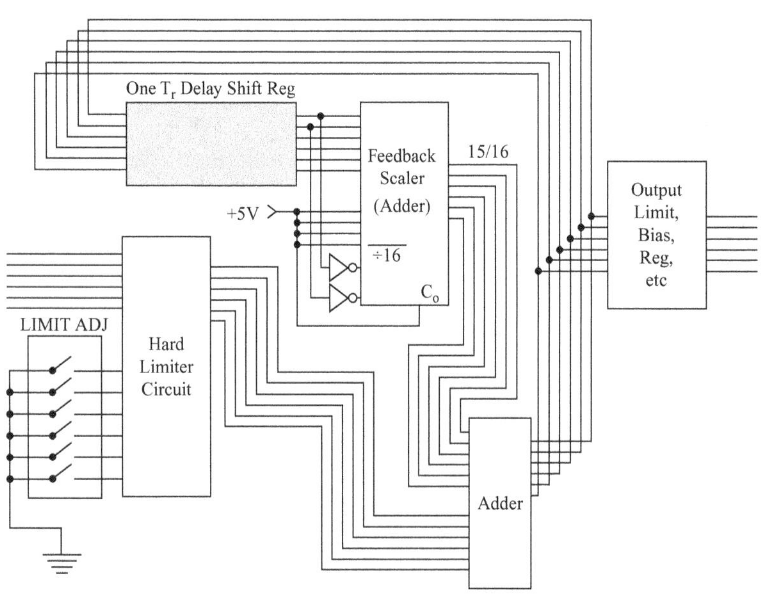

FIGURE 6-21

Abbreviated digital enhancer diagram.

The enhancer does a fine job of eliminating interference, and of creating strong, bright, targets on the ppi display; air traffic controllers will insist on using it when it is made available. However, it is not without shortcomings. The manner in which a target is enhanced over several T_r's is illustrated in the lower portion of Figure 6-22. Note that the first hit, limited below visibility, does not provide an output, and that the decay of the feedback loop continues to produce a target long after echoes have ceased to be present. This causes the center of the target to appear to be further clockwise than it actually is, and the degree of error is variable with target ***run length,*** the number of hits or azimuth width. If enhanced data is used as an input to a digitizer, the azimuth-sliding-window process used to calculate the azimuth center will be in error.

In synthesis systems, the time of a range cell is related to the coho frequency; in magnetron systems, another oscillator must be provided, since the coho must be interrupted at each T_r. The ASR-8 uses a 30-MHz coho divided by 14, to produce range cells of 0.467 µs. Later systems use a 31.07 MHz coho divided by 24, to provide 0.7725 µs range cells, which are 1/16 nmi in radar range. This is useful in data processing, as 1/16 is 2^{-4}, and fractions such as 1/8, 1/4, and 1/2 are readily available.

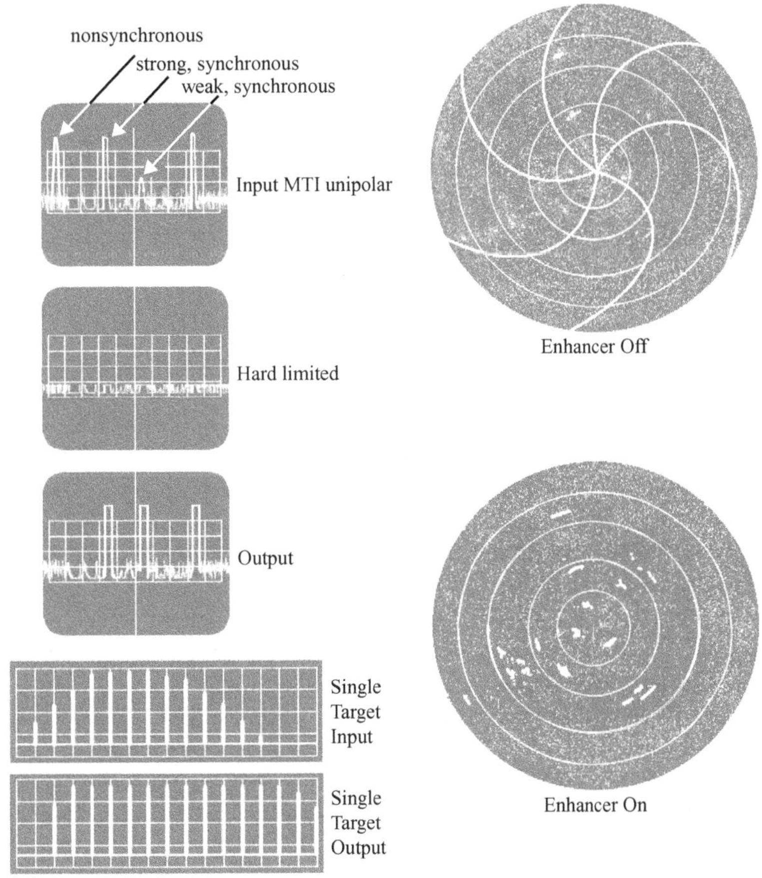

FIGURE 6–22

The effects of an integrator/enhancer.

The enhancer will strengthen the appearance of all synchronous targets; this unfortunately includes mti clutter residue; the ppi presentations in Figure 6-22 illustrate this. A change from enhancer-off to enhancer-on operation may necessitate a change in the ***mti velocity response shape,*** which determines the minimum visible target Doppler.

The Log and Normal Receiver Processing

Were it not for the need to convert it to a form suitable for the enhancer, to make the log ftc video available, and to temporally align it with the mti video, the detected normal video at the output of the normal i-f unit in Figure 6-1 could be used for display. It is routed to a ***log/normal select*** switch, where it will pass, unaltered, unless the ***NORMAL LOG FTC*** switch has been placed to ON.

Differentiating ftc

The output of the log receiver in the generic system of Figure 6-1 is applied to the ***delay-line ftc*** circuit. The term ***fast time constant (ftc)*** is another old carryover from early-day radar, and is not truly descriptive of the operation of this circuit. As early as the 1940s, to remove large blocks of weather from the displays, normal video was applied to a differentiating network with a fast time constant, so that long-duration video would be "canceled" after a duration of time approximately equal to that of the radar system, t_p. The general principle and effect is illustrated in Figure 6-23.

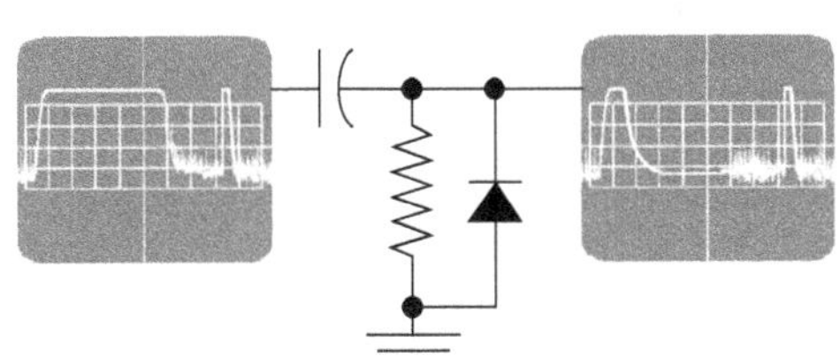

FIGURE 6–23

Large block of video differentiated by a fast time constant.

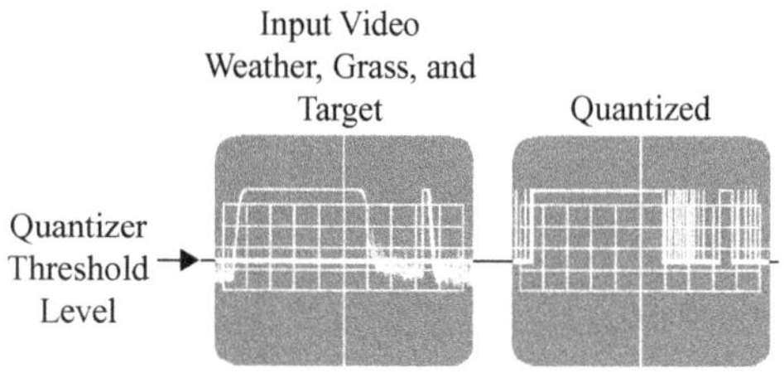

FIGURE 6–24

Quantized video.

The original ftc circuit accomplished the purpose of removing the objectionable blocks of bright weather displayed on the operator's ppi, but it eliminated most of the targets in the weather, also. Normal video rapidly reaches the limit level with increasing signal strength, very roughly speaking, at about 20 dB above system noise. This means, of course, that any target in weather of significant strength would be obliterated by limiting.

IAGC and Soft Limiting

An early attempt to improve upon differentiating ftc, called ***instantaneous automatic gain control (iagc),*** appeared in the ASR-4, -5, and -6. A fast-action agc circuit in the forward i-f amplifier stages caused an abrupt gain reduction; the agc action was delayed, in order that any target not longer in duration than t_p could pass unaffected. By accomplishing signal strength reduction at this point in the amplifier, targets in weather were less likely to be destroyed by limiting, which would ordinarily occur further downstream. A latter-day mti i-f improvement called ***soft limiting*** bears similarity to iagc. Heavy i-f currents caused by strong clutter are used as gain reduction in an effort to improve subclutter visibility.

As radar technology expanded, increasing attention was devoted to improving digitizer performance. To process radar data, the received analog information must be converted to ones and zeros, and ***thresholding*** circuits called ***quantizers*** were employed; any radar video pulse exceeding a threshold level would become a "one," and any that did not would become a "zero," as illustrated in Figure 6-24. Of course, the thresholding would occur on noise pulses, as well as targets, as there was no way for the circuit to recognize the difference. In the earliest days, threshold breaks caused by anything other than radar targets were called ***false alarms,*** as they had the potential to become a target detection. Raising the threshold above all the noise caused weak targets to be lost, and lowering the threshold to detect weak targets caused an unacceptable number of false alarms. The rate at which these occurred was called the false alarm rate, as the rate increased, so also did the likelihood that undesirable threshold breaks would become digital target detections. Obviously, it became important to keep noise and weather at as constant a level as possible, so that threshold breaks could be held to a minimum.

Constant False Alarm Rate (CFAR)

A number of methods to provide a ***CFAR*** have been devised. One technique involved wide-band amplification to provide good radar pulses, followed by hard limiting, as in the enhancer input circuitry, followed again by narrow-band amplification to restore the amplitude of the pulses to a greater degree than the noise. ***Pulse width discrimination*** in the digitizers inhibited the very narrow pulses created by noise threshold breaks. Another technique utilized a logarithmic i-f amplifier, followed by a t_p delay-line cancellation circuit, as shown in Figure 6-25. Many consider the delay-line ftc and its derivations to be synonymous with ***CFAR circuits;*** others consider CFAR circuits to encompass a wider variety, including video integrators.

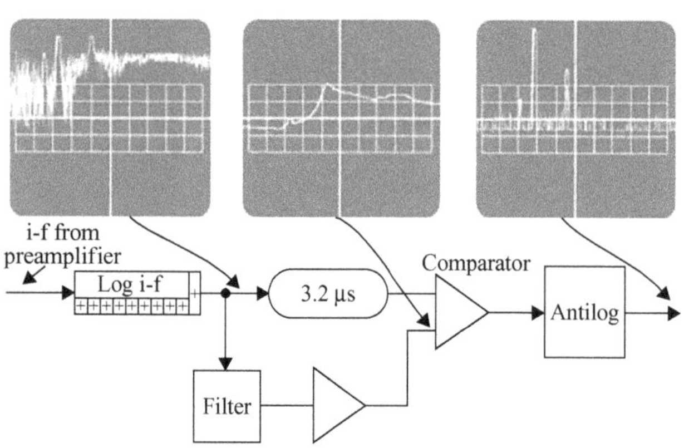

FIGURE 6–25

Delay line ftc.

Log ftc circuits had been in use as an electronic countermeasures device against chaff for quite some time, but first entered the air traffic control realm when introduced in the 1960s as a "Modification to Provide Precipitation Suppression" in GCA radars used in Viet Nam, where heavy seasonal rainfall rendered the X-band PARs nearly useless.

Delay-Line ftc (see Figure 6-25)

The logarithmic video from the log i-f is routed to two places, a low-pass filter, and a delay line. The outputs of the filter and delay line are both applied to a comparator, where the average from the filter is subtracted from the delayed video. The purpose of the delay is to allow time for the filter to build an average before the comparison takes place. Because the video is logarithmic, the lumped averages from the low-pass filter are also logarithmic, and the log average is subtracted from the log video. Recall that if one logarithm is subtracted from another, the antilog of the subtrahend is a divisor of the antilog of the number from which it is subtracted. Since log video is representative of the signal power in deciBels, the comparator subtracts the average in deciBels from the signal in deciBels, effectively attenuating the signal power. The output of the comparator is applied to an antilog circuit, which amplifies the inputs in exactly the opposite manner as did the log i-f, restoring the video to a linear condition. Because of the delay to the logarithmic video, the video would be "late" when log ftc was selected. Other circuitry, not shown, will delay the 6-bit normal/log digital data for proper temporal relationships with the mti data; both are delayed, but the normal is delayed by an additional seven clocks (approximately 3.2 μs).

In mti systems, it is not possible to use a logarithmic i-f amplifier, because signal intelligence needed for phase detection would be distorted. The need for ftc is, nevertheless, still present, but must be accomplished downstream of the canceler. In the ASR-8, a digital log ftc was introduced, made possible by integrated-circuit technology. The canceled mti unipolar video was made logarithmic by a digital arithmetic process, and the average was achieved by a nine-range cell accumulator circuit. More information is available on this circuit in Chapter 13, "An mti Processor".

The Destagger Circuitry

As with most other units, this circuit may go by different names, according to the manufacturer. It may also be called the "realignment unit," "alignment unit," "destagger," or even "destaggerer."

In earlier systems, where quartz delay lines were used for mti canceler delay, it was necessary to "destagger" the video upstream of the canceler (see Figure 6-26). To "destagger" the video is to convert it from two or more different system intervals to a single system interval. Because these earlier systems required a delay line, delay-line carrier, modulator, and other associated circuitry, they were limited to two or three T_r's. By delaying every other modulator trigger, two T_r's were created, and by delaying every third modulator trigger, three T_r's were created.

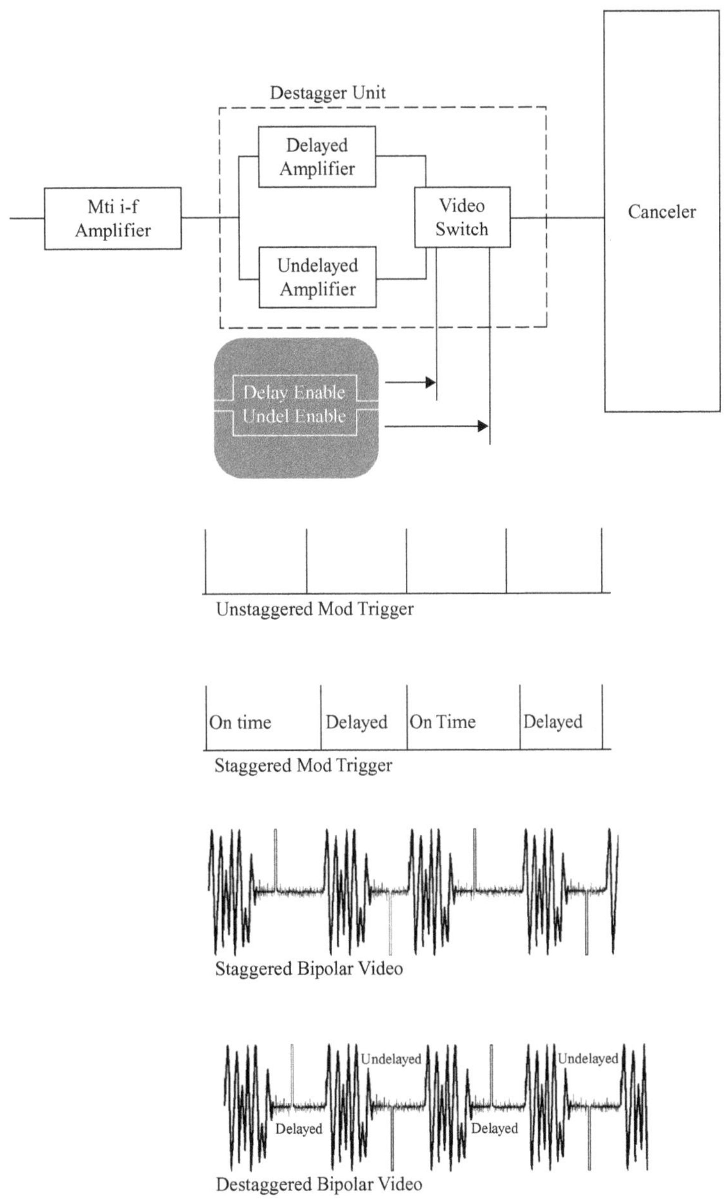

FIGURE 6-26

Destaggering in an analog MTI system.

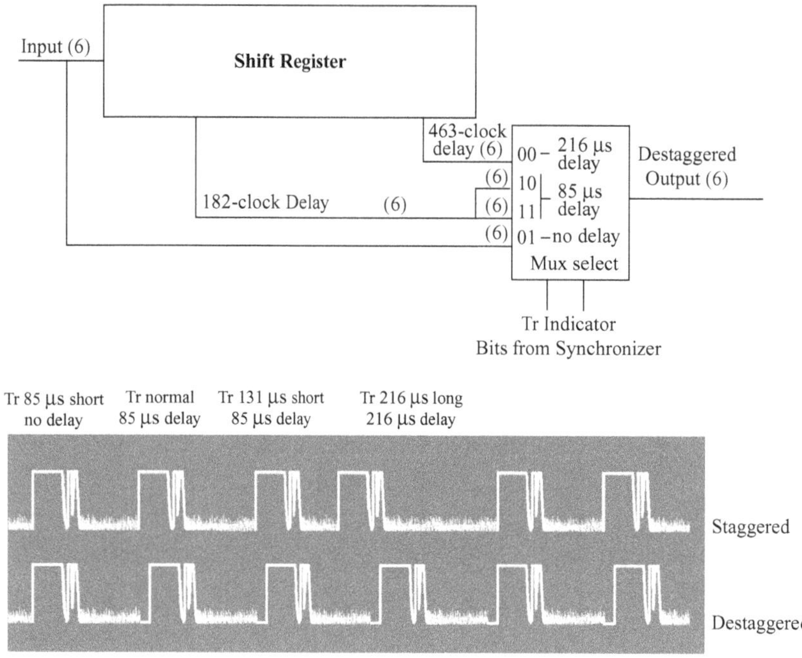

FIGURE 6–27

Destaggering 6-bit digital data.

Once shift registers became the method to delay video in the canceler, the delay became a variable, based on the interruption time between range-cell clocks applied to the shift register. Precanceler destaggering then became unnecessary. In one case (ASR-7), staggered video is actually displayed on ppi's, but triggers to the displays are also staggered, so there is no appearance of multiple videos.

Normal A/D and Enhancer

The normal A/D circuit is necessary to convert the video to a form suitable for processing in the enhancer and destagger circuits. In the ASR-8, and as shown in the generic system of Figure 6-1, the A/D provides 6 bits for the enhancer. The enhancer is identical to the mti enhancer; in the ASR-8, it is even on the same circuit card. Earlier systems contained only one enhancer (called integrator), with an mti/normal gating circuit on the input.

Destaggering in the generic system of Figure 6-1 is accomplished with shift registers (see Figure 2-27). When this technique became possible with the introduction of integration circuits, it made possible the use of more intervals. The generic system shown has four T_r's, as does the ASR-8. The four T_r's are easily implemented, as 2 bits can represent the selected T_r, and operate a four-input multiplexer in the destagger circuitry.

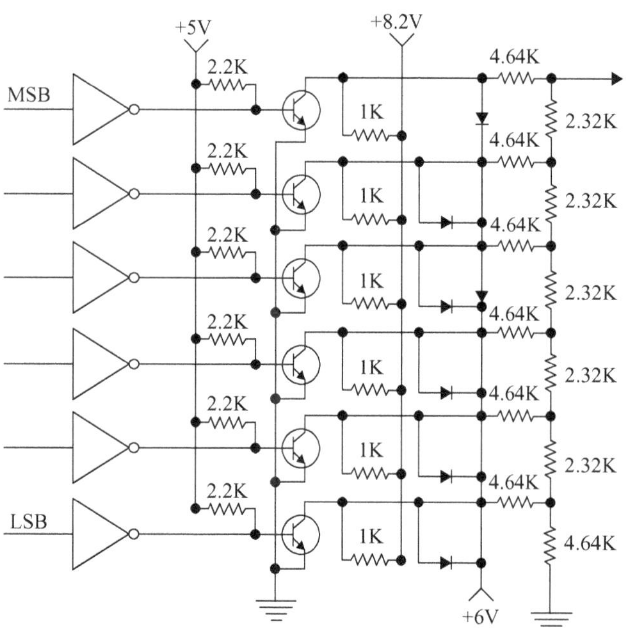

FIGURE 6–28

Digital-to-analog converter.

The Digital-to-Analog Converters

These circuits are known as an R/2R ladder circuit, and closely resemble those used in the ASR-8 (see Figure 6-28). If the input bit is low, the inverter output is a "1," and the associated npn transistor is forward biased by the electron flow from the emitter, through the base and 2.2 K resistor, to the +5-V supply. The transistor then conducts at saturation, and the collector is at near-ground potential, reverse-biasing the associated diode, which has a positive 6 V on the cathode. When an input bit is high, the inverter output goes low to stop current flow through the transistor. The collector voltage rises to maximum, forward-biasing the associated diode, and causing the 6-V reference to be applied to the ladder. Whenever a transistor is cut off, current flow through the ladder includes one 4.64 K resistor and some number of the vertically drawn resistors, to form a voltage divider. The more significant bits cause a voltage drop across more resistance, and the voltage at the output is greater. The resistance values cause the voltage drops to be halved from the msb down.

The ladder voltages created by different set bits are additive. When only the msb is set, the output is half the reference voltage; if only the second msb is set, the output is one-fourth the reference voltage, and if only those 2 bits are set, the output is 3/4 the reference voltage.

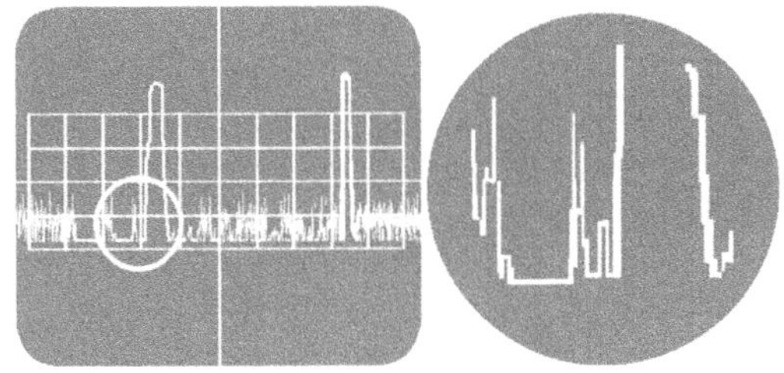

FIGURE 6–29

Steps on video caused by A/D and D/A conversions.

In some applications of D/A converters, the reference voltage may be varied by either adjustment, or by a signal, which is then "modulated" by the digital input.

The effects of A/D and D/A conversion on radar video are worthy of note to the technician (see Figure 6.29). Once the video is produced at the output of the D/A, and subject to final limit and amplifier settings, it is suitable for display. However, close inspection will reveal "steps" from the processing. If this video is to be used in digitizing equipment, which contains quantizers, these steps can cause unstable target leading-edge range declarations; if a step varies in amplitude, the range of the threshold break will also vary.

The Line Drivers

Consistent with their name, these units will provide the necessary current drive for the coaxial lines and balanced pairs connecting the radar site and indicator facility. To make it possible to send a display site pretrigger over the same line with video, therefore eliminating the need for one cable, a trigger is combined with the normal video, but at an amplitude much greater than the video. For example, the trigger may be 40 V, while the video may be only 6 V. This permits the trigger to be "stripped off" the video at the receiving end, accomplished by a circuit biased to inhibit the lower voltage video. A spare coaxial cable is usually provided, and it is common practice to use the spare cable for a separate trigger cable until a failure necessitates the use of the normal video cable for the trigger.

The outputs of the line drivers, video, trigger, and azimuth data are routed to a ***radar cable junction box,*** the final demarcation point for all wiring between the radar and indicator facilities. In addition to the data cables, many pairs may be provided for remote control of the system, communications, "readbacks," and more. In some of the earliest systems, the interfacility connections were accomplished with overhead telephone-type wiring and poles, but all newer systems are underground. Earlier ARSR facilities were equipped with ***radar microwave link (RML)*** systems to transfer data; analog radar information frequency-modulated X-band klystrons, and readback-and-control information was accomplished by subcarriers on a ***voice data multiplex (VDM)*** system.

The Digitizer

Rather than a line driver or RML, ARSR facilities now contain a ***common digitizer,*** so called both because it may be used with any analog ARSR radar data and because it may supply both the FAA and military users with digital radar message data. The common digitizer is a complex equipment in itself and performs many functions, such as automatic quantizer thresholding adjustment, pulse width discrimination, automatic clutter elimination, beacon data processing, beacon/ search correlation, and many more. In many cases, ASR radar data will also encounter a digitizer, although it will most often be located at the indicator facility. In any case, one basic underlying principle will be utilized in nearly all digitizers, a data-processing technique called the ***conventional azimuth sliding window.*** The word "conventional" is used to distinguish the process from ***azimuth centroiding,*** used in the ASR-9, and from ***monopulse azimuth*** determination, now used in latter-day beacon interrogators and fire-control radars. Still further, it is essential for clarity that "azimuth" not be omitted from "azimuth sliding window," to preclude any confusion with "range sliding window," which is an entirely different process, and done as part of a digital mean-level CFAR process.

The actual manner in which this is implemented is equipment dependent, but a simplified and generalized description is illustrated in Figure 6-30. A quantized radar echo becomes a digital "1," called a ***search hit***; the

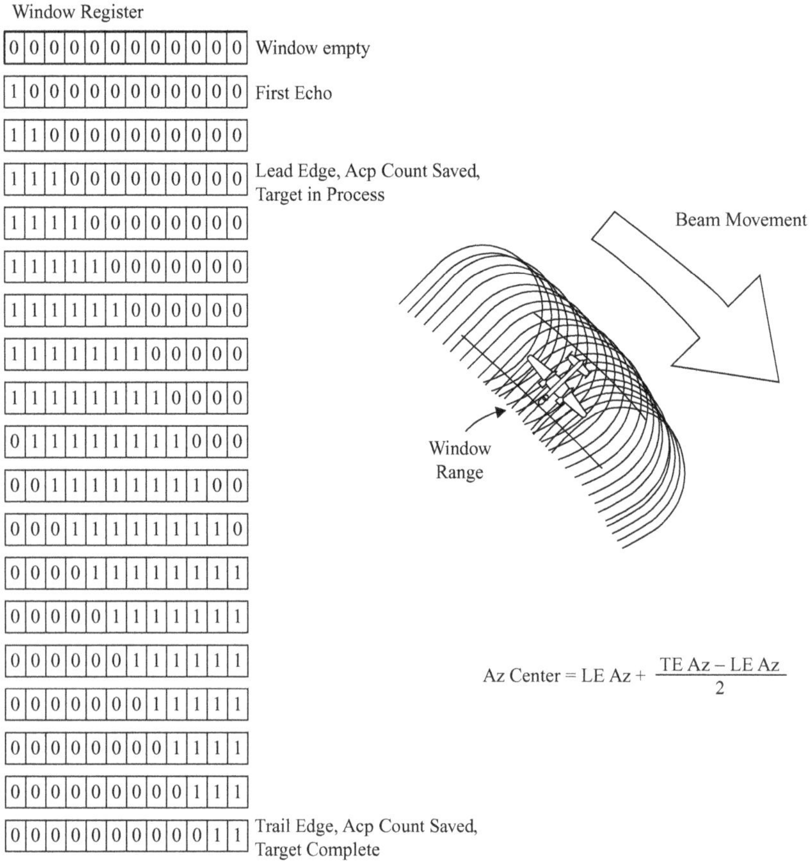

FIGURE 6–30

The azimuth sliding window process.

word "search" is of military origin and is synonymous with "surveillance," as in ASR or ARSR. When a search hit occurs during the radar live time, called ***real time,*** a ***range counter*** output is saved to indicate the presence of a hit. At the end of live time, in ***dead time,*** memory locations are reserved for the ranges of the hits which occurred during the T_r; in the generic illustration, the memory locations are in the form of a shift register; a single location associated with a single range is shown. When hits have accumulated at the end of a T_r, a search of the memory is made before any new location is assigned; if there is already an opened location for a given range, another "1" is shifted into the register. Any target for which a memory location has been opened is said to be ***in process.*** Once a target is in process, the shift register contents will be updated with a "0" if there is no hit for its range. Because of false alarms, there will be several locations which contain only one or two hits, and are then emptied.

Once a predetermined and adjustable number of hits, obtained over several T_r's, are contained in the shift register, a ***target lead edge*** is logically declared; the acp count is then retained in memory, linked with the shift register hit count. As the beam sweeps across the target, the shift register hit count increases; when the beam has passed the target, the shift register hit count declines with each T_r. When (1) the lead edge has been detected and (2) the hit count has decreased to ***target trail edge*** value, the trail-edge acp count is retained for use in the center-of-azimuth computation. Center-of-target azimuth accuracy depends upon proper adjustment of the lead- and trail-edge criteria, and azimuth offsets may also be provided.

The detection of targets as the echoes are received during the T_r is called ***real-time processing;*** real time refers to those conditions under which data occurs at the same rate at which it is received, where 12.36 μs of time elapses for each radar mile. Those radars which display analog data on ppi's may be called ***real-time systems.***

Having developed a group of linked digital data containing the target range, center-of-azimuth, run length, mti/normal, and other information, an output message may be assembled in an ***output buffer*** and transmitted over modems and telephone lines to ARTCCs. To handle a large number of targets, more than one modem channel will be employed with a priority system; if a second target becomes available for transmission while one modem is busy transmitting the first, the target data is routed to the second, or the third, if the second is busy. These output messages have little relationship with radar real time; the time at which they occur may be somewhat "behind" the radar antenna position and will have no relationship with radar real-time range. These conditions are not significant, since the accurate target range and azimuth are contained in the message.

In instances where monopulse azimuth is unavailable, a similar conventional azimuth sliding window may also be applied to beacon transponder codes, where an additional ***code validation*** process assures that the received codes are consistent. Beacon and search targets are subjected to a comparison; when they are in proximity, and within a tolerance, a single ***reinforced*** target is declared. Another ambiguous term is the word ***correlated.*** In some systems, this may refer to a search-reinforced beacon target; in the ASR-9, a correlated target is one upon which a search track has been established and is not necessarily connected to the declaration of a beacon target.

Review Questions

1. On approaching and entering a radar site, the technician should observe ______________.
2. The most frequently used test connector in the radar system is ________________.
3. In the synthesis digital system, ____________ is the basic source of system timing.
   In the digital magnetron system, ___________ is the basic source of system timing.
4. What is the purpose of the "Preset" on the range cell counter?
5. The purpose of staggering the T_r is to______________________.
6. Staggered *fp* causes blind-velocity targets to become visible because____________________.
7. The charging choke in the transmitter modulator _________________ the charge on the pfn.
8. The pulse transformer matches the impedance between ________________ and ____________.
9. The bifilar windings on the pulse transformer are used to _______________________.
10. An inverse current meter reading of zero indicates _____________________.
11. The charging diode was made necessary by __________________________.
12. Briefly explain the initial tuning procedure for a new klystron drift tube.
13. The latest state of the art in transmitter design employs _______________________.
14. State the difference between a "north mark" and "azimuth reference pulse."
15. What is the difference between STANDBY and MAINTENANCE?
16. The circular polarizer is tested by____________________________.
17. The azimuth ring on the antenna is adjusted on installation to indicate_______________, because ______________________.
18. The apg azimuth reference pulse can be precisely set with an ________________________.
19. The double-balanced mixer provides a better mds because _______________________.
20. The exciter may also be called______________________, and its purpose is ________________.
21. Beyond amplification, the rf driver is needed to____________________.
22. After changing the transmitter frequency, the receiver ______________________ must be tuned.
23. The lna is necessary to _____________________________.
24. The receiver "front end" includes _____________ and may also be called ______________.
25. A technician observes the phase-detected output of the mti i-f amplifier and finds no grass or signals. The normal i-f amplifier shows the same effect. The technician should look for _________________and the most likely cause is the _______________.
26. The passive waveguide system is used at __________ranges for the purpose of _________.
27. A high, fluctuating, reverse power is measured at the bidirectional coupler. This may very well indicate ______________________.
28. Circular polarization is used to ________________________.
29. The need to replace a T-R tube might be indicated by _____________or ______________.
30. As seen on an oscilloscope, explain how log video differs from normal. ____.
31. A log i-f amplifier may require a factory-specified test procedure because _______________.
32. Log TC is used to _______________________________.
33. Name five signals that may pass through the rotary joint. ______________________, _____________, ______________, _______________, ______________________.
34. The enhancer exhibits two beneficial and two detrimental effects, ___________, ________________, ________________, and ________________.

35. If a D/A converter has a reference voltage of 10 V, and only the third msb is set, the output will be ______________________________.
36. In an analog mti system, destaggering must be done _______________in a digital system, ______________________________.
37. In a system with two feedhorns, the recovery point of the active path stc would _____________ in relation to that of the passive path.
38. Radar digitizers employ a process called ________________________________.
39. A digitizer may employ more than one modem because ______________________.
40. If a radar and beacon target are detected at the same azimuth and range, the target is said to ______________________________.

Answers to Review Questions

1. On approaching and entering a radar site, the technician should observe ***antenna rotation, an audible modulator, and receiver crystal current.***
2. The most frequently used test connector in the radar system is ***the INCIDENT POWER connector on the directional coupler.***
3. In the synthesis digital system, ***the coho*** is the basic source of system timing. In the digital magnetron system, a ***separate master oscillator*** is the basic source of system timing.
4. What is the purpose of the "Preset" on the range cell counter? ***It establishes the*** f_p.
5. The purpose of staggering the T_r is ***to reduce the effects of mti blind velocities.***
6. Staggered *fp* causes blind-velocity targets to become visible because ***the Doppler remains constant, but a change to the*** $\Delta\phi$ ***is introduced.***
7. The charging choke in the transmitter modulator ***doubles*** the charge on the pfn.
8. The pulse transformer matches the impedance between ***the PA and pfn.***
9. The bifilar windings on the pulse transformer are used to ***prevent the high-voltage pulse from being added to the filament voltage.***
10. An inverse current meter reading of zero indicates that ***the inverse current diode is probably open.***
11. The charging diode was made necessary by ***staggered*** f_p.
12. Briefly explain the initial tuning procedure for a new klystron drift tube.

    The tube is first adjusted according to the manufacturer's chart and then the lowest possible high voltage is applied. The technician views the power, spectrum, and detected pulse shape as he tunes.
13. The latest state of the art in transmitter design employs ***many, small, low-power, transmitters.***
14. State the difference between a "north mark" and "azimuth reference pulse." ***An azimuth reference pulse is produced by the apg. A "north mark" is a strobe to be displayed on a ppi.***
15. What is the difference between STANDBY and MAINTENANCE? ***The transmitter is in dummy load, in both cases. In MAINTENANCE, control of the system has been electrically released by the user.***
16. The circular polarizer is tested by a ***test fixture which enables power measurements in both vertical and horizontal planes.***
17. The azimuth ring on the antenna is adjusted on installation to indicate ***true north, because it will not change.***
18. The apg azimuth reference pulse can be precisely set with ***an oscilloscope or "pulse catcher."***
19. The double-balanced mixer provides a better mds because ***noise pulses from the stalo are self-canceling.***
20. The exciter may also be called ***the buffer mixer,*** and its purpose is to ***synthesize the transmitter frequency by mixing the coho and stalo frequencies.***
21. Beyond amplification, the rf driver is needed to ***properly shape the drive pulse.***
22. After changing the transmitter frequency, the receiver ***preselector*** must be tuned.
23. The lna is necessary to ***provide high gain and low noise, because it is the first active stage in the receiver, and therefore has the greatest effect on noise figure and mds.***
24. The receiver "front end" includes ***everything in the path from the antenna to the preamplifier outputs,*** and may also be called the ***common receiver.***
25. A technician observes the phase-detected output of the mti i-f amplifier and finds no grass or signals. The normal i-f amplifier shows the same effect. The technician should look for ***a loss of crystal current,*** and the most likely cause is the ***stalo or mixer.***
26. The passive waveguide system is used at ***close*** ranges for the purpose of ***reducing clutter strength.***

27. A high, fluctuating, reverse power is measured at the bidirectional coupler. This may very well indicate ***a defective rotary joint.***
28. Circular polarization is used to ***attenuate precipitation echoes.***
29. The need to replace a T-R tube might be indicated by ***excessive recovery time or mds.***
30. As seen on an oscilloscope, explain how log video differs from normal. ***It appears to have a poor signal-to-noise ratio, and the limit level is indistinguishable. The signal-to-noise ratio of normal will be around 4:1, and limit level is obvious.***
31. A log i-f amplifier may require a factory-specified test procedure because ***the bandpass will be distorted by the logarithmic gain, and the dynamic amplitude range is of the greatest interest.***
32. Log ftc is used to ***attenuate rain clutter and noise, and establish a CFAR.***
33. Name five signals that may pass through the rotary joint. ***Active and passive waveguide paths, beacon directional, beacon omnidirectional, and polarizer control and readbacks.***
34. The enhancer exhibits two beneficial and two detrimental effects, ***improved appearance of weak targets, elimination of nonsynchronous interference, increase in clutter residue, and center-of-target azimuth shift.***
35. If a D/A converter has a reference voltage of 10 V, and only the third msb is set, the output will be ***1.25 V.***
36. In an analog mti system, destaggering must be done ***upstream of the canceler;*** in a digital ***system, anywhere upstream or downstream of the canceler, or not at all, if staggered triggers are used for the indicators.***
37. In a system with two feedhorns, the recovery point of the active path stc would be ***at a greater range in*** relation to that of the passive path.
38. Radar digitizers employ a process called ***a conventional azimuth sliding window.***
39. A digitizer may employ more than one modem because ***a single modem cannot produce message outputs fast enough to "keep up" with the radar data.***
40. If a radar and beacon target are detected at the same azimuth and range, the target is said to be ***reinforced.***

CHAPTER 7

Secondary Radar Systems

General

"Radar" is an acronym for "radio detection and ranging." Even though most of us assume radar to be an echoing system, "radio detection and ranging" can be achieved by another means. A radar can also be an "answer-back" system, in which an ***interrogator*** transmits a signal to all aircraft equipped with ***transponders***. On receipt of the interrogation, the transponder transmits a coded reply at a frequency differing from the interrogator's. To differentiate between the answer-back systems and the conventional echoing radars, the latter are called ***primary radars*** and the former are called ***secondary radars.*** Secondary radar systems have been built both with ground interrogators for aircraft transponders and with aircraft interrogators for ground or shipboard transponders.

Secondary Radar is Essential Safety Equipment

The secondary radar system is absolutely essential to safe air traffic control, as it is the only means by which radar controllers may reliably identify aircraft on their displays. ARTS computer data processing now maintains tracks on transponder-equipped aircraft to detect minimum safe altitudes or potential collision conditions. Prior to its introduction, to identify a specific aircraft on a display, radar controllers would issue course-change instructions to the aircraft pilot and then watch for the turning target on their ppi displays. This was hazardous, as another turning aircraft could be misidentified as the one under control, or the aircraft could be turned into another aircraft or ground obstruction if located somewhere other than the controller's estimated position. The secondary radar transponder in the aircraft provides controllers with an identification code and altitude report. Beyond that, a pilot can close a switch to transmit a special "ident" signal for positive identification during communication with the controller.

Advantages and Disadvantages of Secondary Radar

It is far too easy for the civil aviation community to forget that the original intent for radar was to provide early detection and warning of an enemy attack. The secondary radar system has many advantages over the primary system, but above any of its inherent problems, it has one very serious shortcoming. It is totally reliant upon aircraft transponder equipment. Without primary radar, aircraft not containing an operational transponder cannot be detected, whether it be for reasons of equipment failure, or deliberate clandestine operation. For this reason, the secondary radar should always be considered an accessory to the primary system, however necessary it may be. Operation without a primary radar, particularly in a ***terminal radar*** environment, where low-altitude air traffic arrivals and departures are controlled, should be considered hazardous.

Dangers in Excessive Reliance

Any system that relies on active equipment in the aircraft presents some safety or security hazards. As the author writes this, an orbiting satellite was destroyed by a communist Chinese missile. The new jeopardy to ***Global Positioning Satellites (GPS)*** used for aircraft navigation is now clear. ***Instrument Landing Systems (ILS)*** have replaced most ***Precision Approach Radars (PAR);*** GCA controllers using PAR could once guide an aircraft to the runway with"no gyro" approaches. In readiness for emergency deployments, military organizations still use and maintain transportable GCA equipment containing both PAR and ASR equipments. There can be no better example of the naiveté in excessive reliance on secondary radar than the events on September 11, 2001, in which terrorists disabled aircraft transponders to conceal their attack on the World Trade Center twin towers and the Pentagon.

Evolution of the Technology

It is easier to grasp the technology, and it provides much insight, to look at its history. The original system concepts are still in use, have only been expanded upon over six decades, and will remain for years to come. In the following discussion, the author separates the growth into five generations.

A characteristic of the secondary radar system is that echoes from the transmitted interrogation will not enter the receiver, because the interrogator's receiver is tuned to 1,090 MHz, but the transmitter radiates at 1,030 MHz. A tuned circuit called a ***diplexer*** was used for decades to steer the 1,030-MHz transmitted energy to the antenna and to steer the received replies to the receiver. The configuration limits the secondary radar system to being only a data-exchange device, unable to detect echoes from weather, ground obstructions, or those aircraft not equipped with operational transponders.

The First Generation

Air traffic control beacon system technology has passed through a technological progression ever since World War II, when the purpose was to identify friendly or enemy aircraft on radar displays. These early systems were called ***Identification Friend or Foe (IFF),*** and were a carefully guarded secret. Ten years after the end of the war, this author received training on this system. The training, with all equipment and books, was contained in a fenced enclosure, on the second floor of an aircraft hangar, guarded by an armed USAF policeman. Soon after the war, the military also procured AN/FPN-13 transponders for ground and shipboard purposes around the world. The author maintained one of those systems aboard a USAF telemetry picket ship on the Atlantic Missile Range in 1959.

The Second Generation (see Figure 7-1)

The CAA and FAA began using the military interrogator equipment in joint-use military/civilian facilities in the late 1950s. These earliest systems had very limited processing equipment, and transponders could send only 64 possible codes. In many cases, the old wartime practice of reading codes directly from raw pulse trains on PPI displays continued, as production of both military and civilian decoding equipment began in the United States. The name of the civil system eventually acquired the name ***air traffic control radar beacon system (aka "service") (ATCRBS).*** A ground-station interrogator transmits the ***challenge,*** also called an ***interrogation,*** in the form of two 1,030-MHz bursts, called the ***mode pair.*** Aircraft with transponders set to receive the mode pair transmit 1,090-MHz beacon reply code trains back to the interrogator facility. Note in Figure 7-1 that the beacon antenna is simply a set of four dipoles around the primary radar feedhorn. Antennas were a principal source of problems throughout the history of these systems, and finally led to monopulse techniques in the fifth-generation equipment.

The Third Generation (see Figure 7-2)

Beacon decoding systems first appeared in both military and joint-use USAF–FAA systems in the 1950s. These systems obviated the need for increasingly busy controllers to read the raw pulses, and permitted them to see a combination of "slashes" and "blooms." Although only rarely used in equipment-failure emergencies now, some of these equipment still remains in service in the United States. However, countries all over the world now maintain some form of ***secondary surveillance radar (SSR),*** now the internationally accepted description. So, also, is ***primary surveillance radar (PSR).*** Some of these foreign systems remain primitive by today's standards, and there is still a need for some to have knowledge of those earlier systems. By the 1970s, the beacon reply code had been expanded to 4,096 possibilities, and even a potential for 8,192. The new code trains were called ***discrete beacon replies.***

The Fourth Generation (see Figure 7-3)

As air traffic increased dramatically in the 1960s, excessive reliance on fallible human memory presented an obvious hazard. The USAF already was using a vacuum-tube computer, the AN/FST-2, to maintain surveillance

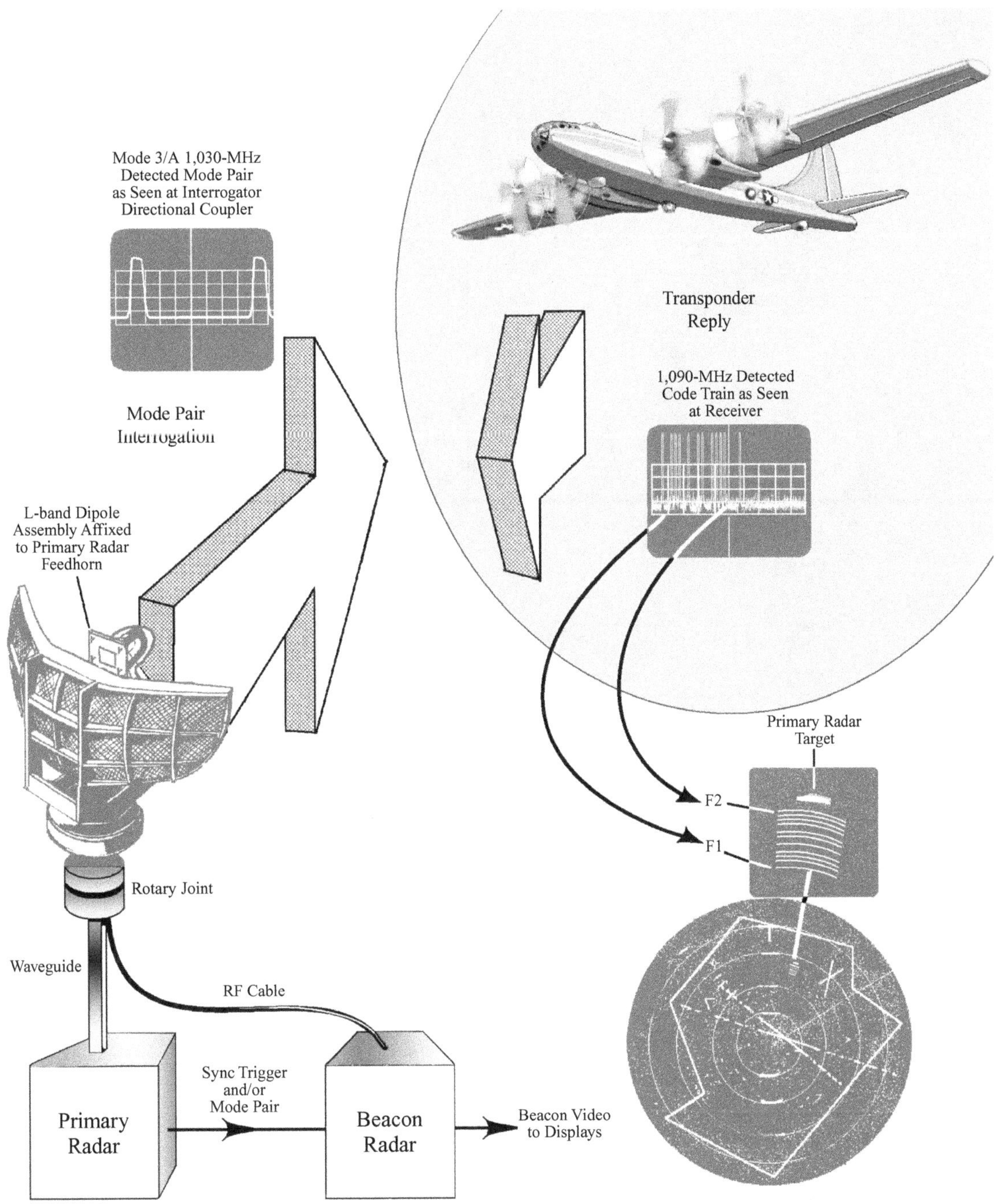

FIGURE 7–1

The second-generation ATCRBS system.

of air traffic for defense purposes. A new system, the ***Automated Radar Tracking (aka "terminal) System (ARTS),*** was developed for airport-area (terminal) beacon systems. The first system was deployed in New York City, and used a solid-state, discrete-component computer already in use by the US Navy for fire-control radar systems. The new systems provided for controller data entry, scan-to-scan tracking, alphanumeric displays, and much more. ARTCCs also went to automation using IBM mainframe computers. The original ARTSI evolved into the ARTSIIIA by the late 1990s, with the final system bearing little resemblance to the first. An ARTSII was produced for use at low-density facilities. It has now evolved to the ARTSIIE, which remains in service now, and for some time into the foreseeable future.

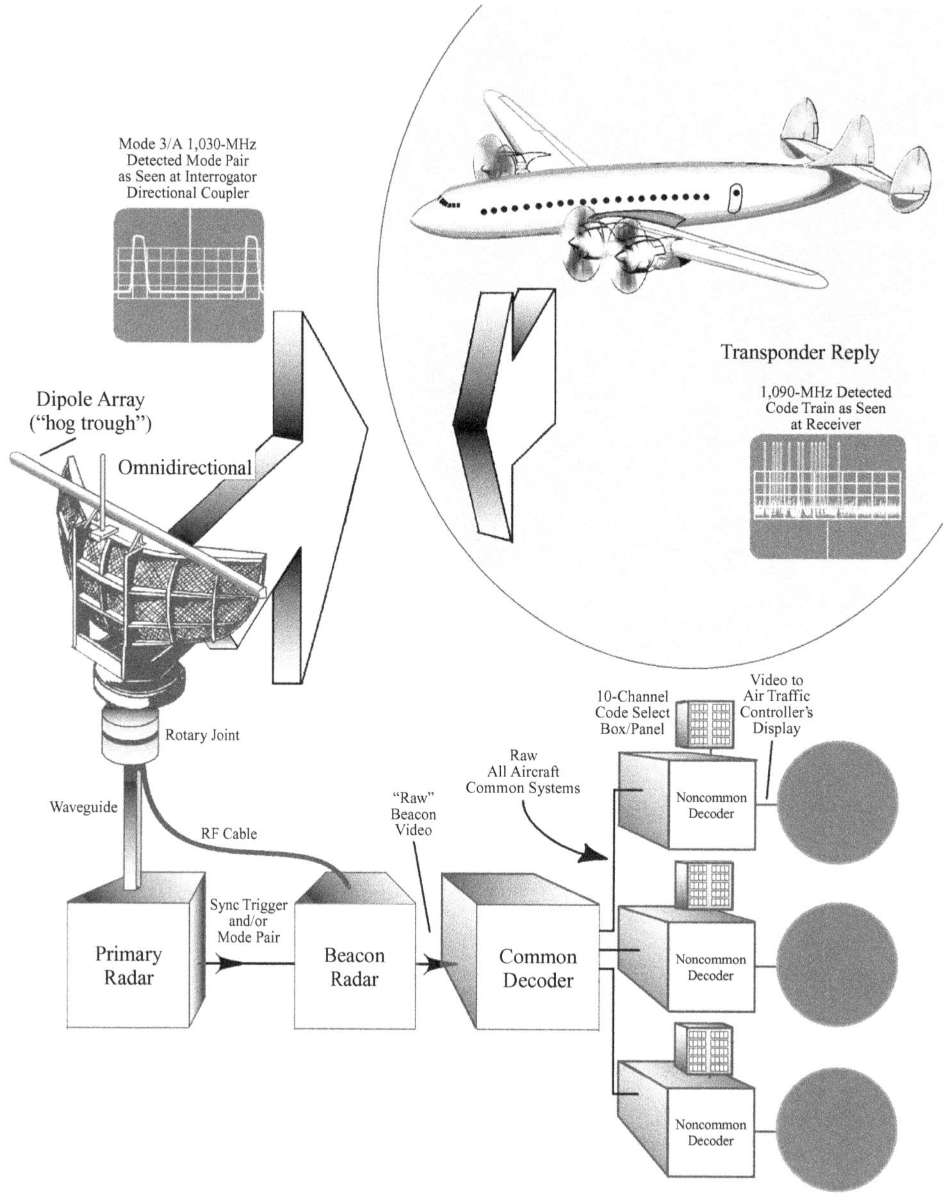

FIGURE 7–2

The third-generation ATCRBS system.

ARTS Data Processing

An ARTS is a computer data-processing system, a computer running on an air traffic control application program. It employs a central processing unit, temporary and permanent data storage, display equipment, input/output hardware to handle peripheral equipment, and much more. Display equipment may be at a TRACON or at remote air traffic control towers. Displays have evolved from the analog ppi to color television equipment.

Data Processing at an ARTCC

The ARTCC is an air traffic control facility devoted to high-altitude flight-following between terminal control areas. ARSR facilities, also called ***enroute or long-range radar,*** provide radar data messaging for ARTCCs. The ARSR may have its own self-contained digitizer. Older systems require separate ***common digitizers,*** so named because they were common to all ARSRs and because the data could be provided to both civil ARTCCs and military Air Defense Command facilities. ***Enhanced ARTS (EARTS) Systems*** are used in unusual air traffic control situations as a combination ARTS/ARTCC. Among the locations are Alaska, Puerto Rico, and Hawaii.

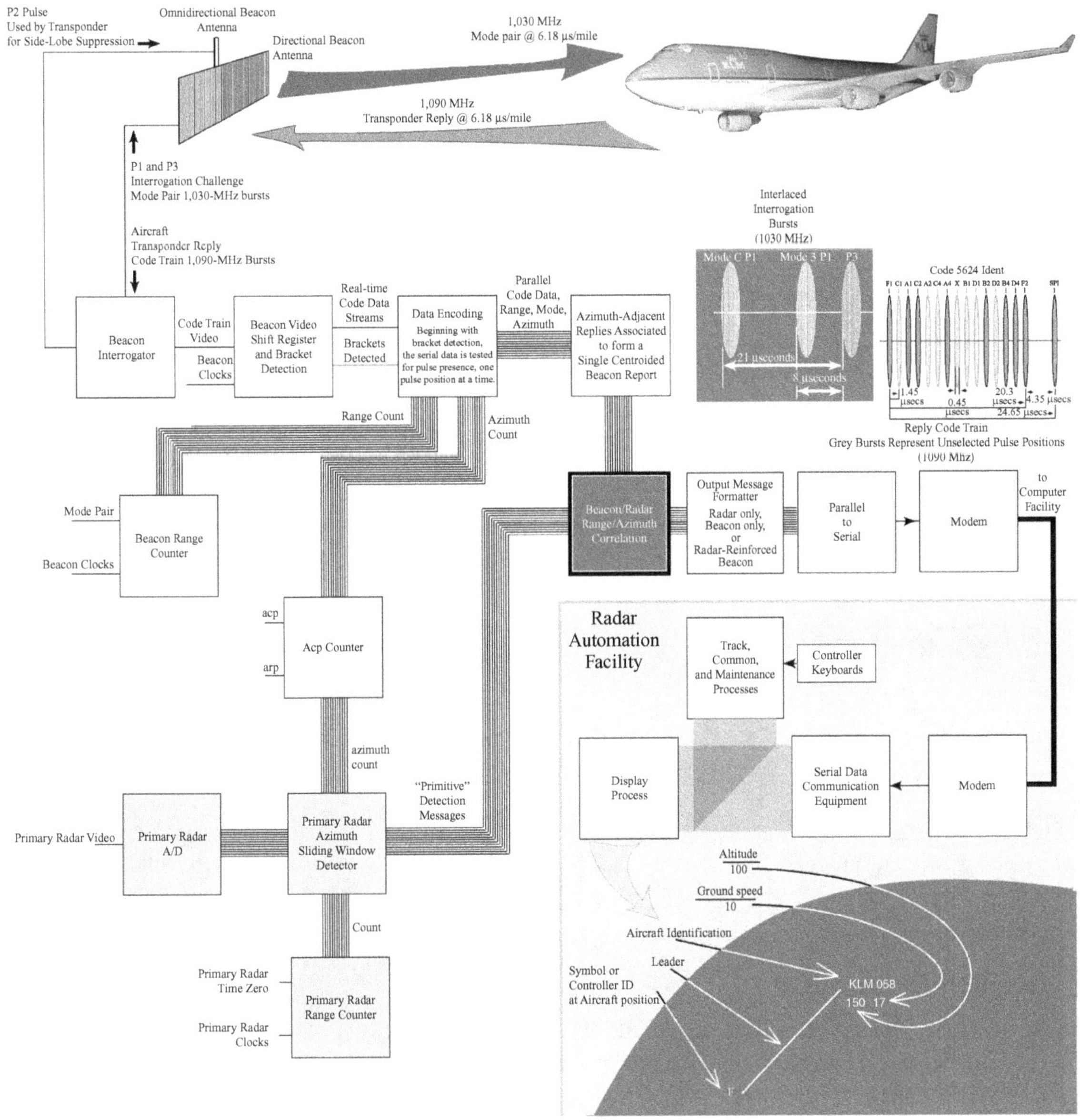

FIGURE 7–3

The fourth-generation ATCRBS/ARTS system.

ARTSIIIA computers expanded for increased input–output (I/O) at EARTS evolved into multi processor systems at all ARTSIIIA facilities. The ARTSIIIA remained in service until the late 1990s, when it was replaced by the ARTSIIIE and ***Surveillance Terminal Automation Replacement System (STARS).***

Data Utilization and Processing

Because the secondary radar system receives ample power from the transponders, and because high transmitter power is unnecessary and undesirable, ***interrogator equipments*** are small, relatively inexpensive in comparison to primary radar systems, and less susceptible to failure. One of the earliest interrogators, the AN/UPX-6, was (very roughly and approximately) 19" × 12" × 30". In comparison to the "downstream" hardware which it supplies with data, the interrogator and its receiver have been relatively simple and inexpensive. The data-processing equipment handling the received transponder replies is quite another story.

At either the radar site or the terminal facility, a single center azimuth, called ***centroid,*** of the group of replies for a single aircraft must be calculated and included in a digital target ***report,*** a data message describing the range, azimuth, altitude, beacon code data, interrogation mode, and more, for input to the computer systems. Until the fifth-generation equipment, centroiding was always achieved with the azimuth sliding-window process (see Figure 7-4). A superior method of azimuth determination was, of necessity, incorporated when Mode Select technology required monopulse operation.

Code Validation

Part of the sliding-window process tests the first replies for consistency, called ***validation***. Another process compares beacon report data to-proximate search report data to improve data confidence in both systems with a "search reinforced" bit in the report. At an ARSR facility, the report data is transmitted to an ARTCC, perhaps hundreds of miles away. At an ASR facility, the data may be sent only a couple of miles, within the same building, or many miles to a centralized ***terminal radar control facility (TRACON)***.

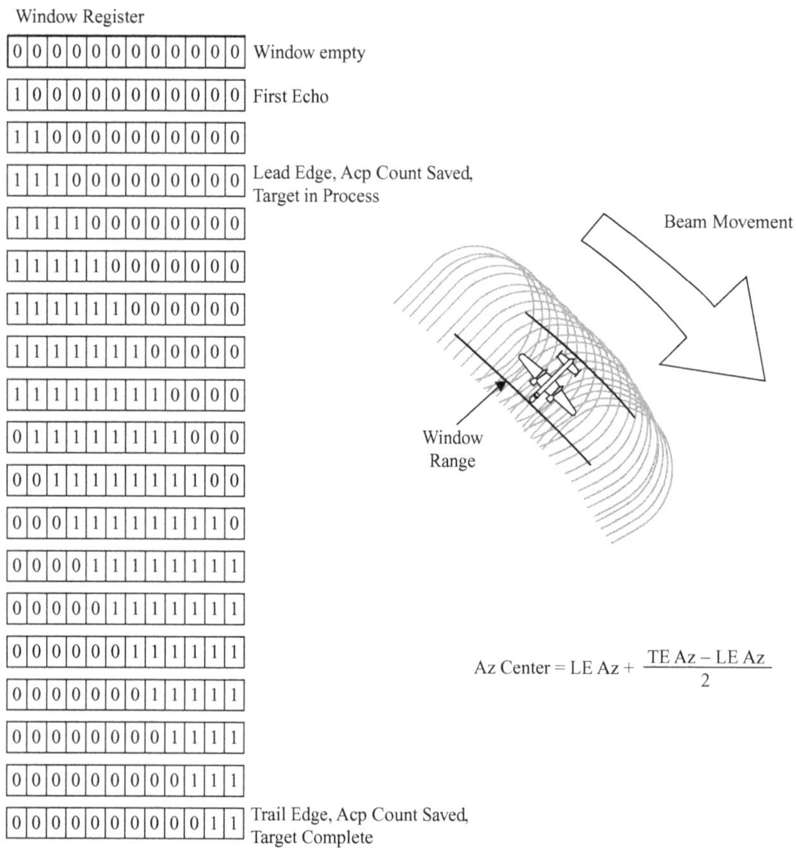

FIGURE 7-4

The sliding-window process.

The Fifth Generation

With increasing air traffic density and speeds, the ATCRBS has been plagued with increasing limitations, and it is gradually being supplemented with a data-exchange system called ***Mode Select (Mode S),*** formerly known as the ***discrete address beacon system (DABS).*** DABS was so named because it provides a means for every aircraft to keep its own precise identity in a 24-bit address, providing 16,777,215 possibilities. The Mode S system is, of necessity, compatible with, and contains, existing ATCRBS technology. It is significantly different in operational theory. It is popular in the United Kingdom and western Europe, where it is called ***Monopulse Secondary Surveillance Radar System (MSSR).*** It is further compatible with a new system mandated by Congress in 1986, the ***TCAS (Traffic Collision Avoidance System)*** in which an interrogator and transponder in an aircraft could exchange data with other aircraft and provide visual cockpit displays. Also integral to TCAS equipment is an interface to a GPS system dedicated to air traffic control. TCAS Mode S transmitters can further emit a "squitter" signal to be received by Mode S ground equipment or other aircraft. The "squitter" concept is borrowed from ***distance measuring equipment (DME)*** and permits the TCAS or MSSR equipment to gather range and identity information.

ARTCRBS Interrogator Basics

Basic ATCRBS Interrogator Principle (see Figure 7-5)

The ATCRBS interrogator must transmit a "mode pair," two 0.8-μs bursts at 1,030 MHz, called "P1 and P3." There are really three bursts transmitted, but the third one, called "P2," is switched to, and transmitted from,

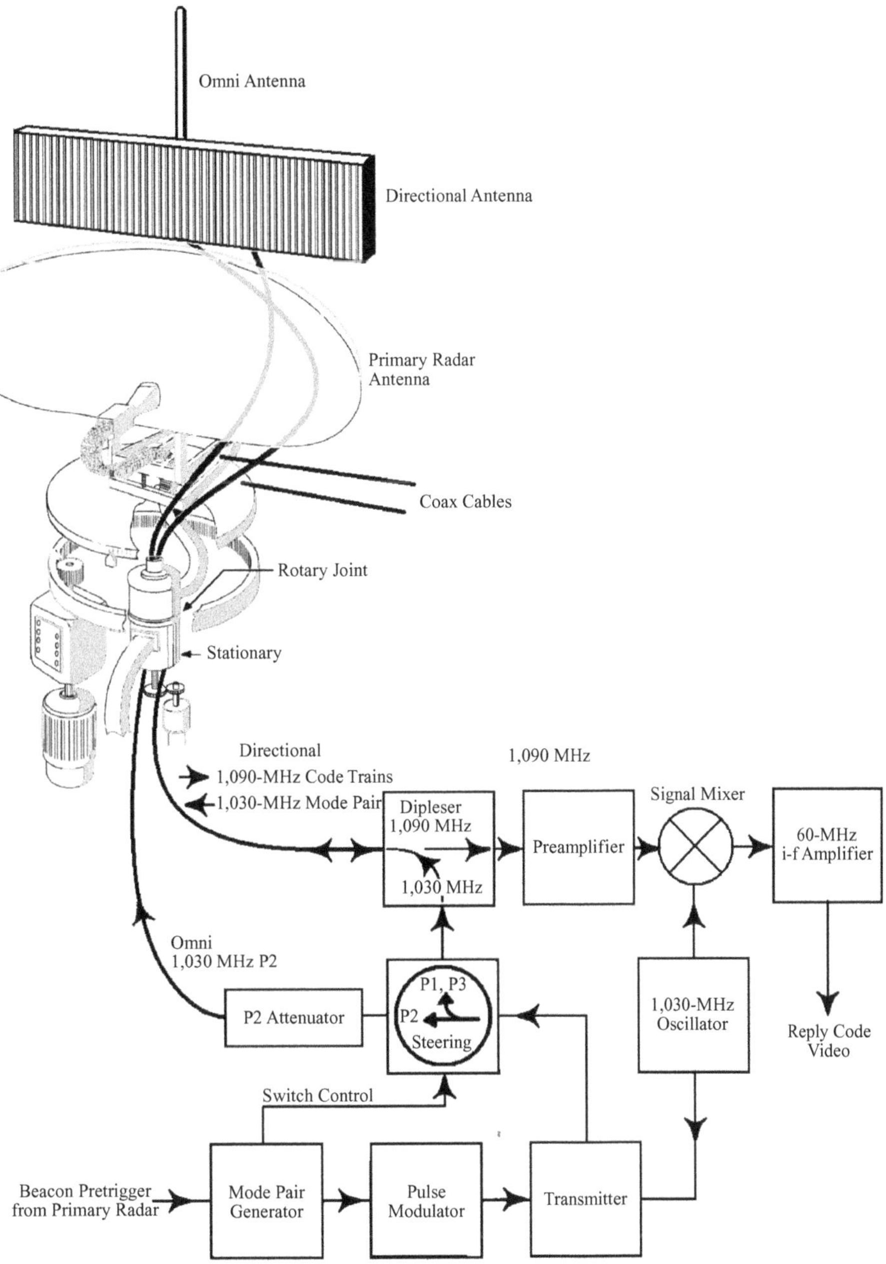

FIGURE 7–5

Simplified block diagram, generic ATCRBS interrogator set.

an omnidirectional antenna for side-lobe suppression. That subject is addressed in another part of this chapter.

The interrogator set also contains a 1,090-MHz receiver. Steering of the transmitted 1,030 MHz and received 1,090 MHz was originally achieved with a tuned-circuit diplexer. It is now done with ferrite circulators, not yet available in WWII, or for many years after. The ATCRBS transponder replies with a series of 0.45-μs bursts in a 20.3-μs code train. For 0.45-MHz pulses, an optimum band-pass of $1.2/t_p$ would be 2.7 MHz, and greater values are used for better pulse shaping. The higher bandwidths do result in more noise, but the high sensitivity of primary radar receivers is unnecessary for these one-way receivers. The minimum discernible signal is usually more (worse) than –90 dBm. The 1,030-MHz frequency used for the transmitter is a convenient local oscillator source, and the resultant intermediate frequency is 60 MHz.

$$P_{r_{pri}} = \frac{P_t G_t A_0 A_e}{16\pi^2 R^4}$$

$$P_{r_{sec}} = \frac{P_{t_t} G_{t_t} A_e}{4\pi R^2}$$

Legend	
$P_{r_{pri}}$	Received echo power in a primary radar.
$P_{r_{sec}}$	Received transponder reply.
P_t	Peak transmitter power of a primary radar.
P_{t_t}	Peak transmitter power of a radar beacon transponder.
G_t	Primary radar antenna gain.
G_{t_t}	Beacon transponder antenna gain.
A_0	Target cross-sectional area.
A_e	Primary radar or interrogator antenna effective aperture.
R	Target range.

FIGURE 7–6

Received signals in primary and secondary radars.

Detection Probability

Because the ground interrogator facility receives a reply transmitted from an aircraft transponder, target detection probability (for transponder-equipped aircraft) is far greater than that in a primary radar dependent upon echoes. See the equations in Figure 7-6. The signal strength of the reply is attenuated only by the one-way spherical surface area equation $4\pi R^2$, instead by $16\pi^2 R^4$ as it is in a primary radar. The signal source is the aircraft itself, not the ground station. The P_t is that of the transponder (here called ***transponder peak power*** P_{tt}**)**, instead of the interrogator. Of course, the interrogator P_t must be sufficient to reach the transponder, and the challenge is also subject to the $4\pi R^2$ attenuation. Difficulty in obtaining transponder replies is unusual; it is far more likely that unwanted replies from distant aircraft will occur; this also causes unwanted data to other radar sites, called ***false replies unsynchronous in time (FRUIT).*** Still further, in a secondary radar, there is no cross-sectional area (A_0) of the target to be considered, but the approximate omnidirectional ***gain*** **(**G_{tt}**)** of the aircraft transponder antenna will be a determining factor. Because the challenge is only attenuated in the ***interrogator-to-aircraft (uplink)*** direction, much lower transmitter power is required than for a primary radar.

Interrogation

The challenge pulses may be as little as 50-W peak. Excessive power creates many problems, such as side-lobe interrogations, excessive replies from targets beyond the intended range, and others. Figure 7-5 shows a second transmitted P2 pulse to an omnidirectional antenna; it is used by the transponder for ***side-lobe suppression (SLS),*** addressed in the ensuing discussions in this chapter. Because of the strength of the transponder reply, the 1,090 MHz receiver requires an stc which recovers to maximum sensitivity more gradually. In primary radar systems, a 12 dB/octave stc recovery rate is used; in beacon systems, a 6 dB/octave curve is used. Some military beacon systems have named the sensitivity time control ***gain time control (GTC).***

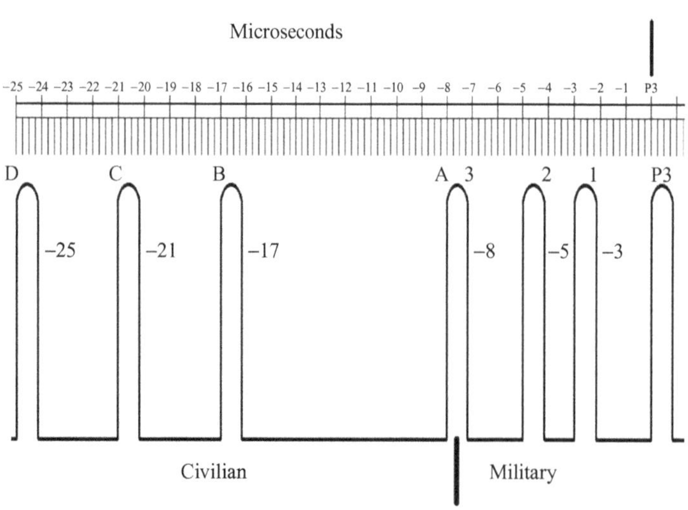

FIGURE 7–7

P1 pulse positions for modes.

Challenge/Interrogation Modes

Each T_r, the interrogator transmits a challenge comprising two 0.8-μs, 1,030-MHz bursts; the two bursts are called a mode pair, and are assigned the identifications ***P1 pulse and P3 pulse***, respectively. See Figures 7-7 and 7-8. The P3 pulse bears a fixed temporal relationship with the primary radar unstaggered triggers, and the P1 pulse may occur at one of six possible times, as illustrated in Figure 7-7, in advance of the P3. The challenge will elicit a response from all transponders set to reply to the mode pair temporal spacing, which may be 3, 5, 8, 17, 21, or 25 μs, for modes identified, respectively, as 1, 2, 3/ A, B, C, and D. There are other military-only modes not addressed in this book.

In the past, the f_p of the interrogator would usually be the same as that of the primary radar at ARSR facilities, but will likely be 1/3 that of the primary f_p at ASR facilities, to reduce ***range-ambiguous (second-time)*** replies from targets beyond the necessary range. Latter-day MSSR systems use a substantially lower f_p, about 80 Hz.

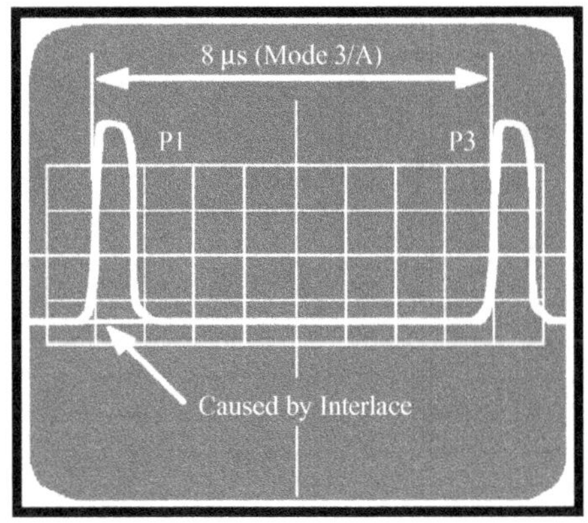

FIGURE 7–8

Mode 3/A pair on an oscilloscope.

Evolution of the Challenge/Interrogation

Originally, the military IFF systems utilized three mode-pair spacings: 3, 5, and 8 μs, designated modes 1, 2, and 3, respectively. When civilian systems were deployed, other modes were needed to permit continued use of military systems for security purposes, and to provide for other civil purposes. To distinguish between civil and military modes, the four civil modes were designated with letters A, B, C, and D. Military traffic is usually under civil control, so one civil mode, A, was chosen as a universal military/civil mode and uses the same 8-μs spacing as does mode 3. This 8-μs mode is commonly called "3/A." Mode B spacing is 17 μs, Mode C is 21 μs, and Mode D is 25 μs. Mode C has an important special purpose; it is used to elicit an altimeter-reporting reply code from the aircraft. Mode B is now used in MSSR systems for antenna pattern measurement.

Traffic Collision Avoidance System (TCAS), MSSR, and Mode S

In the late 1980s, Congress required that large aircraft be equipped with TCAS systems as an added safety feature. These systems are 1,030-MHz interrogators and 1,090-MHz Mode S transponders in aircraft, and can provide the cockpit with a display of nearby aircraft, including type, identification, and altitude. The exchange of such information requires bidirectional data transmission of the same type as that used by air traffic control. The FAA beacon systems, to handle these messages, had to be radically changed to a monopulse system, in which aircraft data could be received and centroided in a single interval. There were many side benefits to ATCRBS from the change.

Mode 4 Challenge (see Figure 7-9)

This latter-day system is used in ARSRs by military entities for national defense. When absolute friend-or-foe determinations must be made, a complex mode 4 challenge containing a special security code is transmitted.

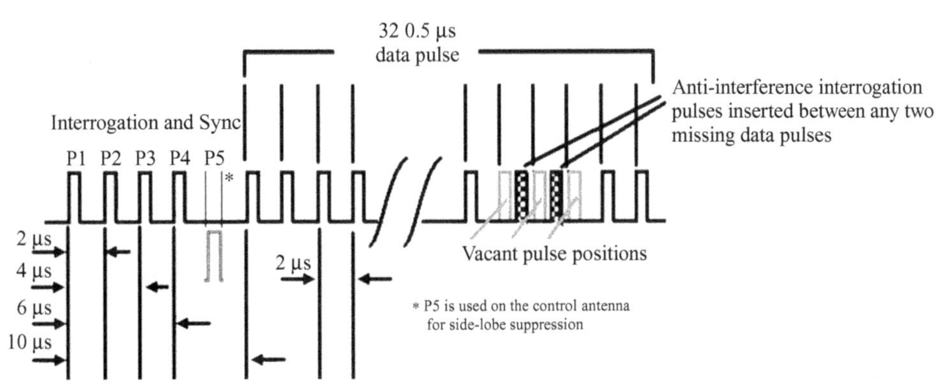

FIGURE 7–9

Mode 4 interrogation.

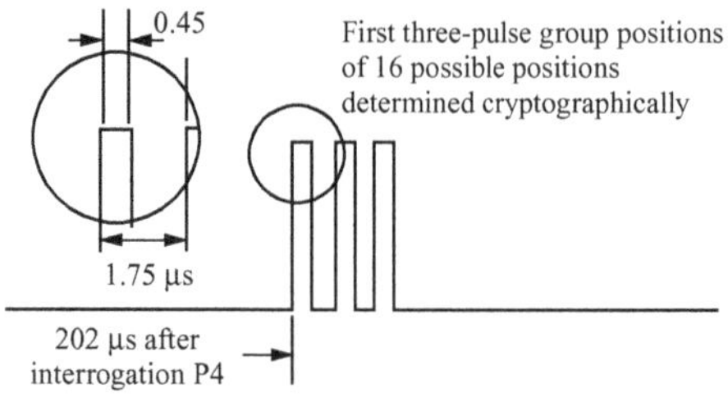

FIGURE 7–10

Mode 4 reply.

In the absence of an appropriate reply, the aircraft becomes a possible foe. The security codes are changed frequently.

Mode 4 Reply (see Figure 7-10)

The mode 4 reply is a simple three-pulse group, but it can be in 1 of 16 possible positions, the position determined cryptographically.

MSSR Basic Interrogator-Receiver Concepts

General

There are currently two MSSR systems used by the US FAA. One is the original 1980s' version, simply called "Mode S System." It is used mostly in conjunction with ASR-9 systems and can be operated in an ***interim beacon interrogator (IBI)*** mode, without monopulse. The other is named the ATCBI-6, and it was intended for use at ARSR facilities, although it can be adapted to ASRs. A significant difference in the ATCBI-6 is that it always operates in monopulse. Most discussions in this chapter will more closely resemble the ATCBI-6, partly because it more likely resembles systems used in other nations, and partly because it is the most current state of the art.

Basic Concepts (see Figure 7-11)

The Mode S system illustrated is abbreviated to show only the transmitter and three logarithmic receivers. More details on a receiver are given further in this chapter. The system uses a monopulse technique, in which an array, phased for 1,030 MHz, is used for transmit to provide a narrow beam. The original 5-foot antenna in the Mode S system contained 32 dipole columns, was center-fed, and the individual columns are spaced by 11.25° at 1,030 MHz. Since there were 16 columns on each side of the center, there is 11.25° × 16, or 180° of phase shift on each side of the feed. The latest antenna, for the ATCBI-6 has 36 columns, spaced by 11.0° at 1,030 MHz and is 7 feet in height. In receive, the same antenna is used, but phasing differs for 1,090 MHz providing different patterns. If the outputs from the two sides are taken from points in phase, the entire array acts as one, forming a narrow beam, called ***sum (E)***. If the outputs of two sides of the center are taken from points 180° out of phase to 1,090 MHz, two separate receive patterns called ***delta (Δ)*** are formed on each side of the boresight. There are two Δ patterns, and the one receiving the stronger reply is identified by comparison of the *E* and Δ replies in a phase detector. The lobe is identified by a sign bit indicating positive or negative outputs from the phase detector. Logarithmic receivers are used to produce unlimited outputs representing signal power. The error is determined by digital amplitude comparison of the logarithmic *E* and Δ receiver outputs, and the difference is used to address an off-boresight angle (OBA) table for azimuth correction to the 16,384 acp count. This method permits a single report of a beacon reply, makes the sliding window obsolete, reduces FRUIT, and provides the necessary time for more complex data exchanges.

Transmitter

The Mode S transmitter is more complex than an ATCRBS transmitter. It must operate in both ATCRBS and Mode S, substantially increasing the duty cycle. Both the P1 and P2 pulses must be transmitted on the directional antenna, so as to appear as side-lobe interrogations and inhibit ATCRBS transponders. The P1–P2 combination is followed by a long (up to 30.5 µs) P6 data burst containing a phase synchronization signal and 56 or 112 bits. These interrogations are in addition to the P1 and P3 pulses for ATCRBS interrogations, and the resultant duty cycle of all is a significant load increase to the transmitter. Additionally, military use of the ATCBI-6 can even further tax the transmitter with mode 4 challenges. The omnidirectional antenna also requires (1) a P2 pulse for side-lobe suppression in ATCRBS, and (2) a P5 pulse coinciding with the P6 phase reversal sync for side-lobe suppression in Mode S. The solution is a "low-duty" P2 and P5 transmitter, and a "high-duty cycle" P1, P2, P3, P4, and P6 transmitter. Further, an Ω receiver is used for ***receive side-lobe suppression (RSLS)***.

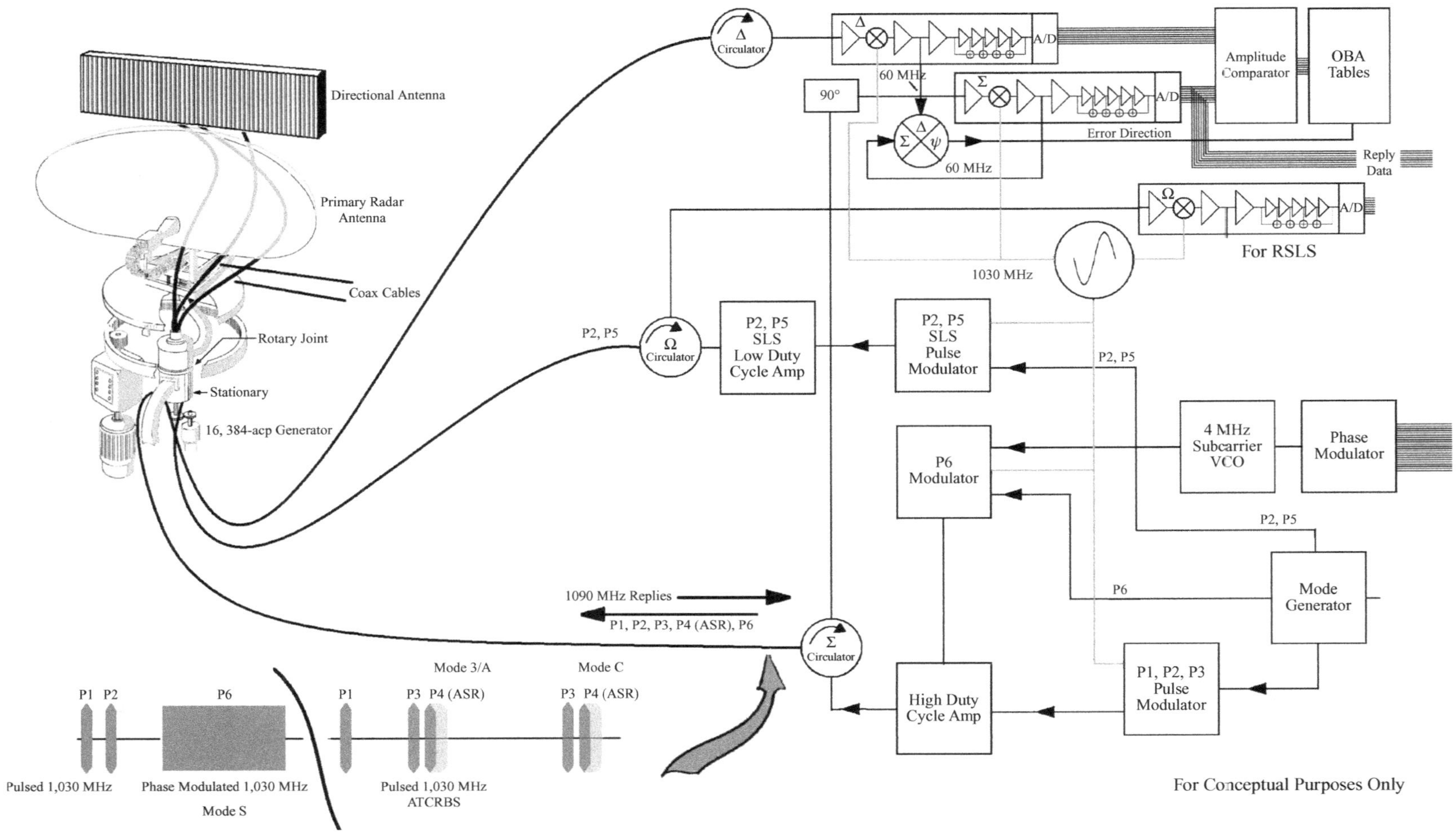

FIGURE 7–11

MSSR simplified interrogator-receiver system.

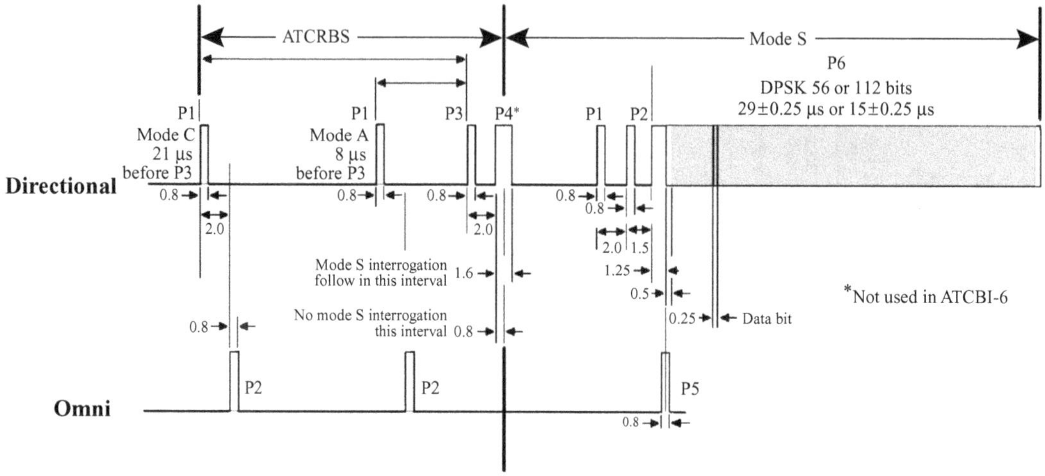

FIGURE 7–12

ATCRBS interlaced with mode S.

Mode S Interrogation (see Figure 7-12)

Mode S has entered the beacon scene over the past three decades. It incorporates automated data exchanges between the interrogator and aircraft transponder. See Figure 7-12. Mode S ***uplink interrogations*** use P1, P2, P4, P5, and P6 bursts. The P6 burst is actually a 4-kHz subcarrier modulated with a digital phase shift keying method called ***Manchester***. There are two basic types of Mode S interrogations, called "***all call***" and "***roll call***." The P4 pulse is necessary with ASR systems to indicate the transmission of both ATCRBS and Mode S in the same interval. The pulse width is 0.8 µs to indicate single-ATCRBS modes and 1.6 µs to indicate both mode types. It occurs 2 µs after the P3.

All Call

The all call is used for surveillance and acquisition. The first 32 bits of the 56-bit data content of the all-call P6 burst will initiate a ***downlink reply*** of identity and altitude from Mode S transponder-equipped aircraft not already stored in a data-memory "roll," an azimuth-ordered list. Once stored, the aircraft will be addressed on subsequent scans with roll calls, by its unique identity in the last 24 bits.

Roll Call

Once aircraft data has been acquired and stored, the system begins a software scan-to-scan track, placing a range-azimuth software "window" around the aircraft, independently addressing it in the next-scan roll-call interrogation, as the antenna position, indicated by the acp count, reaches the counterclockwise lesser-azimuth side, of the track window. At that point, the current and previous-scan range-azimuth rectangular coordinates are compared to calculate the aircraft course and speed. Once the speed and course have been determined, the soft window is shaped and moved in the course direction, and a ***track*** is established.

Uplink P6 Contents

The P6 comprises 56 or 112 data bits modulating a 4-kHz subcarrier, following transmission of the wide P4, the P1–P2 combination, a 1.25-µs P6 leader, and a 0.5-µs phase reversal for sync. The bits are described as "ones" or "zeros," determined by their phase in a ***digital phase-shift keying (DPSK)*** scheme. Each data bit period is 0.25 µs in length. A data "one" is represented by a phase change. For instance, phase reversal sync set to "one," (1) no change, (2) change, (3) change, (4) no change, (5) no change, (6) change, (7) change, would represent 1100110 (first bit in message on right).

Receipt by the Transponder

In the aircraft transponder, the received uplink interrogation DPSK signal is hard-limited, quieting the receiver, and eliminating noise, as in FM radio. The P6 DPSK signal can be a maximum of 30.35 μs, short enough to end before the transponder ATCRBS inhibit ends.

The P6 content may be in several "uplink" formats, and the replies may also be in several "downlink" formats. Of the 56 data bits, a field of five identifies the format type, a second field of 27 is for commands, and the remaining 24 bits are for address and parity. In all call, intended only for acquisition, the last bits are set to all "ones"; in roll call, they contain the address of specific aircraft previously acquired. Additional roll calls for the same aircraft may contain other types of data, as identified by the format data.

P5 Pulse/Burst

The P5 burst is used for side-lobe suppression as in ATCRBS, and it is timed to inhibit the Mode S transponder from locking onto the first phase change in the message. The P1–P2 combination will be treated by the ATCRBS transponder as though the interrogation had been via a side lobe, inhibiting it for 35 μs after both pulses have been received. The P5 pulse is necessary to prevent both Mode S and equal-amplitude P1–P2 ATCRBS side-lobe interrogations from initiating Mode S replies.

ATCRBS Mode Interlace

In latter-day systems, there is no need to synchronize the ATCRBS timing to the radar timing, because the primary and secondary radars each contain independent range counters and each performs independent center-of-azimuth calculation. Completed radar and beacon data reports are compared in a "merge" or "correlate" software process. Figure 7-8 is an oscilloscope presentation of ATCRBS mode pairs prior to fourth-generation systems. A trace appears through the baseline of the P1 pulse; this is because the oscilloscope has been synchronized to the primary radar, and the P1 pulse is not occurring at the same time relationship with the primary radar and P3 pulse in each T_r. There is a need to challenge in more than one mode. The P3 pulse maintains a constant temporal relationship with the radar unstaggered pretrigger and appears at the same point on the oscilloscope each time it is triggered, so the baseline of the P3 pulse appears to be "open." Typical interlace patterns have been 3/A, 3/A, C, or 3/A, 3/A, 2, C up to the fourth-generation. The 3/A interrogations were usually more frequent because their replies have been universally used for aircraft identification and code validation.

At any ground facility, conventional interrogations in both modes 3/A and C are required to provide control and flight following of ATCRBS-only aircraft. At civil-military joint-use facilities, there may also be a requirement to challenge in mode 2. To provide for these combinations, a ***mode interlace pattern (MIP)*** is used, and different modes are interrogated over successive T_r's. Mode S further increased the number of interrogations in the interlace pattern, a major factor in the need for monopulse. Monopulse provides for immediate centroiding and does not require a collection of azimuth-adjacent replies.

Interlace with Mode S

In Mode S systems, a necessary reduction of mode 3/A replies makes conventional azimuth sliding-window centroiding impractical, but a superior and more efficient means of receiving data and determining azimuth from a single reply is made available by ***monopulse azimuth determination.*** A 16,384-acp azimuth pulse generator (as opposed to 4,096 in earlier systems) is used for azimuth accuracy, and a very low challenge repetition rate (80 Hz in the ATCBI-6) is used to permit more time for Mode S data. Interlace patterns may differ for higher rpm antennas, where there is inadequate beam-on-target "dwell time" to complete the interlace pattern.

Interlace Pattern Determined by Antenna Rotation Rate

The Mode S system must process conventional ATCRBS and Mode S replies at ASR rotation speed and beam width. The Mode S interrogations must be interlaced with modes 3/A and C. Although an interlace pattern for

ASRs is available in the ATCBI-6, the usual interlace pattern for the ATCBI-6 at ARSR facilities is 3/A, roll call, C, roll call, all call, roll call. This is called a ***discrete Mode S*** type. The ATCBI-6 was originally intended for use with a 5- or 6-rpm antenna, which permits plenty of time for entire intervals to be devoted to a single mode. In contrast, see Figure 7-12. An ASR antenna rotates nearly three times as fast and requires an ATCRBS/Modes S/all call mode, where a P4 must be used. The P4 pulse may be either 0.8 μs or 1.6 μs, and the wider pulse indicates that both ATCRBS and Mode S interrogations are contained in the same T_r. The Mode S does not respond to the ATCRBS P1 and P2 pulses, because the wide P4 pulse identifies the dual-use interrogation. The Mode S P1–P2 combination does inhibit the transponder long enough (35 μs) for the Mode S receipt of the P6 DFSK signal, which is a maximum of 30.5 μs. The ATCRBS transponder will not respond to the P4 pulse because it occurs 2 μs after P3; during the 3 μs this transponder assembles its reply. To preclude the possibility that an ATCRBS P1–P2 sidelobe interrogation might appear to the Mode S transponder as a Mode S interrogation, a P5 pulse is transmitted on a near-omnidirectional "control" antenna in synchronization with the P6 start to prevent phase synchronization.

Interlace versus Azimuth

Consider a 5-rpm ARSR ATCBI-6 operating at an f_r of 80 Hz, and with an interlace pattern of 3/A, roll call, C, roll call, all call, roll call. Six intervals of 12.5 ms each are needed, and one antenna rotation requires 12 s. One interval uses 0.00104, or 0.104% of the antenna scan time. The six intervals total 75 ms, which is 0.00625, or 0.625% of the antenna scan time. There are 16,384 acp's per revolution, so each interval requires 17 acp's and the interlace pattern requires 102.24 acp's. There are 0.022° per acp, so 0.374° of azimuth rotation occurs during the interlace pattern. In spite of the beam motion during dwell, the azimuthfor each aircraft reply to all six modes in the interlace pattern remains accurate, since the monopulse off-boresight azimuth determination corrects all replies to result in a single azimuth.

Reply Codes

Conventional ATCRBS Reply Codes

The latter-day aircraft beacon transponder sends a 20.3-μs train of 0.45-μs, 1090-MHz rf bursts in one of 4,096 possible combinations, providing a method of identification. Even further, the aircraft pilot can press a switch to cause an additional pulse called the ***ident or special position identifier (SPI);*** the ident pulse occurs 4.35 μs after the F2 pulse of the code train.

Original A-B Code Train

The original code train, illustrated in Figure 7-13, contained ***framing pulses,*** also called ***bracket pulses,*** and a maximum of six data pulses to permit 64 possible combinations. The first three data pulses were designated the "A" pulses and could have eight possible combinations, 0 through 7. The "A" pulses were designated A1, A2, and A4; the code value was obtained by adding the indicated binary values of the pulses present. The second group of three pulses was designated "B" and contained the B1, B2, and B4 pulses. Because the code train uses groups of three pulses to represent numbers 0 through 7, it represents an octal numbering system. When expressed in decimal numbers, the code is expressed as A,B; the value of the A code first, then the value of the B code. In the military systems, the ident was achieved by repeating the code train; the F1 pulse of a second code train occurred 4.35 μs after the F2 pulse of a first code train. The two ident methods are illustrated in Figures 7-13 and 7-14, and 7-15.

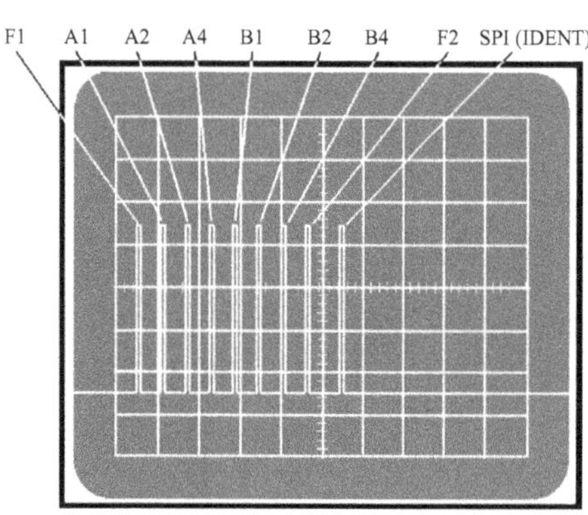

FIGURE 7–13

Beacon code train with A and B pulses.

"Discrete" Codes

As air traffic increased and the state of the art progressed into the late 1960s, ***discrete*** codes, illustrated in Figure 7-14, were implemented to raise the number of possible combinations from 64 to 4,096. The number 4,096 is 2^{12} in binary, 8^4 in octal, and 64^2. To upgrade the code train, "C" pulses were inserted between the "A" pulses, and "D" pulses were inserted between the "B" pulses. The first "C" pulse occurs between the F1 and A1 pulse, and the last "D" pulse occurs between the B4 and F2 pulse. This left a "hole" between the A4 and B1 pulse. In this space, there is provision for a special-purpose "X" pulse. Although it has not been so utilized at this time, one potential future use of the "X" pulse could be used to double the possible number of codes to 8,192.

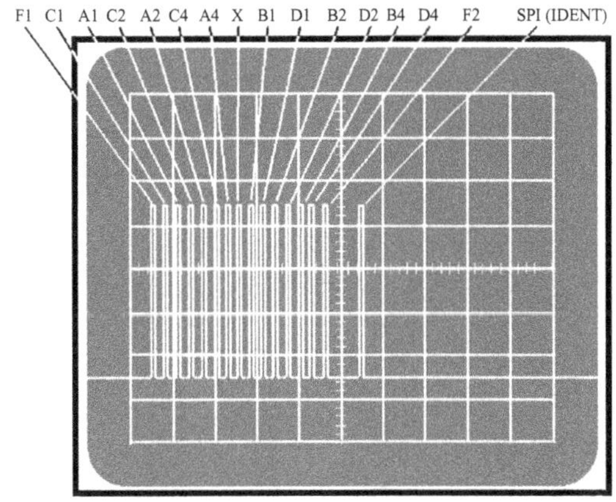

FIGURE 7–14

"Discrete" code train.

Octal Numbering

Because the code train uses groups of three pulses to represent numbers 0 through 7, it represents an octal numbering system. Those familiar with binary and octal numbers might find it confusing that the least significant bit occurs first, and to the left side of an oscilloscope display (A1, B1, etc.). When expressed in octal numbers, the code is expressed in the octal sum of set pulses in the A, B, C, and D value order. For instance, if an A1, A2, B4, C1, and D2 pulse were present, the code would be 3412. Some codes have special meanings, such as 7,700, for emergency, and 7,600 for aircraft communications failure. Processing systems contain special circuits or software for

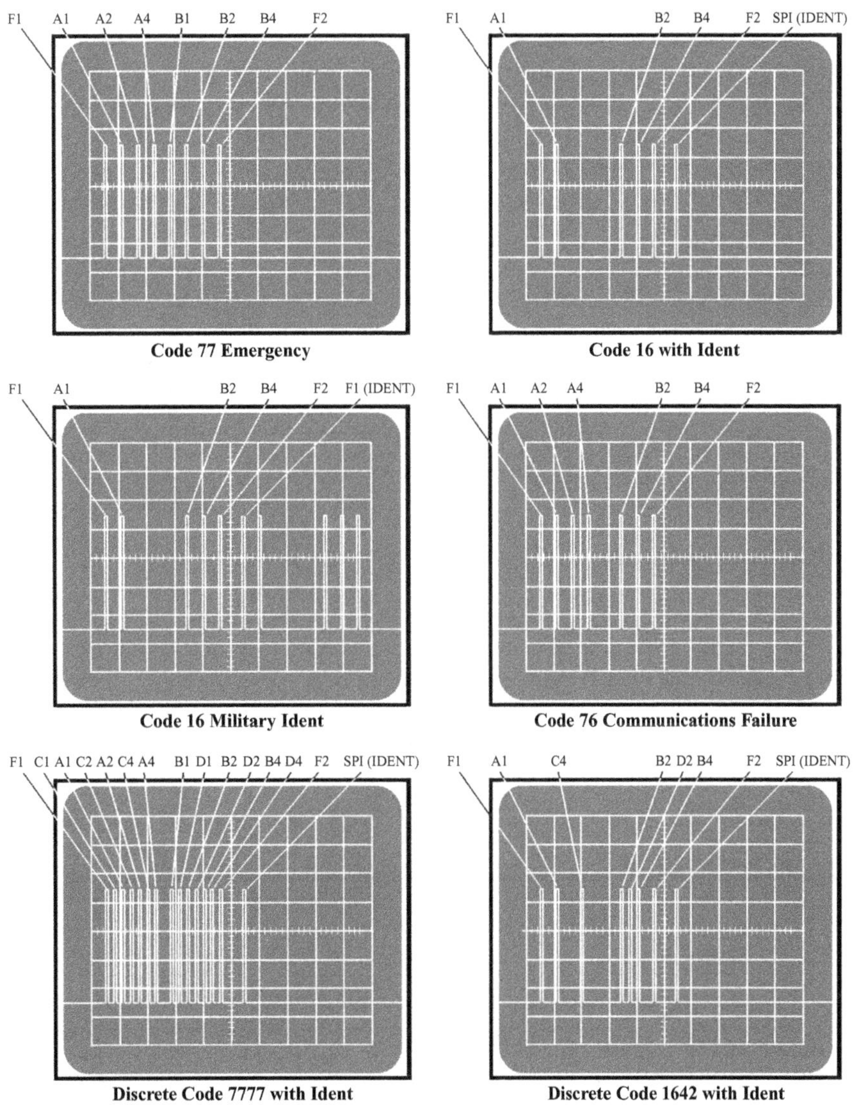

FIGURE 7–15

ATCRBS beacon code trains.

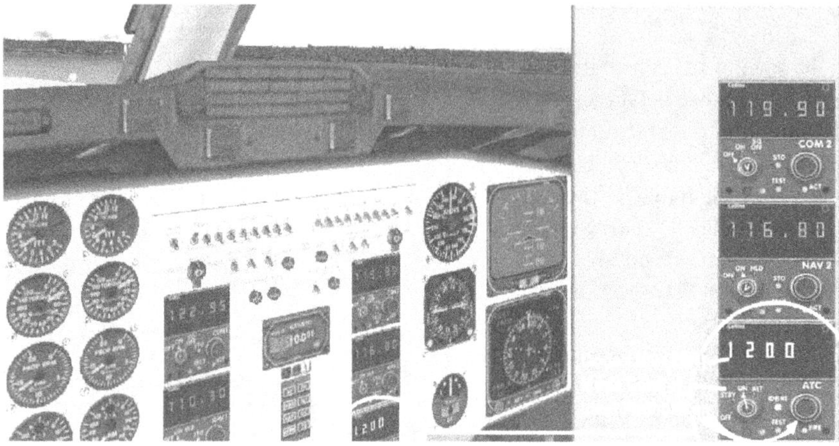

FIGURE 7–16

Beacon transponder controls in an aircraft cockpit.

recognizing these codes and alerting air traffic controllers. Figure 7-15 illustrates several oscilloscope code-train displays.

Civil and Military Idents

In the military systems, the ident was achieved by repeating the code-train; the F1 pulse of the second code train occurred 4.35 μs after the F2 pulse of the first code train. See Figures 7-14, 7-15, and 7-16. To a civilian system, the F1 pulse of the second code train served the same purpose as a ***special position identification (SPI)*** pulse. The SPI pulse is commonly known as ***ident,*** and it is activated with a pushbutton by the aircraft pilot when requested by an air traffic controller, with an instruction such as "Braniff 237 squawk ident."

Mode S Reply Codes (see Figure 7-17)

A Mode S reply may be initiated by an "all call" or "roll call" interrogation with P1, P2, and P6 bursts. The Mode S reply is a four-pulse "preamble" identifying it as such. Unlike the DPSK in the interrogation, it is followed by 56 1-μs, data-bit pulse-position "slots," each containing two smaller pulse-position slots, one for a data "one," and one for a data "zero." The preamble enables mode S receive processing at the radar site, synchronizes that equipment for data receipt, and inhibits mode 3/A and mode C receiving equipment. The contents of P6 may be in several different formats, describing many things such as altitude, aircraft identification and type, GPS coordinates, and much more.

Altitude Codes (see Figures 7-18 and 7-19)

For mode C challenges, the code train reply from the aircraft would appear on an oscilloscope to be identical to the replies for other modes, but the code represents altitudes that have been organized into a ***Gray code.*** The altitude is represented from below sea level to above 100,000 feet in 100-foot increments. The aircraft is equipped with a pressure-altitude digital reporting device, in which the barometric pressure port is plugged at standard pressure.

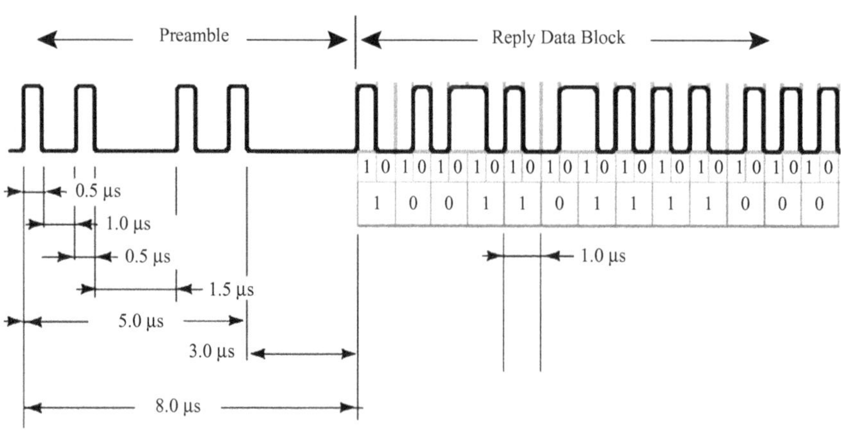

FIGURE 7–17

Pulse position modulation, Mode S reply.

Adjusting for Barometric Pressure

Local altimeter setting corrections are performed in the ATC computer systems to present accurate altitude information to controllers. A direct mechanical encoding of binary or octal data is not practical, because there is a high potential for gross errors. A simple example of such an error-prone device is illustrated in Figure 7-18. It is impossible to arrange the mechanical switches in such a manner that they all transition at precisely the same point. As the number attempts to change from decimal values of 3 to 4, the numbers 2 and 6 are erroneously encoded.

Gray encoding defeats the potential for the gross errors that can be encountered by direct binary encoding; the switch points are overlapped, as shown in Figure 7-19, so that only one switch may be in error near any altitude transition point. Each transition of the 2,000' increment represented by the B2 pulse occurs in the center of the 1,000' increment represented by the B4 pulse, providing a 500' division. The C pulse total weight, instead of being a maximum of seven, is restricted to a maximum of five, dividing the 500' B-code division to 100'. Figure 7-20 illustrates the altitudes from −1,000' to 100,000' in 1,000' steps, with the encoding from 19,000' to 21,000' expanded to 100' steps.

The A, B, and D pulses represent the altitude in 500-foot increments, but inspection of the code train in the usual manner will tell little about the indicated altitude, because the weights are differently assigned to the pulses in the altitude code. To read an octal code, the bits are arranged, msb to lsb, A4, A2, A1, B4, B2, B1, C4, C2, C1, D4, D2, D1. In the gray code, the bits are arranged, msb to lsb, D2, D4, A1, A2, A4, B1, B2, B4, C1, C2, C4. The D1 pulse is not used. The B4 pulse changes state in 1,000-foot increments, the B2 pulse changes state in 2,000-foot increments, and the B1 pulse changes state in 4,000-foot increments. All other pulses, except for the C, bear a similar multiplier-of-two relationship.

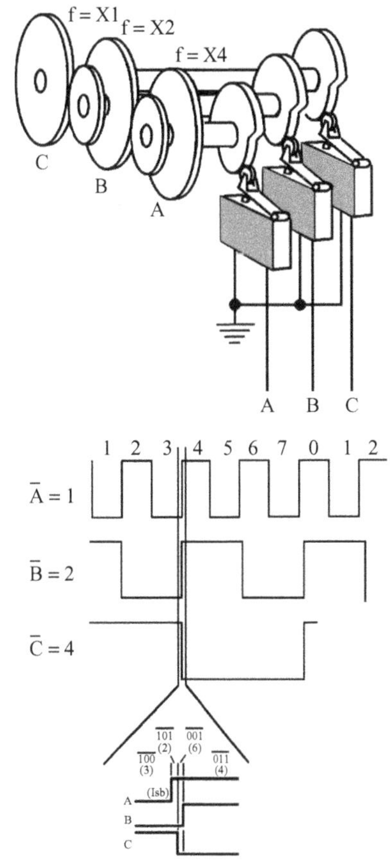

FIGURE 7–18

Electromechanical binary encoding errors.

Matching the Display of Primary and Secondary Radar

Timing Relationships

Figure 7-21 illustrates a mode 3/A and C interlace, and a transponder reply at minimum range; the mixed mode-pair/receiver video is typically available for view at the interrogator's receiver. In older real-time analog systems, it was significant that the P3 pulse remained constant in a specific temporal relationship to the primary radar. When the P3 pulse is transmitted, the complete pair of mode bursts has been radiated from the antenna, and is in route to any transponder which may receive it. The transponder cannot reply until it has received both the P1 and P3 pulses. Three microseconds after having received the P3 pulse, the reply code begins to be transmitted at 1,090 MHz, with the F1 pulse, first. If there were a transponder at the antenna, with zero range, the F1 pulse would then arrive at the receiver 3 μs after the P3 pulse. The P3 pulse might then be considered as beacon range zero −3 μs. Data processing of beacon code trains will be synchronized to the P3 pulse; in latter-day digital radars, a beacon range counter will be set to zero when the P3 pulse occurs. In MSSR, the transponder "turn-around time" delay is 128 μs because of the data-exchange message lengths. That time further explains the long T_r's and low f_p's necessitating the use of monopulse "instant-azimuth" processing.

Radar Mile Remains 12.3552 μs (see Figure 7-22)

After the transponder has received a mode pair challenge, it will produce a code train beginning with an F1 pulse, any number of data pulses from 0 to 13, and a second framing pulse, F2. Assembly and transmission of the output code train requires 3 μs after receipt of the challenge, and 20.3 μs is used between the F1 and F2 pulses. An ident

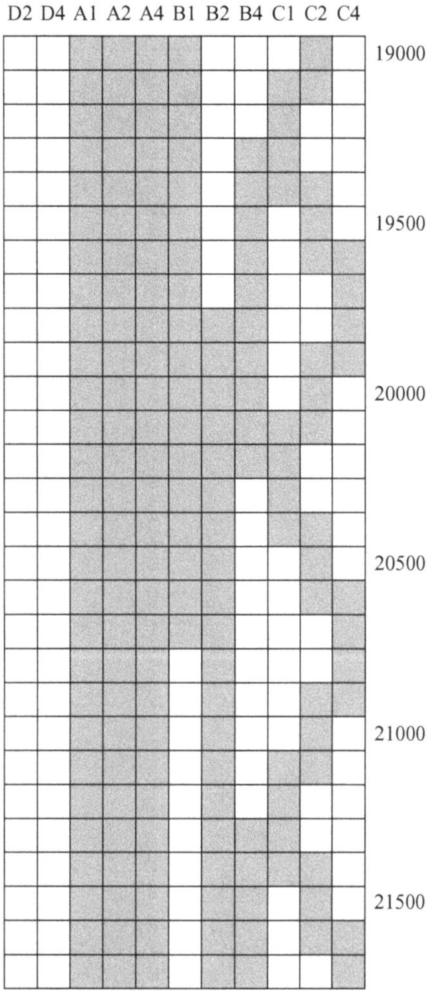

FIGURE 7–19

Gray coding.

pulse can occur 4.35 μs after the F2 pulse. The system must receive the entire code train before it may utilize the data, so beacon range becomes 12.3552 μs per mile, plus 3 μs, plus 20.3 μs, plus 4.35 μs, after the P3 pulse. In short, the range of a processed beacon reply is 12.3552 μs per mile plus 27.65 μs in respect to the P3 pulse. In mode S, the aircraft range remains 12.3552 μs, but range counter time zero differs because of the message length and transponder delay.

Real-Time P3-to-"Main Bang" Timing Relationship

If the interrogator is operated in conjunction with a primary radar, and if the beacon and radar are real-time systems (in which the received information is used and displayed at the same rate it is received, and in which one nautical mile is equivalent to 12.3552 μs), it is necessary to make provisions to ensure that the radar targets and processed beacon reply codes appear at the same place on the radar indicator. Because of the 27.65 μs processing delay, the beacon targets would appear about 2.24 miles more distant than the associated primary radar targets if the P3 pulse were aligned with the primary radar transmitter pulse. This is corrected by timing the mode pair to cause the P3 pulse to occur 27.65 μs in advance of the radar transmitter pulse.

Timing Error Alarms

In older systems, an alarm circuit was usually connected to the radar andbeacon triggers to alert air traffic controllers when the beacon timing advance was more or less than originally set, since this would place beacon information at an incorrect range on the real-time displays. The alarm will appear on air traffic control panels as a red light, labeled RANGE ERROR.

In other than those real-time systems, where the P3 pulse had to precede the radar transmitter burst by a fixed amount for coincidence on the display, timing relationships between primary and secondary radars may be less important. In the newer digital systems, there is a departure from real time. The primary and secondary data are independently made into stowable messages containing the range, azimuth, mode, code, etc. For instance, if an independent beacon range counter begins running at P3 time, or the end of P6 plus delay constants, and beacon data messages are created using the ATCRBS beacon range count, the mode pair relationship to the radar time zero is less insignificant. In the ASR-9, the search primary radar data messages, not in real time, are compared to centroided beacon target data messages, also not in real time. In the ATCBI-6 monopulse system, the f_p is 80 Hz and bears no temporal relationship or synchronization to the primary radar f_p.

Measuring "Reinforcement Rate"

For analysis purposes, automated systems contain a 1/2 nautical mile "offset" to temporarily cause the secondary beacon data to be deliberately disassociated with its related primary radar search target. This is done to permit analysis of the percentage of radar reinforced beacon targets. The beacon-only reports can be separately counted (by automation) during the offset mode and then counted again to determine reinforcement in the operational mode. If the radar reinforcement percentage is low, it may indicate poor primary radar performance or range-azimuth alignment differences between the radar and beacon. It might seem that there could be no

ABCD v Altitude

0020 -1000	2220 13000	1120 27000	7324 41000	4024 55000	4226 69000	7126 83000	1322 97000
0620 sea level	2420 14000	1720 28000	7524 42000	4624 56000	4426 70000	7726 84000	1522 98000
0320 1000	3020 15000	1220 29000	3124 43000	4324 57000	6026 71000	7226 85000	5122 99000
0520 2000	3620 16000	1420 30000	3724 44000	4524 58000	6626 72000	7426 86000	5722 100000
4120 3000	3320 17000	1024 31000	3224 45000	0124 59000	6326 73000	5026 87000	
4720 4000	3520 18000	1624 32000	3424 46000	0724 60000	6526 74000	5626 88000	
4220 5000	7120 19000	1324 33000	2024 47000	0224 61000	2126 75000	5326 89000	
4420 6000	7720 20000	1524 34000	2624 48000	0424 62000	2726 76000	5526 90000	
6020 7000	7220 21000	5124 35000	2324 49000	0026 63000	2226 77000	1126 91000	
6620 8000	7420 22000	5724 36000	2524 50000	0626 64000	2426 78000	1726 92000	
6320 9000	5020 23000	5224 37000	6124 51000	0320 65000	3026 79000	1226 93000	
6520 10000	5620 24000	5424 38000	6724 52000	0526 66000	3626 80000	1426 94000	
2120 11000	5320 25000	7024 39000	6224 53000	4126 67000	3326 81000	1022 95000	
2720 12000	5520 26000	7624 40000	6424 54000	4726 68000	3526 82000	1622 96000	

C Codes

C	total no. A,B,D bits	hundreds of feet
4	odd	7
6	odd	6
2	odd	5
3	odd	4
1	odd	3
4	even	8
6	even	9
2	even	0
3	even	1
1	even	2

Expanded

Dec	ABCD	D2	D4	A1	A2	A4	B1	B2	B4	C1	C2	C4	Altitude
482	7120	0	0	1	1	1	1	0	0	0	1	0	19000
486	7130	0	0	1	1	1	1	0	0	1	1	0	19100
484	7110	0	0	1	1	1	1	0	0	1	0	0	19200
492	7510	0	0	1	1	1	1	0	1	1	0	0	19300
494	7530	0	0	1	1	1	1	0	1	1	1	0	19400
490	7520	0	0	1	1	1	1	0	1	0	1	0	19500
491	7560	0	0	1	1	1	1	0	1	0	1	1	19600
489	7540	0	0	1	1	1	1	0	1	0	0	1	19700
505	7740	0	0	1	1	1	1	1	1	0	0	1	19800
507	7760	0	0	1	1	1	1	1	1	0	1	1	19900
506	7720	0	0	1	1	1	1	1	1	0	1	0	20000
510	7730	0	0	1	1	1	1	1	1	1	1	0	20100
508	7710	0	0	1	1	1	1	1	1	1	0	0	20200
500	7310	0	0	1	1	1	1	1	0	1	0	0	20300
502	7330	0	0	1	1	1	1	1	0	1	1	0	20400
498	7320	0	0	1	1	1	1	1	0	0	1	0	20500
499	7360	0	0	1	1	1	1	1	0	0	1	1	20600
497	7340	0	0	1	1	1	1	1	0	0	0	1	20700
465	7240	0	0	1	1	1	0	1	0	0	0	1	20800
467	7260	0	0	1	1	1	0	1	0	0	1	1	20900
466	7220	0	0	1	1	1	0	1	0	0	1	0	21000

FIGURE 7–20

Altitude codes.

difference in the radar and beacon azimuths, but differences or errors can be introduced in a center-of-azimuth-calculation (centroiding) process employed by digitizing equipments. Until recent introduction of monopulse systems, the azimuth centroid for beacon reports had always been determined by a conventional azimuth sliding window.

The Interrogator Transmitter and Antenna System

The system illustrated in Figure 7-1 is an early type; this is apparent from the antenna system, because there is no omnidirectional antenna shown. In the earliest systems, L-band dipoles around the primary radar feedhorn were the usual form of radiation, and this method is still used in some ARSR systems. However, the ASR S-band antenna reflector did not provide adequate beam shaping, and, in the early 1960s, the ASR feedhorn dipoles were replaced with a linear dipole array mounted atop the S-band reflector; the antenna was given the slang name, "hog-trough," because of its appearance. It also exhibited a number of undesirable characteristics and was replaced in the 1970s by a 5-foot-tall fixed-dipole

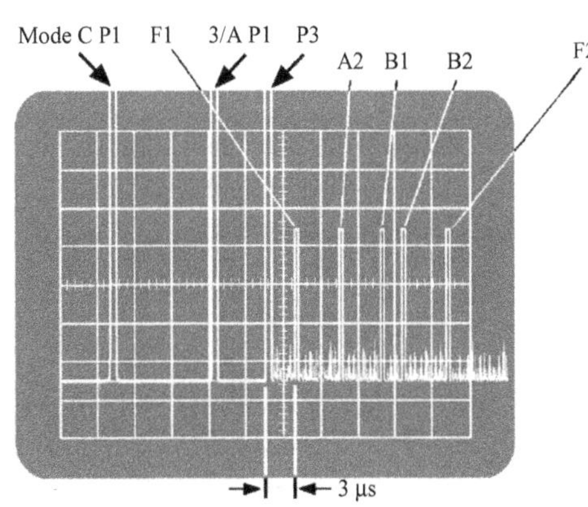

FIGURE 7–21

Modes 3/A and C interlace: transponder at Minimum Range.

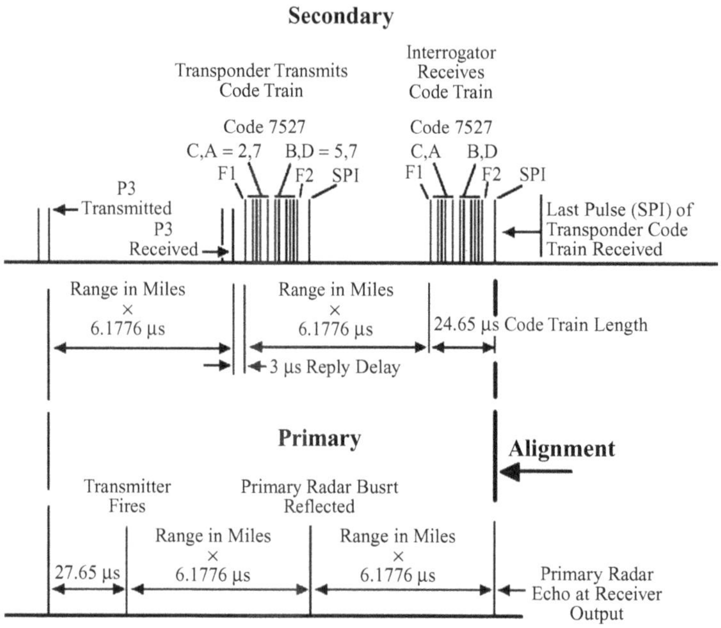

FIGURE 7–22

Range and timing, Secondary vs. Primary.

array, phased in both the vertical and horizontal planes for beam shaping. The antenna is appropriately called the 5-foot array and it is illustrated in Figure 7-23. An even newer beacon antenna resembling the 5-foot array in appearance only, is actually 7 feet in height, is called "the LVA" for "large vertical aperture." It is used in MSSR, which will eventually be the standard type.

Side-Lobe Suppression (SLS) and Improved Side-Lobe Suppression (ISLS) (see Figures 7-24–7-27)

With the increasing volume of air traffic and use of the 1,030- and 1,090-MHz spectra in the early 1970s, beacon propagation problems became very serious. In order to reduce the number of false reply code trains, two "fixes" were employed. One of these was called side-lobe suppression (SLS), the other was called improved side-lobe suppression (ISLS).

Because the transponders must be sensitive to low-level mode pairs, and because of the side and back lobes of beacon radiation patterns, many unwanted replies may be received. To reduce these, an omni antenna is employed to perform, in conjunction with transponder circuitry, as a part of a side-lobe-suppression system. A third 1030-MHz burst, not a part of the mode pair challenge, and called P2, is transmitted from the omni antenna 2 μs after the P1 burst is transmitted from the directional antenna. Because of the directional antenna gain, a transponder in the main beam will receive a P1 burst greater in power than the P2 burst from the omni antenna. If the received mode pair is from a side lobe, the P2 burst will be greater in amplitude than the P1, and the transponder will generate an inhibit to any further reply for 35 ± 10 μs. The principle is illustrated in Figures 7-25 and 7-26. ATCRBS transponders will also be inhibited by Mode S interrogations, where the P1 and P2 are transmitted on the main beam to identify the Mode S message type.

The aircraft transponder contains a threshold circuit to detect excessive-amplitude P2 bursts. It will not respond to 1,030-MHz bursts that are below the threshold, since that is a normal condition when the mode pair is received in a straight path from the directional antenna. Transponder receipt of any initial apparent P1 burst of duration greater than 0.7 μs makes the receiver ready to test the level of the P2 burst. The first burst, presumably P1, desensitizes the receiver by 9 dB plus the received pulse amplitude. After the trail edge of the pulse, the receiver sensitivity recovers at approximately 3.5 dB/μs. When the P2 pulse arrives, the receiver will then have recovered by (2 μs – 0.8 μs) × 3.5 dB, totaling 4.2 dB. If the P2 pulse is substantially greater than the P1 pulse +9 dB, it may have been because of side-lobe interrogation, and it will break the threshold to cause the 35 ±10 μs inhibit. This SLS-inhibit effect was capitalized upon in mode S design, where an equal-amplitude

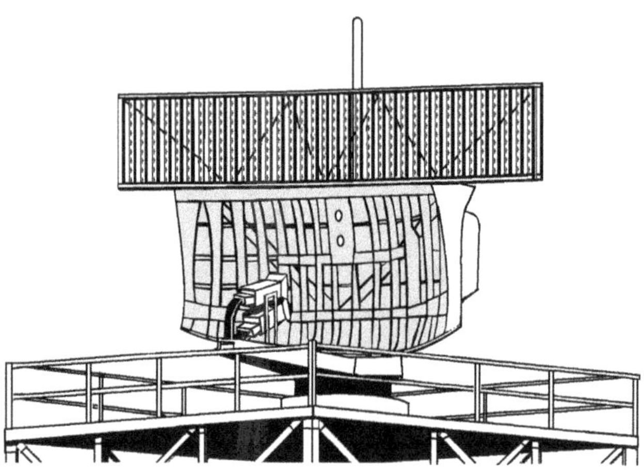

FIGURE 7–23

The 5-foot array on an ASR reflector.

P1–P2 combination indicates a Mode S interrogation while simultaneously disabling any ATCRBS reply.

The interrogator may employ ISLS to correct another problem, called "ground-bounce" interrogation (see Figures 7-25 and 7-27). In this system, both P1 and P2 bursts are transmitted from the omni antenna, but the P2 burst is 3 dB greater in power. If the transponder receives the omni signal in the absence of the directional-antenna mode pair, it will generate the 35-μs inhibit. When both signals are normally received with two P1s occurring in coincidence, the P1 power addition raises the P1 total to the point that the inhibit will not occur. If the mode pair arrives late at the transponder, it will have been because of a ground bounce, and the transponder will already have been inhibited by the P1–P2 omnidirectional bursts.

Location of false targets and the ground objects causing them can be determined by the trigonometric "law of cosines," illustrated in Figure 7-27, both graphically and mathematically. The calculation can be used to incorporate "fixes" for such conditions.

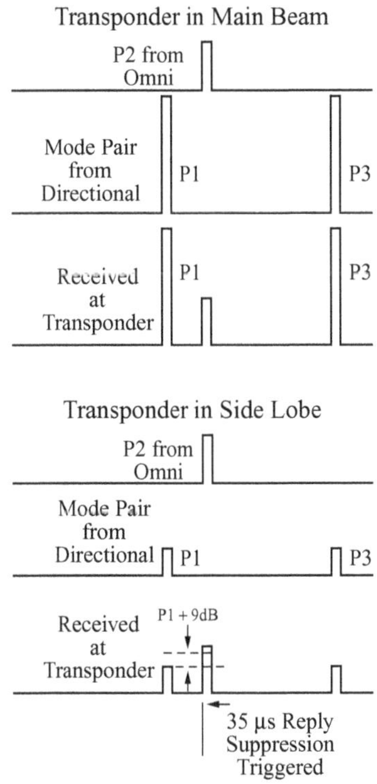

FIGURE 7-24

Side-lobe suppression in the transponder.

SLS and MSSR

Side-lobe returns in MSSRs need additional attention. In these systems, P1 and P2 are both transmitted on the directional antenna and identify the interrogation as Mode S. Transmitted in that manner, the ATCRBS transponder treats the interrogation as though it were a side lobe, and the ATCRBS reply is inhibited for 35 μs. The Mode S transponder recognizes the P1–P2 combination as a Mode S interrogation, and prepares to synchronize its DPSK phase detection at the beginning of P6 time. There is a possibility that the P1–P2 combination really is only an ATCRBS omnidirectional side-lobe interrogation. To prevent the DPSK circuitry from synchronizing on noise, fruit, or a P3 pulse in that condition, a P5 burst is transmitted in coincidence with the P6 pulse, but on a "control" antenna, similar to the omni, but with a pattern that is "notched" around boresight.

Receive Side-Lobe Suppression (RSLS)

The ATCBI-6 MSSR uses the control antenna for a side-lobe receiver to further exclude undesired replies. In addition to side lobes, undesired Mode S replies can also result from TCAS exchanges between aircraft.

ATCBI-3 Hardware Transmitter Block Diagram

Figure 7-28 illustrates a manner in which the P1, 2, and 3, 1,030-MHz bursts are generated in a hypothetical interrogator transmitter; the block diagram closely resembles the ATCBI-3 system. The frequency generation begins with a 42.917-MHz crystal oscillator, multiplied to 128.75 MHz, and then modulated by all three 0.8-μs P1, 2, and 3 pulses in a gated amplifier. These modulating pulses have been generated in the pulse mode generator, which is, in turn, triggered by a beacon pretrigger from the primary radar. At ARSR facilities, the beacon pretrigger will be at the same f_p as the primary radar; at ASR facilities, the pretrigger will usually be at 1/3 f_p of the primary radar. The bursts from the gated amplifier are applied to two more frequency doublers to obtain 515 MHz. The frequency-generation circuitry up to and including

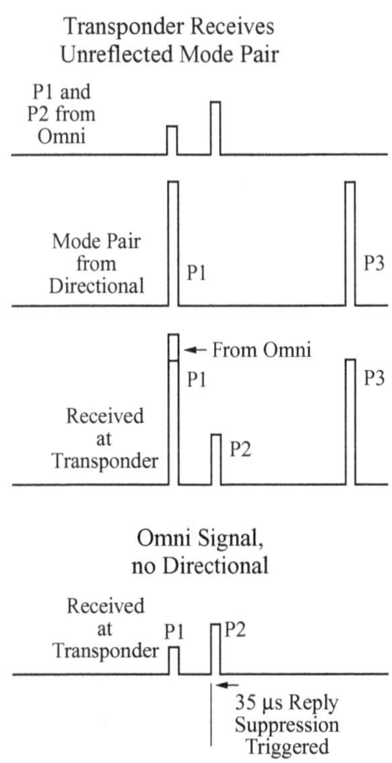

FIGURE 7-25

ISLS. The omni signal creates a suppression gate.

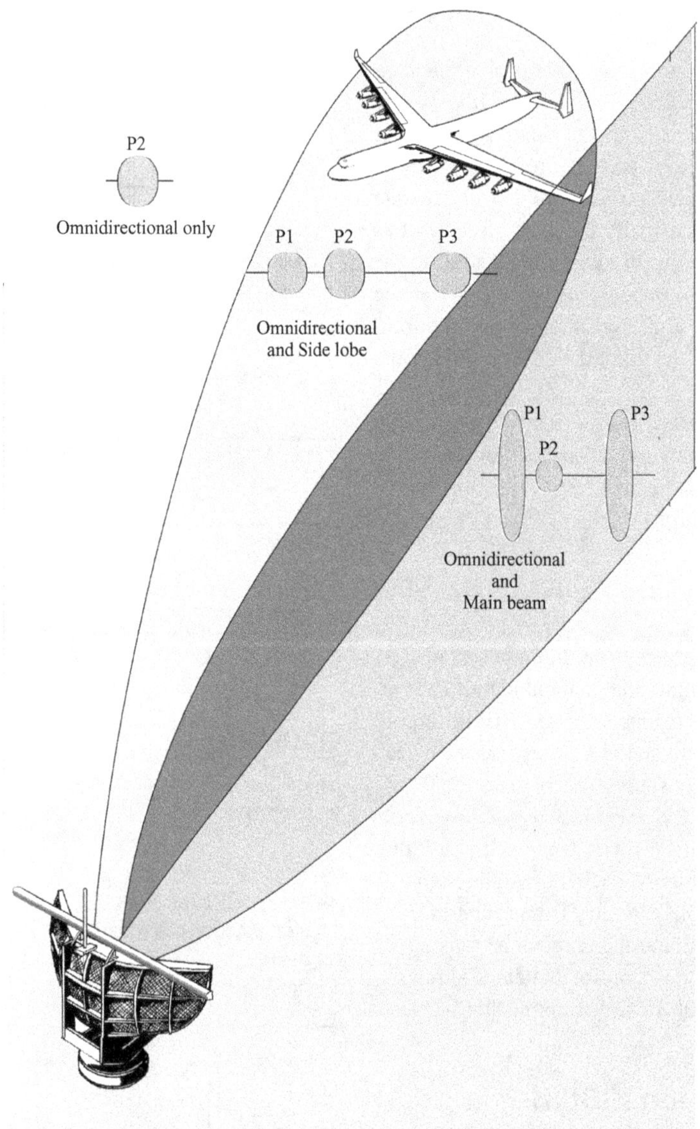

FIGURE 7–26

Side-lobe suppression (SLS).

the 515-MHz output is in a module called the exciter. The exciter output is applied to the four final stages of the transmitter. A frequency doubler produces 1,030 MHz, and three more stages of amplification provide the final power output of all three bursts. To facilitate the amplification of the L-band frequency, the doubler and amplification stages are made of microwave cavities and separation of the P2 pulse from the P1 and P3 pulses is achieved with switching circuitry. A directional coupler serves as a power splitter, providing two outputs, one 3 dB down from the other; this will ultimately provide the means through which a P1 burst 3 dB less than the P2 burst may be used for ISLS. All three bursts from the −2 dB directional coupler output are applied to a diode switch, operated by an SLS gate from the pulse mode generator. The diode switch is used to separate and steer the P2 pulse through the ISLS circuit, and to the omni antenna. In ISLS operation, a gate from the switch driver allows the −3 dB P1 burst from the −5 dB directional coupler port to be placed on the line to the omni antenna along with the P2 burst.

The output of the SLS diode switch is applied to a tuned diplexer, very roughly equivalent to the duplexer in a primary radar. The 1,030 MHz is steered toward the antenna, and received reply codes are steered toward the receiver. The directional coupler following the diplexer provides for power, sensitivity, and other measurements. Channel switches, equivalent to the primary radar waveguide switches, provide routing for a redundant channel, and additional directional couplers provide for connections for monitoring circuits.

Peak power output may be as high as 3 kW, but is more likely to be in the order of 500 W, determined by facility commissioning data. For the directional antenna path, transmitter power measurements are performed in the same manner as in a primary radar, except that the pulse width for duty-cycle calculation is effectively doubled because there are two. The mode pair duty cycle is, therefore, $2t_p/T_r$. In measuring SLS power, the duty cycle is t_p/T_r for a single pulse, exactly as for a primary radar. In ISLS operation, the ISLS P1 burst is half the P2 burst, and the total power in the two-pulse package is 1.5 times the power in the P2 pulse. The average power is mathematically equivalent to that for a single pulse, 1.5 times the peak of the P2 pulse, so the duty cycle for a single pulse may be used with a multiplier of 1.5. Of the total ISLS power, the P1 burst power contributes 1/3, and the P2 burst power contributes 2/3, so the P2 burst peak will be 0.667 times the calculated peak using the duty cycle for a single pulse. A multiplier of 2/3 is −1.76 dB. Mathematically stated, duty cycle and peak power calculations are as shown in Figure 7-29.

The Receiver

The ATCRBS receiver is a simple, superheterodyne, amplitude-detection type, resembling the normal receiver in a primary radar. One familiar with radar frequency synthesis might expect that the receiver local oscillator frequency source would be the same as that of the 1,030-MHz transmitter. However, this is not always the case; the local oscillator in ATCBI-3 was independent and operated at 1,030.5 MHz to avoid a specific interference problem between the receiver and transmitter. For the 0.45-μs transponder reply code pulses, optimum bandwidth would be 2.66 MHz, but more attention is devoted to pulse reproduction than sensitivity, and the receiver bandwidth may be 8 to 10 MHz. The wide bandwidth is achieved with stagger-tuned i-f stages. Receiver sensitivity is measured at a point where the injected test signal is viewed at twice the noise level; this is called ***tangential sensitivity***. This level is used because it is the lowest signal level at which the receiver output may be reasonably expected to break a quantizer threshold in the interrogator, or in the decoding or processing equipment downstream. Tangential sensitivity should be 3 dB greater than mds, since the test signal power is twice the noise level. The maximum acceptable level is specified at −87 dBm, very high in comparison to primary radar mds values in the order of −106 to −110 dBm. The difference is in part because of the wide bandwidth, in part because there is no low-noise front-end amplification, and in part because a double-balanced mixer is not used.

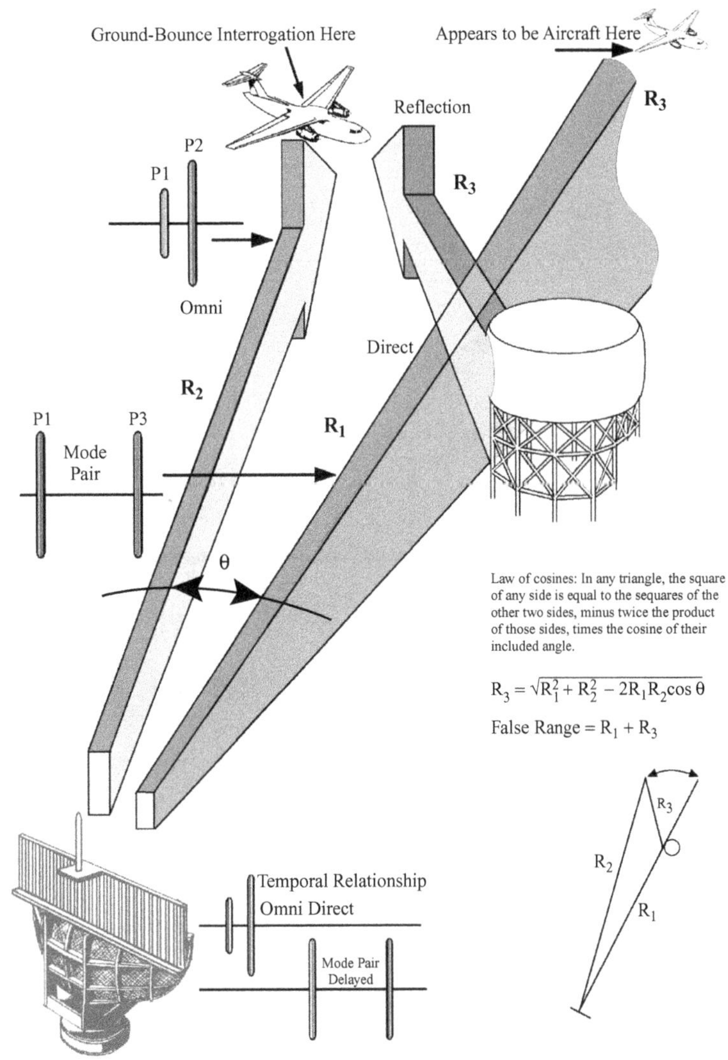

FIGURE 7–27

Improved slide-lobe suppression (ISLS).

STC/GTC

Although both terms are used and considered synonymous, "GTC" may be some what more applicable because the intent is to attenuate transponder replies rather than increase the sensitivity to echoes. In earlier equipments, the stc waveform was originally applied as a negative bias to an i-f amplifier stage; in later versions, the PIN-diode stc was employed. Since most of the noise in the signal is created by the local oscillator, the noise does not appear to be affected by stc in the later models, and downstream quantizers may operate with fewer noise breaks at the greater ranges. The video output of the second detector is clipped, amplified, and limited, before being made available as an output. In some systems, it may be quantized. In others, only raw video may be available at the interrogator output. Although the stc waveform has a very similar effect on the receiver as in primary radar, the effect on beacon system performance may be more pronounced. Because of the strength of the reply code trains, and the potential that they may be the result of side-lobe interrogations and side-lobe

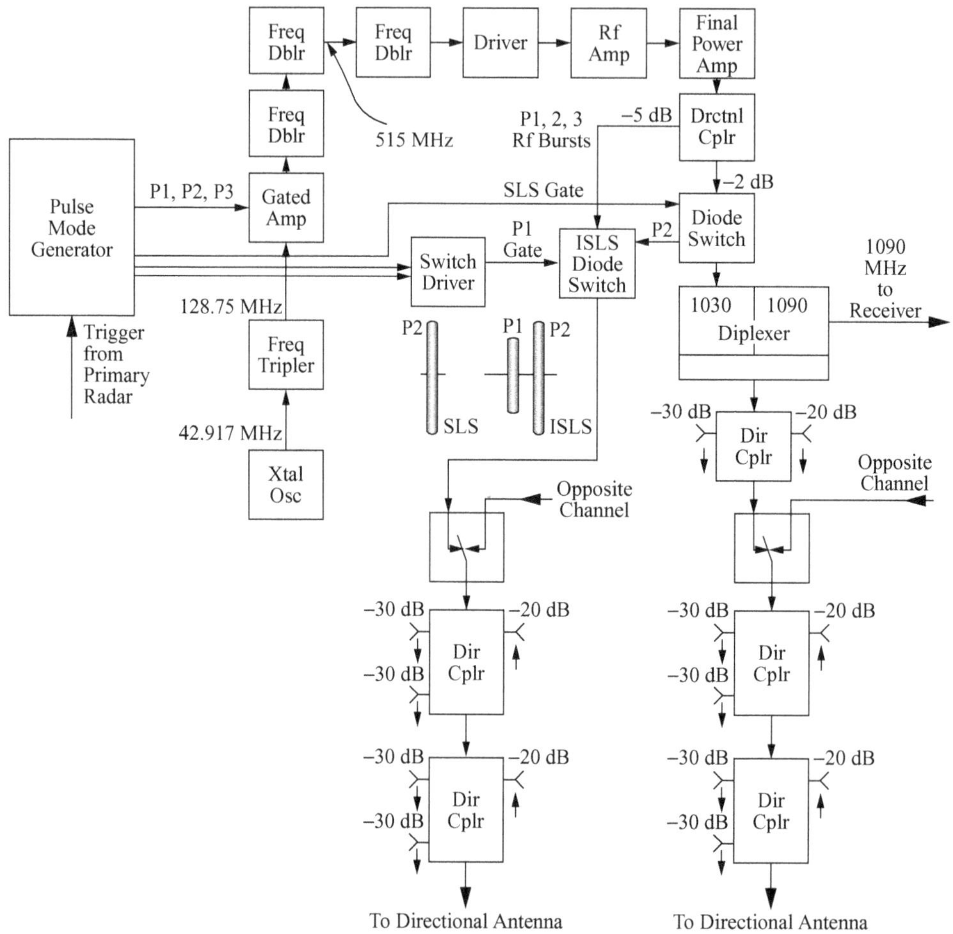

FIGURE 7–28

An early interrogator transmitter.

reply, a condition called ring-around, illustrated in Figure 7-30, may appear in the absence of stc. This was not unusual in generations 2 and 3.

Code Train Data Processing

The received code trains may be used in several manners. The output of the receiver is called raw beacon, and it may be displayed on ppi's, as in Figure 7-34. In the earliest days of air traffic control with beacon equipment, controllers were trained to read the code directly from the raw train displayed on the ppi. Now this raw beacon video is generally not used for air traffic control but is useful to the technician on monitor ppi's. In many cases, the beacon data will be quantized in the interrogator equipment; if not, it will encounter a quantizer somewhere in the processing stream.

$$DC_{directional} = 10 \log \frac{2t_p}{T_r}$$

$$DC_{SLS} = 10 \log \frac{t_p}{T_r}$$

Mode Pair, SLS: $P_t = P_{ave} - DC_{dB} + Atten_{dB}$

ISLS P2: $P_t = P_{ave} - DC_{dB\ single\ pulse} + Atten_{dB} - 1.76\ dB$

Legend:

DC = duty cycle
DCdB = duty cycle in dB
Directional = for directional antenna
Mode Pair: P1 and P3 pulses
P2 = P2 pulse used for Omnidirectional antenna
SLS = Side-lobe suppression
ISLS = Improved side-lobe suppression

FIGURE 7–29

Duty cycle and peak power calculations.

To reject those noise spikes that break the quantizer threshold, as illustrated in Figure 7-31, beacon video quantizers may also contain pulse width discrimination. It is unfortunate that the word "quantizer" now has two definitions. In beacon systems, and many other earlier radar systems, a quantizer is an amplitude-standardization circuit. In other equipments, it has been used to describe A/D converters, where a parallel combination of "1s" and "0s" represents an analog voltage.

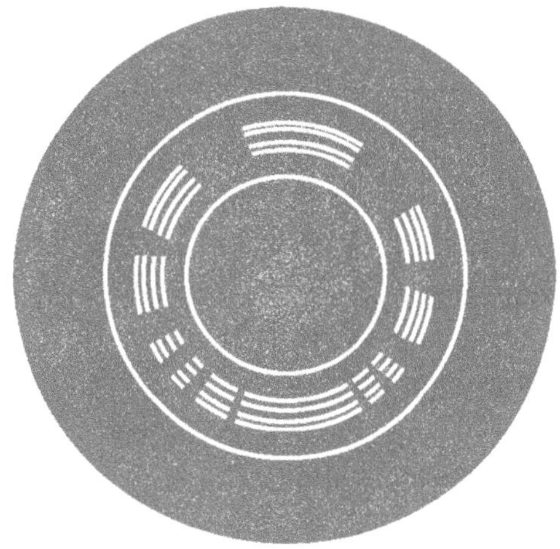

FIGURE 7-30

"Ring-around."

Decoders

Analog decoders were in use by the late 1950s. Today, at airport terminal facilities, automated systems, using beacon code train data, provide controllers with far more information. However, the decoders must still remain available for those times in which the automated systems are out of service.

Figure 7-32 illustrates a hypothetical beacon control panel, as might be found at an air traffic control position. In the earliest beacon systems, air traffic controllers viewed the entire beacon code train on the display, and they were trained to read and calculate the code from the displayed video.At that time, there were only 64 possible codes, using the A and B pulses. This visual code reading was subject to error and required important controller time and attention; very early in the history of IFF and ATCRBS, beacon decoders were manufactured to provide analog display of up to 10 selected beacon codes at an air traffic control position. More codes could be selected at other positions with decoders, so the number of possibilities for a facility was not limited to 10.

The beacon decoder system consists of two major equipment portions. One of these, called a ***common decoder,*** contains a 24.65-μs delay line, illustrated in Figure 7-33. The quantized beacon video is applied to the delay line input and propagates down the line to the termination at the end. The delay line contains a tap for both framing pulses and each data pulse; when the first pulse, F1, of the code train has been delayed by 24.65 μs, it is present on the last delay line tap. When a pulse is simultaneously present on the F1 and F2 delay line taps, a bracket detection, also called ***common system (C/S)*** detection, is made by means of an AND gate. At the time the bracket detection is made, the bracket-detection C/S pulse and all the delay line outputs are enabled and placed on parallel lines, to be made available to several ***non-common decoders,*** each associated with individual controller display positions. Should an SPI/ident pulse follow the F2, it will be present at the delay line input when the bracket detection is made; the first 4.35-μs delay makes this possible.

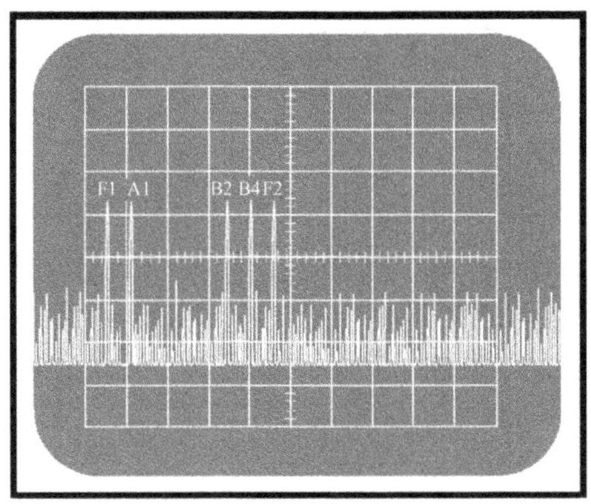

Code 16 Raw

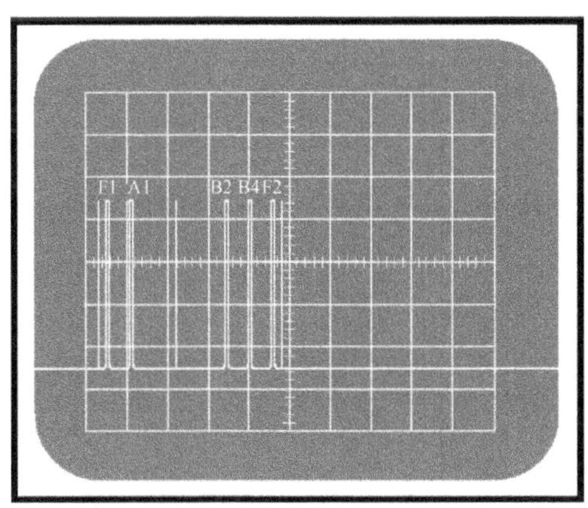

Code 16 Thresholded (Quantized)

FIGURE 7-31

Raw and quantized code 16.

Non-common Decoders (see Figures 7-2, 7-32–7-34)

Each noncommon decoder is controlled by the 10 thumbwheels illustrated in Figure 7-32. Each pair of thumbwheel switches produces logic-circuit inputs; at bracket-detection time, the switches are logically compared to the parallel delay-line outputs from the common decoder.

When a received code train matches the code selected by the thumbwheels, a "double slash" will appear on the display.

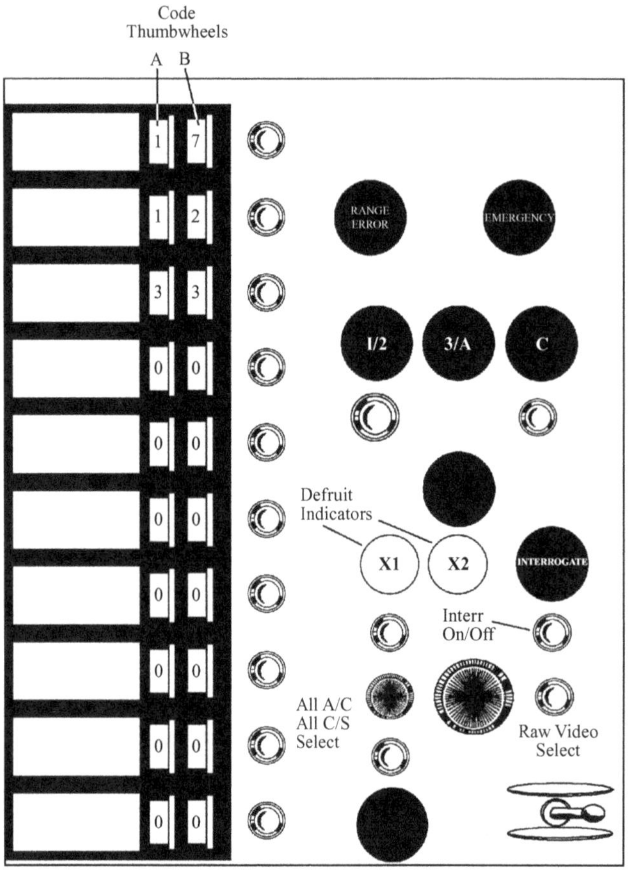

FIGURE 7–32

Beacon control panel.

The push buttons next to each set of thumb wheels allow the controller to enable an ident display for the selected code while he requests an ident from the aircraft pilot; the ident will appear as an intensified "bloom" between the selected-code double slashes. The appearance of raw, decoded, and ident targets is illustrated in Figure 7-34. The common system also performs emergency decoding, and the emergency lights and alarms will appear on all control panels.

When COMMON SYSTEM has been selected, a single-pulse "slash" will be displayed for all bracket detections. The common decoder also performs an "all aircraft" function; when ALL AIRCRAFT has been selected on the control panel, any reply that does not have 20.3-μs brackets will be displayed in its entirety.

Fourth-Generation Processing

Fourth-Generation Digital Data Processing (see Figure 7-35)

To digitally process beacon data, upgraded state-of-the-art techniques were employed to provide parallel code-information data, bearing some resemblance to the delay-line technique employed in the analog decoders. The data-processing objective is to first construct two parallel reply words, containing the range count, interrogation mode, and potential "garble" conditions. "Phantom," "garble," and "interleave" are conditions resulting from the mixing of reply code trains from aircraft at close proximity. Further detail on this subject is provided further in this chapter in a section entitled "Interference between Code Trains."

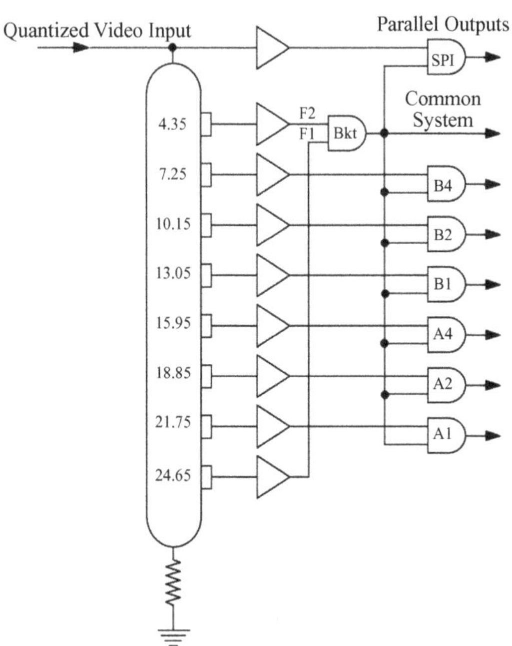

FIGURE 7–33

Delay line decoder.

Delay Lines Replaced by Shift Registers

In later decoding systems, the code data was applied to a serial shift register with a high-speed clock, for instance, 9.5 MHz. When the appropriate stages of the shift register contained "1s," a bracket detection occurred, and code data from the bracket-detection register was placed into code-storage shift registers. To provide for the possibility that interleaved code trains might be progressing through the shift register at the same time, and that two bracket detections could occur while data being clocked down the shift register, the code data would be sent to one of two registers for temporary code storage and conversion to parallel form; they were called X and Y.

Multiple Code Extractors

Later versions of code extractors, such as the one in the ASR-9 beacon reply processor, illustrated in Figure 7-36, employed more channels, higher speed clocks, and longer shift registers, containing more stages. The beacon reply processor contained a 242-stage shift register, clocked at 11.7 MHz. The contents of the shift register were used for bracket detection, and the serial delayed shift-register outputs were selectively steered to one of four channels by a channel-select logic. With this frequency and number of stages, each data pulse should occupy five or six shift register stages, and the 242 stages provide for 20.6 μs of data. When two of the serial shift register stages, 238 stages apart, first contain “1s,” a “lead-edge” bracket detection is made, and when the data from those stages returns to “0s,” a “trail-edge” bracket detection is made. Both the lead edge and trail edge of the brackets are used in initiating a sequential test of each serial shift-register-delayed data pulse for the appropriate pulse position and presence.

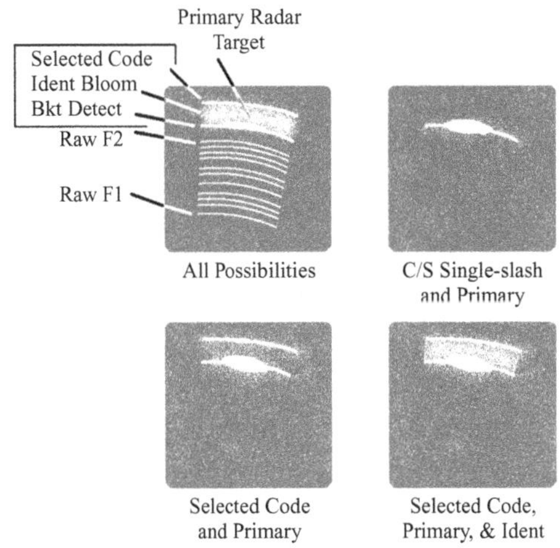

FIGURE 7–34

Raw and decoded beacon.

The “assigned” channel would begin logically testing for the presence of a bit in each data position, ultimately forming a parallel word in the beacon code extractor (BCX) first-in, first-out (FIFO) to describe the code. The code requires 14 bits: 12 for the A, B, C, and D pulses, 1 for the SPI pulse, and 1 for the “X” pulse. Two additional “garble flag” bits will be used to indicate phantom or garble conditions, described in an ensuing paragraph.

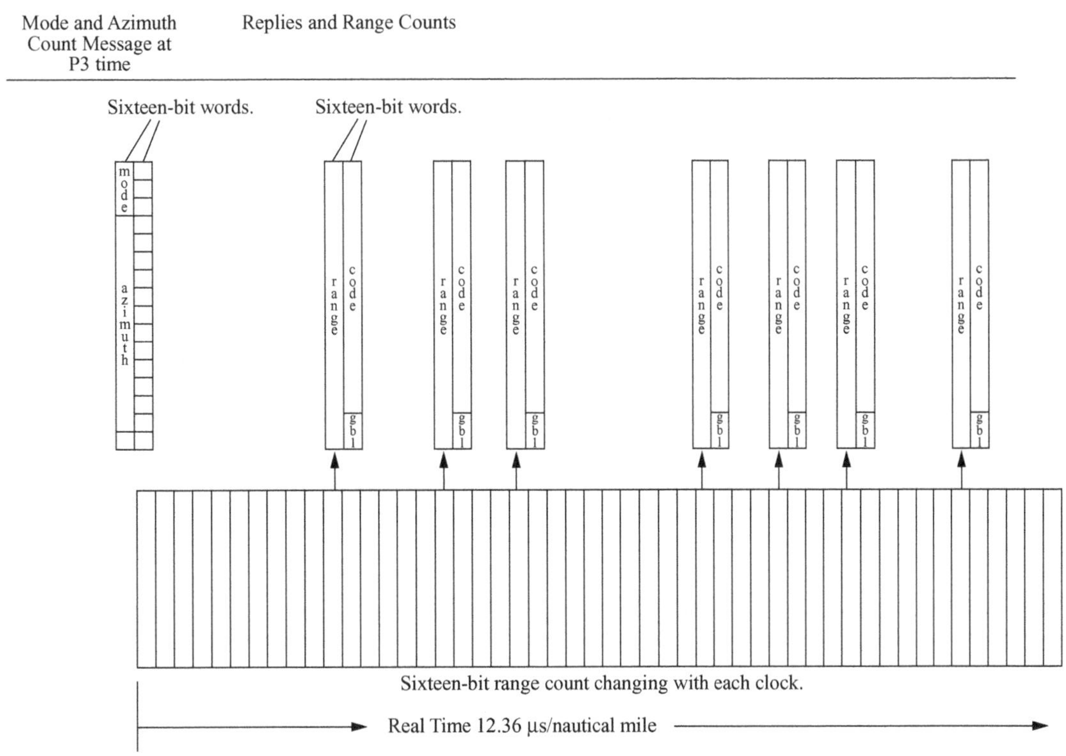

FIGURE 7–35

Real-time data stream into HSIB.

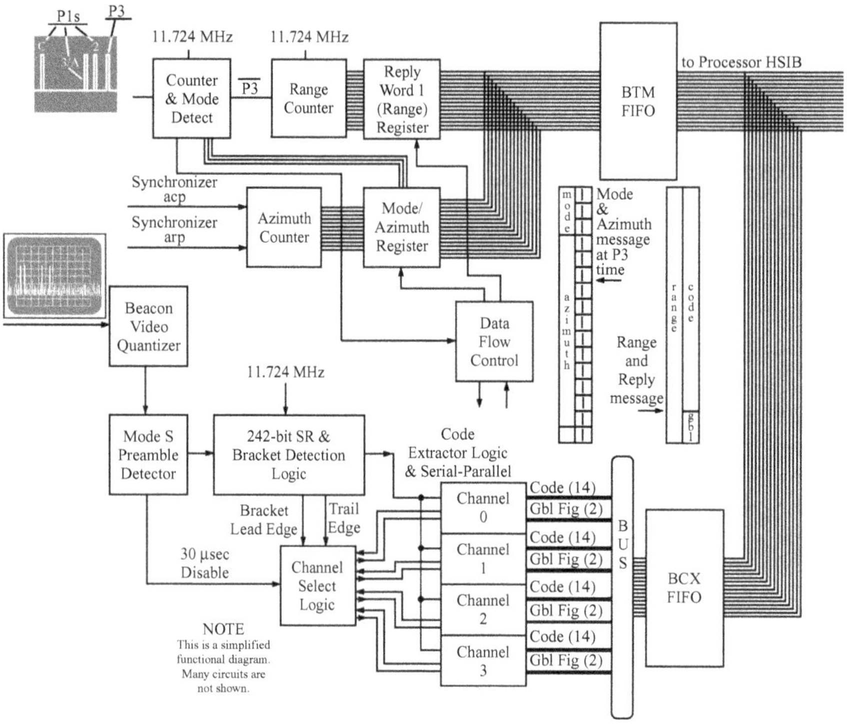

FIGURE 7-36

Simplified block diagram, ASR-9 beacon reply processor.

When the serial testing of data pulses has been completed in the individual selected channel, a series of events is initiated. First, the 16-bit beacon range count is placed on the output data lines by the beacon timing and monitoring (BTM) FIFO. Immediately following, the code-extractor outputs are placed on the data lines. For each beacon reply, then, a two-word message describing range andcode is transmitted to a high-speed interface buffer (HSIB) in the array signal processor (ASP), the computer which will perform the azimuth-sliding-window process and code validation. The reply-code-message output illustrated in figure 7-35 still is very close to real time, but will no longer remain in real time after being stored in the computer's HSIB; the real-time relationship is no longer necessary at that point, since the range count is associated with the code word.

Identifying the Interrogation Mode and Azimuth

The reply message transmitted to the HSIB contains no information to identify the challenge interrogation mode. However, that information, as well as the azimuth count, is made available to the HSIB at the beginning of each beacon T_r. On receipt of a P1 pulse, a counter in the mode detect logic begins running; when the P3 pulse occurs, the count will be used to indicate the mode, and a 3-bit word to describe the mode will be transferred to the mode/azimuth register. At the same time, the current azimuth count is also loaded into that register, forming a word to describe the mode and azimuth of the current T_r. As the range counter runs throughout the T_r. On detection of the P3 pulse, the beacon range counter is preset to a starting value which takes into account the 3-µs transponder delay.

One reason for the increase in the number of shift register stages and clock speeds was to provide for more critical logic testing of the size and position of each framing or data pulse. Refer to a proceeding section entitled "Interference between Code Trains." Many shortcomings in beacon data processing have been attributable to several code-train faults, among these are conditions called phantom, garble, and interleave. Higher scrutiny of the pulse position and size provides greater opportunity to detect these conditions. The ASR-9 code extractor tests bracket and data pulses for tolerances, adjusted via a computer terminal. The tolerances are stored in a memory and are called variable site parameters (vsp). BITE test signals of changing parameters are regularly circulated through the beacon code extractor during deadtime, to ascertain the ability of the extractor to detect an error.

More about the Beacon Azimuth Sliding Window

The second step of the process is to input the replies to an azimuth sliding window, described at preceding points in this chapter, and in this book. The window examines associated replies at the same or proximate ranges and adjacent azimuths to form a single target report from (1) the acp count when replies stop, called "trail edge," minus (2) the acp count when the replies began, called "lead edge." The result is the ***run length***. The lead edge acp count plus half the run length is the centroid, a single acp count describing the target azimuth. Criteria adjustments to the number of "hits" required to declare lead edge, or trail edge, maximum run length, proximity of range and azimuth of primary radar targets needed to set a ***radar reinforced*** bit, and other parameters are provided. A regular performance test of the system is a measurement of the ***reinforcement rate.***

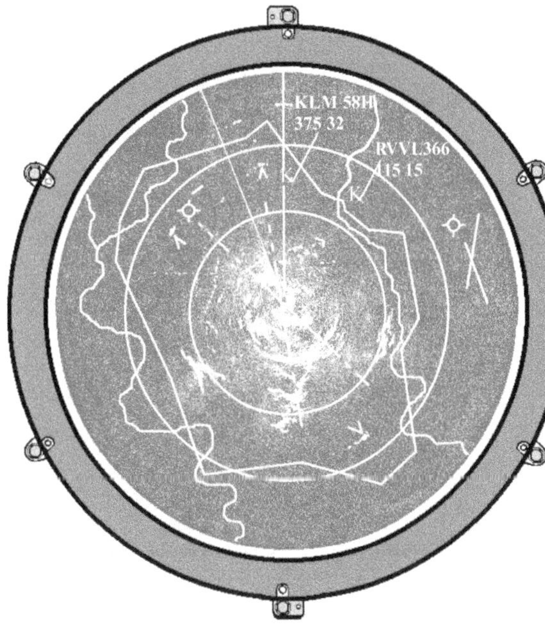

FIGURE 7–37

Alphanumeric data blocks on a controller display. Analog ppi, alphanumerics by the deadtime stroke generator.

Final Data Display (see Figure 7-37)

Whatever the system, the objective of the beacon code extractor is to provide a parallel code-data word for use in conjunction with an azimuth-sliding-window detector. The processing will include a code validation, which verifies that successive mode 3/A replies are consistent. Completed beacon reports are assembled into individual data messages containing mode, code, range, and azimuth. These completed target messages will be further used by data-processing computers such as the ARTS II, ARTS III, EARTS, or those in an ARTCC. Those computers may use this information for association with stored flight plans, and air traffic controllers may enter other data to be associated with the aircraft, including assumption of control, aircraft identification and type, handoffs to other controllers or facilities, and much more. Figure 7-37 illustrates a display of aircraft alphanumeric data blocks from an ATC computer. Near the end of this chapter is a script and illustrations to exemplify the use of the overall system.

Beacon Reply and Mode Pair Reconstitution

In the ASR-9 system, to make the digital-data messages compatible with analog decoders and display equipment, the indicator-site equipment actually reconstitutes a synthetic analog beacon code train, mode pair, radar display trigger, and azimuth data; since the information in the target messages accurately describes the range and azimuth of each target, the synthetic analog data is a faithful reproduction of the original information. The ASR-9 equipment that does this is called the Surveillance Communication Interface Processor (SCIP). The SCIP also provides data conversion to make the data compatible with the ARTS IIIA,ARTSIIIE, or ARTS IIE.

Interaction between Code Trains (see Figure 7-38)

Disregarding a possible ident pulse, the code train length is 20.3 μs between framing pulses, and this is 1.64 nmi in radar range. Clearly, a second or third beacon reply may occur while a first one is still being received. When the code data pulses of one occur in temporal alignment with code data pulse positions of another, a garble condition is said to exist. When the code trains are being received at the same time, but the data pulse positions of the two are not in alignment, an interleave condition is said to exist; this poses no problem in digital processor systems such as the ASR-9, since the separate code extractor channels will operate properly. Whenever there is a 20.3-μs separation between any two pulses of two code trains, a false framing-pulse detection can occur, and a phantom condition is said to exist. In the ASR-9 beacon reply processor, 2 of the 16 bits in the reply message may represent phantom or garble conditions.

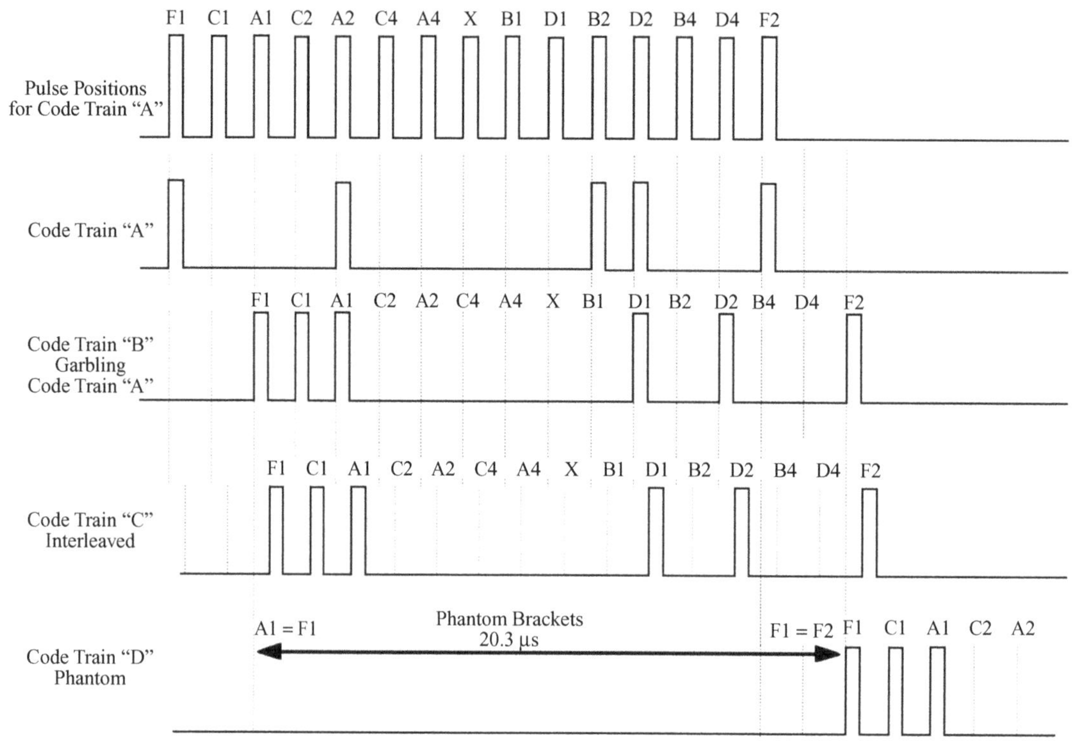

FIGURE 7–38

Phantom, garble, and interleave conditions.

Interference and FRUIT

Because there are so many aircraft with transponders, and because they are interrogated by so many radar facilities, the environment in, around, or even near, any populous area is saturated with replies. One characteristic of the reply makes it possible to separate the reply to the local radar interrogator from all the others; it will repetitively occur at the same range, making it synchronous. Replies to other interrogators will not be in synchronization to the local one, and processing techniques permit their elimination. Replies to interrogators other than the local one are called false replies unsynchronous in time (FRUIT).

FRUIT was eliminated by a beacon video ***defruiter*** from the second into the fourth generations of equipment. The defruiter performs a challenge-mode-sensitive T_r-to-T_r comparison of the code trains; successive replies to the same modes are compared for temporal match, somewhat similar to the primary radar video integrator (aka "enhancer"). In contrast to the enhancer, the beacon video defruiter is simply a coincidence device, without a feedback loop to "build" a larger target. If a code train occurs at the same range in subsequent intervals, it is synchronous. There have been several techniques to accomplish this over the years. Three generations of video defruiting equipment have been used. The first of these used an automatic tracking delay line to maintain the beacon video delay precisely equal to the radar T_r. The tracking delay line was an unacceptable maintenance workload, and the defruiter used vacuum tubes. The second version used storage tubes, written upon and read with spiral sweeps. The storage tubes developed "holes" which were difficult to discover. The third generation is still in limited use and uses digital memory. Because of an azimuth-sliding-window process, defruiting is inherent in a digitizer such as the common digitizer, ARTS, or the digitizers in the ASR-9, ASR-11, ARSR-3, and ARSR-4. The need for defruiters is further reduced by monopulse systems.

Range Ambiguities (Second-Time Replies)

Because of the abundance of power in the reply code, secondary radars are vulnerable to range ambiguity, a condition in which a reply from a distant aircraft is received in the next T_r to provide false range information. For this reason, interrogators are always operated with long T_r's of 3,000 µs or more. This reduces the hits-per-scan (*N*s) (number of replies between 3-dB beam dimensions as the antenna scans across the target) in short-range systems such as airport surveillance radars (ASR), but there is an abundance of transponder replies in most cases, anyway. A beacon system may employ a staggered f_p, where the T_r's are of several different intervals; this causes second-time targets to be nonsynchronous and thereby eliminated by defruiting, or by the sliding-window process. Second-time replies are now further reduced in MSSRs because of their low f_p's and long T_r's.

Test Equipment

Special Test Equipment

Apart from an oscilloscope, the most frequently used equipment at an ATCRBS interrogator-site facility is an elaborate special-purpose ***beacon test set.*** This is actually a two-part unit, containing a video section, and an rf section. It can perform a multitude of test functions. It will generate up to three range-positioned code trains, present only during selected challenges, and called mode-sensitive code trains. It provides swept-frequency and marker outputs, rf detection, and calibrated signal attenuation. Viewing test video is much easier than viewing live.

Graphic User Interfaces (GUIs)

These have been provided with new equipment beginning in the fourth generation. They are a computer, display, and keyboard, running on an application program designed for the system. Latter-day equipment contains online ***built-in testing (BIT)*** and off-line ***fault isolation testing (FIT).*** The commands or results are accessible through a GUI, and much analysis, troubleshooting, and system control can be performed from the terminal. Single-board computers in many systems provide serial interfaces for personal computers with commercial communications software.

Some particular oscilloscope techniques are necessary in observing beacon phenomena, and may require some practice to develop the technician's skill. The following are a few of the most common:

> In older systems, synchronization to the P3 pulse may be readily obtained from the primary radar or beacon pretrigger. Newer systems may well not be in sync with the primary radar.
>
> To observe multiple reply codes, as when measuring video amplitudes, synchronize the oscilloscope internally by the vertical amplifier, and set the timebase to about 5 μs/cm to observe whole code trains.
>
> Synchronizing internally on the oscilloscope vertical amplifier, increase the sweep speed to about 200 nsec/cm to observe pulse widths and shapes.
>
> The mode pair mixed with the video will make it difficult to sync on reply codes only. To keep the distracting interlaced mode pair out of the oscilloscope code-train video presentation, look for a video source where those timing signals have been inhibited.
>
> Synchronizing the oscilloscope externally with the mode pair may produce an unstable presentation because of the interlace pattern.
>
> To observe individual reply codes, as when inspecting the reply from a parrot or beacon test set, synchronize the oscilloscope with the beacon pretrigger on the EXTERNAL TRIG input, then use the DELAYED SWEEP function to place the expanded delayed video on the display.

Test Equipment Evolution

New test equipment has replaced old in many cases. Among the most significant improvements is a new generation of rf power meters. The power-sensing thermistors and bolometers have been replaced by detector diodes, and peak power may be measured by electronic sampling and holding. The MSSR equipment has necessitated an entirely new beacon test set for adaptation to mode S monopulse equipment.

Sliding-Window Detection

For decades, the means of finding the azimuth center of transponder replies has been the conventional ***azimuth sliding window***, and it is still in use at many facilities as this is being written. It has been addressed both in this chapter and in Chapter 6. In ATCRBS systems, the window begins to operate with a ***code validation***, when two azimuth-adjacent replies to mode 3/A interrogations are identical. As the window is in progress, the reply codes are retained in memory. When the window detects "trail edge," the azimuth center is computed. The transponder range and mode C altitude data are already present in the reply-code messages, and all the information necessary for the formation of a complete ***report*** to describe the aircraft, its altitude, and its range. The reports are then compared to proximate centroided primary radar reports to detect "reinforcement" or "correlation" to increase confidence in the beacon report accuracy. In this "correlation" or "merge (ASR-9 term)" process, it has been customary to use the beacon range count and primary radar azimuth in a final report. The beacon report is subject to azimuth inaccuracies, but the beacon range counter is faster and therefore more precise.

Deficiencies in the Beacon Azimuth Sliding Window

Since the introduction of beacon systems into civilian air traffic control radar in the late 1950s, there has been a continual effort to overcome a number of problems. Many of the shortcomings were a result of excessive transmitter power from both interrogators and transponders, from terrain effects on radiation patterns, from the high f_p's of ASR-produced interrogator pretriggers, from unnecessarily wide interrogator radiation patterns, and from the coarse azimuth positional information obtained from the 4,096-acp counter.

As the population of transponder data around radar sites increased dramatically from 1960 until the present, a natural increase in fruit, phantoms, and garbles accompanied the progression. ASR beacon triggering was counted down by three, or sometimes four, times, and interrogator power was substantially lowered. ASR antennas were upgraded to the 5-foot dipole array to better shape the beam and reduce propagation flaws. Since ARSRs operate in L band, the ARSR reflectors are adequate for ATCRBS, although some work was done to improve the beacon dipole assembly used as a feed to the reflector.

Defruiting equipment was used through the 1960s to 1990s, but was made obsolete by the growth of automation equipment and the precipitate increased use of the sliding window. The sliding window was a defruiter in itself. However, environmental fruit or heavy synchronous data could still introduce phantom or garble flags, "zeroing out" code data in the beacon reports.

In the early 1970s, FAA engineers and technicians at many sites began to research the multitude of beacon problems. As part of this effort, the author built a test fixture to isolate, photograph, and count replies from single transponders. The counts varied from less than 10 to 45. The sliding-window centroid for such variations could clearly not be consistent. The environmental fruit created by the high counts was also substantial.

FIGURE 7-39

The LVA antenna used in the ATCBI-6 for monopulse detection.

Monopulse Transponder Detection (see Figure 7-39)

The latest improvement to ATCRBS, the ***monopulse*** antenna, has alleviated many pattern-related deficiencies in beacon performance. Monopulse techniques are not new in radar; earlier systems incorporating multiple feedhorns and single reflectors are traceable to at least the 1950s, and WWII experimentation at the Naval Research Laboratories and/or MIT Radiation Laboratories may have begun as early as 1943. This application as ***monopulse secondary surveillance radar (MSSR)*** began to appear in the late 1980s in the FAA-and ICAO-member mode S systems, and was made necessary by those, because of the addition of new modes in the interlace pattern. The name of the newer system may be misleading where monopulse detection is concerned, because the monopulse azimuth determination may, as in the FAA's ATCBI-6, apply to all forms of operational modes, challenges, and replies, both the old ATCRBS and the new mode S. On the other hand, the FAA system called "mode S" intended for ASRs has an ATCRBS-only mode called "interim beacon interrogator (IBI)," in which the antenna is employed as a conventional planar array, and the data must be centroided by the azimuth sliding-window process in the ASR-9 or other equipment.

History of Dipole Arrays

The FAA mode S equipment uses a 6-foot tall, 32-column array of dipoles, and the ATCBI-6 uses a similar 7-foot, ***large vertical aperture (LVA)*** antenna with four more columns. Dipole arrays have been used in radar since the earliest days. Though crude by today's standards, the SCR-270, which detected the Japanese attack on Pearl Harbor, used such an antenna, as did several other radars. The appearance of those early, low-frequency, large arrays earned them their slang nickname, "bed-spring" (see Figure 7-40).

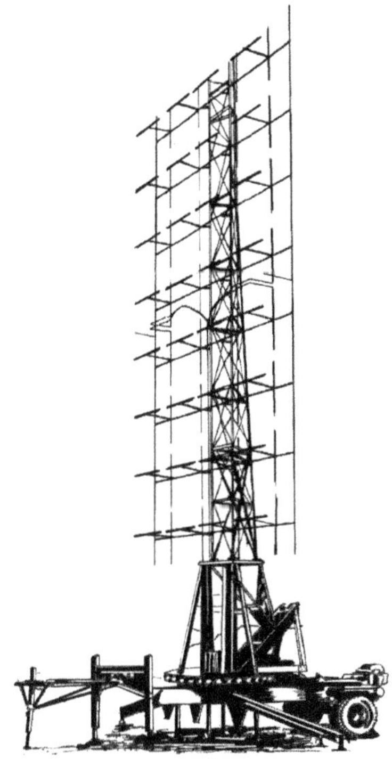

FIGURE 7–40

Fixed dipole array for SCR-270.

Early in WWII, the MIT Radiation Laboratories program to develop a "landing radar" produced the "squeezing waveguide," a single row of dipoles, with probes into a waveguide (see Figure 7-41). The antennas became the precision-approach elevation and azimuth antennas for the AN/MPN-1, the first air traffic control radar, and the antennas were so successful that some remain in use to this very day. The waveguide wall opposite the dipole probes moved back and forth to vary the dipole phasing, and the beam-pointing angle. The number of dipoles in the row determines the pattern beam width, and the main principle is the development of a constant plane of in-phase radiation in front of the radiating dipoles. It is useful to consider this early, perhaps very first, controlled phased array, to fully appreciate and gain insight into latter-day antennas.

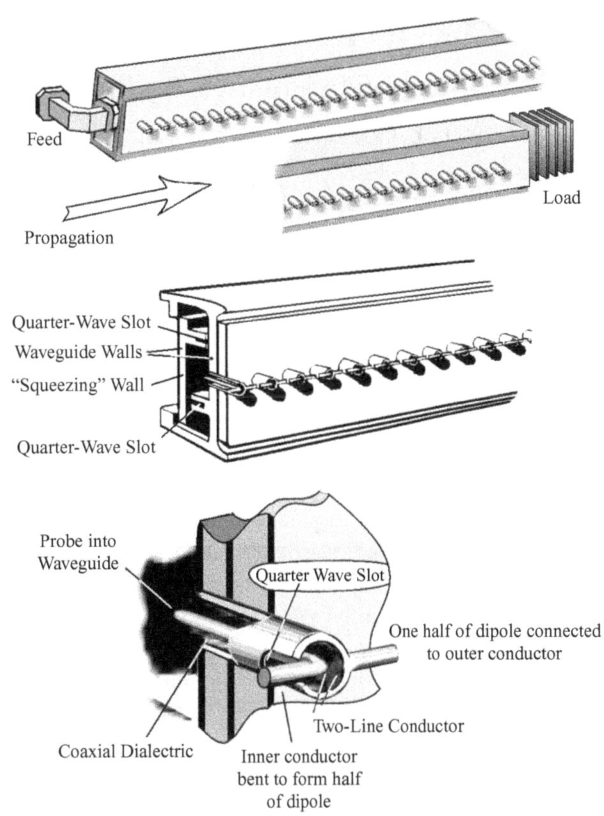

FIGURE 7–41

Squeezing waveguide and dipole construction.

Phasing the Dipoles (or Slots)

The "squeezing waveguide" antennas used in precision GCA radars by the USAAC, USAF, CAA, and FAA since WWII may be more complex than those now used in ATCRBS (see Figures 7-41 and 7-42). Ignoring the mathematics, the general principle of forming a constant phase plane and beam is similar to other directional antennas. The electrical phase of the energy applied to the dipole probe inside the guide determines the initial phase of the radiated field from each dipole. The probes are spaced either more or less than (depending on the antenna) $\lambda/4$ apart in the 9,010-MHz to 9,080-MHz area, but the phasing is varied according to the position of the moving wall. The antennas are designed so that "normal

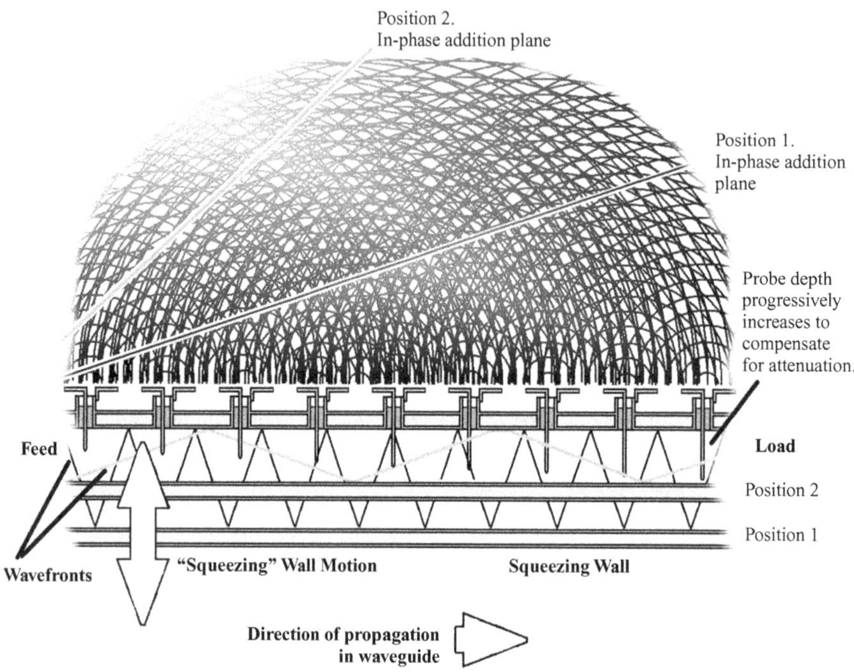

FIGURE 7–42

"Squeezing wall" varied direction of the antenna beam.

(straight ahead)" never occurs, because the beam is subject to splitting. At some point in the space in the front of the array, a straight line may be drawn through all the dipole radiation patterns of the same phase, and the beam and wavefront are formed.

ATCRBS systems in the 1960s and beyond used a single-row dipole array called, in slang, the "hog trough." An array such as the 5-foot tall one developed in the middle 1970s used dipoles phased in both the horizontal and vertical dimensions to shape the pattern in both dimensions, partly to reduce "ground bounce" interrogations. Although dipoles have been the most common method for ATCRBS antennas, slot radiators have also been used for over 40 years in several types of radar antennas.

Beam Control by Frequency Agility

In the late 1960s, ITT Gilfillan, under a USN contract, developed an X-band, raster-scan, phased-array antenna for proposed use as a precision-approach radar (PAR) (see Figure 7-43). It was to be an improvement to GCA and ***carrier-controlled approach (CCA) systems***, some of which used separate azimuth and elevation variable-width-waveguide arrays. The single antenna incorporated a squeezing waveguide, but, rather than dipoles, the outputs were probes, radiating into 109 ***"serpentine"*** vertical antennas. Other ITT Gilfillan serpentine-fed, frequency-scanned radars had already been deployed. The AN/SPS-48 antenna is clearly visible atop many USN ships to this very day. It rotates physically for surveillance, but is flat and rectangular, with a folded serpentine waveguide visible on one side to provide for vertical scanning and altitude determination. The serpentine waveguide feeds a tall stack of horizontal guides with rows of slots to provide radiation. A serpentine waveguide is one which has been folded to provide maximum phase delay in minimum physical distance. The serpentine-feed method in L, S, or C band did not offer the needed precision for X-band use, so the array illustrated in Figure 7-39 was engineered. The vertical antennas were constructed from S-band waveguide with internal precision-machined dividers to produce the serpentine effect. The vertical scanning could be achieved with small frequency changes to the transmitter, and the horizontal scanning with the squeezing waveguide. The required precision machine work could only be performed by a very few selected companies at the time, and the vertical frequency scanning required automatic correction circuitry via a small monitor array to compensate for temperature expansions or contractions. The crossed-slot radiators offered a sine–cosine relationship with the radiated pattern and echo for circular polarization. The radiation pattern was a "pencil beam" for sharp definition in both horizontal and vertical dimensions, each dimension only a fraction of a degree.

Fire-Control Secondary Radar

Phased-array and monopulse technology has grown in sophistication over past decades. New hardware, such as electronically controlled phase shifters, has made many other antenna designs practical. Perhaps the most notable use of the technology is the US Navy's AN/SPY-1 series of systems, used on Aegis-class cruisers and Arleigh

Burke-class destroyers, claimed by some to be the world's most sophisticated radar systems. These are azimuth-elevation monopulse radars. There are no rotating antennas. The primary radar antennas on those ships are unmistakable, resembling four huge stop signs on the four corners of their superstructures. For rapid defensive action, the shipboard electronics equipment must also include secondary radar to point to, and challenge, targets immediately upon SPY primary radar acquisition. It employs a circular phased array for 1,030 MHz and 1,090 MHz. In the late 1970s, the FAA borrowed upon the Navy's Aegis/Arleigh Burke SPY technology in an experiment nicknamed "quick radar," conducted at the Raleigh-Durham, NC airport.

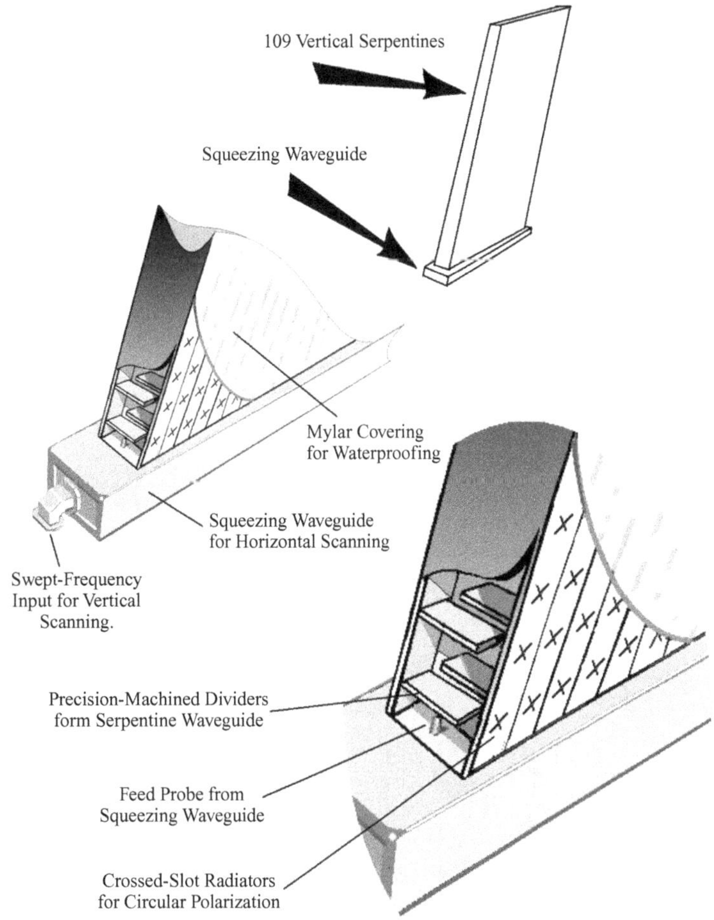

FIGURE 7–43

A raster-scan precision phased-array antenna using a squeezing waveguide and slot radiators. Developed by ITT Gilfillan about 1965.

Monopulse by Amplitude (see Figures 7-44 and 7-45)

Having established that the phasing of dipole or slot radiators by radiator spacing or frequency offers opportunities to shape or point the beam, the potentials for their use become apparent. In the vertical dimension, radiators may be phased to produce a sharp bottom cutoff to reduce ground-bounce interrogations and minimize the "cone of silence" above the radar site.

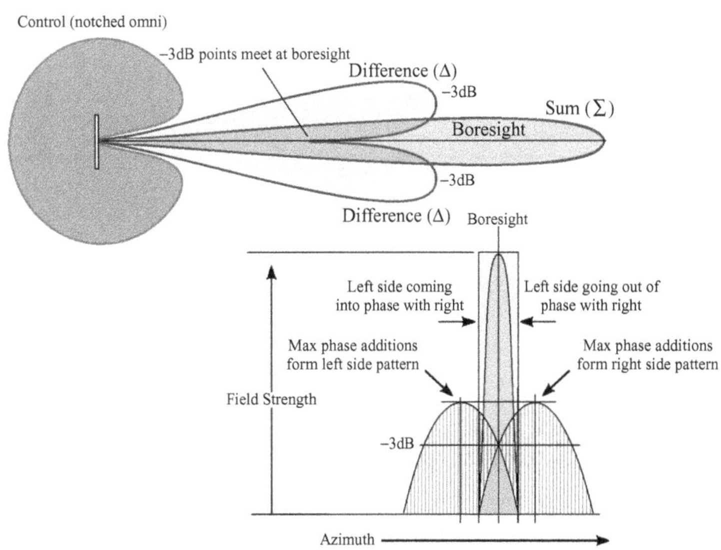

FIGURE 7–44

Monopulse by amplitude.

FAA Monopulse Antennas

The FAA mode S equipment monopulse antenna comprises 32 columns of dipoles, electrically divided into two adjacent 16-column horizontal arrays, fed in the center of the two, and phased for transmitting at 1,030 MHz. The horizontal spacing is 11.25°, which is 1/32 of 360°, and which would provide a 180° phase shift from center across each of the 16 columns. The λ of 1,030 MHz 11.45 inches, and that of 11.25° is

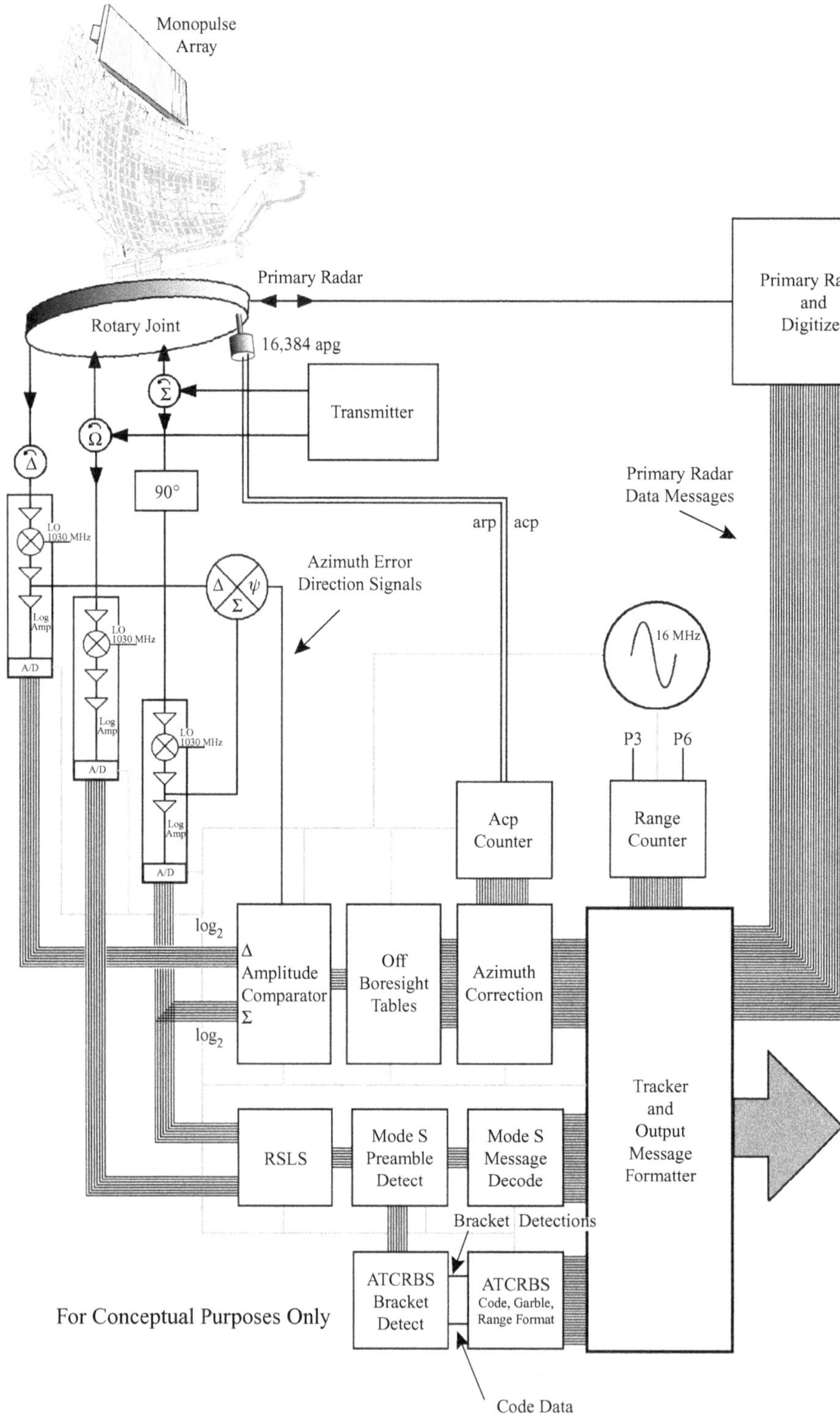

FIGURE 7–45

Simplified partial functional diagram. Monopulse by amplitude.

only 0.3515625 inches. The columns can clearly not be that closely spaced, and are most likely 371.25°, about 11.8078125 inches apart, for 1,030 MHz, providing more electrical change between columns.

Mode S 5' or 6' (tall) array. The energy fed to both sides attenuates as it propagates toward the ends, further shaping the narrow transmit radiation pattern, called "boresight," originating from the use in missile-guidance and weapons-pointing systems. Since the mode S antenna is used in secondary radar, the dipole phasing differs in receiving 1,090 MHz. The λ of 1,090 MHz is 10.83 inches, 0.030 inches per degree. For 1,090 MHz, the 11.8078125-inch column separation is 392.5° (32.5°), and the characteristics of the antenna are so changed that the left and right halves receive transponder replies 180° out of phase at the center, splitting the received pattern into two difference (Δ) patterns. The 3-dB points of each overlap each other at the center of the interrogate boresight pattern. Since the replies are originated by the 1,030-MHz interrogation boresight pattern, most will be in the boresight region, and those 1,090-MHz replies directly on interrogation boresight will contribute equal signal strength of opposite phase from the two sides of the array.

The Δ output provides two separate, wider, receive patterns. The phase comparison between the Δ and Σ replies indicates which of the two arrays is receiving the larger 1,090-MHz reply on the entire array. A characteristic phase detector response is cosinusoidal, but the inputs are phase-adjusted to make it sinusoidal. When the two inputs are in phase, a zero-degree difference provides a maximum positive output. When the two inputs differ by 180°, the output is maximum negative. When Δ is compared to Σ in the phase detector, the polarity of the ϕ output indicates which antenna lobe (leading, clockwise, or lagging, counterclockwise) is producing the Δ output. Comparison in a phase detector produces either a positive or negative output (or none at boresight azimuth), depending upon which of the two patterns produced the stronger signal. The actual amount of algebraic amplitude difference between the Δ and Σ replies indicates the value of the angular difference from boresight. To determine the actual azimuth of the reply, the difference is converted to a digital word to address an "off boresight azimuth (OBA) table," and the table output is added to (or subtracted from) the azimuth change pulse count. The acp generator provides 16,384 pulses per revolution, 45,511 pulses per degree, 0.022° per acp. The corrected azimuth is used in formatting a reply word.

LVA Array

The 7-foot tall ATCBI-6 LVA antenna uses 36 columns spaced 11° and about 11.79886 inches at 1,030 MHz, about 0.00795" closer than the mode S. At 1,090 MHz, the phase spacing for 11.79886 inches is 393.2° or 33.2°. Across the entire array, the difference in the 1,030-MHz phase spacing between 11° and 11.25° is, of course, cumulative, so the 32-column array is 360°, and the 36-column array is 396°.

Beacon Monopulse Operation

General

The monopulse interrogation modes and transponder reply codes were described in preceding paragraphs. This section will deal more closely with the hardware implementation.

Receiver Side-Lobe Suppression

Until the introduction of the monopulse systems, SLS was a function of the transponder. The newer systems have a receiver connected to the "control" antenna. The control antenna is has a similar pattern to omni antennas and is used in the same manner in transmit, but the pattern is "notched" around the main antenna pattern. At all other azimuths all replies received on side or back lobes are attenuated by outputs from the control antenna. This provides for a substantial reduction to fruit and false replies.

Except for the mode S equipment operating in its "IBI" mode, all replies from a single aircraft are processed with monopulse-corrected azimuth. Were it not for the azimuth correction, the mode interlace would create azimuth differences within the series of replies from a single aircraft. With the correction to boresight, which is the acp count, all the reply messages from a single aircraft will contain the same azimuth data. An offset adjustment provides final accuracy. "Parrots" (remote transponders at ground locations) are used to verify azimuth.

In the FAA ATCBI-6, the monopulse azimuth calculation is performed in both mode S and ATCRBS modes. In the FAA "mode S" equipment for ASR radars, the equipment can revert to conventional ATCRBS interrogation synchronized by the primary radar, and requiring centroiding by the sliding-window process. In the mode S/ATCRBS interlaced operation, replies may contain data about the aircraft and flight plan, and that must be saved in memory from one scan to the next. The "all call" invites replies from those aircraft not already established in the memory "rolls." Those aircraft in the rolls are examined on each scan, and their range and azimuth updated each time.The change in rectangular coordinates is then used in calculating the aircraft speed and course, so that the aircraft address identification can be included in the predicted "roll call" at the necessary azimuth on the next scan.

Rotational Rate

The rotational rate of an ASR radar is about 12.5 rpm, and an ARSR is about 5 rpm. A scan, then, takes about 4.8 s or 12 s (secs per rev = 1/rpm × 60), respectively. An ATCRBS system with a repetition rate of 80 Hz has a repetition period of 12.5 ms. For an ASR, 384 periods would occur; for a lower rpm ARSR, 1,000 periods will occur. For an ASR, 42.67 acp's would occur during each interval. For the lower rpm ARSR, 16.384 acp's will occur at each interval. For the ASR, then, the acp count changes by 0.94° per interval, and for the ARSR, the count changes by only 0.36° per interval. Most MSSR arrays have a characteristic beam-width of about 2.5°. If, as in the ATCBI-6, the beam width is 2.4° at the 3-dB points, there is time between those points for 2.55 intervals at 12.5 rpm, and for 6.66 intervals at 5 rpm. There are six intervals in the ATCBI-6 interlace pattern, and that system is intended for use at ARSR facilities. Clearly, the ASR requires a different scheme, such as combined ATCRBS/mode S all-calls in a single interval. The mode pair in these systems may contain a 0.8-μs or 1.6-μs P4 pulse to identify those interrogation as a combination ATCRBS/mode S.

An MSSR Receiver and Processing System (see Figure 7-41)

The MSSR interrogator transmitter was addressed in the interrogator section of this chapter. The 10-MHz P1 and P3 bursts are routed through the Σ circulator to the antenna for use in the ATCRBS mode, and the P2, P3, P4, and P6 bursts follow the same path. The P2 burst is routed to the "control" ω circulator for use in ATCRBS transponder side-lobe suppression, and the P5 burst follows the same route for use in mode S transponder side-lobe suppression. An addition to the new systems is ***receiver side-lobe suppression (RSLS).***

Unlike past ATCRBS receivers, there is a receiver connected to the ω antenna to be used for attenuating or removing "out-of-beam" transponder data. There are two other receivers, the Σ and Δ. All three receivers contain a preamplifier, a signal mixer with 1,030-MHz local oscillator input, a 60-MHz i-f preamplifier, and a 60-MHz i-f logarithmic amplifier. The logarithmic amplifiers are necessary to prevent limiting, and to put the signal into a "deciBel scaling," so the amplitude may be representative of signal strength. The receiver outputs are converted to parallel digital words, clocked out at a 16-MHz rate. I-f outputs of 60 MHz from the Σ and Δ receivers are applied to a phase detector for comparison. Unlike a more critical ***phase or Doppler monopulse*** used in fire-control radar, the only purpose of this phase detector is to determine the "off-boresight" direction, and the only significance is whether it is positive or negative.

See Figure 7-44. Recall that a radiation pattern wavefront and its propagation direction are formed in a constant-phase plane. When transmitting, the interrogator directional power is applied to the two electrical halves of the array, phased so the radiated fields from each will both add and subtract to form a third, narrow, "boresight" beam. While receiving, the signal simply falls upon the aperture, the entire width of the array. However, on receive, the two electrical halves of the array provide outputs of a 180° phase difference. On-boresight replies then provide zero output from the phase detector.

Phase Detector

The characteristic response of a phase detector is cosinusoidal (see Figure 7-46). At a zero degree difference between the two inputs, the output will be maximum positive, and at 180°, the output will be zero. To shift the response 90°, so that it will be sinusoidal, there is a phase shift upstream of the Σ receiver, to provide a maximum positive or maximum negative output to indicate the direction off boresight of each reply.

Off-Boresight Calculation

The Δ and E i-f outputs for each reply were put in logarithmic form to allow nonlimited comparisons in the amplitude comparator. The difference indicates the amount off boresight, but it is a logarithmic number not directly representative of azimuth.The difference result is used to address an OBA table, and that table output represents the OBA value in fractions of an acp, which is 0.021972656°. The acp count is then refined by the correction and sent to the tracker for message formation.

All of the preceding azimuth determination was made for a single reply, and in real time. The interrogation mode, the data in the reply, and the range count must also be included in the assembled message. Range zero may not be the same for mode S and ATCRBS, and some provision must be made for corrections depending on mode type.

The mode S preamble detector separates the mode S and ATCRBS replies, and the ATCRBS begins serially decoding and formatting the parallel replies for input to the tracker and output message formatter.

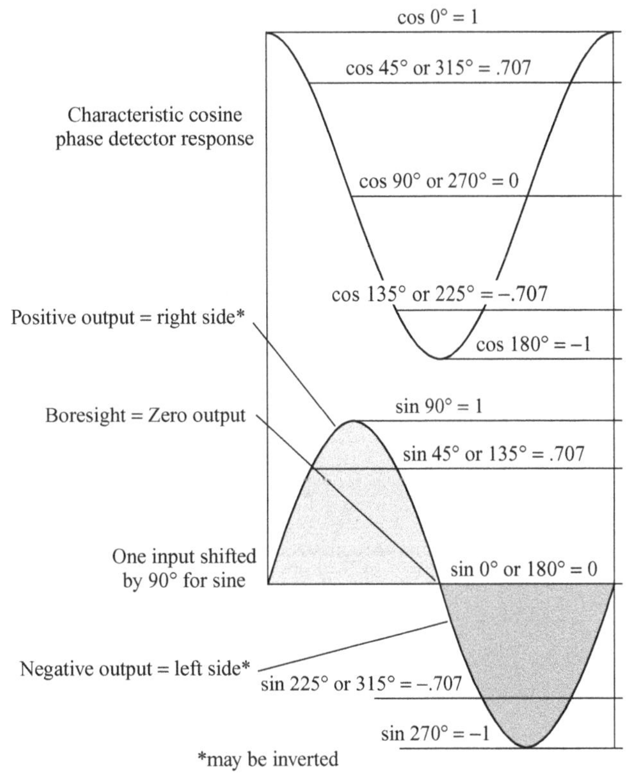

FIGURE 7–46

Phase detector response.

Monopulse by Phase Difference or Doppler (Interferometry)

This method is used in fire-control primary radar systems, which must have instant high-precision information for gun-pointing or missile guidance. The transmitted signal is focused into a beam with the constant-phase plane as any summed linear array, but the receive antenna is divided into halves, with a receiver for each. The phases of the signals from the two separate arrays are directly compared to determine the angle off boresight. There is no phase difference when the target is on boresight, because the wave front reaches both arrays simultaneously. When the target is off boresight, the separate arrays appear as two apertures to the wavefront, and there is a detectable phase delay between the wavefronts received by the two.

Multiple Facilities and Coordination

A fictitious use of ATCRBS systems in the following aircraft departure script illustrates the safety gained by "flight following," continuous "radar" (actually secondary radar) monitoring. Even though the pilot will be flying VFR (visual flight rules, the pilot is doing his own navigation), his aircraft will be continuously identified on radar displays by a series of air traffic control facilities, including an ***Airport Traffic Control Tower (ATCT)*** with ***Remote ARTS Color Displays (RACD)*** and keyboards, a ***Terminal Radar Approach Control facility (TRACON),*** and an ***Air Route Traffic Control Center (ARTCC).*** Note that every transmission is confirmed with the inclusion of the aircraft identification.

NOTE:

This example of air traffic control operations is solely intended to help radar technicians envision the network of air traffic control radar facilities. It is inappropriate and should not be used for the training of pilots or air traffic controllers. Air traffic control operations require very precise procedures and language. This example may contain omissions and variations from the true equipment, facilities, procedures, conditions, and interfacility agreements at the locations described.

The Consolidated Liberator B-24 has taxied to the "hold bars" on the taxiway at the end of runway 17L, Wiley Post Airport, Oklahoma City. He selects Wiley Post tower at 126.9 MHz on his communications radio (see

FIGURE 7–47

Aircraft takeoff. Tower comm and assingned code 1203.

Figure 7-47). (The takeoff direction will be about 170°. On takeoff, the aircraft will enter class C airspace, which permits VFR operation, but requires air traffic control clearances.)

B-24 Pilot: "Wiley Post Tower, Revival 95366 at Runway one seven left. Ready for takeoff with west departure."

ATCT Controller: "Revival 95366 taxi into position and hold. Traffic is Boeing 727 on runway. Squawk 1203."

B-24 Pilot: "Squawk 1203. Taxi and hold 17L. Revival 95366." (The pilot moves his aircraft into takeoff position at the north end of runway 17L/35R.)

FIGURE 7–48

Aircraft cockpit in turn

The B-24 transponder is now communicating with an interrogator located at an ASR-9 site southwest of Will Rogers International Airport, also in Oklahoma City. The beacon data is supplied to an ARTS system at the Will Rogers TRACON. ARTS displays are located at both the TRACON and the ATCT (see Figure 7-48 and 7-49).

The Boeing 727 takes off on 17L. The ATCT controller waits for prescribed separation until the B-24 is safe from wake turbulence.

ATCT Controller: "Revival 95366 cleared for takeoff, west departure approved."

B-24 Pilot: "Cleared for takeoff, west departure. Revival 95366." (Aircraft takes off, climbing at

FIGURE 7–49

Handoff, departure control comn, and reply code 1203.

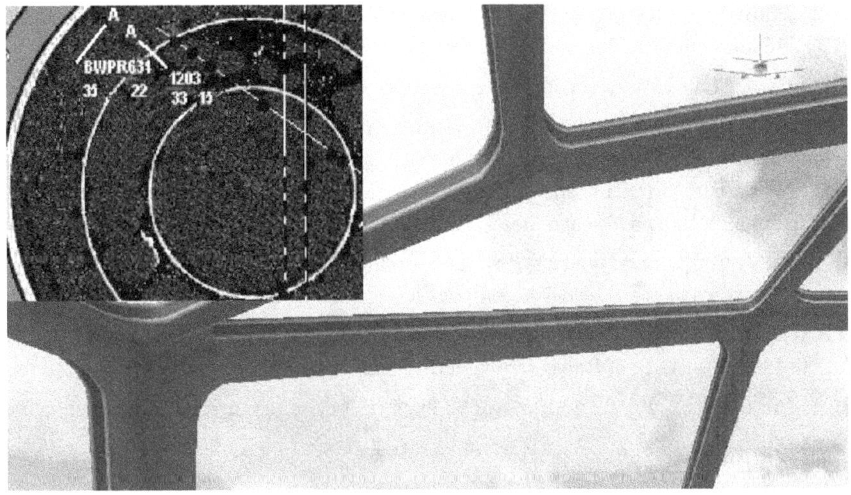

FIGURE 7–50

Boeing 777 at 12 o'clock.

500" per minute, 130 nautical mph, slowly turning toward the west.) Aircraft reaches a point 5 miles from Wiley Post.

ATCT Controller: "Revival 95366, you are leaving my airspace." Frequency change approved.

B-24 Pilot: "Going to 120.45. Revival 95366."

(The pilot changes his com radio frequency to the Will Rogers TRACON west departure control.) The ATCT Controller types a "handoff" entry to the display position for west departure control. The entry goes into the ARTS computer at the Will Rogers Airport TRACON. On his ARTS Color Display in the TRACON IFR dark room, the west departure controller sees a blinking yellow data block for beacon code 1203 to indicate the handoff request.)

B-24 Pilot: "Oke City Departure Revival 95366 Liberator type bravo two-four requests transition through Charlie airspace."

The departure controller in the IFR darkroom has already noted the handoff request and accepts the handoff with a keyboard entry. The data block alphanumerics turn solid white to indicate that his position is now handling the aircraft.

TRACON West Departure Controller: "Revival 95366 Roger. Radar contact 2 miles east of F29 altitude 3,300 feet. Altimeter 29.05, cleared through Charlie airspace." F29 is a navigational "fix" identified on navigational maps, GPS displays, and the radar video map on the departure controller's display. The altitude has been obtained from the B-24 mode C transponder and then automatically adjusted in the ARTS equipment to incorporate local barometric pressure. The local barometric pressure is given so that the pilot can set his analog altimeter. Once this is done, the analog altimeter should agree with the ARTS-corrected altitude.

B-24 Pilot: "Copy 2 miles east of F29 at 3,300 feet. Revival 95366. Cleared through Charlie airspace." (The aircraft altimeter should agree with the controller's report.)

The B-24 has proceeded further to the south than anticipated before completing his turn to the west. At this point, a "conflict alert" generated by the ARTS equipment occurs at the TRACON west departure display position. An alarm sounds at the controller position. Two aircraft data blocks, including the one for the climbing B-24, begin flashing red. The TRACON west departure controller identifies the two aircraft causing the alarm.

TRACON West Departure Controller: "Traffic Alert! Revival 95366. Boeing 777 at 3,500 feet 12 o'clock less than 1 mile (see Figure 7-50). Report traffic in sight." (B-24 Pilot looks for Boeing 777 straight ahead.)

TRACON West Departure Controller: "Traffic Alert! Beawopper 634 heavy. Liberator type bravo two-four 3,300 feet 6 o'clock less than 1 mile. Report traffic in sight." (Both pilots now know the B-24 is behind the 777.)

B-24 Pilot: "Have the Boeing 777 in sight. Revival 95366." (Pilot sees Beawopper 634 heavy Boeing 777, clearly dead ahead, flying in the same direction, but at a greater airspeed.)

Beawopper Airlines Pilot: "Do not have the Liberator bravo two-four in sight. Beawopper 634 heavy." (It's because it's climbing directly behind and below him, but falling back.)
Several minutes pass as the B-24 continues climbing to the west.

TRACON West Departure Controller: "Revival 95366, you are leaving my airspace. Contact Oke City Approach on 116.3."

B-24 Pilot: "Copy. Going to 116.3. Revival 95366." (Changes radio frequency. The new frequency is at another ARTS display position, but still in the Oklahoma City Will Rogers TRACON.)

B-24 Pilot: "Oke City Approach Revival 95366 with you. Request flight following."

TRACON West Area Controller: "Revival 95366 radar contact. Twenty miles west of Will Rogers 9,500 feet." (Controller read 1203 on top line of data block, and 950 (9,500 ft) on second line of data block.) The controller makes keyboard entries to "start a track" on the data block, entering an ***ACID (aircraft identification)*** REV366. The data block turns white and the ACID replaces the beacon code in the data block.

B-24 Pilot: "Copy radar contact. Revival 95366."
Several minutes pass as the B-24 continues to climb.

TRACON West Area Controller: "Revival 95366, you are leaving my airspace. Squawk 1200. Contact Fort Worth Center on 122.55. Radar service terminated." (1200 is a standard code used to identify VFR aircraft.)

B-24 Pilot: "Going to 122.55. Revival 95366." (Pilot changes aircraft radio frequency.)

B-24 Pilot: "Fort Worth Center Revival 95366, Liberator type bravo two-four with you. Request flight following."

The B-24 pilot has changed to the Fort Worth ARTCC, the facility watching traffic in this area. The center is receiving beacon data from an ARSR facility in Oklahoma City. The ARSR has a common digitizer to provide the messages.

ARTCC Sector Controller: "Revival 95366 Roger squawk 5322."

B-24 Pilot: "Copy squawk 5322. Revival 95366."

The B-24 transponder is now being interrogated and detected by an en route ARSR facility, and the data from the ARSR is being transmitted to the ARTCC by microwave links (see Figure 7-51). The ARTCC controller is in Fort Worth, but he is speaking to the B-24 pilot over another microwave link to a remote control air-to-ground (RCAG) near the B-24. The ARTCC controlling a particular airspace is dependent upon which ARSR facilities

FIGURE 7–51

En Route. Flight following by ARTCC.

are supplying it with radar data. Some ARSRs may supply more than one ARTCC, as well as military users. An ARTCC may receive data from many ARSR facilities, to cover thousands of square miles, and major areas of the country.

ARTCC Sector Controller: "Revival 95366 radar contact 4 miles southwest of KODX, flight level (FL) three-one-zero. Altimeter 28.95." (Add two zeros to the FL; 31,000 feet.)

B-24 Pilot: "Copy radar contact. Revival 95366."

Review Questions

1. The first secondary radar systems were the ________________.
2. How many second-generation reply codes were available?
3. A secondary radar system is an______________ ______________ system.
4. The first secondary radars were called ________.
5. Name all six ATCRBS beacon modes and the mode pair spacing for each.
6. ATCRBS code data pulses are ___________ wide.
7. What is the difference between the ATCRBS and Mode S P2 pulse?
8. The F1 and F2 pulses may be called _________ or ________ pulses.
9. Describe the received data in Mode S.
10. ATCRBS mode pulses are ________ wide.
11. The temporal separation between F1 and F2 (lead edge-to-lead edge) is ________.
12. Mode interlace is to ___________________.
13. Name three reasons for using MSSR.
14. In fourth-generation systems, beacon code data appears on controller displays as ____________.
15. Beacon reply code 7700 identifies _____________.
16. Beacon reply code 7600 identifies _____________.
17. Beacon reply code 7500 identifies ____________.
18. A garble condition is _______________________.
19. A phantom condition is _______________________.
20. Mode C interrogations obtain ________________ information.
21. FRUIT is ______________________________.
22. Beacon systems require (longer or shorter) ___________ duration stc than do primary radars because _________________.
23. Compare centroiding in ATCRBS and mode S ________________.
24. SLS is employed to alleviate _________________________.
25. ISLS is employed to alleviate _________________________.
26. RSLS is____.
27. In amplitude monopulse, there are two Δ receive antenna patterns. How does the system recognize which one is receiving data?

Answers to Review Questions

1. The first secondary radar systems were the ***WWII military IFF systems***.
2. How many second-generation reply codes were available? ***Sixty-four***.
3. A secondary radar system is an **answer-back** system.
4. The first secondary radars were called ***IFF***.
5. Name all six ATCRBS beacon modes and the mode pair spacing for each. ***M1, 3 μs; M2, 5 μs; M3/A, 8 μs; MB, 17 μs; MC, 21 μs; MD, 21 μs.***
6. ATCRBS code data pulses are ***0.45*** **μs** wide.
7. What is the difference between the ATCRBS and Mode S P2 pulse? ***It is transmitted on the omni antenna in ATCRBS, but on the directional antenna in mode S.***
8. The F1 and F2 pulses may be called ***bracket or framing*** pulses.
9. Describe the received data in Mode S. ***A preamble, followed by 56 or 112 pulse-position-modulation data bits.***
10. ATCRBS Mode pulses are ***0.8 μs*** wide.
11. The temporal separation between F1 and F2 (lead edge-to-lead edge) is ***20.3*** μs.
12. Mode Interlace is to ***allow transmission of different modes on different intervals.***
13. Name three reasons for using MSSR. ***It provides a way to exchange data with an aircraft. It provides more accurate azimuth information than the "many-hits-needed" sliding window. It is less susceptible to, and creates less, FRUIT than previous systems.***
14. In fourth-generation systems, beacon code data appears on controller displays as ***alphanumeric data blocks.***
15. Beacon reply code 7700 identifies ***emergency.***
16. Beacon reply code 7600 identifies ***radio communications failure.***
17. Beacon reply code 7500 identifies ***hijack.***
18. A garble condition is ***pulse positions of two code trains in coincidence.***
19. A phantom condition is ***20.3 μs*** between pulses other than brackets.
20. Mode C interrogations obtain ***aircraft altimeter*** information.
21. FRUIT is ***false replies unsynchronous in time.***
22. Beacon systems require (longer or shorter) longer duration stc than do primary radars because ***the range equation is 6 dB per octave instead of 12.***
23. Compare centroiding in ATCRBS and Mode S ***ATCRBS in earlier systems used the azimuth sliding window. In MSSRs, both ATCRBS and Mode S obtain single-hit azimuth by mono-pulse.***
24. SLS is employed to alleviate ***replies from side-lobe interrogations.***
25. ISLS is employed to alleviate ***replies from "ground bounce" interrogations.***
26. RSLS is ***Receiver side lobe suppression in the MSSR interrogator equipment.***
27. In amplitude monopulse, there are two Δ receive antenna patterns. How does the system recognize which one is receiving data? ***By the positive or negative output of a sinusoidal-response phase detector, comparing the E and Δ replies.***

CHAPTER 8

Microwave Transmission Lines and Cavities

Origin of this Chapter

The author of this book makes only a limited claim to the originality of this chapter. It has been rewritten from Chapter 10 of US Department of the Air Force Manual 52-8, entitled Radar Circuit Analysis, published June 30, 1951. The entire chapter has been retyped, edited, supplemented, and reformatted. Some of this was done to more closely conform to latter-day technical writing standards, some to add more current information, some for clarification, and some to make it more closely resemble the style of the other chapters of this book. Most importantly, no information has been deleted or diluted, and useful information has been added. All the illustrations have been retouched or redrawn, and all the labeling in the illustrations has been retyped. Some illustrations have been added; those taken from Manual 52-8 are so annotated.

Introduction

One of the most significant areas of knowledge that differentiates the radar technician from others is waveguide and cavity technology. This book is intended for technicians with a foundation in electronics fundamentals, and knowledge of rf transmission-line theory is a necessary prerequisite for a good understanding of this chapter. Whatever the technician's background may be, it is unlikely that he knows much about waveguide and cavity theory if he has not previously been exposed to radar or some other type of microwave equipment.

The Beginning

Early radar systems operated at VHF, and even lower, frequencies. This imposed serious limitations because of the large size required of directional antennas. In Great Britain, in the late 1930s, there was an urgent need to create higher frequencies, and two physicists at the Birmingham University, Randel and Boot, created the multicavity magnetron for use in the first S-band (10-cm wavelength) radar transmitters. Conventional means of conducting energy over cables, through transformers, etc., just would not work at these new microwave frequencies, and physicists at the Massachusetts Institute of Technology developed the entire science of waveguides and cavities, called microwave technology. A waveguide is a "pipe" for transferring high-frequency energy. Cavity resonators are metal containers in which electromagnetic oscillation can exist when the cavity is properly excited.

The source material in Air Force Manual 52-8 provided an analogy to the propagation of energy over a two-wire line, an approach to breach an "understanding gap" between familiar signal flow over wires, and through the space in a waveguide. Although microwave hardware may have evolved to improved techniques, the physical science is still the same; scientific truths forever stand the test of time, and this chapter should provide the radar technician with a good foundation.

Need to Know

Beyond the basic professional obligation for the technician to understand his equipment, a knowledge of microwave theory is necessary for him to maintain or install a radar system. Installation work may require knowledge of waveguide characteristics and test procedures. Poor radar performance may be traceable to waveguide or microwave-component mismatches, and the waveguide antenna rotary joint is a source of occasional major radar failure. Waveguide damage may have serious implications. Cavity tuning is a necessity for the radar technician, and tuned cavities are used as radar test equipment.

Waveguides: Guided and Unguided Electromagnetic Waves

Propagation on Wires and in Space

In general, there are two methods for transferring electrical energy. One of these is by current flow in wires, as the student learns in basic Physics and Electronics courses. Another means of transferring energy is by the movement of electromagnetic fields in space. Electrical energy can be transferred as current flow over two-wire or coaxial lines; in space, it moves as electromagnetic fields, where the energy source is an antenna.

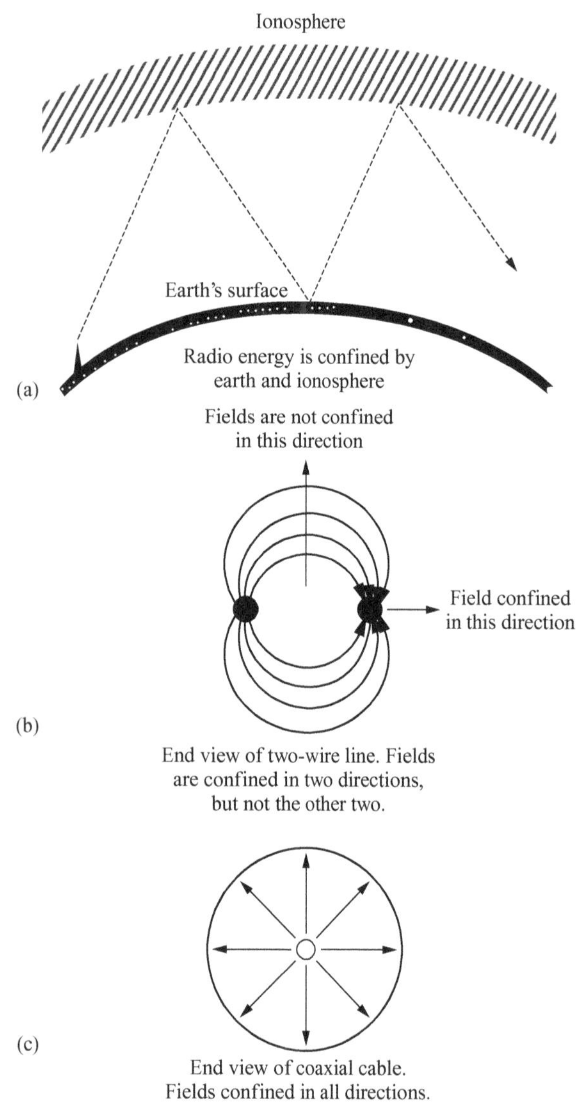

FIGURE 8–1

Guiding waves. From USAF Manual 52-8.

Electromagnetic Waves in Space

Electromagnetic fields, moving in space, are confined largely to the area between the earth and the ionosphere, the thick layer of free ions and electrons which exists at a height of about 60 miles above the earth (see Figure 8-1A). The change in the dielectric constant of this part of the atmosphere is sufficient to reflect electromagnetic fields which strike it. Extremely high-frequency radiation, striking it almost perpendicularly, is an exception; this will penetrate the layer and propagate into space, allowing systems such as satellite communications. An electromagnetic field that starts out in an angular direction neither parallel nor perpendicular to the earth and the ionosphere follows a back-and-forth path in between these two areas and may or may not be reflected back to the earth. This phenomenon is sometimes called "skip" in communications.

First Discoveries

The transfer of energy by electromagnetic fields in space and by currents in wires are related; Michael Faraday (1791–1867) and James Clerk Maxwell (1831–1879) had noted this. Even two-wire lines are elements to guide electromagnetic fields from one place to another; television twin-lead from the set to the antenna is an example. The currents in the wires may even be considered merely incidental to the moving fields.

A two-wire line is a poor guide for transferring electromagnetic fields, because it does not confine the fields in a direction perpendicular to the plane which contains the wires, as shown in Figure 8-1B. Some energy escapes in the form of radiation.

Electromagnetic fields can be completely confined in this direction, when one conductor is extended around the other to form a coaxial cable, as shown in Figure 8-1C. In a coaxial cable, energy transfer takes place by electromagnetic fields, as well as by current flow.

Transmission by a coaxial cable is inefficient at high frequencies, since ***skin effect*** limits the current-carrying area of a conductor to a thin layer at its surface. Also, field formation is limited by the amount of current flow associated with it. When the resistance in a conductor is increased, current flow in the conductor is proportionally reduced, in turn reducing the field magnitude. The surface area of the coaxial cable's inner conductor is much less than the surface area of its outer conductor. The inner conductor thus offers more resistance to the current than does the outer conductor; the efficiency of energy transfer is thereby reduced. If there were a way to remove the center conductor while retaining the fields, energy could be transferred with much less loss; this is what may be achieved with a hollow pipe called a ***waveguide.***

A waveguide does not necessarily have to be circular in cross-section. Waveguides may be square, rectangular, or elliptical. The most common shape is rectangular. Metal walls are also unnecessary because the fields will be reflected whenever they encounter any different dielectric than that substance in which they are traveling. For example, fields can be made to travel through a ceramic rod with little loss of energy. When they encounter the air at the surface of the rod, they are reflected back into the rod.

Waveguides versus RF Lines

There are three types of losses in the rf lines; they are caused by the copper, dielectric, and radiation to the surrounding atmosphere. Copper loss is an I2R loss and becomes appreciable whenever skin effect reduces the conducting area of the lines. Dielectric losses are due to the heating of the insulation between the conductors.

Advantages of Waveguides

Considering the aforementioned losses, waveguides have the following advantages:

1. The copper losses are small. A two-wire line consists of a pair of relatively small conductors, offering significant resistance. The center conductor of a coaxial cable similarly offers significant loss. A waveguide, having no center conductor, has a large surface area. When current flows, the copper losses are less than those in the other types of lines.
2. Dielectric losses are small in waveguides. In two-conductor lines, insulation is necessary between the conductors. The fields which move around this insulator create heat, dissipating power. Since a waveguide has no center conductor to support, there is no need for a dielectric, and only air or gas is contained; these exhibit negligible dielectric loss.
3. Radiation losses are less in a waveguide than in a two-wire line. The fields are entirely contained within the guide, and a negligible amount of energy is radiated.
4. The power-handling capacity of a waveguide is greater than that of a coaxial line of comparable size. Power is a function of E^2/Z_o, where E is the maximum voltage in the traveling wave, and Z_o is the characteristic impedance of the line. E is limited by the space between the conductors. In the coaxial line illustrated in Figure 8-2, this distance is S1. A circular waveguide breakdown potential is determined to be space S2, twice the amount of S1.
5. A waveguide is simpler to construct than a coaxial line. In the waveguide, the center conductor is eliminated completely.
6. The physical ruggedness and durability of a waveguide is greater than that of a coax. The waveguide has no center conductor or insulators, which can be displaced or broken.

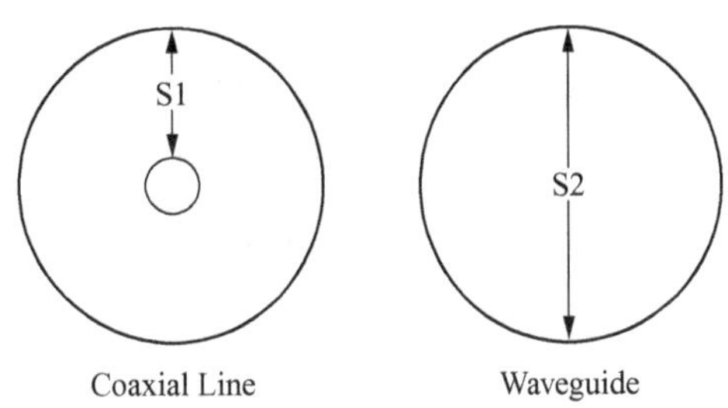

FIGURE 8–2

Spacings, coax versus circular waveguide. From USAF Manual 52-8.

Disadvantages of Waveguide

Waveguides are impractical to use at any but extremely high frequencies, because the dimensions become very large at lower frequencies. The cross-sectional dimensions of a waveguide must be in the order of a half-wavelength for it to contain electromagnetic fields properly. Recall that the wavelength, λ, is equal to *c*/*f*. A waveguide used at 1 MHz would be about 700 feet wide. At 200 MHz, a waveguide would have to be about 4 feet wide, while at 10,000 MHz, it would need to be only 1 inch wide. There have been radar systems employing 2-foot-wide waveguide, but these are rare. As a part of the 1960's frequency-diversity program, the AN/FPS-35 USAF/FAA joint-use radar operated at UHF with very large waveguide, and the AN/FPS-24 operated at VHF with rigid coaxial lines. A waveguide has a cutoff frequency; if the dimension of the guide is a half wavelength or less, energy will not be propagated through it.

Waveguide Theory and the Two Analogies

A thorough mathematical analysis of the fields in a waveguide is beyond the intended complexity of this book. However, it is possible to obtain an understanding of many of the properties of waveguide propagation by using two analogies, to be described in proceeding parts of this chapter. One of these is an analogy to rf on a two-wire line; the other is an analogy to propagation of electric waves as in light, more resembling the theories of Heinrich Rudolph Hertz (1857–1894).

Waveguide as Compared to a Two-Wire rf Line (see Figure 8-3A)

Assume that a waveguide has the form of a two-wire line; there must be some means of supporting the two wires. The support must be a nonconductor, so that no power will be lost by radiation leakage. This line is spaced, insulated, and supported by porcelain stand-off insulators. At communication frequencies, the absorption of power by the dielectric material (insulators) causes them to exhibit a low resistance and capacity; the equivalent electrical circuit at higher frequencies is shown in Figure 8-3B. For frequencies of 3,000 MHz and above, a better insulator than nonconducting porcelain must be used. A superior high-frequency insulator for this purpose is a quarter-wave section of rf line called a ***metallic*** insulator, because it will appear to be an open circuit to signals on the line. Such an insulator is illustrated in Figure 8-3C. As there are no dielectric losses in a quarter wave section of an rf line, the impedance at the open end (the junction of the two-wire line) is very high.

A metallic insulator can be placed anywhere along a two-wire line. Figure 8-4A shows several on each side of a two-wire line. A point to note in this line is that the supports are a quarter wave at only one frequency, limiting the high efficiency of the two-wire line to only one frequency.

The use of several insulators results in the improved conductivity of a two-wire line when the sections are connected together. This connection is made between the two adjacent insulators through a switch, as shown in Figure 8-4B. When the switch is open, both quarter-wave sections are excited by the main line. Under this

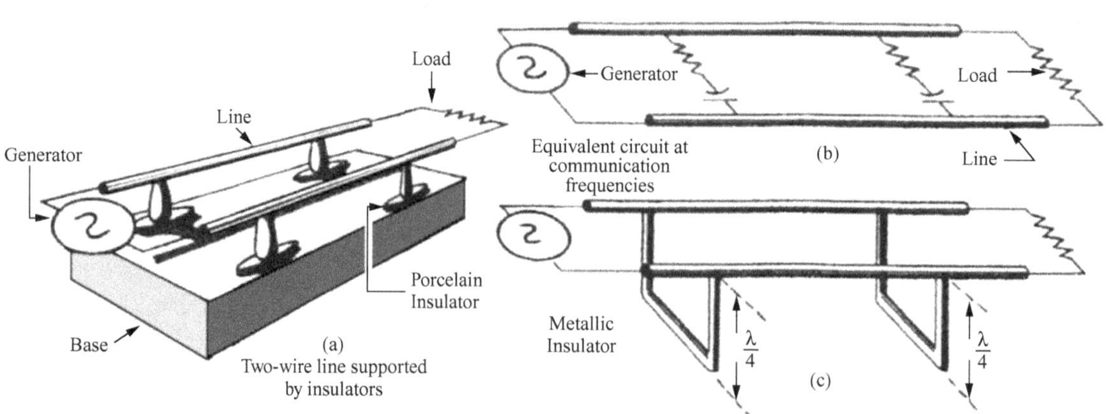

FIGURE 8-3

Insulating the two-wire line. From USAF Manual 52-8.

condition, there will be standing waves on the quarter-wave (λ/4) sections. When the switch is connected to the same place on each section, the relative phase relationship of the voltages at the connection will be the same for each section. The No. 1 section will be excited first by the generator. When the switch is closed, the No. 2 section will be partly excited by the No. 1 section through the switch. Less energy from the main line will be required to excite the No. 2 section. The parallel paths shown cause less resistance to exist along a given length of line, and energy is transferred with less copper loss.

When more and more sections are added to the line until each section makes contact with the next, the result is a rectangular box in which the line is at the center, as shown in Figure 8-4C. The line itself is actually part of the wall of the box. The rectangular box thus formed is a waveguide.

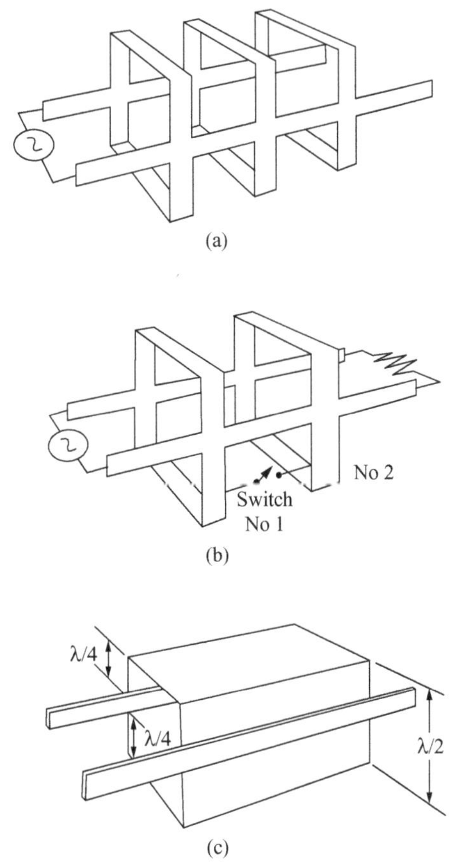

FIGURE 8–4

Development of waveguide by adding λ/4 sections. From USAF Manual 52-8.

Effect of Different Frequencies on a Waveguide

Recall that the quarter-wave section is limited in operation to a certain frequency. In contrast, when a solid wall of insulators is added, the section will operate at other frequencies. The waveguide shown in Figure 8-5A is the one just discussed. When the frequency being transferred by the guide is increased, the quarter-wave section must be shorter. These shorter wavelengths are easily accommodated if you assume this two-wire line is made up of a wide bar or strip in each wall of the guide, as shown in Figure 8-5B. The shorter distance remaining is the shorter quarter-wave section. Thus, the wide bar shown is theoretically well insulated at any frequency higher than the one which creates the near-minimum frequency in Figure 8-5A. There is a practical upper frequency limit at which this analogy is applicable. For example, when the bar is a half-wavelength across, the waveguide will be four quarter-wavelengths across, and may act as though there were two bars instead of one, as shown later in the chapter.

The next consideration is a frequency lower than the original. Lowering the frequency will lengthen the sections and narrow the bar. Beyond some lower frequency, this bar does not exist because the quarter-wave

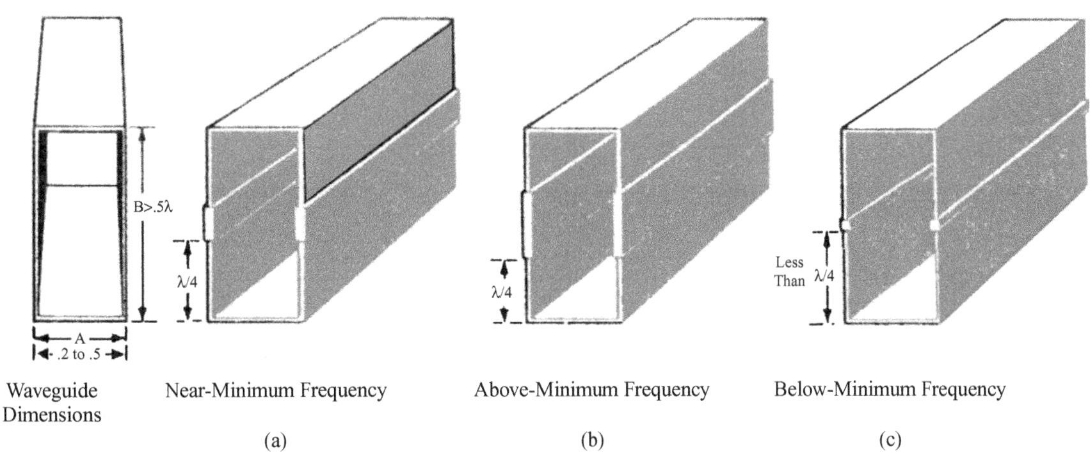

FIGURE 8–5

Effect of different frequencies on the waveguide. From USAF Manual 52-8.

sections meet one another. At a still lower frequency, the sections become less than a quarter wave, as in Figure 8-5C. A section less than a quarter wave is inductive, so the impedance across the place where the conducting bars belong is not a high resistance, but an inductance exhibiting a reactance. The inductive reactance will dissipate the energy in the line very swiftly through high currents which flow back and forth in one place, as though the inductance were taking energy during one half cycle, and then returning it to the line during the next half cycle.

It follows that in a waveguide, there is a low frequency limit, called ***cutoff frequency,*** below which the waveguide cannot transfer energy. The width of the waveguide at the cutoff frequency is equal to one half wavelength, since at this frequency, the two quarter-wave sections touch, and add to one another. Mathematically, the cutoff wavelength is expressed by the equation $\lambda c = 2B$, where λ_c is the cutoff wavelength, and B is the long dimension of the rectangular cross-section.

Because cutoff occurs when the width of a waveguide is below a half-wavelength, most waveguides are made 0.7 wavelengths in the wide dimension, to give a margin between (1) the actual size and (2) the size for cutoff. The other dimension is the distance between conductors, and similarly, as in the two-wire lines, is governed by the voltage breakdown potential of the dielectric, which is usually air. Widths of 0.2 to 0.5 wavelengths are common.

Electromagnetic Fields in a Waveguide

Energy in a waveguide is transferred by the electromagnetic fields, while currents and voltages merely aid in forming these fields. In any waveguide, two fields, ***electromagnetic*** and ***electrostatic,*** are always present.

The Electric Field

An electrostatic (electric) field exists where there is a difference in the number of electrons between two points (see Figure 8-6). An electric field consists of a stress in the dielectric field, conventionally represented by arrows pointing to the more negative potential in diagrams. An electrostatic field forms between the two parallel plates of a condensor as illustrated in Figure 8-6A. When the top plate of the condensor is made positive by a battery, electrons move from the top plate and get deposited on the bottom plate. This immediately sets up a stress in the dielectric between the plates, represented by the arrows, pointing from positive to negative. The amount of stress, called ***magnitude***, is sometimes is shown by the length of the arrow. The *number* of arrows may also be used to indicate the field magnitude. In the case of the condensor, note that the arrows are evenly spaced across the area between the two plates. As the voltage across the plates is the same at all points, the electrostatic lines between the plates are evenly distributed. This electrostatic or electric field is abbreviated ***e-field,*** and the lines of stress are called the ***e-lines.***

In Figure 8-6B, an instantaneous wave of voltage is applied to the two-wire transmission line. This line is equal to one wavelength of the applied signal. Half the wave is positive, and the other half is negative. The instantaneous e-field is the same at the negative and positive points, but the arrows representing each field point in opposite directions. The voltage along the line varies sinusoidally. Therefore, the density of the e-lines varies sinusoidally.

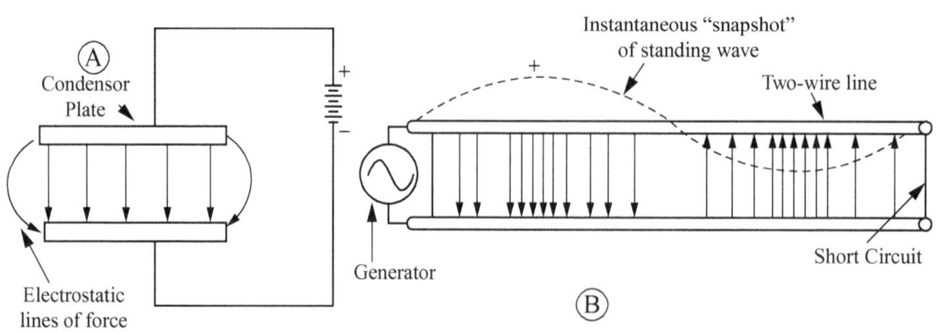

FIGURE 8–6

Electric fields, condensor plates, and a two-wire line. From USAF Manual 52-8.

Development of the e-fields in a waveguide can be shown by expanding upon the two-wire line as illustrated in Figure 8-7. This illustrates a two-wire line which has quarter-wave insulators. The illustration shows the two-wire line previously described with several double-and quarter-wave insulators, or half-wave frames used as insulators.

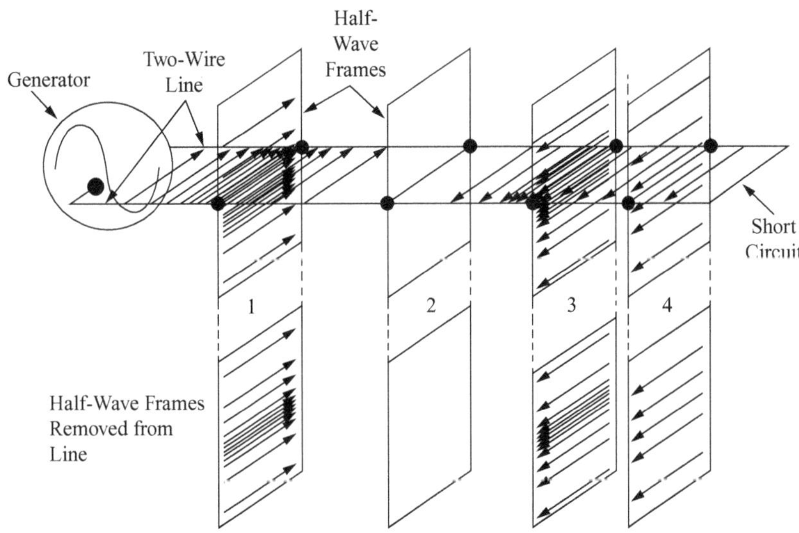

FIGURE 8–7

Magnitude of fields on half-wave frames very with strength of field on mail line. From USAF Manual 52-8.

In Figure 8-7, the e-field on the main line is the same as that in the transmission line illustrated in Figure 8-6. The half-wave frames located at points of strong e-fields will exhibit a strong e-field across them. Frame number 1 in Figure 8-7 is an example of an insulator with a strong e-field across it. Frame number 2 is at a zero voltage point, and has no field on it. Frame number 3 has a strong field of polarity opposite that of frame number 1, and frame number 4 has a weaker field on it, because it is at a lower voltage point on the main line. The illustration is an introduction to development of the three-dimensional aspect of the full e-field in a waveguide, shown in Figure 8-8.

See Figure 8-8, an *instantaneous "snapshot"* of the e-field existing at the time the standing wave is peaked. The voltage magnitudes actually vary at the frequency of the input signal. The field results when an infinite number of quarter-wave sections are connected to the line to form a rectangular box. The e-field is strong at one-quarter and three-quarter distances from the shorted end, but becomes weaker at the sine rate toward the upper and lower walls and toward the ends and center. Again, the phenomenon of wavelength is present, as shown in Figure 8-8A.

Certain ***boundary*** conditions must exist, in order for propagation to occur in a waveguide. One is that there must be no electric field parallel to and against (tangent to) the walls of the guide. This is satisfied when the e-field diminishes to zero at the top and at the bottom of the guide by natural rf line action; at other places, the field is perpendicular to the walls.

The Magnetic Field

An electron in motion creates a circular magnetic force around the electron. A hypothetical electron in motion toward the reader will exhibit an endless clockwise magnetic force, as though from a north to a south magnetic pole. The small magnetic lines of force around all the individual electrons in motion in a conductor will add to make a summed magnetic field around the conductor. The presence of the force is illustrated by the closed loops around the single wire in Figure 8-9A. The magnetic force lines must be a continuous closed loop in order to exist. The magnetic force lines are called ***h-lines,*** and the magnetic field formed by the h-lines is called an ***h-field.*** The h-lines may be depicted in terms of magnitude in the same manner as the e-lines, and they may be shown approaching or departing the viewpoint, with (1) dots to indicate approaching, and (2) ×'s or +'s to show departing. The h-lines point

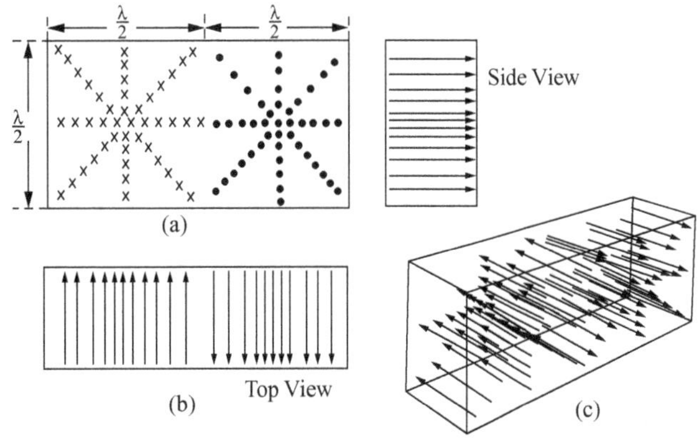

FIGURE 8–8

e-Field in waveguide. From USAF Manual 52-8.

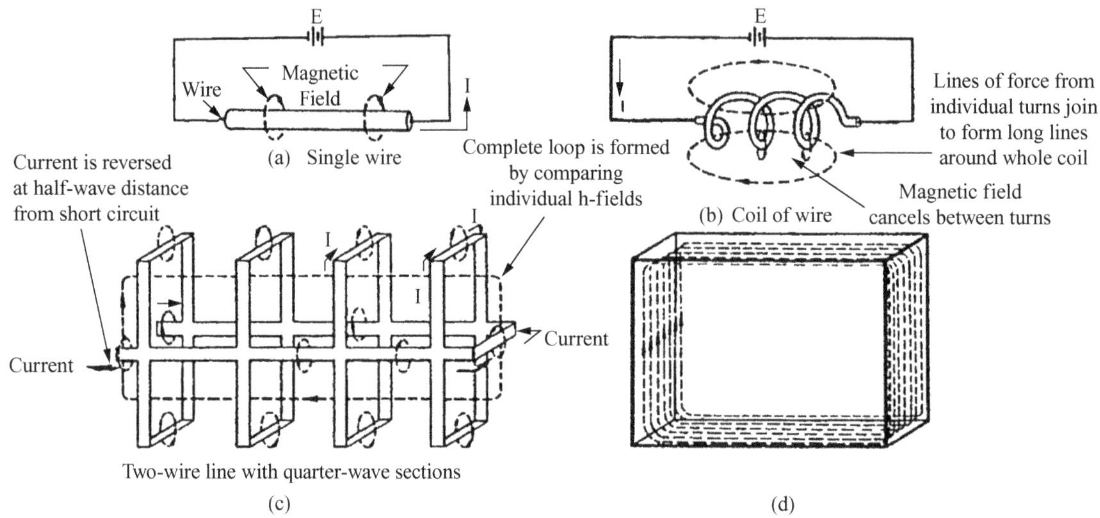

FIGURE 8–9

Developing the waveguide h-field. From USAF Manual 52-8.

from north toward south magnetic poles. The strength of the h-field is directly proportional to the current. In some illustrations in other material, the h-field arrows may be hollow-diamond-shaped, to distinguish them from e-field arrows.

Although h-lines encircle a single straight wire, they behave differently when the wire is formed into a coil (see Figure 8-9B). There, the individual h-lines tend to form around each turn of wire, but in doing so, they take opposite directions between adjacent turns. This causes cancellations, and results in a zero field strength between the turns. However, in the interior and exterior of the coil, the fields combine to form continuous h-lines.

An action similar to the magnetic activity in a coil takes place in a waveguide. A two-wire line with a quarter-wave section is shown in Figure 8-9C. Current flows in the main line and in the quarter-wave section. The current direction produces the individual h-lines around each conductor. When a large number of sections exist, the field cancels between the sections, but their directions are the same, inside and outside the waveguide. At half-wave intervals on the main line, current flows in opposite directions and produces h-line loops in opposite directions. In Figure 8-9C, current at the left end is opposite to the current at the right end. The individual loops on the main line are opposite in direction. All around the framework, they join in such a way that the long loop shown at Figure 8-9D is formed. On the outside of the waveguide, the individual loops cannot join to form a continuous loop, so there is no magnetic field.

Figure 8-10 illustrates a conventional presentation of the magnetic field in a waveguide three half-wavelengths in length. The h-field is strongest at the edges of the waveguide because the current is the highest at that point. The current is lowest at the center of each set of loops because there, the current standing wave is zero at all times.

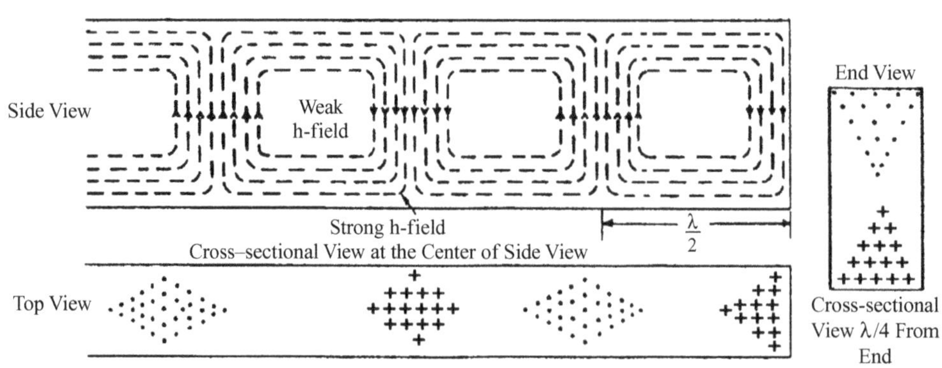

FIGURE 8–10

Magnetic fields in a three-half-wavelength waveguide. Form USAF Manual 52-8.

Remember that Figure 8-10 illustrates an instantaneous condition. During the peak of the opposite half cycle of signal input, all the field directions are reversed.

Recall that the propagation of electric fields requires that there must be no electric field tangent to the walls of the guide. A second boundary condition for electromagnetic fields requires that, at the surface of the waveguide, there be no perpendicular component of the magnetic field. Since all the h-lines are parallel to the surface, this condition is satisfied.

The waveguide e-fields and h-fields obviously exist simultaneously. The h-fields cause currents, which, in turn, cause voltage differences. These cause e-fields, further causing currents and h-fields. As the fields are dependent on one another, the wave is propagated by the continual interdependent development of fields.

Figure 8-11A is a three-dimensional, conventional illustration of both e- and h-fields in the waveguide. Because such an illustration is excessively complex, more simplified illustration techniques are used, such as those in Figures 8-11B–D. In these diagrams, one field is shown with ×'s and dots, and the other with lines.

Figure 8-11 illustrates only one of many waveguide ***modes.*** The ***dominant mode*** is shown. There are higher modes, with different field configurations, that may occur either accidentally or deliberately.

An example of another field configuration is illustrated in Figure 8-12A. If the size of this waveguide is doubled over that of the waveguide shown in the previous illustrations, the cross-section will be a full wave rather than a half wave. The two-wire conductor can be assumed to be $\lambda/4$ down from the top or up from the bottom. The remaining distance to the bottom or top is $3\lambda/4$.

A $3\lambda/4$ section has the same high impedance input as the $\lambda/4$ section. Thus, the two-wire line is properly insulated, and will transfer energy. The field configuration will show a full wave across the wide dimension, as illustrated in Figure 8-12B.

This field configuration can be applied to a circular waveguide. The two conductors shown in Figure 8-12C are assumed to be part of the waveguide wall. The remaining part of the wall forms the $\lambda/4$ sections. The $\lambda/4$ section insulates the two conductors, making it possible to transfer energy with minimum losses. The resulting field configuration in Figure 8-12D is the dominant mode for a waveguide with a circular cross-section.

Waveguide Propagation by Electric Waves

A somewhat different analogy of waveguide signal propagation deals with the field rather than with current and voltages. This analogy is more precise than the two-wire conductor analogy. It assumes that power flow is in the fields, rather than in the conductors.

The fields in a waveguide are the same as those radiated into space by an antenna. Figure 8-13 illustrates a small portion of a field radiated into space by an antenna. The e-lines are parallel to the antenna, and the h-lines are perpendicular to it. The fields move outward at the speed of light. At any given point, the fields are

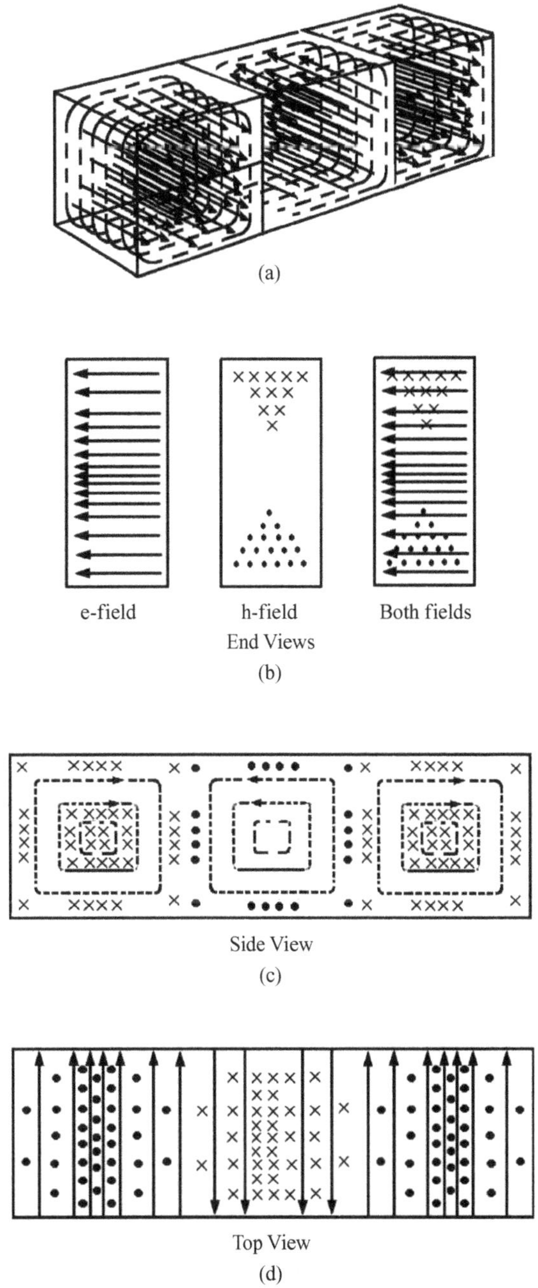

FIGURE 8–11

Both e- and h-fields shown. From USAF Manual 52-8.

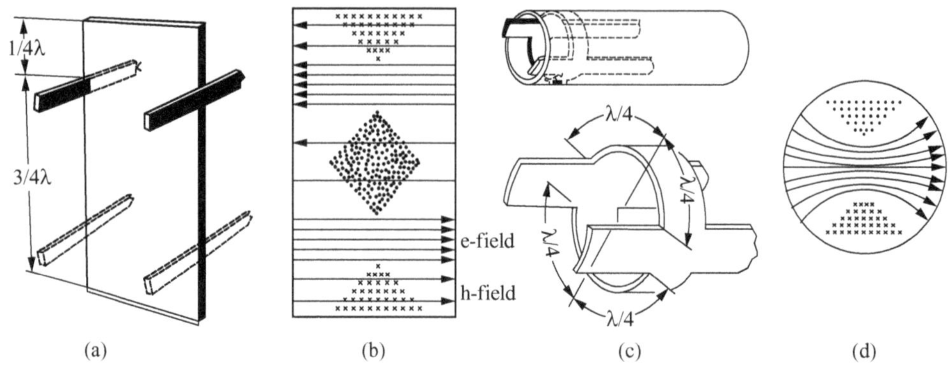

FIGURE 8–12

Another field configuration. From USAF Manual 52-8.

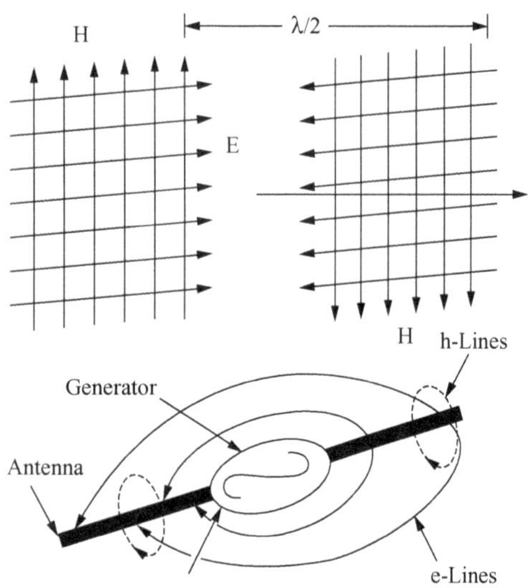

FIGURE 8–13

Field radiated by antenna. From USAF Manual 52-8.

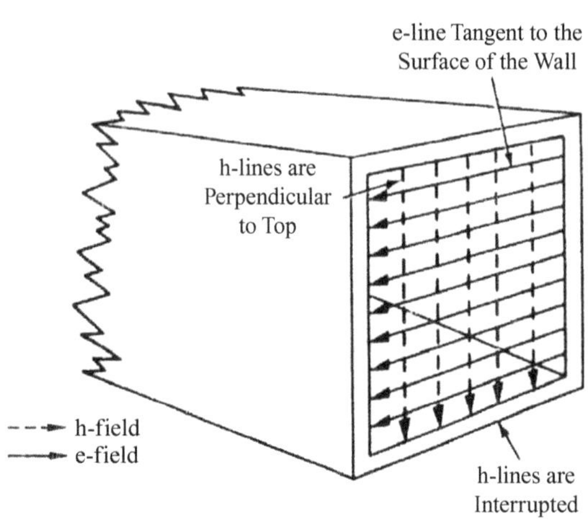

FIGURE 8–14

Fields must satisfy boundary conditions. From USAF Manual 52-8.

changing at the signal frequency. Recall that distance equals rate times time ($d = rt$). If the wave moves at the speed of light, c, and the time required for one cycle of the signal frequency is $1/f$, then the distance traveled by one cycle is

$$d = rt \Rightarrow \lambda = c\left(\frac{1}{f}\right) = \frac{c}{f}$$

$$\lambda_{(\text{cm})} = \frac{3\times 10^{10}}{f} \Rightarrow \lambda_{(\text{inches})} = \frac{1.18\times 10^{10}}{f}.$$

Only a small part of the total field is shown in Figure 8-13. The e-lines and h-lines form huge closed loops after they leave the antenna. Again, the energy propagation through a waveguide and the energy radiated into space by an antenna are the same form of electromagnetic radiation.

The field configuration shown in Figure 8-13 cannot exist in a waveguide, because that configuration does not satisfy the required boundary conditions. First, there cannot be any e-lines tangent to the surface of the walls. Since the e-lines are evenly distributed across the area, some would be across the top and bottom wall. This short-circuits the voltage and causes the e-lines to vanish. Other e-lines, pushed to the wall by mutual repulsion of like charges, are also shorted; the effect is cumulative and removes the entire e-field.

The other boundary condition which must be satisfied is that there must be no component of the magnetic field perpendicular to the wall. Figure 8-14 shows that the h-lines are correctly parallel to the side walls but incorrectly perpendicular to the bottom; so h-lines of this type cannot exist in the waveguide. Still further, an h-line cannot exist unless it is a closed loop.

When a small antenna is placed in the waveguide and excited at an appropriate frequency, a wavefront is radiated and would attempt to expand outward in a circular fashion (see Figure 8-15). The portion of the wavefront traveling in direction "B" proceeds straight down the waveguide and is quickly attenuated because it does not meet the boundary conditions for propagation. However, the part of the wavefront which travels in direction "A" is reflected from the wall. The wall is a short circuit to the energy and causes the wavefront to be reflected in the reverse phase. The wavefront traveling in direction "C" is reflected from the other wall and proceeds in the opposite phase. Thus, the radiation fields are contained in the waveguide.

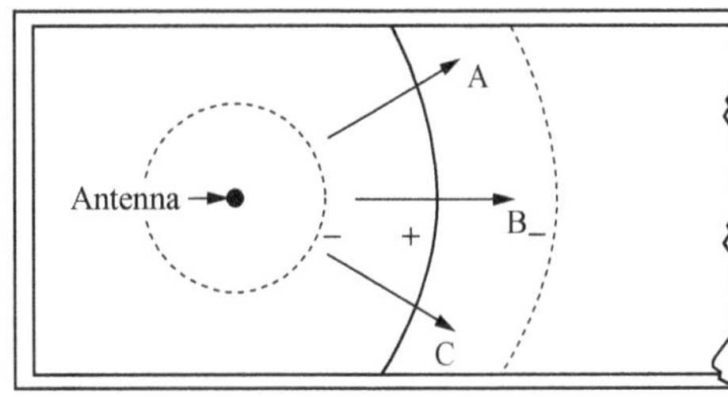

FIGURE 8–15

Fitting a field into a waveguide. From USAF Manual 52-8.

Path of Wavefronts in a Waveguide

A side view of the propagation of wavefronts in a guide is shown in Figure 8-16. Two directions of the wavefront propagation are shown, "A" and "B." The "A" wavefronts are shown by light solid or broken lines, and the "B" wavefronts are shown by heavy solid or broken lines. Wavefront A is always traveling at an upward angle, and wavefront B is always traveling at a downward angle. The solid lines represent positive wavefronts, and the broken lines represent negative wavefronts.

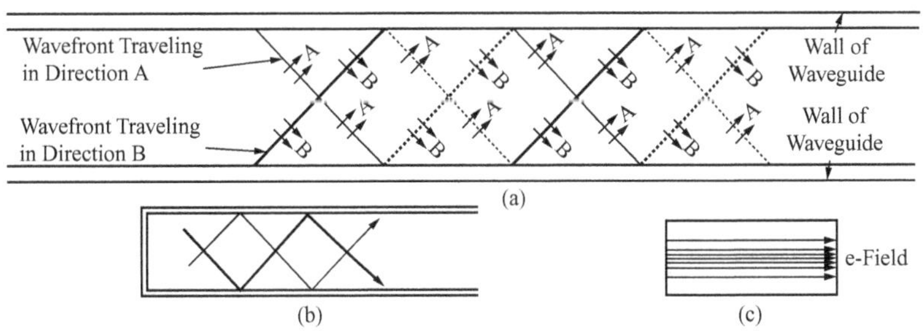

FIGURE 8–16

Wavefront paths. From USAF Manual 52-8.

At the point in the guide where the positive solid-line wavefronts cross, there is a maximum e-field; where the negative broken-line fronts cross, there is a maximum e-field in the opposite direction. Additionally, at those points where the entire wavefronts have arrived at the walls, the opposite polarities cause a cancellation, reducing the e-field to zero. Propagation is therefore possible because there are no e-lines tangent to the walls of the guide; the e-field is shown in Figure 8-16C.

The Wavefront Crossing Angle

The angle at which the wavefront crosses the guide is a function of (1) the wavelength and (2) the guide's cross-sectional dimension (see Figure 8-17). As the frequency increases, the angle of incidence decreases, and the signal travels farther before it reaches the other side. At lower frequencies, the wavefront crosses the guide at more nearly right angles to the walls. At some frequency, the angle is 90°, and the wave travels back and forth across the guide until the energy is dissipated by the resistance of the walls of the guide. At this frequency, called the ***cutoff frequency,*** the distance from side to side is one-half wavelength. At this cutoff frequency, the attenuation is a linear function of length and is very high.

The physical distance between the points of wavefront reflection becomes a major factor in the design of phased-array or frequency-scanned antennas. In some of these, the waveguide dimensions are physically varied, and in others the frequency is varied; in both cases, a precise change of the dimension or input frequency causes a precise change in the length of the wavefront path from one reflection to the next.

The velocity of propagation on a two-wire line is less than its velocity in air, because the capacitances and inductances must charge and discharge as the signal moves. There is also some delay in a waveguide, attributable to the manner in which the field travels. As shown in Figure 8-17C, the path of a wavefront at a relatively low frequency is along the zig-zag arrow at the velocity

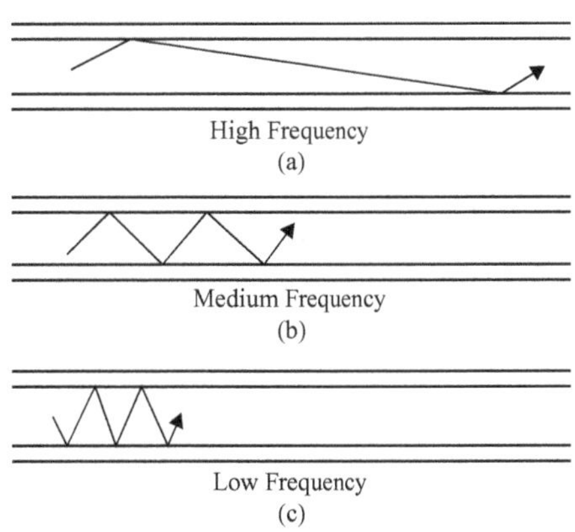

FIGURE 8–17

Wavefront crossing angles. From USAF Manual 52-8.

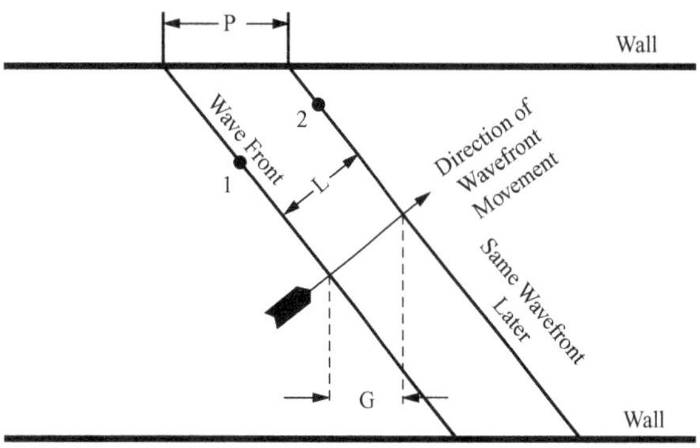

FIGURE 8–18

Relation of phase, group, and wavefront velocities. From USAF Manual 52-8.

of light. However, because of all the trips back and forth across the guide, the wavefront actually travels very slowly along the guide. In Figure 8-17A, the frequency is higher and the wavefront velocity, called the ***group velocity,*** increases, because fewer trips are made back and forth across the guide, and the group velocity approaches the speed of light.

The axial velocity of a wavefront, called the group velocity, has a relationship with the diagonal velocity that causes an unusual phenomenon. The velocity of propagation appears to be greater than the speed of light. During a given time, a wavefront moves from point 1 to point 2, shown as distance *L*, at the velocity of light (*c*) (see Figure 8-18). Due to this diagonal movement in the direction of the arrow, during this time, the wavefront has actually moved down the guide only the distance *G*, which is necessarily a lower velocity. This is called ***Group Velocity (Vg).*** However, if an instrument were used to detect the two positions at the wall, they would be the distance *P* apart. This is greater than the distance *L* or *G*. The movement of the contact point between the wave and the wall is at a greater velocity. Since the phase of the rf has changed over the distance *P*, this velocity is called the ***Phase Velocity (Vp).*** The mathematical relationship between the three velocities is stated by the equation

$$c = \sqrt{V_p V_g}$$

where
c = velocity of light = 3×10^8 m/s
V_p = phase velocity
V_g = group velocity.

This equation indicates that it is possible for the phase velocity to be greater than the velocity of light. As the frequency decreases, the angle of crossing is more of a right angle. In this condition, the phase velocity increases. For measuring standing waves in a waveguide, it is the phase velocity which determines the distance between voltage maximum and minimum. For this reason, *the wavelength measured in the guide will actually be greater than the wavelength in free space.* From a practical standpoint, the different velocities are related in the following manner: if the signal frequency being propagated is sine-wave modulated, the modulation envelope will move forward through the waveguide at the group velocity, while the individual cycles of rf energy will move forward through the modulation envelope at the phase velocity. If the modulation is a pulse, as in radar transmissions, the pulse envelope will travel at the group velocity, while the rf waveshape will move forward within the envelope. Since the standing-wave-measuring equipment is affected by each rf cycle, the wavelength will be governed by the rapid movement of the changes in rf voltage. Since intelligence is conveyed by the modulation, the transfer of intelligence through the waveguide will be slower than the speed of light, as is the case in other types of rf lines. The difference in propagation times of carrier and intelligence may have great implications in the design and tuning of microwave tubes, such as the klystron drift tube in radar transmitters.

Because of the way the fields are assumed to move across the waveguide, it is possible to establish a number of trigonometric relationships (see Figure 8-19). The angle that the wavefront makes with the wall (angle θ) is related to the wavelength and dimension of the guide and is equal to

$$\cos\theta = \frac{\lambda}{2B}$$

where λ is the wavelength in free space of the signal in the guide, and *B* is the inside wide dimension of the guide. The group velocity (V_g) is related to the velocity of light (*c*) as follows:

$$\frac{V_g}{c} = \sin\theta = \sqrt{1 - \left(\frac{\lambda}{2B}\right)^2}.$$

Further, since it is possible to measure the wavelength in the guide (λ_g), the wavelength in space is equal to

$$\frac{\lambda_g}{\lambda} = \frac{1}{\sin\theta} = \frac{1}{\sqrt{1-\left(\frac{\lambda}{2B}\right)^2}}.$$

Solving this for λ, the equation becomes

$$\lambda = \frac{2B\,\lambda_g}{\sqrt{\lambda_g^2 + 4B^2}}.$$

After measuring the wavelength and the inside dimension of the waveguide, it is possible to calculate most other quantities associated with the waveguide.

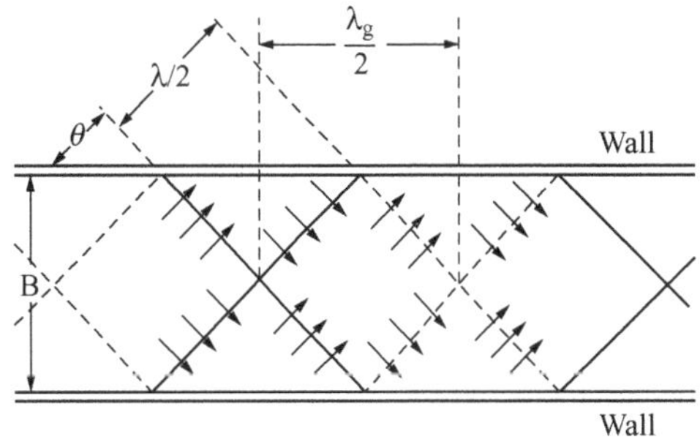

FIGURE 8–19

Trigonometric relationships. From USAF Manual 52-8.

Numbering System of the Modes

The normal configuration of the electromagnetic field within a waveguide is called the ***dominant*** mode of operation. The mode thus far developed for the rectangular waveguide is the dominant mode of operation. The dominant mode for the circular waveguide is also shown in Figure 8-12. A wide variety of higher modes are possible in either type of waveguide. The higher modes in the rectangular waveguide are seldom used in radar, but some of the higher modes in the circular waveguide are useful.

To identify modes, any field configuration can be classified as either a ***transverse electric*** or a ***transverse magnetic*** mode. These modes are abbreviated TE or TM, respectively. In a TM mode, the plane of the h-field is perpendicular to the length of the waveguide. No h-line is parallel to the direction of propagation. This mode is also sometimes called an E-mode. A wavefront in free space, or on a coaxial line, is in a TEM mode, since both fields are perpendicular to the direction of propagation. This mode cannot exist in a waveguide. In addition to the letters TE or TM, subscript numbers are used to complete the description of the field pattern (see Figure 8-20).

Rectangular Waveguides

The first small number of the subscript indicates the number of halfwave patterns of the transverse lines which exist along the short cross-section. The second small number indicates the number of transverse half-wave patterns that exist along the long dimension of the guide through the center of the cross-section.

Circular Waveguides

The first number indicates the number of full waves of the transverse field encountered around the circumference of the guide. The second number indicates the number of half-wave patterns that exist across the diameter.

Counting Wavelengths for Measuring Modes

In the rectangular mode illustrated in Figure 8-20A, all the electric lines are perpendicular to the direction of movement. This makes a TE mode. In the direction across the narrow dimension of the guide, parallel to the e-line, the intensity change is zero. Across the guide along the wide dimension, the e-field varies from zero on the top through maximum at the center to zero at the bottom. Since this is one-half wave, the second subscript is 1. Thus, the complete description of this mode is $TE_{0,1}$.

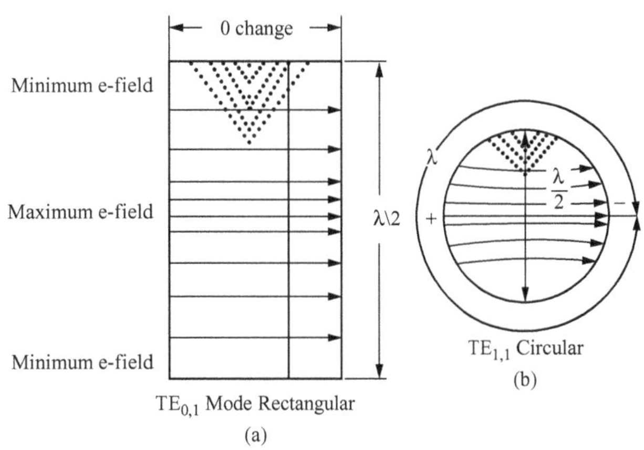

FIGURE 8–20

Counting wavelengths to number modes. From USAF Manual 52-8.

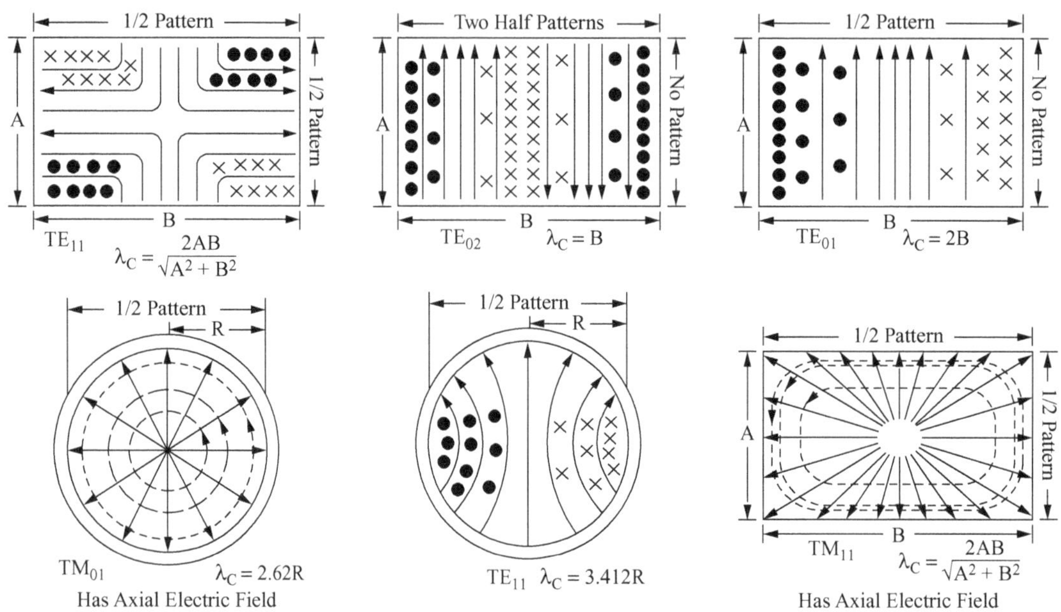

FIGURE 8–21

A variety of modes. From USAF Manual 52-8.

In the circular waveguide in Figure 8-20B, the e-field is transverse and the letters which describe it are TE. Moving around the circumference, beginning at the top, the field goes from zero, through maximum positive (tail of arrows), through zero, through maximum negative (head of arrows), to zero. This is one full wave, so the number is 1. Going through the diameter, the start is from zero at the top wall, through maximum in the center to zero at the bottom, one-half wave. The second subscript is 1. The complete designation for the circular mode becomes $TE_{1,1}$.

Several circular and rectangular modes are possible. In each diagram illustrated in Figure 8-21, note that the magnetic and electric fields are maximum in intensity in the same area. This indicates that the current and voltage are in phase. This is the condition which exists when there are no reflections to cause standing waves. In previous examples in which fields were developed, the fields were out of phase because of a short circuit at the end of the two-wire line.

Introducing Fields into a Waveguide

A waveguide is a single conductor and does not have the two connections of ordinary rf lines. It is necessary to use special devices to apply energy to a waveguide, or to recover it from the guide. A waveguide exhibits ***reciprocity,*** a condition in which energy can be applied or retrieved with equal efficiency. Waveguides may be excited by electric fields, magnetic fields, and electromagnetic fields.

Excitation with Electric Fields

A small probe, or antenna, can be placed in a waveguide. When fed with an rf signal, the current in the probe sets up an electrostatic field as shown in Figure 8-22A. This causes the e-lines to detach themselves from the probe and to form in the guide. When the probe is correctly positioned, a field of considerable intensity will be developed. The optimum location for the probe is in the center, parallel to the guide's narrow dimension and $\lambda/4$ from the shorted end of the guide, as shown in Figure 8-22C. The field is strongest at the $\lambda/4$ point. This is the point of maximum coupling between the probe and the field. Of course, the probe will work equally well in the center of any unidirectional field. For example, a $3\lambda/4$ distance from the shorted end will also be a good location for the probe.

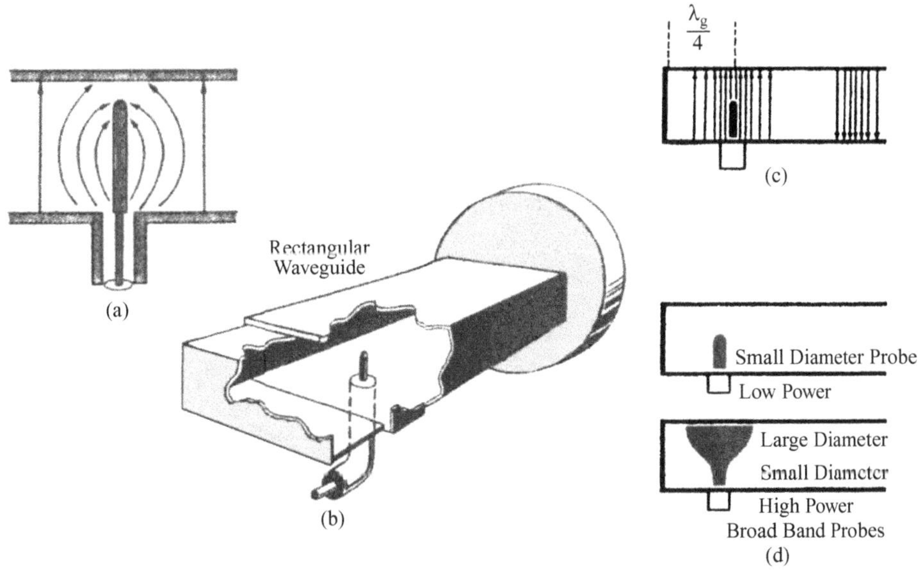

FIGURE 8–22

Waveguide excitation by the electric field. From USAF Manual 52-8.

The probe may be fed with a coaxial cable, when the power level is low enough that arcing will not occur. In comparison with the waveguide, this cable is extremely short, to ensure that the greatest benefit will be derived from the waveguide. Impedance matching between the coax and the guide is accomplished by adjusting the distance from the probe to the end of the waveguide. That distance can be adjusted by moving the shorted end of the guide, and by varying the length of the probe, as illustrated in Figure 8-22B. A mismatch will cause unwanted reflections in the waveguide.

The excitation power can be adjusted by means of the length of the probe, by moving it into or out of the center of the e-field, or by shielding it. Where it is necessary to vary the degree of excitation frequently, the probe is made retractable, and the end of the waveguide is fitted with a movable plunger. Such adjustments may be found in the receiver waveguide, as to adjust the amount of injected local oscillator power for a given receiver mixer crystal current. In many radar systems, the position of the probe and the end piece is often predetermined by the factory and permanently set.

Both the transmitted and received signals in pulse-modulated radar systems comprise a wide spectrum of frequencies because of the harmonic content of the modulating pulse. The waveguide system must be able to pass the entire spectrum, rather than only the carrier center frequency. The waveguide cannot, therefore, be tuned to a single frequency, and must be ***broad-banded*** sufficiently to pass the entire spectrum, and to permit operating frequency changes within the band. To accomplish that, a probe feeding system with a large-diameter, conical, or door-knob-shaped probe is used; such probes also permit the application of high power.

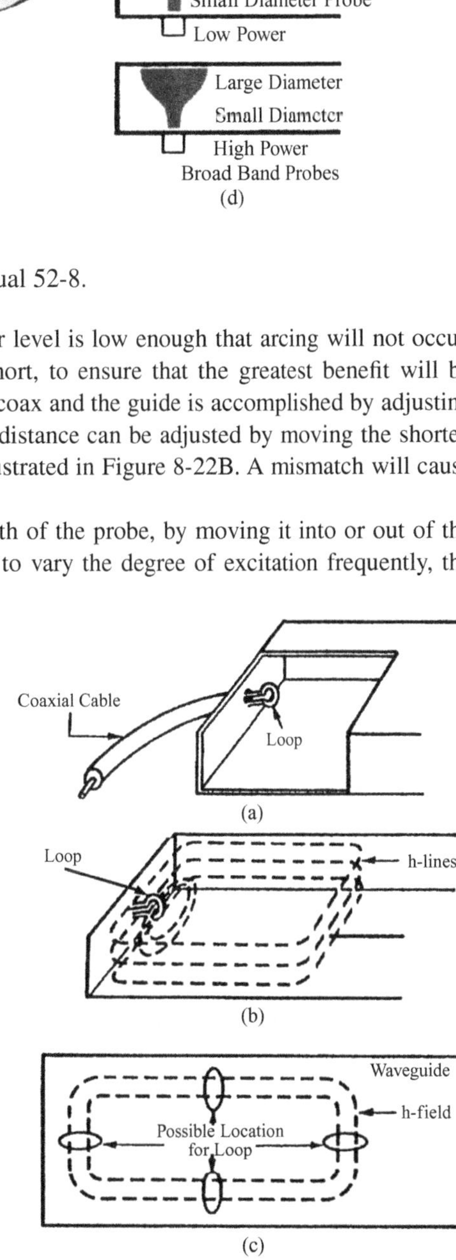

FIGURE 8–23

Magnetic excitation. From USAF Manual 52-8.

Excitation with a Magnetic Field

Excitation can also be achieved by creating an h-field, with a small loop in the waveguide (see Figure 8-23). A magnetic field around the loop is created with a high current at the appropriate frequency. The h-field expands to fit the guide, e-fields result,

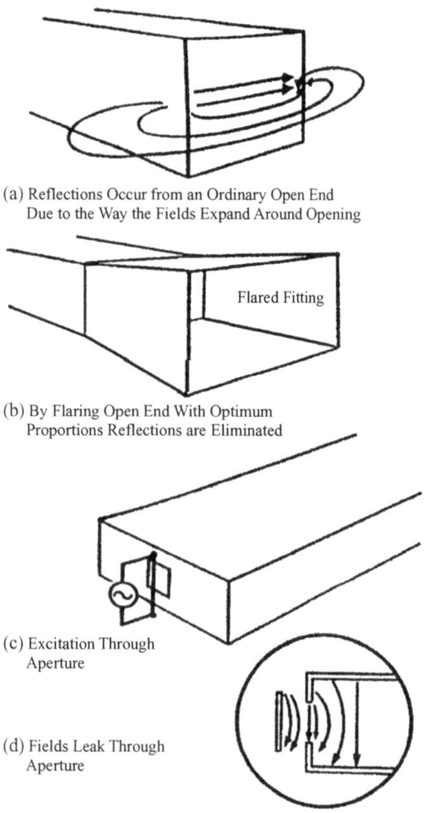

FIGURE 8–24

Electromagnetic excitation. From USAF Manual 52-8.

and propagation begins. Note that the loop is fed by a coaxial cable. The location of the loop for optimum coupling to the guide is at the place where the desired magnetic field will be of greatest strength. To reduce excitation power, the loop can be rotated, or moved, until it encircles a smaller number of lines of force. The location of the loop is often predetermined at manufacture by laboratory tuning.

Excitation with Electromagnetic Fields

An open-ended waveguide will not serve as a means to radiate or receive energy. When energy leaves such a guide, fields form around the end to cause an impedance mismatch, as illustrated in Figure 8-24A. Reflections and standing waves would result. The opening of the guide may be flared, as in Figure 8-24B; it then performs in a manner similar to a V-type antenna. This ***feedhorn*** eliminates reflection by matching the impedance of the waveguide to that of free space. When the feedhorn receives electromagnetic fields, the fields are gradually shaped to fit the waveguide. The feedhorn, as with other waveguide components, exhibits reciprocity and may be used either to radiate energy into the atmosphere, or to receive it from the atmosphere.

Energy of the correct frequency may also be transferred to or from waveguides through slots or openings; this is sometimes employed where very loose coupling is necessary (Figures 8-24C and D). Any device which generates an e-field may be placed near the aperture, and the e-field will expand into the waveguide. A single wire is shown in Figure 8-24D; e-lines are created parallel to the wire by the voltage difference between different parts of the wire. In expanding, the e-lines will first exist across the aperture, and then across the interior of the waveguide.

Bends, Twists, Joints, and Terminations

In order for energy to move through a waveguide without reflections, the size, shape, and dielectric material of the waveguide must be constant throughout its entire length. Any abrupt change in size or shape causes reflections. To prohibit these, any change in the direction or the size of the guide must be gradual. Where an abrupt change is a physical necessity, special devices, such as bends, twists, joints, or terminations, must be used.

Bends

One way to bend a waveguide is to make the bend gradual. A gradual bend must have a radius greater than two wavelengths to minimize reflections. A bend can be made in either the narrow or wide dimension of a guide. The bend shown in either Figure 8-25A or 8-25D is called an ***E-Type bend,*** and the bend shown in either Figure 8-25 B or 8-25C is called an ***H-Type bend.*** Where a sharp 90° bend is required, reflections are minimized by joining two sharp 45° bends, spaced by $\lambda/4$. The interaction of the direct reflection at one bend, and the inverted reflection from the other, will result in cancellation of the reflection.

Radar installation work requires some versatility, and completely rigid waveguide assemblies are subject to stress and damage, as from ship motion, antenna tilt changes, antenna tower motion, vehicle motion, and others. To allow for these conditions, manufacturers provide sections of flexible waveguide, sometimes simply called ***flexguide.*** These sections can be bent or twisted as required, and consist of a spirally wound ribbon of brass, resembling a spring. The brass is imbedded in a flexible, airtight, rubber jacket. As skin effect keeps the current at the inner surface of the waveguide, the inside surfaces are chromium- or silver-plated for maximum conductivity. It is noteworthy, for the technician searching for pressure leaks, that the rubber jacket on the flexguide is a potential source for the leak.

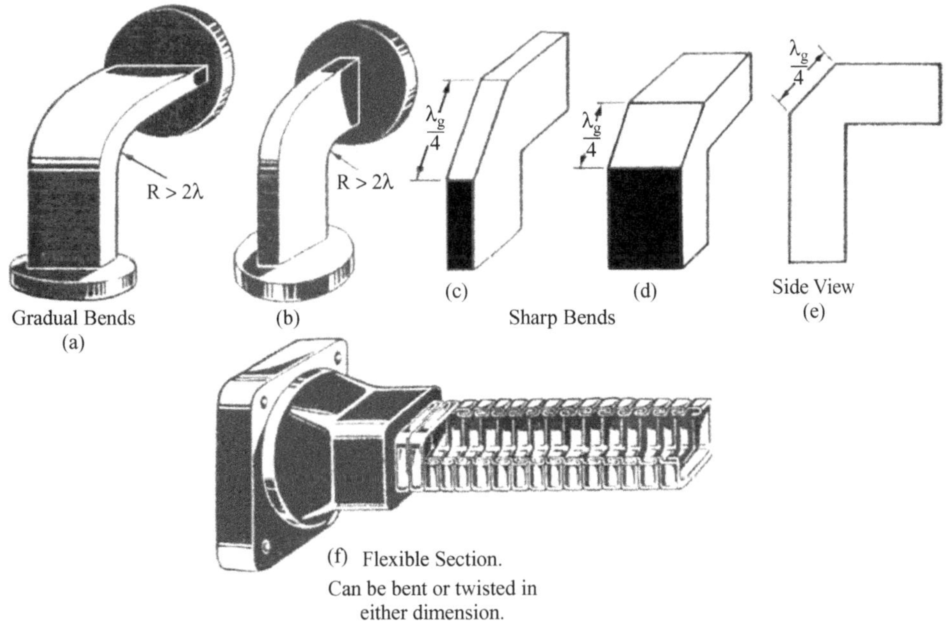

FIGURE 8–25

Waveguide bends. From USAF Manual 52-8.

Rotating the Field

Some components require rotation of the electromagnetic field for a number of reasons (see Figure 8-26). This can be accomplished by twisting the waveguide as illustrated, or by use of flexguide. The twist should be gradual and extend over two wavelengths or more to prevent reflections. Antenna ***circular polarizers*** may use a more complex field rotation technique, which causes two sets of e- and h-fields in the feedhorn, electrically separated by 90° to permit sine–cosine power sharing between horizontally and vertically polarized radiation patterns.

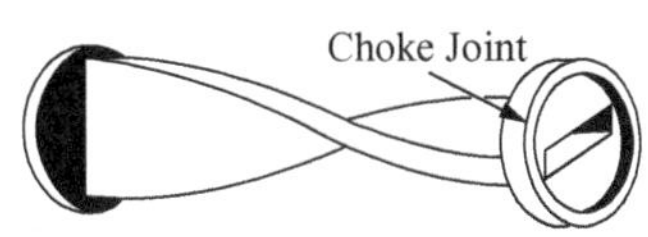

FIGURE 8–26

90° twist. From USAF Manual 52-8.

Joints

A waveguide system is necessarily assembled by sections, connected together by joints. There are permanent, semipermanent, and rotating joints. However, irregularities at the joints can create standing waves, allowing energy to escape. A permanent joint is made at the factory. The waveguide sections are machined within a few thousandths of an inch and then welded to make a hermetically sealed, mirror-finish joint.

It is usually necessary for installation and maintenance that sections must be made for disassembly, and semipermanent joints are manufactured. The most common of these is the ***choke joint.*** Figure 8-27A illustrates a cross-sectional view of a choke joint. It consists of two flanges connected

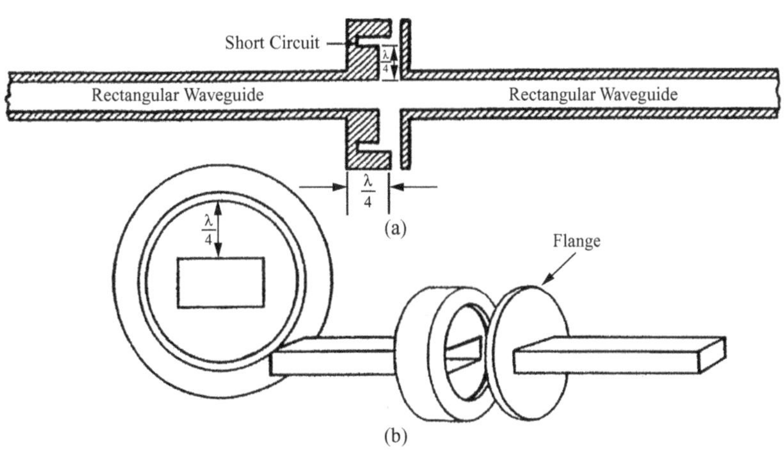

FIGURE 8–27

Choke joint. From USAF Manual 52-8.

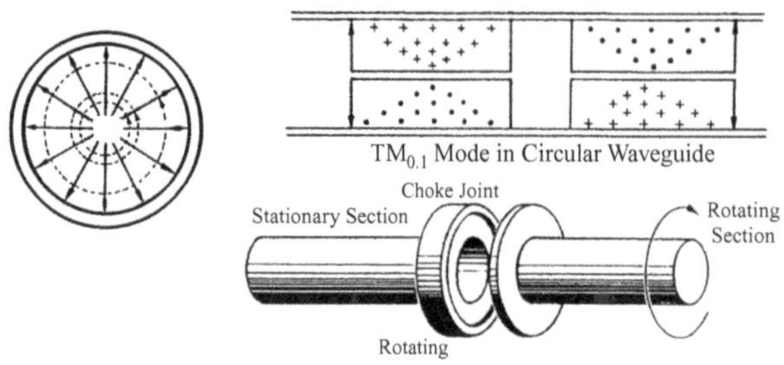

FIGURE 8–28

Circular rotary joint. From USAF Manual 52-8.

to the waveguide at the center. The right-hand flange is flat, and the one at the left is slotted a $\lambda/4$ deep at a distance $\lambda/4$ from the point where the walls of the guide are joined. The $\lambda/4$-slot is shorted at the end. The two $\lambda/4$ recesses total $\lambda/2$ and reflect a short circuit at the place where the walls are joined. The waveguide joint thus appears continuous to signals. The two sections can be separated as much as a tenth of a wavelength without appreciable loss of energy. That tolerance allows space to seal the waveguide with a gasket. Waveguides are often pressurized with compressed air or gas to keep moisture out. The quarter-wave distance from the walls to the slot is modified slightly to compensate for the slight reactance introduced by the short space and open circuit from the slot to the periphery of the flange.

The name "choke joint" is said to come from the similarity between the action of this joint on rf fields and the action of an rf choke in a power supply lead. An rf choke keeps rf in the circuit where it belongs. Similarly, the choke joint keeps the electromagnetic fields in the waveguide where they belong. The loss introduced by a well-designed choke is less than 0.03 dB, while an unsoldered permanent joint, well machined, has a loss of 0.05 dB or more.

Rotary Joints

Rotary joints are necessary in radar systems with physical-motion antennas, because the transmitter is stationary and the antenna must rotate. A simple waveguide rotary joint is illustrated in Figure 8-28. A stationary circular waveguide is mated to a circular rotating waveguide with a choke joint, closely machined and assembled for minimum clearance without contact. The circular waveguides are in the $TM_{0,1}$ mode, and the field does not rotate in the rotating guide.

The circular rotating joint will operate satisfactorily, but places many limitations on design. One is that only one waveguide path is permitted; some radar systems may require that as many as six paths be provided. Another is that waveguide systems are generally constructed of rectangular guide, and transition hardware would be necessary. Therefore, most rotary joints are manufactured with rectangular guide, somewhat as the very simple one illustrated in Figure 8-29. It should be emphasized that the state of the art of rotary joints has now progressed far beyond this simple introduction of the basic idea. Latter-day rotary joints may contain as many as five channels.

For mechanical reasons, the rotary joint itself must be circular and involves the use of a rigid coaxial joint to provide axial symmetry of the fields and the circular cross-section for rotation. In the rotating joint, a probe forms the end of the center conductor of the coaxial joint. The probe takes energy from one waveguide. It is then conducted through the coaxial joint, through another probe, into the other waveguide. The probe in the other waveguide is the other end of the center conductor. The center conductor remains stationary with respect to one waveguide and rotates with respect to the other. To make the rotating electrical connection, the outer conductor can either be fitted with sliding contacts, or with the tubing, by a half-wave slot. The slot, which is shorted at the end, reflects a short at the junction of the two outer conductors. In this method, no mechanical contact is required between the two sections of the outer conductor. The inner conductor of the coaxial cable is supported by insulating washers.

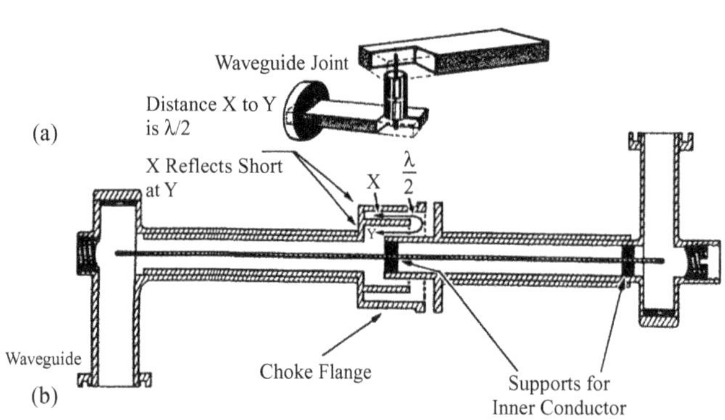

FIGURE 8–29

Rectangular-Guide rotary joint. From USAF Manual 52-8.

T-Junctions

There are many requirements to split a signal for use in two places, or to combine two signals into a single waveguide. This requires a T-junction. It may be connected either in the narrow side, as shown in Figure 8-30A, or in the wide side of the waveguide, as shown in Figure 8-30D. When the T-junction is in the plane of an h-field of a $TE_{0,1}$ mode, it is called an ***H-type junction,*** and when the junction is in the plane of the e-lines, it is called an ***E-type junction.***

The H-type junction is a parallel connection with the main line. For example, when the end of the T-joint shown in Figure 8-30B is short-circuited at a distance $\lambda/2$ from the ***center*** of the waveguide, the result is the equivalent parallel circuit shown in Figure 8-30C. Note that Figure 8-30C shows a $\lambda/2$ section connected to a two-wire line. This section will reflect a short circuit at the line, and will not permit energy to pass. Similarly, in the waveguide itself, a short circuit is reflected to the center, where the e-lines are supposed to be. Since an e-line cannot exist at a short circuit, no energy will pass that point. If the shorted end of the T-section were only a distance $\lambda/4$ from the center of the waveguide, an open section would be reflected there, and the passage of energy would be unaffected.

The E-type joint illustrated in Figures 8-30D–F is a series connection with one side of the main line. In this case, the added waveguide section is $\lambda/2$ in length, acts as a short circuit at the junction, and will allow energy to propagate down the main waveguide. If the added section were $\lambda/4$, the circuit at the junction would be an open one, and propagation down the main guide would be stopped.

A special T-joint of both types, called the magic tee, serves a purpose in the mixer circuits of some microwave receivers (see Figure 8-31). The local oscillator is injected by an H-type junction, and the radar rf echoes are injected by an E-type junction. The local oscillator e-fields undergo a phase reversal as they propagate toward the crystal mixers, but the signal input goes to both mixers in the same phase. The circuit is used for cancellation of noise from the local oscillator and is described in more detail in other chapters of this book.

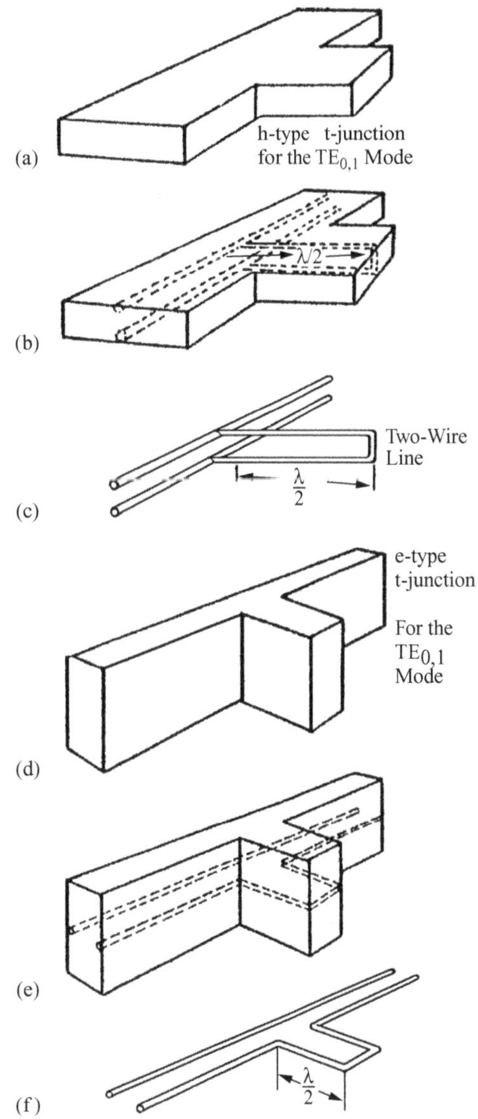

FIGURE 8–30

T-Junctions. From USAF Manual 52-8.

The ATR Switch

One important use of the T-joint has been with a device used to "disconnect" the waveguide path to the transmitter during the "listening time," so as to prevent it from absorbing the signal power of received echoes. This device was originally called an "automatic receive-transmit switch," or "R/T box." In later years, it became known as an ***anti-transmit-receive (ATR) device*** (see Figure 8-32). It was part of a two-way switching circuit called the ***duplexer,*** used to (1) steer transmitter energy to the antenna without damaging the receiver, and (2) steer received echoes to the receiver without "losing" them in the transmitter. Although some ATRs are still in use in certain radar types as

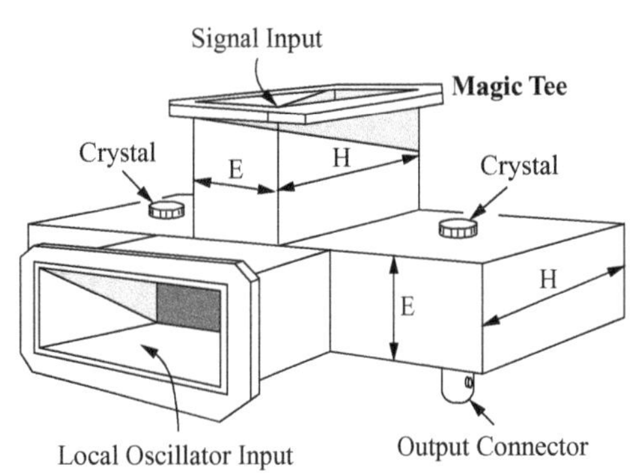

FIGURE 8–31

Magic tee.

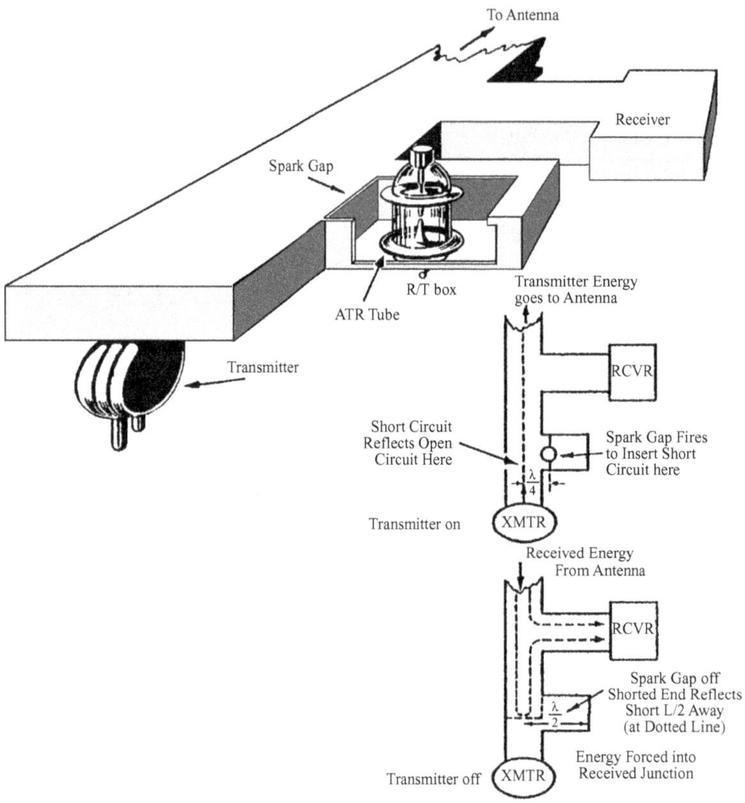

FIGURE 8–32

The anti-transmit-receive (ATR) switch. From USAF Manual 52-8.

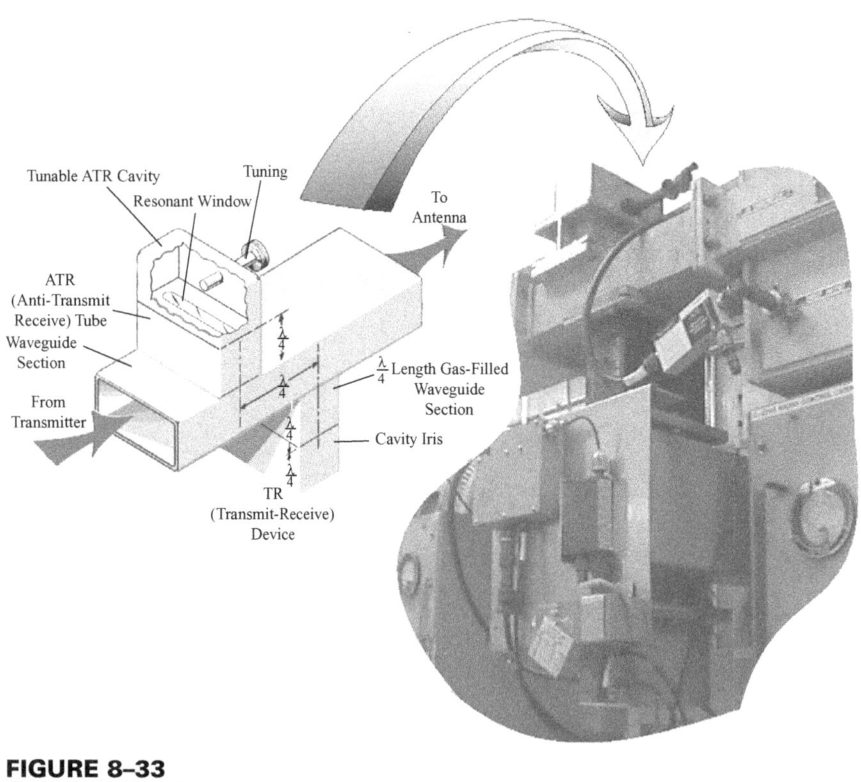

FIGURE 8–33

ATR tube in FPS-20 series ARSR duplexer.

shown in Figure 8-33, the ATR is seldom used in new radar design, as ***four-port circulators*** and ***ferrite load isolators*** have eliminated the need for them. However, a similar device in the receiver waveguide path, called the transmit-receive (TR) device is still necessary, and in use.

In the partial ATR system illustrated in Figure 8-32, an H-type T-junction is connected in the main waveguide. A spark gap is located one quarter wave from the center of the main guide. The junction is shorted at the distance of a half wave from the center of the main waveguide. When the powerful transmitter is turned on, a spark jumps the spark gap. This causes the waveguide branch to short at that point; the short is inverted to an open circuit at the center. Therefore, transmitted energy can pass, unhindered.

Energy received from the radar echo enters in the waveguide from the opposite end, after the transmitter is turned off. This energy is not great enough to cause a spark to jump across the spark gap. This time the shorted end of the branch reflects a short at the center of the waveguide, and reflects the received energy back to another T-junction, which is located at the input to the receiver. Thus, no energy is absorbed by the "off" transmitter. There is a similar arrangement called a TR device in the waveguide leading to the receiver. It offers high impedance to the transmitted pulse, but low impedance to the received echo.

Matching Devices

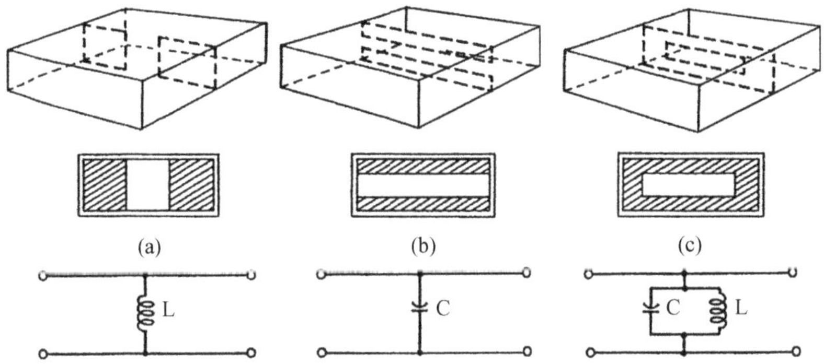

FIGURE 8–34

Reactive plates. From USAF Manual 52-8.

Some devices used in microwave circuitry introduce inductance or capacitance, often deliberately. When inductive or capacitive reactance is present and undesirable, it can be tuned out with small fins or plates in a waveguide.

Figure 8-34 illustrates several reactive plates used to introduce capacity, inductance, or both, in a waveguide. When these plates are employed as shown in Figure 8-34A, they cause oscillations in the higher modes. Since a waveguide is too small for higher modes at the same frequency, these frequencies are not propagated, but remain in the vicinity of the plates. If the edges of the plates are vertical with respect to the plane of the h-field, the modes produced are the TM type. The effect of this on power flow is that of inductance across the two-wire line. This causes reflections and a shift in the standing wave pattern. The wider the space between the plates, the greater the inductive reactance.

When the partitions are arranged perpendicular to the e-field, as in Figure 8-34B, a local e-field and the higher modes of oscillation are created between the edges of the plates. These oscillations cannot be propagated, but change the dominant mode to a TE mode, and introduce capacitive reactance. As with the TM mode, the wider the opening, the greater the reactance.

By combining both types of plates, and leaving a small opening in a large guide, as in Figure 8-34C, a resonant circuit is created.

Terminating a Waveguide

Since a waveguide is a single conductor, it is not as easy to define its characteristic impedance (Z_o) as it is for a coaxial line. The characteristic impedance of a waveguide is approximately equal to (1) the ratio of the strength of the electric field to (2) the strength of the magnetic field for energy traveling in one direction. This ratio is equivalent to the voltage-to-current ratio in coaxial lines on which there are no standing waves.

The lowest Z_o of a circular guide is about 350 Ω. In a rectangular waveguide, it may be any value, depending on the guide dimensions and the signal frequency. In this guide, it is directly proportional to the narrow dimension when the other dimension and the frequency are fixed, and may vary from approximately zero to 465 ohms.

A waveguide cannot be terminated with a fixed resistor, as can a coax. There are several special devices for termination. One is to fill the end of the waveguide with graphited sand, as illustrated in Figure 8-35A. As the fields enter the sand, currents flow, creating heat, dissipating the energy. None of the energy thus dissipated as heat is reflected back into the guide. Such devices are used as "dummy loads," to absorb the power from a standby transmitter. In high-power radar systems, the dummy loads may even be cooled with a circulating water/ethylene-glycol mixture. It is noteworthy that the heat from the load generates noise at the receiver input and can increase (degradate) the measured minimum discernible signal.

Another arrangement, Figure 8-35B, uses a high-resistance rod, which is placed at the center of the e-field. The e-field creates current flow through the high-resistance rod, which dissipates the energy as an I^2R loss. Another method is to use a wedge of high-resistance material as in Figure 8-35C. The plane of the wedge is perpendicular to the h-lines. When the h-lines cut the wedge, a voltage is induced, and the produced current through the high resistance of the wedge produces an I^2R loss. Again, the loss is dissipated in the form of heat, and very little energy reaches the closed end to be reflected.

Each of the preceding terminations is intended to match the guide impedance to ensure a minimum reflection. There are designs where the objective is to reflect all the energy from the end of the waveguide. A common way to do this is to permanently weld a metal plate at the end of the waveguide as shown in Figure 8-36A.

When it is necessary that the end be removable, a plate is attached to the end of the guide. The contact between the guide and end plate must be perfect, so the h-field will not be attenuated when current flows.

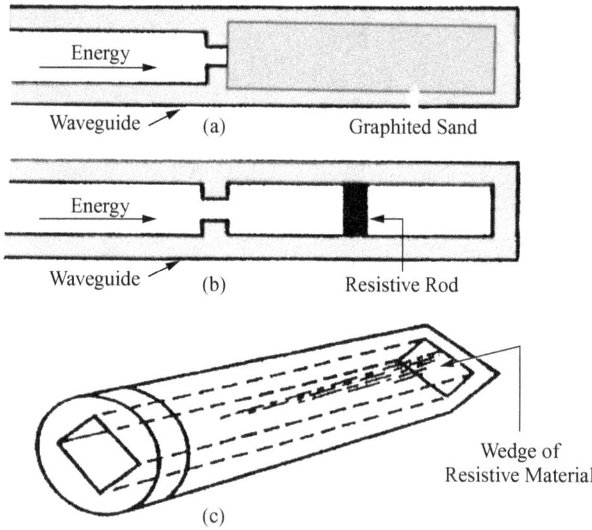

FIGURE 8–35

Waveguide terminations. From USAF Manual 52-8.

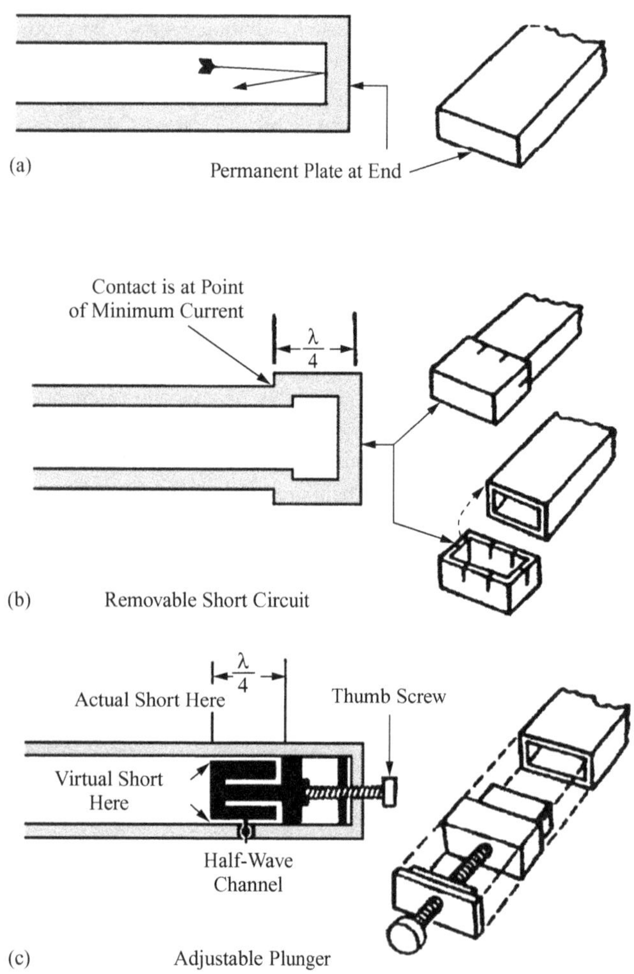

FIGURE 8–36

Reflecting terminations. From USAF Manual 52-8.

Such a perfect contact is unnecessary when the connection is made at a minimum-current point, located at λ/4 from the end. A cup is used instead of the end plate, as illustrated in Figure 8-36B. This cup is λ/4 long and large enough to fit over the end of the guide. The voltage between the opposite sides of the cup opening is high, but as the reflected h-field cancels the incident h-field, the resulting current is small, and reflection is minimum.

When the end is necessary to be adjustable, the contact must be nearly perfect, but that is impossible. An arrangement similar to the choke joint consists of an adjustable plunger, which fits into the guide as shown in Figure 8-36C. The guide walls and plunger form a λ/2 channel. The λ/2 channel is closed at the end and reflects a short circuit across the other end, where a perfect connection is supposed to exist between the wall and the plunger. The actual physical contact is made at λ/4 from the short circuit, where the current is minimum, due to the standing waves. This makes it possible for the plunger to slide loosely in the guide at a point where the contact resistance to current flow is very low.

Cavity Resonators

Conventional low-frequency resonant circuits are based on coils and condensors, connected either in series or in parallel (see Figure 8-37A). To increase the resonant frequency, it is necessary to decrease the capacity, the inductance, or both. As the frequency is increased to ever-higher values, a point is reached where the inductance is a half-turn coil, and the only capacity is that of the stray capacity in the coil. At extremely high frequencies, this resonant circuit would consist of a coil about an inch long and a quarter-inch across. In this circuit, the current-handling capacity and breakdown voltage would be low.

The current-carrying ability of a resonant circuit may be increased by half-turn loops in parallel, without appreciably changing the frequency. It adds capacity in parallel to lower the frequency and inductance in parallel to equally increase the frequency; the frequency remains about the same.

In Figure 8-37C, several half-turn loops are added in parallel. In Figure 8-37D, several parallel λ/4 Lecher Lines are shown in parallel; these are resonant when they are near λ/4. When more and more loops are added in parallel, the assembly becomes the closed resonant box shown in Figure 8-36E. The box is λ/4 in radius, or λ/2 in diameter, and is called a ***resonant cavity.***

A resonant cavity exhibits the same characteristics as a tuned coil-capacitor tank circuit. In the resonant cavity, there are a large number of current paths. The resistance of the cavity to current flow is very low, and the Q of the resonant circuit is very high. While it is difficult to attain a Q of several hundred in a coil of wire, it is fairly easy to construct a resonant cavity with a Q of many thousand. A cavity is as efficient at low frequencies as at higher frequencies, but the large size required at low frequencies prohibits its use at those frequencies. For example, at 1 MHz, a resonant cavity would be a cylinder about 500 feet in diameter. When the frequency is in the vicinity of 10,000 MHz, the diameter of the cavity is only 0.6 inch. This makes the cavity even smaller than a conventional tuned circuit. Cavities used as resonant circuits begin to appear in equipment operating at 1,000 MHz (1 GHz) or above. At the time this is being rewritten, experimental radar systems in the area of 55,000 MHz (55 GHz) are being built; one intended use of these is in automatic brake control radar for automobiles.

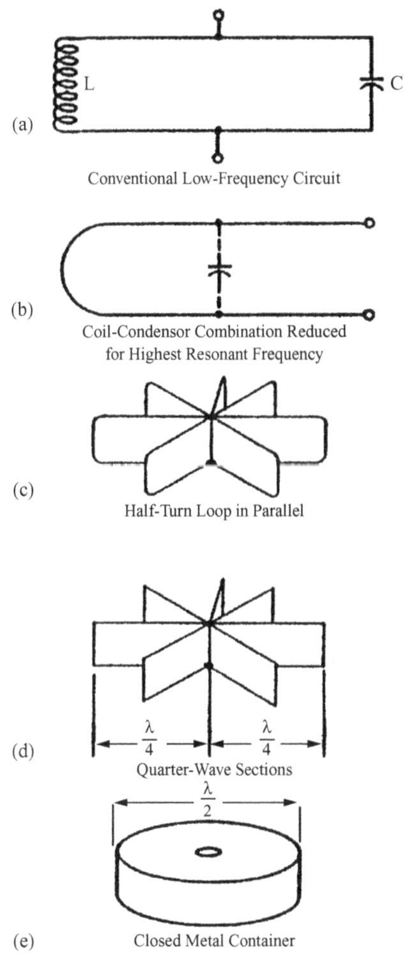

FIGURE 8–37

Developing a resonant cavity from λ/4 sections. From USAF Manual 52-8.

The Fields in a Cavity

A resonant cavity bears considerable similarity to a waveguide, since its operation is best described in terms of fields, rather than voltages and currents. The different field configurations in cavities are also called modes. In Figure 8-38A, the dominant mode of the cylindrical cavity is illustrated. The voltage is represented by e-lines between the top and the bottom of the cavity. The current, due to skin effect, flows in a thin layer on the surface of the cavity. Here, the strength of the current is indicated by graduated arrows. The magnetic field is strong where the current is high, and the strongest h-field is at the vertical walls of the cylinder. The h-field diminishes toward the center, where the current is zero. This is attributable to the standing waves on the λ/4 section. The e-field is maximum at the center and decreases to zero at the edge, where the vertical wall is a short circuit to the voltage. The curves of e-field and h-field density are shown in Figures 8-38B–D to show two types of cavities and their fields.

The modes in a cavity are identified by a numbering system similar to that for waveguides; it differs, in that a third subscript is used to indicate the number of patterns of the transverse field along the axis of the cavity (perpendicular to the transverse field).

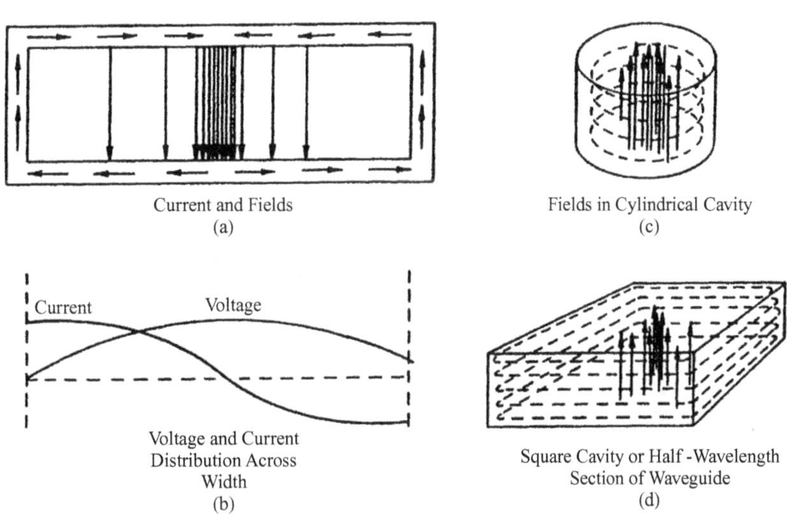

FIGURE 8–38

Voltages, currents, and fields in a cavity. From USAF Manual 52-8.

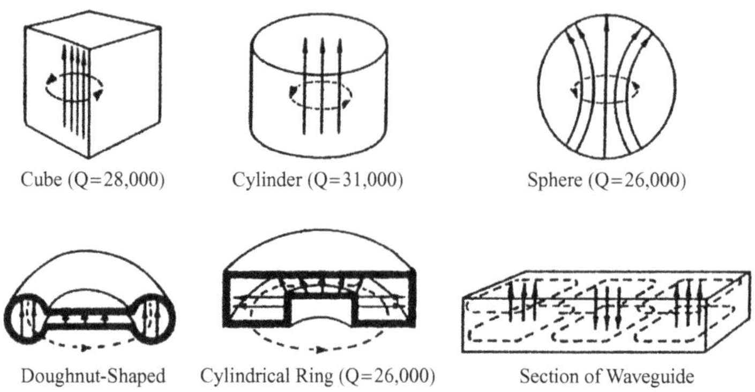

FIGURE 8–39

Some cavity types. From USAF Manual 52-8.

For example, the cylindrical cavity in Figure 8-38C is a form of circular waveguide. The axis is the center of the circle. The transverse field is the magnetic field. Therefore, it is marked TM. Around the circumference is a constant magnetic field. The h-lines are parallel to the circumference, and the first subscript is 0. The distance across the diameter is $\lambda/2$, so the second subscript is 1. Through the center, along the axis, the h-field strength is a constant 0; this makes the third subscript 0. The complete description of the mode is $TM_{0,1,0}$.

When a $\lambda/2$ guide section is closed at both ends to make a rectangular cavity, standing waves are created, and resonance occurs (see Figure 8-38D). The simple mode in this cavity is the same as the dominant rectangular-waveguide mode, $TE_{0,1}$. The third subscript of the mode is determined by the e-field plane, and is 1. The complete description of the illustrated rectangular-cavity simple mode is, therefore, $TE_{0,1,1}$.

Cavities may have a variety of shapes. Any chamber enclosed in conducting walls resonates at several frequencies and produces a number of modes. Figure 8-39 illustrates several types of cavities, indicating the Q of each cavity. Of those shown, the cylinder-type cavity is useful in wavemeters or frequency-measuring devices. The cylindrical-ring-type is used in super-high-frequency oscillators as the frequency-determining element. The diagram of a waveguide section illustrates a device sometimes used as a mixing chamber for combining signals from two sources, as in signal mixers or synthesizers.

Cavity Excitation

The cavity may be excited by a probe, loop, or by electron injection (see Figure 8-40). The current in a probe creates parallel e-lines, and they, in turn, start oscillation. A magnetic loop, placed in the region where the e-field is located, will start oscillation; the currents in the loop start an h-field in the cavity. Either of these two methods can also be used to extract energy from the cavity. A third method, very important to a microwave electron-tube design, uses a cylindrical-ring cavity. The energy is injected into the cavity by clouds of electrons, shot through the holes in the center of a perforated plate. As each cloud goes through, it creates a disturbance in the space inside the cavity until a field is created. The cloud of electrons makes the perforated plate positive, by repelling the electrons away from it, and a current results from the difference in potential. The current creates an h-field. Energy may be removed from the cavity by placing a loop at the outside edge.

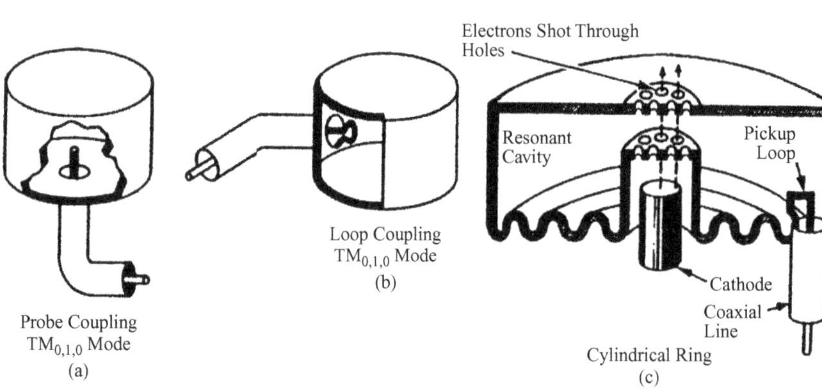

FIGURE 8–40

Cavity excitation methods. From USAF Manual 52-8.

Tuning the Cavity

Figure 8-41 illustrates three methods to adjust the cavity resonant frequency. One method uses a disk in a cylindrical cavity. If the $TE_{0,1,1}$ mode is used, the size of the cylinder may be changed along the axis to adjust the resonant frequency. The frequency increases as the

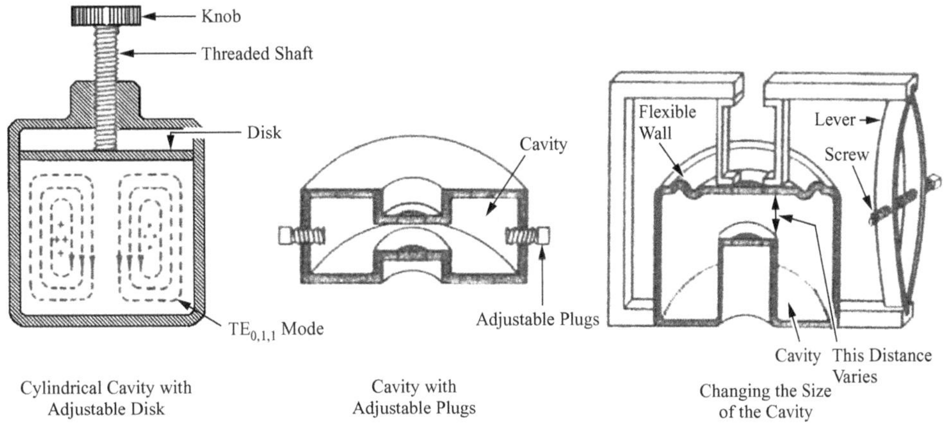

FIGURE 8–41

Cavity tuning methods. From USAF Manual 52-8.

cavity volume is decreased. The movement of the disk may be calibrated in terms of frequency; the use of a micrometer to move the disk is common; a precisely tabulated chart or graph offers the technician a means to convert micrometer depth to frequency.

Another tuning method illustrated in Figure 8-41 employs threaded slugs, inserted in the cavity. The slugs reduce the strength of the magnetic field, which has an effect equivalent to reducing the inductance of a tuned circuit. As the slug is further turned into the cavity, the frequency increases.

A third method of tuning the cavity, illustrated in both Figures 8-41 and 8-42, is used in the ***reflex klystron,*** one of the oldest microwave tubes in radar, and still in use today. The interior of the cavity is part of the interior of a vacuum tube, and is sealed and evacuated. The frequency is changed when the top and bottom of the cavity are moved toward or away from each other. A screw at the end of a lever compresses or expands two external "bows," which move the flexible top of the cavity up or down. As the volume and the ***capacity*** are changed, the frequency changes. Because the change in capacity is the chief result of the tuning, the resonant frequency is inversely proportional to the distance from the top to the bottom; as volume decreases, frequency increases.

Tuning can also be achieved by changing the method of exciting the circuit. This can be done by tuning the exciting loop either capacitively or inductively.

Uses of Cavities

Resonant cavities are used in many ways in microwave equipment. The reflex klystron is illustrated in figure 8-42A. These have been used as local oscillators, as "pump" oscillators in parametric amplifiers, and as transmitter output tubes in microwave link equipment. In the klystron, the cathode emits electrons to pass through the grids toward the plate. When they pass through the cavity area, they disturb the field to excite the cavity. The resonant frequency is changed by the spacing of the grids, which are a part of the top and bottom of the cavity.

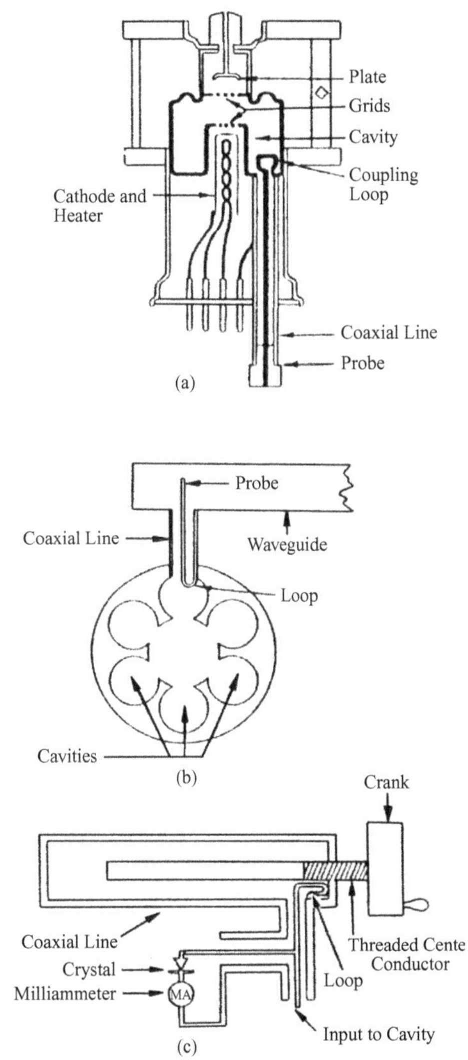

FIGURE 8–42

Cavity applications. From USAF Manual 52-8.

Illustrated in Figure 8-42B is the use of resonant cavities arranged around a circle in the magnetron oscillator. These cavities, coupled to one another through capacity at the openings, are excited by moving electron clouds. The energy from the electrons is passed around the ring from all cavities, to the one with a loop. That cavity serves to transfer energy to the waveguide. From the waveguide, the energy goes to the antenna, where it is radiated into space. Magnetrons are still used in radar transmitters today, even though superior methods have been invented. They are more economical than other methods, and easier to replace than other types. The magnetron has become extremely popular as the main and central device in microwave ovens.

Since the frequency of the transmitter is fixed, there are no frequency-adjusting devices on a waveguide. However, the different speeds at which the electrons travel and the variable impedances in the waveguide may cause either a change in frequency, called ***frequency pulling,*** or oscillations in a mode which differs from the one intended.

Cavities may also be used as wavemeters, devices for measuring frequency. One type of wavemeter consists of a cylindrical cavity with an adjustable disk. Another type operates essentially like the first one but employs a coaxial cavity and operates in a different mode as shown in Figure 8-42C. In the second type, the signal is introduced into the cavity by a loop, which is located at one end of the cavity. The signal is strongest when the loop is at a high current point in the standing wave. The standing wave of the current is maximum at odd multiples of $\lambda/4$ distances from the open end of the guide. When the length of the threaded center conductor is varied by the crank, the distance from the open end to the loop can be made equal to an odd multiple of $\lambda/4$. This causes the current maximum to occur at the loop, which, in turn, causes the input impedance at the loop to be zero. When the distance from the open end of the guide to the short at the closed end is $\lambda/4$, or any odd multiple of $\lambda/4$, the current introduced into the guide will be at the location of parallel resonance.

The threaded shaft illustrated in Figure 8-42C, can be calibrated either in wavelength or in frequency. The input impedance can be indicated by any suitable indicator, such as a crystal rectifier and a dc milli-Ammeter. In operation, when the input impedance is zero, the signal is shorted out. In this case, the meter will read zero. To check the frequency for a zero impedance reading, turn the crank until a dip occurs on the meter and read the frequency on the threaded shaft.

Another use of a cavity is as an impedance matching device, such as the illustration of a small guide connected to a larger one in Figure 8-43. Normally, when two waveguides are connected together, there will be reflections. However, a small section of waveguide can be matched through an intermediate-dimension cavity to a larger section, without reflections. This is provided that plates are used at the junction to cancel the reactive effects resulting from the different dimensions. Thus, although the cavity itself has standing waves in it, there are none on either waveguide.

Another use of a cavity is as a ***ringing circuit,*** which oscillates for a considerable time duration after being started by an external circuit. For example, it may continue to ring as long as 50 μs after receiving a 1-μs burst, similar to a bell, after having been struck.

Figure 8-44 illustrates an "echo box," one of the oldest and most useful of all radar test equipment. The echo-box ringing, displayed on an oscilloscope as detected normal radar video, is illustrated in Figure 8-45. By today's standards, the echo box illustrated is a true antique, but the basic principle is still the same. The echo box illustrated was used in World War II, X-band (3 cm λ) airborne radar systems, and featured an antenna, so that it could be used outside the aircraft without the necessity of cable connection, as illustrated in Figure 8-46.

The echo-box cavity in Figures 8-44 and 8-46 was a 3-inch-diameter, 10-inch-long cylinder. A movable piston in the cylinder determines the volume of the cavity. The piston position was calibrated in MC (now MHz), and it served as a frequency meter. For many years, precision echo boxes were the main tool used to set radar frequency. The cavity is energized by the antenna input through a small aperture. Once excited, the continued ringing is radiated by the antenna, back into the radar system as a simulated large echo.

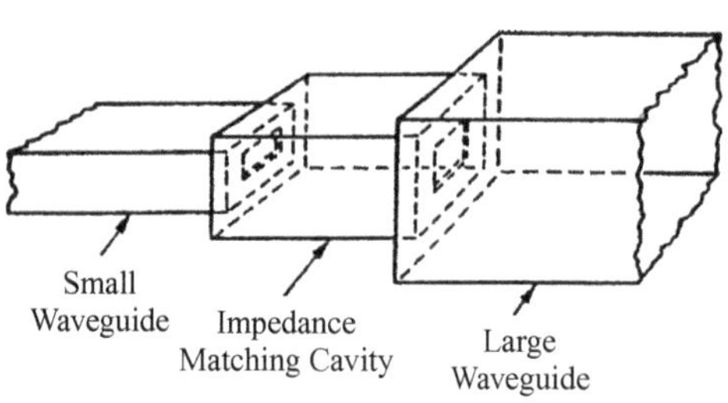

FIGURE 8–43

Z-Matching cavities. From USAF Manual 52-8.

Figure 8-47 illustrates the waveguide system from a 1940's airborne radar. It is useful as an introduction because of the simplicity and absence of many latter-day refinements. The transmitter is a magnetron; in those days, that was the only microwave transmitter available. The TR and ATR devices, each at $\lambda_g/4$ from the center of the main guide, comprise the

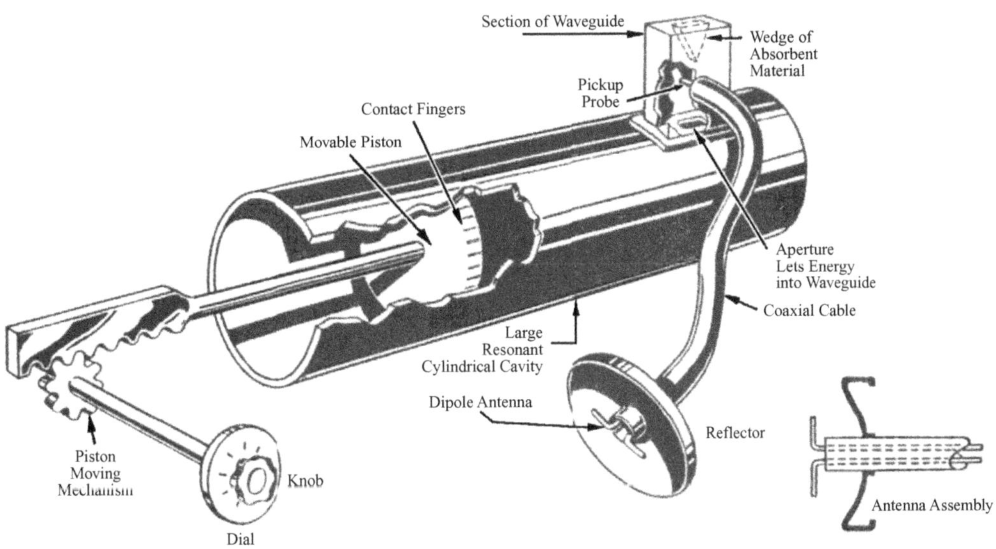

FIGURE 8–44

An early-day echo box. From USAF Manual 52-8.

duplexer. The TR tube, energized during transmit time to protect the receiver, was a gas-filled device, energized to a short circuit by the high-power transmitter burst. A "keep-alive" voltage of about –750 V allowed the TR device to energize easily. Newer TR devices do not require the keep-alive voltage.

The test probes shown were simply direct waveguide probes. By the early 1950s, directional couplers were introduced to provide precisely calibrated waveguide measurement ports. The antenna employed two rotary joints, one for tilt and one for azimuth. The Cutler feed, a cavity which reflected energy through slot radiators to the antenna reflector, is now obsolete, and feedhorns are usually a simple, tapered, expanding, waveguide to match the waveguide impedance to the impedance of free space.

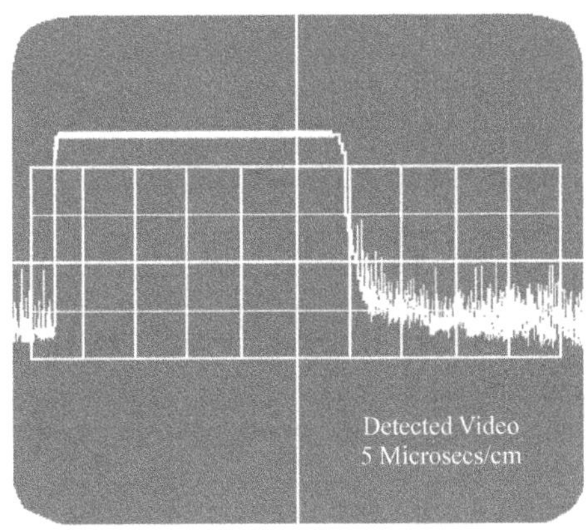

FIGURE 8–45

Echo box ringtime.

Received echoes enter the main waveguide from the antenna; the signals propagate toward the magnetron but are stopped by the ATR device. The length of the T-junction is $\lambda/2$ from the center of the guide, so the closed end of the section reflects a short circuit at the outer center of the main waveguide. This reflects the signal back along the guide to the mixing chamber. During the "listening time" neither T/R or AT/R can be fired by the weak echoes. After the transmitter burst, there is a brief period of time required for the T/R device to recover;

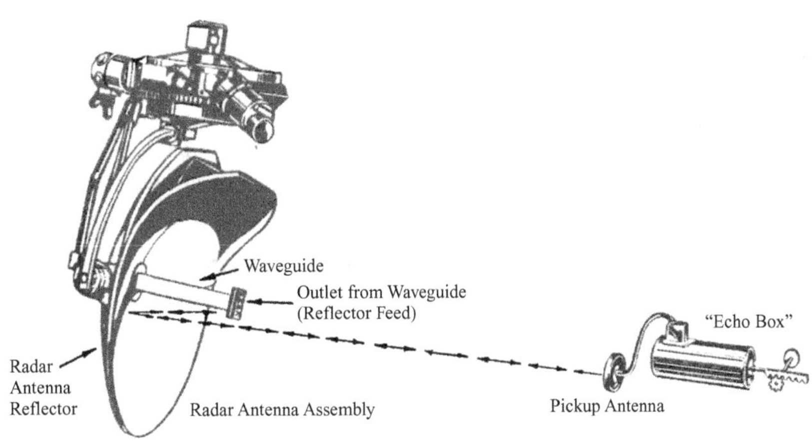

FIGURE 8–46

Echo box and aircraft radar. From USAF Manual 52-8.

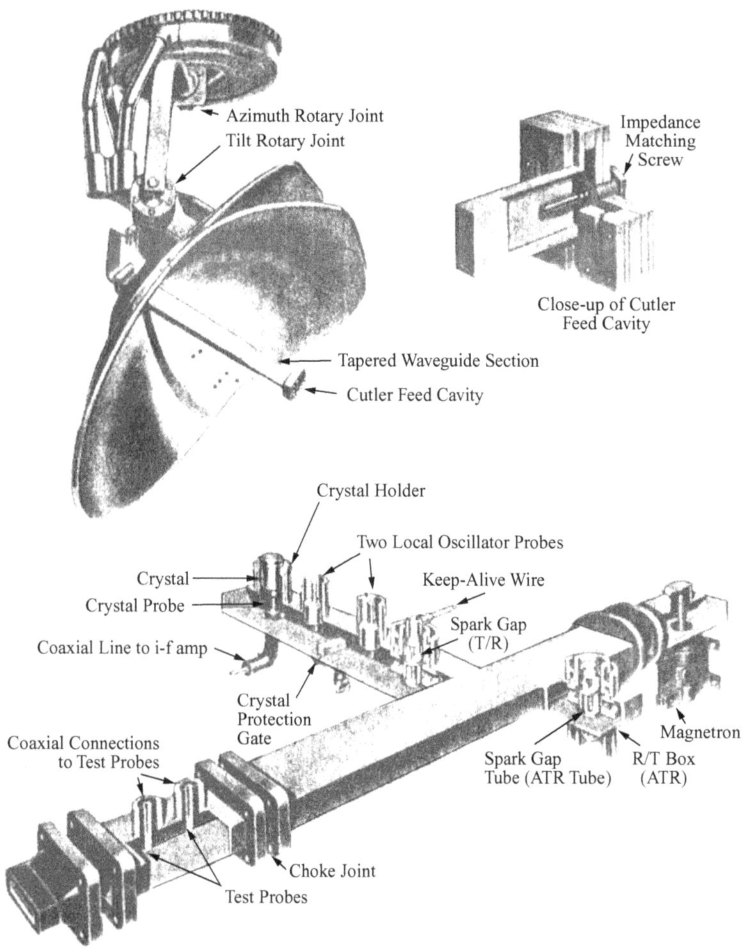

FIGURE 8–47

Early airborne radar waveguide system. From USAF Manual 52-8.

in modern FAA ASR radars, for example, this may be roughly 5–10 μs, depending on the radar model. The widely accepted definition of this ***T/R recovery time*** is the measured time from the leading edge of the transmitter burst at which the incoming echoes are attenuated by 3 dB from maximum. The procedure is described in detail in another chapter of this book.

Echoes entering the mixing chamber strike the end wall, and reflections again occur. The chamber becomes a cavity resonator. The local oscillator and signal inputs both cause current flow through a pick-up probe at $\lambda/4$ from the end of the cavity. That pickup probe is connected to a crystal mixer. The local oscillator operates near the transmitter frequency, but differs by the amount of the desired intermediate frequency (i-f). Although frequencies such as 60 MHz have been employed, the most common i-f has been 30 MHz for many years. There are now several systems that use 31.07 MHz; the new value is related to digital processing techniques. The particular system illustrated shows two local oscillators and two signal mixers; this was a special arrangement in which a different i-f would be produced by a radar beacon transmitter (at X-band, not at 1,030 or 1,090 MHz). This arrangement may still be found in military airborne radars.

The crystal protection gate is a relay-operated shutter to close the waveguide path to the mixers when the transmitter high voltage has been turned off; strong signals, as from an adjacent transmitter, would not energize the T/R device when the keep-alive voltage was off. The FAA ASR-4, -5, and -6 contained such a waveguide shutter. These are not entirely obsolete, but are becoming rare, since the receiver mixer crystal in modern radars is not as directly exposed to incoming signals, and because the newer T/R devices remain "ready" in the absence of high voltage.

The entire waveguide system is assembled with gaskets at the joints to permit pressurization. Where it is necessary to close a section of guide for pressurization, mica windows are used. These do not interfere with rf propagation.

Waveguide Test Equipment

The Early Frequency Meter

A resonant cavity can be used to measure the frequency of the signal in a waveguide. Early echo boxes were not sufficiently accurate for this, and even latter-day models are not as accurate as a frequency meter. Today, the echo box, spectrum analyzers, frequency counters, and absorption-type wavemeters are all used to measure frequency; accuracy varies with test equipment type.

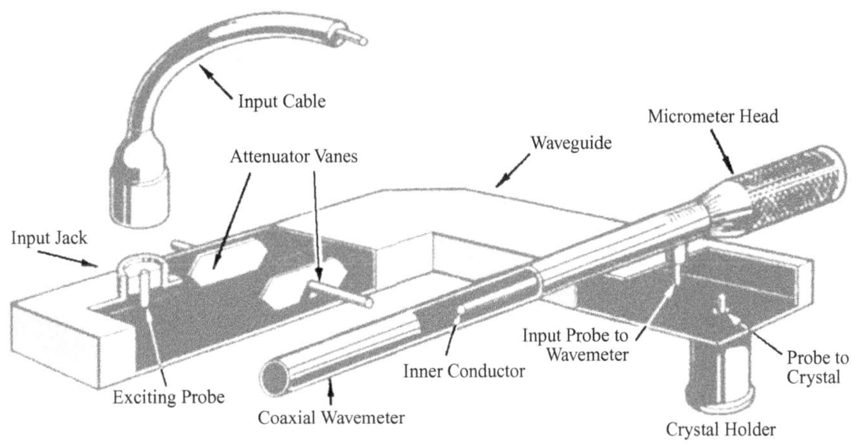

FIGURE 8–48

An early frequency meter. From USAF Manual 52-8.

Figure 8-48 illustrates an early-day frequency meter. There are still operating radars containing devices similar to this. A coaxial cable routes the rf signal to be tested to the wavemeter input connector. A probe sends the signal down the guide through attenuator vanes. When these vanes are at the walls of guide, they have no effect, but when they are moved together, they reduce the wide dimension, until it is less than cutoff λ for the input frequency. The signal bounces back and forth between the vanes, with very little of it getting out of the waveguide proper. It then goes to a coaxial cavity. The cavity is similar to that illustrated in Figure 8-41, but it has a micrometer head to ensure precision. The coaxial line is resonant when it is 15 quarter-wavelengths long. Resonance is indicated by the rectified output from the crystal, which may be applied to a milli-Ammeter. If the input is pulsed, the detected pulse can be viewed on an oscilloscope. When the pulse or milli-Ammeter is maximum, the micrometer is tuned to the input signal, and the micrometer reading is applied to a calibration chart to determine the frequency.

The Slotted Line

Slotted lines are available from test-equipment manufacturers and are regularly used in microwave laboratories (see Figure 8-49). Standing waves on any waveguide system can be measured by inserting a slotted section into the waveguide. A travel probe, usually part of the slotted-line assembly, is used to measure the electric field strength. The travel-probe assembly is a small guide section that fits onto the slotted section, and the probe extends into the slotted waveguide. The energy picked up by the probe is transmitted through the short guide to a crystal, where it is rectified and then carried by a coaxial cable to a meter.

The Directional Coupler

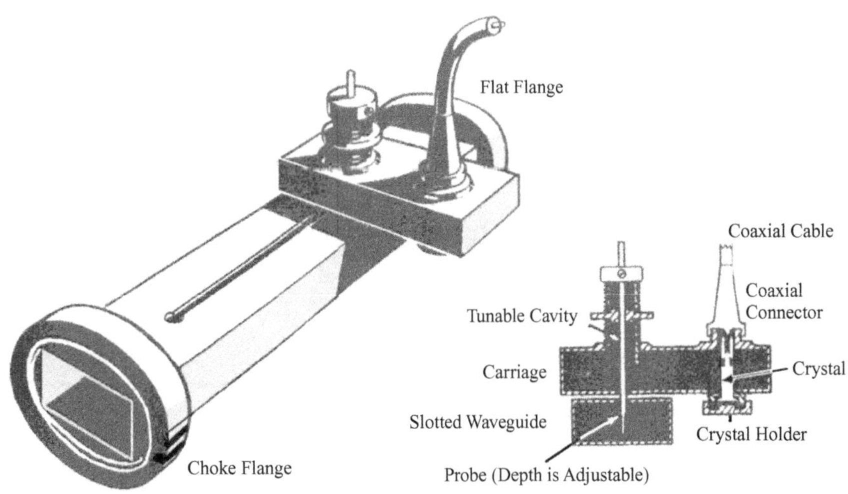

FIGURE 8–49

Slotted line. From USAF Manual 52-8.

The directional coupler is generally a permanent part of the waveguide assembly in a radar system, but nevertheless should be classified as test equipment (see Figure 8-50). It is provided as a test fixture for an isolated and attenuated connection of test equipment to the waveguide. The coupler is affixed to the narrow dimension of the guide, and several

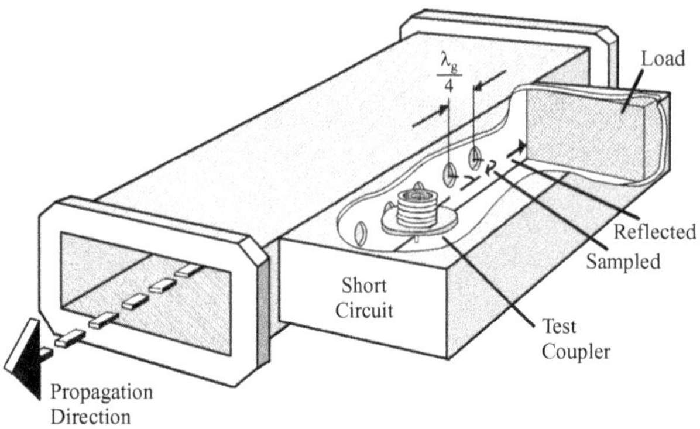

FIGURE 8–50

The directional coupler.

ports, spaced by $\lambda_g/4$ in the waveguide wall, permit energy to propagate into the coupler chamber, and continue in the same direction toward the shorted end; the short then reflects the energy back toward the load, where it is absorbed. The size of the holes and the depth of the test coupler probe cause a precise attenuation at the test coupler. Because of reciprocity, the attenuation will equally affect either (1) an injected test signal or (2) the power of a signal in the waveguide.

Affixed to the coupler assembly is a tag to indicate the precise attenuation at the test port. Because of the hole spacing, the coupler is somewhat frequency sensitive, and the tag will contain a chart or graph, so the technician may precisely determine or interpolate the attenuation for the frequency under test.

The directional coupler may have test chambers on both sides of the waveguide, reversed in respect to one another; such a coupler is called a ***bidirectional coupler*** *(*see Figure 8-51*)*. One of these will receive energy propagated from the transmitter toward the antenna, or it will permit injection of a test signal from the test port back toward the receiver. The other will receive energy that has been reflected from the antenna back toward the transmitter, or it will permit injection of a test signal to be propagated toward the antenna. They may be labeled, respectively, FORWARD and REVERSE, or INCIDENT and REFLECTED. Often, a radar system will not have a bidirectional coupler, but instead, two directional couplers in series. Because the REVERSE coupler is used for the measurement of reflections, and because these reflections will be of very low power in respect to the transmitted power, the REVERSE coupler will exhibit considerably less attenuation than the FORWARD coupler.

The directional coupler, particularly the INCIDENT coupler, is the most frequently used test connector in the radar system. The two couplers are used for the following purposes:

1. ***Measurement and certification of transmitter power.*** FCC regulations regarding power express interest in the amount of power at the transmitter output, without respect to antenna gain. The INCIDENT connector is the accepted port for measurement. In measuring power,

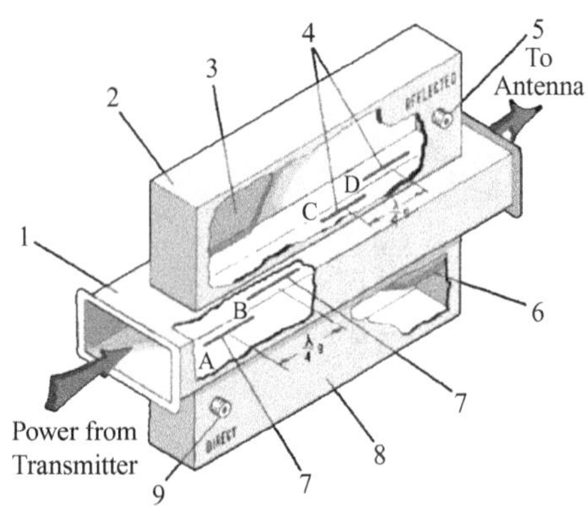

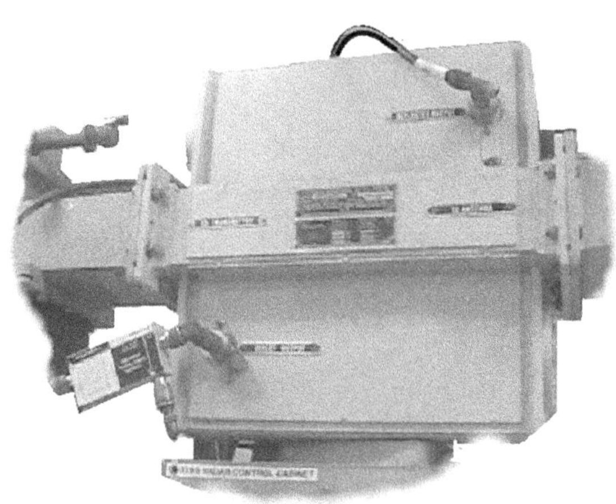

1. Main Waveguide
2. Reflected Energy Waveguide
3. Power Dissipating Load
4. Reflected Energy Coupling Slots
5. Reflected Energy Output Jack
6. Power Dissipating Load
7. Incident Energy Coupling Slots
8. Incident Energy Waveguide
9. Incident Energy Output Jack

FIGURE 8–51

The bidirectional coupler used in the FPS-20 series ARSRs.

use as much added attenuation between the power meter and coupler as possible, as it will increase the accuracy of the reading. Any mismatches will cause reflections, and added attenuation will rapidly reduce the reflections as the energy travels back and forth.

2. ***Measurement of minimum discernible signal.*** An rf signal generator at the transmitter frequency is modulated by a pulse, precisely equal in width to the transmitter burst, is injected at the INCIDENT directional coupler. In the signal generator, a 0-dBm peak burst is applied to a precision variable attenuator, and the generator is connected to the directional coupler through a cable of known attenuation to the test frequency. The test signal is viewed on an oscilloscope, and reduced until it is barely visible; the signal ***in the waveguide*** is the minimum discernible signal. The actual level will be 0 dBm minus the generator attenuator setting, minus the cable attenuation, minus the directional coupler attenuation. The cable can be calibrated by (1) directly measuring power at some convenient point, and at the transmitter frequency, (2) again measuring the power with the cable between the source and power meter, and (3) calculating the difference between (1) and (2).
3. ***Inspection of the transmitter pulse shape with a video detector.*** Care must be taken to ascertain that adequate attenuation is used to protect the detector and prevent saturation. The detector must also be properly terminated, or the pulse will charge the junction, it will be unable to rapidly discharge, and the trail edge of the pulse viewed on the oscilloscope will be stretched.
4. ***Inspection of the transmitter frequency spectrum with a spectrum analyzer.*** In tuning klystron systems, this is absolutely essential. Inspection of magnetron systems is also necessary, but there is usually little that can be done about poor spectra, other than replacing the tube.
5. ***Connection of the echo box for any of the following procedures:***

NOTE:
The directional coupler attenuation must be low enough to accommodate an echo box. If the echo box has not been specified as part of the test equipment for the system, it is possible that this will not be the case, and that the sampled power will be inadequate to usefully excite the echo box.

(a) ***Ringtime measurement.*** If within specifications, this is a fast, albeit rough, indication that (1) the transmitter power is adequate, (2) the stalo is tuned to the proper frequency, and (3) the receiver is operating satisfactorily. Unless otherwise specified, the ringtime is measured from the lead edge of the transmitter burst, called ***main bang*** (WWII slang), to the point where the ringing decays to grass level (see Figure 8-52).

(b) ***Pulse width measurement.*** The meter on the echo box is indicative of the tuning; when it is peaked, the echo box is tuned to the transmitter frequency. When the echo box is tuned away from the peak to the first minimum reading, it is tuned to a null on one side of the main lobe. To determine the pulse width:

(1) Tune to the peak.

(2) Tune to the null on one side, and note the frequency indicated on the dial.

(3) Tune to the null on the other side, and note the frequency.

(4) Calculate the frequency difference between the nulls.

(5) The frequency difference is $2/t_p$. The reciprocal of half the difference is the pulse width.

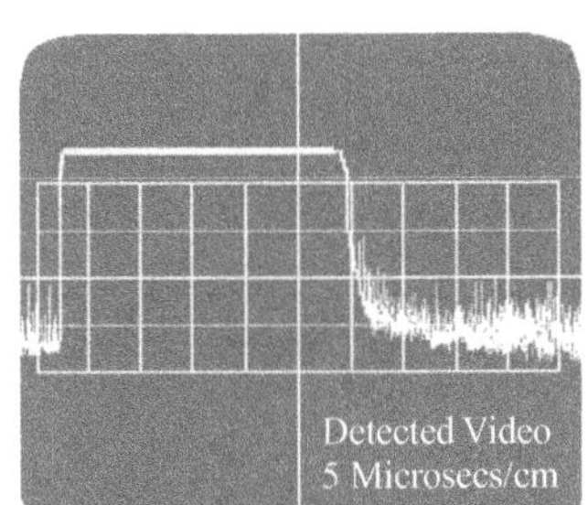

FIGURE 8–52
Echo box ringtime on normal video.

(c) ***Precision indication of stalo tuning in magnetron Mti dystems.*** See Figure 8-51. The echo box is very sharply tuned, and rings back only a small portion of the spectrum. If (1) the echo box is tuned to peak, (2) the stalo is

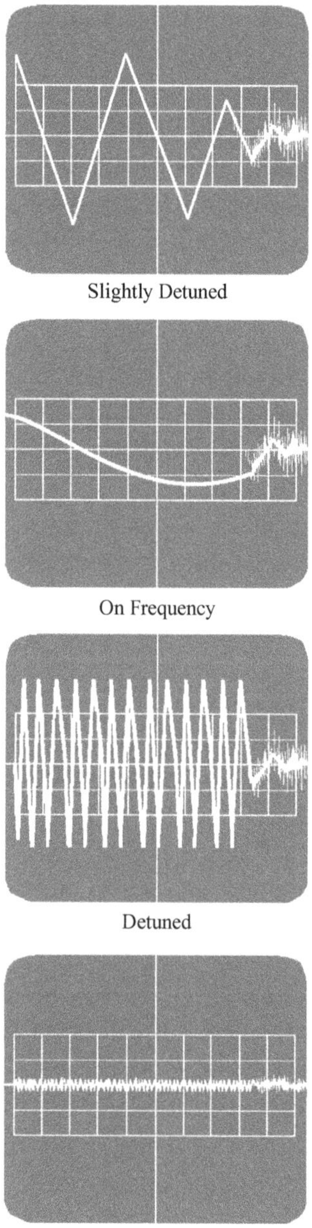

FIGURE 8–53

Echo box ringtime at the phase detector output.

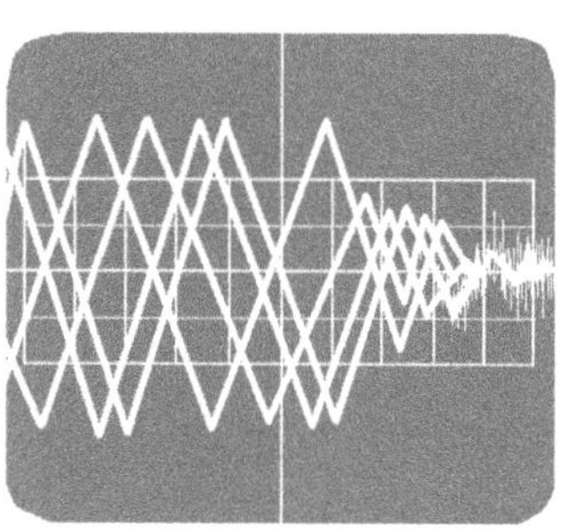

FIGURE 8–54

Echo box ringing, incoherent mti.

precisely on frequency, and (3) the coho is precisely on frequency, precisely 30 MHz will "ring down" through the mti i-f amplifier to the phase detector. If (1) the echo box signal is precisely 30 MHz, and if (2) the coho is precisely 30 MHz, the phase difference between the echo box signal and coho will not change during the entire ringtime, and the output of the phase detector will remain constant throughout the ringtime. If the stalo is off frequency, the i-f will not be precisely 30 MHz, and the phase detector output will vary as the coho and i-f phases differ at the rate of the frequency difference.

(d) ***Phase detector amplitude and balance inspection.*** With the echo box slightly detuned, the excursions may be inspected for equal positive and negative values, and for proper amplitude (see Figure 8-53).

(e) ***System frequency stability and coherence inspection.*** Figure 8-54 illustrates the appearance of the phase detector output in an incoherent mti system.

6. ***VSWR measurements.*** By measuring the power in the waveguide as sampled at the output of both the INCIDENT and REFLECTED couplers, the levels, in Watts, may be entered into the equation to calculate the VSWR:

$$\text{VSWR} = \frac{1+\sqrt{\dfrac{P_R}{P_I}}}{1-\sqrt{\dfrac{P_R}{P_I}}}.$$

Many organizations will provide charts to interpolate the VSWR. The use of these charts is emphatically discouraged by the author, as it will eventually destroy the technician's competence to perform the calculation.

In observing the reflected power of an on line transmitter while the antenna is rotating, note the power variation; this is called ***WOW,*** a term taken from audio-recording technology. Excessive WOW is an indication of a defective rotary joint; the acceptable amount of WOW may be specified in manufacturer's or other engineering data pertaining to the specific rotary joint.

7. ***Automated measurements.*** In newer radar systems, provisions are being made to automate many of the measurements described in this chapter. ***Hardline,*** a miniature, metal, rigid coax, is permanently connected to directional couplers, and the hardline is connected to test circuitry in the radar equipment. For only one of many examples, in the ASR-9, a calibrated test pulse of −106 dBm (in the waveguide) at a programmed Doppler is injected in deadtime. The test pulse output from a point in the mtd system is applied to a calibrated threshold circuit; the rate of threshold break determines the computer-reported mds. For instance, 50% indicates an mds of −106 dBm, and a higher percentage indicates a lower (better) mds. Additionally, the level of a calibrated pulse 5 μs after the transmitter is indicative of TR recovery time.

Review Questions

1. What, if anything, will occur when the wide dimension of the waveguide is less than $\lambda/2$ of the applied frequency?
2. The electrostatic field cannot exist when ________________________.
3. The wavefront crossing angle becomes _______ when the cutoff frequency is reached.
4. The starting phase of the transmitter pulse changes between the final power amplifier and antenna feedhorn. Why?
5. A waveguide is bent in the narrow dimension, across the wide dimension. This is an ___-type bend.
6. The most efficient way to connect sections of waveguide is with a __________.
7. A rotary joint must be used with ________________.
8. A tee-joint in the narrow dimension is an _____-type tee.
9. Two tee-joints of both types combined into one is called a _______________.
10. A plate which narrows the waveguide wide dimension is (capacitive, inductive).
11. A plate which obstructs both dimensions is ________________________.
12. An echo box is ________________________________.
13. A slotted line is used ________________________________.
14. Once in the chamber of the directional coupler, rf continues to propagate (1) toward the load, or (2) in the same direction.
15. A technician has applied a calibrated test signal from an rf signal generator to the INCIDENT power connector on the bidirectional coupler; the generator attenuator has been adjusted to place the signal at minimum discernible. The chart on the directional coupler indicates –55 dBm at the transmitter frequency, the test cable is 2.4 dB, and the dial on the signal generator is 50 dB. What is the mds?
16. The INCIDENT power is 55 dBm, and the reflected power is 17 dBm. What is the VSWR?

Answers to Review Questions

1. What, if anything, will occur when the wide dimension of the waveguide is less than $\lambda/2$ of the applied frequency? ***Propagation will stop, because this is cutoff frequency.***
2. The electrostatic field cannot exist when ***the electrostatic lines of force are tangent to the waveguide walls.***
3. The wavefront crossing angle becomes ***zero*** when the cutoff frequency is reached.
4. The starting phase of the transmitter pulse changes between the final power amplifier and antenna feedhorn. Why? ***The phase and group velocities differ.***
5. A waveguide is bent in the narrow dimension, across the wide dimension. This is an ***E-type*** bend.
6. The most efficient way to connect sections of waveguide is with a ***choke joint.***
7. A rotary joint must be used with a ***physical-motion antenna.***
8. A tee-joint in the narrow dimension is an ***H-type tee.***
9. Two tee-joints of both types combined into one is called a ***magic tee.***
10. A plate which narrows the waveguide wide dimension is ***inductive.***
11. A plate which obstructs both dimensions is ***a resonant circuit.***
12. An echo box is a ***precision tuned cavity.***
13. A slotted line is used ***to locate and measure field strengths in a waveguide system.***
14. Once in the chamber of the directional coupler, rf continues to propagate ***in the same direction.***
15. A technician has applied a calibrated test signal from an rf signal generator to the INCIDENT power connector on the bi-directional coupler; the generator attenuator has been adjusted to place the signal at minimum discernible. The chart on the directional coupler indicates –55 dBm at the transmitter frequency, the test cable is 2.4 dB, and the dial on the signal generator is 50 dB. What is the mds? **0 dBm – 2.4 dB – 55 dB – 50 dB = –107.4 dBm**.
16. The INCIDENT power is 55 dBm, and the reflected power is 33 dBm. What is the VSWR?

$$\text{VSWR} = \frac{1+\sqrt{\frac{P_R}{P_I}}}{1-\sqrt{\frac{P_R}{P_I}}}$$

55 dBm = 316 Watts 33 dBm = 2.0 Watts $P_R/P_I = 0.08$, VSWR = 1.17.

CHAPTER 9

Radar Synchronizers

Introduction

Synchronizers may differ substantially in systems, each producing the unique set of triggers and gates for the radar's purpose and design, and by the hardware available at the time of manufacture. Some fundamental design concepts applicable to many synchronizers will be discussed in this chapter, and one synchronizer necessary for the ASR-9 MTD system will be introduced to illustrate a more complex application. Figure 9-1 is a simple diagram of some basic timing signals required of an early ASR radar. Triggers and gates for use throughout the radar system may be produced by the synchronizer, or by other units which rely upon the synchronizer for timing triggers. Among the various timing signals are the transmitter triggers, the coho gate in magnetron systems, stc triggers, display triggers, live-time gates to enable the receiver outputs for 741 μs (for 60-mile ASR radars) or 2,471 μs (for 200-mile ARSR radars), deadtime triggers to initiate ***built-in-test (BIT)*** functions, and many others. Satisfactory operation of the MTI system places stringent requirements on the synchronizer design and stability that would not otherwise be necessary. In sophisticated tracking and range/elevation/azimuth three-dimensional systems, timing becomes so complex, and subjected to so many variables, that only a computer with several interrupts and complex subroutines can suffice.

The Master Trigger Blocking Oscillator

A blocking oscillator is a vacuum-tube circuit characterized by a transformer between the plate and positive voltage supply. It is used to produce high-amplitude, high-current triggers, and can run with or without a trigger input. The natural period in an MTI system is substantially greater than the desired period. More information on this circuit can be found in Appendix A. The MTI system depends upon T_r-to-T_r, pulse-to-pulse, comparisons of bipolar video to achieve cancellation of zero- and near-zero-Doppler clutter targets in the comparator circuit of the canceler. To prevent ***residue***, which may be described as "uncanceled portions of stationary targets," it is absolutely essential that the one-T_r-delayed target occur in exact time coincidence with the undelayed target. In analog systems, engineering such a system was no small matter. The amount of delay had to be precisely the same as the T_r, and because of the length of time (about 1,000 μs for an ASR, or 3,000 μs for an ARSR), manually adjusting the master timing blocking oscillator to match the delay to the video was simply impossible, because either the delay time or blocking oscillator period would quickly drift away from the critical adjustment. Still further, a blocking oscillator with such a long period exhibited a considerable amount of jitter.

See Figure 9-2 for an example of a timing error. If a 0.8-μs t_p, 1000-μs T_r, ASR radar canceler delay line changed by 0.05%, there would be a temporal (in time) error of 0.5 μs, and there would only be coincidence between the delayed and undelayed pulse for 0.3 μs. For the remaining 0.5 μs of the 0.8-μs echo pulse, there would be no cancellation; there would be residue resulting from the lead edge of the earlier pulse, and there would be residue from the trail edge of the later pulse. The final result of such an error would be that uncanceled MTI display video would appear to be similar to normal radar video with ftc.

The earliest MTI systems used mercury delay lines for video delay; these were very sensitive to temperature changes, and it was necessary to provide some kind of automatic regulation to the master trigger blocking oscillator period, which was T_r, to maintain cancellation. The circuitry which accomplished this was called an ***automatic temporal control***. Except for discussing the broad principles of the automatic temporal control to illustrate the first approach to this problem, there is no reason to discuss it in great detail, as it is now obsolete.

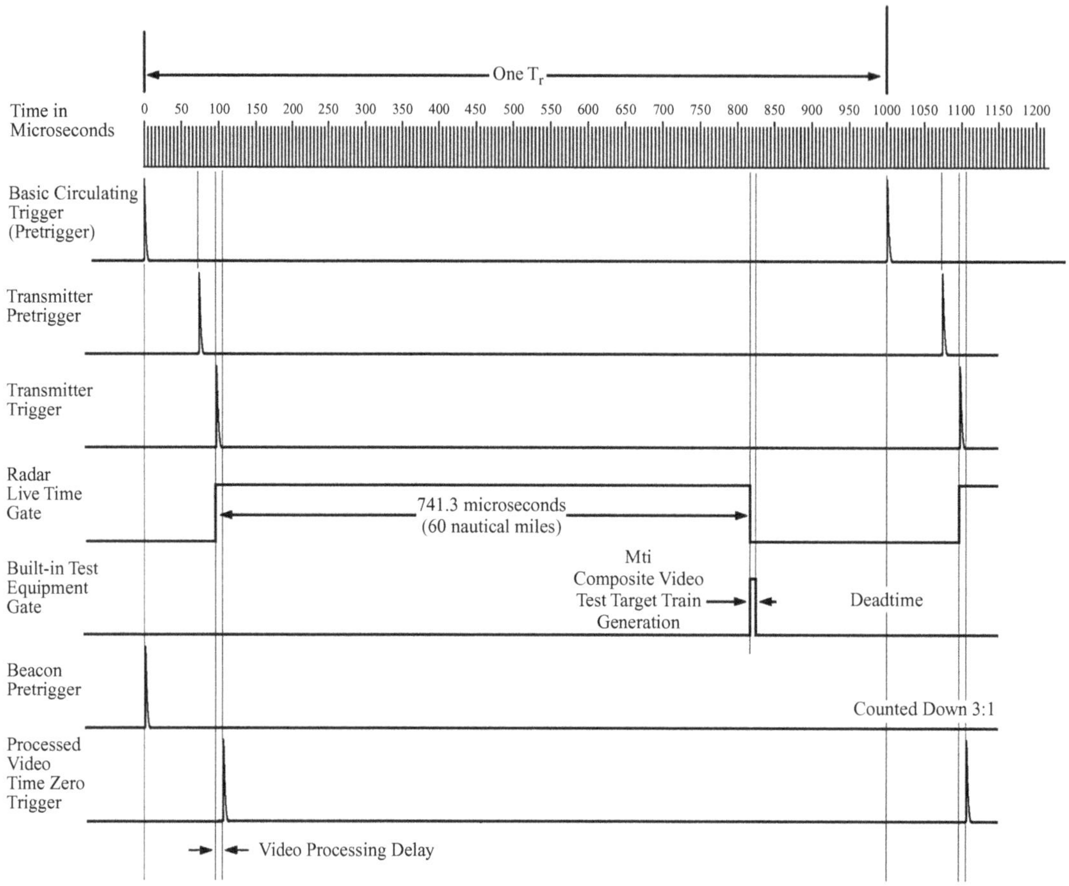

FIGURE 9–1

Typical timing signals in an ASR radar.

The Automatic Temporal Control

Because of their length, and the materials used for their manufacture, MTI delay lines are not capable of satisfactorily delaying pure bipolar video, and require a carrier frequency. Carrier oscillator frequencies of 30 MHz and 20 MHz were commonly used. The carrier would be modulated by bipolar video, applied to the delay line via a driver, and then recovered, amplified, and detected, at the output end of the delay line.

Regulation by Phantastron

In the systems using automatic temporal control circuits, a dc-regulated phantastron gate generator served as the main timing regulator. The gate disabled the carrier during deadtime and was appropriately called the carrier oscillator gate. Since the carrier oscillator gate was delayed by the delay line, the delayed-gate transitions were representative of the amount of delay. Temporal differences between (1) the trail edge of the carrier oscillator gate and (2) the system trigger would be integrated into a correction bias voltage for the phantastron. A fine manual adjustment to the bias, called of course the AUTOMATIC TEMPORAL CONTROL adjustment, provided for precise correction. The phantastron circuit was an exotic milestone design, and more information about it can be found in Appendix A.

The Circulating Trigger Loop

Mercury delay lines were soon replaced with superior quartz lines, and the automatic temporal control was replaced with the ***circulating trigger loop***. This method was used for analog canceler systems in many MTI systems from the late 1950s until the introduction of digital mti in the 1970s. At the time this is being written, there are still old, yet operational, radar systems employing this circuitry. See Figure 9-3, which is an illustration of a single-canceler

MTI system with a circulating trigger loop. The circulating trigger generator is the master trigger generator for the entire system. It is a free-running blocking oscillator with a natural period considerably greater than the T_r. However, under normal operation, it does not run at its normal period, because it is re-triggered by a trigger which has passed through the canceler delay line.

In the rf carrier modulator, bipolar video from the phase detector modulates the 20- or 30-MHz carrier from the carrier oscillator by, typically, 30%. The video-modulated carrier is then applied to the quartz delay line and appears at the output, one T_r later. Because the modulated carrier is severely attenuated by the delay line, it must be applied to a carrier amplifier, and recovered in a manner similar to a received signal. The amplifier resembles an i-f amplifier and contains an amplitude detector at the output. It goes by different names, depending upon the radar manufacturer, and has been called the ***delayed channel amplifier, recovery amplifier, post-delay amplifier***, and others.

There are two outputs from the delayed channel amplifier. One output applies the delayed, usually inverted, bipolar video to the ***comparator***; in these analog systems, the comparator was ordinarily a simple resistive bridge with a balance control. The other output was applied to a ***trigger pickoff unit***, a major functional part of the circulating trigger loop.

When the master trigger blocking oscillator in the circulating trigger generator creates a trigger, the trigger is applied to a separate input of the rf carrier modulator. The trigger causes 100% modulation, so that it appears at a higher amplitude than the 30%-modulated video at the output of the delayed channel amplifier. The trigger pickoff circuitry is a threshold device; the threshold may be broken by the trigger, but not by the video, so that the delayed trigger will be applied to the master trigger blocking oscillator. The delayed trigger is applied to the master trigger blocking oscillator grid, re-triggering it before its natural period has ended. As a result, *the period of the master trigger blocking oscillator is dependent on the delay line time.* For final precise video temporal matching, a small adjustable delay is usually provided in the video line to the comparator.

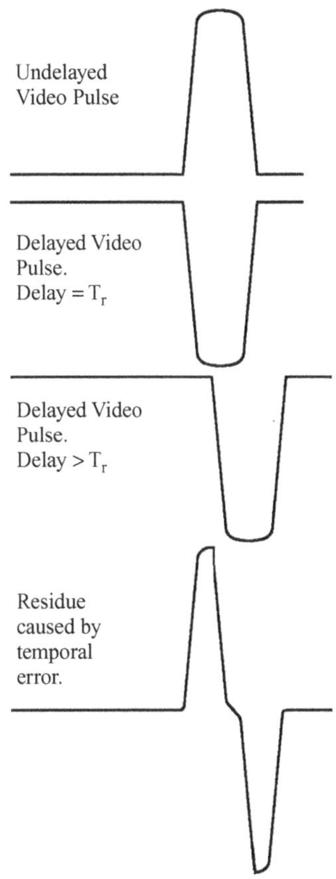

FIGURE 9–2

MTI comparator temporal error.

The circulating trigger generator provides an output that is routed out of the canceler to the synchronizer, which, in turn, develops and distributes triggers and gates to the rest of the radar system. Among these are the transmitter trigger, stc trigger, coho gate, and many others. In many systems, the synchronizer outputs may only be triggers, and the gates are developed by flip-flops in the units that require the gates.

Digital MTI and Elimination of the Delay Line

One of the greatest advantages gained by digital MTI was the elimination of the need for a delay line. Instead of the delay line, the first digital MTI used a many-stage shift register (1,664 stages long × 10 parallel stages wide in the ASR-8) for delay, and the time delay is dependent only on the shift clock period and number of clocks applied to the shift register. Gone was the need for an rf carrier, modulator, delay line, delayed channel amplifier, and all the alignment required of that analog circuitry. A latter-day development, before MTI began to be replaced by Doppler-filter MTD, employed a larger memory that could be written over multiple intervals, and the data from each range cell in all those intervals could be compared by the application of a single arithmetic operation. Such a technique was used in a ***solid-state-receiver and digital-MTI (SSR/DMTI)*** modification to ARSR-1 and -2 and FPS-20 series radars.

Analog-to-Digital Conversion of the Bipolar Video

The digital MTI system is, in many ways, simply a digital replica of the analog system (see Figure 9-4). Its canceler contains a comparator, which is a subtractor circuit, instead of an analog resistive bridge or transistor pair. To provide information in the proper form for the digital canceler, the phase detector output must be converted into parallel binary words, and at a rate which will provide a new word at approximately every $3t_p/4$; this time pe-

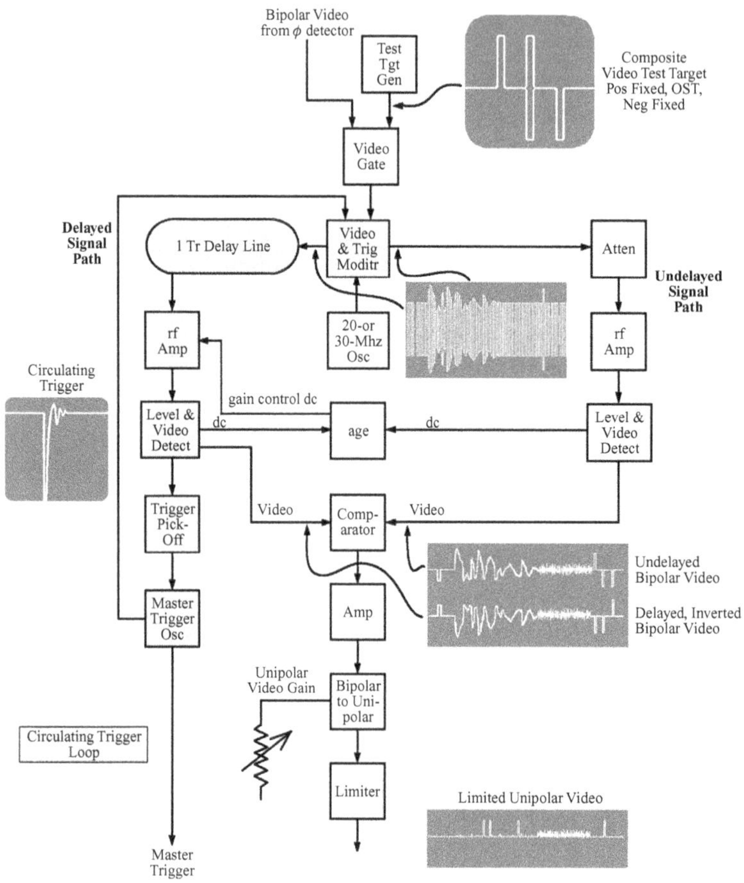

FIGURE 9–3

The circulating trigger loop.

riod is called a ***range cell***. Within the canceler, this binary data must range from a maximum negative to a maximum positive binary number, as illustrated in Figure 9-4.

Basic Timing

Because a number of timing events are taking place in each range cell, the digital MTI system requires higher frequency timing in addition to the once-per-T_r triggers and gates of an analog system. In a synthesis system, the uninterrupted c-w coho provides a convenient source for all of these signals. In a magnetron system with digital circuitry, a master oscillator is the timing source. The coho is counted down to the necessary rates for operation of the digital system. Recall that the coho is compared to the i-f in the phase detector; here lies the explanation for the 31.07 MHz i-f in newer systems. A 31.07-MHz coho frequency can be counted down to 1.2945 MHz, the period of which is 772.5 ns, which, in turn, is 1/16 of a radar nautical mile.

Range-Cell Clock Rate

The rate of the range-cell-rate clock is influenced by a number of factors. As the rate is increased, the digitized video becomes an increasingly better representation of the analog phase-detector output, because the temporal

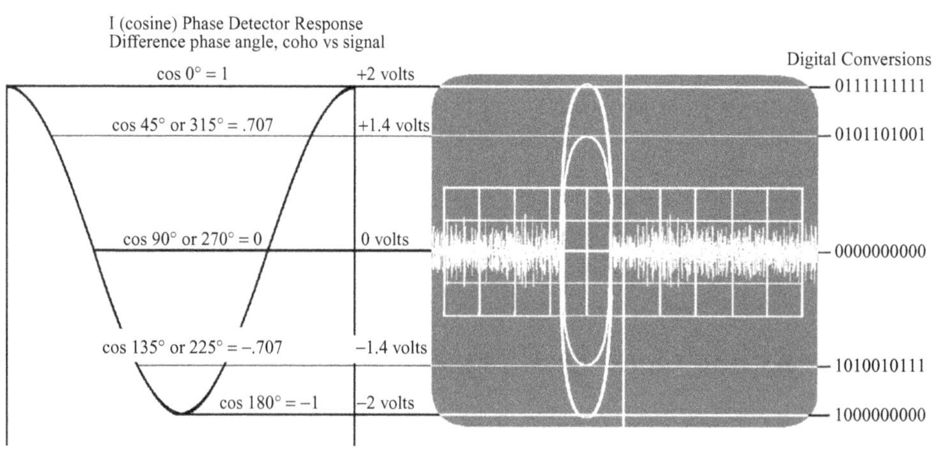

FIGURE 9–4

Phase detector response, bipolar video, and A/D.

resolution increases. The final choice for the range-cell time must be significantly less than the transmitter pulse width (about 75% is a common choice) to preclude "missed" echoes. As the range-cell clock rate is increased, more range cells are present in each system interval, and more storage is required. The maximum clock rate is limited by the ability of the A/D converter(s) (also called "quantizers") to develop a digital word within a specific time frame, by the spectral width of the sampled bipolar video, and by the maximum rate at which the storage medium (usually a shift register) may be updated.

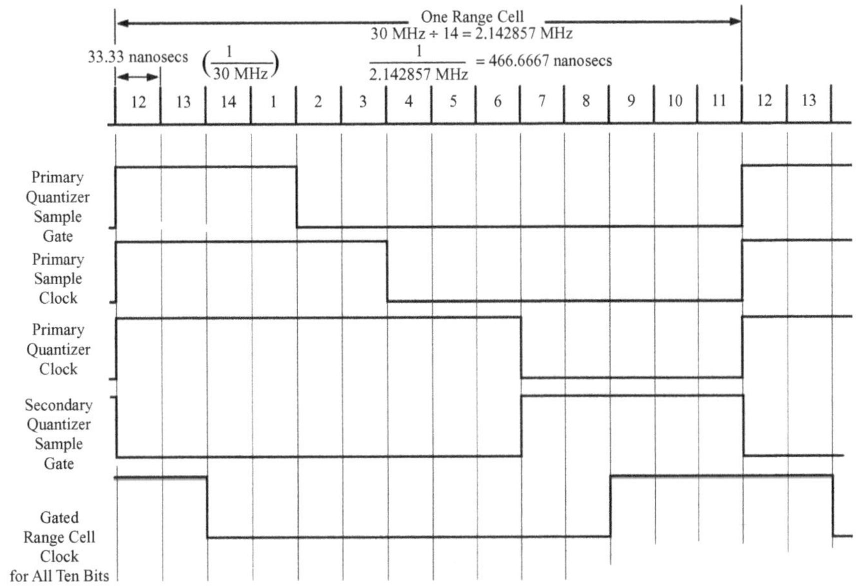

FIGURE 9–5

ASR-8 A/D (Quantizer) timing signals.

Timing Events Within the Range Cell

Within a range cell, some very intricate timing must take place to operate the A/D converters and other parts of the MTI system (see Figure 9-5). One of these necessary timing signals is a bipolar video sampling gate. This gate is used in the A/D converter to bring the charge on a capacitor to the level of the bipolar video from the phase detector. Any pulse-to-pulse time differences in the range-cell clocks may result in sampling errors in the bipolar video, causing cancellation residue to occur. Visualize a block of clutter at a range of 123.456 μs, and a phase of 67°. Adjacent to it at 123.457 μs, the clutter phase is 77°. If the quantizer sampling gate changes by 0.001 μs from one system interval to the next, the canceler could be presented with an apparent 10° phase change ($\Delta\phi$). To preclude such an occurrence, the range-cell clock rate must be very stable. Two cancelers are used in series, typically called ***cascade***; the operation of the second canceler is identical to the first.

Timing Circuitry Block Diagram

Figure 9-6 is an illustration of the timing circuitry in a basic synthesis-type digital MTI system; it will serve as a reference for the proceeding discussion. Although it may bear a considerable resemblance to some systems, it is a very simplified illustration; do not assume it to be representative of any particular radar set. The basic timing source is a 30-MHz coho, divided by 14 to provide 0.467-μs clocks to a basic f_p generator. The A/D timing signals are created in the countdown circuitry. Note that the timing circuitry comprises two groups of blocks which bear a close resemblance; one of these is a basic f_p generator, outlined by dashed lines, and the other is a staggered f_p generator. The operation of the staggered f_p generator is dependent upon the basic f_p generator.

Range Cells versus Shift Register Stages

Whatever be the type of digital system, the range-cell timing must be based on a precise crystal-controlled oscillator to ensure that the range cells are accurate time marks. If the radar's maximum range is 60 miles, and if the range cells are 467 ns, we may be certain that 12.36 × 60/0.467 will indicate that 1,588 range-cell clocks will occur during the radar live time. This number of memory locations will be necessary for the storage of "live-time" digital data

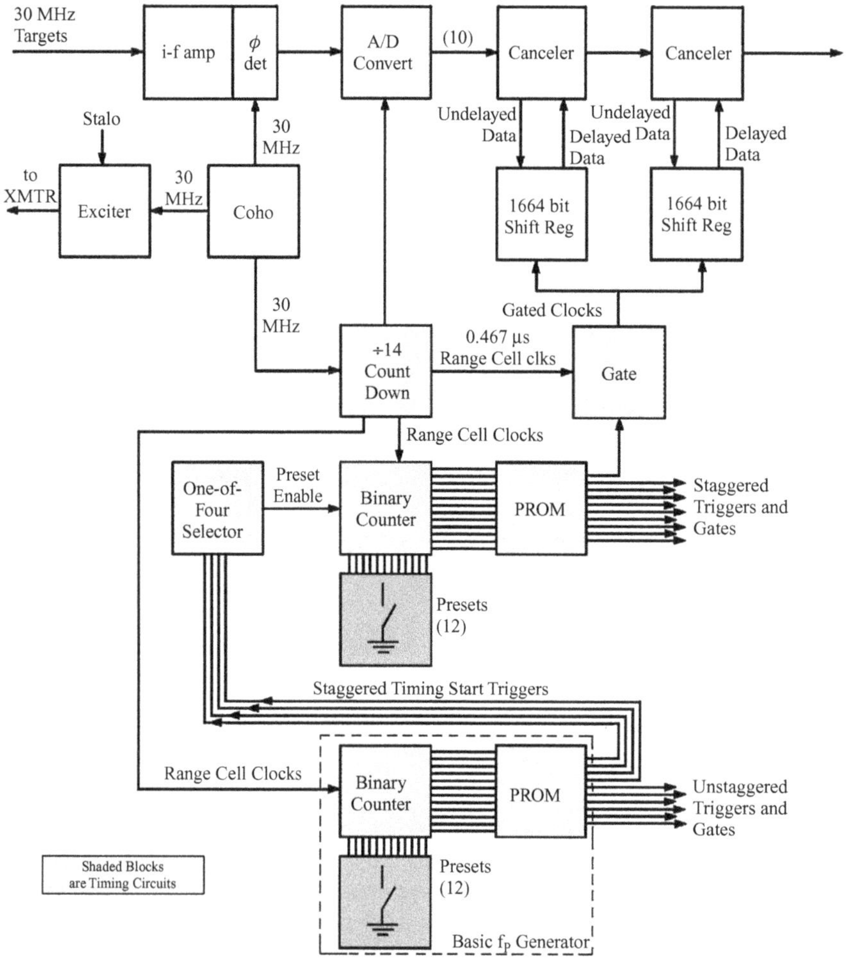

FIGURE 9–6

Basic timing block diagram, digital MTI.

in the first canceler system. Some additional range cells will also be required for special test signals, so a total of perhaps 1,665 memory locations (this is really the number in some systems) may be required for one canceler. This choice is not simply by chance; shift-register integrated circuits are available in 128, 256, and 512 stages. Note that $(512 \times 3) + 128 = 1{,}664$. The 1,665th loca tion is an input register in the canceler.

Range Cell and Trigger Synchronization

Synchronizing the range-cell clocks to the coho only meets part of the timing requirements for the radar system. The clocks to the storage shift registers must (1) be gated on when the transmitter is fired, and then (2) gated off after the precise number of clocks needed for all the memory locations in the storage shift register have occurred. This means that those clocks must somehow be counted, and that they must begin with the transmitter trigger, or with range zero.

The Basic Unstaggered f_p Generator

One circuit area in Figure 9-6 is outlined with dashed lines, to indicate that it is the basic f_p generator. It is operated by range-cell clocks, counted down by 14 from the uninterrupted 30-MHz coho, to produce a period of 0.467 μs. Having established that system timing must be derived from counted range-cell-rate clocks, it should be clear that the count in the counter must be related to the system interval timing. The counter is preset to an adjustable starting point once at each system interval. As the count passes through numbers for which decodes have been programmed, the events for the system interval are initiated; when the count reaches a maximum number, it is forced back to the preset value, and the next system interval begins.

f_p Adjustment

In an analog delay-line system, the f_p can only be changed by replacing all the quartz delay lines in the system; there may be six, or even more. Setting the T_r in a digital system becomes a simple matter of adjusting the counter preset with the 12 dual in-line package (DIP) switches, illustrated in the basic f_p generator in Figure 9-6. The higher the preset, the fewer clocks are required for the counter to reach the number at which it returns to preset.

Since fewer clocks are used between presets, the f_p becomes higher. An example of how a typical preset might be calculated is as follows:

Maximum Count to Counter Reset: 4,097
Range-cell clock period: 467 ns
Desired system interval: 961 μs

$$\text{Number of clocks per } T_r = \frac{961\,\mu\text{s}}{0.467\,\mu\text{s}} = 2{,}057.8$$

$$\begin{gathered} \textit{Maximun count} - T_r = \text{Preset} \\ 4{,}097 - 2{,}058 = 2{,}039 \end{gathered}$$

Staggered f_p

Introducing a staggered f_p complicates the timing system design. The staggered f_p generator in Figure 9-6 operates in a similar manner as the basic generator inside the dashed lines, but the counter is preset at a different time, at each T_r. Decoded triggers or gates from the basic f_p PROM are used to establish the staggered timing through a logical sequencer. All four decoded stagger times are supplied to the sequencer at each T_r. However, the sequencer contains a 2-bit counter and associated logic circuitry; it selects only one of the four stagger timings each T_r. The sequencer outputs preset a second counter in accordance with the selected stagger period; the start time of each staggered T_r is therefore different for each of the four decodes. The staggered counter presets are adjusted to provide the same average f_p as the counter in the basic f_p provides for unstaggered timing.

Digital Timing in Magnetron Systems

The synthesis-system coho is a free-running oscillator which must not be disturbed by switching signals, making it a convenient source for system timing. The magnetron-system coho cannot be used as a basic timing source, because the magnetron's coho must be gated off and re-started each time the transmitter is pulsed. Since (1) the entire synchronizer depends on the coho to run the counter, and since (2) the counter produces all the addresses to the PROM, which (3) produces all basic triggers and gates, then (4) as soon as the PROM produced a gate to shut off the coho, the counter would stop and (5) the synchronizer would be "frozen" so that no further signals could be produced. For that reason, a magnetron system with a digital MTI processor must have an independent master clock oscillator. The independent oscillator is even further necessary, because re-phasing the coho at each T_r would also cause small T_r-to-T_r temporal errors in the comparator.

MTD System Timing

Preface and Introduction

At the present time, there are three major MTD systems in the FAA: the ASR-9, ARSR-4, and ASR-11. There were previous versions on small-scale deployments, and more are being deployed, or are under development, in the United States and other countries. Because it was deployed in the largest numbers, this discussion will be based on the ASR-9. The other systems will have many similarities.

Timing for an MTD system bears some similarity to a digital MTI system, but the MTD timing is many times more complex. As with a digital MTI system, the timing source is the 31.07-MHz coho, and 0.7725 μs (1/16-nmi) range-cell clocks are obtained with a 24:1 countdown. Additionally, the MTD system contains an A/D converter that requires sample gates every range cell.

Because this book is devoted to fundamental principles, rather than specific equipments, the details of the MTD synchronizer will not be discussed here. The ASR-9 synchronizer contains seven circuit cards, none of which can be described briefly. However, certain basic functions of the ASR-9 synchronizer should be understood by the student because they are important parts of the more generalized operational theory of MTD. An entire chapter of this book is devoted to MTD processing. The subject must, however, be lightly touched upon here, in an introduction to the synchronizer timing scheme.

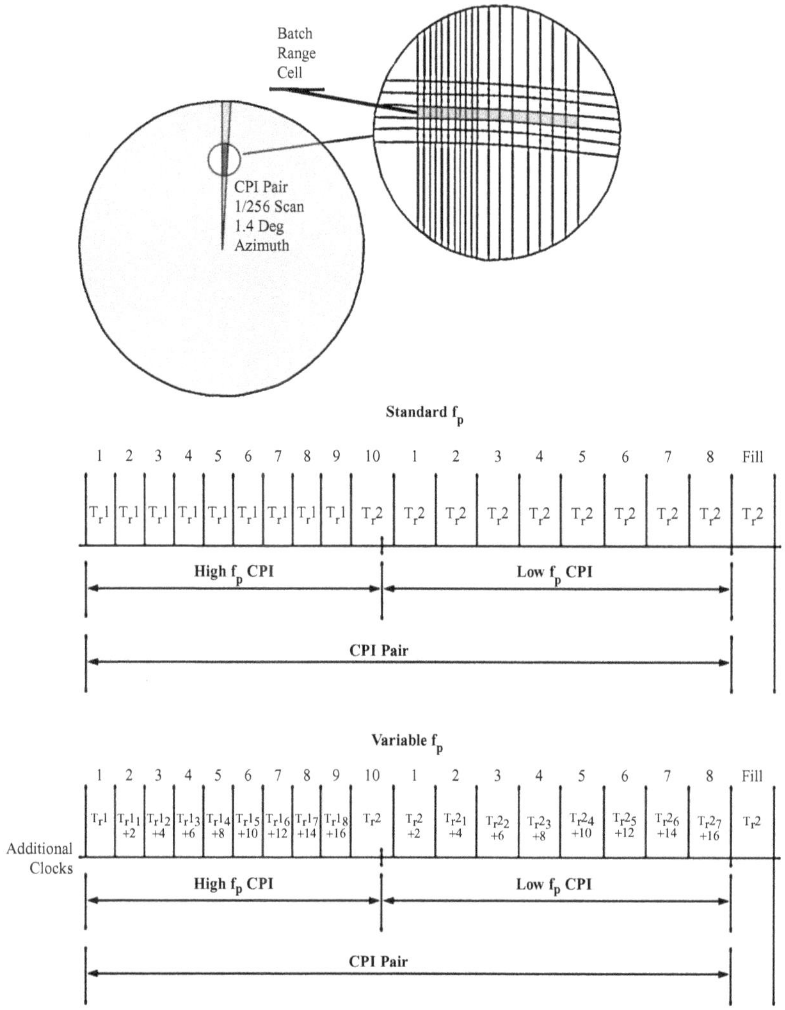

FIGURE 9–7

ASR-9 CPI pair.

Azimuth Data and the Synchronizer

In simpler, earlier systems, most radar timing signals and/or events occurred in real time during the individual system pulse repetition intervals (PRI, PRT, etc.) named ***"interpulse period" (IPP)*** by Westinghouse, the ASR-9 manufacturer (see Figure 9-7). All these events were unrelated to the antenna pointing angle, and the azimuth data was, for the most part, needed only to cause radar data to be displayed at the appropriate angle, in real time (no appreciable delay), from the ppi sweep origin. The MTD system introduced numerous new requirements to synchronize azimuth data and timing.

To begin, the wide azimuth change pulses (over 25 µs) are routed into the synchronizer for lead-edge detection and operation of a 12-bit binary counter, with the msb = 2,048, and maximum count = 4,095. To correspond to aircraft compasses and air traffic control displays, the counter must be referenced to magnetic north, so a "preset" value is obtained from an EEPROM after occurrence of an azimuth reference pulse. The EEPROM may be changed by a keyboard entry at a maintenance terminal.

The Coherent Processing Interval Pair (CPIP)

The entire 360° circle around the radar antenna is divided into 256 coherent processing interval pairs (CPIPs), each containing two coherent pulse intervals (CPI). Each CPIP must occur in less than 1.4, in order that all 256 may be completed before the beginning of a new rotation. Not coincidentally, the antenna beamwidth is 1.4°, at the –3-dB points. Accordingly, each CPI must occupy less than 0.7°, so that each is one-half the CPIP in azimuth. The purpose of this scheme is to permit the operation of digital Doppler filters and to do so at two f_ps, so as to create data that will differ for blind velocities, but not for fixed echoes. Consider the complications of this scheme:

1. Each CPIP must begin at a dedicated azimuth.
2. The rotational rate of the antenna may be accelerated or slowed with wind or power fluctuations.
3. The first of the two f_ps must be at a high rate, and the second at a substantially lower rate, so that a blind-velocity echo ($f_{Doppler} = f_p$) in one will not appear as a blind-velocity echo in the other.

4. The total time required for each CPI in the CPIP must be nearly equal, so that each will be about 0.7° in azimuth width.
5. The user must be provided with a choice of CPIP "IPP sets" to allow for site and area conditions and FCC requirements/assignments.
6. The time required for completion of the CPIP and additional tasks must be less than 1/256 of an antenna revolution, or 16 acp's.
7. The radar transmitter must not be interrupted at the end of the CPIP, because the first echoes in the next CPIP would be lost as the spectrum was under new development.

The solution to all of the aforementioned considerations was to transmit an 18-IPP CPIP, with 10 IPPs in CPIA at a high f_p and 8 IPPs in CPIB at a lower f_p. The f_p choices are such that there is time remaining for additional transmitter "fill pulses" between the end of one CPIP and the beginning of the next. During that time, data is not processed for target-detection purposes, but computer-scheduled built-in testing (BIT) is performed.

The Single-Board Computer

The single-board computer (SBC) is becoming a main component of many latter-day radar systems. Those used in the ASR-9 are mid-1980s technology, and much faster and more uniform "off-the-shelf" designs are now available. One popular new hardware configuration is the ***Versa Module Europe (VME)*** "crate" and standardized bus structure. The advantages to computers in the radar circuitry are numerous. Among the biggest are that equipment operation can be modified with application program software, adjustments can be made with keyboard entries from a personal computer, and access to some or all of the equipment can be made available from remote locations.

Generating all the timing sequences demanded by the MTD processing is beyond efficient utilization of discrete-component, integrated-circuit, or even large-scale-integrated circuitry. The SBC is a circuit card that is connected into the same backplane as the other hard-wired dedicated cards in a variety of ASR-9 cabinets. Because the operation can be modified by program or data changes, an external interface to a personal computer permits the technician or engineer to make adjustments. There are more than a dozen of these SBCs in a complete ASR-9 system, serving a multitude of purposes. The processor and surrounding hardware is essentially the same for all; however, the units are not all interchangeable because firmware program storage depends on the intended function.

Computer Hardware

Brief Description of Computer Components

The functional operational concepts of a computer would increase the size of this radar book beyond the author's intent. However, the latter-day radar technician often has already been trained in computer technology, and enters the radar profession well prepared. Anticipating that, a brief review of elementary concepts is all that will be offered here. In earlier times, air traffic control computers were discrete-component or integrated-circuit design, and the technician needed to be able to troubleshoot to the failed component. The lengthy, detailed training necessitated by that troubleshooting concept gave the interested student an excellent understanding of all those operations that now take place inside a small, sealed processor. He or she who is old enough to have learned such a system as the ARTSIII has a distinct advantage and insight into modern technology.

Program Counter

The program counter is run by the computer clock. As the count increments, instructions are pulled from a random-access memory (RAM). The program counter is reset, and the RAM is loaded by an instruction from a nonvolatile stored "boot" program when power is applied, or when a reset is commanded.

Basic Input/Output System (BIOS)

In personal computers, the ***basic input/output system (BIOS)*** may be stored in a ROM and establishes communications with essential peripherals. A program ROM may be found in single-board computers to serve the same purpose.

Nonvolatile RAM (NVRAM)

Computers often contain an NVRAM with accessible configuration settings. Contents may be changed, but not lost on power failure.

Instruction Repertoire

Many types of instructions cause the processor to perform a multitude of operations, utilizing internal registers. In the ARTSIII as an example, the two main 30-bit data registers were called "A" and "Q." Either could be loaded from RAM, or the contents of either could be stored to RAM. One could be added to the other or added to the contents of a memory location. Division and multiplication could be performed. "OR," exclusive "XOR," "AND," "MASK," "LEFT SHIFT," "RIGHT SHIFT," and many more are available. Beyond A and Q, several "arithmetic" B registers served as "holding places" and could be used as "up" or "down" counters in program loops. Of course, all these operations were entirely in binary "machine code," and modern systems use an "operating system" for the user interface. In the earlier systems, programming was done in machine code. "Assemblers" were used, but the assembly language was only a little more than a shorthand for the machine-code instructions. In the early days of personal computers and MS DOS, "compilers" allowed programmers to write in BASIC or PASCAL, and then "compile" their programs into executable machine-language programs.

Program Execution

Given a simple set of instructions with none to change the program count beyond its single-address incrementing, the counter might simply run to maximum and "fold over" to the beginning count. In actual practice, the starting program address would probably not be zero, and the final program address would contain an instruction to return to the beginning. So, unless something happens, the program runs in an endless loop.

Interrupts

When an "interrupt" occurs on a line into the processor, the program completes the instruction under way, stores the address of the next instruction to be performed, and jumps to an "interrupt handler" address to perform another program. At the end of the handler program, the final instruction recalls the main program address previously stored and resumes.

Interrupt Priorities

Many processors use a seven-level interrupt priority scheme, where level seven has the highest priority. If a level seven interrupt handler is running, a ***"status register"*** indicates that the processor is busy with that task; lower level interrupts cannot be immediately handled, but may proceed when the level 7 status register bit clears. On the other hand, if the level 7 interrupt occurs while a lower level interrupt is in progress, the lower level handler will be interrupted in the same manner as any interrupt to the main program execution.

Status Registers

Status registers may be used to monitor the internal operation of the computer, or essential external "health," which may include power supply voltages and temperature.

Interrupt Tabling

Since low-priority handlers may be easily and frequently interrupted, there must be a way to allow them to be held in "tables" or "stacks" until the status register begins to "unlock" them.

Watchdog Timer

A self-contained timeout and shutdown circuit to halt the processor operation should it fail to execute instructions for a prescribed amount of time.

The ASR-9 Timing Scheme

The ASR-9 Synchronizer SBC hardware and program use a seven-level interrupt scheme to control all the system timing functions necessary to regulate the number of fill pulses, start each CPIP, schedule BIT functions, and much more (see Figures 9-7 and 9-8). Beyond the A/D converter, MTD processing is not at all similar to MTI; there are no cancelers and no delay lines or shift registers. Phase-detected bipolar target information is stored in memory as it is received in ascending real-time ***range order***. When an entire group of data from 18 T_rs has been stored in a memory, the memory is read in ***batch order***, subjected to an arithmetic process in a bank of Doppler filters, sorted according to Doppler, and subjected to a complex thresholding and statistical-analysis target-detection process before being sent to the user facility as digital messages.

CPIP Azimuth

There are 256 CPIPs in a scan, and each CPIP is intended to be associated with a specific azimuth "wedge." For example, the first CPIP occurs at magnetic north, and the 128th occurs at south. The azimuth for any CPIP can be found with ratio and proportion, as shown in the following equation:

$$\theta_{\text{CPIP}} = \frac{(\text{CPIP } no + 1) \times 360}{256}.$$

The engineering necessary to ensure that this occurs correctly was no small challenge, and is one reason for the complexity of the MTD synchronizer program.

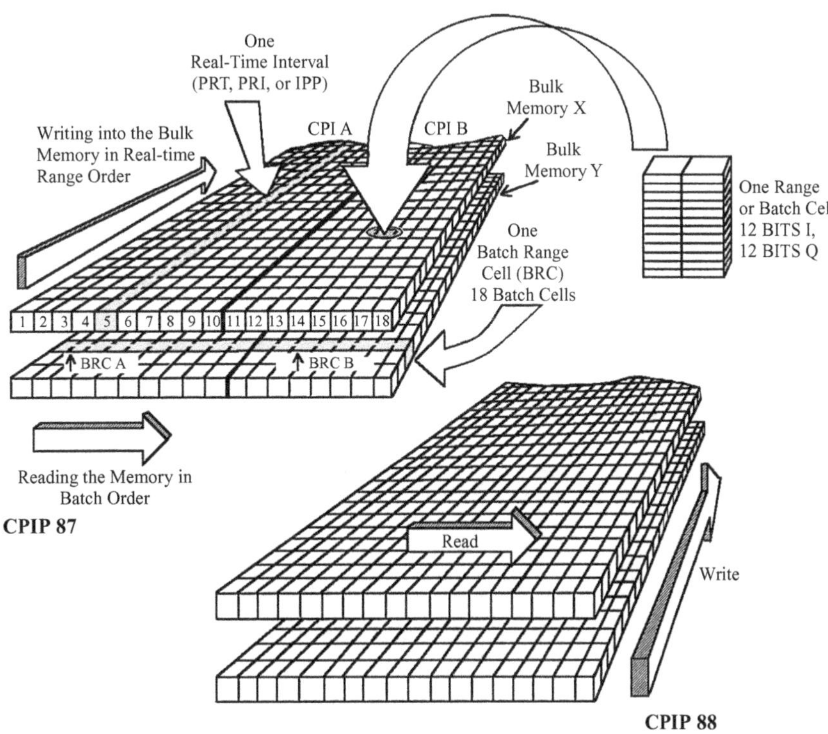

FIGURE 9–8

Range-to-batch conversion.

The MTD Front End

The "front end" of the MTD processing system is called the ***digital signal processor (DSP)***. The DSP contains the Doppler filters and a memory system that is used to supply data to those filters; the filters are actually digital math processors. The timing that is used to read the memory and then operate the filters, and all the rest of the DSP, is called ***batch timing***. There is a multitude of batch timing signals generated in the synchronizer; they occur at 10.35 MHz, 5.175 MHz, 2.5875 MHz, and 1.29 MHz. Batch time is not real time, but it does have a direct relationship with real time. In addition to all the batch-time signals, the synchronizer generates a multitude of real-time triggers. It also generates a multitude of test initiation signals, digital stc information, azimuth data, and more.

CPIP and f_p Stagger

For each 1/16 nmi, from 0 to 60 nmi, the filters in the Doppler filter bank analyze the echoes that have been received from several transmitter bursts, to determine the Doppler frequency of any target that might be present. To facilitate this, the transmitter is pulsed with two groups of triggers, called a ***coherent processing interval pair (CPIP)***, illustrated in Figure 9-7. One group contains 10 triggers and the other contains 8. Each group is called a ***coherent processing interval (CPI)***. The group of 10 triggers occurs at a higher rate than the group of 8, providing different pulse-to-pulse phase changes ($\Delta\phi$) for the same target, and accomplishing an effect somewhat comparable to a conventional staggered f_p in an MTI system.

Complexities Introduced by the CPIP Process

Consider some of the difficulty presented to the technician or engineer by the CPIP scheme. To calculate peak power, the average f_p must be known. To find the average f_p, the technician can use a frequency counter, or he or she may calculate it from the known T_r for each CPI. However, the average f_p will decrease if the antenna rotates at a lesser rate, and more low-f_p fill pulses are used, so the precise scan time must be found. Simply using the reciprocal of 12.5 rpm is not adequate, because the manufacturer's tolerance on rotational rate is ±10%, so the scan time must be measured. To further complicate the problem, a modification to the ASR-9 provided CPIPs in which the intervals between pulses in each CPI progressively increase in length by two range-cell clocks, and this must be accounted for also.

Calculation and Example of ASR-9 Average f_p

A procedure for measuring and calculating the ASR-9 average f_p is necessary for the calculation of peak power. It is as follows:

(a) Determine the CPI pair set in use (assume set 19, for this example).

(b) Determine the actual rotational speed of the antenna. To be precise, measure the time required for several scans, and then calculate an average. For example, if 20 scans required 90 s,

$$T_{scan} = \frac{90\text{ s}}{20} = 4.5\text{ s}.$$

(c) Find the basic T_rs for the standard CPIP set (16 through 28 are variable sets that begin with the T_rs shown for 0 through 15) (see Table 9-1).

(d) For either a standard or variable CPIP set, the total time used for 4,608 T_rs in 256 CPI pairs per scan would be at least as great as shown in the following equation:

$$T_{scan_{CPIP}} = 256 \times [9(T_{r_{high}}) + 9(T_{r_{low}})]$$

Example:

$$T_{scan_{CPIP}} = 256 \times (9 \times 773.3 \times 10^{-6}) + (9 \times 994.25 \times 10^{-6}) = 4.0724\text{ s}.$$

(e) If the CPIP set is variable, there are 144 additional clocks in each CPIP, so for sets 16 through 28, add 28.477 ms, because

$$0.7725\ \mu\text{s} \times 144 \times 256 = 28.477\text{ ms}.$$

(f) Fill pulses at the low f_p will make up the time difference between the scan time and the time used for all the 256 CPIPs. The number of required fill pulses is

$$N_{\text{fill}} = \frac{T_{\text{scan}} - T_{\text{scanCPIP}}}{T_{\text{rlow}}}$$

Example:

$$N_{\text{fill}} = \frac{4.5\text{ s} - 4.072\text{ s}}{994.25 \times 10^{-6}} = 430.03.$$

(g) Determine the average f_p as follows:

$$T_{\text{rave}} = \frac{T_{\text{scan}}}{4608 + N_{\text{fill}}}.$$

Example:

$$T_{\text{rave}} = \frac{4.5\text{ s}}{4608 + 430.03} = 893.2\text{ ms}.$$

Synchronizers in Redundant Radar Channels

FAA radar systems always contain two redundant channels, so that one is available for either maintenance or immediate use while the other is in use. In most cases, then, each radar channel has at least three possible modes of operation, ***ON LINE***, ***STANDBY***, and ***MAINTENANCE***. In the ON LINE mode, the radar channel is in use and under control of the air traffic user facility. In the STANDBY mode, the radar channel is ***immediately available*** to the user facility; a technician should never perform work on a standby radar channel without coordination and agreement by the user. Unless the radar system is a ***diversity*** type, where both radar channels may radiate simultaneously, the standby mode will connect the standby transmitter to a high-power waveguide termination called a ***dummy load***. In the MAINTENANCE mode, the user facility has relinquished control of the radar channel to the technician, and cannot gain use of it at the control facility. The technician then assumes responsibility

TABLE 9–1 CPIP Sets

CPIP set	IPP (T_r)	
	High f_p	Low f_p
0 or 16	757.08	973.39
1 or 17	762.49	980.34
2 or 18	767.90	987.30
3 or 19	773.30	994.25
4 or 20	778.71	1,001.20
5 or 21	784.12	1,008.15
6 or 22	789.53	1,015.11
7 or 23	794.94	1,022.06
8 or 24	800.34	1,029.01
9 or 25	805.75	1,035.97
10 or 26	811.16	1,042.92
11 or 27	816.57	1,049.87
12 or 28	821.97	1,056.82
13	827.38	1,063.78
14	832.79	1,070.73
15	838.20	1,077.68

for the radar channel until it is returned to service, and there is no risk that the user will assume control of a channel under test or adjustment.

Different manufacturers may use different terms to describe these modes of operation, but the three will always exist as a matter of safety and policy. Some examples of different names of the three modes that might be encountered are

ON LINE	STANDBY	UNAVAILABLE
ANTENNA	LOAD	MAINTENANCE
OPERATE	AVAILABLE	UNAVAILABLE
RADIATE	DUMMY LOAD	MAINTENANCE
ON LINE	OFF LINE	UNAVAILABLE
MAIN	STANDBY	MAINTENANCE

In the case of ***diversity*** operation, both radar channels operate at different frequency, and are connected to the antenna through a tuned diplexer; the transmitter bursts are temporally offset by a few microseconds to prevent excessive rf power in the waveguide, which can cause arcing. In this case, both channels can be in the same mode, and the data from both is added together when both are radiating and in OPERATE.

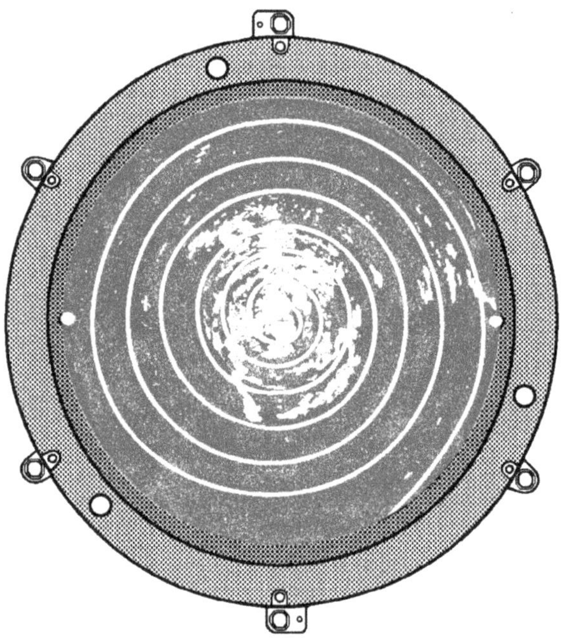

FIGURE 9–9

Cross-channel interference.

Interference between Channels

If two redundant radar channels are operating in the same building, the transmitter burst from one will be received by the other, even if one is terminated in a dummy load (see Figure 9-9). The amount of environmental rf burst energy and pulse harmonics available in the building and waveguide assemblies is so great, in comparison to the minimum discernible signal of the sensitive receiver, that it is highly unlikely that this received ***main bang*** could be hidden from the ppi displays. If the radar channels are completely redundant, each one will contain its own synchronizer. In an analog circulating-trigger-loop system, both channels would use the individual circulating trigger loops within them, and the delay line time would determine the T_r and f_p of each. In a digital system, where clocks are derived from counted-down coho or master oscillator oscillations, the T_r and f_p of each channel will be dependent upon the precise frequency of the coho or master oscillator in the channel. In either case, the probability that the f_p of both channels would be exactly the same is extremely remote, and the probability that the channels could remain at exactly the same f_p for more than a few seconds is even more remote.

Review Questions

1. Name the three evolutionary methods of synchronizing an MTI radar system.
2. State the basic principle of a circulating trigger loop.
3. The ________________ is the basic timing source in synthesis digital MTI systems.
4. A ___________________ is the basic timing source in a magnetron digital MTI system.
5. What might be the purpose of a 0.7725-μs clock in a digital MTI system?
6. What might be the purpose of a 100-ns gate at a 1.29-MHz rate applied to the quantizers (analog-to-digital converters)?
7. The ASR-9 synchronizer produces transmitter triggers in what pattern?
8. The average f_p of an ASR-9 depends upon _______________ and ______________.
9. An ASR-9 antenna takes 4 min, 37 s to make 58 revolutions. The f_p set in use is number 21. The average measured power is 61.6 dBm, and the transmitter pulse width is 1.0 μs. What is the peak transmitted power?
10. A technician has obtained control of the standby channel of a radar system, and placed it to the maintenance mode. Why is this mode provided?
11. A technician has obtained control of the standby channel of a radar system, placed it to the maintenance mode, set the synchronizer to INTERNAL, and energized the transmitter. Why is the telephone ringing?
12. Briefly describe the operation of two radar channels in diversity (both on line and in operate mode).
13. A single-board computer (SBC) is used to control a synchronizer. The Motorola processor can accept interrupts at seven levels of priority. Of the events listed below, which might have the higher priorities?

    Radar Pretrigger
    Azimuth Reference Pulse (arp)
    Azimuth Change Pulse (acp)
    Receiver Sensitivity BIT Test
14. In the preceding question, if the SBC receives an azimuth reference pulse 100 ns before the radar pretrigger, what might one expect the processor to do?

Answers to Review Questions

1. Name the three evolutionary methods of synchronizing an MTI radar system. ***Automatic temporal control, circulating trigger loop, and digital counter.***
2. State the basic principle of a circulating trigger loop. ***Essentially, the circulating trigger is routed through the canceler delay line, so the trigger period and video delay will always be equal.***
3. The ***coho*** is the basic timing source in synthesis digital MTI systems.
4. A ***master oscillator*** is the basic timing source in a magnetron digital MTI system.
5. What might be the purpose of a 0.7725-μs clock in a digital MTI system? ***A 0.7725 μs clock could provide 1/16 nmi range cells in a digital MTI system.***
6. What might be the purpose of a 100-ns gate at a 1.29-MHz rate applied to the quantizers (analog-to-digital converters)? ***The reciprocal of 1.29 MHz is 0.7725 μs, or 1/16 nmi. The 100-ns gate is probably used as the sample gate for the quantizer.***
7. The ASR-9 synchronizer produces transmitter triggers in what pattern?
   The ASR-9 synchronizer produces transmitter triggers in CPIPs, 10 pulses at a high f_p, and 8 pulses at a low f_p.
8. The average f_p of an ASR-9 depends upon the ***f_p set*** and ***antenna speed***.
9. An ASR-9 antenna takes 4 min 37 s to make 58 revolutions. The f_p set in use is number 21. The average measured power is 61.6 dBm, and the transmitter pulse width is 1.0 μs. What is the peak transmitted power?

$$T_{scan} = \frac{4 \text{ min } 37 \text{ s}}{58} = \frac{4.61667 \text{ min}}{58} = 4.77586 \text{ s}$$

$$T_{scanCPIP} = [256((9 \times 784.12 \times 10^{-6}) + (9 \times 1008.15 \times 10^{-6}))] + 28.477 \times 10^{-3} = 4.15787 \text{ s}$$

$$N_{fill} = \frac{4.77586 - 4.15787}{1008.15 \times 10^{-6}} = 613$$

$$T_{r_{ave}} = \frac{4.77586 \text{ s}}{4608 + 613} = 914.74 \text{ μs}$$

$$DC_{dB} = 10 \log \frac{1}{914.74} = -29.6 \text{ dB}$$

$$P_t = 61.6 \text{ dBm} + 29.6 \text{ dBm} = 91.2 \text{ dBm} = 61.2 \text{ dBW} = 1.32 \text{ MW}$$

10. A technician has obtained control of the standby channel of a radar system, and placed it to the maintenance mode. Why is this mode provided? ***The maintenance mode is provided to keep the radar unavailable to the user while it is being worked on. Similarly, the standby mode is provided to keep the radar unavailable to maintenance personnel when it has not been released.***
11. A technician has obtained control of the standby channel of a radar system, placed it to the maintenance mode, set the synchronizer to INTERNAL, and energized the transmitter. Why is the telephone ringing? ***The displays at the control facility are filled with "running rabbits."***
12. Briefly describe the operation of two radar channels in diversity (both on line and in operate mode). ***The two transmitter bursts are both routed to the antenna, but separated by a few microseconds. The receiver video from the two channels is added together.***

13. A single-board computer (SBC) is used to control a radar synchronizer. The Motorola processor can accept interrupts at seven levels of priority. Of the events listed below, which might have the higher priorities?

    Radar Pretrigger

    Azimuth Reference Pulse (arp)

    Azimuth Change Pulse (acp)

    Receiver Sensitivity BIT Test

    The order would likely be as listed. The pretrigger starts all timing events. Real-time azimuth data is essential to accuracy. BIT testing is not immediately essential, and most likely would not be done after the radar "listening" or "live time."

14. In the preceding question, if the SBC receives an azimuth reference pulse 100 ns before the radar pretrigger, what might one expect the processor to do? ***The lead edge of the arp might start an interrupt "handler" program, but the pretrigger would "put that on hold" until the pretrigger handler program finished, at which time the arp handler would resume.***

CHAPTER 10

Radar Transmitters

Introduction

Because of the high power emitted from a radar transmitter, there is a greater potential for failure in this unit than in any other. The radar technician will probably do more repair work on the transmitter, than on all the other units in the system, and a sound foundation knowledge will pay big dividends in system restoration and outage prevention.

The size of this chapter substantially exceeded preliminary planning. Even so, it is limited from what it might have been; the target size of the entire book played a part in this decision.

The fundamentals of common radar-transmitter hardware were addressed in preceding chapters. This chapter will deal in greater detail with several aspects of the pulsed radar transmitter. One topic will be the origin and composition of the emitted energy, a wide spectrum of frequencies. Another subject is the ***modulator,*** which will apply a high-voltage pulse to a third area of interest, the ***final power amplifier,*** which may be a ***magnetron,*** a ***klystron power amplifier tube,*** a ***traveling wave tube (twt),*** or an ***amplitron,*** also called a ***cross-field amplifier (CFA).*** There are still other means to transmit the radar burst. A major component in the modulator is the pulse-forming network, an artificial transmission line; considerable discussion of this device will be contained in this chapter.

Compressed High-Resolution Pulse (CHIRP) Transmitters

In recent years, considerable effort has been devoted to systems which provide a low peak power, but which employ a very wide transmitter pulse; during the pulse, the transmitter frequency changes. In some applications, in the receiver, a frequency-sensitive delay system offers more delay to higher frequencies, causing the pulse to "pile upon itself," to be compressed into a narrow pulse, with a higher peak power. In other applications, there may be several receivers. Wide, frequency-modulated pulses, which overlap in space, may still be separated because of the compression and different frequency content in the overlapped portions; thus, the receiver compression, and narrow-pulse range resolution, yield the accepted name of the expansion-compression technology, ***compressed, high-resolution pulse (CHIRP).*** Whether by clever intent or accident, the comparison to the changing frequency in the transmitter pulse, and the chirping of a bird, is both amusing and fitting.

CHIRP systems have been employed in many variations. In some cases, the frequency must be stepped, from one precise value to another; such a technique may be necessary for phased-array antennas; it also provides for the portions of the pulse to be amplified by multiple receivers. In other cases, the transmitter is linearly frequency-modulated during the burst. An important use of CHIRP in recent years has been in the low-power, solid-state transmitting systems. By combining the output of a multitude of low-power transmitters, over a very wide pulse, the need for dangerous and high-failure-potential units is significantly decreased. The FAA ARSR-4 employs such a technique, transmitting two low-power bursts at each T_r, with t_ps of approximately 60 μs and 90 μs. CHIRP systems may also have an advantage in ***electronic counter-counter measures (ECCM).*** A ***Barker code*** phase-shifts parts of the burst, so that the echoes can be identified as having not come from another source. Received rf which does not match the coding within the transmitted burst may then be inhibited or severely attenuated.

This chapter will deal only with those simple transmitting systems in which there is a constant frequency within the transmitted pulse; this type of system is the most common in the FAA.

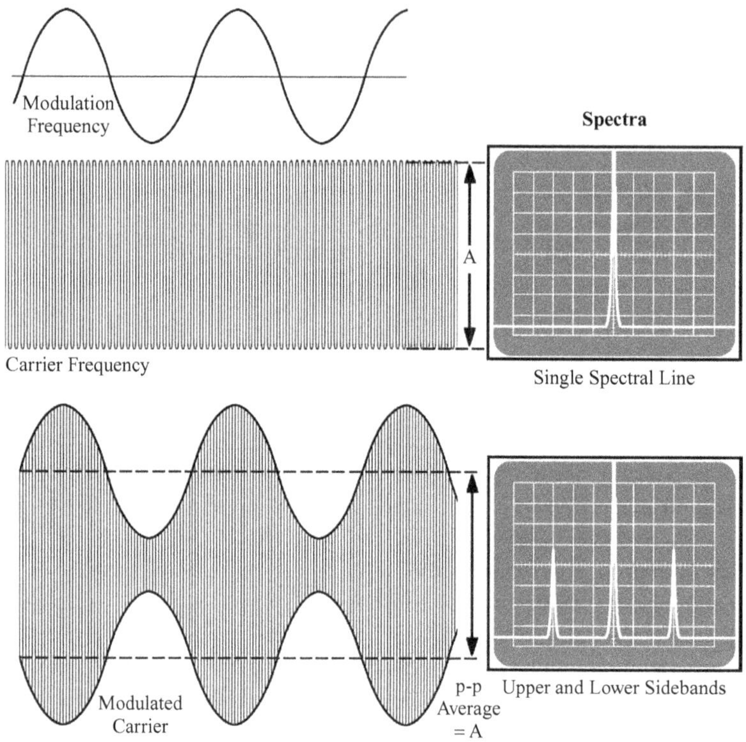

FIGURE 10–1

Amplitude modulation and spectrum.

Modulation

This book is intended for readers with a foundation in basic electronics, which would include ***amplitude modulation (a-m)*** theory. A radar transmitter is pulsed, but the modulation is somewhat similar to amplitude modulation exceeding 100%. In the presence of the pulse, a maximum carrier power is created and, in the absence of the pulse, the carrier power is totally disabled. The character of the transmitted output is a very important consideration; it is determined by a-m principles and by the harmonic content of the pulse. It is all too easy to incorrectly assume that the transmitter only emits a burst of a single frequency, which is impossible; believing such an overly simplistic concept deprives the technician of important knowledge, necessary in tuning and trouble analysis, in both the transmitter and receiver. Accordingly, a review of the rudimentary principles of a-m is in order.

Figure 10-1 illustrates the effect of modulating a carrier frequency with a sine wave. Without the modulation, the c-w carrier power is indicated on a spectrum analyzer as a single spectral line. When the carrier is modulated by the sine wave, two additional frequencies are generated: (1) the carrier plus the modulation frequency and (2) the carrier minus the modulation frequency. The carrier power remains unchanged because the average of the modulation sine wave voltage is zero.

Carrier and Intelligence

It is clear, in Figure 10-1, that the modulation frequency is impressed upon the carrier frequency as ***intelligence.*** The intelligence can be recovered from the carrier, in the receiver, by means of the ***first and second detectors.***

Linear versus Nonlinear Devices

Recall, from basic electronics courses, that a signal may pass through either a ***linear*** or a ***nonlinear*** device. A linear device is one which produces a cosine wave of current change, when a sine wave of voltage is applied. For instance, when the sine wave of voltage is at minimum, at 0° or 180°, the current is changing at a maximum rate because the voltage is rising or falling at its highest rate. A resistor is a linear device. A high-power rf amplifier, which boosts a signal without distorting it, is called "a linear." A "nonlinear" device does not produce a cosine wave of current change with a sine wave of applied signal. A diode is a nonlinear device. So, also, is a class B amplifier, or an overdriven amplifier. Most importantly, an a-m modulator is a nonlinear device.

Modulation Creates Harmonics

When a sine wave is impressed upon a nonlinear device, the signal is distorted, producing ***harmonics,*** other frequencies, mixed with the original, ***fundamental,*** frequency. Figure 10-1 shows the frequencies emitted from an a-m transmitter, indicated on a spectrum analyzer display. In the case of the unmodulated carrier, a pure sine wave of the carrier frequency, unmodulated, is transmitted as a single frequency, but the application of a modulating sine wave creates upper and lower sidebands. The frequency distance of the sidebands, from the center-frequency carrier, is the modulation frequency. The carrier power is not reduced by the modulation because the average of the modulation sine wave is zero.

A closer inspection of individual oscillations of the modulated carrier signal will reveal that the resultant wave is no longer sinusoidal and that the modulation has distorted the carrier wave (see Figure 10-2). By so distorting the carrier, harmonics are created; the number of harmonics are determined by the shape of the modulation signal. In Figure 10-1, the modulation is sinusoidal, and only an upper and lower sideband are shown.

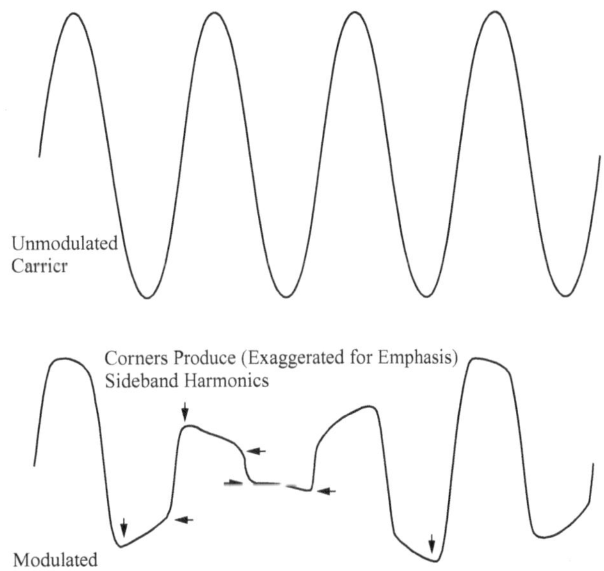

FIGURE 10–2

Distortion of carrier by modulation.

Frequency Mixing

If two frequencies are combined in a circuit, the instantaneous voltage of those summed and combined frequencies, at any single point in the circuit, can obviously be of only one resultant value at any time; as shown in Figure 10-3. In conventional a-m transmitters, complex modulating waveforms are more the rule than the exception, as very few sounds create either single frequencies, or pure sine waves. When complex waveforms of this sort modulate a carrier, additional sidebands are created, as shown in Figure 10-4.

Modulation Percentage

Figure 10-5 illustrates two conditions of carrier modulation, one in which the carrier output reaches zero during the negative modulation peaks; this is 100% modulation. The other illustrates a 200%, overmodulation condition, in which the carrier is totally disabled for an entire half cycle. The overmodulation condition is undesirable for audio purposes, as it will garble the sound. The overmodulation also has another effect, which begins to relate to pulsed radar modulation. At those points where the carrier is disabled, or enabled, the abrupt change creates a multitude of harmonics, causing the output spectrum to "splatter." In a crowded spectrum environment, such as the a-m radio spectrum in a populous city, this could cause serious interference problems.

Pulse Modulation

To this point, the review of amplitude modulation was intended merely as a necessary refresher, and an introduction. The transmitted pulsed-radar signal is a spectrum of frequencies, which is related to a ***Fourier series,*** named for the French physicist and mathematician, ***Jean Baptiste Joseph Fourier (1768–1830),*** a self-made genius who had become an orphan at 8 years of age, and who was educated in a Benedictine military school [1]. The Fourier mathematical series is used to calculate the amplitude and frequency of harmonics, and is used in both mechanical and electrical engineering. The interest in the effect of combining harmonics for a resultant is important in mechanical engineering, as synchronous vibrations can combine to create the type of resultant forces that can destroy structures, such as bridges,

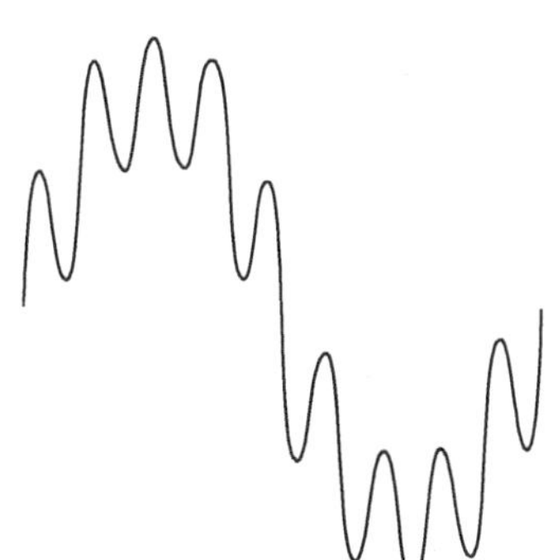

FIGURE 10-3

Frequency mixing and resultant.

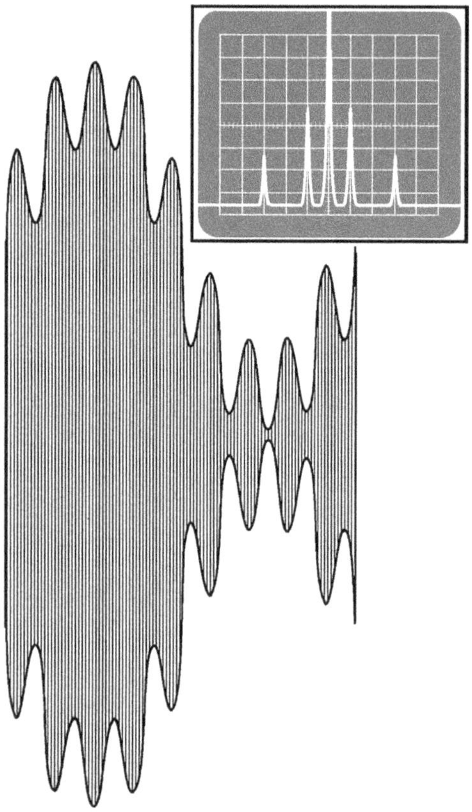

FIGURE 10–4

Additional frequencies and sidebands.

buildings, and towers. In the 1930s, a large suspension bridge in Seattle, nicknamed "galloping Gertie," was destroyed by harmonic oscillations.

Combining Harmonics to Form a Pulse

Figure 10-3 illustrated the waveform resulting from the mixing of two frequencies. To depict the presence of multiple frequencies, writers and illustrators often show them as in the upper portion of Figure 10-6. Although all those frequencies may be present, only a resultant may exist in a circuit or component, simply because two different voltages cannot simultaneously exist at the same time and place; such a resultant is illustrated in the lower portion of Figure 10-6.

In Figure 10-6, the individual frequencies have a special relationship to one another; they are synchronous, and all the oscillations periodically reach a peak at the same point. Additionally, as the frequencies increase, their peak-to-peak amplitudes decrease. Still further, the frequency increases are precise multiples of the original, and they are called ***harmonics.*** Because of these relationships, the resultant begins to assume the shape of a pulse, which approaches a rectangular shape, as more and more harmonics are added. The repetition of the pulse, f_p, is at the same frequency as the lowest frequency harmonic, since that sine wave provides the predominant influence on the additions of all the harmonics. That lowest frequency harmonic, therefore, is called the ***fundamental frequency,*** and the harmonic which is twice the frequency of the fundamental is called the ***second harmonic.*** If enough harmonics of the correct amplitude are added together, the pulse can become perfectly rectangular.

It is very unlikely that a deliberate combination of individually generated frequencies would be attempted, but there is a ***reciprocity*** in repetitive, nonsinusoidal, electrical signals. If a multiple number of synchronous harmonics may be combined to create a repetitive resultant pulse, then a multiple number of synchronous harmonics may also be obtained from a repetitive pulse that has been formed by other means. For instance, it is possible to apply a repetitive pulse to a circuit, sharply tuned to one of the harmonics, and obtain a sine-wave output at the harmonic frequency.

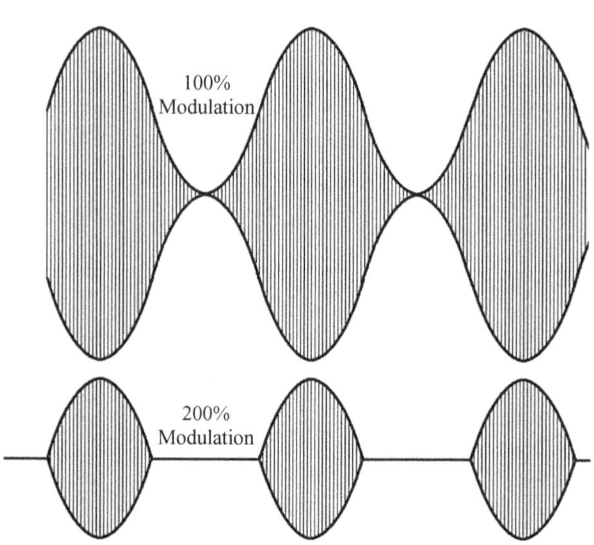

FIGURE 10–5

Modulation percentage. Pulse modulation.

The sin *x*/*x* Amplitude Wave

The instantaneous amplitude and polarity of all harmonic oscillations, mechanical or electrical, have been proven to be principally based upon a mathematical function, in which the sine of an angle, expressed in radians, divided by the numerical value of the radian measure, provides a unique and special curve, as illustrated in Figure 10-7. This sin *x*/*x* curve shall become the principal determining factor in the spectral content of the radar transmitter output. Figure 10-8 illustrates the approximate shape of the curve, to a point where $x = 3\pi$; when the curve is plotted further, it will assume a negative value past π, since the sine becomes negative past 180°.

The shape of a pulse determines the harmonic content; the more closely it resembles a perfectly rectangular shape, the more harmonics it can produce. The precise values of the peak-to-peak amplitude of each harmonic may be computed by a Fourier series [2] as follows.

Instantaneous voltage $e_{resultant}$ at time t_x, referenced pulse center:

$$e_{\text{resultant}} = 2\,E_{\text{pulse peak}}\left(\frac{t_p}{T_r}\right)\left(\frac{\sin\left(1\pi\frac{t_p}{T_r}\right)}{\left(1\pi\frac{t_p}{T_r}\right)}\right)\cos 1(2\pi f_p)t_x$$

$$+$$

$$2\,E_{\text{pulse peak}}\left(\frac{t_p}{T_r}\right)\left(\frac{\sin\left(2\pi\frac{t_p}{T_r}\right)}{\left(2\pi\frac{t_p}{T_r}\right)}\right)\cos 2(2\pi f_p)t_x$$

$$+ \cdots +$$

$$2\,E_{\text{pulse peak}}\left(\frac{t_p}{T_r}\right)\left(\frac{\sin\left(n\pi\frac{t_p}{T_r}\right)}{\left(n\pi\frac{t_r}{T_r}\right)}\right)\cos n(2\pi f_p)t_x.$$

This series allows the physicist or engineer to calculate the value of the resultant at any time, t_x, after the center of the pulse. The term $n(2\pi f_p)$ provides for the phase angle rotation during the time t_x and "n" multiplies the harmonic frequency by the value of "n," the harmonic number.

Until recent years, calculation of the resultant harmonic spectrum was tedious and time-consuming. However, it is no longer necessarily so. A computer program, written in BASIC, with repetitive loops, to increment the harmonic number and t_x, can provide an accurate plot of the entire shape. Other computer programs may provide even simpler means for entering equations and printing or displaying the plots between limits.

Calculation of the Harmonic Resultant at Pulse Center

In many cases, the greatest interest in harmonic amplitude will be at the time of the pulse. Should this be the case, the expression $\cos n(2\pi f_p)$ may be dropped, since the phase angle at the center of the pulse is 0°, and the cosine of 0° is 1. The equation for the amplitude of the resultant at the center of the pulse then becomes

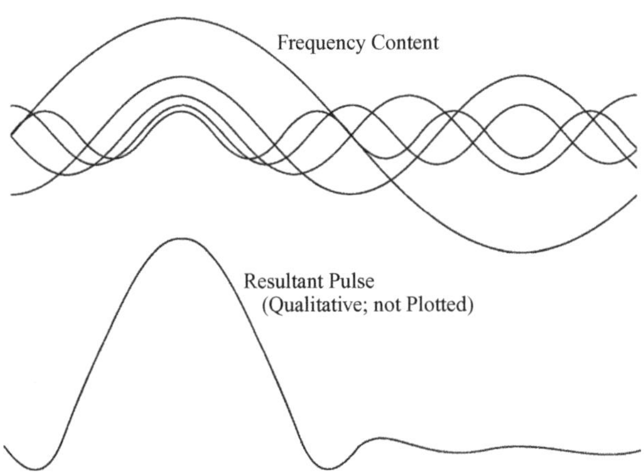

FIGURE 10–6

Harmonics and resultant. Combining harmonics to form a pulse.

$$E_{\text{resultant}} = 2\,E_{\text{pulse peak}}\left(\frac{t_p}{T_r}\right)\left(\frac{\sin\left(1\pi\frac{t_p}{T_r}\right)}{\left(1\pi\frac{t_p}{T_r}\right)}\right)$$

$$+$$

$$2\,E_{\text{pulse peak}}\left(\frac{t_p}{T_r}\right)\left(\frac{\sin\left(2\pi\frac{t_p}{T_r}\right)}{\left(2\pi\frac{t_p}{T_r}\right)}\right)$$

$$+ \cdots +$$

$$2\,E_{\text{pulse peak}}\left(\frac{t_p}{T_r}\right)\left(\frac{\sin\left(n\pi\frac{t_p}{T_r}\right)}{\left(n\pi\frac{t_p}{T_r}\right)}\right).$$

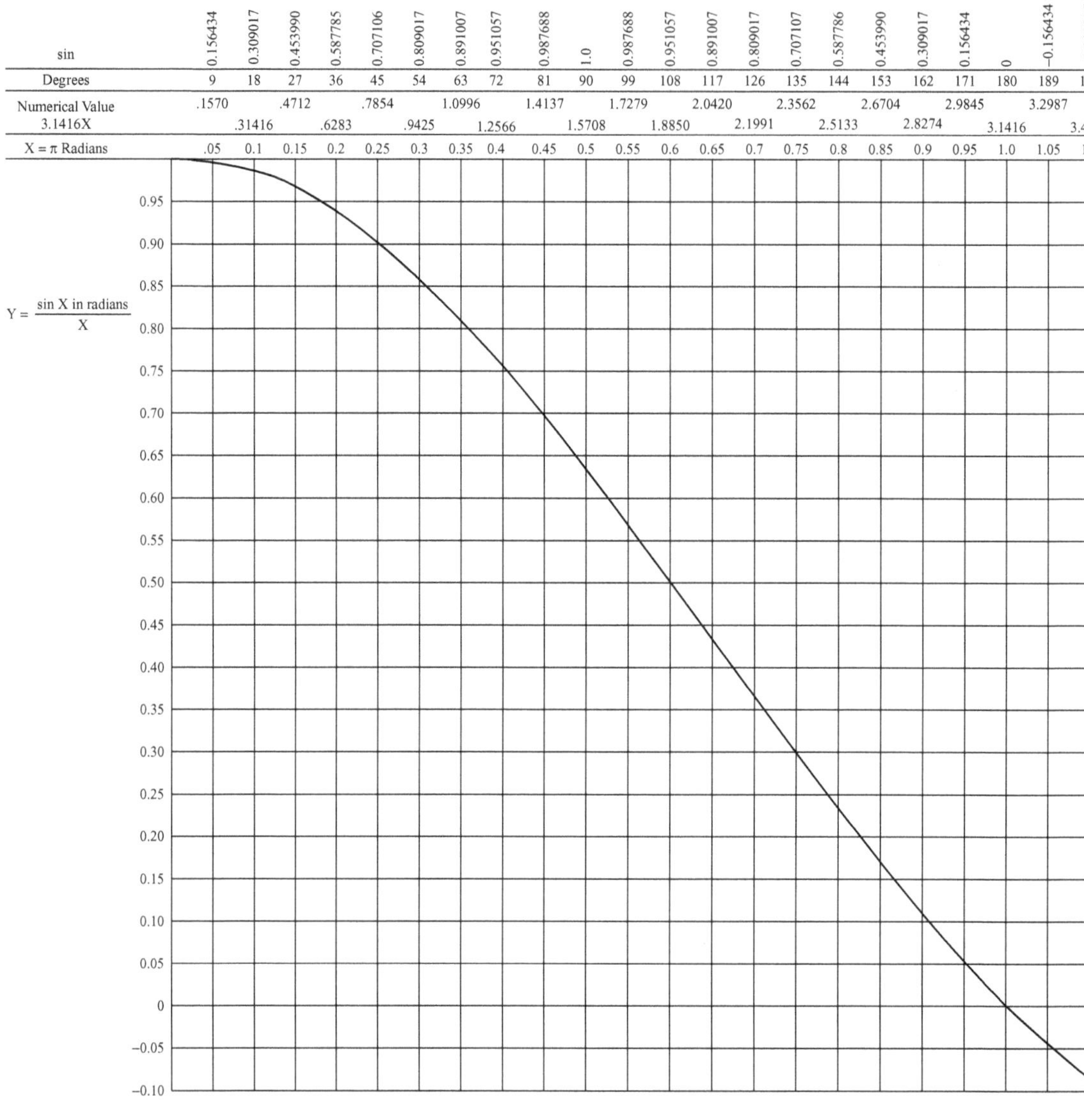

FIGURE 10–7

Plot of sin *x*/*x*, *x* in radians.

Calculation of the Amplitude of a Single Harmonic

The last part of the equation, where "*n*" describes the harmonic number, may be used independently for the computation of the amplitude of any harmonic; this is of significant interest to this discussion of pulse amplitude modulation, because it will determine the transmitted spectrum shape:

$$E_{\text{pulse center}_n} = 2\,E\left(\frac{t_\text{p}}{T_\text{r}}\right)\left(\frac{\sin\left(n\pi\frac{t_\text{p}}{T_\text{r}}\right)}{n\pi\frac{t_\text{p}}{T_\text{r}}}\right).$$

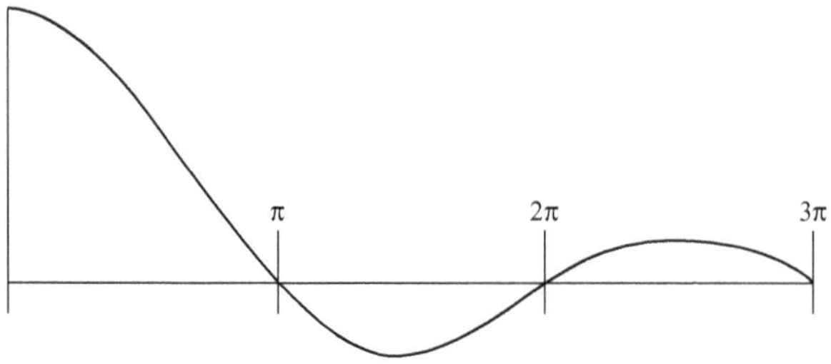

FIGURE 10–8

Sin *x*/*x* to 3π.

One portion of the equation is a sin *x*/*x* function, as in the preceding discussion:

$$\frac{\sin\left(n\pi\frac{t_\text{p}}{T_\text{r}}\right)}{n\pi\frac{t_\text{p}}{T_\text{r}}}.$$

Harmonic Frequency at the Crossover

In the radar pulse Fourier formulas, the sin x/x curve is modified by the duty cycle, t_p/T_r, so that the value of "n" must be increased many times, before attaining that value, $\sin\pi/\pi$, where the amplitude curve reaches zero. That point will occur when $n = 1/t_p$, because

$$\frac{\sin\left(n\pi\frac{t_p}{T_r}\right)}{n\pi\frac{t_p}{T_r}}; \text{ let } n_{xover} = \frac{T_r}{t_p}; \text{ then: } \frac{\sin\left(\pi\frac{t_p T_r}{T_r t_p}\right)}{\pi\frac{t_p T_r}{T_r t_p}} = \frac{\sin\pi}{\pi}.$$

Harmonics occur at multiples of f_p, $f_p = 1/T_r$, so:

$$n_{xover} = \frac{T_r}{t_p} = \frac{\left(\frac{1}{t_p}\right)}{\left(\frac{1}{T_r}\right)} \Rightarrow f_{xover} = \frac{1}{t_p} \Rightarrow n_{xover} = \frac{\left(\frac{1}{t_p}\right)}{f_p}.$$

Number of Harmonics in a Radar Pulse

Now, consider the number of harmonics between the fundamental, and the first crossover, sin π/π, for a typical ASR-8 transmitter pulse, $t_p = 0.6$ µs, $T_r = 877$ µs. If the pulse had perfectly square corners, a plot of all the harmonics to 3π would resemble that in Figure 10-9:

$$n_{crossover} = \frac{\left(\frac{1}{t_p}\right)}{f_p} \Rightarrow n_{crossover} = \frac{\left(\frac{1}{0.6\times10^{-6}}\right)}{\left(\frac{1}{877\times10^{-6}}\right)} = 1461.$$

(A fraction of a harmonic is impossible.)

The Actual Transmitter Spectrum

Because the pulse contains all these frequencies, each frequency will create an upper and lower sideband when the pulse is used to modulate a transmitter. For the pulse harmonic content shown in Figure 10-9, the transmitted spectrum would then be as shown in Figure 10-10. The side lobes are in a positive direction, because the squaring

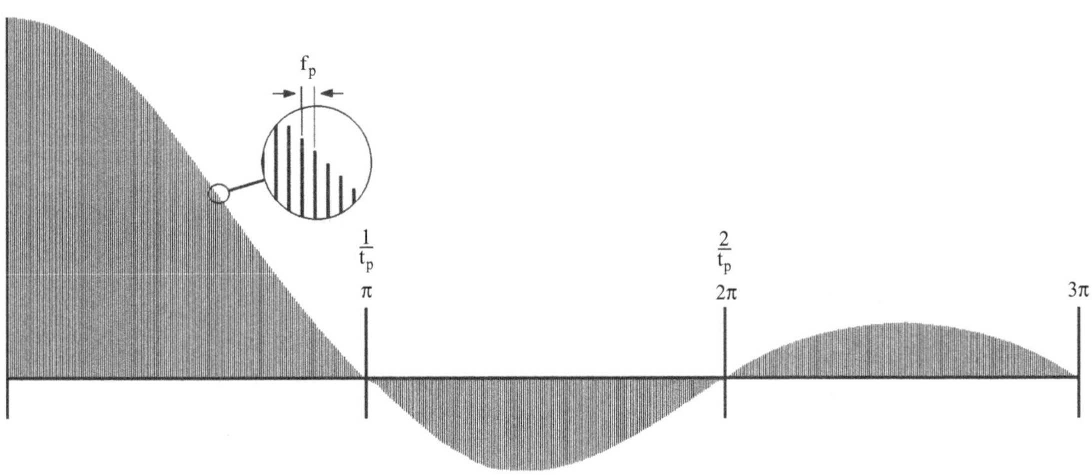

FIGURE 10–9

Harmonic amplitude to 3π.

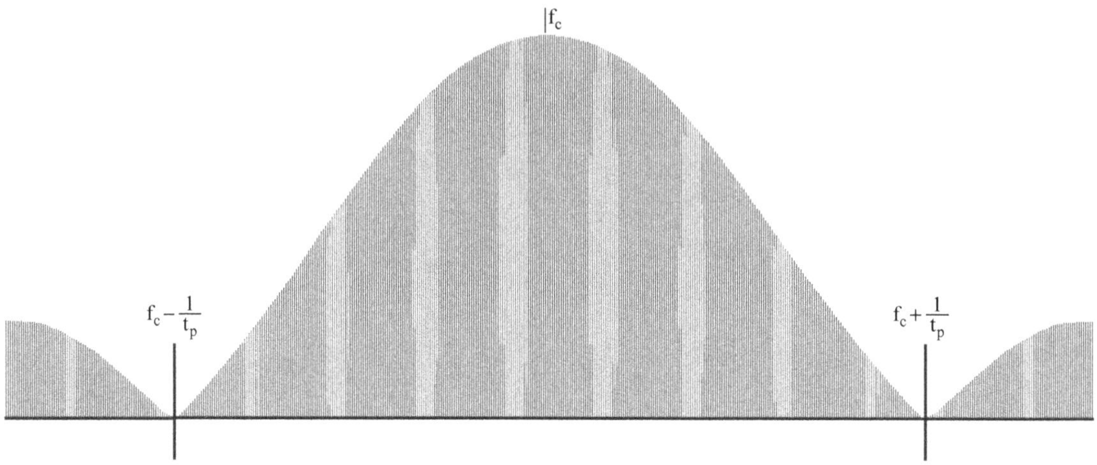

FIGURE 10–10

Pulse-modulated transmitter spectrum.

used in power formulas eliminates any negative numbers; in more practical terms, there just isn't any such thing as negative power. In actual practice, the pulse will not be perfectly rectangular, and the side lobes will be less than shown.

The Spectrum Analyzer

The spectrum analyzer is a special test set, originally developed purely for radar purposes. Today, it is used for many other purposes, but remains a major part of the test equipment collection at any radar facility. The spectrum analyzer is particularly necessary in the maintenance of klystron power-amplifier systems, as tuning of the klystron and driving circuits directly affects the transmitted spectrum.

The spectrum analyzer is a ***frequency-domain*** device, which means that the display has no direct relationship to the radar system timing, and may not be directly interpreted in terms of radar range, or echo time in microseconds (see Figure 10-11). Some spectrum analyzers do have a time-domain feature, but this is an auxiliary function, and not the major purpose.

The major purpose of the spectrum analyzer is to serve as a superheterodyne, swept-frequency receiver. In many uses of the analyzer, it is actually connected to antennas for the purpose of analyzing environmental electromagnetic radiation, and for locating transmitters. It provides a display, from left to right, of a selected band of frequencies, the presence of which will cause a vertical deflection. When used in the maintenance of the radar system, the analyzer is connected to the INCIDENT waveguide directional coupler. Since the analyzer is a delicate and sensitive receiver, considerable front-end attenuation, ordinarily, must be inserted before the connection is made. Once connected, and properly adjusted, the spectrum created by the pulse modulation, described in the preceding explanation, will be displayed.

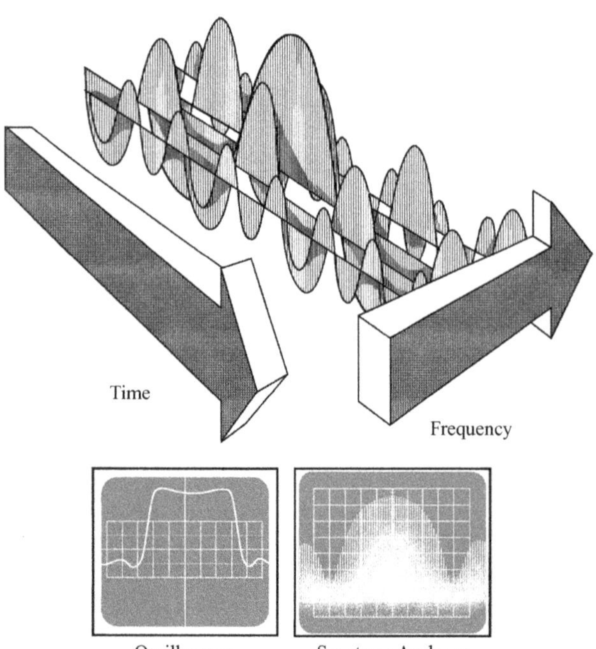

FIGURE 10–11

Time versus frequency domains.

Because the spectrum analyzer is such an important part of the radar technician's work, some basic understanding of its operation is essential, and the fundamental principles will be dealt within the proceeding paragraphs. Some manufacturers regularly conduct short classes on spectrum analysis, and it has been this author's experience that these can provide excellent, valuable information and literature. Attendance at such classes, or acquisition of manufacturer's literature, is highly recommended.

Spectrum Analyzer Functional Description

Figure 10-12 illustrates the front-panel controls of a hypothetical, generic spectrum analyzer. Below the front-panel illustration is a block diagram, greatly simplified, but adequate to offer an understanding of the basic principles.

Central to the entire operation of the spectrum analyzer is the ***sweep generator,*** which creates a sawtooth voltage for use in (1) moving the trace, from left to right, across the face of the cathode-ray-tube display, and (2) providing the voltage-controlled local oscillator with a changing voltage. The cathode-ray tube is generally the electrostatic-deflection type, common in test equipment. The sweep is free running and unsynchronized, totally unrelated to the radar system timing. The sweep speed is adjustable; the choice of speed is dependent upon the user's objective, and the signal to be viewed. Generally, the slowest speeds would offer the best display, but the persistence of the cathode-ray tube makes them difficult to observe, and thus, undesirable. Over the years, direct-view storage tubes and memory devices have been used to overcome the undesirable effects of decaying brilliance.

The crt face is covered by a transparent graticule. The same sawtooth used for the crt sweep is modified before application to the ***voltage-controlled local oscillator.*** One control, the ***FREQUENCY/DIVISION*** adjustment, is effectively a gain control for the sawtooth. As the sawtooth voltage rise is increased, the rate of change in frequency applied to the first detector will also increase, and the local oscillator will span a wide range of frequencies as the trace moves across the crt. Similarly, a slowly rising sawtooth will allow the local oscillator to remain within a narrow band of frequencies as the trace moves across the crt. The dc baseline of the local-oscillator sweep sets the central operating point, the center frequency of the display.

The i-f Amplifier and Resolution Bandwidth

As the input signal is compared to the local oscillator, a changing difference frequency is presented to the ***i-f amplifier.*** When that difference frequency is within the bandpass of the i-f amplifier, an output will be rectified by the ***second detector,*** providing video for use by the display.

The operation of the i-f amplifier plays an important part in the appearance of the display. The i-f bandpass determines the duration of time, for which the difference intermediate frequency for a single input frequency will continue to provide an output. Envision the i-f bandpass as a stationary, open "window," with the band of frequencies "sliding" past it, at the sweep-speed rate, allowing detection of those intermediate frequencies while they are within the "window." If the "window" is wider, more adjacent frequencies may be simultaneously detected, as they "slide" past the "window," and the i-f will provide an output, for a single frequency, for a longer portion of the sweep.

If the bandpass is very narrow, in comparison to the frequency span from the left to right edges of the crt, the output will occur only briefly, and a narrow "spike," or pulse, of video will occur. This pulse has a Fourier harmonic content, and the ability to reproduce and display it will be determined by the bandpass of the video amplifier. Should the video pulse be too narrow for the video amplifier bandpass, the displayed video will be attenuated, and it will not be an accurate representation of the power input.

If the i-f bandpass is very wide, an output will be produced for a wide span of intermediate frequencies, a wider pulse will be presented to the video amplifier, and attenuation attributable to high-frequency losses is less likely, so the representation of input power is more accurate. Additionally, adjacent frequencies tend to "run together," smoothing the video amplifier output.

The i-f bandwidth then determines the ***frequency resolution*** of the spectrum analyzer, and is, accordingly, called ***RESOLUTION BANDWIDTH.*** In order that the resolution bandwidth be precise, it is set with filters, selected by the resolution bandwidth control circuitry.

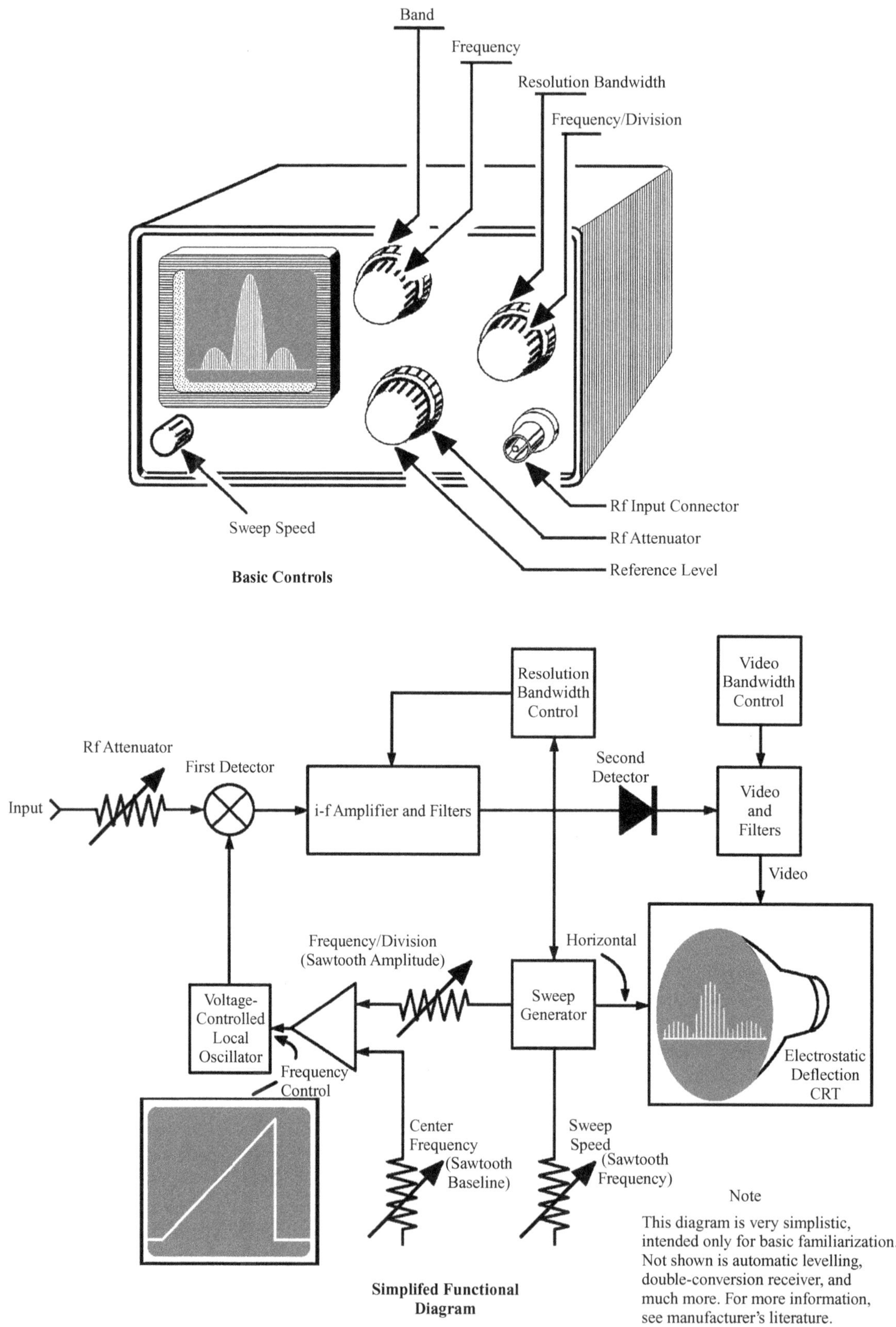

FIGURE 10–12

Spectrum analyzer, simplified illustration.

Sweep Speed versus Resolution Bandwidth

Note in the block diagram shown in Figure 10-12 that there are connections between the resolution bandwidth circuitry and the sweep generator. As the sweep speed increases, the "dwell" time on a given frequency decreases, and the video pulse becomes more narrow. Thus, there is a provision to operate the resolution bandwidth in a fashion which automatically widens the i-f bandpass, as the FREQUENCY/DIVISION is decreased. There may also be a warning indication to the operator, when the RESOLUTION BANDWIDTH adjustment is inadequate for the FREQUENCY/DIVISION setting.

The Display

Figure 10-13 illustrates several spectrum analyzer displays. The display video may be either linear or logarithmic. The i-f amplifier has a logarithmic gain, so that the input will not be limited, and so that its output may be representative of power in deciBels. Antilog circuitry provides for a linear display, as shown in Figure 10-13A. In linear operation, small changes in the spectrum may be more obvious, and the linear mode is, therefore, more desirable to view while tuning the transmitter. On the other hand, measurements of the spectrum are expressed in dB, and the logarithmic mode is used for measurements. The video amplifier output is calibrated to provide a vertical deflection of 10 dB per division in the log mode.

Unlike the display on an oscilloscope used to vertically measure precise voltage differences between two points, the spectrum is measured "from the top down." Recall that dBm or dBW expresses power in relation to 1 mW or 1 W, respectively; however, a dB is only a multiplier or divisor. The spectrum analyzer display then expresses only the power relationships between different points on the display. In Figure 10-13B, the peak of the main lobe is 20 dB below the top graticule line, and the first side lobes are 10 dB less than the main lobe. Latter-day, and more expensive, spectrum analyzers may provide a power-reference "dot," in dBm, and the approximate power may be measured in relation to the reference. Nevertheless, the spectrum analyzer should never be considered an optional means to measure power; even if the center of the main lobe could be accurately measured, the total power includes the whole spectrum, and all the frequencies.

A radar spectrum might easily contain more than a thousand harmonics between the center of the main lobe, and the null at the first crossover point, where $f_{xmtr} = f_{center} + 1/t_p$. These harmonics will all be apparent if (1) the resolution bandwidth is sufficiently narrow, and (2) the sweep speed is sufficiently slow. Such a condition is illustrated in Figure 10-13E. Such a display might be undesirable, because the sweep must move so slowly that the persistence of the cathode ray tube will not allow the entire spectrum to be visible at once; the "older" portions of the display may have significantly decayed in intensity. Further, a slow sweep makes it difficult for the technician to observe the effects of adjustments he may make.

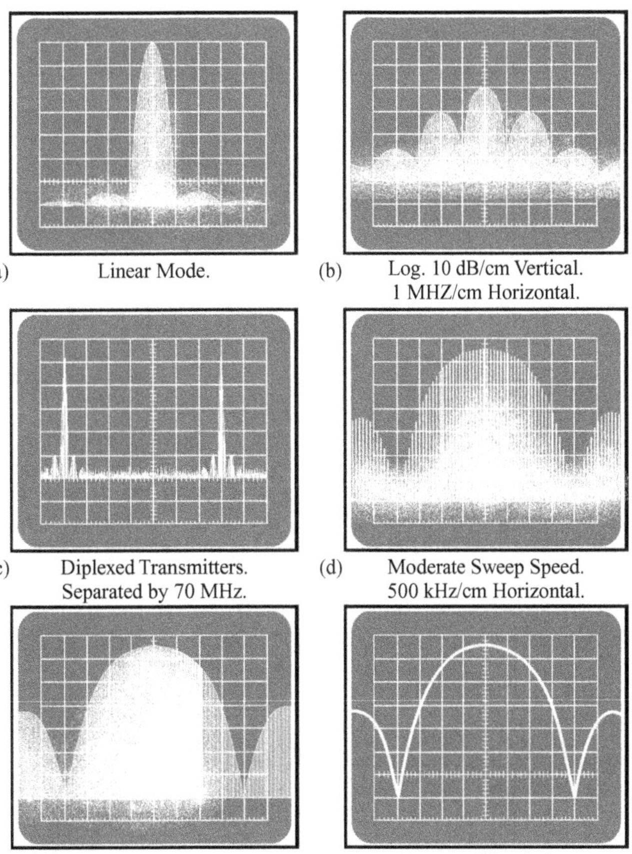

(a) Linear Mode.

(b) Log. 10 dB/cm Vertical. 1 MHZ/cm Horizontal.

(c) Diplexed Transmitters. Separated by 70 MHz.

(d) Moderate Sweep Speed. 500 kHz/cm Horizontal.

(e) Slow Sweep Speed.

(f) Increased Resolution Bandwidth.

FIGURE 10–13

Spectrum analyzer displays, 1-μs transmitter burst.

If the sweep speed is increased above the value in Figure 10-13E, the spectrum will begin to appear as shown in Figure 10-13D. It would appear that there are fewer harmonics; what has happened is that many of the harmonics are being "lost," because the i-f output for those harmonics does not fall within the i-f "window" when the transmitter is pulsed. The likelihood that this will occur increases, as the sweep speed is increased.

Figure 10-13F shows the effect of increasing the resolution bandwidth. The spectral lines have run together, and only the envelope of the spectrum is apparent.

In Figure 10-13C, the spectrum analyzer has been connected to a directional coupler common to two diplexed radar transmitters, and the FREQUENCY/DIVISION has been increased to 10 MHz.

Common Uses

The most-frequent uses of the spectrum analyzer by the technician will be (1) measurement of the difference between side lobes and the main lobe, (2) observation for spectral instabilities, which may cause mti clutter residue, (3) monitoring the spectrum while tuning the transmitter, and (4) precision measurement of the transmitter burst width. Since the Fourier series first crossover is at $1/t_p$, and there are first crossovers at both sides of the center frequency, the reciprocal of half the frequency distance between the nulls at each side of the main lobe is the pulse width. For example, if the main lobe spans 4.4 MHz, the pulse width is 1/2.2 MHz, which is 0.4545 μs.

Video Bandwidth

Some spectrum analyzers may also provide a means to control the video amplifier bandwidth, so as to reduce noise. Obviously, there is a relationship between the video bandwidth and the resolution bandwidth. As the resolution bandwidth is narrowed, producing more narrow pulses with a wider spectral content, the video bandwidth must be made wider.

Consider a single line in the transmitter spectrum, with a 500-kHz/DIVISION display, a sweep speed of 70 μs/DIVISION, and RESOLUTION BANDWIDTH set to 100 kHz. The sweep is scanning 5 MHz in 700 μs, and the resolution bandwidth is 2% of the scan. The individual spectral line will, therefore, produce a video output for 2% of 700 μs, producing a pulse of 14 μs, once every 700 μs. The harmonics of the video pulse are now based on a 700-μs T_r, and a 14-μs t_p, and the video bandpass must be sufficient to reproduce an adequate number of harmonics for good pulse reproduction. If it was determined that all the harmonics to the first crossover were required, the video bandpass should range from zero past $1/t_p$, or at least 71.4 kHz. In many cases, a 100-kHz video bandpass selection is available. Should the video bandwidth be significantly less than this, the amplitude of the display will be decreased, and the relationships in the spectrum will be unreliable.

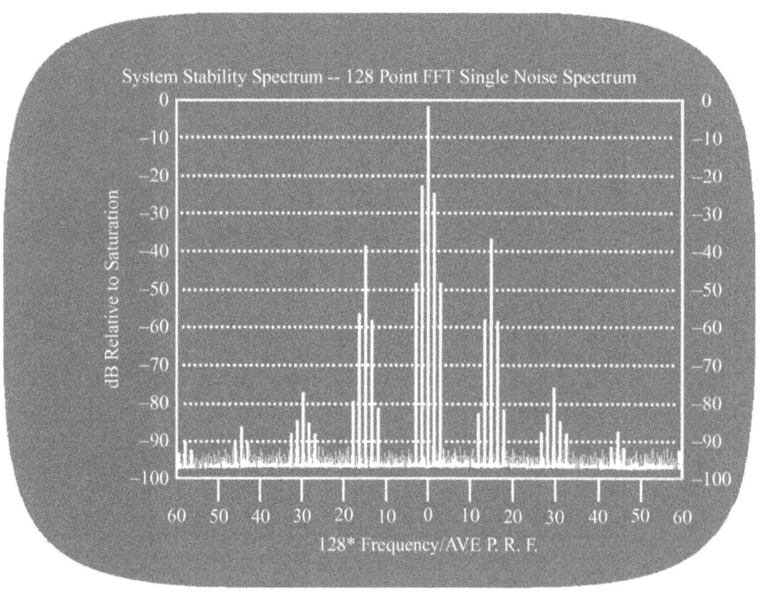

FIGURE 10-14

System stability display, received spectrum.

Fourier Transforms

In recent times, integrated circuits called ***multiplier-accumulator circuits (MAC)*** have been employed in Doppler-filter banks to determine Doppler shift in mtd systems (see Figure 10-14). The Chapter 14 on mtd, in this book, explains the operation of these filters in some detail. The Doppler-filter operation can also be performed by computer software, provided a representation of the transmitter output has been retained in a memory, so that there are no time limitations. In the ASR-9, a special ***system stability test*** provides a means to observe the spectrum of an echo, in the receiver, to verify that the spectrum is repetitive and consistent. A "window" permits the receiver to operate only at a given range, so that the only received data is from a target of interest. The phase-detector digital data for that "window" is then stored in memory for several "hits." The information may then be sequentially read from memory, as data for a computer program, which will test the data for all harmonics within the expected spectrum band. If the spectrum is unstable, it is most likely an indication of poor transmitter performance; the crystal oscillators used for the stalo or coho are rarely unstable.

The Klystron Power Amplifier

General

The klystron power amplifier is the transmitting tube in most high-power synthesis systems. It is common that references to such systems are as "klystron" systems, rather than "synthesis" systems, but the distinction, "synthesis," encompasses all possible methods, including traveling-wave tubes, "hard tubes" (such as tetrodes, in below-microwave radar), twystrons, and solid-state transmitters, as those used in the ARSR-4. The klystron power amplifier is often called the "klystron drift tube," but the proceeding discussion will reveal that the "drift tube" is really only a part of the klystron power amplifier.

Although this author believes it is important that the technician understand the basic internal function of all components, it may be even more important for the klystron-radar technician to understand his transmitting tube, than for the magnetron-radar technician to understand his. The klystron requires tuning, and tuning demands an understanding of rudimentary theory, to prevent damage to the very expensive tube.

Construction

The illustration shown in Figure 10-15 is of a hypothetical three-cavity klystron, the type used successfully, and very reliably, since the 1950s, in the joint-use USAF-FAA AN/FPS-20, and its descendants, including the FPS-64, FPS-65, FPS-66, and FPS-67. The major components are (1) a cathode and filaments, (2) an input-and-"buncher" cavity, (3) an intermediate cavity, (4) an output cavity, and (5) a collector. Latter-day klystrons contain more cavities to provide for both higher power and greater bandwidth. Not shown is a large, external ***focus coil(s),*** or ***focus solenoid,*** essential to the klystron operation. The tube will create and utilize an electron "beam," and, without the magnetic field from the coil, the like charges on the electrons will cause the beam to disperse, dilate, and "defocus."

The Electron Gun

The cathode, filaments, and anode comprise an electron gun, distantly similar to that in a cathode-ray tube. A high negative pulse on the heated cathode causes emission toward the accelerating anode. The accelerating anode is positive, in respect to the cathode, and usually at ground potential. It propels the electrons into the

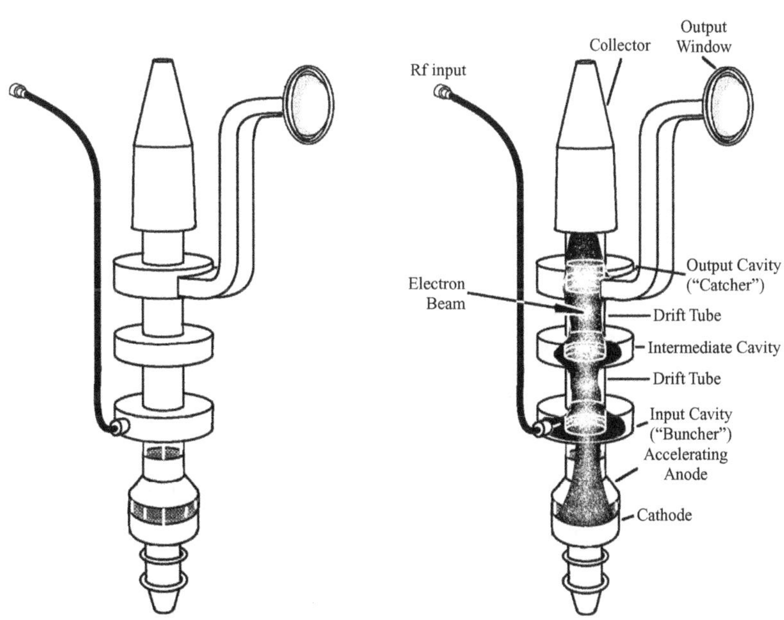

FIGURE 10–15

A three-cavity klystron.

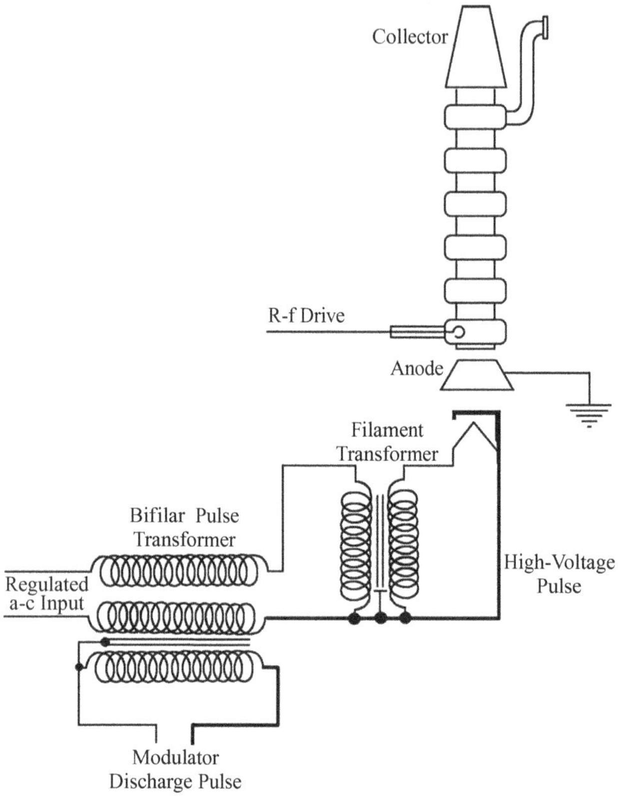

FIGURE 10–16

Bifilar pulse transformer and filament transformer.

gap, surrounded by the input "buncher" cavity. The accelerating anode additionally serves as a "lens" in reducing the diameter of the beam.

There are klystrons utilizing dc filaments, but ac filaments are more common in FAA radars (see Figure 10-16). The filaments require high current, and the most convenient way to achieve it is by reducing a regulated ac voltage with a filament transformer in the pulse tank. It is applied through bifilar windings of the pulse transformer to provide a means to apply both the filaments and high-voltage pulse to one side of the cathode, while providing isolation between the two; the polarization of the windings causes cancellation of the high-voltage pulse in the filament circuit.

The filament current must be precisely regulated, and, although it may be ac, it is often not sinusoidal because of regulation methods. In the ASR-9, for instance, the filament ac is created by a rectangular wave, synchronized to the system timing, in such a manner that the current through the filaments is at a steady, unchanging value at the time when the drive pulse is applied. In the ASR-8, the ac is made irregular and nonsinusoidal, by SCR regulation, and calibration of the filament voltage requires a special ***true rms Voltmeter,*** connected to the filament transformer. The klystron is generally mounted vertically, with the socket immersed in insulating oil. The insulating oil is contained in a tank, often called ***the pulse tank.*** The insulating oil may be circulated through an external cooler. Also to be found in the pulse tank are the filament transformer and pulse transformer. Depending upon the system, still more components may be found in the pulse tank; in the ASR-8, a bank of capacitors, and inductive components, are in the tank.

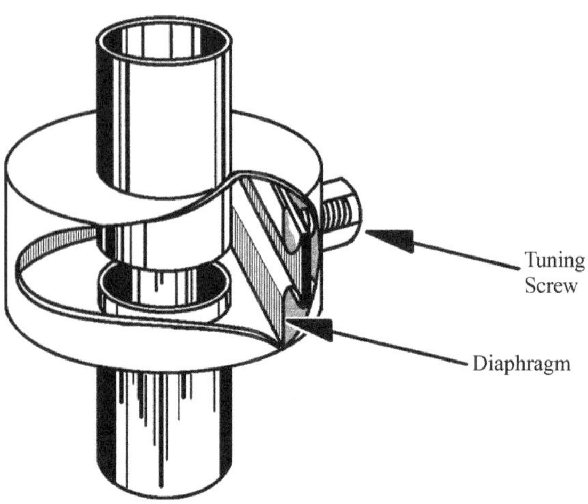

FIGURE 10–17

Tuning diaphragm.

The Input Cavity

A carefully shaped burst of rf drive is applied to an input probe of this cavity. During the negative half-cycles, this drive opposes electrons in the beam, decreasing their velocity. During the positive half cycles of the rf drive, the electrons are accelerated. The input cavity is tuned to the drive frequency, and rings at its natural frequency, to further create fields, which further "modulate" the electron beam. The cavity is tunable. In some tubes, a screw-driven diaphragm, shown in Figure 10-17, provides for this. The alternate slowing and acceleration of electrons creates electron "bunches," or "groups," to proceed into the "drift tube," the conductor to the intermediate cavity. This process is formally called ***velocity modulation*** in engineering and scientific literature. A slightly more detailed description shall follow.

The Intermediate Cavity

As an electron group approaches the intermediate cavity, the negative charge repels free electrons in the cavity space, causing those electrons to flow toward the opposite side of the cavity (see Figure 10-18A). When the electron group is centered over the opening, as in Figure 10-18B, current flow in the cavity is stopped, and when the electron group is passed to the other side of the opening, as in Figure 10-18C, the repulsion induces cavity electron flow in the opposite direction. The electron groups, having been created by the input drive, are occurring at the frequency of the input drive, and the cavity is tuned at, or near, that frequency, by the tuning diaphragm. The cavity then rings of its own accord, in addition to the excitation by the beam, and the e- and h-fields created by the natural ringing of the cavity further cause (1) electron slowing, as the group enters the space, and (2) electron acceleration, as the group leaves the space. The end result of the electron group passing through the space is that it is compressed and better defined.

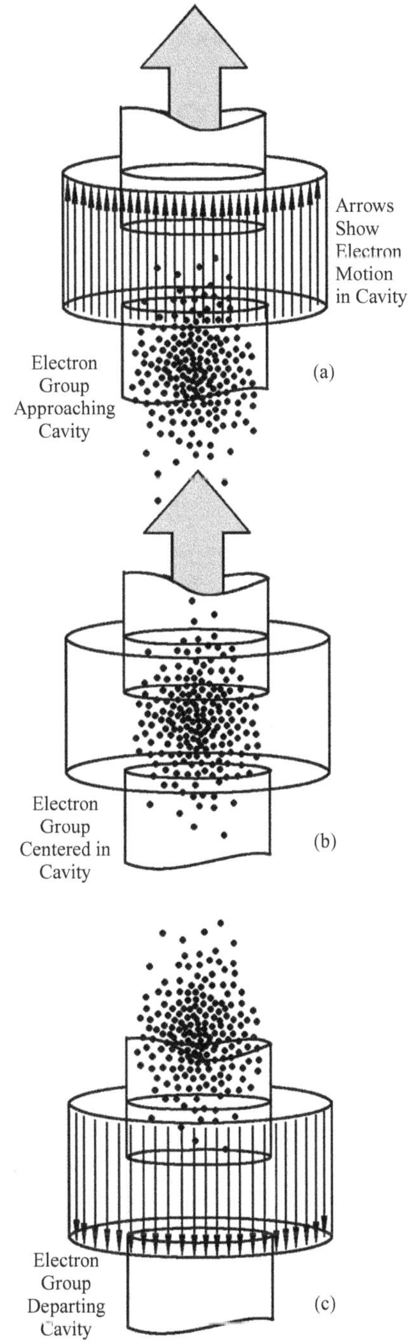

FIGURE 10-18

Electron group passing through cavity.

The Output Cavity

Electron groups pass from the intermediate cavity to the output cavity opening through still another drift tube. The same electron grouping phenomenon again occurs around the "catcher gap" opening to the output cavity. At this point, the groups have reached a very concentrated state, and high energy is created in the output cavity. An opening, covered by a thick glass window, permits most of the energy to be propagated into the waveguide and toward the duplexer. The load on the output cavity, caused by the output window, absorbs energy from the catcher gap, slowing electrons in the beam. In the FPS-20 series radars, air flow is maintained across the output window, to keep it cool. In that system, dried, compressed air is injected in close proximity to the window; a relief valve at the antenna feedhorn allows outward flow and creates the flow across the window. The system also pressurizes the guide to keep moisture from entering. As in the other cavities, the output cavity is tunable.

The Collector

The remaining electrons in the beam proceed to the collector, most of them at a decreased velocity. There are some "out-of-phase" electrons between the groups, and these may actually be accelerated by the polarity of the field in the catcher gap, at the time of their arrival. The collector, normally at ground potential, is often made of copper, to facilitate rapid transfer of the heat caused by the bombardment of electrons. In higher powered systems, as the FAA ARSRs, the collector is cooled by circulating liquid through a jacket. ASRs employing klystrons use air cooling.

Drive Power

The drive burst shape and amplitude are critical to the performance of the klystron. If the pulse that modulated the burst contained exceptionally sharp corners, the drive input would contain too many high-amplitude harmonics. If the burst is low in amplitude, the klystron power output will be equivalently low, and if the burst is excessively

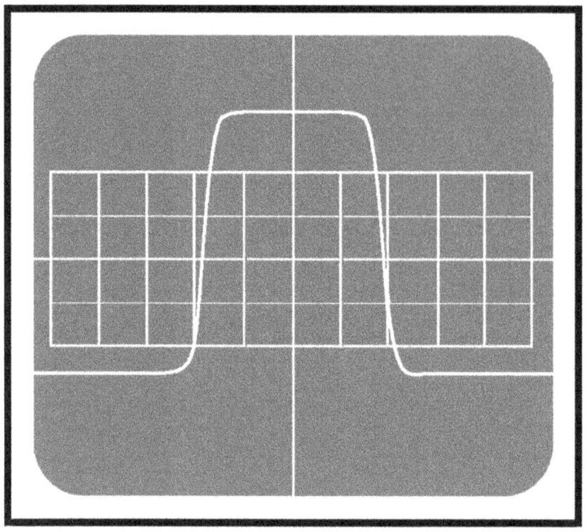

Detected Pulse, Klystron Properly Tuned

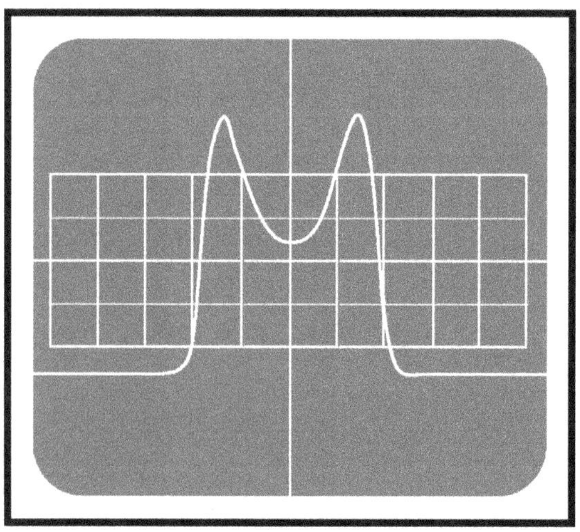

Detected Pulse, Excessive Rf Drive

FIGURE 10–19

Detected bursts.

high, the input cavity will be "saturated," causing the output burst, viewed with a detector on the waveguide directional coupler, to "sag" in the center. Most systems provide a small directional coupler on the drive input to provide for power measurements and/or connection of a video detector, for viewing.

The saturation condition is only roughly comparable to limiting in conventional amplifiers; saturation in the power klystron is a peculiar phenomenon, in which the klystron ceases to perform as designed (see Figure 10-19). Beyond the input cavity, the power klystron will depend on further "bunching" in the intermediate and output cavities. The design is such that the bunches will reach their best form, containing the greatest potential, when they are formed in the output cavity. When the input cavity is overdriven, the bunches are prematurely formed to their highest potential state, and become expanded and dilated by the effects of the intermediate and output cavities; the deformed bunches decrease the power output. The reason for the appearance of "sag" in the detected output pulse is that the dilation of the electron bunches does not begin to occur until the drive burst approaches maximum amplitude.

Body, Collector, and Beam Current

Some of the electrons in the beam strike the drift tubes, creating heat and inducing current flow. This current is called ***body current*** and, if the tube is not properly aligned, can cause enough heat to melt the drift tubes. In some of the higher powered systems, the "body" is cooled by different lines than the collector, and the liquid temperature may be monitored. A current flow is also created by the collector. The sum of the body and collector currents is the ***beam current*** [3]. Some amount of body current is desirable, because a beam of too small a diameter does not provide sufficient coupling into the cavities from the gaps; such a condition could be caused by excessive focus-coil (also called "solenoid") current. Power klystrons may be as little as 30% efficient [3], so as much as 70% of the applied power may be dissipated as heat in the body and collector. An ASR klystron transmitting a 1-μs, 1-MW peak pulse might, therefore, be creating 3 kW of heat that must be removed.

Self-excitation

It is noteworthy for the technician that the klystron may exhibit low levels of self-excitation in the absence of drive, and some close-range echoes may occur. Do not totally dismiss the possibility that the exciter or driver stages may have failed, when the displays indicate exceptionally poor echo performance.

The Ion Pump

Gasses within the klystron may exist for a number of reasons, including long storage and arcs. A "getter," supplied by a low-current, high-voltage power source, removes ions from the tube, and remains in continuous operation.

Radiation Hazard

Wilhelm Roentgen (1845–1923) invented the X-ray, using electron beams to expose photographic film; he died of cancer. Since that time, the danger of X-rays has become better known, and it is well established that they are created by high-energy electron beams. Even cathode-ray tubes operating under exceptionally high voltages produce X-rays. The klystron most certainly produces X-rays, and must be adequately shielded. Periodic monitoring to detect X-ray leakage is an essential safety practice.

Additional Cavities

Most latter-day klystrons have several intermediate cavities; the ASR-8 and ASR-9 use a similar tube, each with four intermediate cavities. As with amplifier stages, the cavities may be "stagger tuned," broadening the bandwidth of the tube. It is essential to the performance of the receiver that the transmitted spectrum contain ample harmonics in the main lobe; the receiver will "reconstruct" the echo spectrum into a pulse.

Further, the expanded capability to tune provides a means to "tune down" the spectrum side lobes; those side lobes may cause deleterious effects on the receiver, particularly in mtd systems. Consider the Fourier content of a pulse, and that harmonic energy exists at times other than during the duration of the pulse. Even still further, concentration of a higher percentage of the total power in the main spectral lobe will increase the efficiency of the overall transmit–receive echoing process.

Tuning

Power klystron manufacturers perform a rigorous set of laboratory-condition tests on each tube, and develop a data form to accompany the tube on shipment. Among the information provided are filament voltage, focus-coil(s) current, and the settings of all tuning mechanisms for frequencies in the band, often in 20-MHz steps.

CAUTION

This information is introductory in nature and should in no way be interpreted as a tuning instruction or procedure. Never attempt to install and tune a power klystron without careful adherence to the radar system manufacturer's instructions. These tubes are very expensive and easily damaged.

In the ASR-8 and ASR-9, an assembly affixed to the focus coil contains mechanical "counters," somewhat similar to the mechanical odometer in an automobile (see Figure 10-20). Each counter is associated with, and physically aligned with, one of the tuning mechanisms in the klystron tube. The counter is turned by a beveled gear, and the beveled gear is driven by another beveled gear on a tuning shaft. When the tuning shaft is inserted, through a passage hole in the focus coil(s), the end engages the tuning mechanism, and the beveled gear on the shaft engages the counter gear. All the parts in the tuning mechanism are made of brass, because of the magnetic field created by the focus coil(s). In the ASR-8, one removable brass tuning "screwdriver," with a knurled handle, is provided. In the ASR-9, six shafts are permanently installed and each is turned by a knob on the tuning assembly.

After installing a new tube, the tuning counters must be zeroed. The tuning mechanisms are first turned either fully clockwise, or fully counterclockwise, depending upon the system. Some klystrons require that care be taken in zeroing the tuning mechanism; if the screw is turned too firmly against its stop, it may jam, necessitating complete removal of the klystron and associated hardware, requiring much labor, consuming much time, and incurring appreciable expense. Once the tuning mechanism has been zeroed, the counters are disengaged and then turned, by the technician's fingers, to zero. The klystron and counters have then been readied for initial adjustment. The technician then consults the manufacturer's data, interpolating as necessary, and turns each tuning mechanism to the appropriate value for the assigned operating frequency.

The tube is first energized with minimum high voltage. The technician observes the spectrum, power, and detected pulse shape, and then gradually increases the voltage, re-tuning as necessary, until the tube is operating

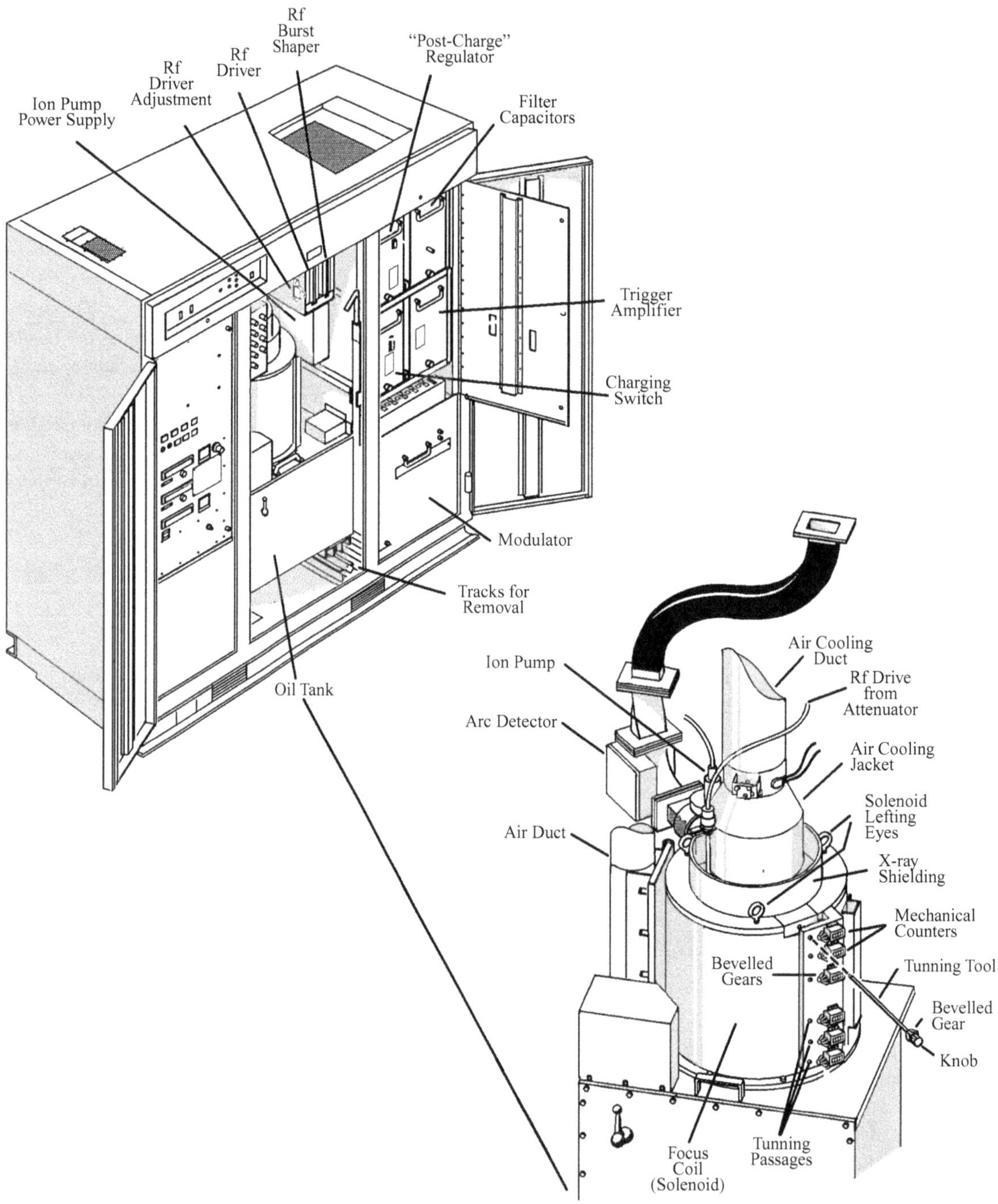

FIGURE 10–20

ASR-9 transmitter. Based on FAA TI 6310.25.

at full power. Follow the manufacturer's instructions for tuning, never "cheat" interlocks, never immediately apply full high voltage to a newly installed tube, and never ignore recommended warmup times. Avoid adjusting any of the tuning mechanisms radically away from the chart figures. Never adjust only for maximum power output, and never adjust without monitoring the detected transmitted burst.

Where no data is available, as in the manufacturer's laboratory, the klystron may be tuned by a procedure in which (1) a minimum high voltage is applied, (2) the directional coupler for the drive input is connected to measure reflected power, (3) the input cavity is tuned for a "dip" in the input power, (4) the output cavity is tuned for a peak, and (5) the intermediate cavities are tuned for a peak. Such a procedure, and the resultant tuning, is called "synchronous" tuning, since all the cavities are tuned to the same frequency. In a six-cavity klystron, the cavities would be tuned in the order, (1) input cavity 1, (2) output cavity 6, (3) next-to-input cavity 2, (4) next-to-output cavity 5, (5) cavity 3, and (6) cavity 4. During the tuning, the drive is continuously lowered to avoid saturation. The first cavities have the highest Q and sharpest response; the fourth and fifth cavities are more broadly tuned.

Synchronous tuning is only a starting point; the tube must be "broad banded," mostly achieved with the intermediate cavities. Should a synchronous-tuned tube be placed in operation with full high voltage, there is a high probability that damage will occur. In the laboratory, the rf drive is supplied by a microwave swept-frequency generator, and the klystron is tuned for the appropriate response at selected frequencies. The tuner positions are then recorded to develop the chart. At a radar facility, the broad banding can be achieved by viewing the spectrum and detected pulse, while tuning the intermediate cavities.

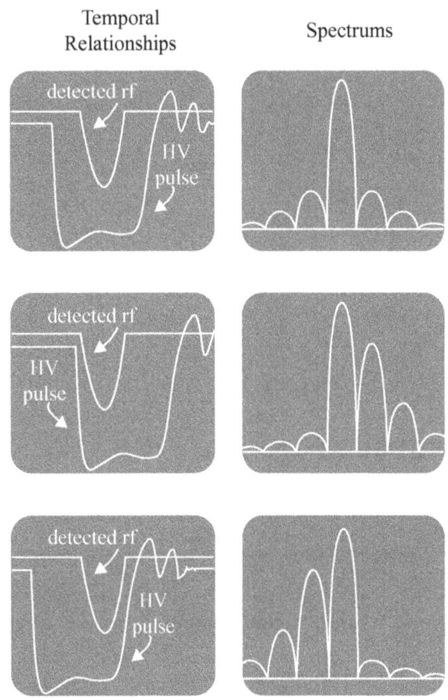

FIGURE 10-21

HV versus drive burst relationship.

Drive Pulse Timing

The high-voltage cathode pulse is of a longer time duration than the drive burst; it begins before the burst, ends after it, and envelopes it (see Figure 10-21). This is done in order that the beam is immediately available for bunching when the drive is applied, and so that variations in the high voltage do not have a deleterious effect on the spectrum. Timing of the high-voltage pulse in relationship to the drive burst must be adjustable, as the high-voltage pulse is subject to variable component delays. In some radars, a manual adjustment is provided; in others, the adjustment may be automatically regulated. As a general rule, the high-voltage pulse, not the drive burst, is adjusted. Temporal changes to the drive burst will alter all the relationships to radar time zero, introducing range errors throughout the system, and in user equipment. The most accurate way to finally adjust the timing relationship between the burst and high-voltage pulse is to monitor the spectrum for balanced side lobes.

The rf driver stage permits adjustment of the drive burst shaping and spectrum. Excessively abrupt rise or fall times cause unnecessarily broad transmitter frequency spectra, and the drive burst is carefully shaped to preclude this. Radar equipment manufacturers may provide precise information regarding the necessary shape of the modulating pulse for the drive burst.

Klystron Interlock Circuits

At this point, it should be clear that the klystron may be easily damaged if all associated equipment is not operating within established tolerances. Excessive collector or body coolant temperatures can indicate mistuning, or cooling system failures. Excessive focus-coil current can prohibit good power development, and cause high collector current. Inadequate focus-coil current can permit beam dilation and high body current. Inadequate filament current can cause cathode damage. Waveguide arcs, if not stopped, can move to the output widow to cause serious damage. Loss of air flow across the output window may promptly cause damage to the glass window, even to the extent of melting. Interlock circuitry is provided to protect against all these conditions. Air and coolant flow meters, air-movement vane switches, temperature-sensitive switches, and more, may all be a part of a klystron transmitter. Arcs are detected by photoelectric devices, to sense light, in the ordinarily dark waveguide.

Out-of-Band Filters

A klystron is rich in harmonics of the fundamental microwave spectrum, and many systems will contain a waveguide filter to prohibit these from being radiated.

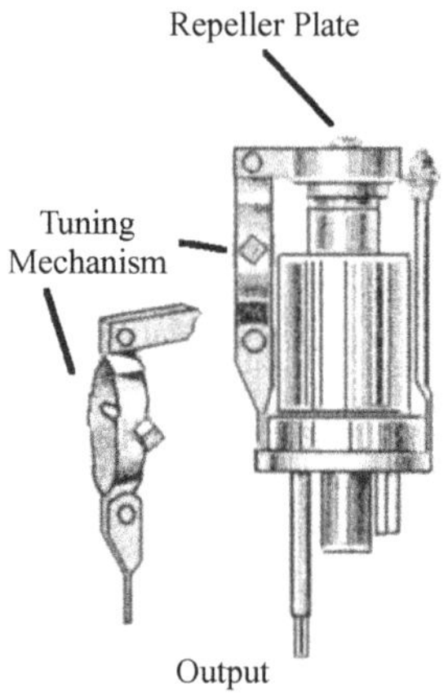

FIGURE 10–22

The reflex klystron. Redrawn from USAF Manual 101-8.

The Magnetron

A Predecessor, the Reflex Klystron

Work on klystrons may well have preceded the magnetron; the brothers Russell and Sigurd Varian [3] of the United States had developed a two-cavity S-band klystron oscillator as early as 1937. However, these early klystrons were not high-power devices, and were not widely used as transmitter tubes. Their development, though, led to the ***reflex klystron*** oscillator, which served as the local oscillator in many early microwave radars. A reflex klystron is illustrated in Figure 10-22. Reflex klystrons are still in use today, particularly in microwave link relay equipment. Klystron high-power amplifiers appeared on the radar scene in the 1950s.

Application by MIT Radiation Laboratories

In 1940, Great Britain provided the United States with the multicavity magnetron, a 10-cm (S-band) high-power oscillator developed by British physicists Sir John Randall and Henry Albert "Harry" Boot at the Birmingham University in England. The magnetron promptly became the transmitter tube in microwave radars for years to come, and is still in use. Compared to the power klystron, it is simple in construction and supporting equipment, and it has the advantages of low cost and economical maintenance. It has the disadvantage of lower power capability, and because it is both an oscillator and transmitter, it is not coherent with the other oscillators in the radar system.

Physical Construction

Figure 10-23 illustrates the construction of a magnetron built in the 1940s. Newer magnetrons are still very similar. The cathode is located in the center of the tube and the anode is at ground potential. Within the anode are eight cylindrical cavities, which are sized to oscillate at the intended frequency. Figure 10-22C illustrates wire straps between the anode segments; these straps were introduced early in magnetron development to prevent "mode skipping"; the straps also affect the operating frequency. For the sake of simplicity, a tuning adjustment is not shown, but tuning may be accomplished by some mechanism to alter the microwave electrical characteristics of the cavity spaces. One method incorporates a movable, circular, piece, which fits over the straps, and which may be moved in position, relative to the straps. Another means incorporates the adjustable insertion of posts into the cavities. There have been some magnetrons which relied upon an external "reference" cavity for frequency determination.

The magnetron illustrated is the original type, called the "hole-and-slot." There are now other forms to provide specific advantages, but the fundamental principles are very similar.

The extensive mathematical and physical principles behind the magnetron operation are beyond the scope of this book; however, there is a need for the technician to have, at least, a qualitative understanding of the manner in which the magnetron functions as a microwave oscillator. Such an understanding will be an aid to fault recognition, and toward an understanding of adjustment procedures. Even beyond those sound reasons, the professional technician should have a natural curiosity and fascination for such information.

The magnetron cannot function without the application of a magnetic field, supplied by an external permanent magnet, as illustrated in Figure 10-23B. A high-potential negative pulse is applied to the cathode, and electrons are propelled toward the grounded anode. However, many factors influence the behavior and trajectory of those electrons, such as the electron velocity, the mass of the electron, the charge on the electron, the external magnetic field, and the e-fields created by the natural ringing of the cavities. These combined phenomena create rotating "clouds" of electrons, which produce a high-energy microwave output.

The "Bottle" Analogy

Even the foremost physicists in the United States initially had difficulty in understanding the magnetron, when it was first given to the MIT Radiation Laboratories in 1940. Probably the leading influence in that project, Isadore

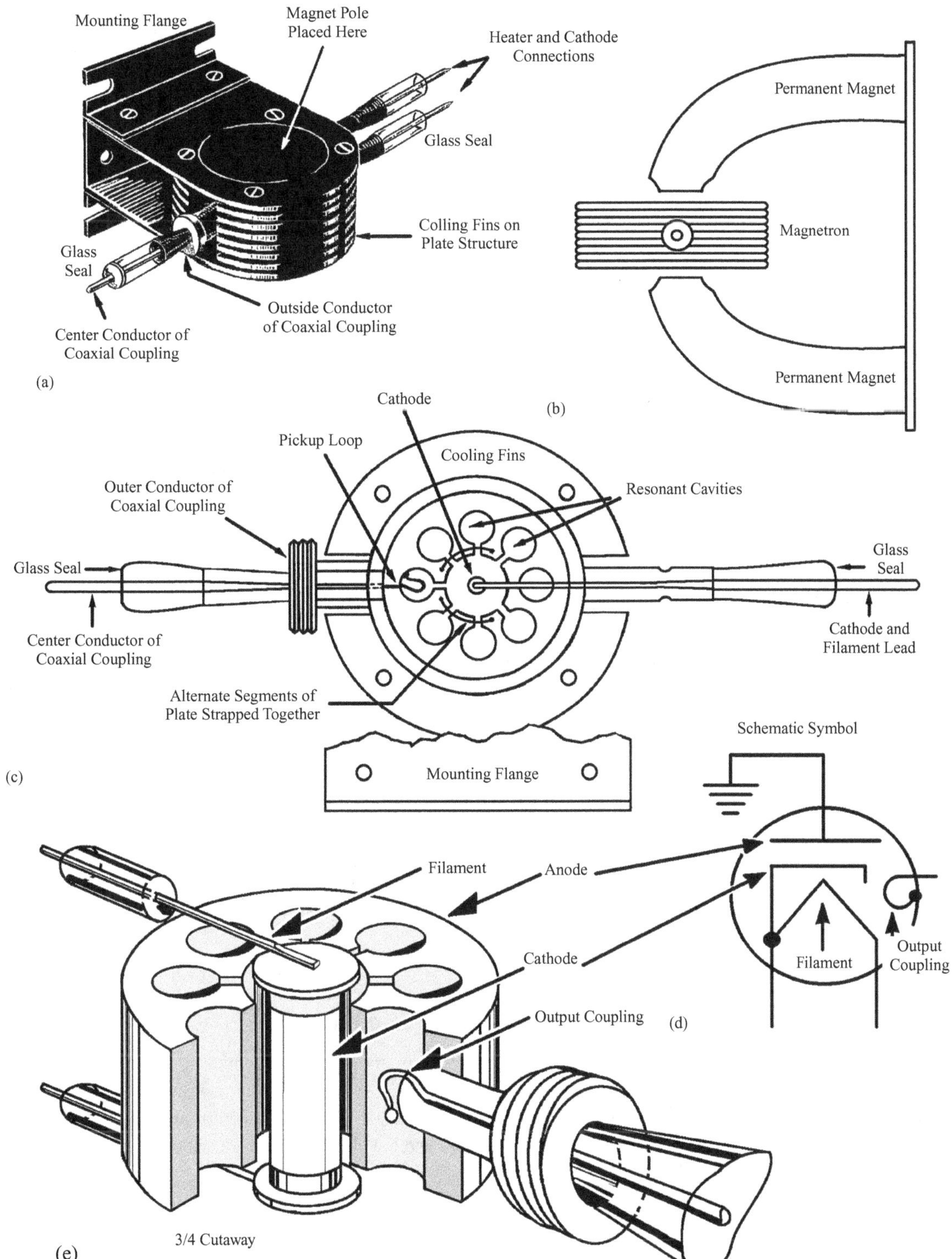

FIGURE 10–23

The magnetron. Some parts from USAF 52-8.

Isaac Robbe [4], in explaining it to others, used an analogy of one blowing across the opening of a bottle to create sound; that analogy became the principle teaching tool in military radar courses. Figure 10-24 is an illustration of the type used in Army Training films during, and after, the war; "GI Joe" is shown creating an audio frequency by blowing across the opening of a soft-drink bottle; the film continues with a simplistic explanation of

FIGURE 10–24

"Bottle" analogy from WWII. Based on a U.S. War Department Training Film Army Service Forces T.F. 11 1385.

the electromagnetic principles of the magnetron. The bottle analogy bears only a vague similarity, but, at least, it provides a good starting point for one to begin forming a concept. Imagine electrons, at high velocity, passing cavity openings to induce microwave oscillations. The microwave oscillations in the cavity then create electrostatic fields which "modulate" the electron motion, forming electron groups.

Underlying Principles

Before any further attention is devoted to the magnetron itself, some major principles must be addressed, to form a foundation for the reader. One of these is the behavior of electrons in magnetic fields, and another is the formation of "cycloids"; these formations occur when a moving electron is subjected to perpendicular magnetic and electric fields. Without some information about these, it would not be possible to understand the operation of the magnetron.

Effect of a Magnetic Field on an Electron in Motion

Figure 10-25A illustrates the behavior of an electron in motion, when subjected to a hypothetical, single-plane magnetic field. The electron has a continual, circular magnetic field about it because its motion constitutes an electrical current. Where that small, magnetic field is in an opposite direction to the external field, there is an attraction, and the electron is physically pulled in that direction. Where the electron's magnetic field is in the same direction as the external field, there is a repulsion, and the electron will be forced away.

Some academic electronics courses teach the use of the "left-hand rule" to remember these force directions. Simply memorizing the three force directions in this illustration will permit the reader to mentally rotate the conditions shown, and adapt them to other field or electron directions. Just remember "approaching, south pole right, force down."

Circular Electron Travel in a Uniform Magnetic Field

Figures 10-25B–C illustrate the electron passing through a uniform magnetic field, which continues to exist after the electron has been forced downward by 90°, as shown in Figure 10-25A. The same attraction and repulsion will continue to affect the direction of the electron, and to reverse it. Once the electron attempts to go in the reverse direction, the fields about it cause an upward motion, and the electron returns to its original direction. In summary, since the magnetic field is uniform, the force directing the course of the electron remains constant, and the electron travels in a circle.

Magnetic Fields

In the science of Physics, different measurement systems have been widely used; the choice has been determined by the size and scale of the quantities normally involved, and by the particular branch of science. The most familiar measurement system in the United States has, until recent decades, been the ***English foot–pound–second*** system. In many countries, and in the more precise sciences, the ***centimeters–grams–seconds (cgs)*** system was used for many years. In electronics, the ***meters–kilograms–seconds (mks)*** system has become nearly universal, partly because of an international agreement in the 1930s, and further, by another 1970s international agreement,

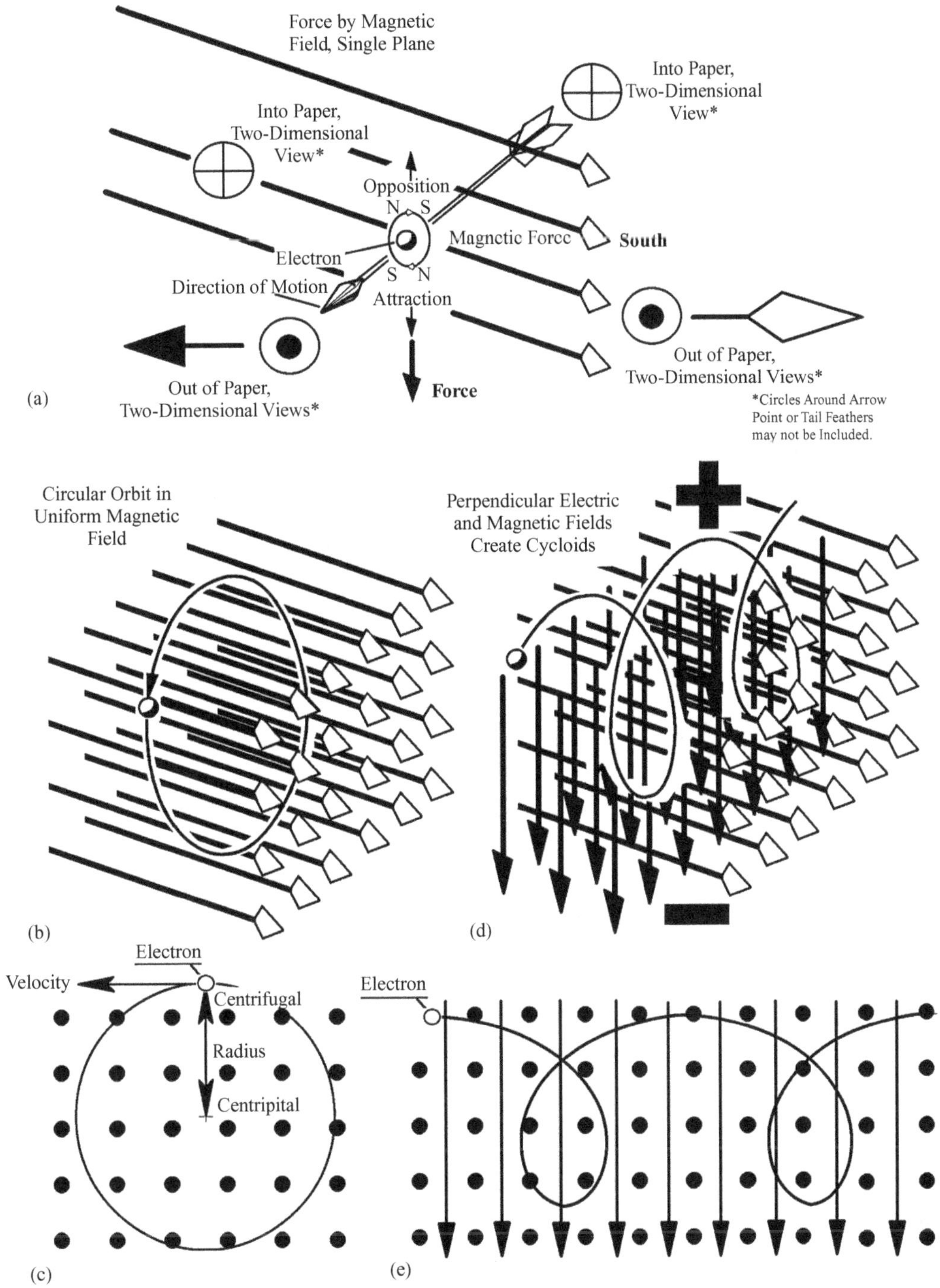

FIGURE 10–25

Behavior of an electron in motion, in magnetic and electric fields.

which created the ***SI system,*** named for Système International d' Unités. The relationship between, for example, the Coulomb and the Ampere demonstrates the necessary tie to mks; a Coulomb is a charge equivalent to 6.28×10^{18} electrons, and the Ampere is the movement of one Coulomb, over a hypothetical pair of "perfect" wires, 1 m in length.

Two German physicists, Karl Friedrich Gauss (1777–1855) [6] and Wilhelm Eduard Weber (1804–1891) [5], worked extensively with the measurement of magnetic fields. ***The Gauss*** is equal to one line of magnetic force per centimeter, and is in the cgs system. ***The Weber*** is a flux measurement in the mks system. A magnetic flux, that changes at the rate of one Weber per second, produces 1 V, and a Weber is a volt-second. One Weber per square meter is equal to 10,000 Gausses [2]. It is significant to note this conversion number; there is 1×10^5 difference (cm to meters and grams to kg) between cgs and mks. Although the mks system, and its successor, the SI system have become the international standard, the Gauss is still the conventional expression of magnetic field intensity, in contradiction to the standards. The Gauss is used on magnetron performance charts and on information regarding the magnets. The term ***Tesla*** is sometimes used to express density in Webers per square meter.

Further involved with electron and magnetic physical mathematics are

The dyne, unit of force, in the cgs system
The Newton, unit of force, in the mks system
1 Newton = 1×10^5 dynes
The mass of the electron, m, 9.11×10^{-28} g
The charge on the electron, e, 1.59×10^{-19} C

The following equations quantify the circular electron movement. The centrifugal force, caused by the electron's mass, opposes the electron's circular course; the centripetal force, caused by the electron's charge, opposes the centrifugal force. The two opposing forces determine the radius of the circle, at which point those forces are equal and opposite. The time required for the electron to make one trip around the circumference will finally be related to the magnetron frequency, and is as shown in the equations. The magnetic field is measured in Gauss, so it may be converted to Webers for use with the mks system. Recall that the Gauss is equal to 1×10^{-4} Webers, but the relationship between mks and cgs is 1×10^{-5}. This must be taken into account if actual numerical results are to be sought. The following equations use the cgs system:

$$F_{\text{centrifugal}_{\text{dynes}}} = \frac{mV^2_{\text{cm/s}}}{R_{\text{cm}}} = \frac{(9.11\times10^{-28}\text{ g})V^2_{\text{cm/s}}}{R_{\text{cm}}}$$

$$F_{\text{centripital}_{\text{dynes}}} = eB_{\text{Gauss}}V_{\text{cm/s}} = (1.59\times10^{-19}\text{ coulombs})\,B_{\text{Gauss}}V_{\text{cm/s}}$$

$$F_{\text{centrifugal}} = F_{\text{centripital}}.$$

Therefore

$$\frac{mV^2_{\text{cm/s}}}{R_{\text{cm}}} = eB_{\text{Gauss}}V_{\text{cm/s}}$$

$$mV^2_{\text{cm/s}} = R_{\text{cm}}\,eB_{\text{Gauss}}V_{\text{cm/s}}$$

$$R_{\text{cm}} = \frac{mV^2_{\text{cm/s}}}{eB_{\text{Gauss}}V_{\text{cm/s}}}.$$

Dividing numerator and denominator by V, and substituting known values, we get

$$R_{\text{cm}} = \frac{9.11\times10^{-28}V_{\text{cm/s}}}{1.59\times10^{-19}B_{\text{Gauss}}}.$$

Simplifying it, we get

$$R_{\text{cm}} = \frac{5.72\times10^{-9}V_{\text{cm/s}}}{B_{\text{Gauss}}}$$

$$\text{Circumference} = 2\pi R_{\text{cm}} = \frac{2\pi(5.72\times10^{-9})V_{\text{cm/s}}}{B_{\text{Gauss}}}.$$

Now,

$$d = rt;$$

therefore

$$\text{Circumference} = V_{\text{cm/s}}t_{\text{seconds}}$$

$$\frac{3.59\times10^{-8}V_{\text{cm/s}}}{B_{\text{Gauss}}} = V_{\text{cm/s}}t_{\text{seconds}}.$$

Now divide both sides by $V_{cm/s}$. The time is dependent only on the mass and charge, with respect to the magnetic field:

$$\text{Time}_{\text{seconds}} = \frac{3.59 \times 10^{-8}}{B_{\text{Gauss}}}.$$

The Cycloid

Figures 10-25D and E illustrate one possible phenomenon that may occur when an electric field is applied perpendicular to the magnetic field. The formerly circular electron motion is made irregular by the opposition or attraction of the electric field. Where the electron trajectory is toward the positive potential, the electron is accelerated, and, where the trajectory is toward the negative potential, the electron is repulsed and slowed. Scientific literature will indicate that the force toward the positive potential is due to the electrostatic attraction, and that the force toward the negative potential is due to the magnetic field. It would be interesting to know if Randel and Boot were so brilliant as to deliberately incorporate all this into the magnetron design, or if the design were more empirical, and more of the phenomenon was envisioned afterward. It has been this author's observation that the genius of scientists of that era was incredible, and the probability that they may have envisioned the entire process may be very high.

Figures 10-26A and B show that the shape of the cycloid is dependent upon the initial direction of the electron; however, most importantly, the points at which the cycloids repeat do not change. The phenomenon is similar to the movement of a point on a rotating wheel, shown in Figure 10-26C. On the illustration of the wheel, the point at the center of the wheel moves at a constant velocity, but points on the radius of the wheel may accelerate and decelerate in respect to the velocity of the center. It might be said that those points on the radius tend to "dwell" in certain areas. Where the point is on an extended radius, past the rolling circumference, the motion reverses at the point where the radius is perpendicular to the rolling surface; the action forms a "loop" in the cycloid. This loop is also the path traveled by an electron, the initial velocity of which is opposite the direction of travel caused by the electric and magnetic fields. In the case of the rolling wheel, all the points slow to a minimum "forward" velocity where the radius is perpendicular to the rolling surface. In the case of electrons, they also slow to a minimum velocity at a similar point. In so slowing, because of the magnetic field, the electrons give up energy to an electric field, creating a concentration of power. Thus, at given points in the motion of electrons in the magnetron, the electrons are concentrated into slowed "clouds," and a concentrated field is created. The electrons all travel in different cycloidal paths, because of the random direction of their emission from the cathode. Where the electrons are being accelerated and diffused, energy is consumed, reducing the field.

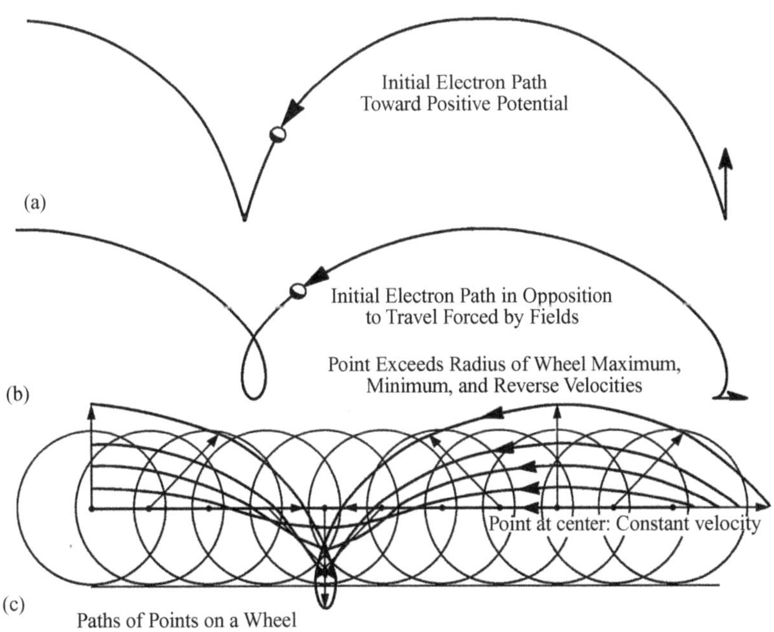

FIGURE 10–26

Cycloidal paths.

Formation of Synchronous Electron Clouds

What is even more significant about the cycloidal rotation of the electrons in the magnetron is that the electric field is not constant and continual, and is only partially caused by the high-voltage cathode. Recalling the bottle analogy, when the electrons

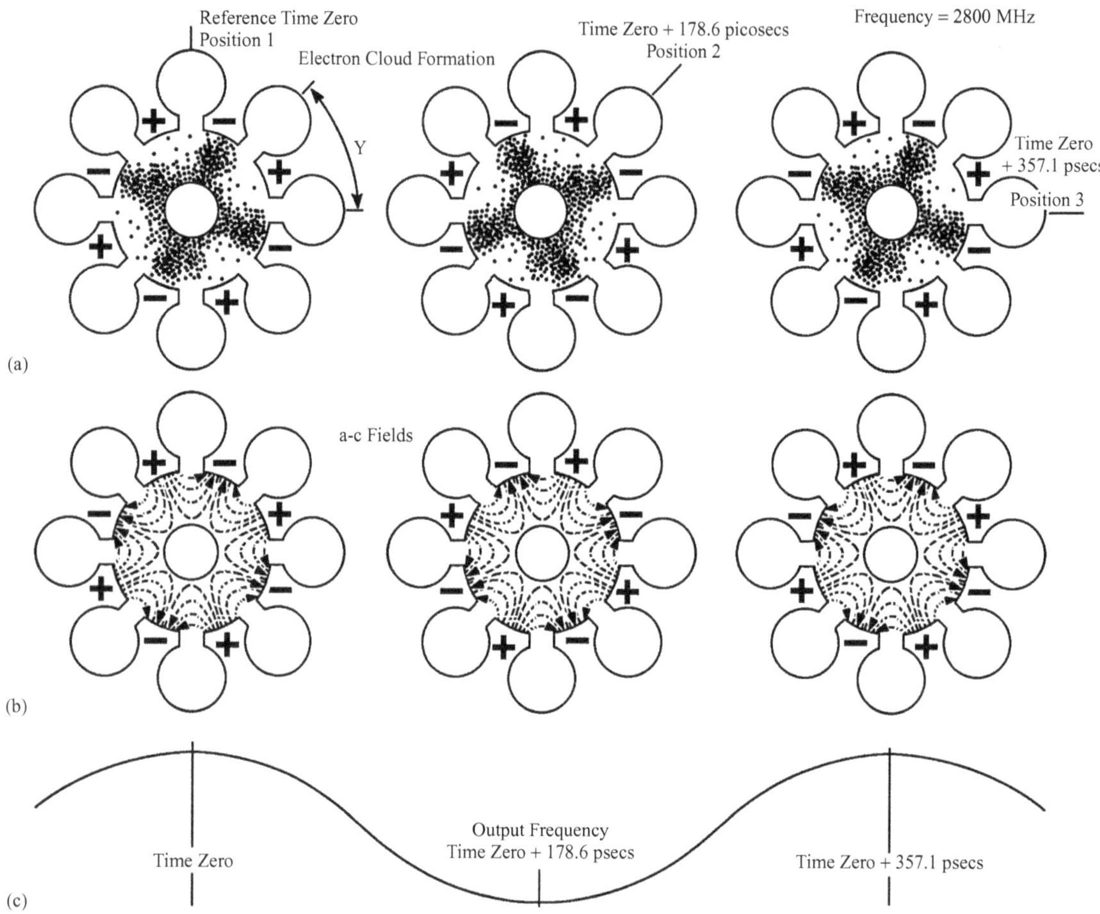

FIGURE 10–27

Magnetron operation.

pass a cavity opening, the fields associated with their motion induce electromagnetic rf activity in the cavity, which, in turn, has the same effect as instantaneously polarizing the anode segments surrounding the cavity inlets.

Since the anode segment polarization is determined by the cavity oscillations, it may be reversed, with each half cycle. Figure 10-27 shows the hypothetical polarization of the anode segments at three successive points, each separated, by a half cycle, from the one adjacent. Because of the oscillations in the cavities, the alternating anode segment polarization, and the magnetic and dc electrostatic fields, the electron motions become synchronized to the cavity oscillations, and the electron clouds gather and dissipate at the magnetron cavity frequency. Many of the electrons do not follow the desired and intended path, and return to the cathode, causing bombardment heating. For this reason, the filament voltage on a magnetron is automatically reduced by a relay, when the high voltage is applied, because the tube does its own heating. Also, because of those electrons which do not contribute to the output power, the magnetron is only somewhere in the area of 30–60% efficient. Efficiencies in the 50–60% region are very common.

It is interesting that much of the magnetron theory literature does not deal with the instantaneous fields in the cavities, assuming the reader should know about these from waveguide and cavity theory. The reader should understand that they are oscillating, and will produce the ac fields in the ***interaction space.*** Much of this literature also attributes this subject as too complex, and beyond the intent of the authors. The information provided here should also be regarded as very simplistic. For instance, it considered only a single frequency, while an entire spectrum exists. Certainly, the technician maintaining a magnetron transmitter needs only a qualitative recognition of the operation.

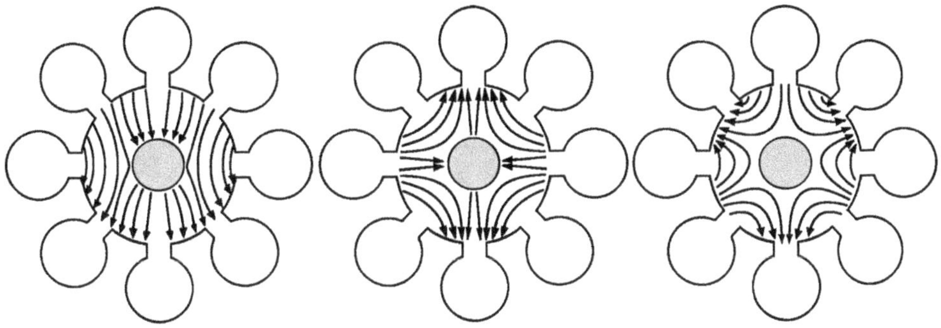

FIGURE 10-28

Magnetron modes of operation.

Modes of Operation

The magnetron illustrated in Figure 10-27 is operating in the desirable "pi" mode, so named because the phase of the cavities are separated by π radians, or 180°. Should the fields be less than optimum, as when the cathode voltage is changed, the anode-segment polarity may cease to alternate as intended, and the magnetron will begin to operate in a "lesser" mode. These other modes, ***π–1, the third mode, π–2, the second mode,*** and ***π–3, the first mode,*** are illustrated in Figure 10-28. The choice of magnetic field strength is based partly on an intent to prevent moding; calculation of the time for a circular trip of an electron in comparison to the intended operating frequency would reveal that the magnetic field would appear to be significantly greater than indicated.

Strapping

To prevent the magnetron from changing modes, wire straps are connected between anode segments, forcing those segments strapped together to be at the same phase (see Figure 10-29). There are several strap configurations. Straps are not practical in the K band, and the "rising sun" magnetron was created to prevent mode "jumping" in K-band oscillators. In the rising-sun magnetron, there are two π-mode frequency responses, and the rectangular shapes of the cavities reduce their "*Q*," to provide broader responses, which overlap (1) the upper response of the lower frequency, and (2) the lower response of the higher frequency. Since all cavities may resonate only at the point of overlap, moding is prevented.

Performance Characteristics

The preceding discussion should have made it obvious that the magnetron depends on electrostatic and electromagnetic fields, and may not operate properly, if the strength of these fields does not fall within certain limits. Figure 10-30 illustrates the relationships between (1) the magnetic field in Gauss, (2) the applied high voltage and current, (3) the efficiency, calculated from power applied versus rf power output, (4) frequency pulling under varied operating parameters, and (5) the peak burst output power in kiloWatts. Note that the deviation beyond parameter limits may force the magnetron into other modes and create poor frequency spectra.

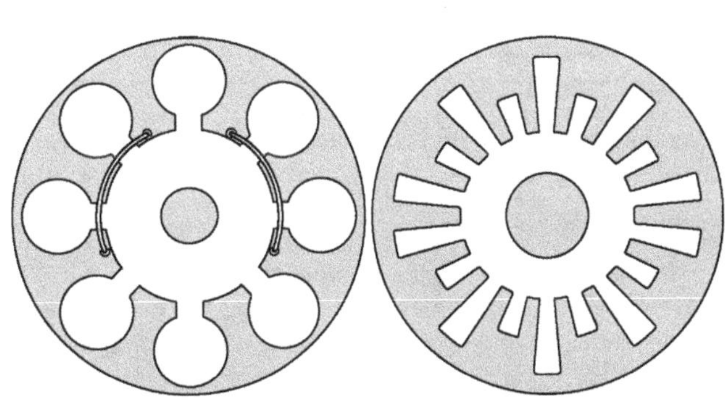

FIGURE 10-29

Strapping and alternate cavity sizing.

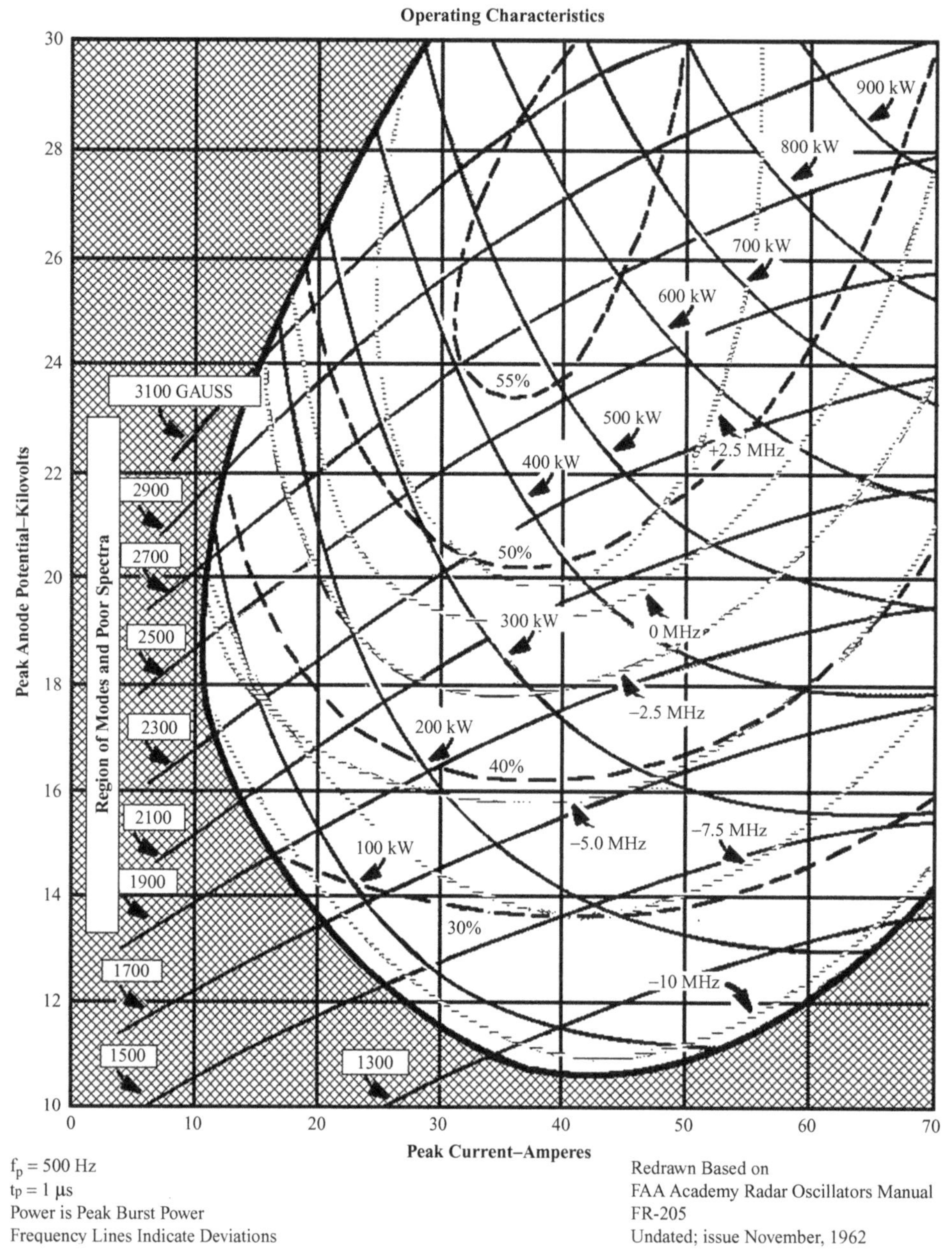

FIGURE 10–30

Operating parameters and characteristics. Redrawn. Based on FAA Academy Radar Oscillators Manual FR-205. Undated: issued November 1962.

The appearance of the spectrum on an analyzer is determined both by the operating fields in the magnetron and by the shape of the modulating pulse applied to the cathode. Since the shape of the modulating pulse is usually determined by fixed components, the odds are very high that a poor spectrum is caused by a magnetron.

The Amplitron, aka Crossed-Field Amplifier

The amplitron may have been so named, because it is an amplifier, yet is somewhat similar to a magnetron (seet Figure 10-31). The amplitron was introduced in the late 1950s or early 1960s as a final power amplifier device, serving two somewhat different purposes in FAA and FAA/USAF radar systems. In the AN/FPS-35 synthesis-type radar system, which operated in the UHF region, an amplitron boosted the power from a series of tetrode amplifiers, which were, in turn, supplied by a frequency synthesizer. In FAA ARSRs, the amplitron was introduced to supplement the power output of a magnetron, and the radar still operated as a magnetron-type system, without all the coherence and stability advantages of synthesis. The amplitron, which may exhibit gain from 3 to 20 dB, is no substitute for the power klystron, which may exhibit gains in excess of 50 dB. Of course, amplitrons may be serially connected for additional gain, and do have a significant advantage in greater bandwidth.

The amplitron's similarity to the magnetron is in the application of perpendicular magnetic and electric fields, cycloidal electron movements, and cloud formations. It does not, however, contain the ring of cavities which cause the magnetron to oscillate. Instead, it utilizes a circular rf transmission line, which provides the cloud-forming rf synchronization fields for the interaction space. Superior to the klystron in one respect, the amplitron is considered a wide-band device, and responds to a wide range of input frequencies; it simply changes the number of cloud formations with input frequency changes. This explains its former use in the FPS-35, a frequency-agile radar, capable of changing frequency, within a wide band, with each transmitter pulse, at programmed azimuths, or randomly. This also explains the amplitron's use in agile-beam, phased-array radars.

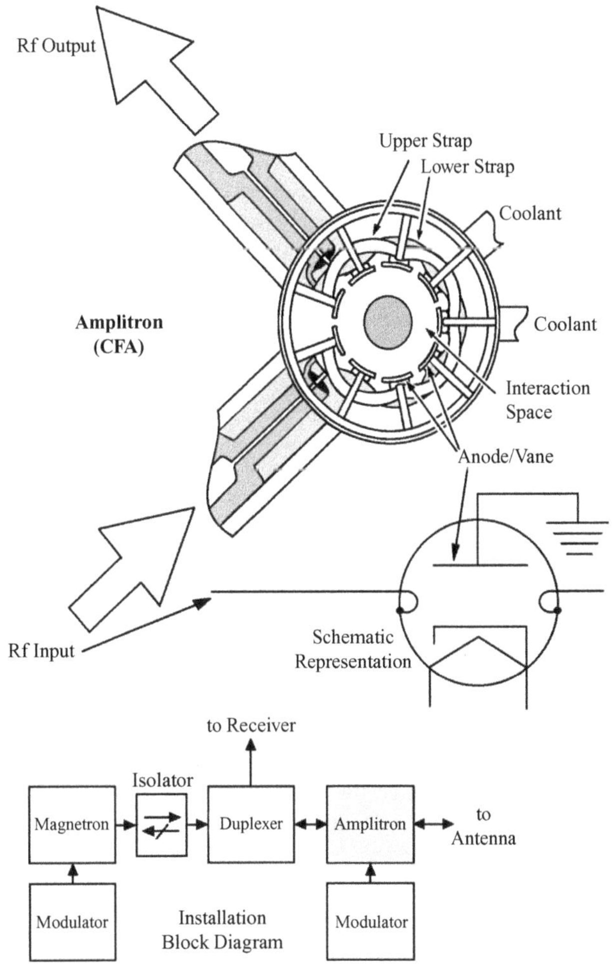

FIGURE 10–31

The amplitron.

Obviously, to amplify the entire burst without distortion, the amplitron must be in full conduction when the drive is applied, and must not cease to conduct until after the drive has ended. Thus, the high-voltage pulse must be wider than that used for the driver, and must "bracket" the drive burst. The amplitron operates in saturation, and, provided there is adequate input power, the output is dependent mostly upon the level of the applied high-voltage pulse. It may be located on the antenna side of the duplexer, since the TR device will have been energized by the lower power burst, and because there is negligible reflection by the amplitron. If reflections do occur, the TR device protects the receiver, and an isolator protects the driving device. In the absence of a high-voltage pulse, received echoes pass, essentially unaffected, through the transmission line in the amplitron. If the amplitron is not energized, the driving device power becomes the radiated output. If the amplitron is energized without drive, a high-power, wide-band noise burst is transmitted.

The Modulator

At one time, pulsed radar modulators were so similar that once a technician gained a thorough understanding of the modulator in one system, nearly all others were so similar, that his main task in becoming familiar with a new system was only to learn the location of components and note design variations. As radar design evolved into larger and more powerful transmitters, and as designers attempted to eliminate the high-failure components in the modulators, new and different concepts appeared. Given the "target" size of this book and chapter, and the variety of modulator types now in use, it is necessary to limit this discussion to the original and fundamental design, with only brief mention of others.

Requirements

In a magnetron system, the magnetron will oscillate, as long as the negative-going, high-voltage pulse remains on the cathode. Obviously, then, the presence of high voltage on the tube at times other than the "main bang" is unacceptable. It would also be unacceptable, in the case of the amplitron, because of the high-power noise that would be emitted. In the case of the power klystron, emission is negligible in the absence of drive; however, continual application of dc voltage would cause excessive average body and collector currents, overheating and damaging the tube. So, in all pulsed-radar cases, the transmitter power amplifier tube requires a high-voltage pulse, and continuous dc is unacceptable.

Beyond the requirements of the high-power tubes, the high voltages and currents necessary for the modulating pulse give rise to many more complexities. It is neither practical nor necessary to construct high-voltage power supplies, capable of continual dc voltages, equal to the cathode voltages of the power-amplifier tubes; if for no other reason, the failure possibilities offered by such enormous voltages would be very high. Since the high voltage is only needed at "main bang" time, the system and components can be made smaller, and more manageable, by allowing the power supply most of the system interval to reach its full potential.

Fundamental Pulse Modulator Principle

To achieve the desired high voltage, earlier modulators employed, among other components, a charging inductance, a pulse-forming network (pfn), and a pulse transformer (see Figure 10-32). The use of these components introduces reactances and all the accompanying implications of ac theory. The modulator then becomes, principally, a circuit which builds a high-voltage charge on the pfn, over the T_r; the pfn is rapidly discharged through a switch, with a high current through the primary of the pulse transformer, which produces the cathode pulse for the power amplifier. Newer modulators use different components and switching arrangements; however, the basic concept of allowing most of the T_r to reach a full potential, and then using a switch to complete a high current discharge path through the pulse transformer, remains in all but the very latest systems.

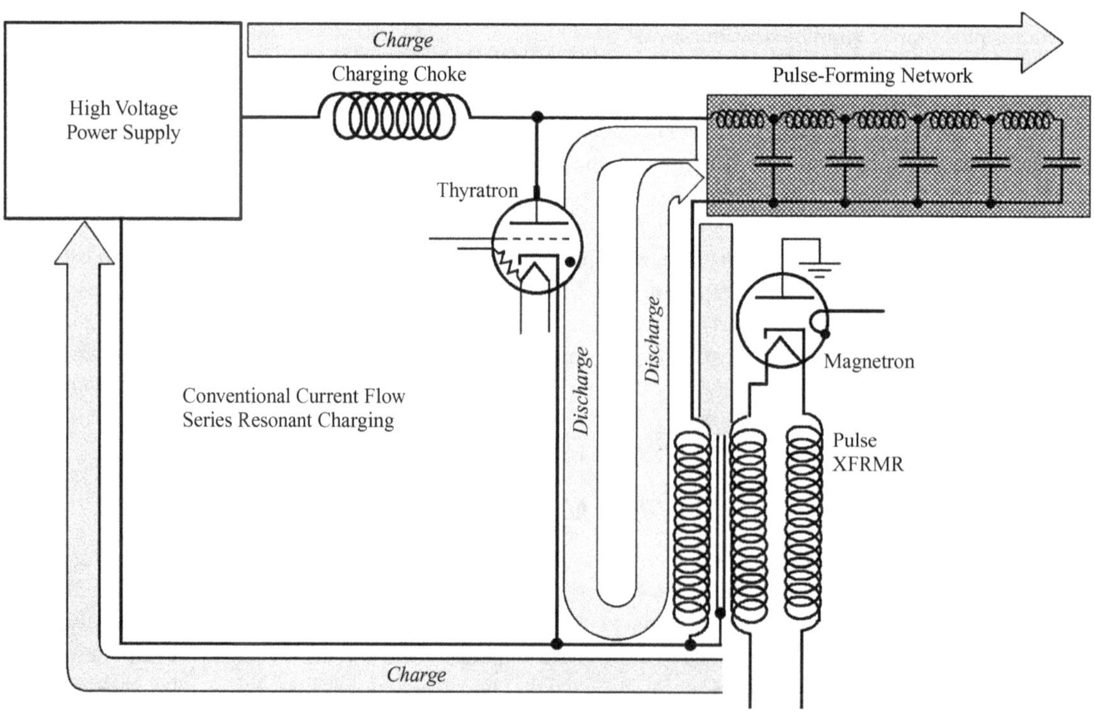

FIGURE 10-32

Series resonant charging, thyratron switch.

The ASR-4, ASR-5, and ASR-6, as an Example

Figure 10-33 is an abbreviated and composite schematic diagram of the ASR-4/5/6 modulator. It will be used for an example of the basic and original principle.

Interlocks

At the upper-left corner of Figure 10-33 is a group of interlock switches, provided both for personnel safety and for protection of the equipment. The opening of any switch in the chain will, in some manner, result in a high-voltage shutdown. If the modulator or high-voltage compartment interlocks are opened, relay K2 will immediately de-energize, and the contacts will cause the high-voltage power supply (HVPS) to be discharged through

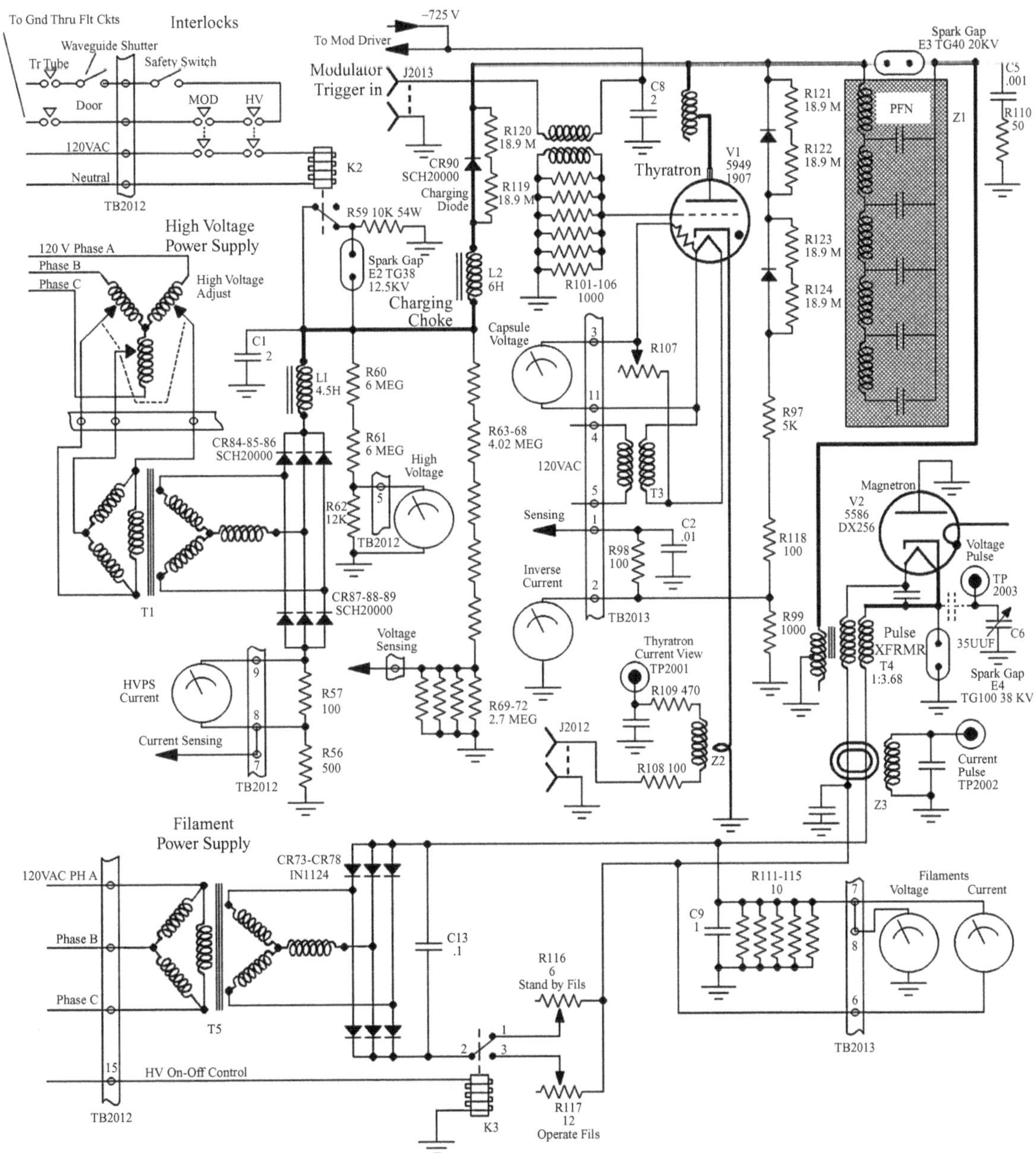

FIGURE 10-33

ASR-4/5/6 modulator.

R59 to ground. Should any of the other interlocks be opened, the 120 V on the coil of K2 will be removed by fault circuitry. Interlock chains such as this are common in transmitter circuits.

The High-Voltage Power Supply

The HVPS uses three-phase ac to provide the necessary current supply. A manually variable inductive component provides for high-voltage adjustment, and a delta-wye transformer supplies a three-phase rectifier circuit. L1 and C1 serve as a power supply filter; L1 also is a part of the charging inductance. A metering circuit, connected to the junction of L1 and C1, indicates the supply voltage. Resistors R57 and R56 serve as a shunt and divider for the power supply current meter. In the event that the high voltage should, for some reason, reach an excessive value of 12.5 kV or more, the spark gap E2 will arc, and the charge will be routed through R59 to ground.

The Charging Chokes and Diode

Inductor L2 is commonly called the ***charging choke*** in most modulators of this type. In this case, both L1 and L2 serve as the charging inductance. The ***charging diode*** was an addition that was necessitated by the introduction of staggered f_p. Once the pulse-forming network has reached maximum charge, it will attempt to discharge in the opposite direction; the charging diode prohibits that action. Before the introduction of staggered f_p, the values of the pfn and charging choke were chosen to cause the pfn to reach full charge just as the modulator trigger occurred.

Charging the pfn to Twice the HVPS Voltage

Except during the pulse, the hydrogen thyratron V1 is not conducting, and the high voltage begins to charge the pfn. During the charge cycle, the pfn, in comparison to the charging inductance, is predominately capacitive. Hence, there is a series resonant circuit comprising the charging inductance and the capacitance of the pfn. Once the pfn has been charged to the HVPS voltage, through the charging inductance, there is no further potential difference at the input and output terminals of the inductance. At that point, however, the current through the coil had created a magnetic field about it, and the collapse of that field sustains the current flow to charge the pfn to twice the high-voltage power supply voltage.

The current path through the pulse-forming network is continued through the primary of the pulse transformer to ground, completing the HVPS load circuit. Compared to the inductance of the pulse transformer, the frequency content in the slowly changing current rise during the charge cycle is insufficient to create enough reactance to induce a voltage on the magnetron cathode, and it would be the wrong polarity, even if it were significant.

The Hydrogen Thyratron

The thyratron, now approaching extinction, is a hydrogen-filled switch tube. Once the hydrogen is ionized by a high-voltage trigger on its grid, and from the modulator driver, often called the ***baby modulator,*** it will conduct until the source voltage is depleted. A capsule in the thyratron maintains the hydrogen supply, and the hydrogen production may be adjusted with voltage. Capsule voltage is ordinarily about 5 V. Inadequate capsule voltage causes the tube to exhibit a reddish glow, and excessive voltage causes a blue appearance. The desirable condition is between the two extremes, a violet color. Improper adjustment will cause transmitter faults, usually shutting the transmitter down by fault interlock circuitry.

In solid-state circuitry, the SCR bears great similarity to the thyratron. When forward biased and triggered, it will continue to conduct until it is no longer forward biased.

The Modulator Driver

A modulator driver is necessary to trigger the thyratron because conventional circuits cannot supply the 600- to 700-V trigger required for the thyratron. The modulator driver is little more than a miniature modulator, frequently called, in slang, the ***baby modulator.*** The input transformer for the main modulator is the pulse transformer for the modulator driver.

Discharging the Pulse-Forming Network

When the thyratron is ionized by the modulator trigger, the pfn is discharged through it, and high current flows through the pulse transformer primary. The turns ratio of the pulse transformer serves two related purposes. Of course, it increases the high voltage by the value of the ratio, but it also matches the impedance of the magnetron to the impedance of the pfn. Recall that the impedance through a transformer changes with the square of the ratio. If the impedance of the magnetron were 650 Ω, the impedance at the transformer primary would be reduced by the square of 1/3.68, which is 0.0738. The 650-Ω impedance of the magnetron then appears to be 47.99 Ω at the primary. The impedance of the pfn will also be in the neighborhood of 50 Ω, and the match will cause a maximum transfer of energy with minimum reflections. Actually, a small mismatch is deliberately introduced to ensure that, at the end of the pulse, the thyratron plate and cathode are, respectively, negative and positive; this assures that the thyratron will be de-ionized at the end of the pulse. Reverse currents, usually called inverse currents, from the pulse-forming network, will flow through an inverse diode circuit, R97, R118, and R99 to ground. The drop across R99 provides a meter indication of inverse current.

As the pfn discharges, the reactance of the pulse transformer, and the impedance of the pfn, being nearly equal, become a voltage divider, with approximately half the pfn charge dropped across each. The pulse on the pulse transformer primary is therefore nearly equal to the high-voltage power supply, but reduced by efficiency losses in the charging system. The pfn is an ***artificial transmission line,*** and the discharge is a voltage wave, the duration being determined by the component values in the network, thus determining the burst width.

The combined operation of the pfn, thyratron switch and pulse-transformer circuit requires a deeper analysis for understanding. That will follow this general discussion. The pfn, the manner in which voltage and current waves move from one end to the other to determine the discharge time, and the width of the transmitter pulse require some analysis of transmission lines.

Magnetron Filament Circuit

The magnetron is supplied with dc filaments. Transformer T5 provides three-phase ac to a full-wave rectifier, and the rectifier outputs are connected to the magnetron filaments through the bifilar windings of the pulse transformer. The pulse transformer bifilar windings are polarized to cancel the transmitter pulse on the filament lines. Relay K3 is energized when the high voltage is energized, so that R117 decreases the filament current. Recall that cathode bombardment maintains magnetron heat during operation, and the filaments may be decreased in operation, because of that bombardment heat. Metering circuits provide for monitoring of both filament voltage and current.

Test Points

For technician observations, TP2001, TP2002, and TP2003 provide for viewing of, respectively, the thyratron current pulse, magnetron current pulse, and magnetron voltage pulse. The magnetron voltage monitor is simply a proximity device, with no direct connections to the high-voltage circuitry.

Spark gaps are provided to preclude damage in the event of excess voltages. They are connected both across the pfn, and on the magnetron cathode.

Real and Artificial Transmission Lines

Real Transmission Lines

A real transmission line is familiar to most nonradar technicians. It is one constructed for the principal purpose of moving signal information, and an ***artificial transmission line*** is a unit constructed to exhibit electrical characteristics resembling a real transmission line. The artificial line does not require great physical length, and may be constructed to withstand high voltages, but is inferior in frequency response, which can be an advantage. The most familiar real transmission line is the coaxial cable, and those characteristics of greatest interest to the technician are (1) the characteristic impedance, (2) inductance per unit of length, (3) capacitance per unit of length, (4) loss per unit of length, (5) a frequency at which the line becomes reactive, (6) voltage rating, and (7) the velocity factor, a decimal number or percentage used to calculate and increase the signal propagation time from the speed of light, 1.015 ns per foot.

Some examples of common real transmission lines are listed in the following table. A very few rough and general rules of application are as follows: (1) near-50-Ω cables are generally used for rf transmission, (2) near-75-Ω cables are generally used for video transmission, and (3) cables with larger center-conductor diameters may be used for receiver circuits to reduce losses, or for transmitter circuits to increase power-handling capability.

Cable Type	Velocity Factor	Z_o (Ω)	C per ft (pf)	Diameter (in)
RG-8	.66	52	29.5	.405
RG-8 foam	.80	50	25.4	.405
RG-11	.66	75	20.6	.405
RG-11 foam	.80	75	16.9	.405
RG-58	.66	53.5	28.5	.195
RG-58 foam	.79	53.5	28.5	.195
RG-59	.66	73	21.0	.242
RG-59 foam	.79	75	16.9	.242
RG-62	.86	93	13.5	.242
RG-62 foam	.79	95	13.4	.242
RG-214 double shield	.66	50	30.8	.425
RG-223 double shield	.66	50	30.8	.212
300-ohm twin lead	.82	300	5.8	

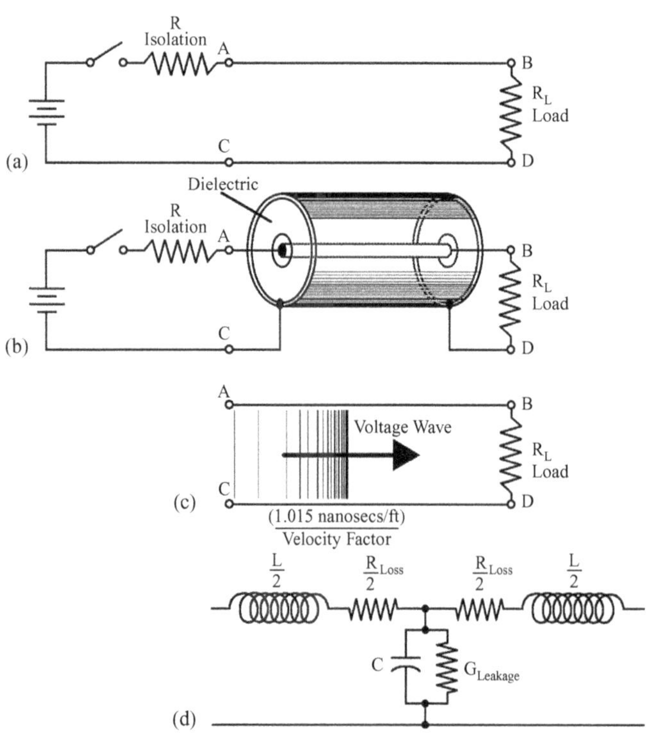

FIGURE 10-34

Voltage propagation on a line.

Common Transmission Line Characteristics

Artificial lines have characteristics similar to real transmission lines, but the characteristic of greatest interest in transmitter design will be equivalent to the propagation time required for a signal to travel between two points on the line. That time will determine the radar modulator pulse width.

Signal Propagation on a Transmission Line

When a voltage is applied, through some form of isolation as the illustrated resistor, to the transmission line, between points A and C, a specific time is required for it to reach the end, the termination at points B and D (see Figures 10-34A and B). The applied voltage moves down the line as shown in Figure 10-34C; the velocity is the speed of light multiplied by the velocity factor, or 1.015 ns/foot divided by the velocity factor. Figure 10-34D shows that the line exhibits the characteristics of many infinitesimal, individual, "tee" sections of resistance, capacitance, and inductance. The capacitance is a result of the spacing between the two conductors, and the dielectric between them. Leakage through the dielectric provides the conductance value *G* in ohms. Electrons flowing through the wire create a magnetic field, and the wire becomes an inductance; the inductance for each side of one tee is shown as "*L*/2," when two adjacent tees are connected, that inductance becomes "*L*." The series resistance represents line loss.

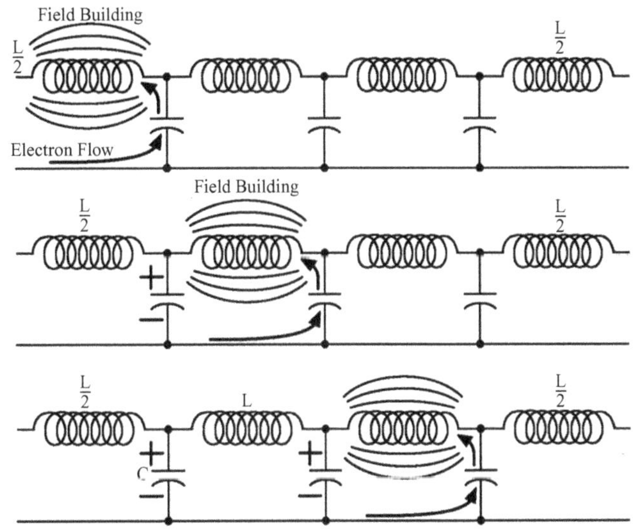

FIGURE 10-35

Charging the line.

Temporal Delay

The reason for the propagation delay to a voltage wave on the line becomes clear when the effect of the individual sections are scrutinized (see Figure 10-35). As the voltage is first applied, the first inductance opposes initial current flow, and the first capacitance is a virtual short circuit. As current begins to flow through the inductance, the capacitor simultaneously begins to charge. By the time the capacitor is fully charged, determined by the values of both the capacitance and inductance, the inductance opposes a reduction in current, and the next section has begun a like operation. The capacitance is charged, and the inductance begins current flow, one section at a time. Since the time it takes each section to charge is determined by the inductance and capacitance, those values determine the propagation rate, as follows:

$$t_{\text{unit}} = \sqrt{L_{\text{unit}} C_{\text{unit}}}.$$

Once the time per unit has been determined, the velocity factor can be calculated. If the velocity were at the speed of light, the signal would propagate at 1.015 ns/foot, so that value divided by the actual time per foot will provide a ratio which expresses the reduction in velocity:

$$VF = \frac{1.015\,\text{ns/ft}}{t_{\text{foot}}}.$$

Because series inductance is additive, and so is parallel capacitance, the ratio between the two is the same, no matter what the length of the line. Obviously, the load upon an applied signal will be determined by the inductive and capacitive reactances. And, because these values are independent of line length, the transmission line will exhibit a signal load which is independent of line length. For lower frequencies, this load is purely resistive, and is called the ***characteristic impedance*** $\mathbf{Z_o}$ of the line. The Z_o is also independent of frequency, provided the frequency is below a value of f_x, the point at which Z_o becomes reactive.

In actual practice, the values of *L* and *C* will be per foot or meter; however, even if the infinitely small values of a single section were known, the equation for Z_o would still be the same:

$$Z_o = \sqrt{\frac{L}{C}}.$$

Manufacturer's information may provide only the VF and values of *C* and Z_o, but, clearly, the value of *L* may be found from those.

Frequencies above f_x and Reactance

When the frequency exceeds f_x, the line begins to exhibit capacitive or inductive reactances. At that point, the characteristic impedance becomes

$$Z_o = \sqrt{\frac{R + j\omega L}{G + j\omega C}},$$

where $\omega = 2\pi f$. Once Z_o becomes reactive, the load presented to the line is variable, and the reactance, X, determines it as follows:

$$Z_o = R \pm jX.$$

Recall from the preceding discussion on frequency spectra, in this chapter, that a pulse may contain even thousands of significant harmonic frequencies, based on a Fourier series, in turn, based on a sin *x/x* curve, at an amplitude-variation rate, which is dependent upon the pulse width and period. It should be apparent that the reactive, higher frequency Z_o of a transmission line, may have a significant effect on the higher harmonics of a pulse, affecting the pulse shape. For this reason, video transferred over long cables is often amplified by a ***line compensator,*** which may provide additional amplification to higher frequencies, to restore pulse shapes.

Frequencies where Z_o becomes reactive, for most coaxial cables, are often in the neighborhood of 50 MHz, and become severe above 100 MHz. A short cable used to connect 3-GHz test equipment might easily attenuate rf signals by as much as 2 or 3 dB.

Artificial Transmission Lines

The similarity between real and artificial lines is so great, that there is considerable overlap in the manner in which the two behave. In fact, real transmission lines are sometimes used in the development of extremely short-pulse, low-power modulators, for extremely short-range radars. An artificial line is simply made of discrete inductors and capacitors, to achieve time delay in a short physical space. Since these lines do not contain the infinitesimally small inductances and capacitances of a real line, their frequency response is severely limited; this is actually desirable in radar modulators, because it limits the Fourier content of the pulse, reducing the amplitude of transmitted side lobes. The effect upon the spectrum differs between synthesis and magnetron systems. A common form of artificial transmission line uses "L" sections, as shown in Figure 10-36. This figure shows a charge resistance, R_{ch}, used to isolate the supply from the artificial transmission line and its termination, R_L.

As capacitance in the line is charged with the applied voltage, energy is given up by the traveling voltage wave, and the charge is left on the capacitance. Thus, the ***total storage capacitance in the line C_{st}*** is significant. When a long cable is used to transfer information over a real transmission line, a ***line driver*** must be used to provide adequate energy for C_{st}. The relationship between C_{st}, L_{st}, and Z_o remains the same as for individual sections, because, again, the series inductances and parallel capacitances add, but maintain the same ratio, shown in the following equation:

$$Z_o = \sqrt{\frac{L_{st}}{C_{st}}}.$$

The values of inductance and capacitance in the individual sections determine the frequency response, and the number of sections determine the modulator pulse width. As fewer sections, containing larger components, are used, the frequency response decreases. The two-way travel time on the line from source to R_L, and then back to the source, ultimately will become the pulse width, t_p.

Charging the Line Capacitance

Academic analysis of the line charge cycle usually begins with ***resistance charging,*** as illustrated in Figure 10-36. The line is usually charged through an inductance; that method allows the line to be charged to twice the HVPS

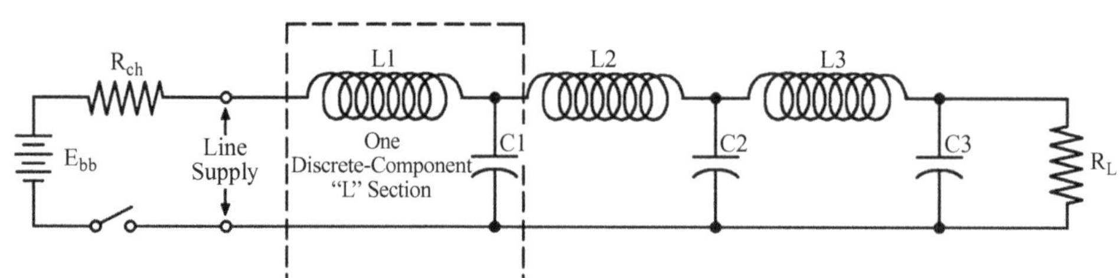

FIGURE 10–36

Artificial line.

potential and makes the charging system nearly twice as efficient. The analysis of resistance charging is useful in preparing the reader for the analysis of the somewhat less-obvious resistance discharging. It is noteworthy that there are some modulators, in low-power radars, that employ resistance charging. Resistance discharging through the pulse transformer is, in fact, what takes place in the conventional modulator, used until the introduction of the very most recent systems.

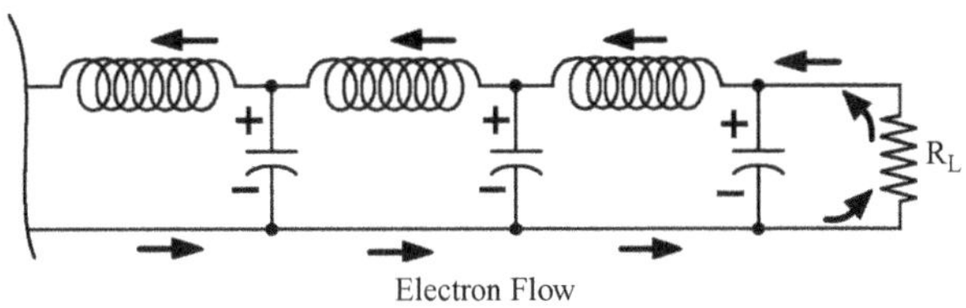

FIGURE 10–37

Terminated in a Z_o match.

The first, incident, voltage wave from the source toward the end of the line, called in some textbooks V_1^+ (subscript "1" for first and superscript "+" for incident), charges each individual capacitance to a voltage, which is a division between any series isolation resistance, R_{ch}, between the source and line, and Z_o of the line. For instance, if the source were 100 V, R_{ch} 40 Ω, and Z_o 60 Ω, the line would be charged to 60 V by the V_1^+ wave. In most cases, the most desirable value of R_{ch} would be that where it is equal to Z_o.

Termination R_L

Once the V_1^+ wave reaches R_L at the end of the line, one of the following five things, described in the proceeding paragraphs, may occur:

1. If R_L is equal to Z_o, current continues to flow through all the inductance, and the load becomes a voltage divider with R_{ch} (see Figure 10-37). Because the voltage wave encountered a load resistance no different than the impedance of the line, there is no change in the current through the inductance, no building or collapsing of the magnetic fields on the coils, and no discharge of the capacitors. The capacitors are charged to the value determined by the R_{ch} and Z_o division, and R_{ch} and R_L are series resistors, connected across the supply.
2. If R_L is zero, a short, the zero voltage across the short will be reflected back down the line, toward the source, removing the charge from the capacitors (see Figure 10-38). On reaching the short, the line current goes to maximum, and the high current through the final inductance raises its magnetic field to twice the value of those preceding it. The voltage source for this doubled current is from (1) the supply and (2) the charge on the capacitor to the left of the inductance. When the capacitance has been discharged, the inductance can no longer sustain the higher rate of current, and the field collapses, sustaining current only at the lesser rate. The next capacitance to the left is now loaded with a short, and the current through next inductance is doubled. The short thus travels back down the line, discharging the capacitance. When all the capacitance has been discharged, R_{ch} is the only load on the supply. Some textbooks may describe this as a reflected voltage

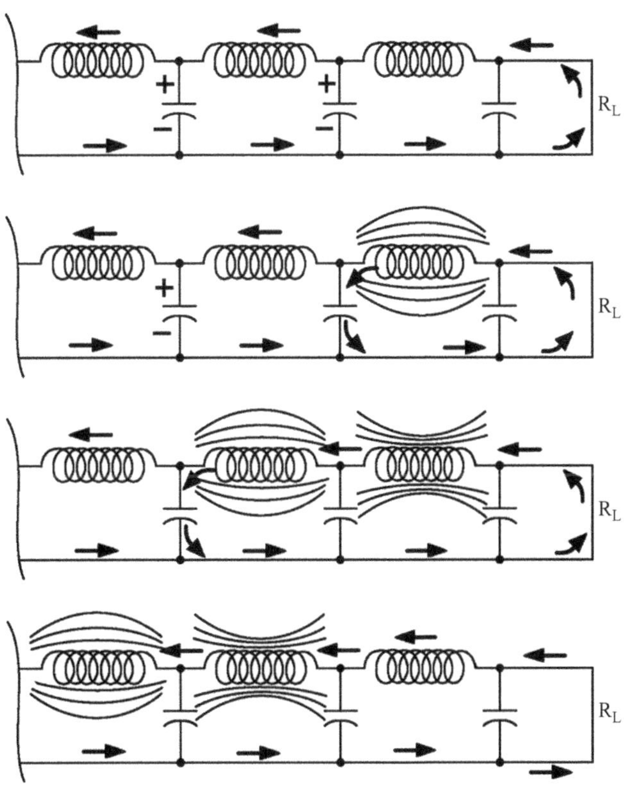

FIGURE 10–38

Shorted termination.

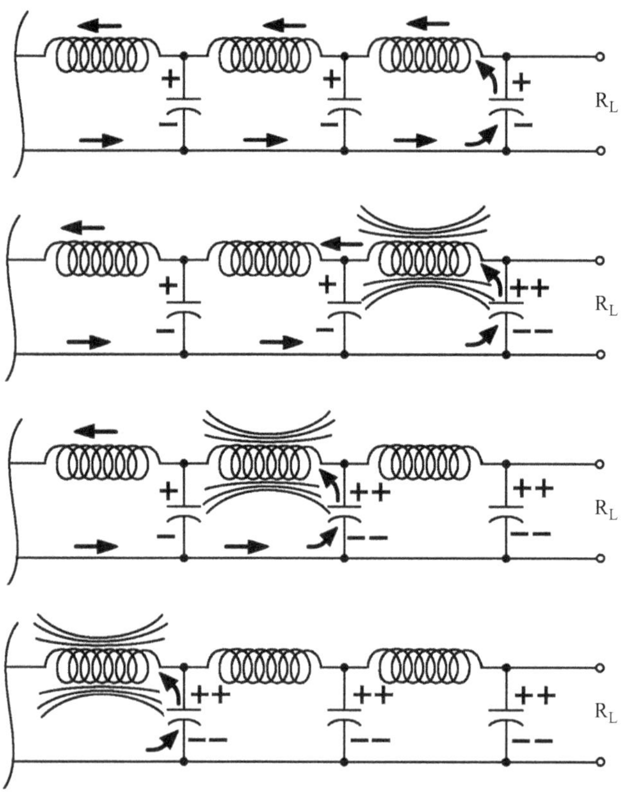

FIGURE 10-39

Termination in an open circuit.

wave V_1^-, opposite in polarity to the incident; this is not incorrect, as the reflected wave is conceivably of equal and opposite polarity as the incident, thereby discharging the capacitance.

3. If R_L is infinite, an open, the voltage may be reflected back down the line, toward the source, at the same polarity as the incident wave (see Figure 10-39). This wave is, again, designated V_1^-, and charges the C_{st} to twice the V_1^+ value. On reaching the open, the current through the T final coil cannot be sustained, and the magnetic field collapses in an attempt to sustain the current; this charges the final capacitor to twice the supplied voltage. Once the last capacitance is charged, its current flow ceases, and the field about the next-left inductor collapses, to charge the next-left capacitance to twice the applied voltage. Once the wave has returned to the supply end, the capacitors are charged to twice the line-supply voltage, and there is no current in the line.
4. If the R_L value is between a match and an open, a like-polarity reflection of calcu lable value will occur. The amount of reflection is determined by the reflection coefficient, Γ_τ (pronounced "gamma sub tau"). Substitution of infinity ∞ for R_L in the equation yields a Γ_τ value of +1, indicating a reflection of the entire voltage wave, as in the preceding discussion.

$$\Gamma_\tau = \frac{R_L - Z_o}{R_L - Z_o}.$$

5. If the R_L value is between a match and a short, a reverse-polarity reflection of calculable value will occur. The same equation applies as in (4). Where the line is terminated in a short, substitution of zero for R_L in the equation will yield a Γ_τ value of −1, indicating a reflection of the entire voltage wave in reverse polarity, as in the preceding discussion.

Reflection at the Sending End of the Line

When the reflection from R_L, V_1^-, reaches the sending end of the line, it will again encounter a discontinuity, and another reflection, back toward the load, will occur. This reflection will further algebraically add to the charge on the line capacitance. The amplitude and polarity of this reflection are determined by another reflection coefficient, Γ_s (pronounced, "gamma sub ess"), calculated in a similar manner as Γ_τ, but determined by R_{ch} and Z_o:

$$\Gamma_\tau = \frac{R_{ch} - Z_o}{R_{ch} - Z_o}.$$

Wave Computations

The actual calculation of several reflections is illustrated in Figure 10-40. Each reflected wave algebraically adds to the voltage V_{st} stored in the capacitance C_{st}. The V_1^+ wave results from the voltage division of the supply voltage

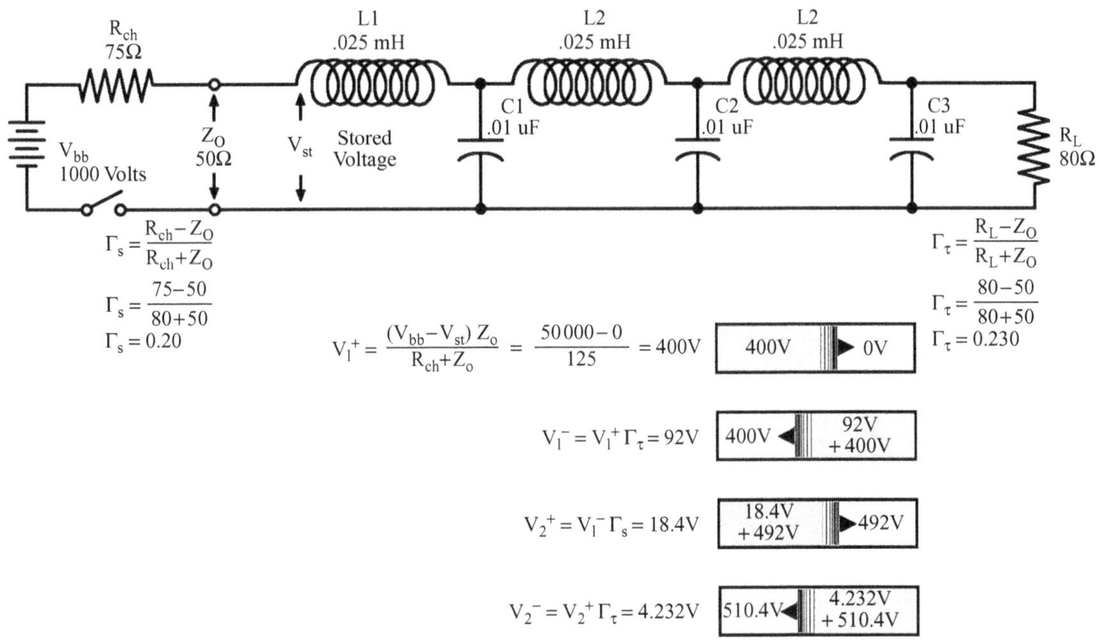

FIGURE 10-40

Reflections when charging a mismatched line.

V_{BB} between R_{ch} and Z_o, and can be computed by calculating the $V_{BB}/R_{ch} + Z_o$ current and then multiplying that result by Z_o. The following equation places this in a form for a single-step computation. In most cases, the V_{st} will be zero:

$$V_1^+ = \frac{(V_{BB} - V_{ST})\,Z_o}{R_{ch} + Z_o}.$$

See the example in Figure 10-40. After V_1^+ has been computed, multiplication by Γ_τ will yield V_1^-, then, multiplication of V_1^- by Γ_s will yield V_2^+, and so on. Each wave adds to the V_{st}, so the charge on the pfn, at any given time, then becomes

$$V_{st} = V_1^+ + V_1^- + V_2^+ + V_2^- + \cdots.$$

Discharging the Line

Figure 10-41 illustrates inductive charging, and the R_{ch} has been replaced with L_{ch}. This makes no difference in considering the discharge path, because the thyratron switch is then energized to become nearly a short circuit; the charge circuit is then of little consequence.

Most FAA radar modulators use open-ended pfns, so that the line will charge to twice the voltage supplied at its input. If R_{ch} is equal to Z_o, there are no further reflections after the first voltage waves to the end of the line and back, called V_1^+ and V_1^-. When the line is so charged to twice its applied voltage, it is ready for discharge, and use in forming the modulator pulse.

If resistance charging were employed, and if R_{ch} and Z_o were equal, they would form a voltage divider, which causes equal drops across R_{ch} and the Z_o of the line input. The V_1^+ wave is one-half the supply voltage, and the V_1^- wave charges the line to a value equal to the power supply voltage. In discharging the line through the primary of the pulse transformer, now called R_{xfrmr}, a new $V_{1_{disch}}^+$ wave of decreasing voltage appears on the line.

When the thyratron switch in Figure 10-41 is energized (closed), all the capacitors are provided a discharge path, but the left-most inductance, being the least inductive reactance in the path, will be the first to permit current flow. A magnetic field about that left-most inductor builds, as the left-most capacitance discharges first. When the

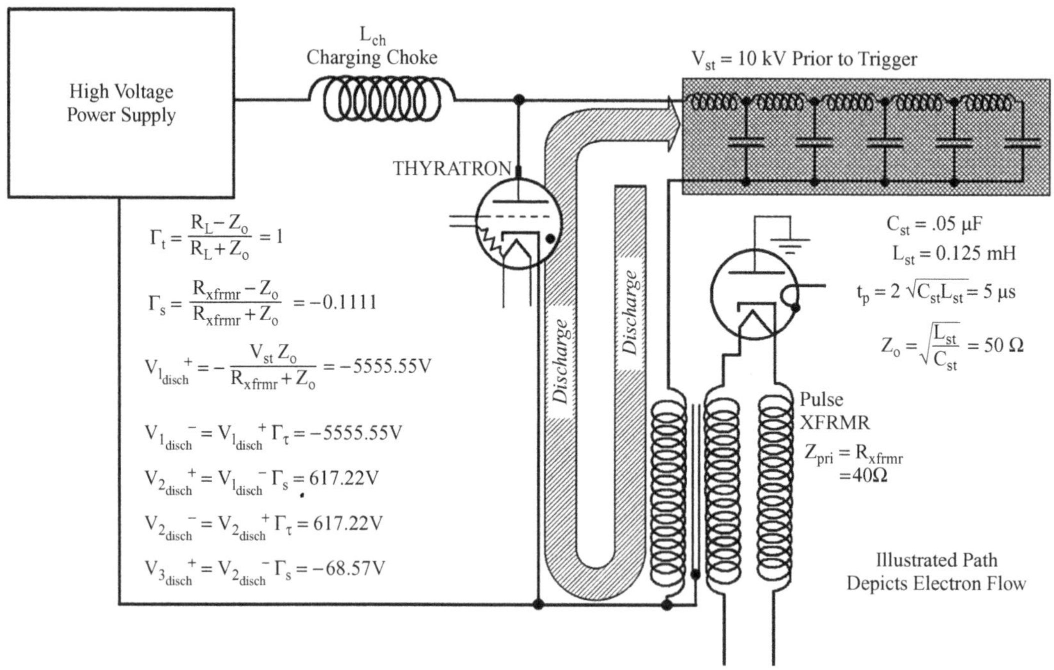

FIGURE 10–41

Discharging a mismatched, open-ended line.

first capacitor is discharged to half its charge, and because of the L and C values, the next capacitor to the right begins to discharge, current flow begins in the next inductor, and the current flow is sustained through the first inductor. The process continues, section for section, until the discharge wave reaches the open end of the line.

When the discharge wave reaches the open end of the line, the last capacitor becomes a short circuit to the last inductor, and the field about that inductor collapses in sustaining current flow through the capacitor; the capacitor discharges completely. Following that, the same action takes place on the next capacitor and inductor to the left, then the next, and so on.

Provided R_{xfrmr} is equal to Z_0, when the discharge wave reaches the end of the line loaded with the pulse transformer primary, all the capacitor charges and inductor fields are exhausted, and no further action can take place until the switch is opened, to allow the line to be re-charged. This will occur inherently, because the thyratron de-ionizes, once its supply is exhausted. A solid-state SCR will operate in a similar fashion.

Forming the Modulator Pulse

During the time of both th$V$e $V^+_{1_{disch}}$ wave toward the open end and the $V^-_{1_{disch}}$ wave to the pulse transformer end, current flows through the pulse transformer primary, creating the high-voltage pulse. The pulse then depends on the time required for both the $V^+_{1_{disch}}$ and $V^-_{1_{disch}}$ waves. Clearly, the time is dependent upon the total capacitance C_{st} and total inductance L_{st}, so the two-way travel t_p is

$$t_p = 2\sqrt{C_{st} L_{st}}.$$

Pulse Shape

Because the artificial line uses discrete components, it will exhibit a cutoff frequency, f_{co}, and pulse harmonics above that frequency will not pass; this is useful in limiting spectrum side-lobe power. The following relationship between sections and pulse width exists:

$$f_{co} = 0.637 \frac{N_{sections}}{t_p}.$$

During the discharge of the line during t_p, the Z_o and R_{xfrmr} form a voltage divider. Provided they are of equal values, there is an equal drop across each, so that one-half the pfn charge is across the pfn, and one-half across the pulse transformer. If R_{ch} and Z_o were equal, causing the pfn to charge to twice the value made available by the R_{ch} and Z_o voltage division, the voltage on the pulse transformer primary would be one-half of the supply voltage, and the system would be only 50% efficient, because of the drop across R_{ch}.

The Pulse Transformer

The pulse transformer serves two major purposes. One is obviously to step up the high voltage to provide ample voltage for the transmitting tube. Another is to match the impedance of the transmitter tube to that of the pfn, to minimize reflections in the discharge path. Recall that the impedance across a transformer varies with the square of the turns ratio, as

$$Z_{primary} = \frac{Z_{secondary}}{N_{ratio}^2}.$$

For example, a common impedance for magnetrons is 677 Ω, and the ASR-4, 5, 6 use a 1:3.68 turns ratio. This indicates a primary winding impedance of 50 Ω, which matches the 50 Ω impedance of the pfn.

Reflections

If R_{xfrmr} is not equal to Z_o, reflections will occur, just as in the charging of the line. Where Γ_τ is the reflection coefficient of the far (open) end of the line, and Γ_s is the reflection coefficient of the supply end of the line:

$$\Gamma_r = \frac{R_L - Z_o}{R_L + Z_o} = \frac{\text{infinity} - Z_o}{\text{infinity} + Z_o} = 1,$$

$$\Gamma_s = \frac{R_{xfrmr} - Z_o}{R_{xfrmr} + Z_o}.$$

Calculating the First Discharge Wave

The discharge wave $V_{1_{disch}}^+$ is the voltage removed from the pfn; the pfn is serving as the voltage source. A discharge equation, similar to the charge equation, may be used; however, to ensure that the calculation of $V_{1_{disch}}^+$ does reduce the pfn charge, a negative sign must be employed:

$$V_{1_{disch}}^+ = -\left(\frac{V_{st}\, Z_o}{R_{xfrmr} + Z_o}\right).$$

Following the $V_{1_{disch}}^+$ wave, the effect on the pfn charge V_{st} may be calculated with reflection coefficients, as in charging. Where a $V_{x_{disch}}^+$ wave is positive, the reflected voltage is added back to the V_{st}.

Inductive Charging

When, as in Figure 10-41, an inductor, commonly called the charging choke, is used in the charging circuit, there is no voltage division between the pfn and R_{ch}. Once the pfn has been charged to the power supply voltage, the field about the charging choke collapses, and the pfn is charged to twice the high-voltage power supply potential.

The inductance L_{st}, in relation to L_{ch}, the inductance of the charging choke, is insignificant, and the $L_{ch} - C_{st}$ circuit is series-resonant (see Figure 10-42). In early radar systems without f_p stagger, the ***resonant frequency*** f_r of the $L_{ch} - C_{st}$ circuit was chosen to be precisely half that of the f_p, so that the pfn charge would be at its peak when the thyratron was triggered. Where the f_r is equal to half the f_p, as described, the charging system is called ***optimum charging.*** If f_r is lower, and a half-cycle of the resonant charging waveform requires more time than the radar system interval, T_r, the condition is called ***linear charging*** because the field about the charging choke is not allowed to totally collapse before the switch tube is triggered, and the pfn waveform appears to be more linear. If f_r is above optimum, the pfn attempts to "ring back," through the charging choke, and there is a lower voltage available when the transmitter is triggered. Such a condition is called ***postoptimum charging.***

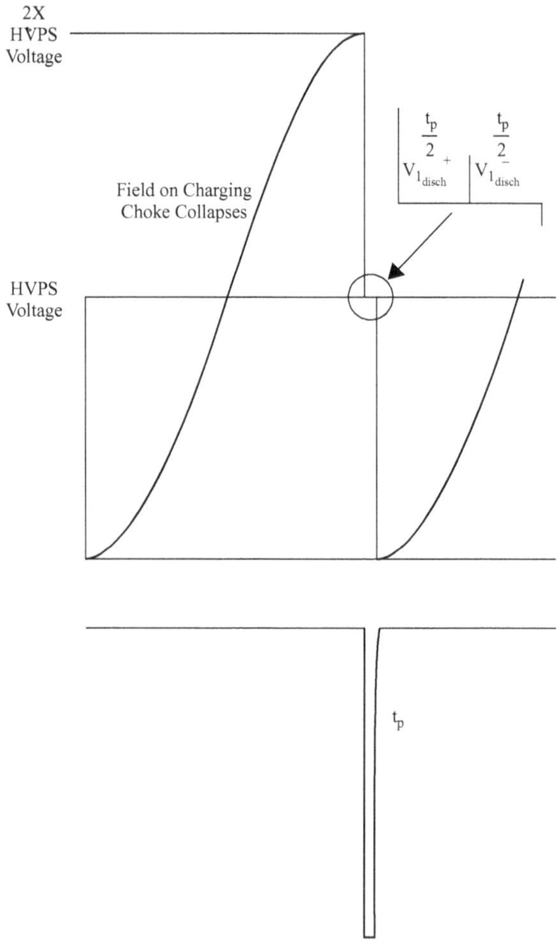

FIGURE 10–42

Inductive charge, resistance discharge, pfn voltage.

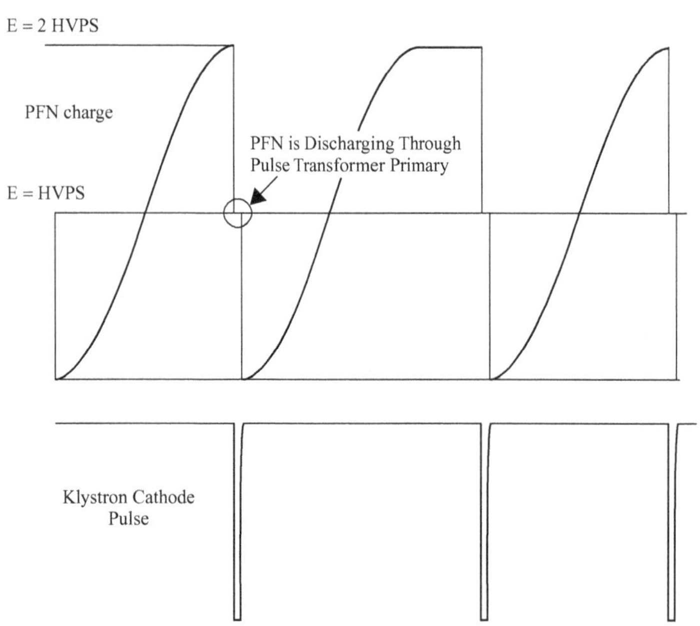

FIGURE 10–43

Staggered charging.

See the pfn waveform illustrated in Figure 10-43. Most systems now utilize postoptimum charging, but a ***charging diode*** prohibits reverse current to the power supply. Such a system provides for the varied intervals in staggered f_p.

Other Modulator Types

Until the widespread deployment of high-power synthesis systems, series-resonant modulators, of the type thus far described, were common, and they are still in use, because of their relative simplicity. One of the major shortcomings of these systems has been the short life of the hydrogen thyratron. Further, latter-day radar systems employ final transmitting tubes which demand a greater high voltage than is easily and safely available than that obtained via the series-resonant system described. Once high voltage reaches excessive potentials, arcing and component breakdowns become an intolerable obstacle.

To overcome the aforementioned inadequacies, manufacturers have incorporated a variety of new designs, employing solid-state devices. One common thread runs through most of these, the use of multiple modules in parallel, to distribute current through several components, so as to develop a large total current, without excessive high voltage. For instance, the ASR-8 uses twelve SCR-triggered, parallel modules, to provide high current into a capacitor bank; the discharge of the capacitor bank produces the high-voltage pulse via a pulse transformer. Because current is steered with saturable inductances, that modulator is sometimes called a ***magnetic modulator.***

The ASR-9 and ARSR-3 use a ***flyback modulator,*** employing ***reverse blocking diode thyristors (RBDTs),*** devices which conduct until the source is depleted, once a reverse "breakdown" potential is exceeded. RBDTs are PNPN devices, which block current in the reverse direction, as would a diode. However, when the reverse voltage exceeds a given value, the RBDT conducts in the reverse direction, until current drops to a minimum value. The RBDTs are employed in "stacks" to reduce the power, heat, and failure likelihood, of each. To ensure proper timing of the discharge, SCR stacks operate, in conjunction with the RBDTs, so as to place a rapid, properly timed voltage change on the RBDT anodes.

References

1. No author credit given. "Jean Baptiste Joseph Fourier," *Collier's Encyclopedia.* UK and Canada: Crowell-Collier Publishing Co., 1964.
2. J. F. Reintjes and G. T. Coate, *Principles of Radar,* MIT Technology Press. New York, NY: McGraw-Hill Book Co., Inc., 1952.
3. B. Nelson, Varian Associates, *Introduction to Klystron Amplifiers,* April 1963.
4. L. Garmon, *Echoes of War* television production for Public Broadcasting System.
5. I. Behrens, "Wilhelm Eduard Weber" Contribution to *Collier's Encyclopedia.,* UK and Canada: Crowell-Collier Publishing Co., 1964.
6. G. W. Dunnington. "Karl Friedrich Gauss" Contribution to *Collier's Encyclopedia.* UK and Canada: Crowell-Collier Publishing Co., 1964.

Review Questions

1. The single-frequency pulsed radar transmits (one, several, or hundreds?) of frequencies.
2. In describing radar system types, the word "synthesis" may be used in preference to "klystron" because ___________.
3. The sidebands of a transmitted radar pulse are caused by_________________.
4. The spectral lines of the radar transmitter pulse are separated by ___________.
5. A radar transmitter is operating with a pulse width of 0.6 μs and f_p of 875 Hz. How many harmonics are contained in the main lobe of the spectrum?
6. The nulls at the sides of the radar transmitter main spectrum lobe are at 9,070 MHz and 9,090 MHz. What is the transmitter pulse width?
7. The RESOLUTION BANDWIDTH control on a spectrum analyzer adjusts the bandwidth of ____________________ (functional area).
8. Adjustment of the VIDEO BANDWIDTH control on a spectrum analyzer has what effect?
9. Give at least one reason that measurement of radar transmitter power should not be attempted with a spectrum analyzer.
10. A "linear" circuit or device is one in which_____________________________.
11. Any nonlinear device produces _________________________.
12. Describe the spectrum analyzer in the most brief and basic terms possible.
13. The most "thorough" display of a spectrum occurs when the spectrum analyzer is adjusted _______________________________.
14. A radar transmitter spectrum is calculated to have approximately 10 times as many frequencies as are visible on the spectrum analyzer. Explain.
15. Name three major radar uses of the spectrum analyzer.
16. Explain the internal power klystron effect when the tube is overdriven.
17. The power klystron is tuned while observing (list three).
18. Bifilar transformers are used so that (dc, ac, either) may be connected to __________.
19. A procedure in which all power klystron cavities are tuned for maximum power is called___________.
20. A new power klystron has just been installed, and the manufacturer's instructions for zeroing the tuning mechanism direct you to turn them fully clockwise. What hazard may be involved?
21. After zeroing the power klystron tuning mechanisms, the next step is to_____________.
22. A "sag" in the detected power klystron output pulse is caused by_______________.
23. Should a tuning chart be unavailable, the power klystron input cavity can be tuned by _______________________.
24. While tuning the klystron, the side lobes appear to be unbalanced. What corrective action should be taken?
25. Define "body," "collector," and "beam" currents.
26. Waveguide arcs are detected by _________________.
27. In what regard the amplitron is superior to the power klystron?
28. In the absence of high voltage, what is the amplitron output?
29. What effect does the amplitron have upon received signals?
30. Both the amplitron and power klystron have advantages over the magnetron. Name at least two.
31. List at least three major parts or signals required for magnetron operation.
32. Explain "π–1" magnetron operation.
33. A poor magnetron spectrum most likely indicates ______________________.
34. "Rising Sun" magnetrons are used to _____________ in ______________ radars.

35. What is the source of the inductance and capacitance in a real transmission line?
36. The velocity factor of a cable is 0.667. How long will it take a 5-μs pulse to travel 300 feet on the cable?
37. A cable has a per-foot capacitance of 35 pF and a Z_o of 50 Ω. What is the inductance?
38. A cable has a per-foot capacitance of 30 pF and a Z_o of 75 Ω. What is the time delay per foot?
39. A cable has a per-foot capacitance of 29 pF and a Z_o of 91 Ω. What is the velocity factor?
40. Two pulse-forming networks provide 2-μs pulses, but one contains six sections, while the other contains three. What differences might appear?
41. The resonant frequency of L_{ch} and C_{st} in a series-resonant modulator is higher than $2f_p$. Is this normal or abnormal, and why?
42. In a resistive charging circuit, R_{ch} is 75 Ω, the pfn Z_o is 50 Ω, and R_L is 25 Ω. The power supply is 1,000 V. Compute $V_1^+, \Gamma_\tau, \Gamma_s, V_1^-, V_2^+, V_2^-, V_3^+$.
43. What is the pulse width of a seven-section pfn, where each section contains a 0.001 μF capacitor, and has a 1 mH inductance?
44. A pfn with a very high number of sections may be undesirable. Why?
45. What is the cutoff frequency of a five-section line, where the sections have 0.002 μF capacitors, and 5 mH inductors?
46. A pfn is charged to 17 kV. The Z_o is 50 Ω, and R_{xfrmr} is 60 Ω. Calculate the discharge waves V_{1disch}^+, V_{1disch}^-, V_{2disch}^+, V_{2disch}^-. What is the charge on the pfn after V_{2disch}^-.
47. A magnetron impedance is 677 Ω, and the turns ratio of the pulse transformer is 1:3.75. What is the impedance of the primary?

Review Answers

1. The single-frequency pulsed radar transmits (one, several, or hundreds?) of frequencies. ***Hundreds.***
2. In describing radar system types, the word "synthesis" may be used in preference to "klystron" because ***it is all-inclusive, and covers all high-power amplifiers.***
3. The sidebands of a transmitted radar pulse are caused by ***the pulse harmonics, whose value is determined by the Fourier series.***
4. The spectral lines of the radar transmitter pulse are separated by ***a frequency equal to*** f_p.
5. A radar transmitter is operating with a pulse width of 0.6 μs and f_p of 875 Hz. How many harmonics are contained in the main lobe of the spectrum?

$$\text{First xover} = \frac{1}{0.6\ \text{ms}} = 1.666666\ \text{MHz}$$

$$\text{Width of main lobe} = 3.333333$$

$$\frac{3.333333\ \text{MHz}}{875} = 3{,}809\ \text{harmonic frequencies}$$

6. The nulls at the sides of the radar transmitter main spectrum lobe are at 9,070 MHz and 9,090 MHz. What is the transmitter pulse width?

$$9{,}090\ \text{MHz} - 9{,}070\ \text{MHz} = \frac{2}{t_p} = 20\ \text{MHz},$$

$$t_p = \frac{1}{10\ \text{MHz}} = 100\ \text{ns.}$$

7. The RESOLUTION BANDWIDTH control on a spectrum analyzer adjusts the bandwidth of ***the i-f amplifier.***
8. Adjustment of the VIDEO BANDWIDTH control on a spectrum analyzer has what effect? ***Permits a "smoothing" of the spectral display, eliminating spikes and noise.***
9. Give at least one reason that measurement of radar transmitter power should not be attempted with a spectrum analyzer. ***The power is distributed over all the spectra, and is not measurable at any single point.***
10. A "linear" circuit or device ***is one in which a sine wave of voltage causes a cosine wave of current change.***
11. Any nonlinear device produces ***harmonic frequencies.***
12. Describe the spectrum analyzer in the most brief and basic terms possible. ***It is a swept-frequency receiver, which offers a frequency-domain display.***
13. The most "thorough" display of a spectrum occurs when the spectrum analyzer is adjusted for ***a very slow sweep, with a very narrow resolution bandwidth.***
14. A radar transmitter spectrum is calculated to have approximately 10 times as many frequencies as are visible on the spectrum analyzer. Explain. ***When the transmitter is fired, only the frtequencies produced during the pulse time may be of adequate power to be received by the spectrum analyzer; the instantaneous position of the spectrum analyzer sweep then determines what frequencies are displayed. If the spectrum analyzer sweep is moving too rapidly, the probability that it will receive all the frequencies is reduced.***
15. Name three major radar uses of the spectrum analyzer. ***Tuning, side lobe measurements, and pulse width measurements.***
16. Explain the internal power klystron effect when the tube is overdriven. ***Final bunching occurs ahead of the "catcher" cavity, and the beam becomes dilated. This does not occur until the drive pulse reaches a maximum amplitude, so the effect of "sagging" is seen only in the center of the detected pulse.***

17. The power klystron is tuned while observing (list three). ***Power output, detected pulse shape, and spectrum.***
18. Bifilar transformers are used so that (dc, ac, either) may be connected so that ***either ac or dc may be connected to the filaments.***
19. A procedure in which all power klystron cavities are tuned for maximum power is called ***synchronous tuning.***
20. A new power klystron has just been installed, and the manufacturer's instructions for zeroing the tuning mechanism direct you to turn them fully clockwise. What hazard may be involved? ***Care should be taken to avoid jamming the mechanisms against their stops.***
21. After zeroing the power klystron tuning mechanisms, the next step is to ***zero all the mechanical position counters.***
22. A "sag" in the detected power klystron output pulse is caused by ***input drive saturation.***
23. Should a tuning chart be unavailable, the power klystron input cavity can be tuned by ***monitoring the input power, and then tuning for a "dip."***
24. While tuning the klystron, the side lobes appear to be unbalanced. What corrective action should be taken? ***The high-voltage pulse timing versus drive pulse relationship needs adjustment.***
25. Define "body," "collector," and "beam" currents. ***Body currents are caused by the portion of the beam that conducts to the drift tubes. Collector currents are caused by the portion of the beam that reaches the collector. The sum of the body and collector currents are the beam current.***
26. Waveguide arcs are detected by ***photoelectric devices.***
27. In what regard the amplitron is superior to the power klystron? ***It is a broad-band tube, and can accept a wide range of inputs; little tuning is required, and it is useful in frequency-agile transmitters.***
28. In the absence of high voltage, what is the amplitron output? ***Essentially the same as the input.***
29. What effect does the amplitron have upon received signals? ***It is essentially only a transmission line, with very little loss.***
30. Both the amplitron and power klystron have advantages over the magnetron. Name at least two. ***They can produce greater power, and can be used in a synthesis system.***
31. List at least three major parts or signals required for magnetron operation. ***Magnetic and electric fields at right angles, and cavities, to sustain and synchronize oscillations.***
32. Explain "π – 1" magnetron operation. ***The normal "π mode" operation is one in which the electrical phase distance between cavity outlets is one-half cycle, π radians, 180°. When the tube begins "moding," the phasing is no longer synchronized to the cavity outlets; "π – 1" is the first such condition.***
33. A poor magnetron spectrum most likely indicates ***a need to replace the tube.***
34. "Rising Sun" magnetrons are used to ***prevent moding*** in ***K-band*** radars.
35. What is the source of the inductance and capacitance in a real transmission line? ***Electrons simply flowing through a wire create magnetic fields, and cause the inductance. The capacitance is caused by the spacing between the two conductors.***
36. The velocity factor of a cable is 0.667. How long will it take a 5-μs pulse to travel 300 feet on the cable?

$$\text{Time} = \left(\frac{1.015 \text{ ns / ft}}{0.667}\right) 300 = 456.52 \text{ ns.}$$

37. A cable has a per-foot capacitance of 35 pF and a Z_o of 50 Ω. What is the inductance?

$$Z_o^2 = \frac{L}{C} \Rightarrow 2{,}500 = \frac{L}{35\times10^{-12}} \Rightarrow L = 0.0875\,\mu\text{H}.$$

38. A cable has a per-foot capacitance of 30 pF and a Z_o of 75 Ω. What is the time delay per foot?

$$Z_o^2 = \frac{L}{C} \Rightarrow 5{,}625 = \frac{L}{30\times10^{-12}} \Rightarrow L = 0.168$$

$$t = \sqrt{LC} = 2.25 \text{ ns.}$$

39. A cable has a per-foot capacitance of 29 pF and a Z_o of 91 Ω. What is the velocity factor?

$$Z_o^2 = \frac{L}{C} \Rightarrow 8281 = \frac{L}{29\times10^{-12}} \Rightarrow L = 0.24\ \mu\text{H}$$

$$t = \sqrt{LC} = \sqrt{(29\times10^{-12})(0.24\times10^{-6})} = 2.638\ \text{ns}$$

$$VF = \frac{\text{time per ft at c}}{\text{time per ft}} = \frac{1.015\times10^{-9}}{2.638\times10^{-9}} = 0.3847 = 38.47\%.$$

40. Two pulse-forming networks provide 2-μs pulses, but one contains six sections, while the other contains three. What differences might appear? ***The six-section pfn will provide a more nearly square pulse, and higher spectrum side lobes than the three-section pfn.***

41. The resonant frequency of L_{ch} and C_{st} in a series-resonant modulator is higher than $2f_p$. Is this normal or abnormal, and why? ***It is normal. It assures that the pfn may reach full charge before being discharged. The charging diode causes the pfn to hold its charge, and allows for staggered*** f_p.

42. In a resistive charging circuit, R_{ch} is 75 Ω, the pfn Z_o is 50 Ω, and R_L is 25 Ω. The power supply is 1,000 V. Compute V_1^+, Γ_τ, Γ_s, V_1^-, V_2^+, V_2^-, V_3^+.

$$\Gamma_\tau = \frac{R_L - Z_o}{R_L + Z_o} = \frac{25-50}{25+50} = -0.333$$

$$\Gamma_s = \frac{R_{ch} - Z_o}{R_{ch} + Z_o} = \frac{75-50}{75+50} = 0.2$$

$$I = \frac{E}{R} = \frac{1{,}000\ \text{V}}{125\ \Omega} = 8\ A$$

$$V_1^+ = 8\ \text{A} \times 50\ \Omega = 400\ \text{V}$$

$$V_1^- = 400\ \text{V} \times -0.333 = -133.2\ \text{V}$$

$$V_2^+ = -133.2 \times 0.2 = -26.64$$

$$V_2^- = -26.64 \times -0.333 = 8.87$$

$$V_3^+ = 8.87 \times 0.2 = 1.77$$

$$V_{st}\ \text{after}\ V_3^+ = 400 - 133.2 - 26.64 + 8.87 + 1.77 = 250.8\ \text{V}.$$

43. What is the pulse width of a seven-section pfn, where each section contains a 0.001-μF capacitor, and has a 1-mH inductance?

$$t_p = 2\sqrt{C_{st} L_{st}} = 2\sqrt{[7(0.001(1\times10^{-6}))][7(1\times10^{-3})]} = 14\,\mu\text{s}.$$

44. A pfn with a very high number of sections may be undesirable. ***Why? The pulse would be very square, creating many harmonics, which would increase the transmitter sideband power. Much of that power would be wasted, because it would be outside the receiver*** Δf.

45. What is the cutoff frequency of a five-section line, where the sections have 0.002-μF capacitors, and 5-mH inductors?

$$t_p = 2\sqrt{C_{st} L_{st}} = 2\sqrt{[5(0.002\times10^{-6})][5(5\times10^{-3})]} = 31.6\,\mu\text{s}$$

$$f_{co} = 0.637\left(\frac{N_{\text{sections}}}{t_p}\right) = 0.637\left(\frac{5}{31.6\times10^{-6}}\right) = 100{,}791\ \text{Hz}.$$

46. A pfn is charged to 17 kV. The Z_o is 50 Ω and R_{xfrmr} is 60 Ω. Calculate the discharge waves $V_{1\text{disch}}^+$, $V_{1\text{disch}}^-$, $V_{2\text{disch}}^+$, and $V_{2\text{disch}}^-$. What is the charge on the pfn after $V_{2\text{disch}}^-$? $\Gamma_\tau = \dfrac{R_L - Z_o}{R_L + Z_o} = +1$

$$\Gamma_s = \frac{R_{\text{xfrmr}} - Z_o}{R_{\text{xfrmr}} + Z_o} = \frac{10}{110} = 0.09$$

$$V^+_{1_{\text{disch}}} = -\left(\frac{V_{\text{st}} Z_o}{R_{\text{xfrmr}} + Z_o}\right) = -\left(\frac{(17)(1 \times 10^3)(50)}{60 + 50}\right) = -7{,}727 \text{ V}$$

$$V^-_{1_{\text{disch}}} = -7{,}727\Gamma_\tau = -7{,}727\text{V}$$

$$V^+_{2_{\text{disch}}} = -7{,}727\Gamma_s = -695.43$$

$$V^-_{2_{\text{disch}}} = -695.43\Gamma_\tau = -695.43.$$

$$V_{\text{st}} \text{ after } V^-_{2\text{disch}} = 17 \times 10^3 - 7{,}727 - 7{,}727 - 695.43 - 695.43 = 155.14 \text{ V.}$$

47. A magnetron impedance is 677 Ω, and the turns ratio of the pulse transformer is 1 : 3.75. What is the impedance of the transformer primary winding?

$$Z_{\text{primary}} = \frac{Z_{\text{secondary}}}{N^2_{\text{ratio}}} = \frac{677\Omega}{3.75^2} = 48.14\Omega.$$

CHAPTER 11

Radar Receivers

Amplifying the Echo

In terms of science, the radar receiver may be the greatest accomplishment in the system; its surface-apparent simplicity unfairly diminishes the scientific and engineering endeavors behind it. Its purpose is simply to amplify and detect radar echoes, but that is no small challenge. Echoes are very low in power, because they have been severely attenuated by, at least, $16\pi^2R^4$, from their transmission to reception, in accordance with the equation

$$P_r = \frac{P_t G_t A_o A_e}{16\pi^2 R^4}$$

P_r = received power
P_t = transmitter peak power
A_o = target cross-sectional area
A_e = antenna effective aperture
R = range.

The Limitation by Noise

The greatest limitation to the receiver's ability to amplify and detect an echo is that there exists noise which may obscure the weaker echoes. The noise comes predominately from two major sources: (1) the movement of electrons through active circuits in the receiver and (2) the radiation from the sun. Other contributions come from the molecular activity of heated gasses in the atmosphere, and molecular activity caused by the heating of the antenna reflector. In any case, there is a direct relationship between heat, molecular activity, and noise. The greatest noise source of all is the receiver local oscillator; since it is an active circuit, it creates noise while generating its output frequency, and that noise is injected into the signal mixer. In fact, the presence of noise, often called grass, on radar video is a good indicator that the local oscillator is running.

The quantity P_a represents the noise injected into the receiver for amplification. Its value depends upon the receiver bandpass Δf, at a standard input termination temperature of 290° Kelvin. Kelvin is used because it begins at absolute zero, as does molecular activity. Kelvin is Celsius temperature plus 273°, and Celsius is Fahrenheit temperature minus 32°, divided by 1.8.

$$P_a = \kappa\, T_t\, \Delta f = (4.002 \times 10^{-21}\ \mathrm{W})\,(\Delta f) = -174\ \mathrm{dBm} + 10 \log \Delta f.$$

κ = Boltzman's constant, 1.38×10^{-23} J/° Kelvin
T_t = standard input termination temperature in degrees Kelvin = 290°
Δf= receiver bandpass in Hz.

Bandwidth Requirement of the Receiver

Recall that the transmitter emits a broad spectrum of frequencies, sidebands created by the Fourier content of the modulating pulse (see Figure 11-1). The receiver must recover enough of this spectrum to "reconstruct" the pulse. Many of the pulse harmonics are of little significant use, and it is not necessary for the receiver to recover them all. The overall receiver bandpass, finally determined by the i-f amplifier, will, in most cases, be only of sufficient width to recover a major portion of the main spectral lobe. Should the bandpass be made wider, the input noise will be more significant than the improvement to the pulse shape, and the signal-to-noise ratio declines. Figure 11-2 illustrates the signal-to-noise ratio, and Figure 11-3 illustrates the signal-to-noise ratio versus bandpass.

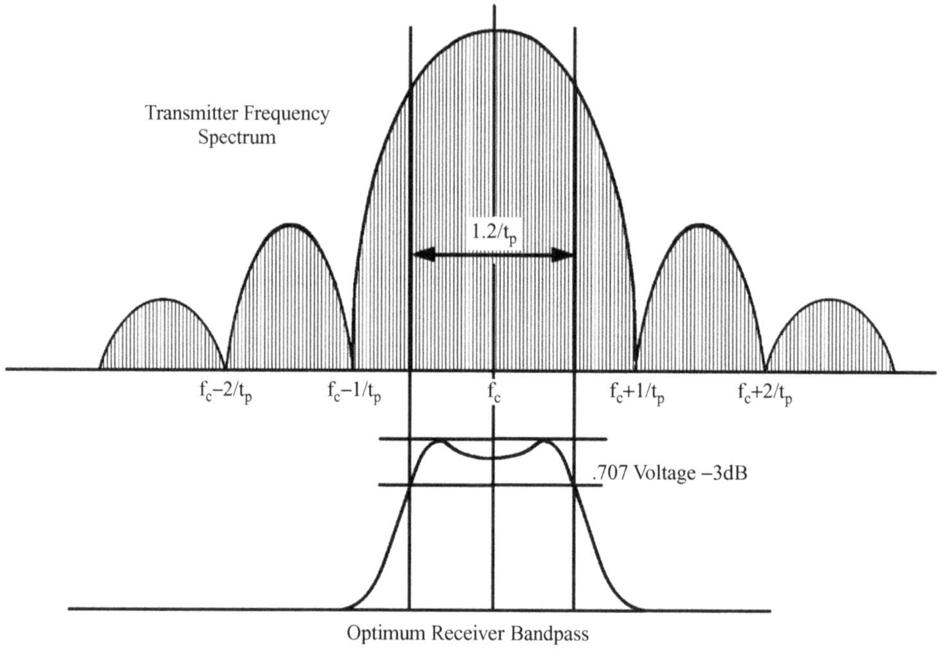

FIGURE 11–1

Spectrum versus bandpass.

Figure 11-3 illustrates (1) that noise power increases linearly with Δf, (2) that no additional signal power is gained after the Δf exceeds $1.8/t_p$, and (3) that the greatest signal-to-noise ratio can be achieved at $1.2/t_p$.

mds Measurement

The ***minimum discernible signal (mds)*** is the smallest signal, $P_{r\ min}$, that can be recovered from the noise (see Figure 11-4). The usual radar mds is measured with an rf test burst, injected into the INCIDENT connector, on the waveguide directional coupler; the test signal may be viewed with an oscilloscope at various points in the system. It may be viewed at the i-f output, or further downstream, so that processing circuitry may be included. It has been a practice to view the test signal on a ppi, to perform an all-inclusive ***overall mds.*** At optimum Δf of $1.2/t_p$, the mds should be

$$P_{r\ min} = -174\ \text{dBm} + 10 \log \Delta f + F_{dB}.$$

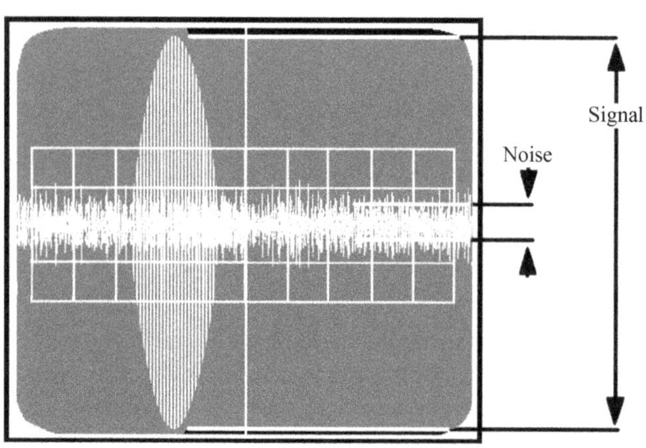

FIGURE 11–2

Signal-to-noise ratio.

"Tangential" mds

In some cases, mds may be measured at a point where the test pulse is at twice noise level, since that is the lowest level at which the test signal may reasonably be expected to "break" a quantizer threshold in digitizing equipment. For such a measurement, –171 dBm is used in the $P_{r\ min}$ equation. This measurement is called a ***tangential mds;*** the origin or rationale behind the definition is unknown.

Noise Figure

The "F" in the $P_{r\ min}$ equation represents the ***noise figure*** of the receiver system. Individual stages or units within the receiver system have noise figures as well, and these add in a specific manner to total F_T, ***the overall noise figure of the system.*** The capital "F" is used to avoid confusion with the "f" used to represent "frequency." For the same reason, the capital "F" should not be used to represent frequency.

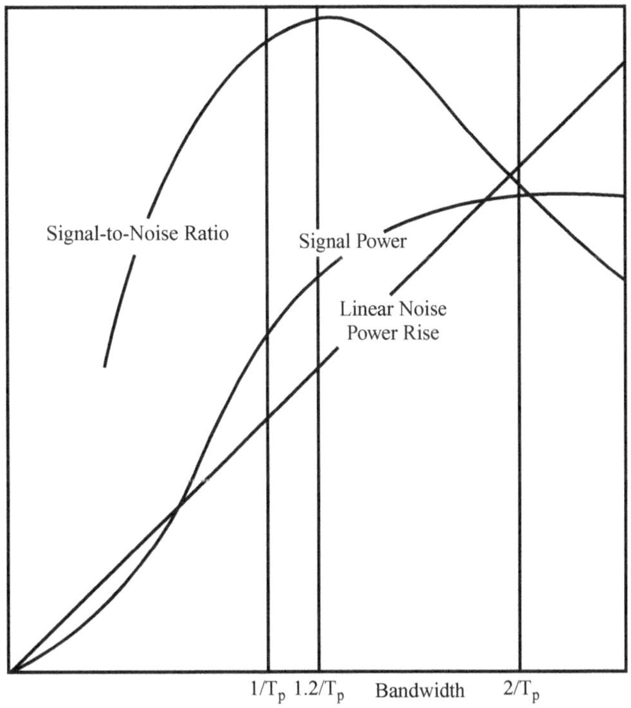

FIGURE 11–3

Noise, signal, S/N ratio versus Δf.

Noise Figure in DeciBels

Noise figure may be expressed either as a whole number, or in dB. In most literature, it will be the responsibility of the reader to recognize that it is expressed in dB, simply by the "dB" in the expression of quantity. In this chapter, noise figure expressed in dB will be further shown with a subscript to preclude any erroneous conclusions, as in "F_{dB}." F_{dB} is simply 10 log F; this conversion is permissible because F is a multiplier in the equation.

Signal-to-Noise Ratio Degradation

In simplest terms, the noise figure is a ratio of two other ratios: (1) the signal-to-noise ratio at the receiver input and (2) the signal-to-noise ratio at the receiver output. If the receiver were perfect, or ideal, there would be no noise added in the amplification of the signal, and the two ratios would be identical, making F equal to 1:

$$F = \frac{S_1 / N_1}{S_0 / N_0}.$$

Noise figure is intended to express the percentage of total noise contributed by the system. Signals are simply amplified by the gains of the stages; noise also increases by amplification by the gains of the stages, but, further, with the injection of electrical-current circuit noise. The signal can, therefore, be discounted, and only noise need be considered. If the receiver were perfect, the noise figure could be expressed as a ratio between the actual noise output, in contrast to what the output would be, were the receiver ideal, and free of any noise contributions:

$$F = \frac{N_{out}}{N_{ideal}}.$$

In accordance with the equation for available noise, P_a, the noise applied to the receiver is $\kappa T_t \Delta f$, and it will be amplified by a gain G, so an output will be $\kappa T_t \Delta f G$, plus any internal noise created by the receiver. If the receiver were perfect, or ideal, a noise figure of "1" would be equal to

$$F_{ideal} = 1 = \frac{N_{out}}{N_{ideal}} = \frac{N_{ideal}}{N_{ideal}} = \frac{\kappa T_t \Delta f G}{\kappa T_t \Delta f G}.$$

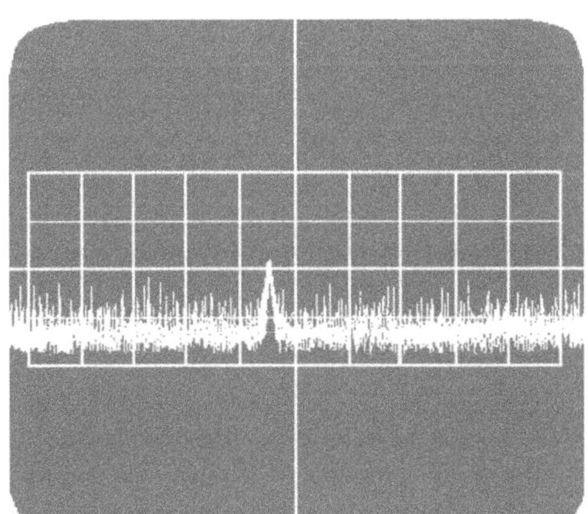

FIGURE 11–4

Test signal just above mds.

However, when internal noise is contributed in the receiver, or in any receiver stage, the noise figure of the receiver, or stage, will be raised from "1" to a value determined by the

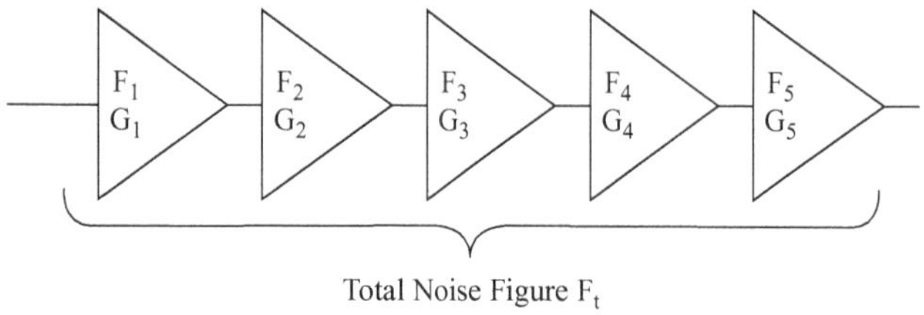

FIGURE 11-5

Noise figure and gains.

contributed internal noise, compared to the ideal noise:

$$F = 1 + \frac{N_{\text{internal}}}{\kappa T_t \Delta f G}.$$

Overall Noise Figure F_t

The effect of unit, or stage, gain, upon the noise figure, may play a dramatic part in the receiver performance; the location of the stage determines the degree of effect it might have (see Figure 11-5).

In Figure 11-5, the first stage, G_1, will contribute some amount of noise to the input, and that contribution will be multiplied by the gains of G_2, G_3, G_4, and G_5. The second stage, G_2, will also contribute noise, but that G_2 contribution will be less significant, because (1) the signal had been amplified by G_1, because (2) the G_2 noise contribution is less significant, in relation to the signal, than was the G_1 noise contribution, and because (3) the G_2 noise will only be amplified by G_3, G_4, and G_5. In short, the closer to the "front end" a stage may be, the more noise it will insert, relative to the amplitude of the signal.

Since the minimum noise figure is "1," it might be said that the "1" represents 100% reproduction of the noise input, and any higher number represents that 100%, plus any internal noise. Thus, an $F = 1.5$ represents a condition in which 50% as much noise was added by the stage. So, if it is necessary to consider the noise contribution only, F minus 1 of any stage would accomplish that. The overall noise figure F_t for the stages in Figure 11-5, then, would be

$$F_t = F_1 + \frac{F_2 - 1}{G_1} + \frac{F_3 - 1}{G_1 G_2} + \frac{F_4 - 1}{G_1 G_2 G_3} + \frac{F_5 - 1}{G_1 G_2 G_3 G_4}.$$

In considering the overall noise figure, it should be apparent from both stated principle and the equations that the first stage of a receiver system will have the greatest effect on noise figure, and subsequently, mds. In early radars, that first stage was the crystal signal mixer, with an F of about 6, and a gain of about 0.8. Those early systems incorporated a special low-noise i-f preamplifier, with a gain of about 10, and a noise figure of 2. Consider the effect on F_T:

$$F_T = F_1 + \frac{F_2 - 1}{G_1} + \frac{F_3 - 1}{G_1 G_2} + \frac{F_4 - 1}{G_1 G_2 G_3}$$

$$F_T = 6 + \frac{2 - 1}{0.8} + \frac{F_3 - 1}{0.8 \times 10} + \frac{F_4 - 1}{0.8 \times 10 \times G_3}.$$

In the preceding equation, the summation value of the first two most significant quantities, F_1 and $F_2 - 1/G_1$, is 7.25. Now, assume that a hypothetical, low-noise, high-gain, rf amplifier is incorporated, upstream of the mixer. The amplifier has a gain of 30, and a noise figure of 1.5. The values in the equation would thus be changed:

$$F_T = F_1 + \frac{F_2 - 1}{G_1} + \frac{F_3 - 1}{G_1 G_2} + \frac{F_4 - 1}{G_1 G_2 G_3}$$

$$F_T = 1.5 + \frac{6 - 1}{30} + \frac{2 - 1}{0.8 \times 30} + \frac{F_4 - 1}{0.8 \times 30 \times 10} + \frac{F_5 - 1}{0.8 \times 30 \times 10 G_4}.$$

Now, in the preceding example, the value of $F_1 + F_2 - 1/G_1$ is 1.67. The difference in the values in the preceding two examples of multiple stages should make it clear that the installation of the hypothetical stage would make a dramatic difference in F_T. In the 1950s and 1960s, considerable effort was devoted to developing such an amplifier. One of the new amplifiers was called a ***maser,*** the most nearly noiseless of all microwave amplifiers. It was very expensive and depended upon very cold operating temperatures to reduce noise; that necessitated refrigeration. Somewhat the lesser in performance, but, nevertheless, impressive, was the parametric amplifier, shown in Figure 11-6.

The Parametric Amplifier (Paramp)

This fascinating device, even though now approaching obsolescence, earned a permanent place in radar history. It relied on a pump oscillator, operating at considerably higher frequency than the radar, for example, 33 GHz for an S band (2,700 –2,900 MHz) ASR. Earlier versions used a reflex klystron for the pump; later ones employed a Gunn diode. The Paramp also contained a circulator, three cavities, and a ***varactor diode,*** really the heart of the amplifier.

The signal inputs entered through a circulator into a cavity tuned to the signal frequency, f_s. In the absence of power to the parametric amplifier, the signal would exit through the circulator, back into the conventional path, essentially unaffected.

With power applied to the amplifier, the signal would be "pumped up" by the action of the varactor, a p-n junction diode. The junction capacitance is a result of the instantaneous applied voltage, and varies at the rate of the applied microwave frequencies, the pump frequency, the signal frequency f_s, and a resultant "idler" frequency f_i. When the diode is reverse biased, the depletion region widens and the capacitance decreases. When the diode is forward biased, the opposite is true. If there is a charge on a capacitor, and the capacitance is increased, the charge must increase; this is the source of the "pumping" action. The parametric amplifier was successful, because it contributed very little noise in amplifying the signal, and it did so because it did not contain resistances. Current through resistance creates heat, which in turn increases molecular activity and noise.

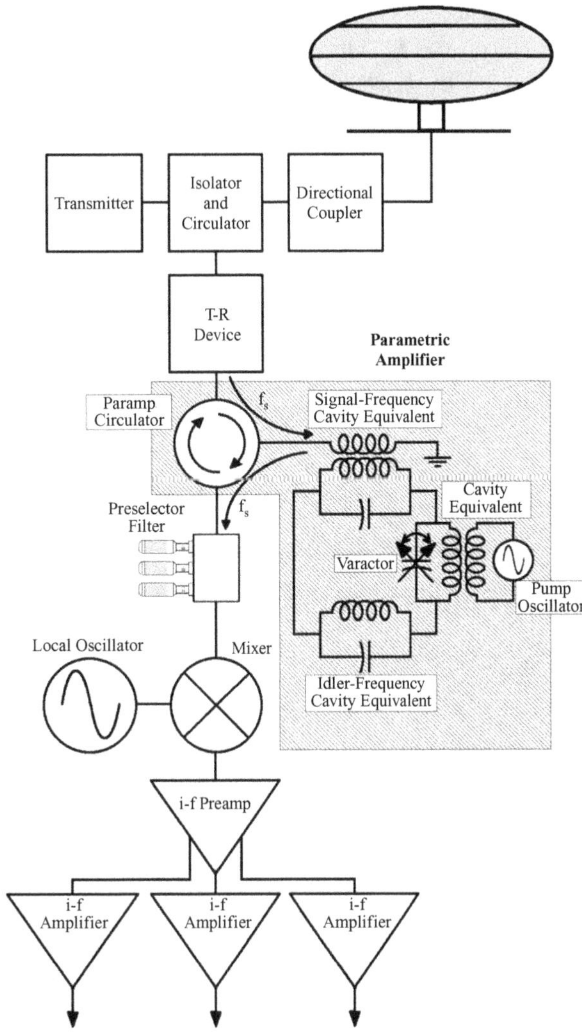

FIGURE 11-6

The parametric amplifier.

Effective Noise Temperature

In recent years, another expression of receiver noise performance has been increased in popularity because of its "fine-grained," precise description of satellite receiver quality. This expression, ***effective noise temperature T_{eff},*** is an input temperature which would account for the noise gained in the circuit, stage, or unit. It is related to F and the standard temperature of 290°K as follows:

$$F_T = F_1 + \frac{F_2 - 1}{G_1} +$$

$$T_{eff} = (F - 1)\ 290°\text{K}.$$

Should the device be perfect, with F equal to 1, the T_{eff} is equal to absolute zero temperature. If F equals 2, the device is producing as much noise as is available at the input, so the T_{eff} is equal to the standard temperature, 290°K. Therefore, between noise figures of 1 and 2, there is a T_{eff} spread of 290°K, allowing for a very fine resolution of the expression of noise performance.

For one example of T_{eff}, consider an F equal to 1.5. In relation to signals, the device is producing half again as much noise at the output, as at the input, and T_{eff} equals 145°K. For another, consider a typical radar noise figure of 4.2 dB; T_{eff} would be calculated:

$$F = \text{antilog}\left(\frac{F_{dB}}{10}\right) = \text{antilog}\left(\frac{4.2}{10}\right) = 2.63$$

$$T_{eff} = (F-1)290°\text{K} = (2.63-1)290°\text{K} = 472.7°\text{K}.$$

To entirely remove F from the T_{eff} equations, substitute F equivalents:

$$T_{eff} = \left(\left(\frac{\frac{S_{input}}{N_{input}}}{\frac{S_{out}}{N_{out}}}\right) - 1\right) 290°\text{ K} = \left(\frac{N_{internal}}{\kappa T_t \Delta f G}\right) 290°\text{ K}.$$

Noise Figure Monitoring

Recall that there is a direct relationship between noise figure and MDS, as expressed in the equation

$$P_{r\,min} = -174\text{ dBm} + 10\log\Delta f + F_{dB}.$$

Observation of mds requires that the technician monitor detected video on an oscilloscope. The characteristics of the video appearance, signal generator construction, and individual subjectivity are such that small variations are not easily detected. Smaller variations may be observed with a ***noise figure monitor,*** test equipment which measures only the noise performance of a receiver. Noise figure monitors have been included as ***built-in test equipment (BITE)*** in several radar systems, but most of these proved less than desirable, principally because of calibration drift. Satisfactorily manufactured test equipment for noise figure monitoring is uncommon, but, nevertheless, available. Such equipment was provided for use with the ASR-7 and ASR-8. The ASR-9 employs precision, digital, automated, mds-measuring BITE, and mathematically converts the mds to noise figure with software.

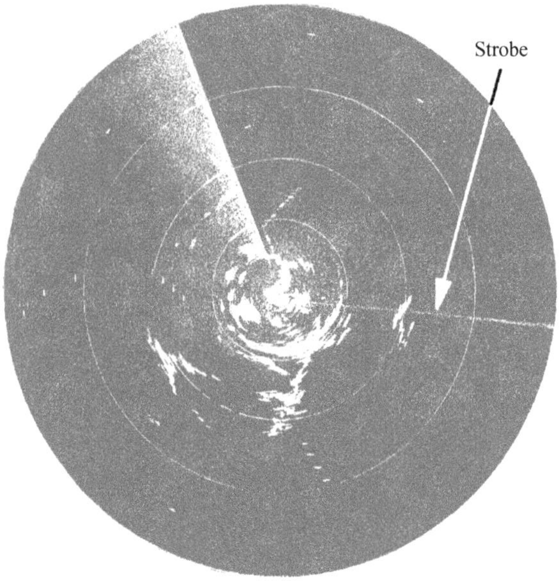

FIGURE 11–7

Solar noise strobe on ppi.

Noise figure test equipment incorporates (1) a calibrated gas-tube noise diode, (2) a directional coupler in the waveguide, normally 20 dB, (3) an output monitor at the i-f frequency, and (4) gating circuitry, synchronized to the radar f_p. The noise figure monitor compares (1) receiver noise power without injected noise to (2) receiver noise power plus the injected test noise power. By determining the gain to the test noise, the relative receiver noise can be calculated, thus providing the comparison between N_{ideal} and N_{out}. The latest noise figure monitors provide a digital display of noise figure to tenths of a dB, and the readings are consistent and repeatable.

Antenna Noise and Solar Strobes

At dawn and dusk, the rotating antenna is pointed into darkness most of the time; it is briefly illuminated by the sun, and a particular phenomena occurs (see Figure 11-7). When the sunlight strikes the reflector, electromagnetic disturbances from the sun, atmospheric thermoelectric molecular activity, and thermoelectric molecular activity in the reflector, all radiate noise into the feedhorn. The effect causes a noise strobe, which can be displayed on a ppi. The precise location of the sun, from a known longitude and latitude, for any particular day of the year, may be found by several means. This ***sun strobe*** then offers an accurate azimuth alignment test. Specialized test equipment for this now may be used to determine the antenna radiation patterns.

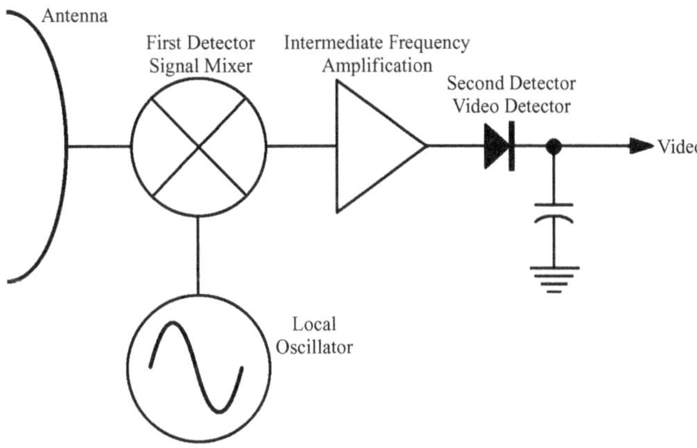

FIGURE 11-8

Basic receiver components.

The Superheterodyne Receiver

The basic components of a generic ASR/ARSR radar receiver were described in Chapter 6, and this book is intended for those with a foundation in the general principles of superheterodyne receivers and amplitude detection. Very simplistically, the most fundamental superheterodyne principles are all there is to radar reception, but microwave considerations, and necessary expansions to the fundamental principle, complicate the design of the radar receiver. First of all, recall the most basic components of a simple, superheterodyne, amplitude-detection receiver shown in Figure 11-8. A local oscillator is beat against the received rf and rectified in a device called the ***first detector.*** The resultant ***intermediate frequency (i-f)*** is further amplified and then reduced to usable intelligence in a ***second detector.***

The Low-Noise Amplifier and Preselector Filter

In the earliest days of radar, a tube-type, low-noise, ***i-f preamplifier,*** usually a design called ***Wollman Low-Noise Amplifier,*** was incorporated at the mixer output, connected by coaxial cables of the shortest practical physical length (see Figure 11-9). A tunable waveguide ***preselector*** was also included in early ASRs to reduce the front-end Δf to about 10 MHz; it is still included in the newest systems. In restricting the front-end bandpass, the preselector reduces noise; it also precludes the possibility that interfering rf frequencies on the "wrong side" of the stalo frequency may be converted to an i-f. The preselector may be located either upstream or downstream of the ***rf amplifier.*** The high-gain, low-noise rf amplifier first appeared as the parametric amplifier, and has since been replaced by off-the-shelf ***GAsFET Amplifiers*** from satellite-receiver technology. The GAsFET amplifiers are commonly called ***low-noise amplifiers (lna).***

Double Conversion

The use of two local oscillators may be employed for several reasons (see Figure 11-10). For instance, a spectrum analyzer uses double conversion; the first local oscillator selects the frequency band, and the first i-f amplifier Δf is broad enough for the entire band. The second local oscillator is swept,

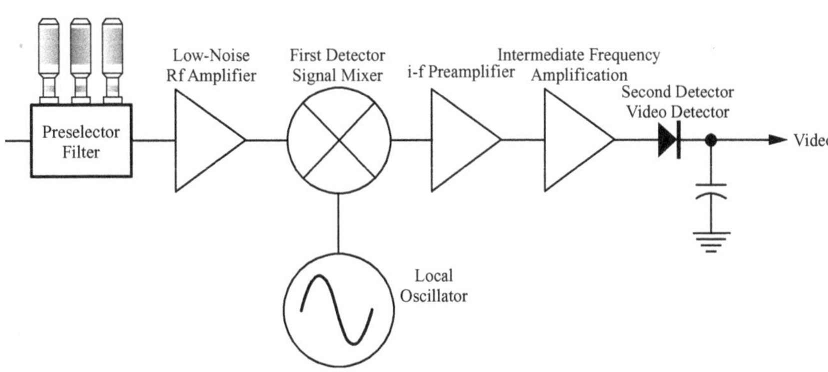

FIGURE 11-9

Preselector, rf amplifier, and i-f preamplifier added.

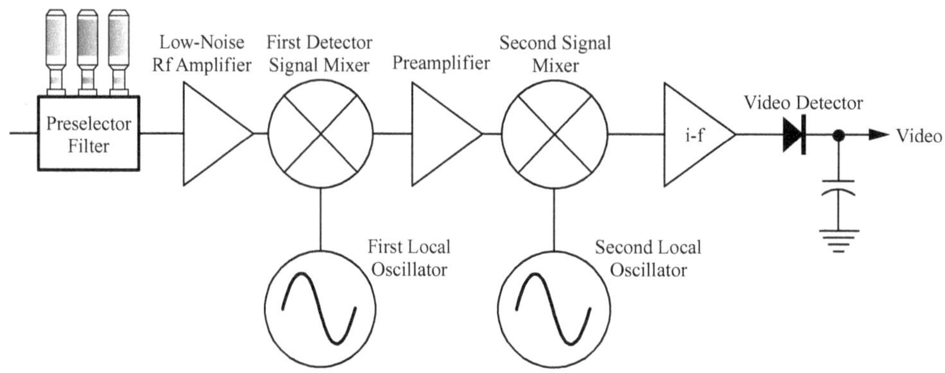

FIGURE 11–10

Double conversion.

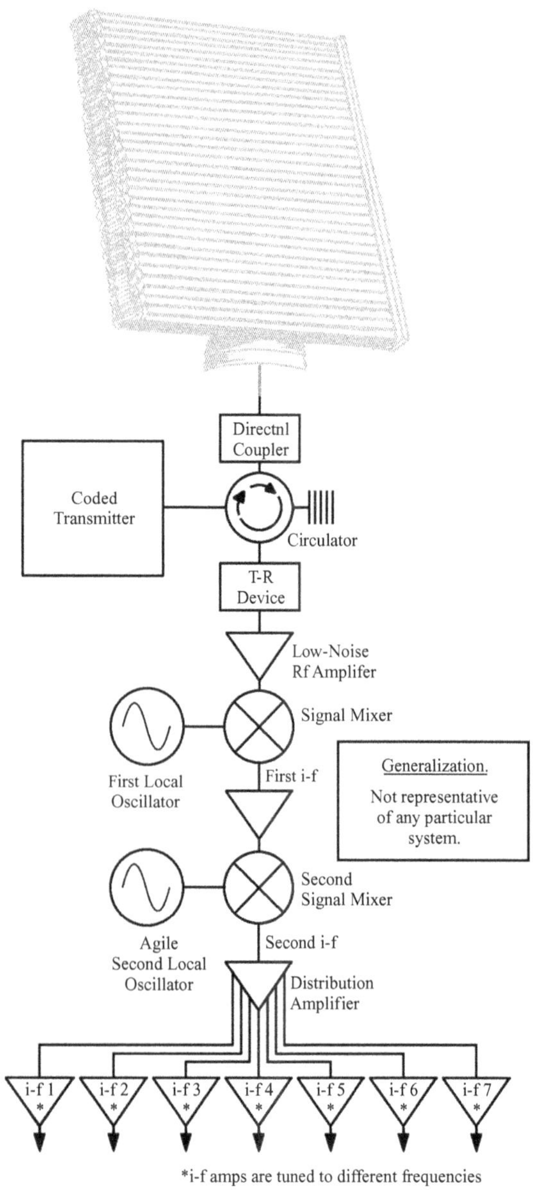

FIGURE 11–11

Multiple i-f signals from coded transmitter burst.

and frequency-agile. There are several applications of double conversion in more sophisticated radar systems, as with phased-array antennas and chirped transmitters.

Multiple i-f Amplifiers

Other schemes may be employed to provide for the simultaneous receipt of different frequencies, as in coded-pulse systems. In the system illustrated in Figure 11-11, the transmitter burst contains different frequencies to cause different beam elevations from the frequency-scanned antenna array. Echoes from those different beams may be separated by i-f amplifiers that have been tuned to different frequencies.

Phase Detection

Each of the simplified systems illustrated in Figure 11-10 employed final ***amplitude detection,*** the simplest form of detection, roughly similar to a-m radio. The detection process becomes much more sophisticated when ***phase detection*** is employed, as in Figure 11-12. In single-conversion amplitude detection, the signal mixer has been called the ***first detector,*** and the ***second detector*** was the diode and capacitor. In phase detection, the signal mixer becomes the ***first phase detector,*** and the phase-detector circuit becomes the ***second phase detector.*** The phase ϕ (phi) of the i-f echo is a result of the phase difference ψ (psi) between the rf echo and the local oscillator. The amplitude and polarity of the second-detector output then are a result of the difference ψ between the ϕ of the i-f echo and the ϕ of the reference oscillator, called the ***coho.***

The phase detector circuit is discussed in detail in Chapter 12. Its cosine response to a variety of inputs of different phases is illustrated in Figure 11-13. Phase detection is necessary in ***coherent*** systems, used to detect Doppler shift for ***moving-target-indicator (MTI)*** processing, or for ***Doppler weather velocity*** detection.

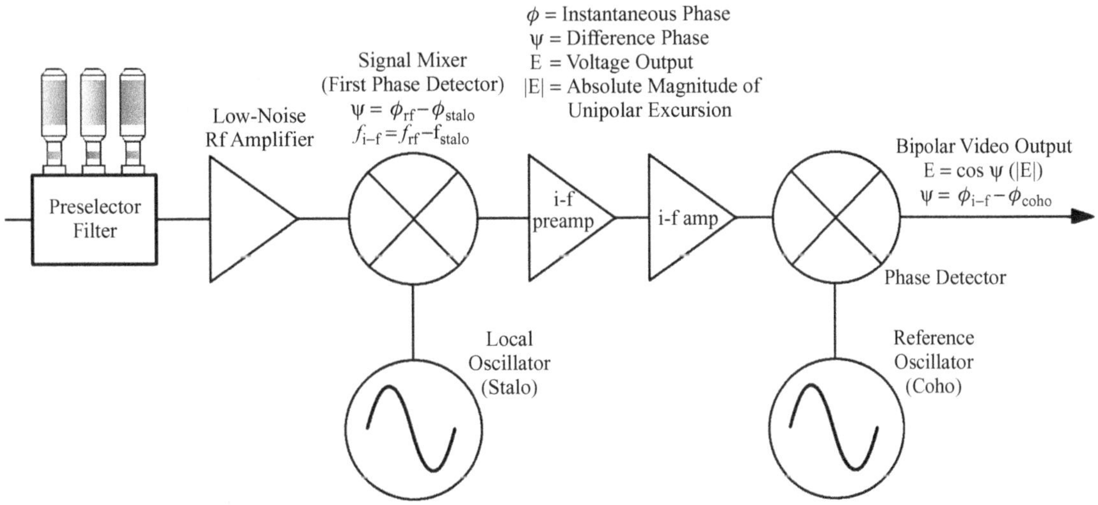

FIGURE 11–12

Superheterodyne phase detection.

Quadrature Phase Detectors

All early MTI systems used a single phase detector, but there was a disadvantage to the system. The phase-detector circuit exhibited an ambiguity; for any value of ψ, another value of ψ would produce the same voltage output. Possibly borrowed from color television circuitry, two phase detectors, called ***in-phase (I)*** and ***quadrature (Q),*** were employed. The *I* phase-detector response is the cosine of ψ, and the *Q* response is the sine of ψ. Based on the Pythagorean theorem, and the trigonometric identity $\sin^2\theta + \cos^2\theta = 1$, the two phase-detector outputs together describe ψ without the ambiguity. The full implications and applications of quadrature phase detection are multiple, and are discussed in Chapters 12–14.

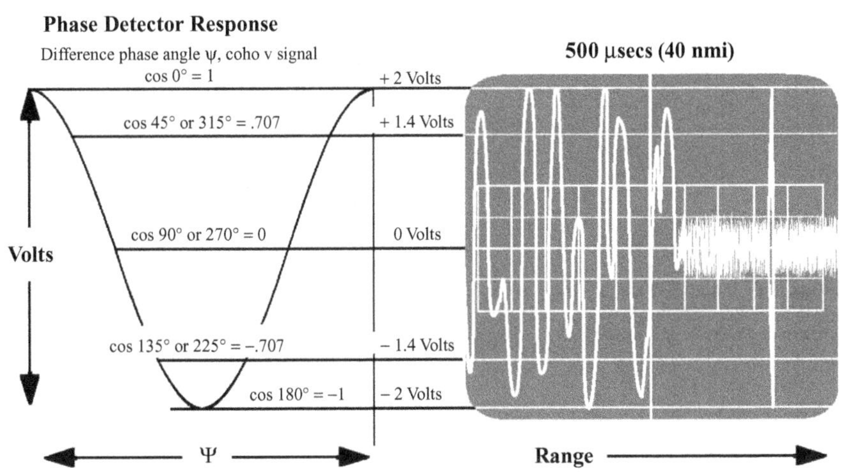

FIGURE 11–13

Phase-detector response and radar bipolar video.

The First Detector/Signal Mixer

When two frequencies are beat together at the same electrical point, they are sometimes in phase, and they add to each other. At other

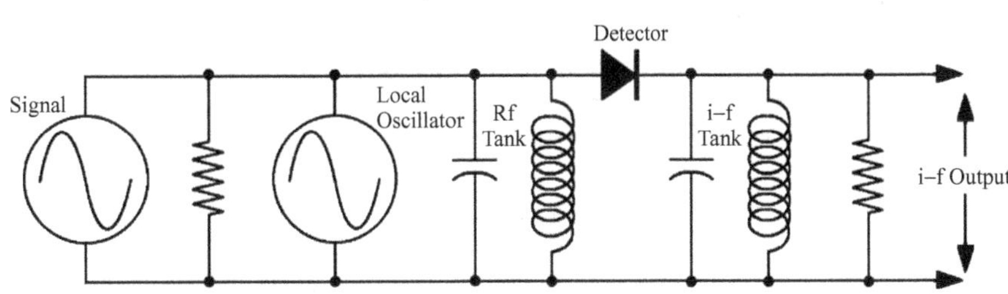

FIGURE 11–14

Equivalent circuit of a signal mixer.

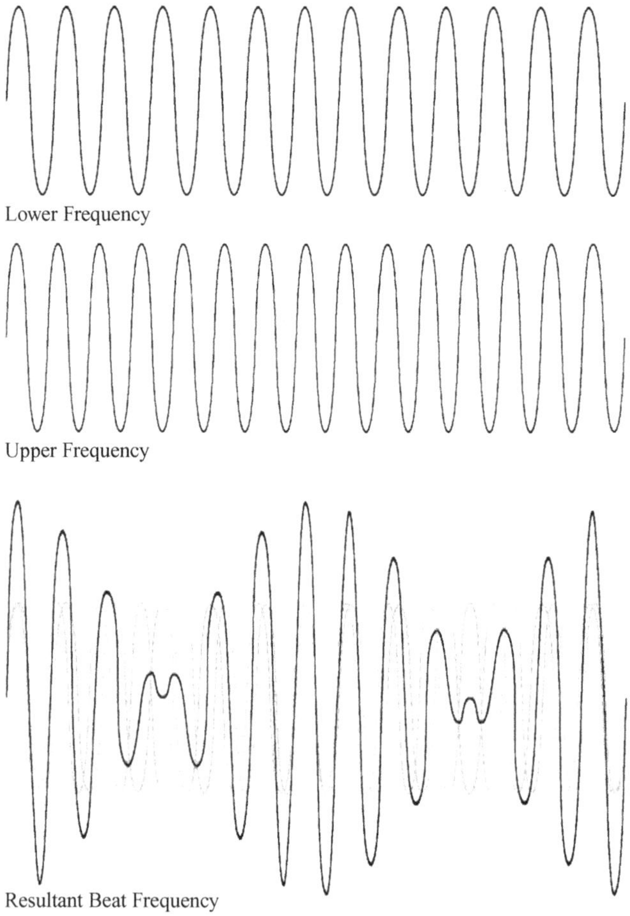

FIGURE 11–15

Beating two frequencies for a resultant.

times, they are of opposite phase and tend to cancel each other. Between those extreme points, there are lesser additions and subtractions. The time between maximum additions is the period of a resultant beat frequency. That beat frequency is the frequency difference between the two applied signals. Once the beat frequency is rectified, a spectrum of harmonic frequencies is created. One of those harmonic frequencies is the difference between the two original frequencies, which will become the intermediate frequency. The average voltage in the resultant frequency is zero, and it must be rectified by the detector to provide a means to recover the intelligence (see Figures 11-14 and 11-15).

The Detector Crystal (see Figure 11-16)

The detector is a latter-day embellishment of one of the oldest devices in electronics. This author can recall building an a-m radio "crystal set" receiver for a Cub Scout project in the 1940s; his father advised him on it, based on his own radio experimentation in the 1910s and 1920s. In the crystal set, enough signal power existed to provide adequate earphone power with no external applied voltage. A sharp metal "cat whisker" contact, indicated by the diode-symbol arrow, is pressed against a crystal slab, the "flat" part of the symbol. The sharp-point contact keeps capacitance at a minimum, to permit high-frequency operation. As an example of but one of many early-day mistaken assumptions, the diode symbol is actually reversed. Conventional current flow is positive to negative, and the arrow portion of the symbol indicates the ability of the diode to conduct from positive to negative and whisker to crystal. In reality, the electron flow is more likely in the whisker-to-crystal direction. In summary, the symbol depicts favored conventional current flow, and the whisker and crystal are reversed.

The radar crystal detector, through improved materials and close-tolerance manufacturing processes, operates satisfactorily in the microwave regions. There is a ***conversion loss*** to the received echo, which may be as much as 6 dB, and considerable noise is injected into the mixer by the local oscillator. The noise injection may be reduced by the use of a ***double-balanced*** mixer, a circuit which causes input noise spikes to be self-canceling.

The Double-Balanced Mixer

Figure 11-17 illustrates a double-balanced mixer, a device incorporating a waveguide ***magic tee hybrid,*** developed in the 1940s' explosion in radar technology. By injecting the local oscillator on the h-type tee, and the signal on the e-type tee, there is a phase reversal in the beat frequency at the two detectors. The detectors then provide half cycles which may be recombined to provide a beat with an i-f baseline. An i-f-tuned circuit will remove the higher frequency components. Noise from the local oscillator will appear on both the detector outputs, then be of reverse polarity, to cancel. The magic tee and double-balanced mixer illustrated rely on achieving a single resultant i-f from equivalent transformer action in the waveguide assembly. In the ASR-8, both detector outputs are routed, on separate cables, to the preamplifier, where the combination takes place, after capacitive coupling, at the connection to the base of a transistor.

Mixer Current

In the absence of echoes, the detector(s) continue to rectify the local oscillator signal, even though no beat frequency is produced. Since the local oscillator runs continuously, the detector outputs, filtered by a capacitor, may be used as a dc voltage to represent the amount of crystal current, which also is indicative of the local oscillator input. Except for the very latest systems, the dc voltage deflected a CRYSTAL CURRENT meter movement (see Figure 11-18). Values of the crystal current are typically in the 0.5- to 0.7-mA range.

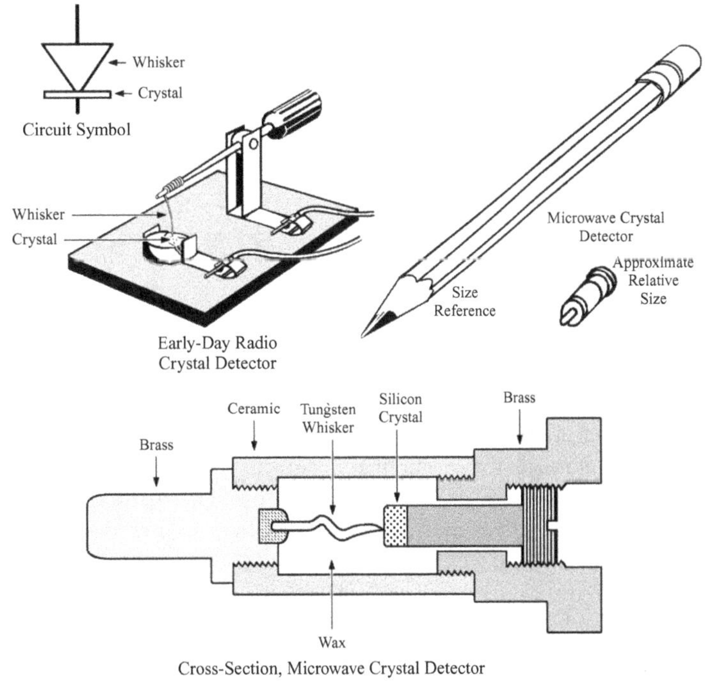

FIGURE 11–16

The detector crystal.

Fault Indications

The technician should be in the habit of observing the crystal current immediately on entry to the radar site; it should be as instinctive as checking for antenna rotation, listening for the f_p from the modulator, and looking at a ppi for displayed video. Most of the noise that is present as "grass" on receiver video is the result of the local oscillator and mixer, and the absence of that grass should immediately draw the technician's attention to the crystal or local oscillator.

The Local Oscillator

When MTI was developed, special attention was devoted to local-oscillator design, since high stability was essential to coherence. The new, exceptionally stable local oscillator was then named ***stabilized local oscillator (stalo),*** and the term is still in use. In early radars, the local oscillator frequency could be achieved only by means of microwave tubes, such as reflex klystrons and "lighthouse" tubes. Early-day radar courses

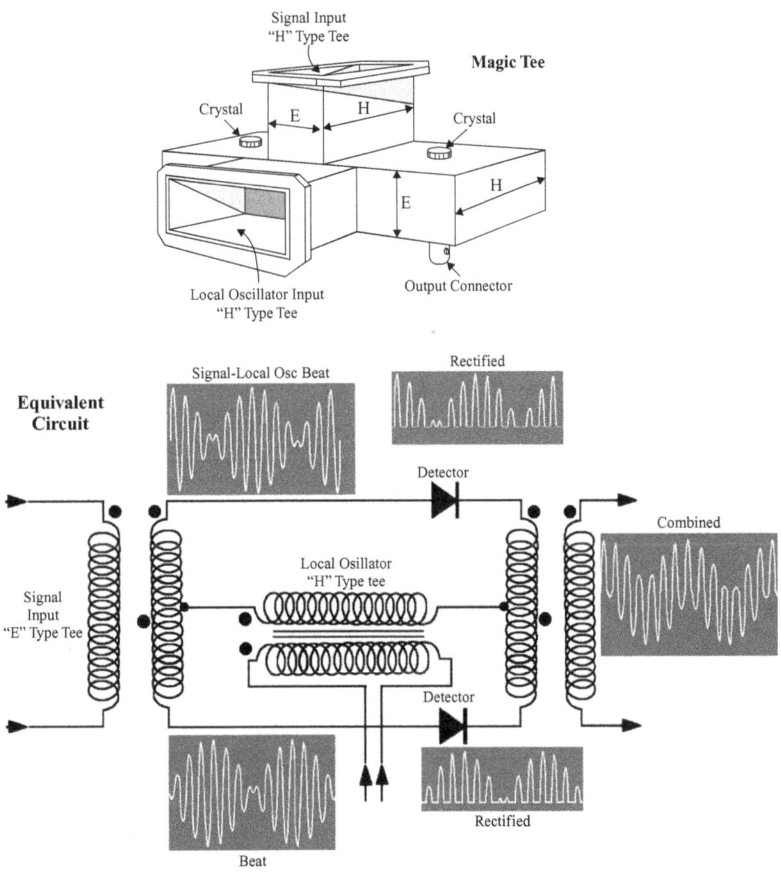

FIGURE 11–17

The magic tee double-balanced mixer.

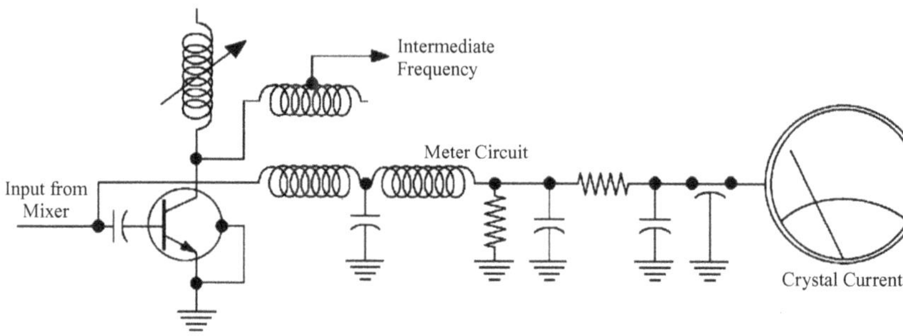

FIGURE 11-18

Crystal current meter circuit in preamplifier.

devoted substantial time to the internal operation of these. Latter-day systems have other means to develop these frequencies with solid-state devices. For instance, the ASR-8 local oscillator begins with a crystal-controlled oscillator in the 95- to 102-MHz region. The generated frequency is applied to a diode, which creates a wide spectrum of harmonic frequencies. A tuned cavity permits passage of the 28th or 29th harmonic into a solid-state amplifier.

Magnetron systems require some form of tuning to ensure that the frequency separation between the stalo and transmitter is precisely the same as the frequency of the ***coherent oscillator (coho).*** In the 1950s, ***automatic frequency control (afc)*** systems were incorporated to accomplish this. The most common technique was one used in other microwave equipment; it was formally called the ***reactance tube modulator,*** but might more descriptively be called the ***swept-bandpass circuit.*** It utilizes the i-f lock pulse as an indication of i-f frequency. The lock pulse was first intended to establish coherence in MTI systems, and was thus already available.

The main object of afc is to convert transmitter frequency error into mechanical frequency tuning correction. Generally, those systems originally designed to incorporate afc contain an electrical motor drive to the stalo cavity tuning. Systems modified to incorporate afc may apply the mechanical tuning correction to the magnetron. One example of this may be found in magnetron version ASR-8s purchased by the USAF; the same stalo used by the klystron synthesis version was used. While this book was in progress, a modification to the ASR-7 upgraded the entire coho–stalo–afc system to incorporate a digital frequency synthesizer. This is, no doubt, a prediction of afc systems of the future.

The Swept-Bandpass afc

Variations on this circuit have been successfully used for years, even in circuitry other than radar. The ingenuity of principle alone is interesting. An abbreviated diagram is illustrated in Figure 11-19, and Figure 11-20 illustrates the waveforms involved in the principle. In tube-type systems, an electron-tube oscillator was mounted in a frequency-determining cavity, and changes to the cavity dimensions would alter the frequency. Typically, there were two frequency adjustments: one coarse, another, fine. The coarse adjustment was a manual cavity-length control, and the fine adjustment was driven by a special, two-phase, ac motor. The motor windings are separated by 90°, and one winding is powered

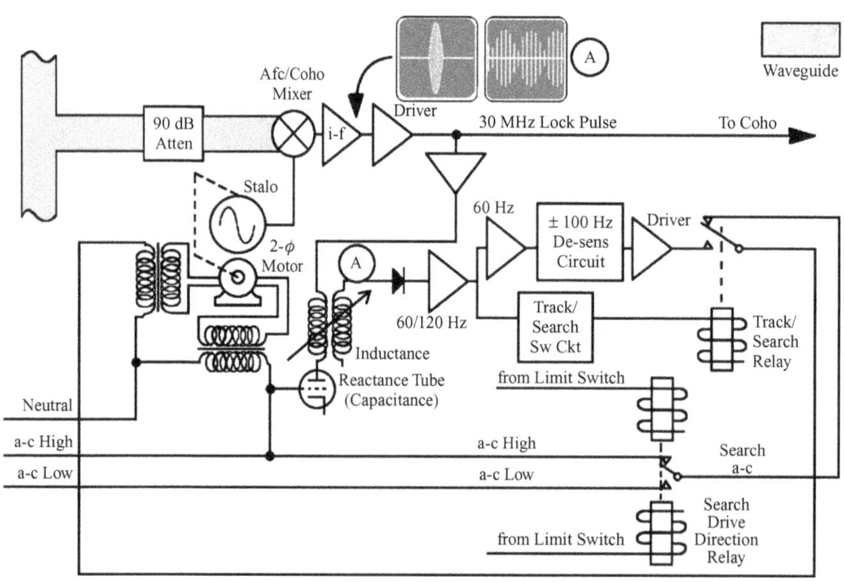

FIGURE 11-19

Generic swept-bandpass afc circuit.

by a constant ac reference. The other winding is powered by the necessary ac drive voltage, the phase of which determines the direction of rotation. The circuit uses a 220-V input, so two ac lines of opposite polarity in respect to neutral are available.

In a magnetron MTI system, a waveguide tee couples rf to a large attenuator, such as the 90-dB attenuator illustrated in Figure 11-19. The attenuator (1) reduces the transmitter burst to a low level, comparable to what an echo might be, and (2) reduces true echoes to such a low level as to be essentially insignificant in the circuitry downstream. The first purpose of this path was to generate the coho lock pulse for MTI coherence, but it also offered a convenient source of tuning information for the afc circuit. When properly tuned, the difference between the stalo and transmitter burst should be precisely 30 MHz, so the lock pulse should be 30 MHz.

Two waveforms are shown in Figure 11-19. The waveform to the left is a single 30-MHz i-f lock pulse, and the one to the right, marked "circle-A," illustrates the output of a transformer in the plate circuit of the reactance-tube-modulator; that tube is really the active central element. The time base of the two waveforms is substantially different; the first illustration of the lock pulse shows a single, 0.8- to 3-µs, 30-MHz burst; the "circle-A" waveform depicts a much slower oscilloscope sweep and shows many of those 30-MHz bursts, modulated by a "sweeping bandpass" in the reactance-tube modulator circuit.

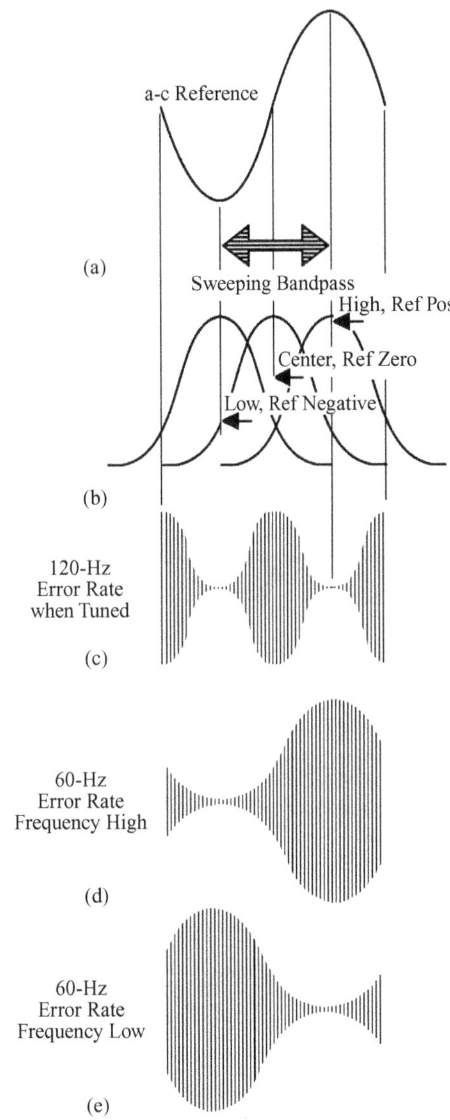

FIGURE 11-20

Swept-bandpass principle.

Sweeping the Bandpass

In the reactance-tube-modulator circuit, the capacitance of the tube circuit and the inductance of the coupling transformer comprise a single-frequency resonant circuit. The grid voltage on the modulator, shown in Figure 11-20A, determines the capacitive reactance of the tube; since the grid is supplied with 60-Hz ac, the capacity in the single-tuned circuit varies at a 60-Hz rate. The bandpass of the single-tuned circuit then varies at that 60-Hz rate, as shown in Figure 11-20B. When the transformer is properly tuned, the circuit bandpass is centered on 30 MHz while the grid voltage is at the two zero baseline points of the ac waveform. When the grid voltage is more positive, the capacity is less, and the bandpass is centered on a higher frequency; when the grid voltage is more negative, the bandpass is centered on a lower frequency.

When the input lock-pulse frequency is 30 MHz, the sweeping bandpass passes through that frequency twice, for each ac cycle, providing two maximums, so that the transformer output has a 120-Hz component, as shown in Figure 11-20C. Figure 11-20D shows a 60-Hz component, while the input is above 30 MHz, and perfectly at the higher frequency extreme. Figure 11-20D shows a 60-Hz output component of the opposite polarity, when the input is at the lower frequency extreme.

Figures 11-20C–E show precise conditions, which are rare. In most instances, the transformer output will have both 120-Hz and 60-Hz components. Clearly, the presence of a 60-Hz component indicates stalo tuning error, and the phase of the error in respect to the reference sweeping voltage indicates the direction of error.

Search Operation

The output of the reactance-tube-modulator transformer is rectified, and then filtered and amplified, to produce an ac waveform of 120 Hz or 60 Hz. That waveform is applied to a track/search switching circuit, which is ultimately

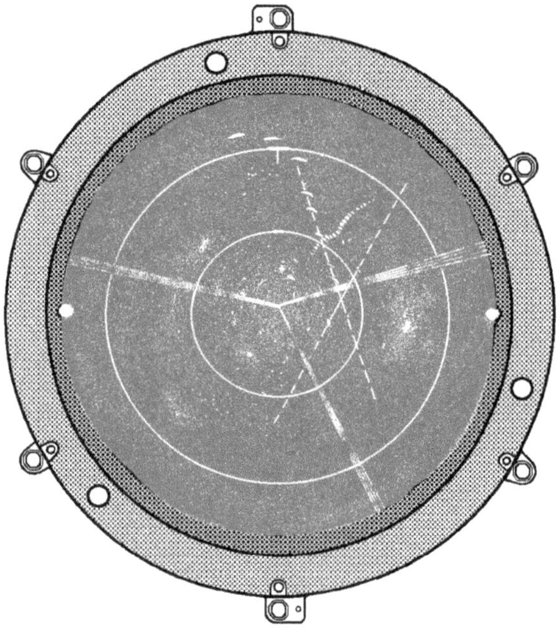

FIGURE 11–21

afc corrections appearing on ppi.

a Schmitt trigger threshold device. In the absence of either 60 Hz or 120 Hz, the track/search relay is de-energized, and maximum ac voltage is applied to the stalo tuning drive motor. When the drive mechanism reaches a limit, the ac polarity is reversed by the search drive direction relay, and the motor rotates in the opposite direction. If the i-f lock pulse frequency is within the approximate 28-MHz range between limits, the search operation will end when the correct frequency is approached, and the 60-Hz error begins to rise. At that point, the track-search relay will energize, and the motor drive becomes dependent upon the output of the error-signal driver circuit. Not shown is an interlock circuit, which prohibits search operation when the transmitter is off.

Track Operation

The track portion of the circuitry contains an amplifier, which is tuned to pass the 60-Hz error, but discriminate against the 120-Hz on-frequency signals. It is undesirable to correct the stalo tuning frequently, since, each time this occurs, the system becomes incoherent, and MTI clutter residue will appear on displays. This effect is illustrated in Figure 11-21. A Schmitt-trigger threshold circuit prevents any error-Voltage output until the error reaches 100 kHz. The track circuit operates over an approximate 3-MHz range, between those points at which the search circuit "takes over."

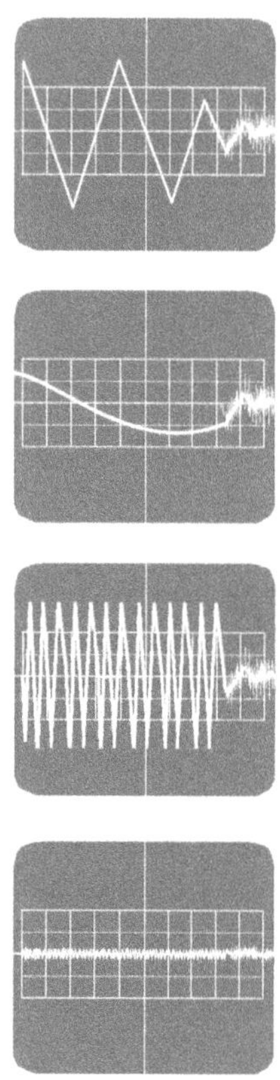

FIGURE 11–22

Precise tuning with an echo box.

Adjustment

This procedure was described in the echo-box discussion in Chapter 8, but the illustration is repeated here, in Figure 11-22, for the reader's convenience. The tuning of the transformer in the reactance-tube-modulator circuit sets the center operating frequency of the afc and stalo. The greatest precision can be obtained with an echo box, and an oscilloscope connected to the MTI phase-detector output. If (1) the coho is operating at precisely 30 MHz, (2) the echo box is precisely tuned to the center of the transmitter spectrum, and (3) the resultant echo-box i-f is precisely 30 MHz, then (4) there should be a zero beat in the phase detector.

Discriminator-Type afc

This variation was used in the ASR-7. A frequency discriminator contains two tuned circuits, one below a center frequency, and the other, above. The combination of the two responses yields the familiar discriminator curve, shown in Figure 11-23. The simple use of the discriminator on the afc lock pulse to achieve correction does not suffice; instead, a sample of the coho is injected into the discriminator at a different time than the i-f lock pulse, and the off frequency indication is achieved by the voltage difference between the two resultant pulses from the discriminator. The voltage difference is then utilized to operate control circuitry for a two-phase drive motor, as in the swept-bandpass afc. Track/search and desensitization circuits are also included.

A major advantage of this type of circuit is that the coho is the reference signal. The main tuning objective is to ensure that the i-f is precisely the same frequency as the coho.

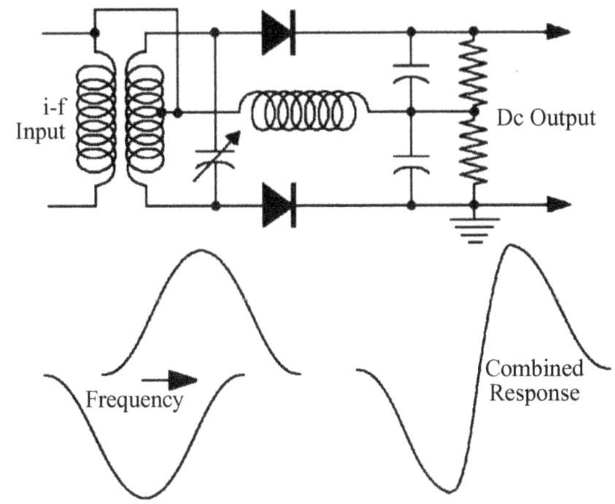

FIGURE 11–23

Fundamental discriminator principle.

i-f Amplifiers

The i-f amplifier is the most significant factor in determining the overall receiver bandpass $\Delta f_{overall}$. As it is likely that there may be more than one i-f amplifier, and they most likely will have different characteristics, any system may have different values of $\Delta f_{overall}$, F_t, P_a, and $P_{r\ min}$. Since this is true, each signal-flow path to a detected i-f output is considered an ***independent receiver system.*** Of course, much of the receiver system is common to all i-f amplifiers, and that portion is commonly called the ***front end.*** The "front end" ends, and the independent receiver begins, at that point in the pre-amplifier where the i-f signal is amplified for distribution. The front end has been called the ***common receiver*** by Texas Instruments, manufacturer of many ASR radars; however, that definition is peculiar to that manufacturer.

Although there are other i-f amplifier types, this section of the chapter will describe only the four major types of i-f amplifiers used in FAA radar systems. These are (1) normal, (2) log, (3) MTI, and (4) MTD. This information may be somewhat repetitious; much of it was introduced in Chapter 6. Each of these i-f amplifiers has its own characteristics and design intent, and differs radically from the others. i-f amplifiers are also often referred to as ***i-f strips,*** as they are a long "strip" or chain of amplifier stages. Conventional expression of "i-f" is the hyphenated, lowercase abbreviation used here. The lowercase is used in technical literature to avoid confusion with reference to controls, testpoints, and mnemonics, and the hyphen is used to avoid confusion with the word "if."

Among the greatest significance in any i-f amplifier are bandwidth, gain, noise figure, type of detection, and characteristics of the output information. Because of the dramatic effect of front-end units on the overall noise figure, the gain may seem less significant than the Δf, because the Δf of the i-f amplifier becomes the Δf in the P_a and $P_{r\ min}$ equations. However, there may be more than 80 dB of gain in an i-f amplifier, and attention to contributed noise is still an important consideration, particularly in the first stages. The i-f amplifier gain is necessary, principally, to bring the signals to a useful level for the second detector operation, whatever type it may be. For many reasons, the i-f Δf may be something other than $1.2/t_p$ to satisfy use-and-purpose intentions.

Because of the need to achieve a wide Δf, the stages of the i-f amplifier are only rarely all tuned to the center i-f frequency. Such a tuning scheme is called ***synchronous tuning.*** Where synchronous tuning is employed, each successive stage in the i-f strip further narrows the overall Δf, since the responses to the edges of the bandpass are de-emphasized, and the response to the center is emphasized. To achieve a sufficient bandpass to amplify the harmonics of the pulse, stages may be tuned to different frequencies in pairs, triples, or quads (see Figure 11-24).

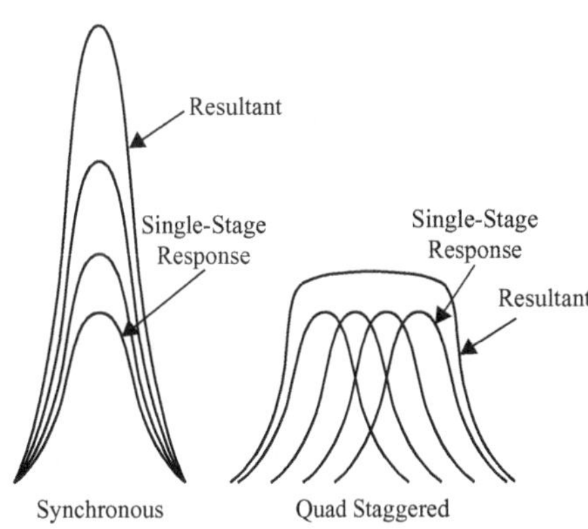

FIGURE 11–24

Responses synchronous versus quads.

The Normal i-f Amplifier

This was the original radar i-f amplifier. It is a standard, amplitude-detection i-f amplifier. It was not called "normal" until the introduction of MTI systems in the late 1940s; then, it became necessary to differentiate the two receiver systems. The word "normal" was a natural term, as that type of receiver had been

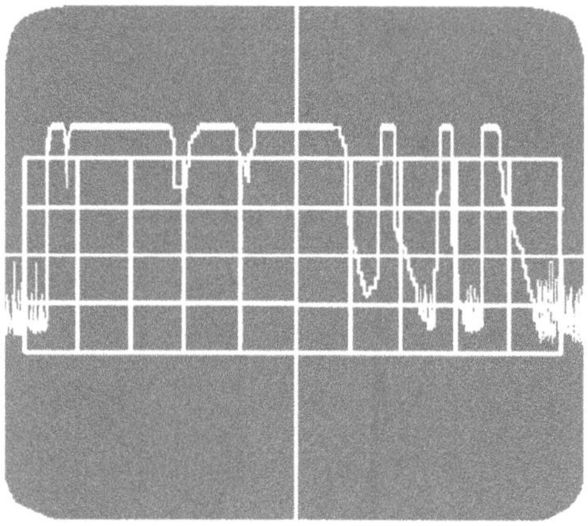

FIGURE 11-25

Normal video pn an oscilloscope.

the "normal," or usual, type, before MTI. The normal i-f amplifier ordinarily has a Δf close to $1.2/t_p$. See Figure 11-25. Most close-in echoes reach limit, and detected normal video is characterized by a "clipped" appearance. A typical normal i-f amplifier output will exhibit an approximate 4:1 signal-to-grass ratio, and the grass will have a fine, dense appearance.

The normal i-f provides the best mds, and its use should be encouraged. There is always an inclination, on the part of users, to entirely rely upon MTI video. Although MTI removes clutter from the displays, it has many disadvantages, such as decreased sensitivity, blind-phase, blind-speed, dim-speed, and tangential effects. Some radar systems have featured "clutter gating"; MTI video was only enabled at those specific ranges where clutter echoes appeared on normal video.

Normal i-f Ancillary Circuits

In early versions of the normal i-f amplifier, particularly the vacuum-tube types, additional circuitry may be found. A connection to a grid circuit(s) in the first stage(s) would allow application of an stc waveform. This method was inferior to the latter-day PIN diode stc, and has now been abandoned. One of the disadvantages was that it eliminated close-in grass, caused by the mixer, causing a range-variable false alarm rate, and a "washed-out" video appearance on real-time displays. Older systems may have incorporated an ftc circuit at the normal i-f detected output; it was simply a differentiating network. In the late 1950s, an instantaneous automatic gain control (IAGC) was introduced in the ASR-4; the circuit would detect and delay the echo by approximately 1 t_p, and then use the delayed video as a negative bias on an upstream stage. The effect was similar to ftc, but offered greater target detection probability, because targets in extended clutter might not be obliterated by limiting.

Special Features

Some normal i-f amplifiers may contain an i-f test connector, at a point upstream of the detector. Although rare, there have been sensitivity test procedures in which the i-f output would be monitored with an rf voltmeter while adjusting a c-w output from the rf signal generator. This connector may also be used for the 30-MHz input to a noise figure test set.

Special Uses

The normal receiver may also be used as a test device for special procedures, as it may serve to indicate signal strength. Two of these procedures are in developing clutter-strength records, and in performing subclutter visibility measurements. Information on these procedures is given on page 264 (Section "Gain and Bandpass Testing").

Testing and Alignment

The normal i-f performance may be verified by conventional gain and bandpass tests. The principle adjustments will normally include an I-F GAIN and VIDEO GAIN control.

The Logarithmic i-f Amplifier

Logarithmic gain has many uses, but those can all be summarized in the universal purpose of placing the output data in a deciBel scaling. Logarithmic data may be used in CFAR circuits for weather cancellation, in weather radars for weather-level detection, and in spectrum analyzers, to provide video representative of power in deciBels related to a reference. Figure 11-26 illustrates the general configuration of a log i-f amplifier. The log i-f performs amplitude detection at the output of each stage, rather than only at the output of the last stage, as would the normal i-f. As signal strength increases from minimum, the last stage, because of all the gain preceding it, will

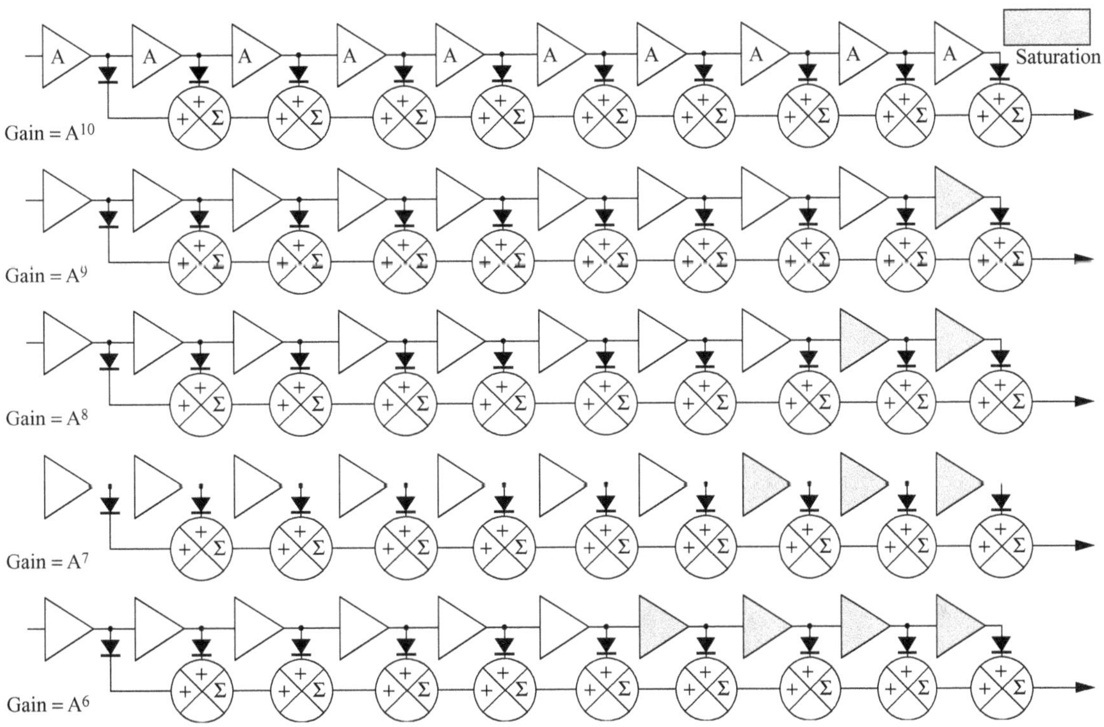

FIGURE 11-26

The logarithmic gain principle.

be the first to reach the limit. In the normal i-f, no further signal-strength changes are seen at the output, but in the log i-f, changes can be further detected. The output of each log i-f stage, after detection, is applied to a summation circuit. Once the last stage reaches saturation, the overall gain of the entire amplifier is reduced, because the last stage no longer amplifies. When the next-to-the-last stage is saturated, the gain of the entire amplifier is again reduced. As the signal continues to increase, more stages saturate, and the gain continues to decrease.

In the older, tube-type, log i-f amplifiers, the individual stage detectors were diodes, as shown. They are not as obvious in newer types, as these may use base-to-emitter transistor junctions as the detector diode.

Video at the log i-f amplifier output would appear to (1) have a very poor signal-to-noise ratio, (2) have no apparent limiting, and (3) have a very irregular and coarse grass. This is a result of the high gain to low-level signals, and the lower gain to high-level signals. Figure 11-27 illustrates the relationship between logarithmic video and the equivalent linear video.

Test Procedures

Special test procedures are required of the log i-f amplifier. With linear amplifiers, the bandpass is checked with a swept-frequency signal generator, and the frequency distance between the bandpass −3 dB, or 0.707 voltage points, is the standard measurement. This procedure is unacceptable for a log i-f, because the gain increases for lower level signals, distorting the bandpass. Manufacturers may publish special test procedures which include a "dynamic amplitude range test" at several frequencies. The ***dynamic range*** of a log i-f is that amount of test signal level change that can be applied between a minimum distinguishable signal and limit level.

The greatest purpose of log i-f amplifiers in FAA ASRs and ARSRs is to provide deciBel-scaled analog information to a delay-line ftc circuit; the i-f amplifier and ftc circuit together are called a ***log ftc*** system. Refer to page 87 (Figures 6-26 through 6-28) for more information on delay-line log ftc.

The MTI i-f Amplifier

A fundamental design objective of this unit is to amplify i-f echoes for phase detection, and to do so in such a manner that phase differences between superimposed echoes will be detectable. Probably the most severe limitation

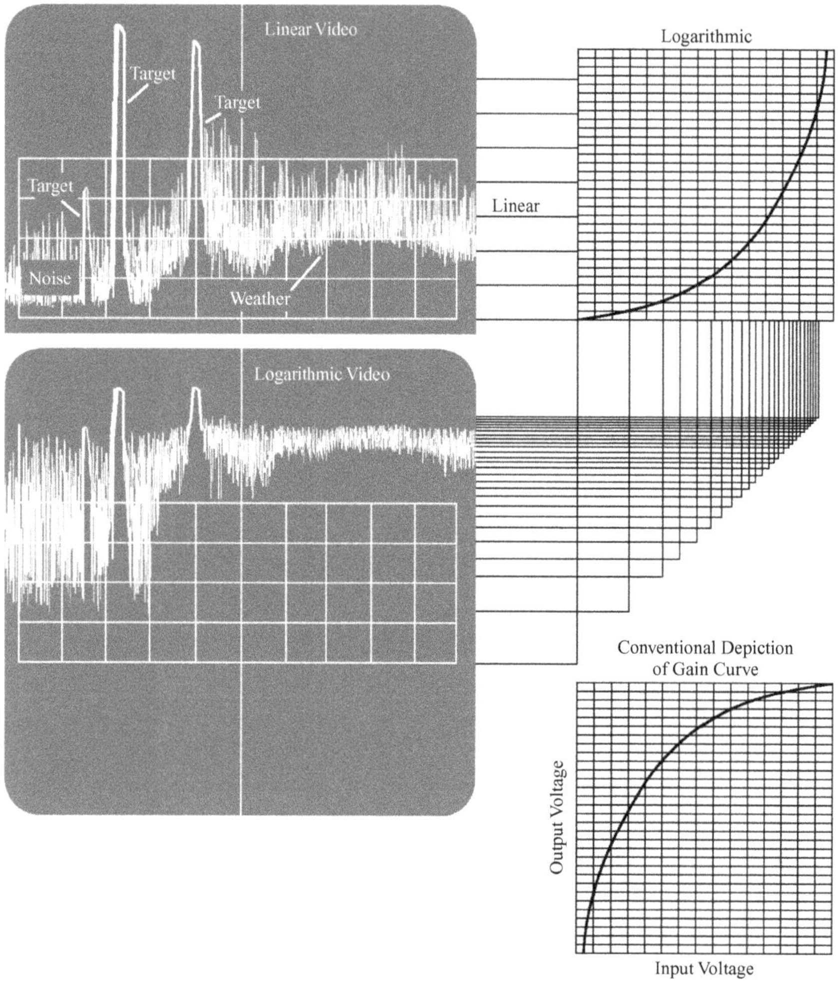

FIGURE 11–27

Logarithmic and linear video.

to MTI performance is ***subclutter visibility,*** a measurement of the ability of the system to detect moving targets superimposed on clutter echoes. This subject is explained in detail on pages 12-30 through 12-35.

The MTI i-f amplifier severely ***hard-limits*** the i-f echoes, so that there is minimal signal amplitude variation at the input to the second detector, the phase-detector circuit. The voltage out of the phase detector must be representative only of the phase difference ψ between the signal and coho; if amplitude variations are allowed, the pulse-to-pulse voltage output for a fixed target will be variable, even though the ψ does not change. This would cause MTI ***clutter residue.*** On the other hand, the depth of limiting plays a major part in subclutter visibility, and more severe limiting worsens the subclutter visibility.

To enhance the amplifier's ability to detect complex superimposed echoes, a wider bandpass, to process more spectral harmonics, is used. Bandpasses in the order of $3.5/t_p$ are not unusual. This, of course, increases the P_a and $P_{r\,min}$ to derogate target detection capability.

Performance Testing

The MTI i-f performance may be verified with conventional gain and bandpass tests, but special in-circuit connections, upstream of the hard limiting, will probably be necessary. See the manufacturer's instructions, to be sure you duplicate the factory test procedure.

Adjustments

There may be multiple adjustments to an MTI i-f amplifier. Probably the most significant, most-used control is the I-F GAIN control. That control is one of two ***MTI payoff controls,*** critical adjustments to MTI performance. The payoff control adjustment is involved and explained in detail on pages 308 and 309. In addition to the I-F GAIN control, the MTI i-f may also contain an I-F LIMIT control, to set the severity of limiting, and it may also incorporate controls to adjust the phase-detector response and output amplitude. These may include VIDEO GAIN, BIPOLAR VIDEO BALANCE, BASELINE BALANCE, and PHASE, an adjustment to the phase shift to the coho where two phase detectors in quadrature have been employed. Of course, the names of these controls may differ somewhat with manufacturers, but their purpose should be obvious to the informed technician.

Video Output

The principles of phase detection and Doppler shift are complex, and are addressed in Chapter 12. For the purpose of this chapter, the technician should only recognize that it does occur, that the cosine phase-detector response represents the ψ between coho and echo, and that the final output "video" is bipolar. Individual fixed targets will produce an unchanging phase-detector output voltage, and individual moving targets will produce a variable phase-detector voltage output; these moving targets have been assigned the slang term ***butterfly*** since the 1940s (see Figure 11-28).

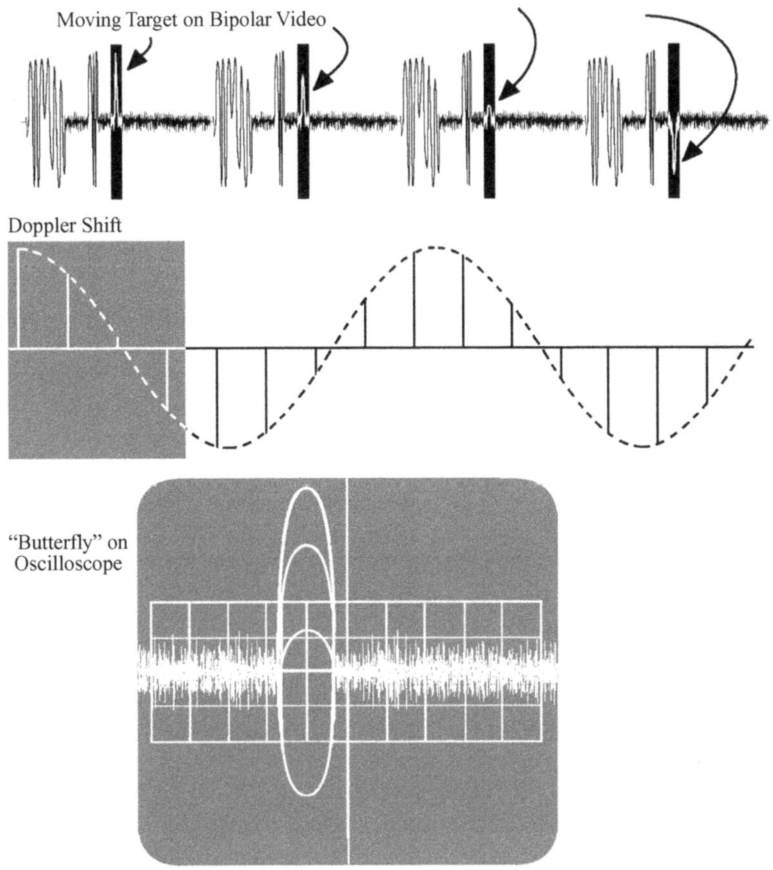

FIGURE 11-28

The "butterfly."

The phase-detector response in latter-day systems is cosinusoidal, and there are ordinarily two phase detectors in quadrature; the quadrature phase detector will exhibit a sinusoidal response. In older systems, quadrature MTI was not practical because of hardware costs, and a single phase detector was employed. Most of these single phase detectors exhibited a modified-cosine triangular response, as illustrated in Figure 11-29. There is also a discussion in Chapter 6.

The MTD i-f Amplifier (ASR-9)

The ASR-9 has only one i-f amplifier; it utilizes quadrature phase detectors, but differs radically from the MTI i-f amplifier (see Figure 11-31). It is very narrow in bandwidth, far less than $1.2/t_p$. The bandwidth is 923 kHz ***at the 6 dB points;*** the ASR-9 pulse width is slightly greater than 1 μs. The narrow bandpass is achieved with a filter. There is no hard limiting, as in the MTI i-f, and the amplifier has a dynamic amplitude range

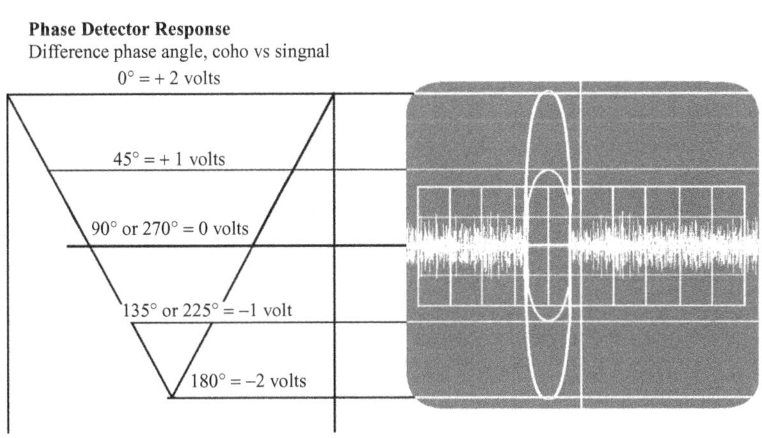

FIGURE 11-29

Triangular phase-detector response.

of 63 dB. A "soft limiter" is incorporated only to prevent exceptionally strong clutter from exceeding the maximum desired phase-detector response excursions of ±3.5 V.

MTD processing bears little similarity to MTI; there are no cancelers, and the processing utilizes the ***rectangular coordinate data*** from the quadrature detectors. Rather than making pulse-to-pulse comparisons of amplitude differences, MTD processing determines Doppler shift through automated analysis of the rotation of target vectors described by the rectangular-coordinate data. The intermediate frequency is 31.07 MHz, popular in latter-day systems. The i-f must be the same frequency as the coho, and 31.07-MHz cohos are used, because that frequency may be conveniently divided to provide 1/16 nmi range cells, a binary power of 2^{-4}, in respect to 1 nmi.

Because one MTD objective is to improve subclutter visibility, worsened by limiting, and because of a digitizer azimuth centroiding technique, based on signal strength, limiting is intolerable in the i-f amplifier, necessitating the wide dynamic amplitude range. However, logarithmic gain is not used to avoid limiting, and the i-f amplification is linear. Therefore, the noise, in relation to the maximum signal, is almost indistinguishable on an oscilloscope. Because it is necessary for the technician to be able to view real-time ppi video at the radar site, there is a special auxiliary amplifier, upstream of the phase detectors, to provide for log detection. The only purpose of this log amplifier is to decrease the signal-to-noise ratio toward a more conventional value, more similar to normal video; without the log amplifier, the ppi display would appear to be noiseless, and weak targets would not be observable.

Obviously, the very narrow bandwidth degrades the mds. However, a phenomenon called ***coherent integration*** occurs downstream in the Doppler filter bank and recovers the loss. The end result of the combined effect is an acceptable mds with less noise to cause digitizer threshold breaks. Overall mds measurements cannot be made with conventional off-the-shelf rf signal generators, because the test signal is incoherent and exhibits a "random Doppler"; such a signal cannot pass through the MTD Doppler filter bank. The ASR-9 contains built-in test equipment to produce a continual −106.5 dBm mds test signal; the test signal contains a programmed pulse-to-pulse phase shift, to cause it to pass through one of the Doppler filters at maximum amplitude.

Although the phase detectors operate in the same fashion as the MTI phase detectors, they may be called ***synchronous detectors*** because the input is not limited and because a single detector voltage does not, by itself, indicate ψ. The two detectors together provide ***rectangular coordinate data.*** Since they are in quadrature, the Q represents the sine of an angle, and the I represents the cosine. Because of the trigonometric identity $\sin^2\theta + \cos^2\theta = 1$ and the closely related Pythagorean theorem, the two values describe both phase and signal magnitude. Chapter 14 describes this in detail, and other information is contained in Chapters 6, 12, and 13. Figure 11-30 illustrates quadrature responses as rectangular coordinates; this illustration is generic, and not necessarily representative of the ASR-9.

Adjustments

The ASR-9 manufacturer, Westinghouse, used the term ***receiver calibration*** in reference to the adjustment of the receiver. As the MTD is latter-day state of the art, the system was designed to automatically make adjustments to the i-f amplifier as necessary. Self-contained ***Single-Board Computers*** govern much of the system operation. The system employs ***servo loops,*** self-maintaining circuitry, in which a ***Doppler test tone*** is circulated through the i-f amplifier. The term "Doppler test tone" implies that audio is injected into the i-f amplifier; this is not the case. The Doppler test tone is a special 31.07-MHz burst created by a ***Doppler Test Tone Generator module,*** and injected into the amplifier in deadtime. It exhibits a precise, programmed T_r-to-T_r phase shift, relative to the coho phase, thus creating a predictable ψ. The degree of test-tone phase shift is commanded by a "receiver calibration" bus. The receiver calibrate information source is a RAM in the synchronizer; a ***Receiver Control module*** provides data steering to the appropriate D/A converters. Very simplistically, the system "knows" what ψ should result from the Doppler test tone, tests for the appropriate voltages, and then adjusts to correct for errors. Sampling of the Doppler test tone is downstream of the i-f amplifier.

Gain and Bandpass Testing

The ability to test an i-f amplifier bandpass and gain should be a basic requisite of all radar technicians; there is no other way to conclusively verify that an i-f amplifier is operating properly. The equation for $P_{r\,min}$ is invalid if the i-f center frequency is incorrect, or if the bandpass is less than $1.2/t_p$. And, if the gain is inadequate, it is likely that the amplifier's noise figure is poor.

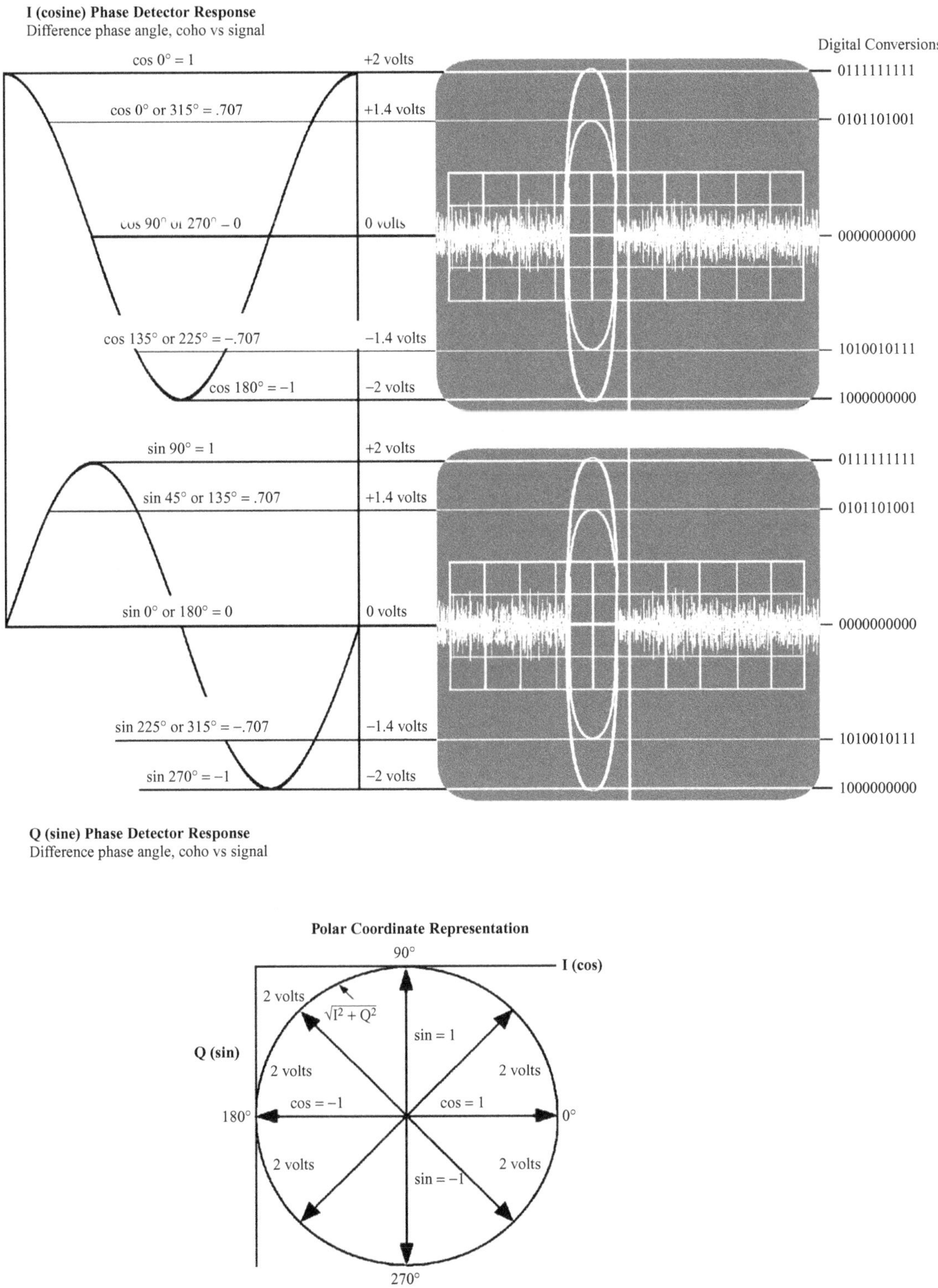

FIGURE 11–30

Rectangular coordinate data.

It is always the best practice to carefully follow the manufacturer's instructions for gain/bandpass testing and receiver adjustments, and adjustments to the i-f should be a rare procedure. Often, the field technician will only exchange the i-f amplifier for another from a central facility; this does not relieve him of the responsibility to perform tests. He must have a method to determine that such an exchange is necessary; except in cases of

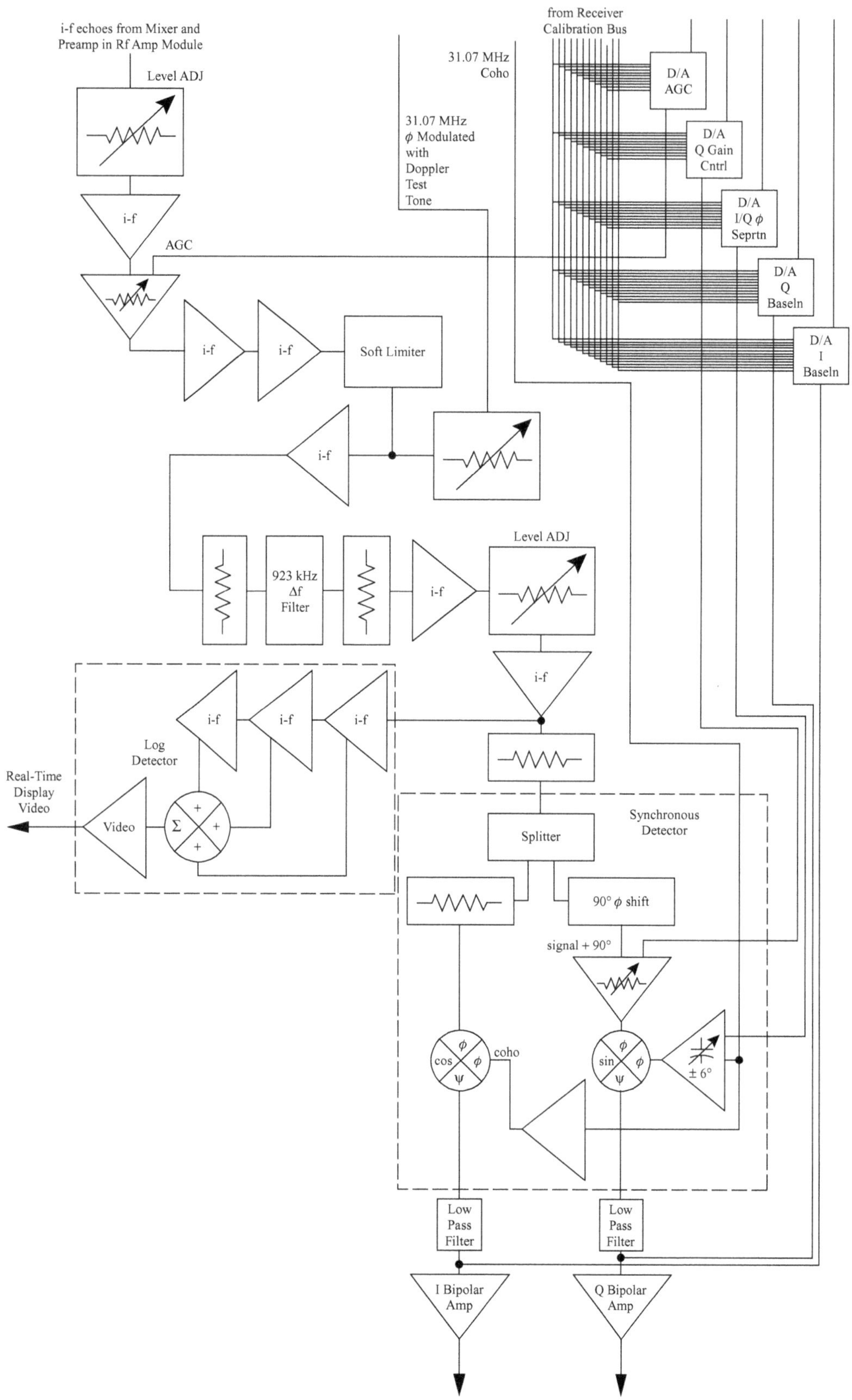

FIGURE 11–31

The ASR-9 i-f amplifier, abbreviated.

catastrophic component failures, only gain-and-bandpass testing can provide the necessary scientific data. No harm can be done by only sweeping and observing the bandpass, and it just might save some expense, or lead the technician toward the *real* problem.

Procedures for connection of the test equipment to the unit under test may vary significantly. Always follow the manufacturer's instructions where possible, and never make adjustments, except under all conditions prescribed by the manufacturer. Often, the manufacturer's instructions will include the fabrication of a special test fixture for matching to the unit's input, output, or both. Testing is not necessarily performed at the final output of the amplifier; often, a connection will be required upstream of the second detector, or of limiting circuits. Alignment may also call for the temporary installation of jumpers, resistors, or other components.

In addition to the necessary adherence to the equipment manufacturer's instructions, the technician must also give consideration to the test equipment manufacturer's instructions. A variety of signal generators have been produced, and both the internal and external configurations of these may differ significantly.

Figure 11-32 illustrates a simplified, functional, generic, swept-frequency signal generator; it is unlikely that it exactly resembles any available signal generators and is incorporated only to provide a discussion basis. In other electronics fields, such test equipment is assigned the slang description "sweep generator." In radar, care should be taken not to use that term, because a "sweep generator" is a main component in display systems.

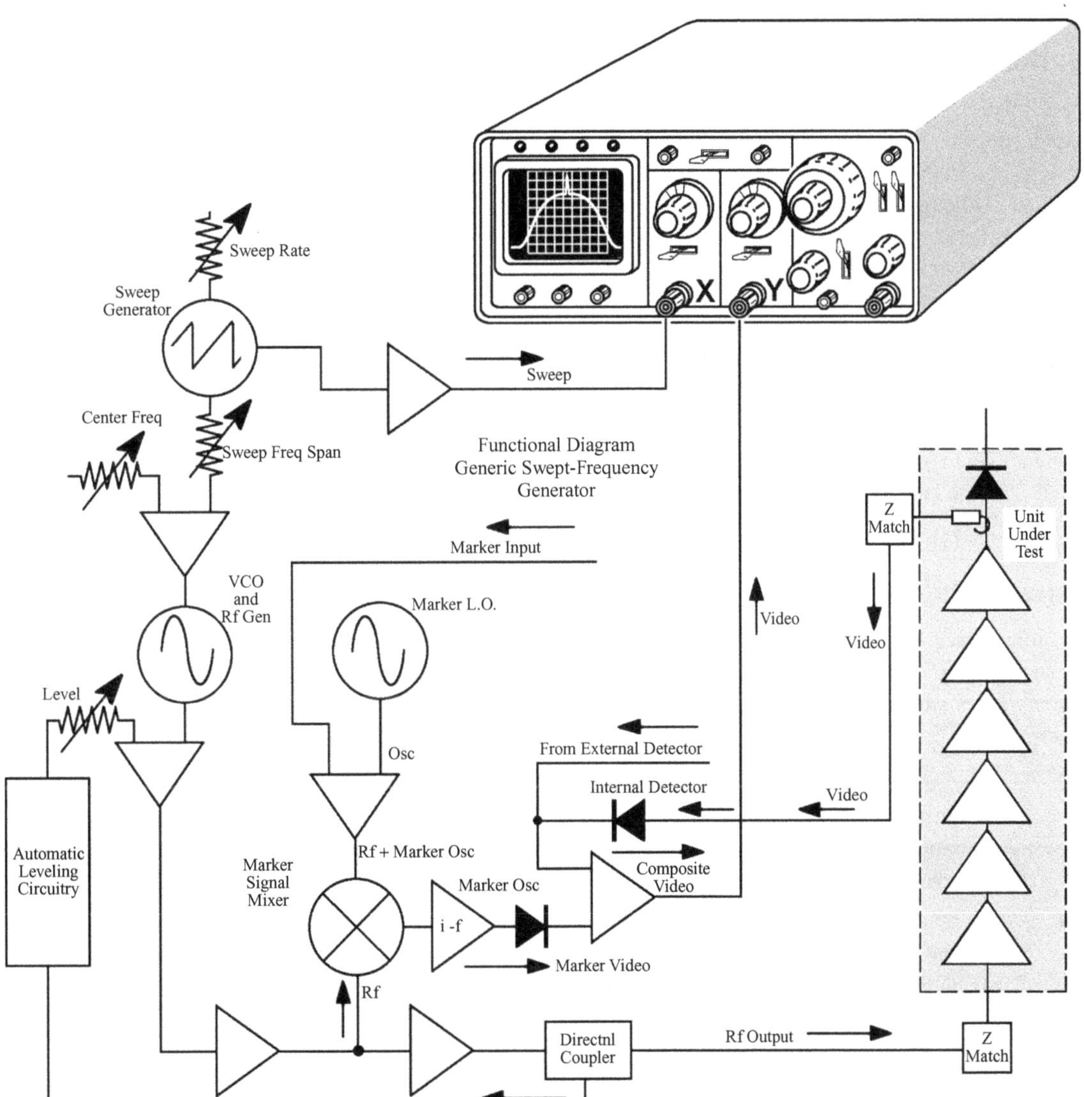

FIGURE 11–32

"Sweeping" the receiver.

The central function of the generator is a voltage-controlled oscillator (VCO); the output may be multiplied to a higher rf band. The VCO output is varied by a ***sawtooth*** (also called ***ramp***) input. Some generators may simply use a 60-Hz sinusoidal waveform for this, and others may use a triangular waveform. A sweeping voltage from the same source is used as the horizontal input for the oscilloscope. The sawtooth to the VCO may be varied in amplitude to increase or decrease the spectrum of frequencies spanned. Adjusting the sawtooth baseline will adjust the center frequency of the swept spectrum. As the frequency increases from low to high, the oscilloscope trace moves from left to right. Obviously, if a triangle or sinusoid is used for the sweep, the oscilloscope sweep will retrace from right to left, while the frequency decreases from high to low.

The rf generator output is amplified for use as the final output drive, but passes through a directional coupler. The directional coupler is used to provide a "sense" output to an ***automatic leveling circuit.*** As the frequency spectrum is scanned by the sawtooth, the VCO output amplitude may vary, and the load to the different frequencies by the test unit may vary. Were this not corrected, the displayed bandpass would be unrepresentative of the amplifier characteristics, because of the input variations. The leveling circuitry ensures that the output level is consistent throughout the spectrum.

The output from the amplifier is routed back into the signal generator for detection and then combined with the marker for display. An external detector, or the amplitude detector in the amplifier, may also be used.

Markers

Providing a band of swept frequencies is of little value if the test equipment operator cannot be sure what those frequencies are. Some sort of "marker" capability must be incorporated. There have been several schemes for these. The earliest generators contained markers spaced by a known crystal frequency, such as 10 kHz, 50 kHz, etc. Although helpful, it was difficult to determine *which* 10 kHz or 50 kHz marker was under observation, and therefore difficult to determine the bandpass center frequency. Later models incorporated marker generators, a variable-frequency marker, and a provision for an external marker. At the radar facility, use of the coho oscillator as an external marker at 30 MHz or 31.07 MHz is the best source to indicate the i-f center frequency.

In Figure 11-32, no marker generator is shown, but a provision for an external marker is incorporated. A marker local oscillator, at the marker i-f, is mixed with the external marker input. In the marker mixer, the swept frequency is beat against the marker-plus-marker i-f frequency; when the marker is equal to the swept frequency, the i-f results, and a detected pulse occurs at the marker detector output. The marker video is then mixed with the bandpass video to provide a composite video output for display on the oscilloscope.

General Rules

The importance of following manufacturer's instructions cannot be overemphasized, particularly where adjustments are to be made. However, it may sometimes be necessary to observe an amplifier's approximate characteristics where the prescribed test fixtures and information are unavailable. Following are some general principles:

1. Impedance mismatches can distort the response. Mismatches cause reflections, which may add to, or subtract from, the true signal. These can be reduced with attenuation, as each reflection is then twice reduced by the two-way travel through the attenuator. Use as much attenuation as possible between the signal generator input and the unit under test.
2. Very often, the i-f input termination will be 50 Ω, and the detected output will be 75 Ω. Check the equipment cable types for verification.
3. If the unit is tested in its normal in-circuit configuration with high-impedance connections to the test equipment, it may closely resemble its true operating characteristics.
4. Take care to ascertain that the swept-frequency spectrum is not saturating the amplifier, and use as low a level as is practical.

Gain Testing

Again, use manufacturer's instructions if available. A simple method of measuring gain is as follows:

1. Connect a variable attenuator between the i-f amplifier output and signal-generator detector input; set the attenuator to zero.
2. Place the signal generator in c-w at the center frequency.
3. On the oscilloscope, observe the dc level from the detected i-f amplifier output. Lower the level enough to ensure that the test signal is not saturating.
4. At the detector input connection on the signal generator, remove the cable from the amplifier output attenuator, then connect the rf output directly to the detector, and note the dc level on the oscilloscope.
5. Connect the rf output back to the i-f amplifier input, and the cable from the i-f amplifier attenuator back to the detector.
6. Use the variable attenuator to bring the oscilloscope dc level back down to the previous level noted. The attenuation added in the variable attenuator is the gain.

Review Questions

1. Identify the four oscilloscope presentations of receiver video illustrated in Figure 11-33.

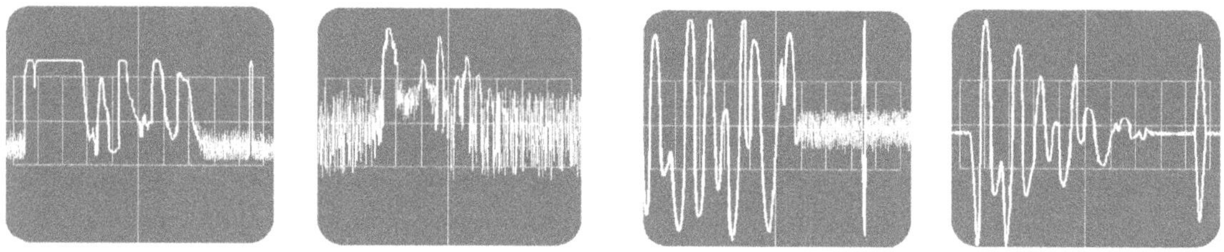

(A) (B) (C) (D)

FIGURE 11-33

Oscilloscope presentations, four receiver videos.

2. Explain "optimum bandpass."
3. The bandpass of an i-f amplifier is 4 MHz, and the transmitter pulse width is 300 ns. The overall noise figure is 3.4 dB, and the mds is –114 dBm. Are these figures believable and reasonable?
4. A radar system exhibits an excessively high mds. Testing reveals that the overall noise figure is 2 dB high, and the gain of the MTI i-f amplifier is 2 dB low. Where should the work begin?
5. The magnetron operating frequency has just been changed by 50 MHz, and the stalo and afc have been properly adjusted to the new frequency. There is only grass on ppi displays. What must be done?
6. Transmitter power measurements indicate maximum allowable power, but the transmitter spectrum shows very high side lobes. Some weak targets normally seen on normal video are not present, but mds and noise figure measurements are normal. What might be the problem?
7. Noise figure test equipment has a display labeled RELATIVE NOISE FIGURE. Why is the word "relative" used?
8. A radar receiver exhibits a noise figure of 4.5 dB. What is the effective noise temperature?
9. The effective noise temperature of a low-noise amplifier is listed as 224°K. What is the noise figure?
10. Name the two major components, and three frequencies, present in a parametric amplifier.
11. How may a receiver be used as a precision azimuth verification?
12. Why is the paramp or GasFET low-noise amplifier so effective?
13. When two frequencies are mixed, what is the resultant beat frequency?
14. Basic electronics courses all teach that a mixer provides the sum, difference, and original frequencies. What causes these?
15. The double-balanced mixer improves noise figure because ________________.
16. An ASR radar manufacturer's instruction book requires that the stalo operate above the magnetron frequency. What might be the reason for this, and what might occur if the stalo were tuned below the magnetron frequency?
17. A magnetron system with afc incorporates mechanical correction drive to the magnetron tuning mechanism. This probably indicates that?
18. The display for a magnetron system appears as in Figure 11-34. What is the most likely cause?

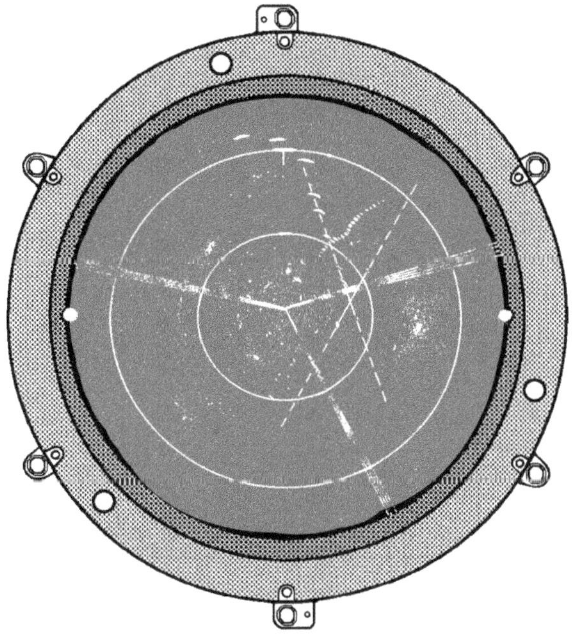

FIGURE 11–34

Strobes of clutter residue.

19. A receiver system contains a multitude of i-f amplifiers, all with the same type of detection. What might be the reason for this?
20. A technician arrives at the radar site and observes no receiver crystal current. What is the most likely cause?
21. The principal purpose of logarithmic gain is ____________.
22. An MTI i-f amplifier exhibits a bandpass of $3.5/t_p$. What might be the reason?
23. Explain the difference between phase detection and synchronous detection.
24. An MTD i-f amplifier output shows very little noise, and mds measured, in the conventional manner, at the log detector output is greater than −100 dBm. However, the displayed performance data indicates an mds of −107 dBm. Explain.
25. An MTD amplifier provides "rectangular coordinate data." Why is this name assigned to the outputs?
26. Explain "synchronous tuning" of an i-f amplifier.
27. Name the usual adjustments provided on an MTI i-f amplifier.
28. The most usual input and output i-f amplifier terminations are ____________.
29. The most important rule in tuning an i-f amplifier is ____________.
30. In observing the i-f bandpass, the best source for a marker at an FAA radar site is ____________.

Answers to Review Questions

1. Identify the four oscilloscope presentations of receiver video illustrated in Figure 11-33.

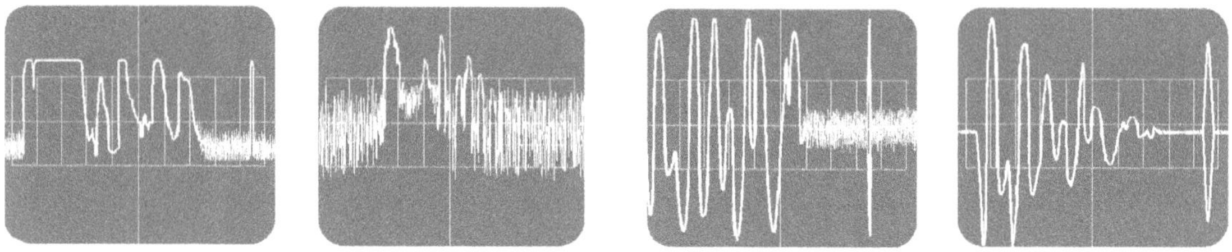

FIGURE 11–33

Oscilloscope presentations, four receiver videos.

"A" is normal video, identified by limiting; by the fine, dense grass; and by the signal-to-grass ratio of 4:1 to 6:1.

"B" is log video. A clear limit level is indistinguishable, and the signal-to-grass ratio is very low.

"C" is MTI bipolar video. The maximum excursions of most of the ground clutter are consistent, because of the hard-limiting upstream of the phase detector.

"D" is the output of a synchronous detector in an MTD system. The ground clutter maximum excursions decay with range, because there is no hard-limiting upstream of the phase detector. The grass level is very low, because of the wide dynamic amplitude range, and because of the very narrow bandpass.

2. Explain "optimum bandpass."

   It is that bandpass which provides the best signal-to-noise ratio and occurs where the bandpass is 1.2/t_p.

3. The bandpass of an i-f amplifier is 4 MHz, and the transmitter pulse width is 300 ns. The overall noise figure is 3.4 dB, and the mds is –114 dBm. Are these figures believable and reasonable?

$$\Delta f_{\text{opt}} = \frac{1.2}{t_{\text{p}}} = \frac{1.2}{300 \times 10^{-9}} = 4\,\text{MHz}$$

$$P_{\text{r min}} = -174\ \text{dBm} + 10\ \log \Delta f\ F_{\text{dB}}$$

$$P_{\text{r min}} = -174\ \text{dBm} + 10\ \log\,(4 \times 10^6) + 3.4\ \text{dB} = -104.5\ \text{dBm}.$$

   The mds measurement is clearly unreasonable, very possibly indicating uncalibrated test equipment.

4. A radar system exhibits an excessively high mds. Testing reveals that the overall noise figure is 2 dB high, and the gain of the MTI i-f amplifier is 2 dB low. Where should the work begin?

   The two 2-dB figures are unrelated because of the equation for overall noise figure. Further, the MTI i-f is seldom used in noise figure testing. The "front end" is the first place to begin looking for troubles.

5. The ASR magnetron operating frequency has just been changed by 50 MHz, and the stalo and afc have been properly adjusted to the new frequency. There is only grass on ppi displays. What must be done?

   The preselector filter is probably only 10 MHz wide at the 3 dB points and must be retuned.

6. Transmitter power measurements indicate maximum allowable power, but the transmitter spectrum shows very high side lobes. Some weak targets normally seen on normal video are not present, but mds and noise figure measurements are normal. What might be the problem?

Much of the measured transmitter power is in the side lobes, outside of the receiver bandpass, and of no useful value.

7. Noise figure test equipment has a display, labeled, RELATIVE NOISE FIGURE. Why is the word "relative" used?

   The noise figure expresses the amount of noise output, in relation to what it would be if the receiver were ideal.

8. A radar receiver exhibits a noise figure of 4.5 dB. What is the effective noise temperature?

$$F = \text{antilog}\left(\frac{4.5\text{ dB}}{10}\right) = 2.8$$

$$T_{\text{eff}}\,(F - 1)\ 290°\text{K} = (1.8)\ 290°\text{K} = 522°\text{K}.$$

9. The effective noise temperature of a low-noise amplifier is listed as 224°K. What is the noise figure?

$$F = 1 + \frac{T_{\text{eff}}}{290°\text{K}} = 1 + \frac{224°\text{K}}{290°\text{K}} = 1.77 = 2.48\text{ dB}.$$

10. Name the two major components, and three frequencies, present in a parametric amplifier.

    The varactor and pump-frequency oscillator are the two major components, and the three frequencies are the signal, pump, and idler frequencies.

11. How may a receiver be used as a precision azimuth verification?

    By observing the solar strobe at sunrise and then comparing its azimuth to meteorological data.

12. Why is the paramp or GasFET low-noise amplifier so effective?

    Because their physical location in the signal-flow stream is very near to the antenna, their noise figure is F_1 in the F_T equation.

13. When two frequencies are mixed, what is the resultant beat frequency?

    The difference between the two original frequencies.

14. Basic electronics courses all teach that a mixer provides the sum, difference, and original frequencies. What causes these?

    When the beat frequency is rectified by the nonlinear mixer diode, a spectrum of harmonic frequencies is created. Among them are the sum, difference, and originals.

15. The double-balanced mixer improves noise figure because

    In the recombination of the signals from the two diodes, noise spikes from the local oscillator will be of equal and opposite polarity, and thereby, canceled.

16. An ASR radar manufacturer's instruction book requires that the stalo operate above the magnetron frequency. What might be the reason for this, and what might occur if the stalo were tuned below the magnetron frequency?

    The phase of the afc error signal is dependent upon the direction of separation between the magnetron and stalo; if the stalo were tuned below, the phase would be reversed, and the stalo correction drive would be in reverse of the necessary direction. Under some conditions, the afc might drive the tuning to the "other side." Some systems have provisions to reverse the error correction for "other-side" tuning.

17. A magnetron system with afc incorporates mechanical correction drive to the magnetron tuning mechanism. This probably indicates that?

    The afc was a modification.

18. The display for a magnetron system appears as in Figure 11-34. What is the most likely cause?

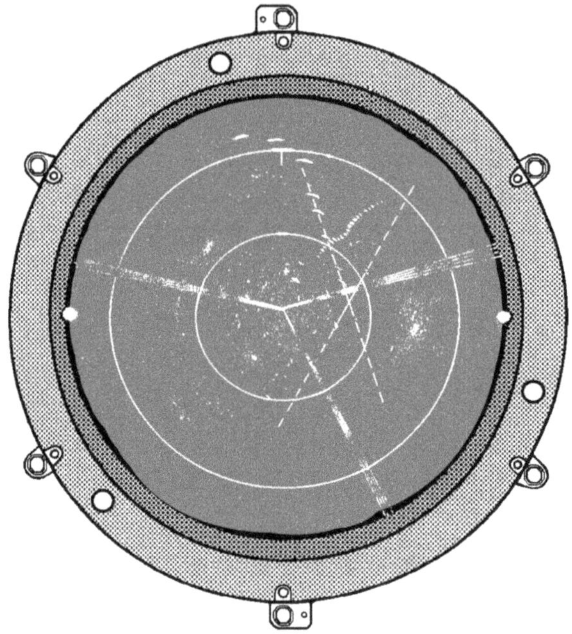

FIGURE 11–34

Strobes of clutter residue.

Excessively frequent afc correction. This is a classic indication of a failing stalo oscillator tube.

19. A receiver system contains a multitude of i-f amplifiers, all with the same type of detection. What might be the reason for this?

    The system may be employing a coded transmitter pulse, containing several different frequencies.

20. A technician arrives at the radar site, and observes no receiver crystal current. What is the most likely cause?

    The first possibility is the stalo and the second is the mixer diode itself.

21. The principal purpose of logarithmic gain is

    To place the signal data into "deciBel scaling," where the voltage represents signal-strength power.

22. An MTI i-f amplifier exhibits a bandpass of $3.5/t_p$. What might be the reason?

    The wider bandpass is desirable in recovering phase differences between superimposed echoes. This is a major factor in subclutter visibility, the greatest limitation to MTI performance.

23. Explain the difference between phase detection and synchronous detection.

    In phase detection, the voltage out of the phase detector is representative of the phase difference ψ between the signal and reference, and the phase-detector input is hard-limited to ensure that the signal strength will have no effect. In a synchronous detector, two phase detectors operate in quadrature, describing both the phase angle and signal magnitude with rectangular-coordinate data.

24. An MTD i-f amplifier output shows very little noise, and mds measured, in the conventional manner, at the log detector output is greater than –100 dBm. However, the displayed performance data indicates an mds of –107 dBm. Explain.

    The i-f amplifier bandpass is very narrow, and the "raw" mds of more than –100 dBm is understandable. The –107 dBm is achieved with built-in test equipment, producing a test signal with a known pulse-to-pulse phase change. The signal will be "pumped up" by coherent integration in the Doppler filter bank.

25. An MTD amplifier provides "rectangular coordinate data." Why is this name assigned to the outputs?

    The two phase detectors operate in quadrature; the "I" is a cosine function, and the "Q" is a sine function. When used together, they describe magnitude and phase angle.

26. Explain "synchronous tuning" of an i-f amplifier.

    When all stages are tuned to the center frequency. Contrasts with "stagger tuning."

27. Name the usual adjustments provided on an MTI i-f amplifier.

    I-f gain, phase-detector baseline adjust, bipolar video balance, bipolar video gain.

28. The most usual input and output i-f amplifier terminations are

    Input, 50 Ω, output, 75 Ω.

29. The most important rule in tuning an i-f amplifier is

    Follow the manufacturer's instructions and duplicate his conditions.

30. In observing the i-f bandpass, the best source for a marker at an FAA radar site is

    The coho.

CHAPTER 12

Moving Target Indicators and the Doppler Effect

Introduction

When this author entered the radar field in 1955, moving target indicator (MTI) systems had been in existence for less than 10 years and were still viewed as new. The non-MTI AN/MPN-1 was still the world's most extensively used air traffic control radar. Of all radar subject material, technicians and engineers in 1955 saw MTI as the greatest mental challenge. With that challenge to his senses of responsibility and competence, the author has devoted the years since 1955 to learning about, writing, and teaching MTI theory. The approach in this chapter is purely his own, and represents years of work and study.

The Frequency and Time Domains

Most understandings of radar theory are based on the time relationships between events that occur during a system interval. And, this is logical because it results from a good grasp of the radar ranging concept. Further, the most-used test equipment at a radar site is usually an oscilloscope, which has to be synchronized by a system trigger to provide an intelligible display. The technician then works and thinks purely in the time domain most of the time. However, absolute, tunnel-vision reliance on these purely temporal concepts may make it difficult to grasp concepts in the frequency domain, which is a major portion of the basis for Doppler theory and MTI principles. For but one example of the importance of the frequency domain, and something not considered in the time domain, imagine how a radar set might operate if only a single pulse were transmitted, just once, and only the time after that single pulse was considered. Would there be a transmitter spectrum? Could the single pulse be reproduced by the i-f amplifier without all the harmonics? A radar system depends upon pulse repetition because of the effect of that repetition in the frequency domain. This chapter will deal with a mixture and interrelation of both the time and frequency domains. The reader may find it necessary to broaden his acceptance of ideas.

Christian Doppler

In 1842, Christian Johann Doppler proved the mathematical relationships between the velocity of sound, the velocity of a vehicle emitting sound, and the apparent frequency shift at a stationary receiving site. Doppler's work began by measuring the pitch of musical instruments on a moving train. His formulas later proved valid for electromagnetic waves propagated through space, and were recognized by Heinrich Rudolph Hertz in his famous papers on propagation. In the earliest days of propagation research, the echoes from objects reflecting energy disappeared from the receiver outputs when the objects moved. Now, the Doppler effect is the basis for radar ground-clutter-reduction systems; a moving object reflects a different frequency than does a stationary object and MTI utilizes this frequency difference to attenuate the stationary echo.

Upward Doppler Shift

If a moving vehicle, containing some frequency-generating device, approaches a receiving site, it moves into the wave it has emitted, thus shortening the wavelength (see Figure 12-1). Since the wavelength is shortened, the receiving site perceives a higher frequency than that of the oscillating device that produced it. The amount of wavelength compression depends upon the wave frequency, the velocity of propagation of sound, and the velocity of the vehicle toward the receiving site.

FIGURE 12–1

Upward shift, whistle on approaching train.

Downward Doppler Shift

In a manner similar to the upward shift, but in an opposite fashion, if the vehicle is transmitting sound waves as it moves away from the receiving site, it moves out of its emitted wave, stretching the wavelength to make the receiving site perceive a lower frequency (see Figure 12-2).

Basic Doppler Equation

Physics textbooks express the Doppler shift in terms of wavelength. The following equation is somewhat simplified; however, it is used extensively in both textbooks and practice, and is accurate for the velocities of aircraft and for microwave propagation:

$$f_D = \frac{V_r}{\lambda}.$$

"Apparent Doppler"

The term f_D describes the Doppler shift as a frequency, academically called the apparent Doppler frequency. The word "apparent" is used by purists because f_D does not actually exist; it, instead, is only a difference between the transmitted and received frequencies, and is technically only "apparent." In actual practice, the shortened terms "Doppler frequency" and "Doppler shift" are widely used. The velocity in this equation is the radial velocity of the vehicle, or the rate at which it moves directly toward, or away from, the receiving site.

FIGURE 12–2

Whistle frequency decreases as train departs.

Radial Velocity

The radial velocity of the vehicle is related to the true velocity, but modified by the cosine of the angular difference between the two; the cosine is used in trigonometric solution of a right triangle (cos θ = adjacent/hypotenuse), where the aircraft ground speed is the hypotenuse and the side adjacent is a resultant velocity from the aircraft to the radar (see Figure 12-3). For example, if an aircraft is flying at 100 nmi/h on a course of 225°, but a receiver is located at 180° from the aircraft, the receiving site can only detect a velocity of 70.7 nmi/h:

$$V_{\mathrm{r}} = V_{\mathrm{ground}}[\cos(\angle_{\mathrm{rcvr}} - \angle_{\mathrm{course}})].$$

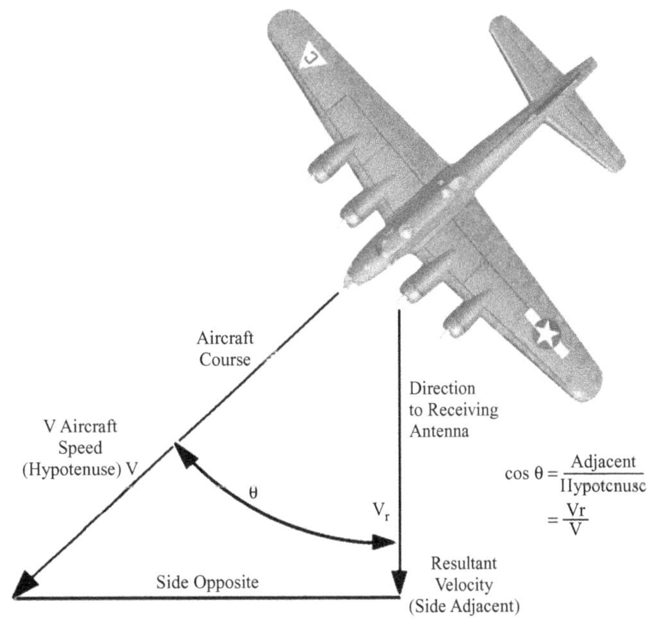

FIGURE 12-3

Radial velocity.

Radar Doppler Shift

When calculating the Doppler shift to radar echoes, the two-way travel time must be considered. The radar Doppler shift occurs twice; the speed of the aircraft causes the aircraft to move either into, or out of, the radar's transmitted rf burst, compressing or stretching waves at the target. Next, the energy reflected back to the receiving site is again subjected to the Doppler shift, further compressing or stretching the waves at the receiver. The Doppler shift occurs twice, and the Doppler-shift equation applied to radar becomes

$$f_{\mathrm{Dradar}} = \frac{2V_{\mathrm{r}}}{\lambda}.$$

Radar Doppler Equation with Transmitter Frequency

In microwave calculations, it is common practice to use frequencies rather than the wavelengths. Substitute the velocity of light and the transmitter frequency for wavelength in the equation, and the radar Doppler, f_{D}, equation in terms of transmitter frequency, f_{xmtr}, becomes

$$f_{\mathrm{Dradar}} = \frac{2V_{\mathrm{r}}}{\left(\frac{c}{f_{\mathrm{xmtr}}}\right)} = \frac{2V_{\mathrm{r}} f_{\mathrm{xmtr}}}{c}.$$

Now, this equation can be rearranged into a simple ratio and proportion, where the Doppler frequency bears the same relationship to the transmitter frequency as does $2V_{\mathrm{r}}$ to the velocity of light, c:

$$\frac{f_{\mathrm{Dradar}}}{f_{\mathrm{xmtr}}} = \frac{2V_{\mathrm{r}}}{c}.$$

Bidirectional Wave Travel and Speed of Light

Use of the Doppler equation will usually involve the use of a known radial velocity, so it is more convenient to use the radial velocity without the "2" multiplier and change the constant for the speed of light; this is permissible because the ratio is unchanged. Reducing the speed of light by a factor of 2 might further be construed acceptable, because the two-way reflection travel causes twice as much time to elapse for energy to traverse a given distance, and $r = d/t$. Reducing the right-hand side of the ratio-and-proportion equation by a factor of 2,

$$\frac{f_{\mathrm{Dradar}}}{f_{\mathrm{xmtr}}} = \frac{V_{\mathrm{r}}}{\left(\frac{c}{2}\right)}.$$

Speed of Light in Nautical Miles per Hour

Since aircraft velocities are expressed in nautical miles per hour (nmi/h), use 582.75 × 10⁶ nmi/h for the speed of light in the wavelength equation. Although some minute rounding was done, this number is based on the most accurate known figures, such as the international standard 6,076.11549 feet/nmi and 299,792,500.6 m/s for the speed of light:

$$\frac{f_{\text{D}_{\text{radar}}}}{f_{\text{xmtr}}} = \frac{V_{\text{r}_{\text{nmi/h}}}}{\left(\dfrac{582.75 \times 10^6 \,\text{nmi/h}}{2}\right)},$$

$$\frac{f_{\text{D}_{\text{radar}}}}{f_{\text{xmtr}}} = \frac{V_{\text{r}_{\text{nmi/h}}}}{291.375 \times 10^6 \,\text{nmi/h}}.$$

The two-way speed of light, 291.375 × 10^6 nmi/h, is 80,937.5 nmi/s, and the reciprocal of that number is 12.3552 μs, the precise time required for rf energy to be reflected by an object 1 nmi from a radar transmitter. This "radar mile," once rounded to 12.4 μs, is now usually rounded to 12.36 μs, but the increased accuracy to 12.3552 μs is occasionally necessary in some precise calculations.

Now, the equation can be rearranged to solve for any unknown quantity within it, and the Doppler shift may be expressed as

$$f_{\text{D}_{\text{radar}}} = \frac{V_{\text{r}_{\text{nmi/h}}} f_{\text{xmtr}}}{291.375 \times 10^6}.$$

Incorporating the Radial Velocity Calculation

It is possible that the technician at a radar facility might be confronted with an MTI problem in a specific part of an air traffic control pattern. One use of the Doppler frequency equation would be for determining the actual Doppler of specific-case targets in known areas. To complete the equation so that it may be singularly used for this purpose incorporate the solution for radial velocity V_r into the radar Doppler equation. For a one-way Doppler shift, the solution was

$$V_{\text{r}} = V_{\text{ground}}[\cos(\angle_{\text{rcvr}} - \angle_{\text{course}})].$$

The two figures available for this calculation will be (1) the course of the aircraft and (2) the aircraft's bearing from the radar antenna. Since these angles are based on opposite points of reference, a 180° difference must be incorporated:

$$V_{\text{r}} = V_{\text{ground}}[\cos(\angle_{\text{target}} - \angle_{\text{course}} + 180°)].$$

Therefore

$$f_{\text{D}_{\text{radar}}} = \frac{\left(V_{\text{ground}_{\text{nmi/h}}}[\cos(\angle_{\text{target}} - \angle_{\text{course}} + 180°)]\right) f_{\text{xmtr}}}{291.375 \times 10^6}.$$

The Tangential Effect

From the equation incorporating radial velocity, it is seen that an aircraft course at an angle of 90° with respect to the radar azimuth has a V_r of zero and an f_D of zero. This is called "the tangential effect" because the target course is at a tangent to a circle around the radar.

The Doppler Effect and Phase Detection

The earliest MTI systems used self-oscillating magnetrons as the power output stage of the transmitter, and the system design required to achieve coherence obscured the relationships among the pulsed transmitter, receiving system, and the Doppler effect. Most latter-day MTI radar systems are of the synthesis class, which, in theory, is much simpler than the magnetron type. In the synthesis system, the transmitter burst is derived from uninterrupted c-w oscillators, and coherence is inherent. A simplified block diagram of a synthesis system, also shown in Chapter 5, is illustrated in Figure 12-4. Because the synthesis system operation is now the usual, and because it is

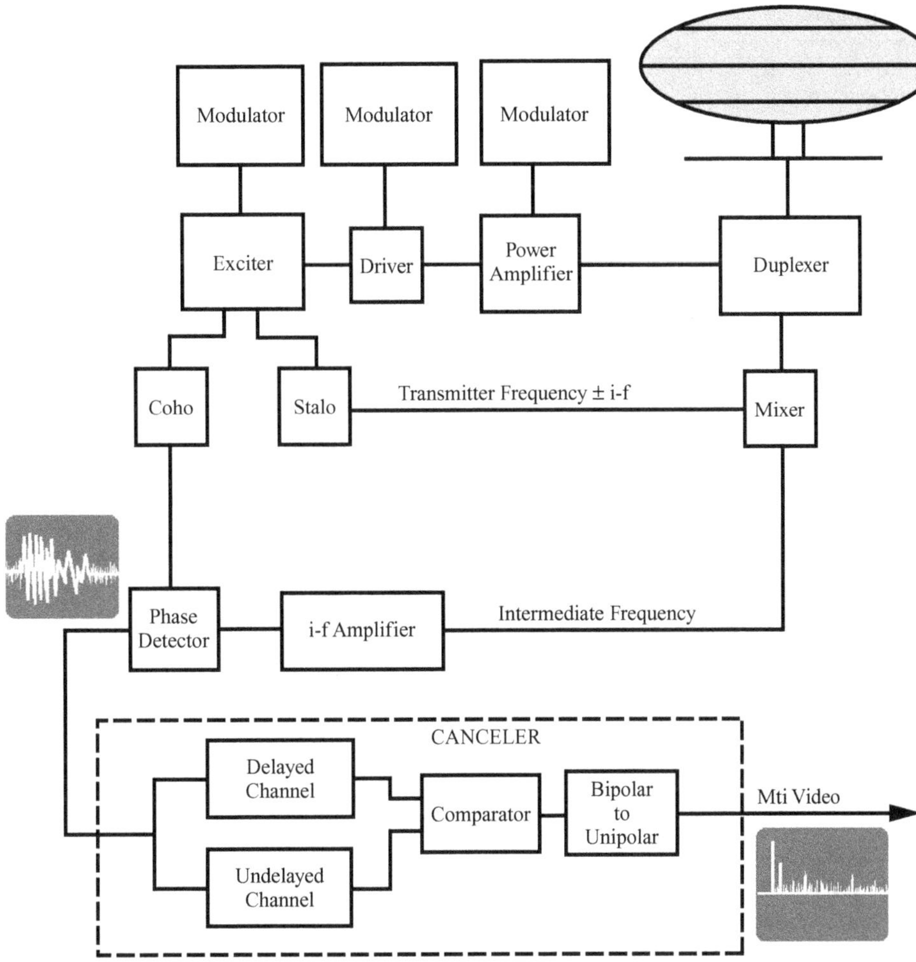

FIGURE 12–4

Synthesis system block diagram.

more obvious to the student, the initial discussion of MTI and Doppler will use the synthesis system as an initial example for discussion. Application of the principle to the more complex magnetron systems will follow.

The synthesis system is so named because the transmitter frequency is directly "synthesized" from the frequencies used in the receiving system. Usually, the receiver contains a stalo (stable local oscillator) and a coho (coherent oscillator). Both of the terms, "coho" and "stalo," come from early MTI system applications. Although the terms were once very descriptive of the purpose and design of the oscillators in magnetron systems, they are now antiquated; they are still applicable to the dwindling number of magnetron systems, and remain in use for the sake of established convention. The stalo is used for a frequency-and-phase reference input to the signal mixer (first detector), and the coho is used for a frequency-and-phase reference input to the phase detector (second detector). The stalo is an rf oscillator, and the coho operates at i-f (ordinarily 30 MHz or 31.07 MHz). In the synthesis system, the stalo and coho are synthesized to compose the transmitter frequency.

First and Second Phase Detectors

The most significant difference in the MTI receiver, as opposed to normal and log receivers, is that it operates on a principle of phase detection, rather than amplitude detection. In the signal mixer (first detector), the stalo is compared to the receiver inputs from the antenna; when an echo is received an i-f output is produced. The phase difference between the stalo and r-f echo determines the phase of the i-f echo. When the i-f output is compared to the coho in the phase detector, the phase-detector output represents the cosine of the phase difference between

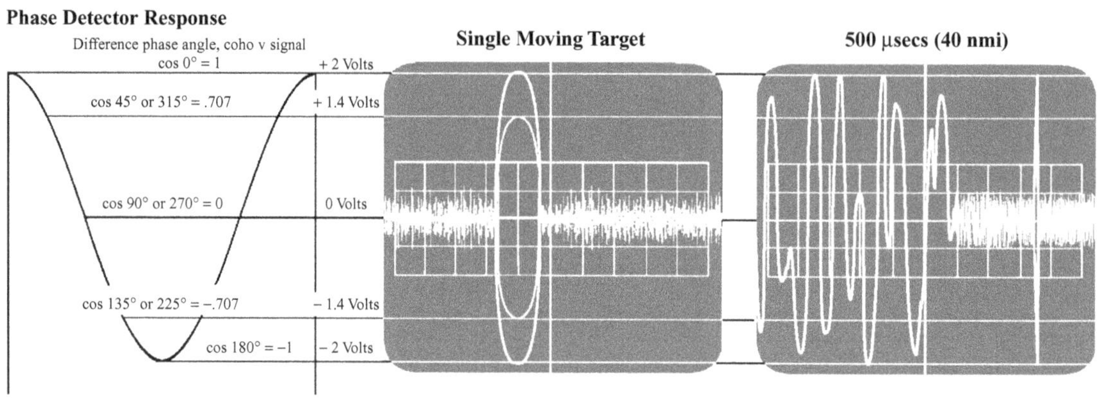

FIGURE 12-5

Phase-detector response and bipolar video.

the i-f echo and coho. Because the two inputs to the phase detector are "coherent," or synchronized, scientific literature, such as Merrill Skolnik's *Radar Handbook*, will sometimes call this device the "synchronous detector." This author will reserve that definition for those applications in which the input from the i-f is not "hard limited." Conventional (canceler-type) MTI systems require this hard limiting, but moving target detector (mtd) systems require that it does not occur, so the mtd phase detectors may appropriately be called "synchronous detectors."

Phase-Detector Response

Many who are familiar with MTI and mtd concepts would say that the phase detector is the "heart" of the system and of the target-detection process (see Figure 12-5). Because of the manner in which it is constructed, the characteristic phase-detector output will represent the cosine of the instantaneous phase difference between the hard-limited i-f echo and the coho. Since the phase detector has a cosinusoidal response, its outputs are called bipolar video, because those signals may be either positive or negative. Depending upon the manufacturer, the bipolar video may also be called coherent video, MTI coherent video, phase-detector video, bipolar coherent video, or coherent bipolar video. Figure 12-5 illustrates the phase-detector response, a single moving target, and 500 µs of video. The 500-µs display shows how ground clutter will appear as bipolar video. The following paragraphs will explain the reasons for the response and video appearance.

Phase Response of the Signal Mixer

The phasing of an i-f signal in any superheterodyne receiver is determined by the phase relationship between the incoming signal and the local oscillator (see Figure 12-6). In amplitude-detection receivers, this phase relationship is generally insignificant, but it is of great importance in the MTI receiver. In Figure 12-6, the phase relationship between the stalo and the rf echo presented to the signal mixer establishes the starting phase of the i-f echo. In a pulsed system, all rf echoes presented to the signal mixer, during any T_r, are of the same initial phase, because they are all reflections of the same transmitted pulse. However, the stalo phase continuously changes during the "listening time," because the stalo runs continuously. In a superheterodyne amplitude-detection receiver, the signal mixer has been called the first detector since the earliest days of radio. In a phase-detection system, the signal mixer is the first phase detector.

The Phase-Detector Operation

The drawings shown in Figures 12-7 and 12-8 illustrate the operation of the phase detector used in the ASR-8. This particular phase detector is contained in an inaccessible circuit package; earlier radars used discrete-component circuits that operated in a similar fashion. Consider first the condition illustrated in Figure 12-7, which shows that two in-phase inputs will produce a continuous maximum negative charge on capacitor C44. Then, consider the condition in Figure 12-8, which illustrates that two inputs with a 180° phase difference will produce

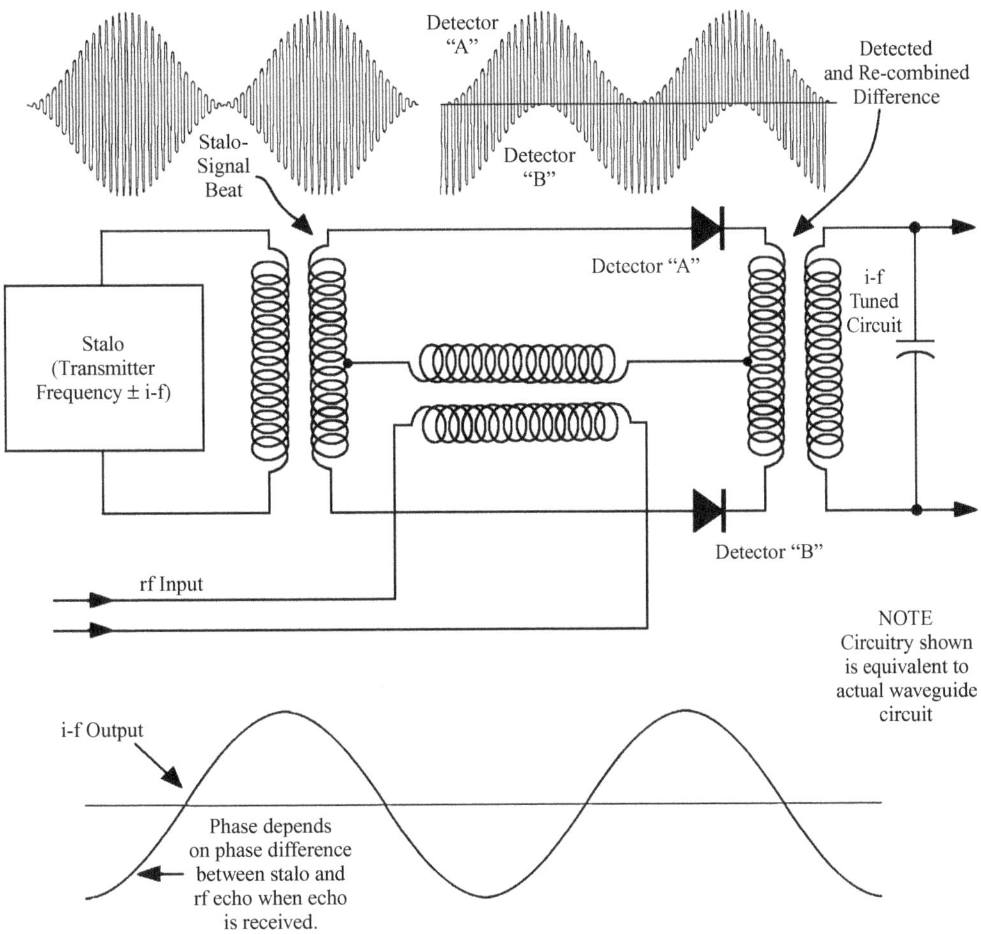

FIGURE 12-6

i-f echo phase determined by signal versus stalo phases.

a maximum positive charge on C44. When the input phase differences are equal to 90° or 270°, the diodes are alternately forward biased during each half-cycle, so that C44 is charged equally in the positive and negative directions, yielding a net charge of zero. Of course, only the most extreme conditions of 0° and 180° were shown, and phase differences between those so far described will yield average C44 charges somewhere between maximum negative and maximum positive. Provided the signal input is sinusoidal, the response to all inputs from 0° to 360° difference becomes the complement of a cosine function, and the response is said to be cosinusoidal. If inverted by an amplifier, then the phase-detector response is a function of the cosine of the phase difference between the signal and coho inputs. The polarity may also be reversed by inversion through the transformers or by reversal of all four diodes.

Comparing Different Frequencies in a Synchronous Phase Detector

Figure 12-9 illustrates that, when two different frequencies are compared, the cosine of their phase difference will vary at the rate of the frequency difference. Since the phase detector has a cosinusoidal response, its outputs are called "bipolar video," because those outputs may be either positive or negative. Therefore, two different frequencies applied to a phase detector will produce a bipolar video which varies at the rate of the frequency difference. This direct relationship between frequency differences and phase differences is fundamental to the principles of MTI and mtd, and is absolutely essential to an understanding of the relationship between the Doppler effect and the phase-detector outputs.

Figure 12-9 illustrates a radical difference between two c-w frequencies. If those two frequencies were very close, such as 30.000 MHz versus 30.001 MHz, a low-frequency detector output of 1 kHz would result. 1 kHz

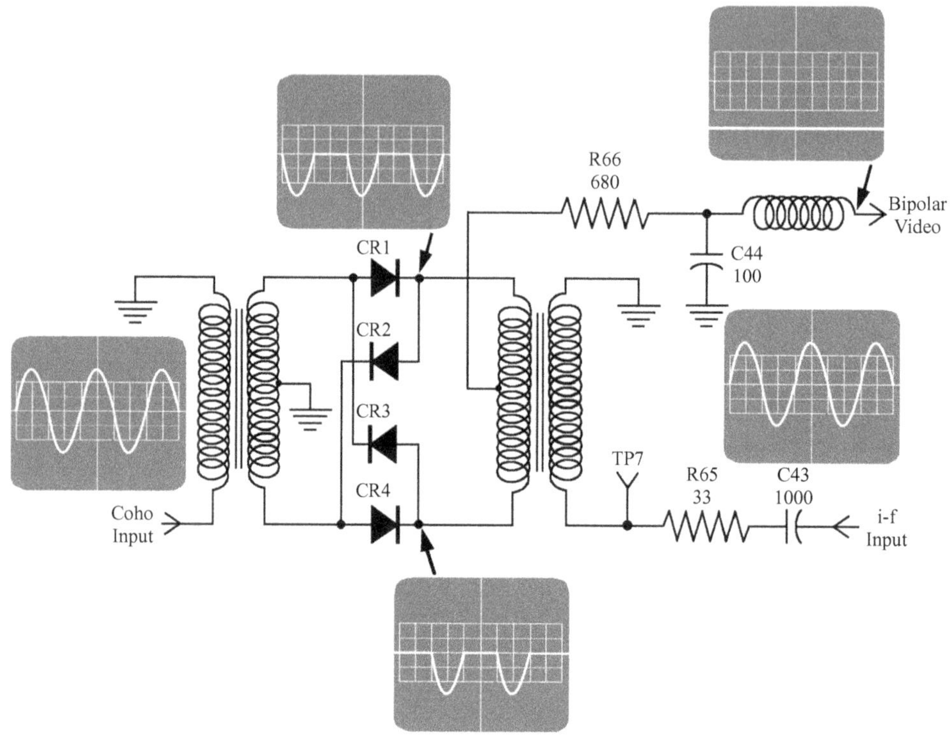

FIGURE 12–7

No phase difference, coho versus i-f.

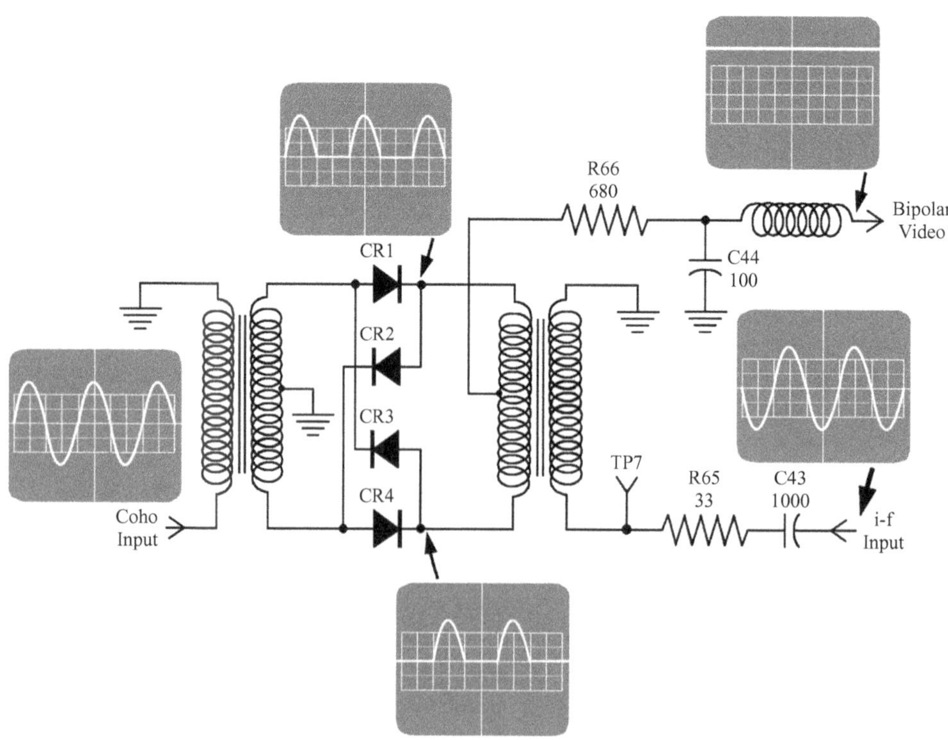

FIGURE 12–8

180° phase difference, coho versus i-f.

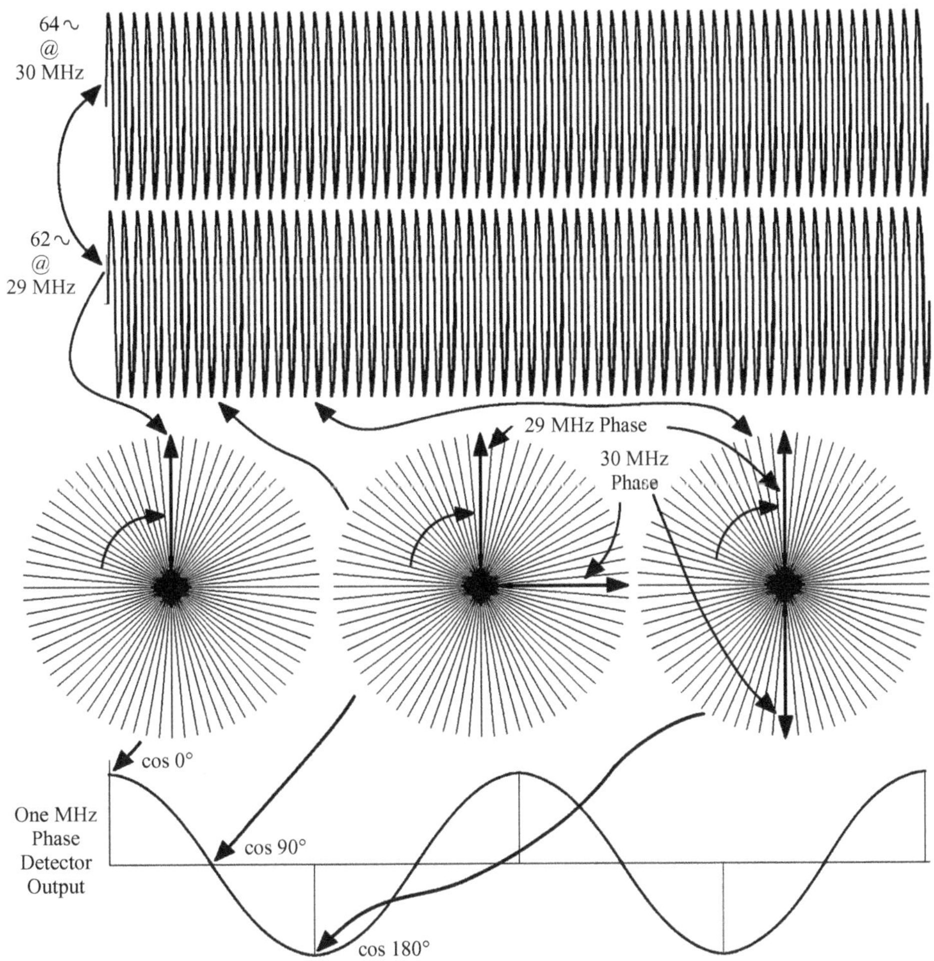

FIGURE 12–9

Phase difference between two frequencies.

is a typical Doppler frequency; in an ASR radar, it would represent a target radial velocity of about 100 nmi/h. How such a target might appear in short-duration bipolar pulses will be described in a paragraph to follow.

Doppler Shift Representation at the Phase-Detector Output

See the synthesis system block diagram in Figure 12-10. This illustration has appeared earlier in this chapter, and in a preceding chapter; the reader should already be acquainted with it. The transmitter frequency is a synthesized combination of the stalo-and-coho sum or difference. Since the rf echoes were compared to the stalo in the signal mixer, and the i-f echoes were compared to the coho in the phase detector, the frequency of the phase-detector output represents the difference between the transmitted and received frequencies. The only difference there can be between the transmitter and receiver frequencies is the Doppler shift. Provided the stalo is tuned to a lesser frequency than the transmitter, the Doppler shift will be upward for approaching targets, and downward for departing targets; however, the shift may be reversed if the transmitter frequency is obtained from the stalo minus coho difference (because received echoes for approaching targets will provide less transmitter–stalo difference). For the remainder of this chapter, it will be assumed that the transmitter frequency is the stalo-and-coho sum, and the Doppler shift is in conventional, expected directions. This may very well be reversed in actual practice, as many magnetron systems require the stalo to be above the transmitter frequency because of the automatic frequency control design.

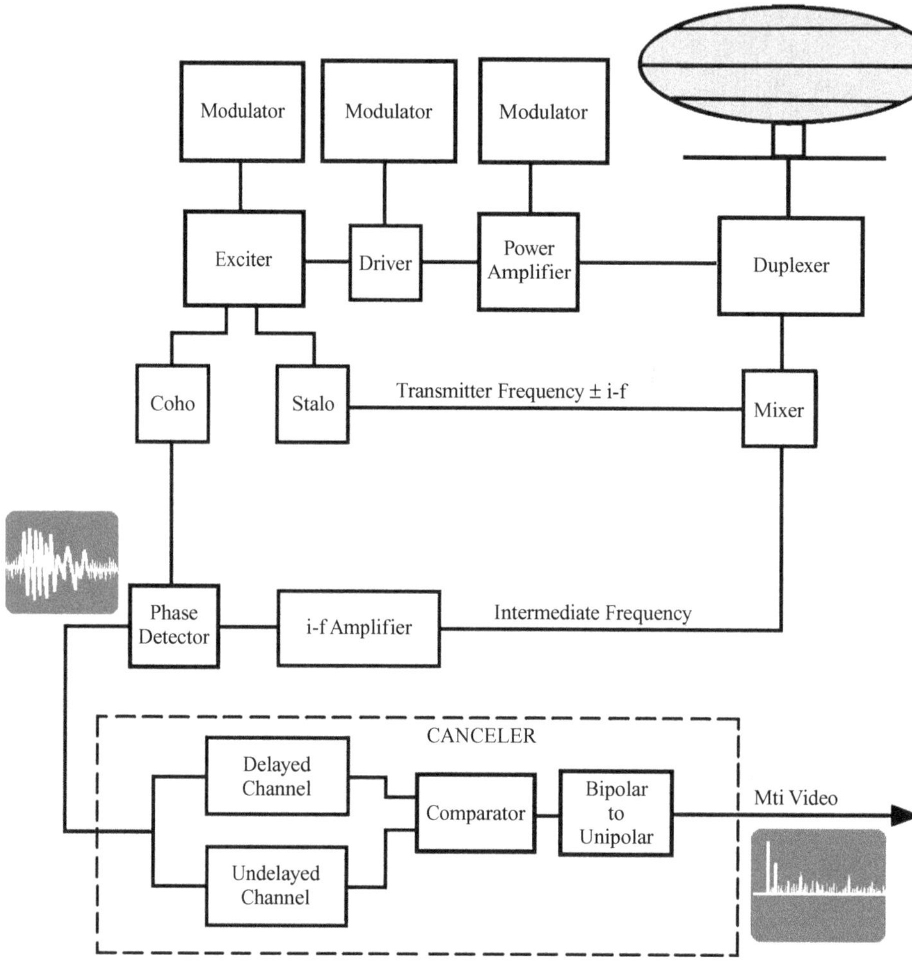

FIGURE 12–10

Synthesis MTI system.

The Doppler Effect in Pulsed Systems

The Echo-Marking Approach

In the 1940s and 1950s, the complex circuitry required of the magnetron system to achieve coherence so obscured the Doppler effect that some training programs even erroneously discounted it, arbitrarily repealing an established law of physics. To put the complex magnetron coherence theory into an understandable, purely time-domain explanation of MTI, instructional material incorporated an "echo-marking" method, which ultimately employed all the same variables used in the Doppler calculations, (1) target radial velocity, (2) transmitter frequency, and (3) velocity of light. In the echo-marking explanation, the V_r is converted into a pulse-to-pulse change in range, the velocity of light is expressed in terms of radar echo time, f_D is expressed in a pulse-to-pulse phase change ($\Delta\phi$), and the transmitter frequency is incorporated by counting the number of transmitter-frequency cycles from the transmitter burst until receipt of the echo. A detailed analysis will mathematically prove that the echo-marking approach is only a circuitous means of calculating Doppler shift, but the discussion would be of no value in this chapter. There will be no further reliance upon the echo-marking approach in this book, but Figure 12-11, and the following discussion of the synthesis system, will make it very clear that there is no difference between the c-w and pulsed Doppler shifts.

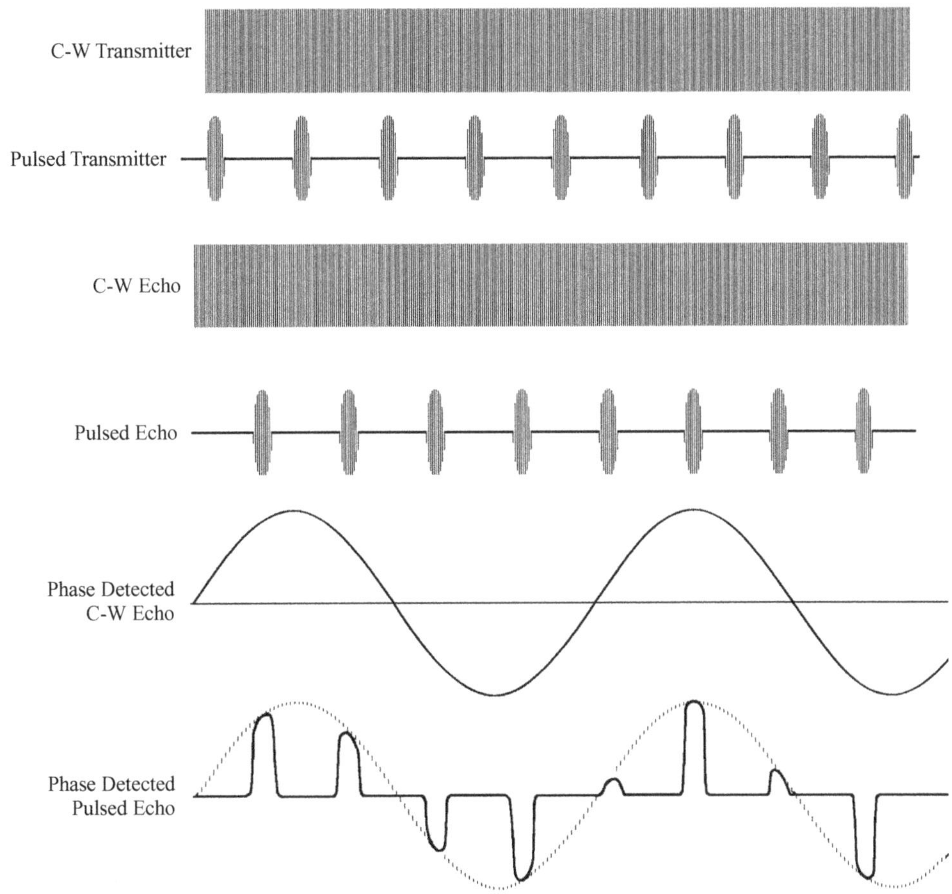

FIGURE 12–11

c-w and pulsed Doppler shifts.

The Identical Doppler Effect in c-w and Pulsed Echoes

As shown in Figure 12-9, the phase-detector output for two inputs of different frequencies varies at a rate equal to the Doppler shift, in a cosinusoidal fashion. If one of those two phase-detector inputs had been gated on and off, but the oscillator producing the frequency never interrupted, the phase detector will perform exactly as if two c-w inputs were applied, except that the phase-detector output will only appear when the oscillations are enabled by the gate. It is important to note, in Figure 12-11, that there is very little Doppler shift exhibited during the duration of a single pulse, and that several pulses must be examined to make the Doppler shift obvious. Such an effect is precisely what occurs in either a synthesis or magnetron system.

Synthesis System Block Diagram, Incorporating the MTI Principle

To expand upon previous discussions, consider the entire two-way propagation of the transmitter burst from the final power amplifier, to the antenna feedhorn, to a target, back to the antenna, and to the signal mixer (see Figure 12-10). After the transmitter burst leaves the final power amplifier, the starting phase continues to change, because of the differences in waveguide phase and group velocities explained in the chapter on waveguide. However, once the burst leaves the feedhorn, and is no longer contained in waveguide, the back-and-forth, "zig-zag" reflections no longer exist to cause phase delay, and the entire burst maintains the same starting phase as it moves through space at the velocity of light. Obviously, the same is true of the "return-trip" echo, back to the feedhorn. On entering the waveguide system, the echo is further subjected to phase delay up to the signal mixer. From the time the transmitter is pulsed, until the time the rf echo arrives at the signal mixer, the stalo and coho have been running continuously. And, since the transmitter burst was synthesized from the stalo and coho, there is an intelligible relationship among all three frequencies, called coherence.

The waveguide phase delays are the same for every transmitter pulse, and because they are constant, are inconsequential in considering the effects on a series of echoes from a target. It is important that the transmitter burst ceases to change phase once it has departed from the feedhorn, and it remains at the same phase until it returns as an echo. It may be correctly argued that characteristics of the target may alter the echo phase, but this will also be most often a constant, and therefore, also inconsequential. Again, remember that, while the burst is traveling through space, the stalo and coho continue to change phase, while the burst does not. The time that it takes for the burst and echo to get to the signal mixer, therefore, determines the phase of the i-f echo, and the amplitude and polarity of the phase-detector output. If the target does not move, the output from the phase detector will be the same from one pulse to the next. If the target does move, the phase-detector output will vary. Most importantly, the rate at which the phase-detector output varies is the Doppler shift.

Cancellation

MTI attenuates ground clutter when the bipolar video is compared to the bipolar video from the previous T_r; this comparison occurs in the canceler, which routes the video through delayed and undelayed paths to a comparator circuit. Depending upon the manufacturer and age of the system, the comparator can be a resistive bridge, a transistor pair, a digital subtractor circuit, or any other circuit that permits one input to be subtracted from the other. The method of delay in the delayed channel may be a quartz delay line, digital shift register, or memory, depending upon the state of the art at the time of manufacture. In any case, the delayed input to the comparator arrives there precisely one T_r late, so that a target of a specific range will be compared to whatever existed at that precisely same range on the previous T_r. If an echo has no Doppler shift, the phase-detector output for that specific target will be the same for each T_r, and the comparison in the canceler will "cancel" the target. On the other hand, if the echo does have a pulse-to-pulse Doppler shift, the comparison will result in a difference, which will ultimately become the canceler output.

The Difference between Radar Doppler and T_r-to-T_r Echo Phase Change

Figure 12-12 illustrates the appearance of a single moving target, at the phase-detector output, over many T_rs, as if there were no other echoes present. Since the cancellation process is based on T_r-to-T_r comparisons, the canceler output for any specific target becomes dependent upon the T_r-to-T_r change in (1) the phase angle of the echo versus (2) the phase angles of the stalo and coho. Since the first days of MTI, this pulse-to-pulse change in phase angle has always been called delta phi ($\Delta\phi$). The Greek letter delta (Δ) is universally used in mathematical subjects for "change," and the Greek letter, phi (ϕ), is regularly used in electrical engineering to describe phase. There is a definite relationship between $\Delta\phi$ and $f_{D\text{radar}}$, but there is also a great difference in what they represent. The radar Doppler, $f_{D\text{radar}}$, is purely a frequency-domain quantity, dependent only upon the transmitter frequency f_{xmtr} and target radial velocity V_r; however the $\Delta\phi$ has implications in both the time and frequency domains. The $\Delta\phi$ could most simply be described as a result of sampling the Doppler at an f_p rate. For instance, if the f_p is precisely equal to the f_D, then each time the transmitter is pulsed, the received echo will be 360° different than the echo from the previous pulse. Similarly, if the f_p is twice the f_D, each received echo

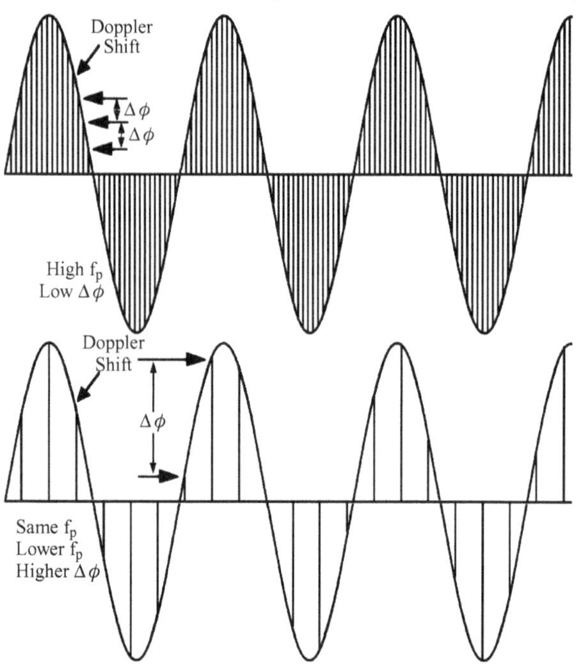

FIGURE 12-12

Doppler frequency versus $\Delta\phi$.

will be 180° different than the echo from the previous transmitter pulse.

The "Butterfly"

Figure 12-12 illustrates the hypothetical appearance of a single moving target over many T_rs. It would be extremely difficult to see such a target on an oscilloscope display, as it would be obscured by all the ground clutter and other echoes. To view a single target on an oscilloscope, the technician would use the oscilloscope delayed-and-expanded function, as illustrated in Figure 12-13. The moving target would appear as a single pulse, bobbing up and down as the wings on a butterfly. Such a target has been assigned the slang description "butterfly," since the 1940s.

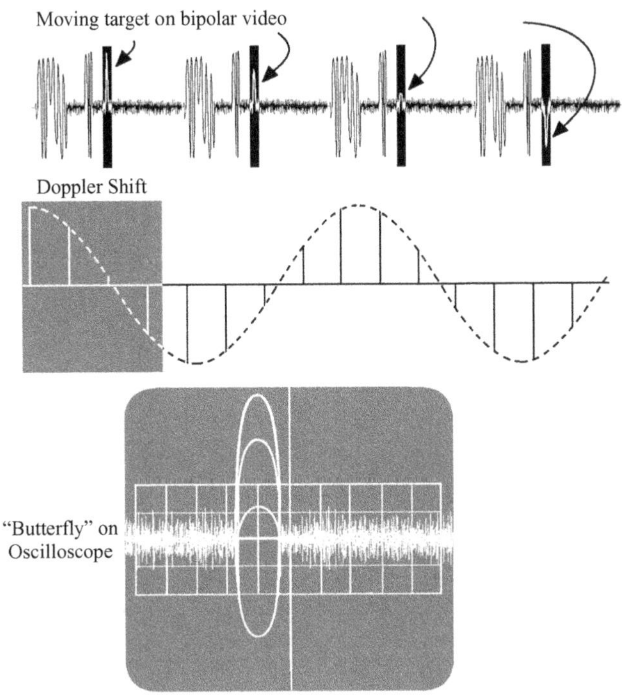

FIGURE 12-13
The "butterfly."

Blind and Optimum Velocities

When the echoes change phase by 360° from one pulse to the next, there does not appear to be any change in the phase-detector output, and the canceler will attenuate the target, just as if it were not in motion; this condition is called the first MTI blind velocity V_{b1} (see Figure 12-14). A second blind velocity occurs when the f_D is twice the f_p, a third occurs at three times the f_p, etc. When f_D is half the f_p, and the $\Delta\phi$ is 180°, it is possible to obtain the greatest output from the canceler, so this condition is called the optimum velocity (V_{opt}). If echoes at optimum velocity happen to be occurring at the peaks of the phase-detector response curve, the target is said to be at optimum velocity, optimum phase, since this condition will provide the maximum possible T_r-to-T_r amplitude difference in the canceler comparator. Optimum velocities occur only at odd multiples of the f_D of the first, since even multiples would be at blind velocities. For example, typical optimum velocities for either an ASR or ARSR radar could occur at f_D values of 500 Hz, 1,500 Hz, 2,500 Hz, etc. Blind velocities for the same system would occur at 1,000 Hz, 2,000 Hz, 3,000 Hz, etc.

Blind Velocity Calculation

A blind velocity occurs when the Doppler shift equals the f_p or any multiple (see Figure 12-14). To calculate the blind velocity, substitute the f_p for the Doppler f_D in the radar Doppler equation, change V_r to V_{b1} for the first blind velocity, and rearrange

$$f_D = \frac{V_r f_{xmtr}}{291.375 \times 10^6}.$$

Therefore

$$V_{r_{b1}} = \frac{(291.375 \times 10^6) f_p}{f_{xmtr}}.$$

It is informative to compare the blind velocities of ASRs and ARSRs; calculations will reveal that these blind velocities will occur in the same general regions, because the ratios of (1) the lower L-band ARSR frequency and f_p versus (2) the S-band ASR frequency and f_p provide (3) a similar first blind velocity. For example, the following calculations for (1) an L-band ARSR with an f_p of 360 and f_{xmtr} of 1,120 MHz and (2) an S-band ASR with an f_p of 900 Hz and f_{xmtr} of 2,800 MHz yield (3) similar first blind velocities of about 94 nmi/h:

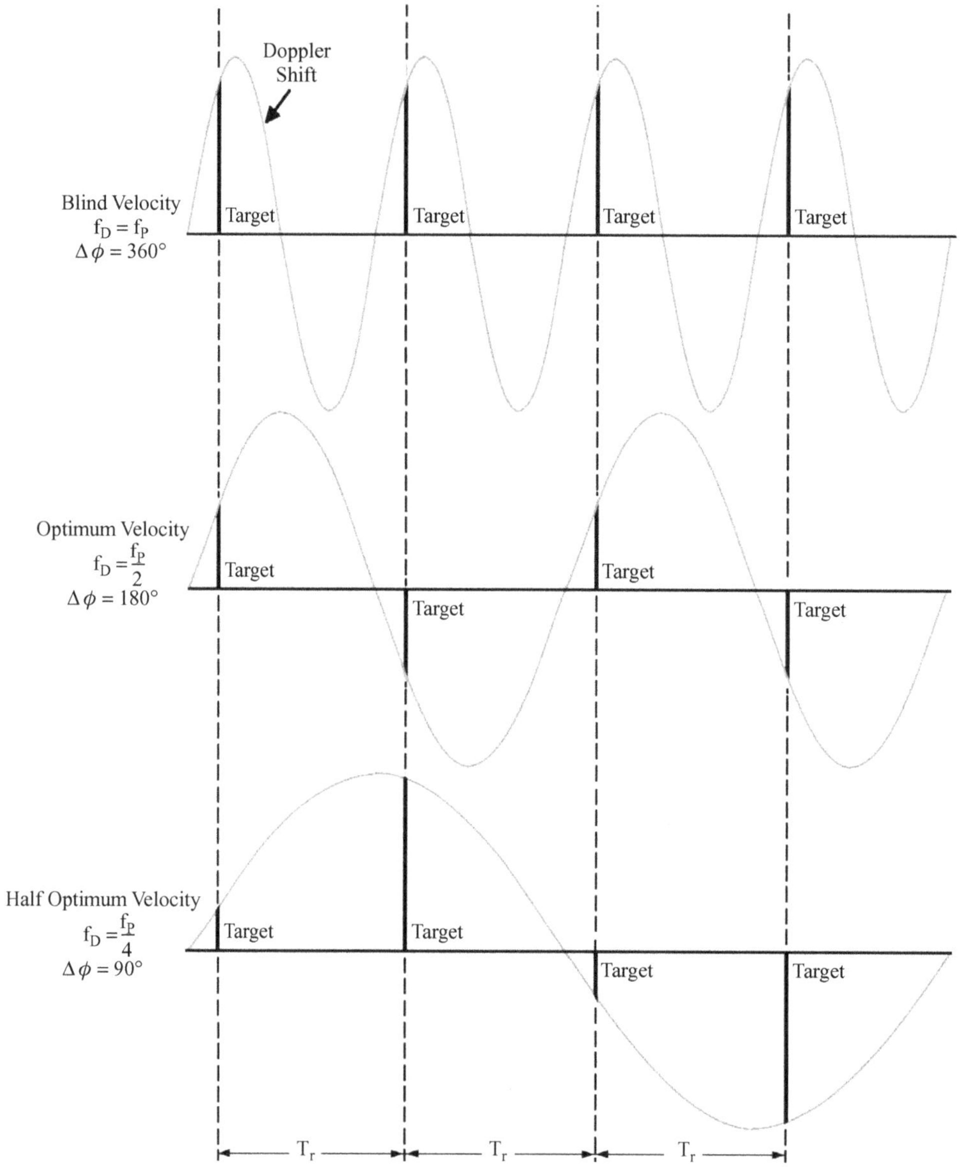

FIGURE 12–14

Blind and optimum velocities.

$$\text{ARSR:} \quad V_{rb1} = \frac{(291.375 \times 10^6)360}{1120 \times 10^6} = 93.7\,\text{nmi/h}.$$

$$\text{ASR:} \quad V_{rb1} = \frac{(291.375 \times 10^6)900}{2800 \times 10^6} = 93.6\,\text{nmi/h}.$$

Blind Velocity and Staggered f_p

To reduce the effects of blind velocities, the staggered f_p introduces T_rs of varied length, so the Δϕ is different from one T_r to the next (see Figure 12-15). This has been described as a Δ(Δϕ), a concise term that actually describes the entire principle. When a target velocity is in the blind speed region, the Δ(Δϕ) reduces the probability of loss by cancellation. The Δ(Δϕ) has no effect on clutter cancellation, because clutter has little, if any, Doppler, and, therefore, a Δϕ of near zero. Obviously, where the Δϕ is zero and the f_D is zero, the Δ(Δϕ) will still be zero.

Optimum Velocity Calculation

Again, the first optimum velocity will occur at one-half the first blind velocity, and at any odd multiple. The first optimum velocity for either an ASR or ARSR is in the general region of 50 nmi/h, which is about 57.5 US statute mph, very close to the usual freeway speed of automobiles. Should a freeway orientation be such that it places traffic on a direct course toward, or away from, the radar antenna, much of the freeway traffic would be at optimum velocity. If the freeway has sufficient elevation to be in the receive antenna pattern, the traffic may cause a long "string" display of targets on air traffic control displays, as illustrated in Figure 12-16. This can be hazardous to an aircraft flying above the freeway, as it cannot be distinguished from the vehicles. mtd systems contain a geocensor map to alleviate such conditions by a "clutter map memory," updated with low-Doppler data from successive antenna scans.

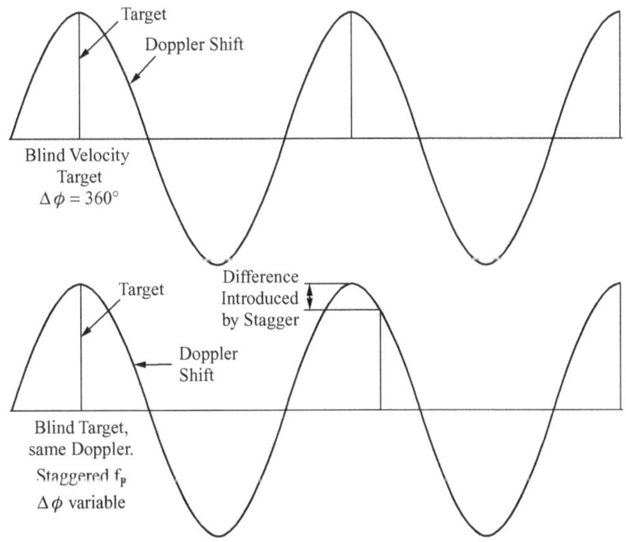

FIGURE 12–15

Staggered f_p introduces $\Delta(\Delta f)$.

The Blind Phase Effect

An optimum velocity condition can exist at any initial phase, and does not necessarily produce a maximum canceler output (see Figure 12-17). As stated at a preceding point in this chapter, the maximum output occurs only when the initial i-f echo-versus-coho phase difference is 0° or 180°, and causes a maximum positive or maximum negative phase-detector output; this condition is called optimum velocity, optimum phase. As shown in Figure 12-17, an optimum velocity can occur where the ϕs are 90° and 270°, providing the canceler with no useful input. Such a condition is called a blind phase condition. The optimum velocity offers the highest probability of a sustained blind phase condition, but the condition may occur briefly or intermittently at any Δϕ, weakening or obliterating the displayed target. A blind phase condition occurs whenever two subsequent echoes from the same target produce phase-detector outputs of the same amplitude and polarity. A blind phase condition could occur, for example, for ϕs of 135° and 225°; the cosine of each is −0.707. The blind phase condition is caused by an ambiguity in the phase detector; there are always two different ϕs which may provide the same amplitude and polarities.

Quadrature Phase Detectors

Probably derived from the *I* (in-phase) and *Q* (quadrature) demodulators in color television receivers, quadrature phase detectors were introduced in the 1970s to eliminate the blind phase effect. The coho reference to a second phase detector is phase-shifted by 90°, so that the second phase-detector response is sinusoidal instead of cosinusoidal. The second sinusoidal phase detector is called the *Q* phase detector, and the first cosinusoidal phase detector is called the *I* phase detector. Each phase detector supplies a complete canceler system with bipolar video, and the outputs of the *I* and *Q* canceler systems are added in a quadrature combiner. (The words "canceler system" are used to distinguish the *I* and *Q* canceler channels from the number 1 and 2 cascaded cancelers in each path.) The

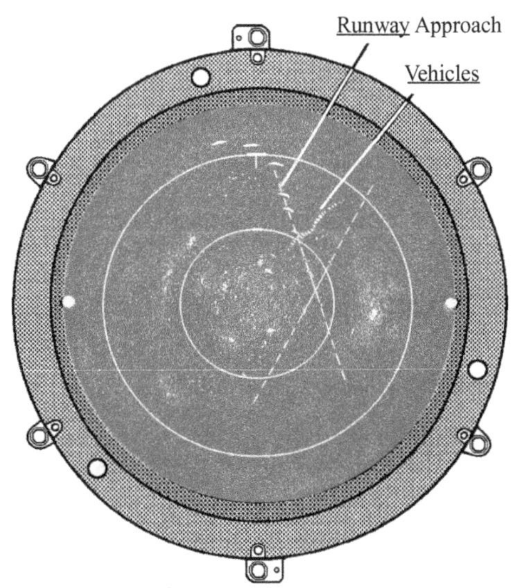

FIGURE 12–16

Freeway traffic on a ppi.

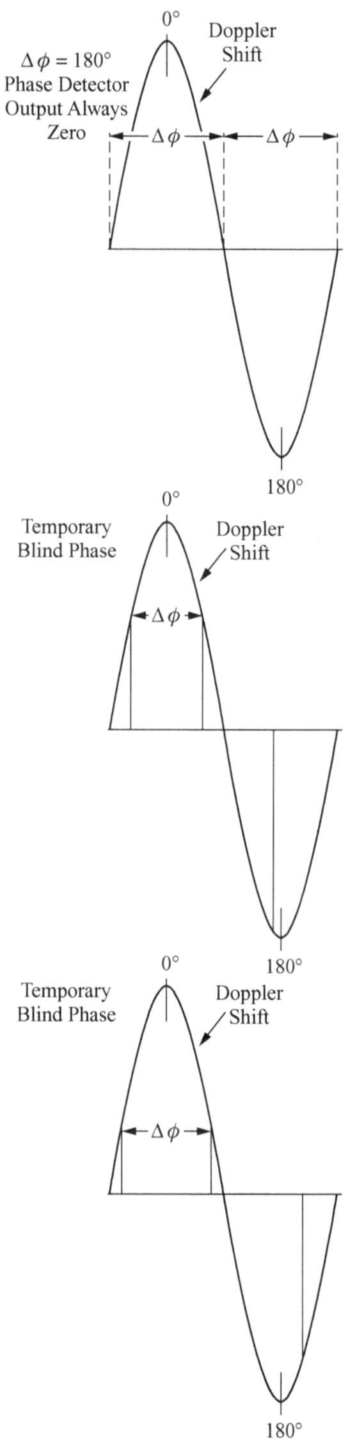

FIGURE 12-17

Blind phases.

quadrature phase-detector responses are illustrated in Figure 12-18, and the manner in which the cancelers are connected is illustrated in Figure 12-19.

Because $\sin^2\theta + \cos^2\theta = 1$, the signal power at either (1) the squared-and-summed phase-detector outputs or (2) the squared-and-summed quadrature combiner output is unaffected by phase angle. Any $\Delta\phi$ which causes a zero change in one phase-detector output will cause a significant change in the other. In mtd systems, quadrature synchronous detectors are used as rectangular-coordinate devices to describe the phase angle and magnitude of the echo, for an entirely different objective than the blind-phase solution.

The Triangular Phase-Detector Response

Before the introduction of quadrature phase detectors, it was necessary to modify the cosinusoidal phase-detector response to a triangular shape. Doppler-shift targets with a low f_D and small $\Delta\phi$ will not provide an adequate pulse-to-pulse difference to provide a discernible MTI output when the ϕ is near a peak of the phase-detector response. The differences near the response peaks are greater when the response is triangular, as shown in Figure 12-20. The triangular shape may be obtained simply by limiting both inputs to the phase detector; comparison of square waves cannot provide a cosinusoidal shape. The triangular shape is not useful in quadrature operation, because the sinusoidal shapes are necessary for implementation of the trigonometric identity, $\sin^2\phi + \cos^2\phi = 1$. Because of the rounded shape of the peaks of the I and Q phase detectors, use of single-phase operation in quadrature systems should be considered inadvisable; targets will "break up" when the ϕs are in the regions of the phase-detector response peaks.

The Velocity Response Shape

Assume there is a single, cosinusoidal phase detector, supplying a single canceler with bipolar video (see Figure 12-21). The canceler can produce (1) a maximum output where the Doppler shift $f_{D_{opt}}$ is half the f_p or any odd multiple of $f_{D_{opt}}$, or (2) a minimum where the Doppler shift $f_{DV_{b1}}$ is equal to the f_p or any multiple of $f_{DV_{b1}}$. Therefore, the canceler output is a function of the T_r-to-T_r change of an input pulse. The canceler output can be plotted in terms of (1) f_D, $\Delta\phi$, T_r-to-T_r change in distance (ΔD), or V_r, in the horizontal plane, and (2) amplitude, in the vertical plane. This might be called a Doppler bandpass; it is commonly called a velocity response shape. Provided the input is sinusoidal or cosinusoidal, the precise shape is a plot of $\sin(\Delta\phi/2)$.

Because of the Doppler bandpass represented by the velocity response shape, it is correct to say that the MTI system canceler is a notched Doppler filter, and it is sometimes so called in scientific literature. Other radar systems use other types of Doppler filters; mtd systems use many narrowband Doppler pass filters to sort the target data according to Doppler. If the phase-detector response is triangular, the "real" velocity response is triangular. The word "real" is used because swept-audio test equipment producing sine-wave outputs will make the response appear to be sinusoidal.

Improvement to the Velocity Response Shape

A single canceler provides a $\sin(\Delta\phi/2)$ velocity response for a sinusoidal input. This may well be inadequate for ground-clutter cancellation, particularly when the antenna rotates at high speed; most stationary ground clutter then presents a small Doppler and $\Delta\phi$ because of the antenna motion, and a true zero-Doppler target is rare.

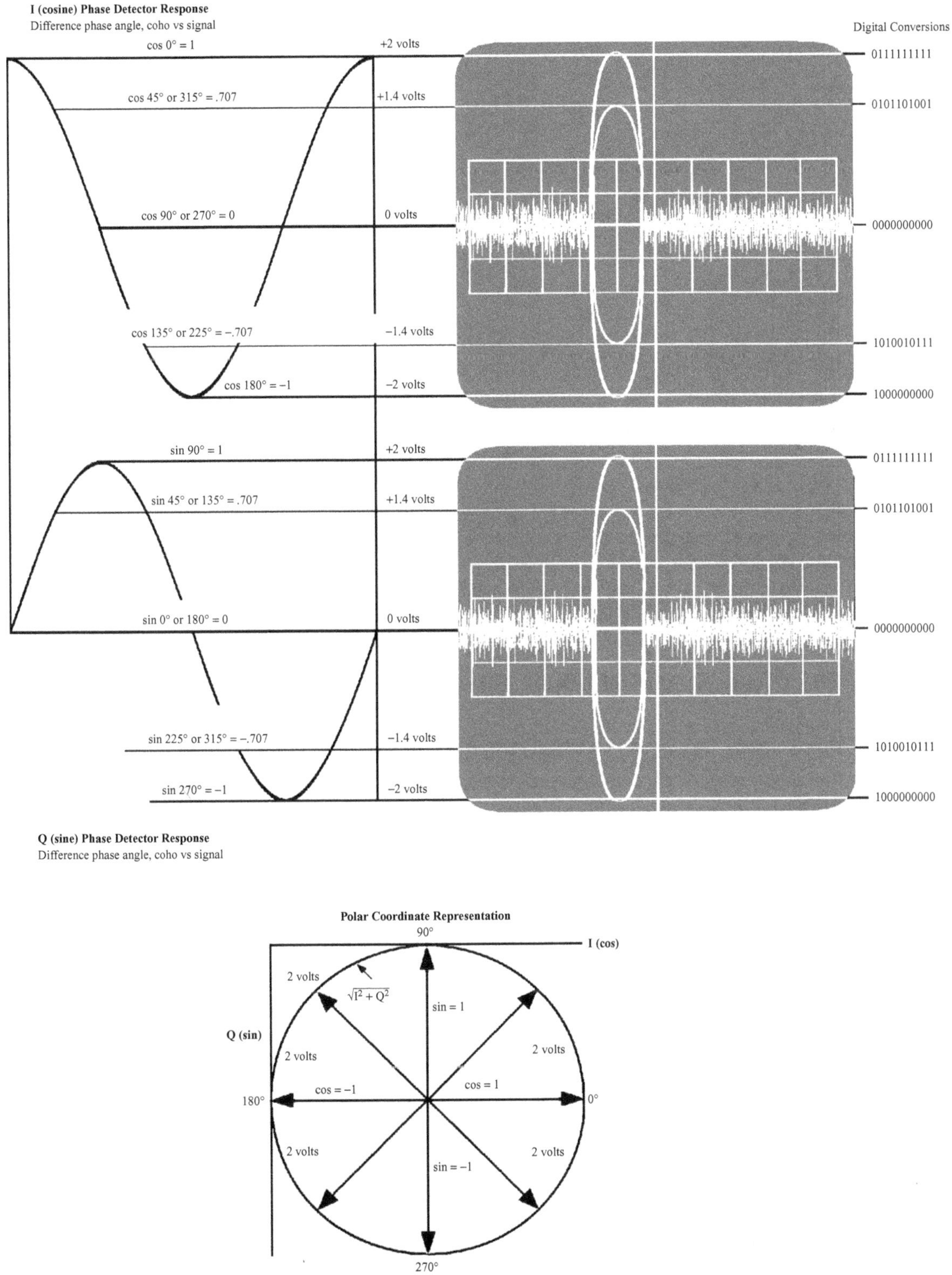

FIGURE 12–18

Quadrature phase-detector responses.

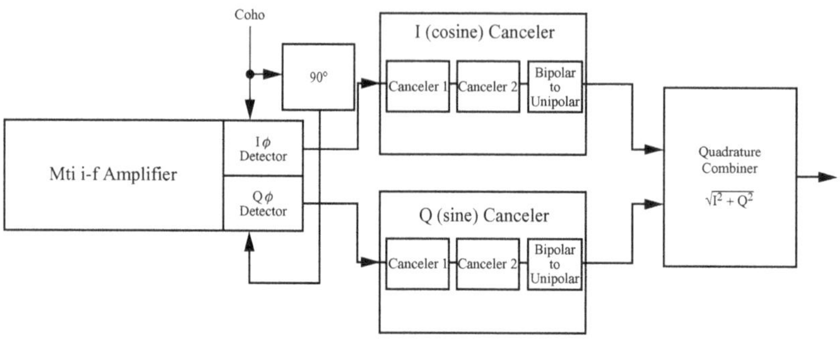

FIGURE 12–19

Quadrature cancelers and combiner.

Clutter residue will result from those stationary targets which present a significant Doppler from antenna scanning effects. A re-shaping of the velocity response can reduce this effect.

Cascaded Cancelers and Feedback

A canceler is a special-purpose amplifier with a bandpass to Doppler audio frequencies (see Figures 12-22 and 12-23.). If two amplifiers of a gain *A* are connected in series, the gains are multiplied, and the overall gain becomes A^2. Similarly, if two cancelers are connected in series (cascade), the total gain becomes $\sin^2(\Delta\phi/2)$. For over 40 years preceding the introduction of Doppler-filter mtd, MTI systems used two cascaded cancelers and feedback loops to alter the response shape; these feedback loops may be adjusted to provide different velocity response shapes. In the ASR-4, ASR-5, ASR-6, ASR-7, and ASR-8, four feedback shapes were made available. As with simpler amplifiers, feedback broadens the bandpass. Because the video within the canceler system is bipolar, the feedback may be either regenerative or degenerative, depending upon the $\Delta\phi$, and some Dopplers are emphasized, while others are de-emphasized.

The choice of velocity response shapes is determined by the clutter characteristics around the radar. Cascade operation without feedback provides the most residue-free display, but also offers the worst moving-target detection. In contrast, the highest feedback levels provide the best moving-target detection, but also the greatest residue; the good target detection is of no value if the user is unable to find the target in the clutter residue.

Bipolar and Unipolar Velocity Response Shapes

The illustrated velocity response shapes in this chapter are of unipolar video, after the canceled bipolar video had been converted. A bipolar velocity response shape might also be observed upstream of the bipolar-to-unipolar converter within the canceler. There are a multitude of circuits to accomplish bipolar-to-unipolar conversion. One such circuit is shown in abbreviated form in Figure 12-24.

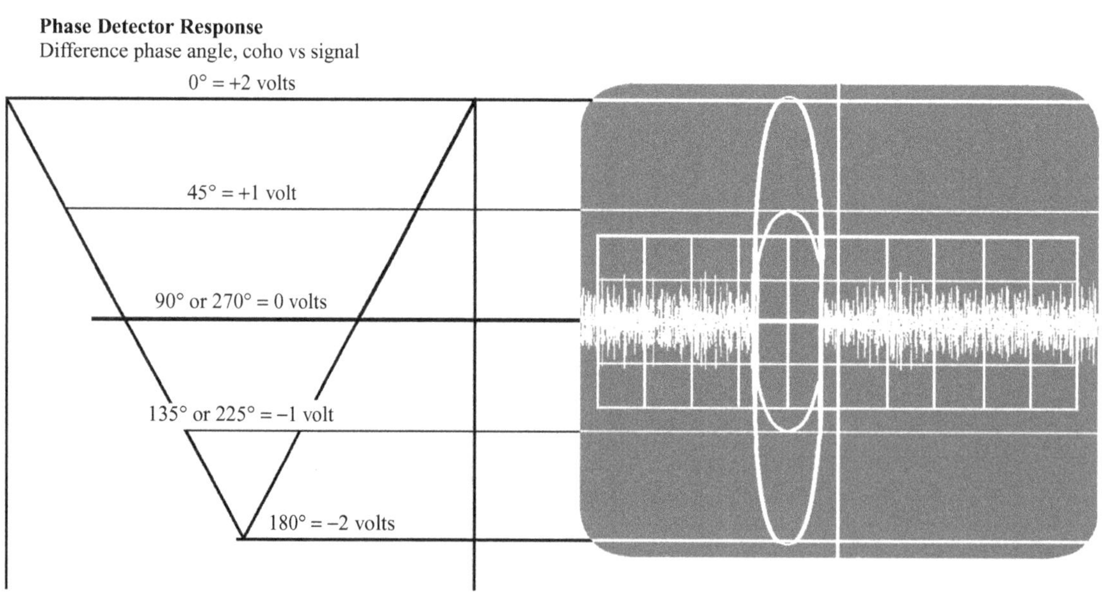

FIGURE 12–20

Triangular phase-detector response.

Sweeping the Canceler with Audio

The velocity response shape may be displayed on an oscilloscope by substituting a swept audio for the phase-detector input to the canceler, and using the audio-oscillator sweep for the oscilloscope horizontal. The swept audio is used to simulate a spectrum of Dopplers, between two f_p multiples, at the canceler input. Figure 12-25 illustrates the manner in which the increasing frequency of the audio becomes unipolar video.

Figure 12-26 shows how the audio will be canceled when its frequency is equal to $f_{DV_{b1}}$, and how a maximum unipolar video output may be obtained when the audio is equal to f_{Dopt}. The swept audio may simulate a wide variety of Dopplers, and also a multitude of blind phase conditions.

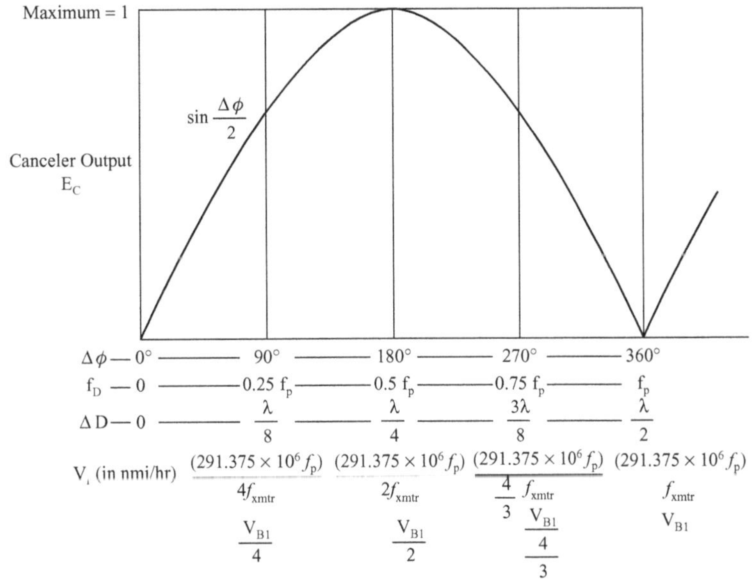

FIGURE 12–21

Single canceler velocity response envelope, sinusoidal input.

Figure 12-27 is an illustration of how an actual velocity response shape for a single-phase-detector canceler system might appear on an oscilloscope display. If (1) the system is quadrature, and if (2) the swept audio into the quadrature canceler bears the same 90° relationship to the in-phase canceler as do the *I* and *Q* phase detectors, then (3) the combined velocity response shape at the output of the quadrature combiner will appear as in Figure 12-28. The simulated blind phase conditions cause smaller notches within the shape, as shown in Figure 12-27. In a quadrature system, these simulated blind-phase notches at the outputs of the *I* and *Q* cancelers will exhibit a sin–cos relationship to each other, and the $\sin^2\theta + \cos^2\theta$ addition in the combiner will yield a result of unity, which is the envelope of the velocity response shape.

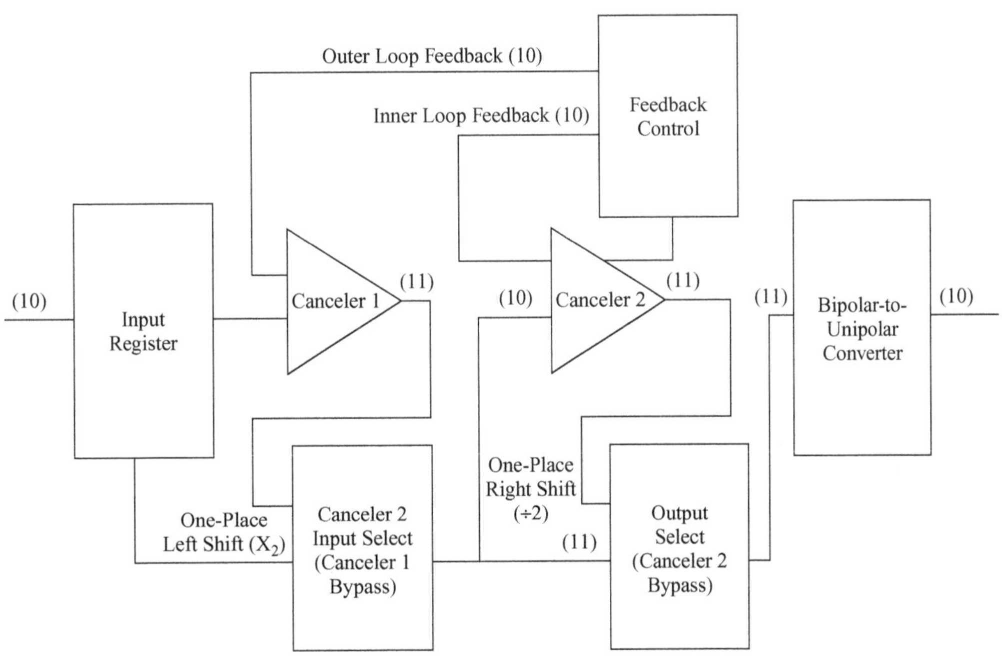

FIGURE 12–22

Cascaded cancelers (digital MTI system).

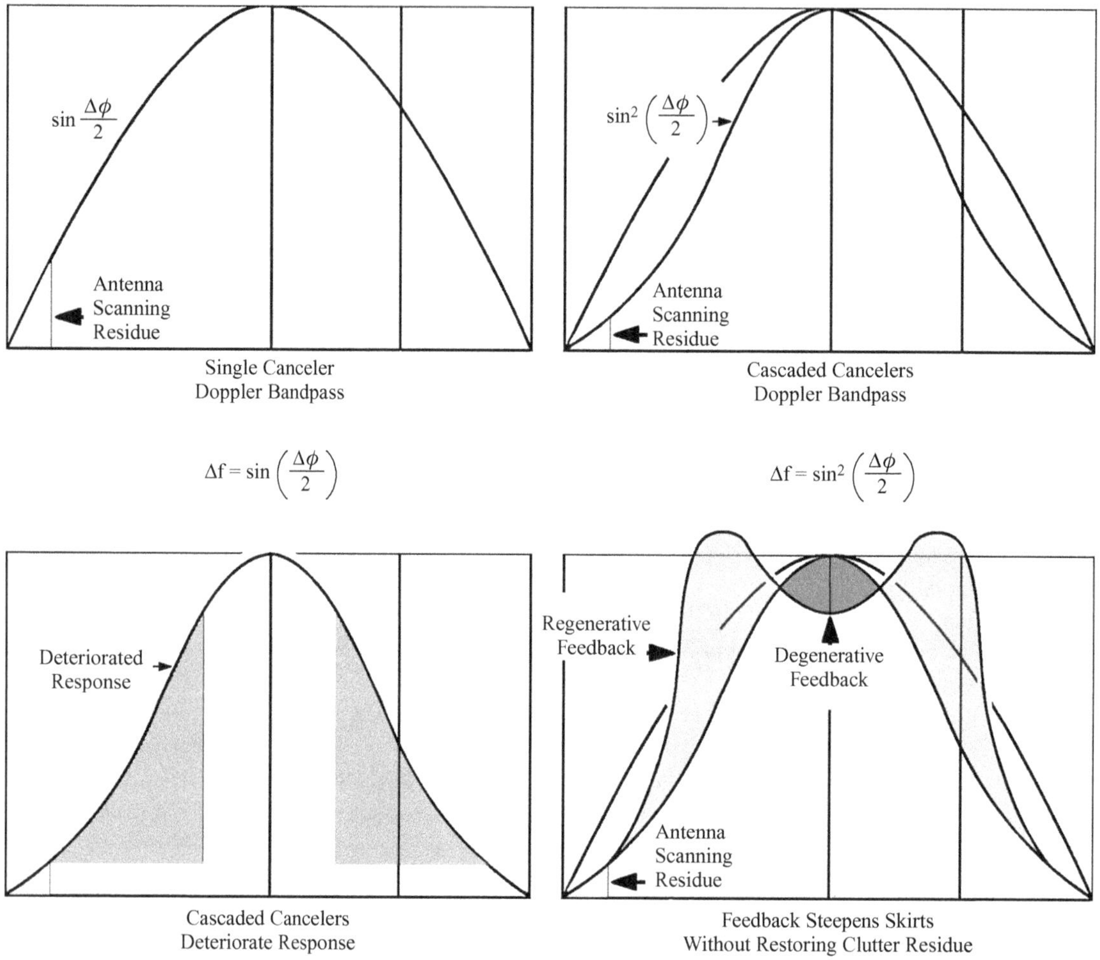

FIGURE 12–23

Adjusting the shape with cascade and feedback.

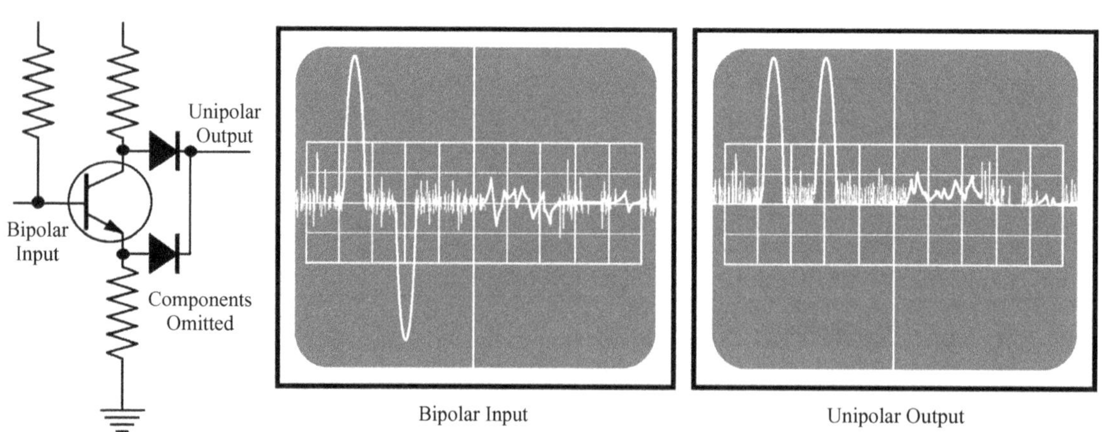

FIGURE 12–24

Bipolar to unipolar conversion.

Mathematical Relationships from the Velocity Response Shape

Figure 12-21 shows that values of f_D, $\Delta\phi$, and V_r are all proportionally related. The first blind velocity occurs where the Doppler shift equals the f_p, and the $\Delta\phi$ equals 360°; therefore

$$\frac{\Delta\phi}{360} = \frac{f_D}{f_p}.$$

The equation can be rearranged to solve for either the Doppler or $\Delta\phi$:

$$f_D = \frac{f_p(\Delta\phi)}{360}$$

$$\Delta\phi = \frac{f_D(360)}{f_p}.$$

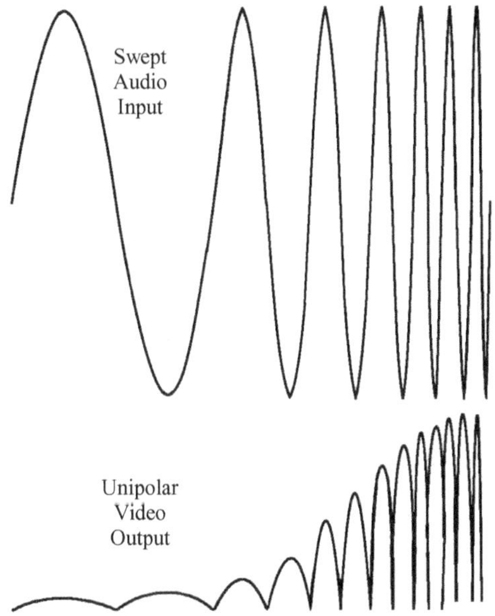

FIGURE 12-25

Swept audio in and out of a canceler.

Finding $\Delta\phi$ from f_D and f_p

Now, sufficient information is established to provide a means to find the $\Delta\phi$ when the Doppler is known:

$$\Delta\phi = \frac{360 f_D}{f_p}.$$

Finding the $\Delta\phi$ from V_r, f_{xmtr}, and f_p

Now, substitute the Doppler solution below for f_D in the solution for $\Delta\phi$:

$$f_D = \frac{V_r f_{xmtr}}{291.375 \times 10^6}$$

$$\Delta\phi = \frac{\dfrac{360 V_r f_{xmtr}}{291.375 \times 10^6}}{f_p} = \frac{360 V_r f_{xmtr}}{291.375 \times 10^6 f_p}.$$

Simplify by dividing the numerator and denominator by 360,

$$\Delta\phi = \frac{V_r f_{xmtr}}{8.09375 \times 10^5 f_p}.$$

Finding the $\Delta\phi$ from Aircraft Course, Radar Bearing, f_p, and f_{xmtr}

Substitute the solution for radial velocity for V_r:

$$\Delta\phi = \frac{[V_{ground}(\cos(\angle_{target} - \angle_{course} + 180°))] f_{xmtr}}{(8.09375 \times 10^5) f_p}.$$

T_r–T_r Radial Target Movement

Since the Doppler equation was originally based on wavelength, and because wavelength and frequency are inversely proportional, there is also a proportional relationship between blind and optimum velocities and rf wavelengths. Consider the radar Doppler equation in terms of λ; for blind radial velocity, substitute V_{b1} for V_r and f_p for f_D, as shown in the following equation. Recall that $d = rt$, where d is the distance traveled in wavelength, r is V_b, and t is T_r. Substitute $1/T_r$ for f_p, and the equation can be rearranged to $d = rt$:

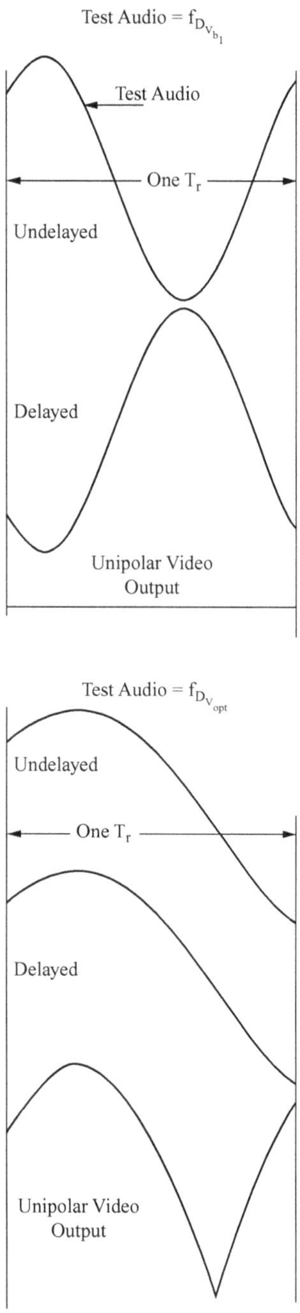

FIGURE 12–26

Sine waves applied to canceler.

$$f_D = \frac{2V_r}{\lambda}, \qquad f_p = \frac{2V_{b1}}{\lambda}$$

$$d = rt$$

$$f_p = \frac{1}{T_r}.$$

Therefore

$$\frac{1}{T_r} = \frac{2V_{b1}}{\lambda},$$

$$\frac{\lambda}{2} = V_{b1} T_r.$$

Note that

$$\frac{\lambda}{2} = d, \quad V_{b1} = r, \quad T_r = t.$$

Therefore

$$\frac{V_{b1}}{V_r} = \frac{0.5\lambda}{\Delta D}$$

$$\Delta D = \frac{0.5\lambda V_r}{V_{b1}}.$$

In short, at the first blind velocity, a vehicle moves by a half-wavelength during a system interval. Since the first optimum velocity is half the first blind velocity, a vehicle at the first optimum velocity moves by a quarter wavelength during a system interval. All other velocities are similarly proportional. For this reason, an MTI reflector can simulate an optimum-velocity target by alternately adding and removing a $\lambda/4$ in its waveguide feedhorn.

The MTI Reflector

An MTI is a passive device, except for a ΔD introduced in the feedhorn assembly (see Figure 12-29). A multivibrator, operating at approximately $f_p/2$, alternately biases microwave diodes to induce or remove a microwave short in the feedhorn. The box on the pole contains a battery to supply power for the multivibrator; power may also be supplied by solar cells. The MTI reflector may be located from less than 1 mile of the radar, up to 10 miles, or even more. At airport terminal facilities, the MTI reflector is necessary to provide a reference for alignment of the runway displays created by radar video mapping equipment. At a terminal facility employing an ASR-9 with an mtd system, multiple MTI reflectors are absolutely essential for radar azimuth verification, because reference permanent echoes, such as mountain peaks or water towers, viewed on normal video, on earlier ASRs, are not available.

Because of the narrow beamwidth of the radar and the directivity of the MTI reflector, alignment to provide a useful echo is critical and requires coordination between technicians at both (1) the display or radar site and (2) the reflector location. The ASR-9, which is operated and controlled via computer terminals, utilizes a target performance window, which permits the technician to "zero in" on an MTI reflector, to obtain precise target strength and Doppler information. Reflectors are frequently damaged by aircraft or vehicles, and may be vandalized or even stolen, even though they are of no use, elsewhere. The reflector may become inoperative because of phenomena ranging from severe weather to growth of vegetation.

The Magnetron MTI System

The Doppler and $\Delta\phi$ relationships of the synthesis system equally apply to the magnetron system, but they are obscured by the additional circuitry required to achieve coherence. There are many dictionary definitions for

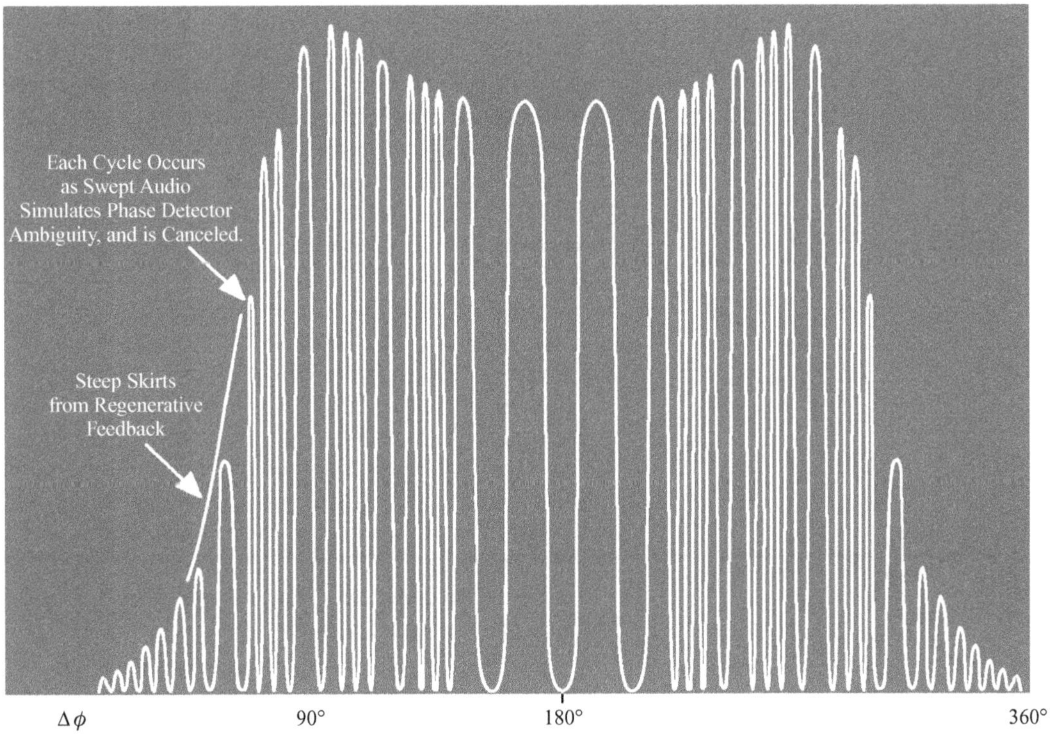

FIGURE 12–27

Single phase-detector velocity response shape.

"coherence" which shed light on its meaning in MTI systems; among these are "to stick together," "to be connected naturally or logically," "to be logically integrated, consistent and intelligible," and others. A synthesis system is inherently coherent, simply because the two oscillators, stalo and coho, both run without interruption, and because the transmitter rf burst is developed from those two oscillators. A magnetron begins oscillation at an unpredictable phase, and bears no intelligible relationship to the stalo or coho. Without additional circuits, the phase-detector output would be incoherent, as illustrated in Figure 12-30. A close inspection of the clutter bipolar video in this illustration will reveal that the incoherence is causing it to differ from one T_r to the next, prohibiting cancellation.

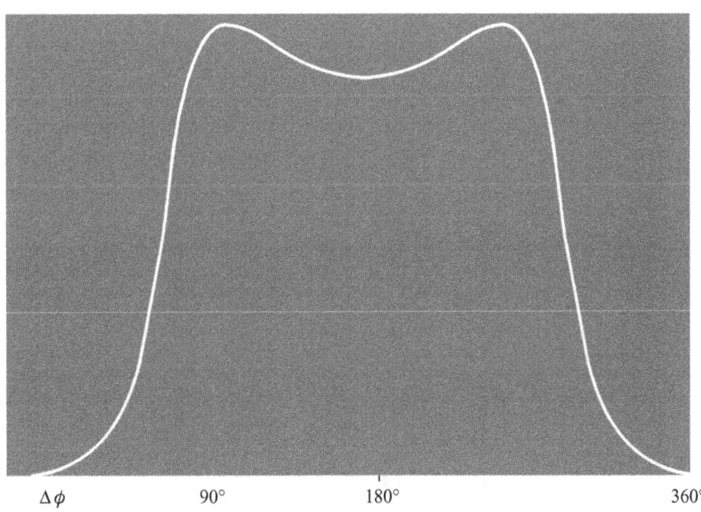

FIGURE 12–28

Quadrature velocity response.

Achieving Coherence

Since there is no way to control the starting phase of the magnetron, there must be a means to adjust the phase of the signals in the receiver to cause the resultant bipolar video to appear exactly as it would in a synthesis system (see Figure 12-31). To accomplish this, the coho is phase-shifted by an i-f representation of the magnetron burst at the beginning of each T_r. A mixer, very similar to the receiver signal mixer, is used to achieve this i-f lock pulse. Because the stalo is used to achieve the lock pulse, and because the same stalo is also used in the signal mixer to beat against reflections of the same magnetron burst, there is coherence between the i-f lock pulse and the i-f echoes. The coho is gated off a few microseconds before the i-f lock pulse occurs, and is re-started just as the lock pulse is applied to the oscillator; the starting phase is then determined by the magnetron's starting phase, and the coho is coherent with the i-f echoes. This original use of the coho, to adjust the system into coherence, is the origin of its name. USAF air traffic control radar schools once taught that the purpose of the coho was "to remember the phase of the magnetron." The phrase may have been a little unsophisticated, but it was very accurate and descriptive, and easy to remember.

There are three frequency sources in this system, the magnetron, stalo, and coho. Only one of these sources, the stalo, runs without interruption. For that reason, the stalo becomes the central reference signal in achieving coherence. The very word, "stalo," is an acronym for "stabilized local oscillator," and was chosen in the earliest days of magnetron MTI in the 1940s. Because the stalo in a magnetron system is the single coherence link between the transmitter and receiver, its design in those early days required great care to achieve the highest attainable degree of stable and consistent phase and frequency.

In the waveguide assembly, a magnetron system contains a coupling circuit used to obtain an attenuated sample of the transmitted rf burst. This magnetron rf burst is then applied to a mixer, similar to the signal mixer, where it is compared to the stalo to produce the i-f difference signal, called the i-f lock pulse. Depending upon the manufacturer, the i-f lock pulse might also be called the coho lock pulse, afc lock pulse, or coho/afc lock pulse. Its original and primary function is to provide an i-f reference burst which is phase-related to both the magnetron and stalo (although the lock pulse is now also used for afc, the original MTI systems in military air traffic control radar had no afc).

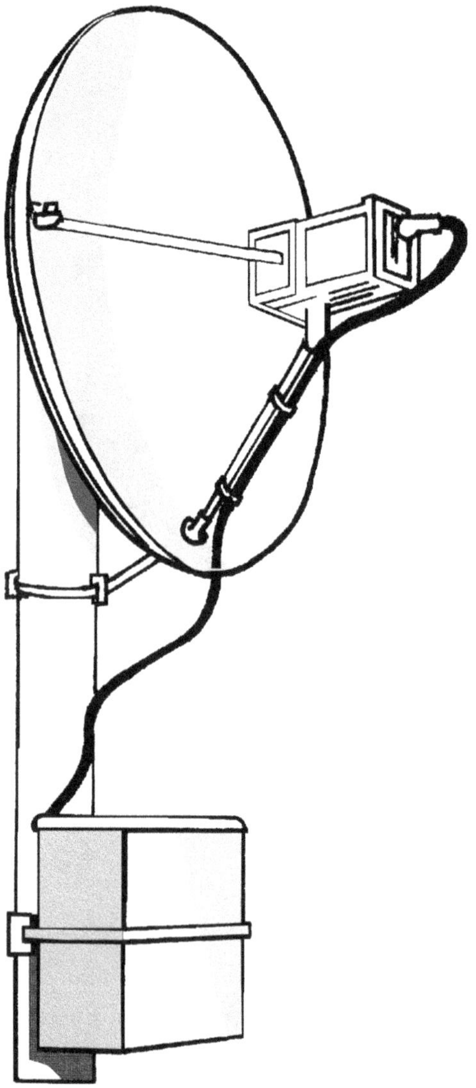

FIGURE 12–29

An MTI reflector.

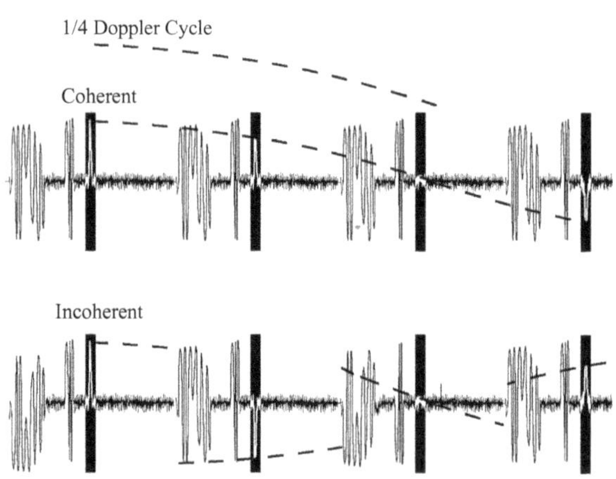

FIGURE 12–30

Coherent versus incoherent phase detector outputs.

Summary of Magnetron Coherence

Coherence has been achieved because:

1. The phase of received rf echoes at the signal mixer is determined by the magnetron burst and the waveguide phase delays.
2. The phase of received i-f echoes is determined by a comparison of the stalo to the magnetron rf echoes in the signal mixer.

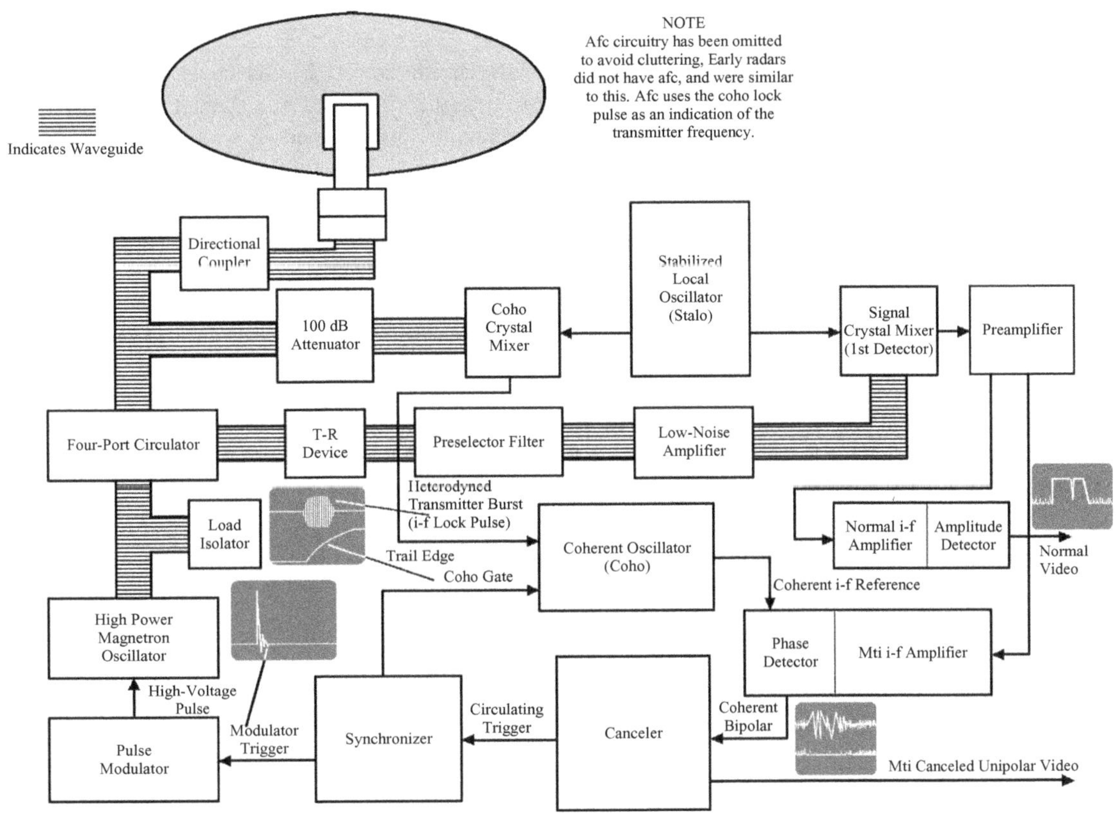

FIGURE 12–31

The magnetron system.

3. The phase of the coho is set by the i-f lock pulse, created at T_0, by comparing the magnetron rf burst to the c-w stalo in the afc mixer.
4. The phases of the i-f and coho inputs to the phase detector are both derived from two common sources, the magnetron and stalo.

Magnetron System Maintenance Techniques

Chapter 8 on waveguide contains a discussion on use of the echo box. This device is extremely useful in observing the "front-end" performance of the coherence circuitry. If the echo box is connected to the directional coupler, and slightly detuned, the frequency in the i-f will be slightly different than the coho, and the phase-detector response can be viewed at the phase-detector output. If the system is coherent, the phase-detector response will be identical with each magnetron burst, and will appear as a synchronized waveform, illustrated in Figure 12-32. If coherence does not exist, the phase-detector response will differ with each magnetron burst.

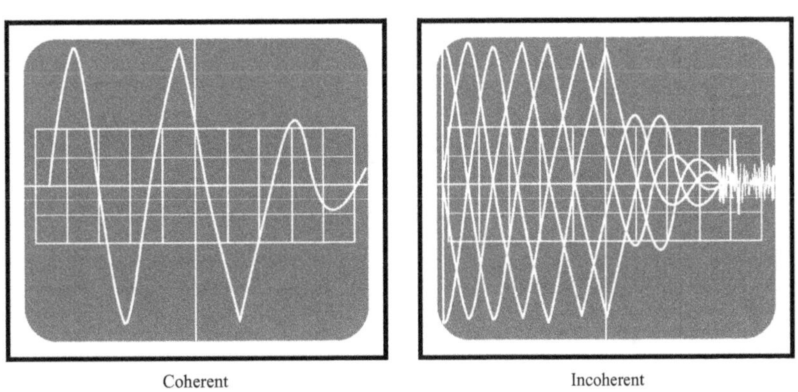

FIGURE 12–32

Echo box phase-detector video.

The Locked Test Pulse Generator

Most magnetron systems are equipped with a locked test pulse generator (LTPG), illustrated in Figure 12-33. The i-f output from the preamplifier, destined for the MTI i-f amplifier, must first pass through switch contacts in the LTPG. If the LTPG MODE switch is in the OFF position, the preamplifier i-f is routed directly to the MTI i-f. If the MODE switch is in position 1, the MTI i-f is supplied with a train of i-f test bursts coherent with the magnetron and stalo. If the mode switch is in position 2, the MTI i-f amplifier is supplied with i-f test bursts derived from a test oscillator; this mode may be necessary when the transmitter is off, or for fault-isolation purposes.

In either normal operation or test mode 1, the coho is phase-locked by the i-f lock pulse as previously described. In test mode 2, the output of the test oscillator is gated for approximately t_p at the beginning of the T_r, and that burst is used to phase-lock the coho.

In either test mode, a 30-MHz burst will be amplified for application to a ringing delay-line circuit. The burst will travel "down" the delay line, but upon reaching the ground, will be reflected back through the delay line. The signal will continue to travel back and forth through the delay line until the circuit attenuation diminishes it to zero, and a long train of test bursts is created, as illustrated in Figure 12-34.

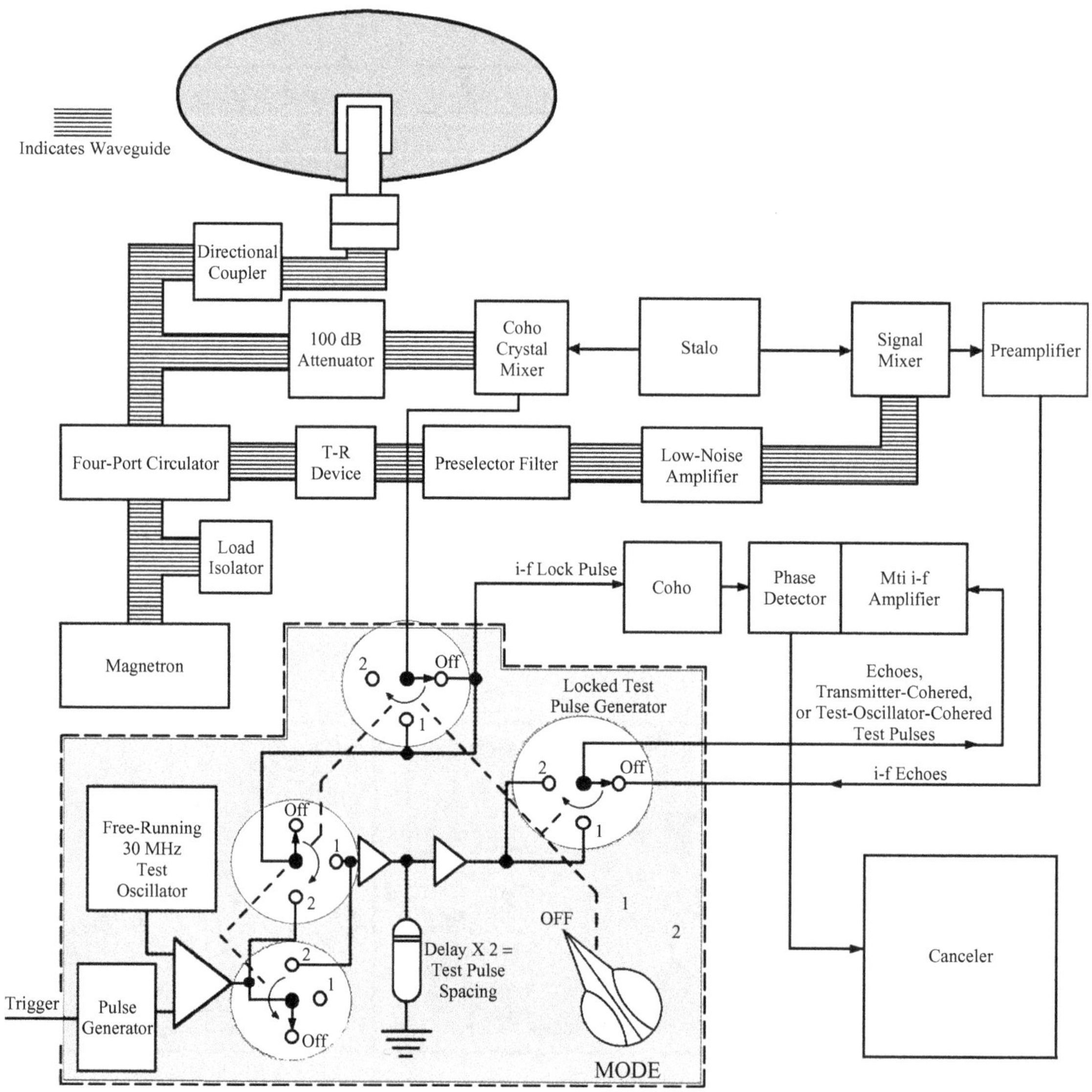

FIGURE 12–33

Locked test pulse generator.

The lock test pulses are useful in isolating system faults. If there is significant clutter residue in all positions of the LTPG MODE switch, it is possible the problem is in the canceler or coho; a video cancellation ratio test of the canceler can eliminate that unit. If (1) there is residue in normal operation, or with the MODE switch in position 1, but (2) the cancellation is normal with the MODE switch in position 2, a coherence problem involving the stalo, magnetron, or coho mixer is indicated. Because of the coho locking each T_r, magnetron frequency drift will not cause residue, but arcing or misfiring may do so. Among the most common causes of high clutter residue is excessive afc correction; each time the stalo is driven back to the "tuned" position, coherence fails. This condition will usually be evidenced by residue strobes on the ppi display.

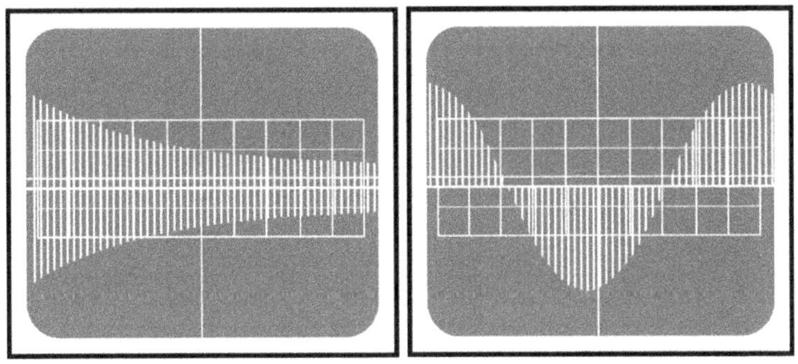

i-f Bursts from LTPG   Phase Detector Output

FIGURE 12-34

Locked test pulses.

i-f Test Pulse Generators in Synthesis Systems

Although it is not possible for a synthesis system to become incoherent, instabilities in the transmitter rf amplification system, stalo, or coho can all contribute to clutter residue, and many synthesis systems contain a test pulse generator to aid in the isolation of these. Additionally, there may be a need for an incoherent i-f test signal for adjustment to the phase detectors, or to produce a test target with a synthetic Doppler. In latter-day synthesis MTI systems employing digital shift registers for the canceler delay, the most likely source of excessive clutter residue is the transmitter, often in high-voltage pulse instabilities. Neither latter-day stalos nor cohos have exhibited a high potential for instability.

The Composite Video Test Target Generator

To independently test the canceler, and isolate it as a source of excessive clutter residue, many systems employ a special test signal, switched onto the bipolar video path to the canceler during dead time (see Figure 12-35). By far the largest number of FAA MTI radars have been built by Texas Instruments, and that manufacturer called this test signal the composite video test target train (cvtt). The cvtt comprises three test targets, each near the t_p of the radar system. One of these is a positive pulse, and occurs once, each T_r. Another is a negative pulse, also occurring once, each T_r. The positive and negative pulses, called the positive fixed target (PFT) and the negative fixed target (NFT), should be canceled. A third pulse, usually positioned between the positive and negative pulses, is called an optimum speed test target (OST); the OST is maximum positive on one T_r, and maximum negative on the next, simulating an optimum-speed, optimum-phase moving target. The OST may be attenuated by 40 dBv (100×) with the activation of a switch. The technician may be wise to leave this attenuator engaged, as the maximum-amplitude OST will be present on the MTI video output to the user facility, and may contribute to confusion, even though it is outside the maximum range.

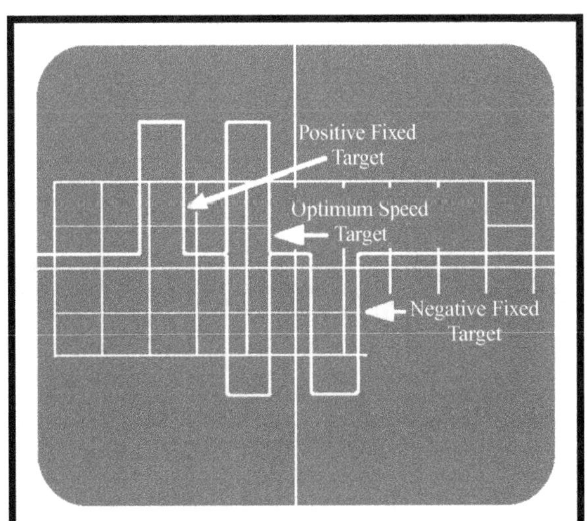

FIGURE 12-35

Composite video test target train.

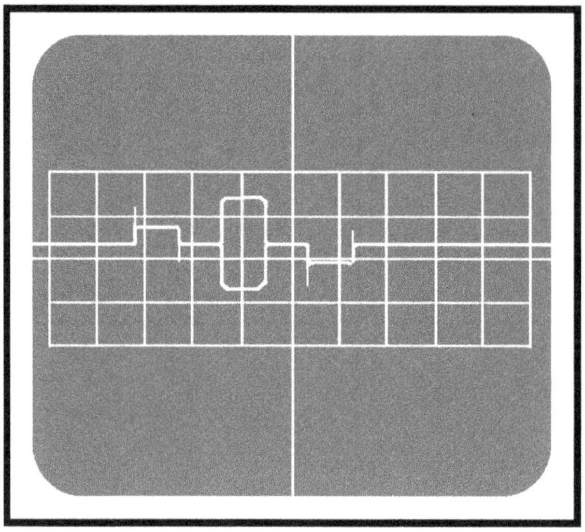

FIGURE 12-36

Canceled cvtt train.

Figure 12-36 illustrates the canceled bipolar cvtt; the 40-dB attenuator switch has been engaged. The main purpose of the cvtt is to ensure that the attenuation to the positive and negative fixed targets is at least 40 dBv for each canceler. When the technician views the cvtt residue on an oscilloscope, the video cancellation ratio exceeds 40 dB if the fixed test target residue is less than the –40-dB OST.

The original purpose of the cvtt was to provide a test signal for viewing during adjustment to the delay time and amplitude balance of analog cancelers. In digital systems, no such adjustment is necessary, and the cancellation ratio test is only necessary to verify that there are no bit failures in the parallel processing of the binary words representing the bipolar video for each range cell.

Live Video

The cvtt is generated by ***built-in test equipment (BITE)***. It appears in dead time, past the maximum normally displayed radar range, sometimes in what is called "BITE time." Active videos will contain bipolar clutter at the phase detector, and canceled "holes" where it had been attenuated by cancellation (see Figure 12-37).

Output Limiting

To make discernible the weakest targets on the ppi, it is necessary to adjust the display to make system noise visible. So that excessive target amplitude does not cause "blooming" on the displays, it is necessary to restrict the video signal-to-noise ratio to approximately 4:1 or 6:1 before application to the display equipment. The signal-to-noise ratio at the phase-detector output is generally very high, sometimes as great as 40:1. The canceler will integrate some of the noise and attenuate some of it, so the canceled video will exhibit a less-dense noise with higher peaks; the "noise texture" is thus changed. Nevertheless, the signal-to-noise ratio will remain very high, and will change with different canceler modes, such as single, cascade, or feedback.

To provide the desired video signal-to-noise ratio, any MTI system will somewhere contain a limiter for the output video or data. In older analog systems, the limiter was a part of the canceler, and was very obvious. In latter-day systems, such as the ASR-8, it was necessarily considerably downstream of the canceler and obscured by digital processing circuitry.

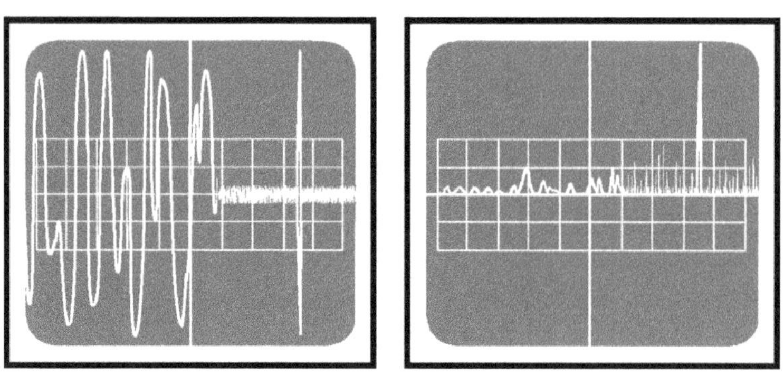

FIGURE 12-37

"Live" phase detector and canceled videos.

The effect of limiting upon the velocity response shape is illustrated in Figure 12-38. Since this is done, the irregularities at the top of the shape, caused by feedback, are all leveled, and are of little consequence.

Figure 12-39 is a simple cascade-with-feedback canceler system block diagram, incorporating an output gain and limit circuit. The gain control, prior to limiting, provides for the video signal-to-noise ratio adjustment. This control may go by several names, according to the equipment manufacturer, but the most common, and most descriptive of purpose, is the UNIPOLAR VIDEO GAIN control. The UNIPOLAR VIDEO GAIN control is the second of two final adjustments to the MTI system; the controls have been assigned the slang description, payoff controls, since the earliest days of MTI. The other payoff control is the MTI I-F GAIN control. Adjustment of the payoff controls will be described in a following paragraph in this chapter.

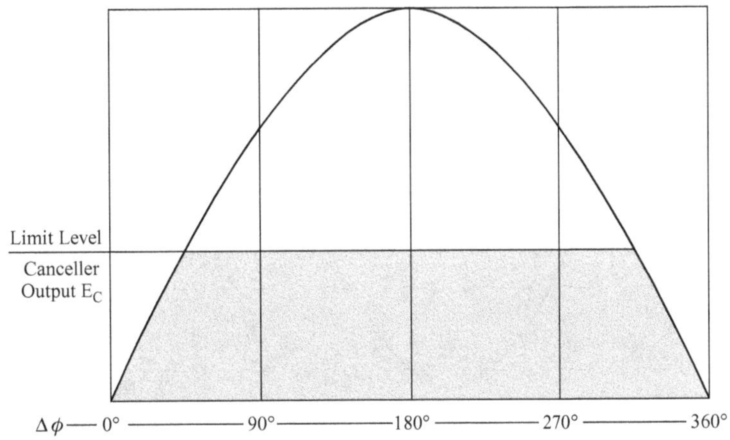

FIGURE 12–38

The limited velocity response shape.

Complex Doppler Echoes and Subclutter Visibility

This chapter has so far been dedicated only to an explanation of radar Doppler shift, and the processing of single-vehicle targets "in the clear," which means, "not over ground clutter." The principle purpose of an MTI processor is not to pass moving targets in the clear, but to make them visible over clutter, which has been attenuated. Target detection may actually be degraded by MTI in clear areas because of blind or dim speeds, blind phase effects, tangential effects, and a higher (poorer) mds caused by the wider bandpass of the MTI i-f amplifier. To ensure maximum detection probability in clear areas, some radar systems have incorporated a clutter-gating threshold circuit, in which MTI video is only enabled when clutter is present on normal video; normal video is enabled in all the clear areas.

Simultaneous Multiple Dopplers

Just as the instruments in an orchestra may all be heard simultaneously, Doppler audios from different sources may occur simultaneously. An aircraft may provide one Doppler-shifted echo from the fuselage, but a wide spectrum, or scattering, of Doppler shifts from propellers, turbofans, or helicopter rotor blades. Ground clutter may provide a small Doppler shift because of antenna rotation, and an aircraft over that ground clutter will also simultaneously provide another Doppler shift to be mixed with that from the clutter.

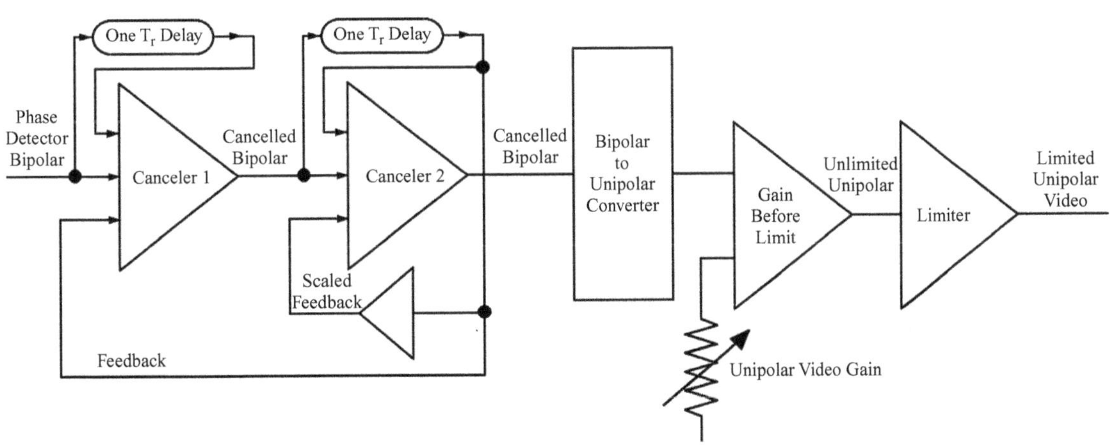

FIGURE 12–39

Unipolar video gain control.

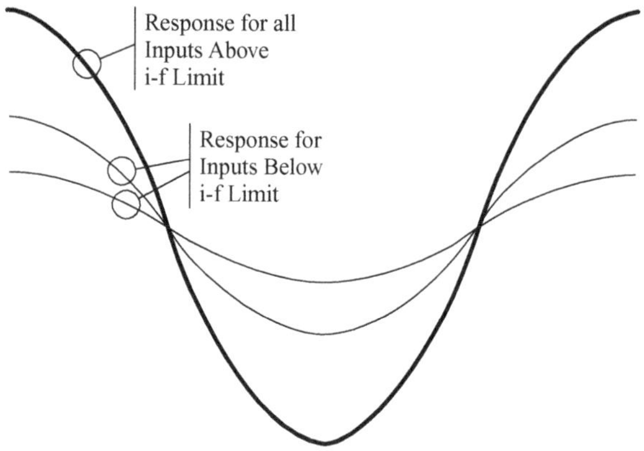

FIGURE 12–40

Phase detector response, limited i-f versus unlimited i-f amplifier.

The MTI i-f amplifier must be "hard limited"; that limiting destroys much of the signal information that could be used to separate mixed Dopplers. A major improvement in mtd systems is the incorporation of a nonlimiting, wide-dynamic-amplitude-range i-f amplifier which could not have been used in a canceler-type MTI system.

i-f Limiting for a Consistent Phase-Detector Response

Because MTI cancellation depends upon T_r-to-T_r pulse comparisons, and because zero-Doppler clutter will present a constant-phase echo, cancellation of zero-Doppler clutter is dependent upon a constant-amplitude output from the phase detector for that zero-Doppler clutter. To ensure this, the phase-detector output must not vary with signal strength; therein lies the necessity for hard limiting. Figure 12-40 shows that the phase-detector response curve is consistent for all echoes, except those which are of insufficient amplitude to reach i-f limiting. Any clutter echoes insufficient in power to reach i-f limit will cause phase-detector output amplitude variations. Those amplitude variations will subsequently cause clutter residue.

Echo Phase "Pulling," Competition, and Resultant

Because the i-f is hard limited, all amplitude differences between two simultaneous limiting echoes are obliterated, and the i-f phase becomes a resultant; that resultant is dependent on both the phase and power of each echo (see Figure 12-41). The larger echo will have the greater "pulling" effect upon the resultant phase. A moving target over clutter then will exhibit a lesser T_r-to-T_r phasedetector output voltage change (ΔE) than it would, were it in the clear. As the clutter signal power increases, the ΔE for the target decreases, and the canceler target output decreases.

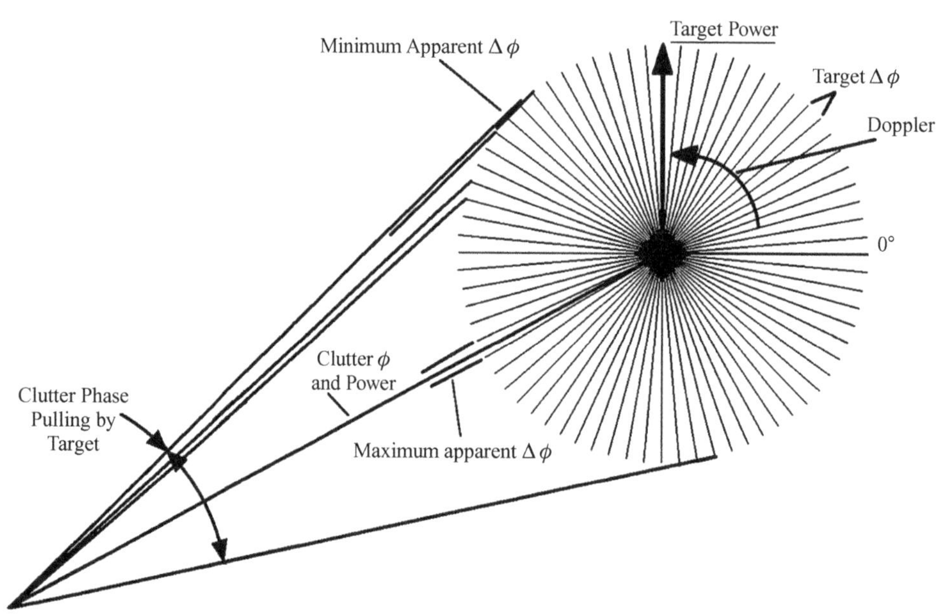

FIGURE 12–41

Phase pulling in i-f amplifier.

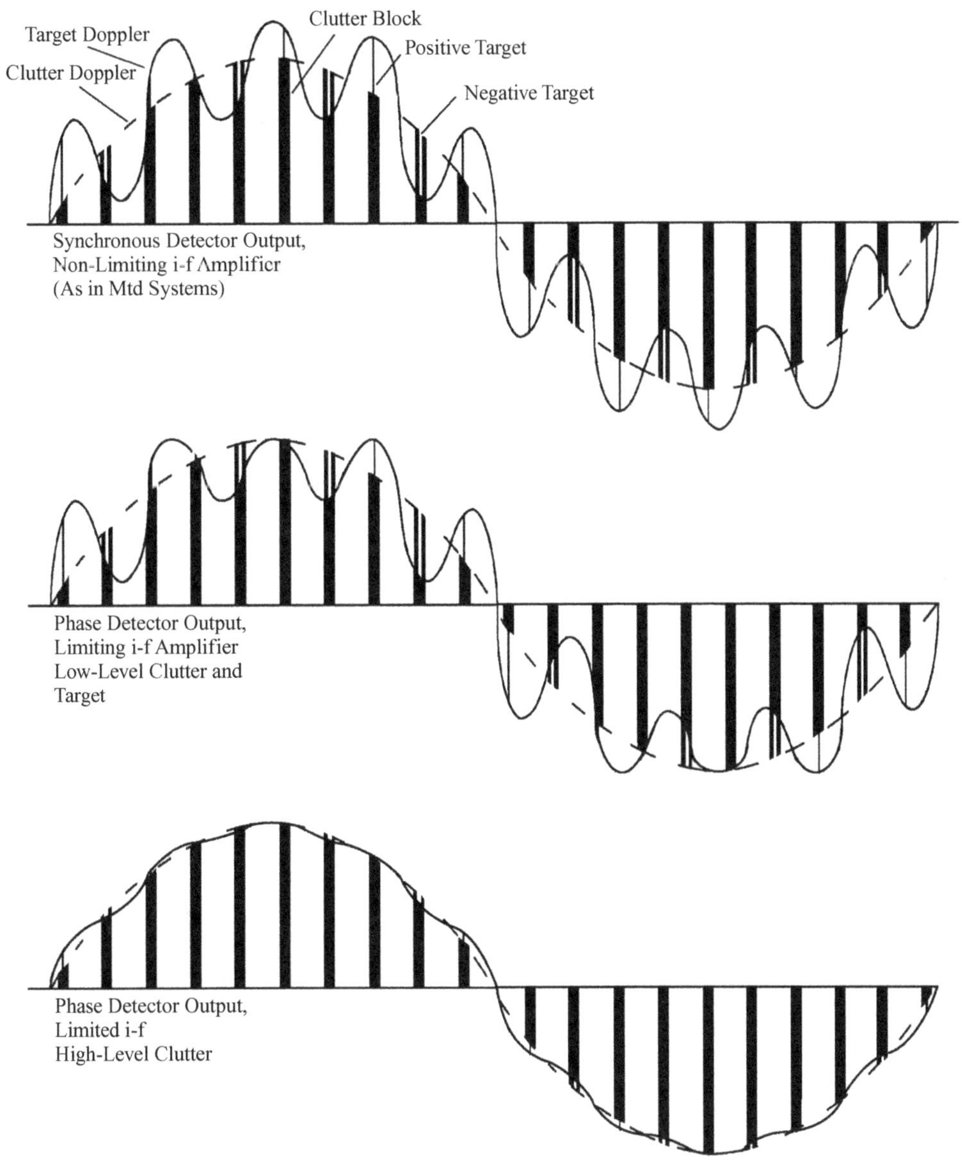

FIGURE 12–42

Complex doppler echoes.

Figure 12-42 is an artist's representation of the effect of phase pulling in terms of Doppler shift. In the first illustration, there is no i-f limiting, and the target Doppler "rides atop" the clutter Doppler. This is the manner in which signals are affected in an mtd i-f amplifier. It would also occur in an MTI i-f amplifier, if both signal powers were insufficient to reach a limit level. The second illustration shows the phase-detector output for a relatively low-level clutter and target. The third illustration is of a target over strong clutter; the target is overwhelmed by the clutter power, and there is insufficient phase pulling to make the target detectable.

Those Phase-Detector Outputs Variable with Signal Power

In order that target Dopplers may be detected over clutter, the minimum practical degree of limiting should be obtained. In contrast, increasing the limit level to reduce the degree of limiting will cause the phase-detector outputs for the weakest clutter to vary with signal strength. Nearly all clutter will be of such a level as the beam first moves onto it, and last moves away from it. Figure 12-43 shows how any block of clutter begins at a low level. This is a major cause of antenna scanning clutter residue.

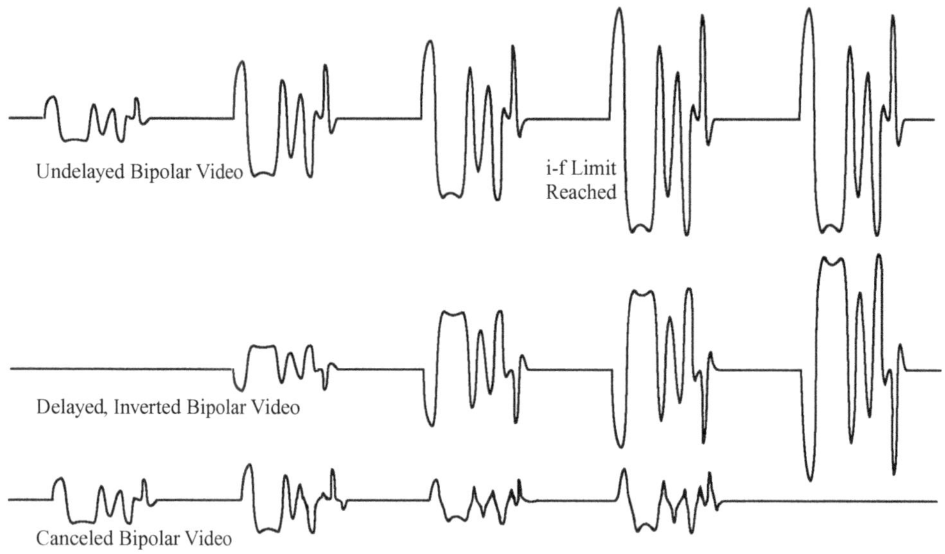

FIGURE 12-43

Residue when antenna first scans onto a clutter block.

Optimum i-f Amplifier Adjustment

The optimum operation of the MTI system can be obtained where (1) the i-f gain has been adjusted so that most of the antenna scanning residue cannot be seen on the ppi, but (2) the degree of limiting has been reduced to a minimum acceptable level, and (3) there is an acceptable mds measurement. Generally, the limit will be fixed, and all adjustment is accomplished with the MTI i-f gain control. This adjustment has always been called the payoff adjustment, because its proper setting will finally determine the satisfactory or unsatisfactory performance, no matter how well the entire system may be working. The payoff adjustment also requires adjustment to the UNIPOLAR VIDEO GAIN control. Any adjustment to the i-f gain will change the noise level out of the phase detector, subsequently requiring restoration of the unipolar video signal-to-noise ratio with the UNIPOLAR VIDEO GAIN.

Payoff Adjustment Procedure

Always consult the manufacturer's instruction book, since any system may have some peculiarities. The general procedure is as follows, and the ppi and oscilloscope presentations at the phase-detector and canceler outputs are illustrated in Figure 12-44.

1. View the bipolar phase-detector and unipolar canceler video outputs on an oscilloscope, and view the MTI display video on a ppi.
2. Increase the UNIPOLAR VIDEO GAIN to make the grass level and clutter obvious on the ppi.
3. Adjust the MTI I-F GAIN so that the clutter residue is of approximately the same brilliance and texture as the grass.
4. Restore the unipolar signal-to-noise ratio with the UNIPOLAR VIDEO GAIN.

The first illustration in Figure 12-44 might appear to be an excellent MTI display, but "black holes," and a total absence of clutter residue is undesirable. Black-hole video can be obtained by operating the MTI i-f with excessive gain, which subsequently causes excessive grass at the canceler output. When the UNIPOLAR VIDEO GAIN is lowered to obtain the correct signal-to-noise ratio, many targets over clutter are reduced below the threshold of visibility. The second illustration is of a properly adjusted system. The third illustration is of operation with inadequate i-f gain to achieve limiting; although this might provide a better moving target over

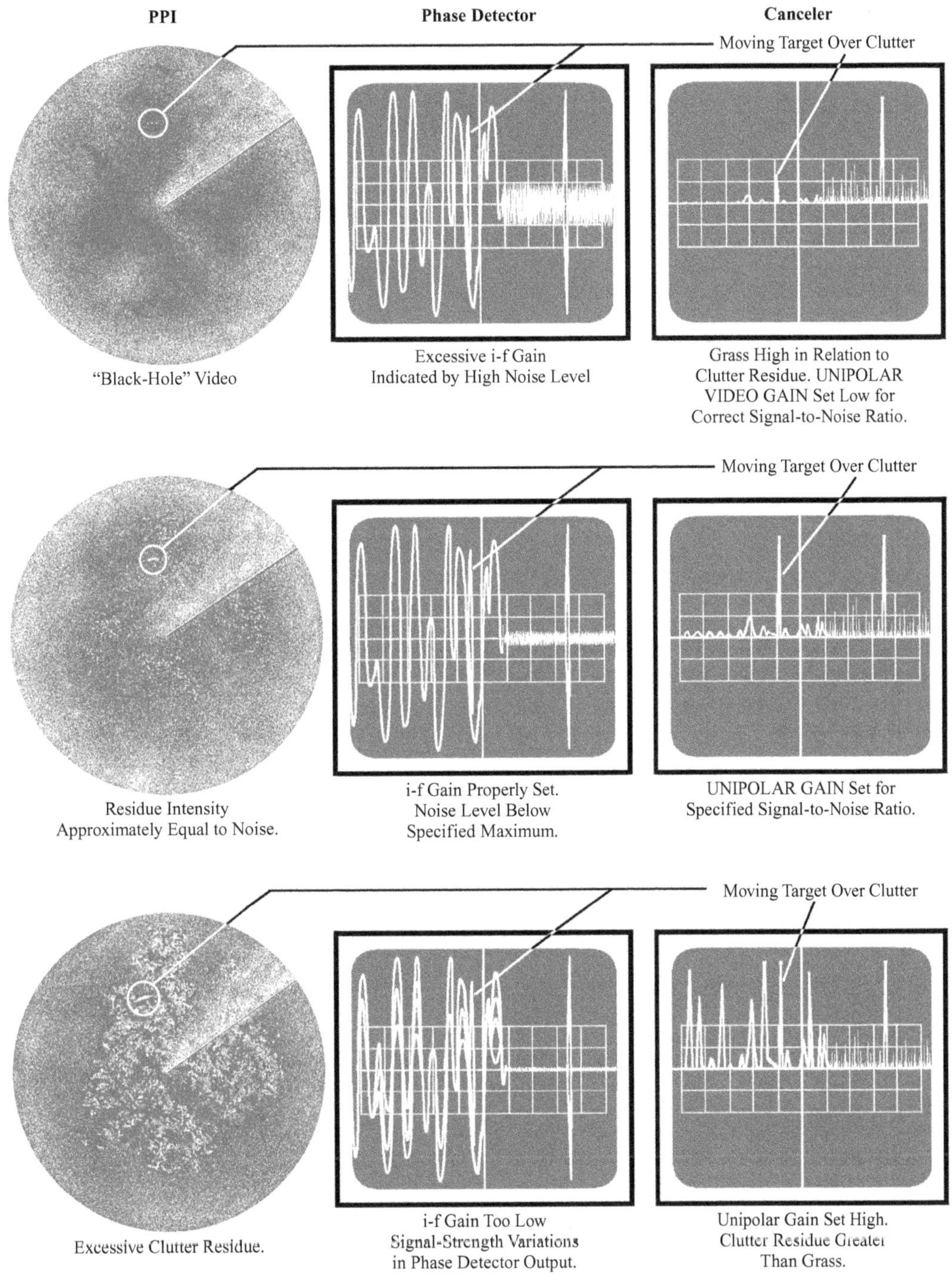

FIGURE 12–44

Payoff adjustment.

clutter, it, of course, is of no advantage, because the clutter residue will obscure many targets, defeating the purpose of MTI.

Maximum i-f Gain

In most latter-day systems, the manufacturer will specify a maximum amplitude for the noise level at the phase-detector output; if that level is excessive after the payoff adjustment, it is an indication that clutter cancellation performance is poor, and that it has been necessary to increase the i-f gain to conceal excessive residue.

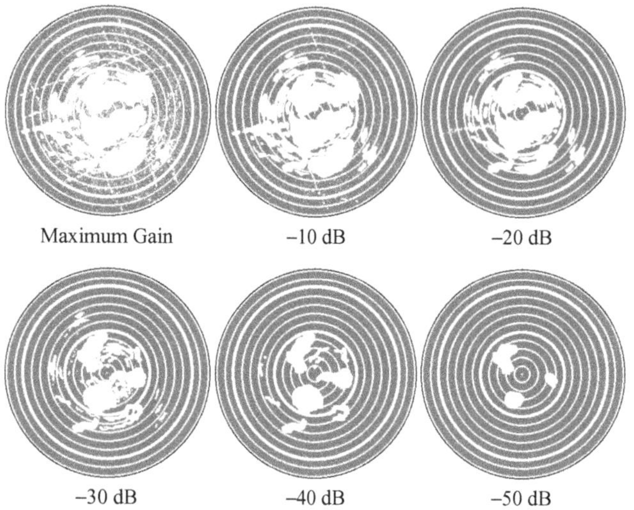

FIGURE 12–45

Clutter strength photos.

Clutter Strength Photographs

The technician will always be confronted with user complaints of system performance. A radar system is not perfect, and total radar coverage of an entire area will never be possible. In determining whether the system is performing adequately, knowledge of the local clutter pattern is necessary. One way to accomplish this is to establish a record of area clutter strength by taking a series of photographs of normal video displayed on a ppi (see Figure 12-45); each photograph in the series is taken with additional attenuation to the normal receiver input.

Subclutter-Visibility Measurement

The ability of the MTI system to detect a moving target over clutter can be quantitatively expressed. The procedure for accomplishing this is one of the most difficult, and demanding of skill, of all those an air traffic control radar technician must do. Experience and practice are essential to obtain consistent measurements.

In the 1950s and 1960s, subclutter-visibility measurements were made in a procedure that required the antenna to be stopped, while pointed at an established piece of clutter. A signal-generator test target of a strength equal to the clutter would be measured, placed over the clutter, and then attenuated until it was twice the clutter residue at the canceler output. Because the signal generator is not cohered to the radar system, it produces a random-speed moving target. The difference between the beginning signal-generator reference level and the attenuated setting at twice the clutter residue level was the subclutter-visibility (SCV) measurement. That "live clutter" procedure is rarely performed, now, because user demands on air traffic control radar systems will not tolerate the time required. However, a very similar procedure, described in the following paragraph, is now performed on the off-line channel using the echo box to produce a simulated block of clutter.

Echo Box SCV Procedure

Perform the following procedure on the off-line, dummy-loaded radar channel (see Figure 12-46):

1. Ascertain that the payoff adjustment has been set while the channel was on line, and then place it in dummy load. Before turning the transmitter on, place the synchronizer to INTERNAL, and then warn the users that "running rabbits" may appear on their displays.
2. Connect the two inputs of a dual-trace oscilloscope to the normal and MTI i-f amplifier outputs. Synchronize the oscilloscope with a radar fp trigger.
3. Connect an echo box and rf signal generator to the INCIDENT directional coupler port of the radar channel. Depending upon the system, this may require a tee connector or 3-dB power divider; some systems may provide a second directional coupler.
4. Calibrate and tune the echo box and signal generator. Tune the echo box for a few bipolar video cycles, as shown in Figure 12-46A. Adjust the signal-generator delay to place the test signal "in the clear," also as shown in Figure 12-46A.
5. For an accurate measurement, the test signal must be exactly 10 dB into the MTI i-f limit; this determination is the most likely source of error and requires care. View the bipolar signal-generator test target and then attenuate the target until it is clearly well below limit. Carefully increase the signal until the bipolar peaks just reach the same amplitude as the echo-box peaks. Note the signal-generator attenuation and

then remove an additional 10 dB. Jot down the attenuator reading, hereafter called S_{ref}.

6. The normal i-f output will be used for signal power references. At this point in the procedure, the signal-generator test signal may be limited on the normal video. Should this be the case, reduce the normal i-f gain until the signal is considerably less than the echo box ringing, as shown in Figure 12-46A. Note the point on the trail edge of the echo box ringtime at which the amplitude is equal to the signal-generator test pulse, as shown by the dotted line in Figure 12-46A, and then adjust the signal-generator delay to place the test pulse at that point on the echo box ringtime, as shown in Figure 12-46B. The test signal is now 10 dB above the MTI i-f limit, and over simulated clutter of equal amplitude.
7. To ensure that the test signal will not be partially distorted, accurate measurement also requires that the test signal be at an optimum phase-detector output; this is achieved by ensuring that it is riding upon a 0-V clutter output, as shown in Figure 12-46B. If not, slightly adjust the echo box.
8. Connect one oscilloscope channel to the canceler unipolar video output. Note the echo-box clutter residue amplitude and then attenuate the test signal until it is twice the clutter residue. The new signal-generator reading is S_{2res}. The SCV is typically around 30 dB and is calculated as

$$SCV = S_{ref\,dbm} - S_{2res\,dbm}.$$

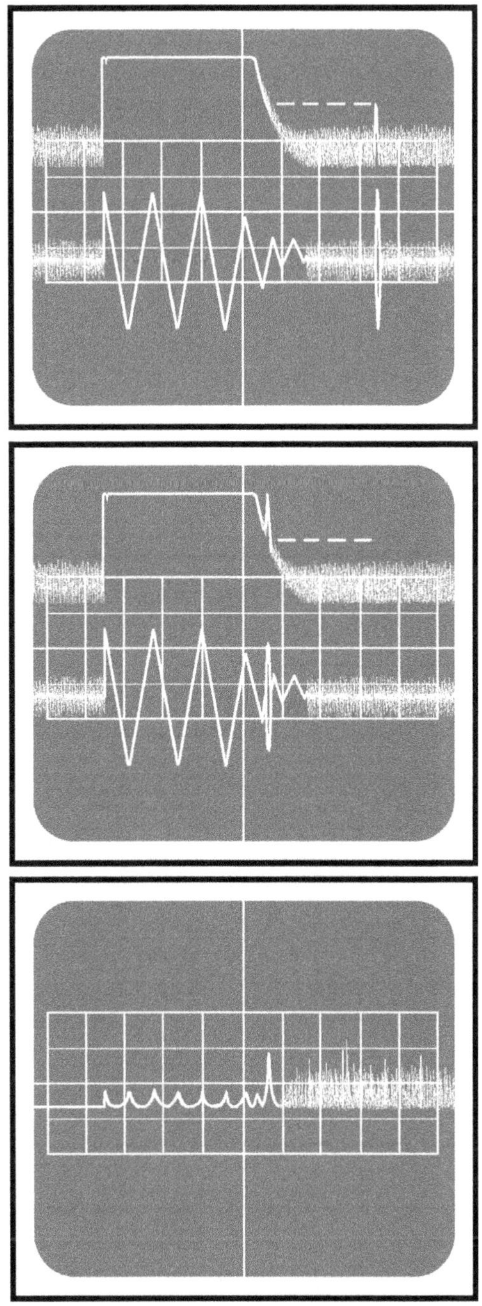

Finding the point on the ringtime where the echo box signal and target are of equal amplitude. (a)

Moving the test target ro the reference point. (b)

Reducing test signal to twice the amplitude of the clutter residue. (c)

FIGURE 12–46

Subclutter-visibility measurement.

The greater the difference in S_{ref} and S_{2res}, the higher the SCV, and the more likely it is that the MTI system can detect a target over clutter. If the system is made to limit the i-f more severely, or if the clutter residue increases for any reason, the SCV will decrease. Increased canceler feedback will increase residue. With the ASR-4, ASR-5, ASR-6, ASR-7, and ASR-8, Texas Instruments labeled the feedback modes in terms of SCV; from minimum to maximum feedback, the canceler modes were labeled 40 dB SCV, 35 dB SCV, 30 dB SCV, and 25 dB SCV.

Any frequency instabilities will make it necessary to increase the i-f gain, subsequently increasing limit severity and decreasing the SCV. Cancellation errors will increase residue and decrease SCV.

The Passive High-Beam Receiver System

The most common means to lower the effect of strong clutter is the use of a passive, high-beam feedhorn, to be used for the receive path at close ranges. This was discussed in Chapter 6.

Isolating Performance Derogation

An MTI system is far from perfect, and many of the user complaints may be related to normal phenomena, such as tangential effects, blind-phase conditions, subclutter-visibility and the effects of strong clutter, anomalous propagation, and blind or dim speeds. The technician's job is to scientifically determine the true source and then correct it, or explain the truth. This is a tremendous responsibility; if he dismisses a real trouble as a normal phenomenon, he is negligent; however if he attempts to fix a nonexistent trouble, he may waste a lot of time that could better be devoted to other work. The answer is to methodically and conclusively prove the existence or nonexistence of a problem. The following paragraphs will detail some approaches.

Poor Cancellation

First, check the payoff adjustment. If an increase to the MTI I-F GAIN seems to solve the problem, check the grass level at the phase-detector output; if it is abnormally high, above manufacturer's specifications, a performance deterioration is indicated. If you are working on a magnetron system, and an echo box is available, an echo-box cancellation ratio is a good place to start. A synthesis system may not provide for an echo box; the "Test Equipment Required" documentation may not specify it, and the directional coupler attenuation may be too great for echo-box excitation. Should an echo-box not be available, perform an i-f cancellation ratio.

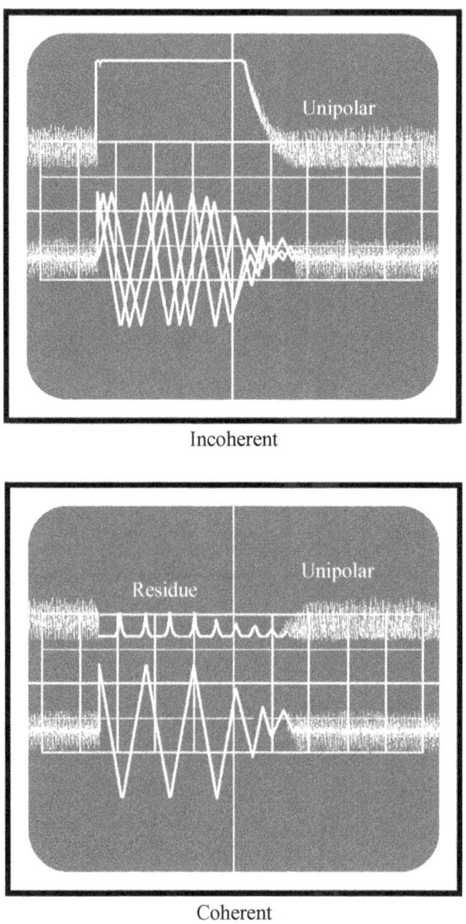

FIGURE 12–47

Echo box cancellation ratio.

Echo-Box Cancellation Ratio

Since the echo box is excited by the transmitter burst, an examination of the quality of cancellation from its ringing signal will perform a test of the entire system (see Figure 12-47):

1. Ascertain the payoff adjustment has been satisfactorily set while the channel was in operation.
2. Place the channel OFF LINE, since live video will obscure the measurement. Place the synchronizer mode to INTERNAL. Ascertain the user will expect to see "running rabbits," and then energize the transmitter.
3. Connect the echo box to the directional coupler INCIDENT connector.
4. View the echo-box residue in the canceler. It is more accurate to find a testpoint prior to output limiting, but using a limited testpoint will make the measurement appear poorer than actual, rather than better.
5. Make the system incoherent by whatever means are available. Disconnecting the coho gate or lock pulse inputs to the coho may achieve this.
6. Measure the amplitude of the echo-box residue, hereafter called U_{uncanc}. To be precise, do this at a point on the unlimited trail edge of the ringtime.
7. Restore coherence.

8. At the same point where U_{uncanc} was measured, measure the amplitude of the echo-box residue, hereafter called U_{canc}.
9. The cancellation ratio is

$$CR_{echobx} = 20\log\left(\frac{U_{uncanc}}{U_{canc}}\right).$$

10. If the echo-box cancellation ratio is within established specifications, the entire system is performing satisfactorily. If the ratio is satisfactory, and the phase-detector output had been excessively noisy, there may be a receiver front-end problem; this could be verified or dismissed by testing the normal receiver mds and/or noise figure. Should the front end prove to be satisfactory, a gain-bandpass test of the MTI i-f is warranted. If the echo-box cancellation ratio is unsatisfactory, proceed with a video cancellation ratio and/or i-f lock pulse cancellation ratio for further isolation.

Video Cancellation Ratio

If the system is digital, it is very unlikely, though not impossible, that the poor cancellation is being caused by the canceler.

Performing this measurement will verify that the canceler is operating satisfactorily. All the Texas Instruments systems, ASR-4, ASR-5, ASR-6, ASR-7, and ASR-8, utilize a cvtt, and use of this was previously explained in this chapter. It will be necessary to perform a video cancellation ratio on each canceler; to attempt it on both in cascade will provide too little residue. Depending upon the system, it may be necessary to use an external video pulse generator. In any case, a fixed positive or negative target should be attenuated at least 40 dBV (100 times) by a single canceler.

Echo-Box Ratio Poor, Video Cancellation Ratio is Satisfactory

If the echo-box cancellation ratio were unsatisfactory, and the video cancellation ratio were within tolerance, a frequency instability would be indicated. In a magnetron system, the most likely source is a loss of coherence, caused by i-f lock pulse failure. The afc and stalo can also cause cancellation failure; this would often be indicated by frequent automatic stalo tuning, exhibited by "wedges," "spokes," or "strobes" of residue. The latest state of the art in stalos for magnetron systems uses a digital synthesizer, and is far less likely to fail in this manner; this is now used as a modification in the ASR-7.

i-f Cancellation Ratio

This procedure is done in a similar manner as the echo-box cancellation ratio, except that the i-f lock test pulses are used (see Figure 12-48). If the system is a magnetron type, and contains the lock test pulse generator, proceed as follows:

1. Synchronizer to INTERNAL. Advise the user that "running rabbits" may appear, before energizing the transmitter.
2. View the canceler output unipolar video, preferably prior to limiting.
3. Place the LTPG MODE switch to "1" (lock test pulses generated from coho lock pulse).
4. Interrupt the coherence by removing the lock pulse input to the coho.
5. Measure the lock test pulse residue U_{uncanc} at the canceler output.
6. Restore coherence.
7. Measure the lock test pulse residue U_{canc} at the canceler output.
8. The i-f cancellation ratio is
9. $CR_{i\text{-}f} = 20\log\left(\frac{U_{uncanc}}{U_{canc}}\right)$. Perform the same measurement with the LTPG MODE switch in position "2" (lock test pulses generated by test oscillator, and coho locked by test oscillator).

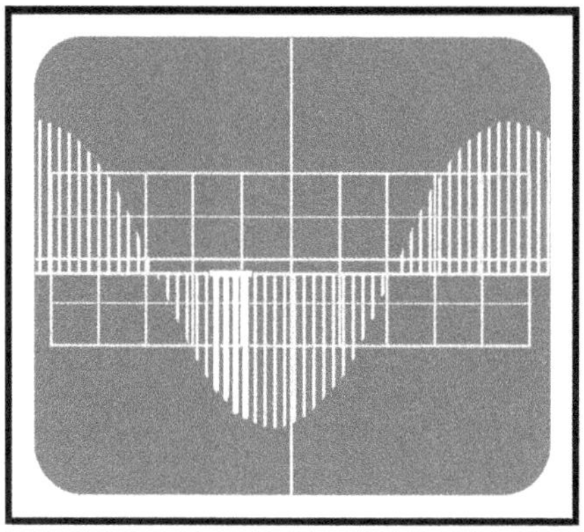

Bipolar Lock Test Pulses

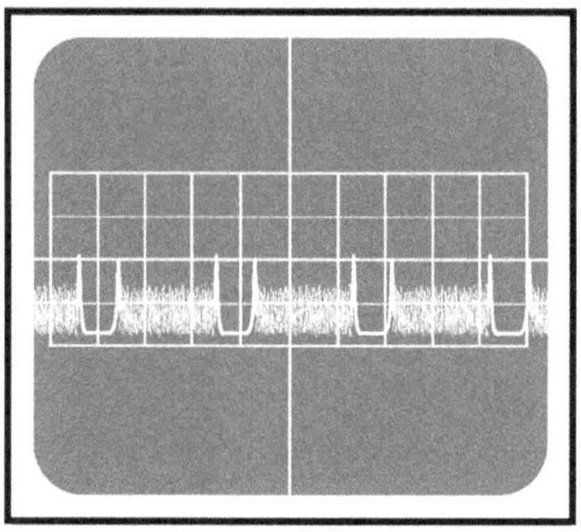

Canceled Lock Test Pulse Residue

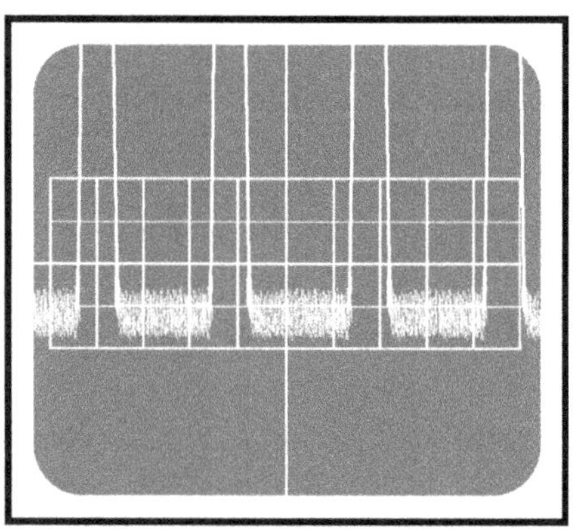

Canceled Lock Test Pulses
When Incoherent

FIGURE 12–48
i-f cancellation ratio.

If both tests yield a satisfactory cancellation ratio specified in the manufacturer's instruction book, the frequency-generation units are eliminated as a problem source. If test mode 2 is satisfactory, but test mode 1 is not, a coherence problem involving the stalo or lock pulse is indicated. If both test modes are unsatisfactory, and the video cancellation ratio is satisfactory, a coho instability is indicated.

If the system is a synthesis type, the only possible i-f cancellation test may be to measure the amplitude of the lock test pulse residue, to verify that it is within specifications described in the manufacturer's instruction book.

Causes of Cancellation Failures

Many of these have already been discussed, but this paragraph is to provide a convenient list of possibilities for a quick reference.

1. ***Trigger Generating Circuitry.*** Missing triggers, such as on alternate T_rs, will cause the bipolar video to be zero on alternate intervals, yielding high residue. This problem will be indicated by the meters on the transmitter, and by frequent transmitter faults.
2. ***Magnetron Misfiring.*** Also indicated by transmitter meters.
3. ***Transmitter High-Voltage Pulse Amplitude Instability.*** In the ASR-8, this is the most likely cause of cancellation residue. A "de-q'ing" circuit is intended to prevent pulse-amplitude variations, but may easily become improperly adjusted.
4. ***Analog Canceler Adjustments.*** These include the delay temporal adjustment, amplitude balance adjustment, or bandpass mismatch between the delayed and undelayed channels. The problem may be determined by the appearance of the cvtt residue. Spikes on the edges of the pulses may be caused by temporal or bandpass mismatches, and low-amplitude pulses will result from amplitude mismatches.
5. ***Klystron Tuning.*** Improper cavity tuning can cause parts of the output burst to "tear," or "jitter." Close attention to the spectrum and detected pulse are necessary.
6. ***Stalo Frequency Jitter.*** Highly unlikely in latter-day systems, more possible in tube-type stalos.
7. ***Coho Frequency Jitter.*** Even more unlikely, but nothing is impossible.
8. ***Excessive afc Correction, Magnetron System.*** This problem causes "wedges" or "spokes" of clutter residue. Most systems have been designed to correct the stalo frequency when the tuning error reaches 100 kHz; a maladjustment can reduce this tolerance, or a bad stalo tube can change frequency excessively.
9. ***Second-time Clutter.*** This can happen in either magnetron or synthesis systems. In a synthesis system, it may not be apparent in stagger-off operation. The problem may also be intermittent, and associated with beam bending (refraction), a form of anomalous propagation. Where staggered f_p is in use, the second-time targets for each T_r occur at different times in relation to T_0, and a multiple clutter display results.
10. ***Antenna Tilt.*** Excessive downward tilt will cause high clutter levels, poor detection of targets over clutter, and high residue from scanning effects. There have been cases in which the tilt adjustment worked downward while the system was in operation; be certain the tilt adjustment screw is unquestionably secure.

11. ***Anomalous Propagation and Refraction.*** The word "anomalous" means "inconsistent with what would normally be expected," and "propagation" refers to the movement of rf energy through space. Inconsistencies in the atmosphere may cause echoes, or beam bending. An unstable downward bend of the beam will cause unstable clutter echoes, which may not cancel. False echoes from atmospheric irregularities, called "angels," may also appear. At some facilities, anomalous propagation may be expected regularly at certain times of the day, particularly at sunrise or sunset.

Further Implications of the Velocity Response

Figure 12-49 illustrates the sin(Δϕ/2) velocity response shapes for two f_ps with a constant f_{xmtr}. This is very informative in many respects, and the following conclusions may be drawn from it.

Doppler Ambiguity

If the canceler output amplitude were to be used as a representation of target velocity, any target velocity above V_{opt} would produce the same output as a target at another velocity below V_{opt}. For instance, the canceler has no provision to distinguish between a Δϕ of 190°, and a Δϕ of 170°. This is insignificant in MTI radars, but becomes significant in Doppler weather radar, specifically used for velocity measurements. Where velocity measurements are the objective, the radar will operate at a very high f_p. However, increasing the f_p decreases the T_r, subsequently increasing the potential for range ambiguity, more commonly called "second-time-around" conditions, where echoes from a transmitter burst arrive at the receiver after the transmitter has been pulsed a second time. These conditions are the source of two major radar class definitions found in Merrill Skolnik's *Radar Handbook*. A Pulsed Doppler Radar is principally designed to measure velocity and to avoid Doppler ambiguities; an MTI Radar is principally designed to reduce clutter, and to avoid range ambiguities. Doppler weather radars use a combination of high and low f_ps to, respectively, (1) determine velocity and (2) remove range-ambiguous information.

Increasing the f_p Decreases Antenna Scanning Residue

It is obvious from the two velocity response shapes in Figure 12-49 that a lower f_p will increase clutter residue. From another viewpoint, it should be equally clear that Δϕ, and clutter residue, will be reduced if the target is allowed less time to move between "hits." Similarly, if the antenna speed is reduced, the ΔD of a fixed target with an oblique surface in respect to the radar antenna beam will also decrease. Still further, a slower antenna rotation decreases the phase-detector ΔE for unlimited clutter, and more of it will be canceled. A slow antenna rotation and high f_p will provide very low clutter residue and a high hits-per-scan (N_s) value. However, the low antenna speed will reduce the display "refresh" rate, important to terminal air traffic control safety, and the high f_p will create range ambiguities. One would find that radars near mountainous areas, particularly in the western continental United States, may have been installed with lower f_ps than in other areas. One might also find that some systems, particularly older ones, offer a choice of antenna speed-reduction gear ratios.

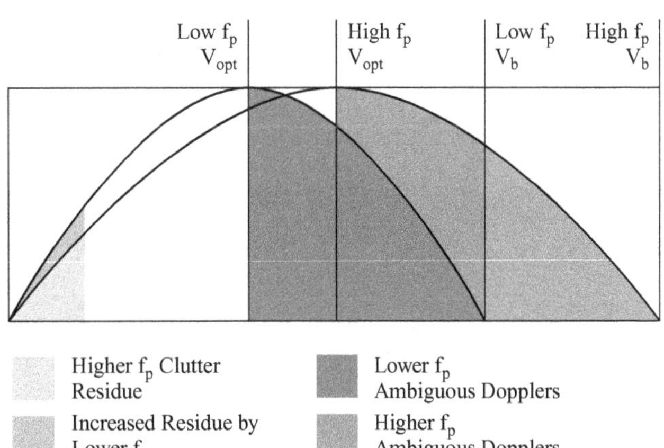

FIGURE 12–49

Velocity response for two f_ps.

Range Ambiguities, Magnetron versus Synthesis

Second-time clutter echoes are particularly troublesome in magnetron systems, because the second-time target will not be coherent with the new magnetron burst, and cannot be canceled. Second-time clutter can be canceled in synthesis systems, provided that the f_p has not been staggered to cause temporal mismatches between the delayed and undelayed canceler videos. Second-time clutter with staggered f_ps is particularly bothersome, since the clutter will appear in as many locations as there are T_rs. As a fix for the noncancellation of second-time clutter in staggered synthesis systems, the ASR-8 contains a RAG Stagger circuit to disable the staggered f_p at programmed azimuths. However, the cancellation of second-time clutter does not remove it as a problem because of subclutter-visibility considerations. For an example of the potential severity of second-time, synthesis-system, subclutter-visibility problems, consider a condition under which aircraft targets over flat farm fields disappear from the display because of a large mountain 75 nmi (86 US statute miles) from the radar.

Different f_ps Provide Different Blind Velocities

Figure 12-49 illustrates the difference in blind velocities for two f_ps, and it should also be obvious in the V_b equation:

$$V_{b1_{\text{nmi/h}}} = \frac{(291.375 \times 10^6) f_p}{f_{\text{xmtr}}}.$$

Staggered f_ps and Canceler Velocity Response

When staggered f_ps are employed, the real velocity response becomes a summation of the velocity responses of the reciprocals of $1/T_r$ of all the T_rs, as if there were multiple f_ps. Actually, for most radar systems, the average f_p for staggered operation is usually the same as the f_p for unstaggered operation.

The combination of velocity response shapes for different f_ps yields a resultant, as illustrated in Figure 12-50. This particular velocity response shape resembles the one for the ASR-4, -5, and -6, which uses only two T_rs. On alternate T_rs, the modulator trigger is delayed, so there is a long and short T_r, and a low and high f_p. On overlapping the two velocity response shapes, it becomes evident that there will be some point in the response where the low and high f_p blind speeds are equal. For this reason, earlier FAA courses once made a point that staggered operation did not eliminate blind speed, but only raised it. This is really "de minimis," as the first blind speed notch in staggered operation is far above the speeds of all but the very fastest military aircraft, and even those do not fly at such speeds inside the continental United States.

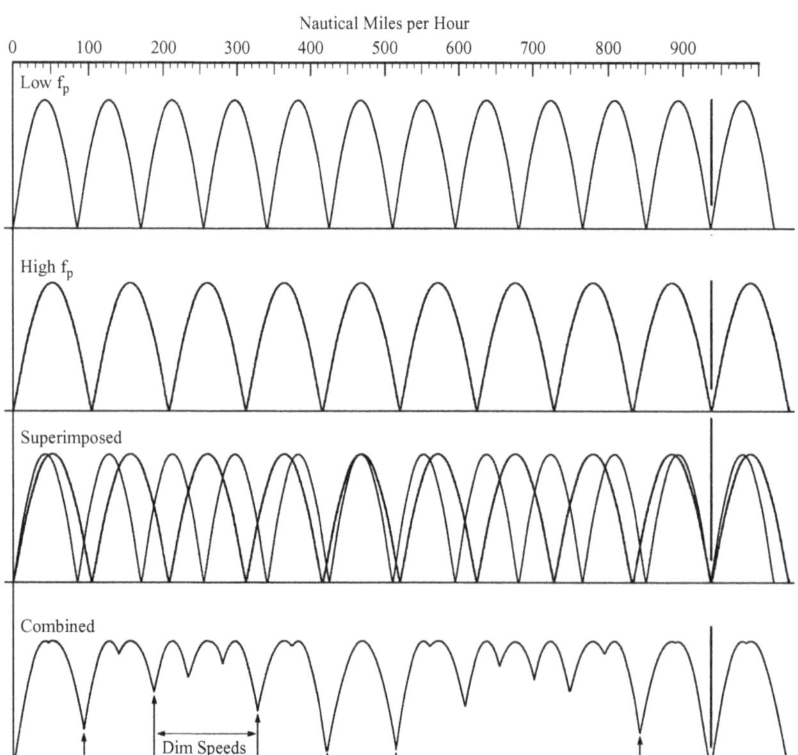

FIGURE 12-50

Effects on the velocity response shape with staggered f_ps.

The staggered responses in Figure 12-50 reach the common blind speed where the low f_p has caused eleven repetitive shapes, and the high f_p has caused nine repetitive shapes. This 9:11 ratio is called the stagger ratio. Later systems employed more complicated ratios, involving several numbers in the teens and twenties. Although these complex ratios raised the blind speed beyond the speeds of existing aircraft, the intent was to smooth the envelope of the resultant curve in an effort to lessen the dim speed effects shown in Figure 12-50.

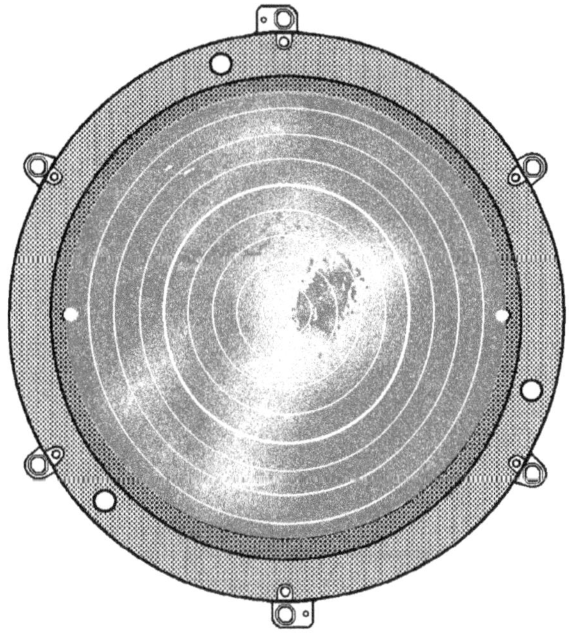

FIGURE 12-51

Weather over canceled clutter.

MTI and Weather

Radar echoes from weather are a collective result of all the raindrops or other forms of precipitation in that weather, and the Doppler shift of those echoes will be over a wide band of frequencies, dependent upon the wind velocities in the weather. How the MTI system responds to the weather depends upon the Doppler shift, and may vary all the way from black holes, caused by cancellation of a gentle rainfall, to an optimum response, for violent weather. Weather echoes can become much stronger than the echoes from either ground clutter or targets, because the precipitation may be occupying a larger portion of the beam. Subclutter-visibility effects may cause black holes in the weather display above strong clutter (see Figure 12-51). Latter-day systems contain digital log ftc to attenuate weather. This will be discussed in the Chapter 13.

The SSR/DMTI Modification for Early ARSR and FPS Series Radars

All MTI systems bear many similarities, and there is no need to discuss all the peculiarities and design nuances. The ASR-8 MTI circuitry is addressed in the next chapter because it was the largest deployment, and because many systems were sold to European and other governments, and may remain in use for years to come. However, this SSR/DMTI modification introduced a few new worthwhile improvements that are worthy of mention.

Current ("Soft") Limiting in the MTI i-f Amplifier

When heavy ground clutter in the i-f signals creates higher currents, a feedback circuit begins to reduce the signal into the phase detectors. This technique was also used in the ASR-9 mtd system. The objective is to improve subclutter visibility by raising the probability that a target echo may not be destroyed by i-f amplifier saturation.

Multiple Staggered f_p

This is not new, but the options have increased. Fifty-seven options of five-T_r sets have been provided. See Table 12-1. Any one of the five may be selected for unstaggered operation, or all five may be used in stagger. The average staggered blind speed is raised to such a high value that it is very unlikely to be reached by any aircraft. The first ***"dim speed (lowered response but not zero)"*** notches occur in the 2,000 nmi/h to 2,700 nmi/h region, depending on the transmitter frequency and selected T_r set. In unstaggered operation, the blind speed is in the general region of 70–100 nmi/h and all whole-number multiples.

Four-Pulse Canceler

All previous discussions have been of delay-line cancelers, in which a delayed bipolar video would be compared to the bipolar video from the phase detector. Using the logic in naming a canceler "three-pulse" or "four-pulse,"

TABLE 12-1 ARSR/FPS SSR/DMTI Modification, T_r Sets

Average number	H/L	Average prf (Hz)	Pulse rep. int. (µs)					S121 settings							
			PRI1	PRI2	PRI3	PRI4	PRI5	8	7	6	5	4	3	2	1
0	L	279.88	3,115	3,939	3,207	3,481	4,123	X	1	0	0	0	0	0	0
1	L	281.06	3,102	3,923	3,193	3,467	4,105	X	1	0	0	0	0	0	1
2	L	282.33	3,088	3,905	3,179	3,451	4,087	X	1	0	0	0	0	1	0
3	L	283.53	3,075	3,889	3,165	3,437	4,069	X	1	0	0	0	0	1	1
4	L	284.82	3,061	3,871	3,151	3,421	4,051	X	1	0	0	0	1	0	0
5	L	286.12	3,047	3,853	3,137	3,405	4,033	X	1	0	0	0	1	0	1
6	L	287.36	3,034	3,837	3,123	3,391	4,015	X	1	0	0	0	1	1	0
7	L	288.68	3,020	3,819	3,109	3,375	3,997	X	1	0	0	0	1	1	1
8	L	289.94	3,007	3,803	3,095	3,361	3,979	X	1	0	0	1	0	0	0
9	L	291.29	2,993	3,785	3,081	3,345	3,961	X	1	0	0	1	0	0	1
10	L	292.65	2,979	3,767	3,067	3,329	3,943	X	1	0	0	1	0	1	0
11	L	293.94	2,966	3,751	3,053	3,315	3,925	X	1	0	0	1	0	1	1
12	L	295.33	2,952	3,733	3,039	3,299	3,907	X	1	0	0	1	1	0	0
13	L	296.65	2,939	3,717	3,025	3,285	3,889	X	1	0	0	1	1	0	1
14	L	298.06	2,925	3,699	3,011	3,269	3,871	X	1	0	0	1	1	1	0
15	L	299.49	2,911	3,681	2,997	3,253	3,853	X	1	0	0	1	1	1	1
16	L	300.84	2,898	3,665	2,983	3,239	3,835	X	1	0	1	0	0	0	0
17	L	302.30	2,884	3,647	2,969	3,223	3,817	X	1	0	1	0	0	0	1
18	L	303.67	2,871	3,631	2,955	3,209	3,799	X	1	0	1	0	0	1	0
19	L	305.16	2,857	3,613	2,941	3,193	3,781	X	1	0	1	0	0	1	1
20	L	306.65	2,843	3,595	2,927	3,177	3,763	X	1	0	1	0	1	0	0
21	L	308.07	2,830	3,579	2,913	3,163	3,745	X	1	0	1	0	1	0	1
22	L	309.60	2,816	3,561	2,899	3,147	3,727	X	1	0	1	0	1	1	0
23	L	311.04	2,803	3,545	2,885	3,133	3,709	X	1	0	1	0	1	1	1
24	L	312.60	2,789	3,527	2,871	3,117	3,691	X	1	0	1	1	0	0	0
25	L	314.17	2,775	3,509	2,857	3,101	3,673	X	1	0	1	1	0	0	1
26	L	315.66	2,762	3,493	2,843	3,087	3,655	X	1	0	1	1	0	1	0
27	L	317.26	2,748	3,475	2,829	3,071	3,637	X	1	0	1	1	0	1	1
28	L	318.78	2,735	3,459	2,815	3,057	3,619	X	1	0	1	1	1	0	0
29	L	320.41	2,721	3,441	2,801	3,041	3,601	X	1	0	1	1	1	0	1
30	L	322.06	2,707	3,423	2,787	3,025	3,585	X	1	0	1	1	1	1	0
31	L	323.62	2,694	3,407	2,773	3,011	3,565	X	1	0	1	1	1	1	1
32	L	325.31	2,680	3,389	2,759	2,995	3,547	X	1	1	0	0	0	0	0
33	L	326.90	2,667	3,373	2,745	2,981	3,529	X	1	1	0	0	0	0	1

Average number	H/L	Average prf (Hz)	PRI1	PRI2	PRI3	PRI4	PRI5	8	7	6	5	4	3	2	1
34	L	328.62	2,653	3,355	2,731	2,965	3,511	X	1	1	0	0	0	1	0
35	L	330.36	2,639	3,337	2,717	2,949	3,493	X	1	1	0	0	0	1	1
36	L	332.01	2,626	3,321	2,703	2,935	3,475	X	1	1	0	0	1	0	0
37	L	333.78	2,612	3,303	2,689	2,919	3,457	X	1	1	0	0	1	0	1
38	L	335.46	2,599	3,287	2,675	2,905	3,439	X	1	1	0	0	1	1	0
39	L	337.27	2,585	3,269	2,661	2,889	3,421	X	1	1	0	0	1	1	1
40	L	339.29	2,571	3,251	2,647	2,873	3,403	X	1	1	0	1	0	0	0
41	L	340.83	2,558	3,235	2,633	2,859	3,385	X	1	1	0	1	0	0	1
42	L	342.70	2,544	3,217	2,619	2,843	3,367	X	1	1	0	1	0	1	0
43	L	344.47	2,531	3,201	2,605	2,829	3,349	X	1	1	0	1	0	1	1
44	L	346.38	2,517	3,183	2,591	2,813	3,331	X	1	1	0	1	1	0	0
45	L	348.31	2,503	3,165	2,577	2,797	3,313	X	1	1	0	1	1	0	1
46	L	350.14	2,490	3,149	2,563	2,783	3,295	X	1	1	0	1	1	1	0
47	H	352.61	2,647	2,836	2,761	2,609	3,327	X	0	0	0	0	0	0	0
48	H	354.48	2,633	2,821	2,595	3,310	3,310	X	0	0	0	0	0	0	1
49	H	356.38	2,619	2,806	2,731	2,582	3,292	X	0	0	0	0	0	1	0
50	H	358.29	2,605	2,791	2,716	2,568	3,275	X	0	0	0	0	0	1	1
51	H	360.23	2,591	2,776	2,702	2,554	3,257	X	0	0	0	0	1	0	0
52	H	362.19	2,577	2,761	2,699	2,540	3,239	X	0	0	0	0	1	0	1
53	H	364.17	2,563	2,746	2,673	2,526	3,222	X	0	0	0	0	1	1	0
54	H	366.17	2,549	2,731	2,658	2,513	3,204	X	0	0	0	0	1	1	1
55	H	368.19	2,535	2,716	2,643	2,499	3,187	X	0	0	0	1	0	0	0
56	H	370.23	2,521	2,701	2,629	2,485	3,169	X	0	0	0	1	0	0	1

the original cancelers would have been called "two-pulse," because the information from two transmitter pulses is compared. The earliest cancelers relied on precise delay lines to compare incoming analog video to delayed analog video. In the ASR-7 and ASR-8, the delay lines were replaced by long shift registers, and the bipolar analog video was converted to parallel digital words for input to the shift register. Use of the clocked shift register insured accurate range-cell timing and reduced the possibility of temporal differences between delayed and undelayed data.

In the four-pulse canceler, the data from three stored previous intervals is compared to the incoming digitized data in an arithmetic operation, where a "weighted" summation of data from three intervals, previously stored in three RAMs, is compared to the incoming data. As the new data is first stored, a sequence word accompanies it to identify its order in the stagger sequence. Data from "even" intervals is made positive, and data from "odd" intervals is made negative in "weighting" the sum. Since there are five intervals, the data in the sum is "rolling" to provide velocity response shaping.

Review Questions

1. What is the significance of the numbers 291.375×10^6 and 80,937.5?
2. What is the relationship between aircraft groundspeed and radar Doppler shift?
3. What are the first and second detectors in an MTI receiver system?
4. Why is the "coho" so named?
5. When an aircraft is moving toward the radar antenna, will the echo frequency be increased, or decreased, or unchanged?
6. Can a Doppler shift be detected in a single pulsed radar echo from a target?
7. What is the phase relationship between the burst from the final transmitter tube and the burst radiated from the antenna?
8. What effect does the waveguide system have upon Doppler shift?
9. What is the phase relationship between the outbound radiated burst when it is (1) 1 mile from the antenna, and (2) 2 miles from the antenna?
10. What variables determine the phase relationship between an rf echo and the stalo?
11. What characteristic response does a single phase detector with sinusoidal inputs have?
12. In terms of the frequency domain, what is the difference in the Doppler shift for continuous wave and pulsed radars?
13. How much delay must be applied to bipolar video in a canceler?
14. What is the purpose of quadrature phase detectors in a canceler-type MTI system?
15. State the trigonometric identity behind the principle of quadrature phase detectors.
16. An aircraft is flying at 275 nmi/h on a course of 300° at a bearing of 100° from the radar. The transmitter frequency is 2,800 MHz and the f_p is 1 kHz. What is the f_D and $\Delta\phi$?
17. Why are cascade cancelers used?
18. What is the purpose of canceler feedback?
19. Explain how a high-altitude moving target, 20 miles from the radar, over a flat agricultural area, might disappear from the displays. The system is a synthesis type, with a high-power transmitter.
20. Why might S-band frequencies not be suitable for an ARSR?
21. Two sinusoidal c-w signals are applied to the two inputs of a phase detector, one at 30 MHz, and one at 30.001 MHz. Describe the output.
22. Twenty coherent echoes at 30.001 MHz are applied to a phase detector with a 30.0000 MHz coho. Describe the output.
23. Describe how coherence is obtained in (a) a synthesis system and (b) a magnetron system.
24. How does a stationary MTI reflector provide a visible target?
25. Name two signals in a magnetron MTI system that do not appear in a synthesis system.
26. In a tube-type magnetron MTI system, MTI targets occasionally "break up" at unpredictable locations on the display. All MTI performance tests are satisfactory. What might be the cause?
27. What effect would increasing the f_p have on residue?
28. What effect would increasing the antenna rotation speed have on residue?
29. A DC-3 is flying at a radial velocity that provides a Doppler shift equal to the f_p. What may you expect?
30. What establishes the signal-to-grass ratio of the MTI video used for display?
31. The echo-box cancellation is poor, but the video cancellation is satisfactory. What might this indicate?
32. To know where strong clutter exists, a radar facility can maintain________.
33. Why should the use of MTI in clear areas be discouraged?
34. A magnetron system is exhibiting wedges of clutter residue. What is the most likely cause?

35. Excessive MTI i-f gain will result in ________.
36. Inadequate MTI i-f gain will result in ________.
37. The "payoff" controls have been properly adjusted when ________.
38. A synthesis system, employing a klystron transmitter, and digital MTI processor, is exhibiting clutter residue. What is the most likely source?
39. Some clutter residue will always be present, and is caused by ________.
40. The technique of using normal video to switch MTI video in or out is called ________.
41. In an SCV measurement, for what purpose normal video is used?
42. In an SCV measurement, the rf signal generator creates a ________.
43. In an SCV measurement, the signal-generator target is first set to ________.
44. What will happen to the subclutter visibility if canceler feedback is increased?
45. What equipment decreases the loss of targets over clutter?
46. A radar site is troubled with clutter residue on the right-hand side of the ppi each morning. What might be the cause?
47. State the principle of staggered f_p and its effects on MTI blind speed.
48. Why are multiple T_rs used in stagger?
49. The subclutter-visibility measurement places a quantitative value on ________.
50. What might be the effects of (a) a gentle rainfall and (b) a thunderstorm on an MTI system?
51. What are radar "angels"?

Answers to Review Questions

1. What is the significance of the numbers 291.375 × 10^6 and 80,937.5?

   291.375 × 10^6 is the two-way speed of light in nautical mph.

   80,937.5 is the two-way speed of light in nautical miles per second.

2. What is the relationship between aircraft groundspeed and radar Doppler shift?

   The groundspeed will create a radial velocity, which will, in turn, create a Doppler shift.

$$f_{D_{radar}} = \frac{V_{ground_{nmi/h}}[\cos(\angle_{target} - \angle_{course} + 180°)]}{291.375 \times 10^6}.$$

3. What are the first and second detectors in an MTI receiver system?

   The first detector is the signal mixer, and the second detector is the phase detector.

4. Why is the "coho" so named?

   It is an acronym for "coherent oscillator," given the name because it was used to provide a phase reference to the magnetron burst.

5. When an aircraft is moving toward the radar antenna, will the echo frequency be increased, or decreased, or unchanged?

   Increased, because of the Doppler shift.

6. Can a Doppler shift be detected in a single pulsed radar echo from a target?

   Not with conventional MTI equipment. Assume a 1-μs transmitter pulse, and a Doppler shift of 1 kHz for a 100 nmi/h V_r. In 1 μs, there would be 30 cycles of 30 MHz, and .001 cycles of Doppler shift.

   Repetition is necessary to detect Doppler shift.

7. What is the phase relationship between the burst from the final transmitter tube and the burst radiated from the antenna?

   There is a phase shift caused by the difference between waveguide group and phase propagations.

8. What effect does the waveguide system have upon Doppler shift?

   None, because waveguide phase delay is constant.

9. What is the phase relationship between the outbound radiated burst when it is (1) 1 mile from the antenna, and (2) 2 miles from the antenna?

   There is no difference in the phases at the two points; once the burst is radiated, the pulse package remains unchanged.

10. What variables determine the phase relationship between an rf echo and the stalo?

    Instantaneous target range and Doppler shift during the receipt of repetitive echoes.

11. What characteristic response does a single phase detector with sinusoidal inputs have?

    The cosine of the phase difference of the two inputs.

12. In terms of the frequency domain, what is the difference in the Doppler shift for continuous-wave and pulsed radars?

    There is no difference.

13. How much delay must be applied to bipolar video in a canceler?

    Precisely one T_r.

14. What is the purpose of quadrature phase detectors in a canceler-type MTI system?

    To eliminate blind-phase conditions.

15. State the trigonometric identity behind the principle of quadrature phase detectors.

    $\sin^2\theta + \cos^2\theta = 1$.

16. An aircraft is flying at 275 nmi/h on a course of 300° at a bearing of 100° from the radar. The transmitter frequency is 2,800 MHz, and the f_p is 1 kHz. What is the f_D and $\Delta\phi$?

$$f_D = \frac{V_{\text{ground}_{\text{nmi/h}}}[\cos(\angle_{\text{target}} - \angle_{\text{course}} + 180°)]}{291.375 \times 10^6} = 2483.27\,\text{Hz}$$

$$\Delta\phi = \frac{360 f_D}{f_p} = 893.9° = 174°.$$

17. Why are cascade cancelers used?
To decrease the response to antenna scanning clutter.
18. What is the purpose of canceler feedback?
To improve upon the sin²(Δφ/2) cascade response.
19. Explain how high-altitude moving targets, 20 miles from the radar, over a flat agricultural area, might consistently disappear from the displays.
The system is a synthesis type, with a high-power transmitter. Subclutter visibility by a second-time mountain.
20. Why might S-band frequencies not be suitable for an ARSR?
Too many blind and optimum velocities; excessive response to low-Doppler targets.
21. Two sinusoidal c-w signals are applied to the two inputs of a phase detector, one at 30 MHz, and one at 30.001 MHz. Describe the output.
A 1-kHz cosine wave.
22. Twenty coherent echoes at 30.001 MHz are applied to a phase detector with a 30.0000 MHz coho. Describe the output.
Twenty pulses, varying at a 1-kHz rate.
23. Describe how coherence is obtained in (a) a synthesis system, and (b) a magnetron system.
The synthesis system is inherently coherent; the magnetron system is made coherent when the i-f lock pulse phase-locks the coho.
24. How does a stationary MTI reflector provide a visible target?
It appears as an optimum speed target, because the echo changes by λ/4 each T_r.
25. Name two signals in a magnetron MTI system that do not appear in a synthesis system.
The i-f lock pulse and coho gate.
26. In a tube-type magnetron MTI system, MTI targets occasionally "break up" at unpredictable locations on the display. All MTI performance tests are satisfactory. What might be a cause?
This is a classic indication of blind phase effects; because it is a tube-type system, it probably has only one phase detector.
27. What effect would increasing the f_p have on residue?
Decrease, because of a decrease in Δφ.
28. What effect would increasing the antenna rotation speed have on residue?
Increase, because of increased Doppler shift.
29. A DC-3 is flying at a radial velocity that provides a Doppler shift equal to the f_p. What may you expect?
A jet might disappear because of the tangential effect, but the propellers on the DC-3 might well provide a good MTI target.
30. What establishes the signal-to-grass ratio of the MTI video used for display?
The postcancellation gain-before-limit control.
31. The echo-box cancellation is poor, but the video cancellation is satisfactory. What might this indicate?
There is some sort of a frequency instability, and it can be further isolated by use of the LTPG.
32. To know where strong clutter exists, a radar facility can maintain ***a series of clutter photographs***.

33. Why should the use of MTI in clear areas be discouraged?

    MTI targets can be lost to blind speeds, blind phases, and tangential effects. Further, the MTI receiver is less sensitive than the normal receiver.

34. A magnetron system is exhibiting wedges of clutter residue. What is the most likely cause?

    Excessive stalo correction by afc.

35. Excessive MTI i-f gain will result in ***black-hole video and poor visibility over clutter.***
36. Inadequate MTI i-f gain will result in ***excessive clutter residue.***
37. The "payoff" controls have been properly adjusted when ***the clutter residue resembles the grass in brilliance and texture.***
38. A synthesis system, employing a klystron transmitter, and digital MTI processor, is exhibiting clutter residue. What is the most likely source?

    The transmitter.

39. Some clutter residue will always be present, and is caused by ***antenna scan motion.***
40. The technique of using normal video to switch MTI video in or out is called ***clutter gating.***
41. In an SCV measurement, normal video is used for what purpose?

    A means to indicate relative signal power.

42. In an SCV measurement, the ***rf signal generator creates a random-speed target.***
43. In an SCV measurement, the signal-generator target is first set to ***10 dB above the MTI i-f limit.***
44. What will happen to the subclutter visibility if canceler feedback is increased?

    The increased residue will decrease the SCV.

45. What equipment decreases the loss of targets over clutter?

    The passive high-beam receiver.

46. A radar site is troubled with clutter residue on the right-hand side of the ppi each morning. What might be the cause?

    Refraction, caused by the sunrise.

47. State the principle of staggered f_p and its effects on MTI blind speed.

    The Doppler for blind speed targets is unchanged, because it is in the frequency domain, but the stagger alters the time-domain $\Delta\phi$.

48. Why are multiple T_rs used in stagger?

    To reduce "dim speed" effects.

49. The subclutter-visibility measurement places a quantitative value on ***the ability of the system to detect a target over clutter.***
50. What might be the effects of (a) a gentle rainfall and (b) a thunderstorm on an MTI system?

    The gentle rainfall may provide Dopplers within the velocity response notch, causing "black holes." The more violent weather will show up as a large, strong, moving target.

51. What are radar "angels"?

    False targets caused by reflections from atmospheric discontinuities.

CHAPTER 13

An mti Processor

Introduction

Moving target indicator (mti) theory was explained in detail in the preceding chapter. For practical application, the technician needs an exposure to an actual system. All systems bear great resemblances to each other, and a good knowledge of one makes all others understandable. The largest deployment of latter-day FAA canceler-type mti systems was the ASR-8, and that system will be used as a basis for the sample system to be described in this chapter. This sample system will differ somewhat from the ASR-8 in that auxiliary, or accessory, circuits, not essential to the processing of the target data, have been omitted.

Timing

Radar synchronizers have been addressed in preceding chapters. Figure 13-1 is an abbreviated block diagram of part of the processor and the associated timing circuitry. The system to be described here is a digital type, employing a synthesis transmitter. The basic timing source is a 30-MHz coho, and the range-cell rate is 1/14 the coho, which is 2.143 MHz; this creates range cells of 0.467 μs, 0.0378 radar nautical miles, 230 radar feet. Within each range cell, 133.3 ns are used to sample the bipolar video in the analog-to-digital converters, hereafter called ***quantizers***. There will be 1,665 range cells in each T_r; the number is dictated by the size of storage-shift register integrated-circuit packages; the odd number arises from the 1,665th stage, a final register in the canceler. The 1,665 range cells, times 0.467 μs, provides 777 μs of processing time, allowing 741.3 μs for 60 nmi of live video, and the remainder for injection of the composite video test target train.

Circuitry to the Canceler Input

The mti Video Conditioner

The outputs from the phase detector are bipolar, ranging from +2 V to −2 V on a 0-V baseline. The quantizer design requires that all inputs be above 0 V; to accomplish that, the phase-detector video is placed on a +2-V baseline in a circuit card called "the mti video conditioner" (see Figure 13-2). The name of the card is misleading, as most of the circuitry on the card is used for test functions. One circuit generates the ***composite video test target train (cvtt)***. In normal operation, the "steering and switching circuits" enable the bipolar video during the 60 miles of "live time," and then enables the cvtt during ***built-in-test-equipment (BITE)*** time.

The Swept Audio Generator

The mti video conditioner can also be operated in test configurations, provided the channel has been placed in the MAINTENANCE mode. One of these is a swept audio, ranging from 300 Hz to 3 kHz, for the purpose of velocity-response-shape testing. An r-c circuit provides an approximate 90° phase shift to a second swept audio output, so that quadrature Dopplers are simulated for the two channels *I* and *Q*.

The Ramp Generator

Another test signal, provided purely for digital testing purposes, is called a ***ramp*** (see Figure 13-3). In earlier times, it would have been called a sawtooth. The intent of the ramp is to substitute the input from the phase

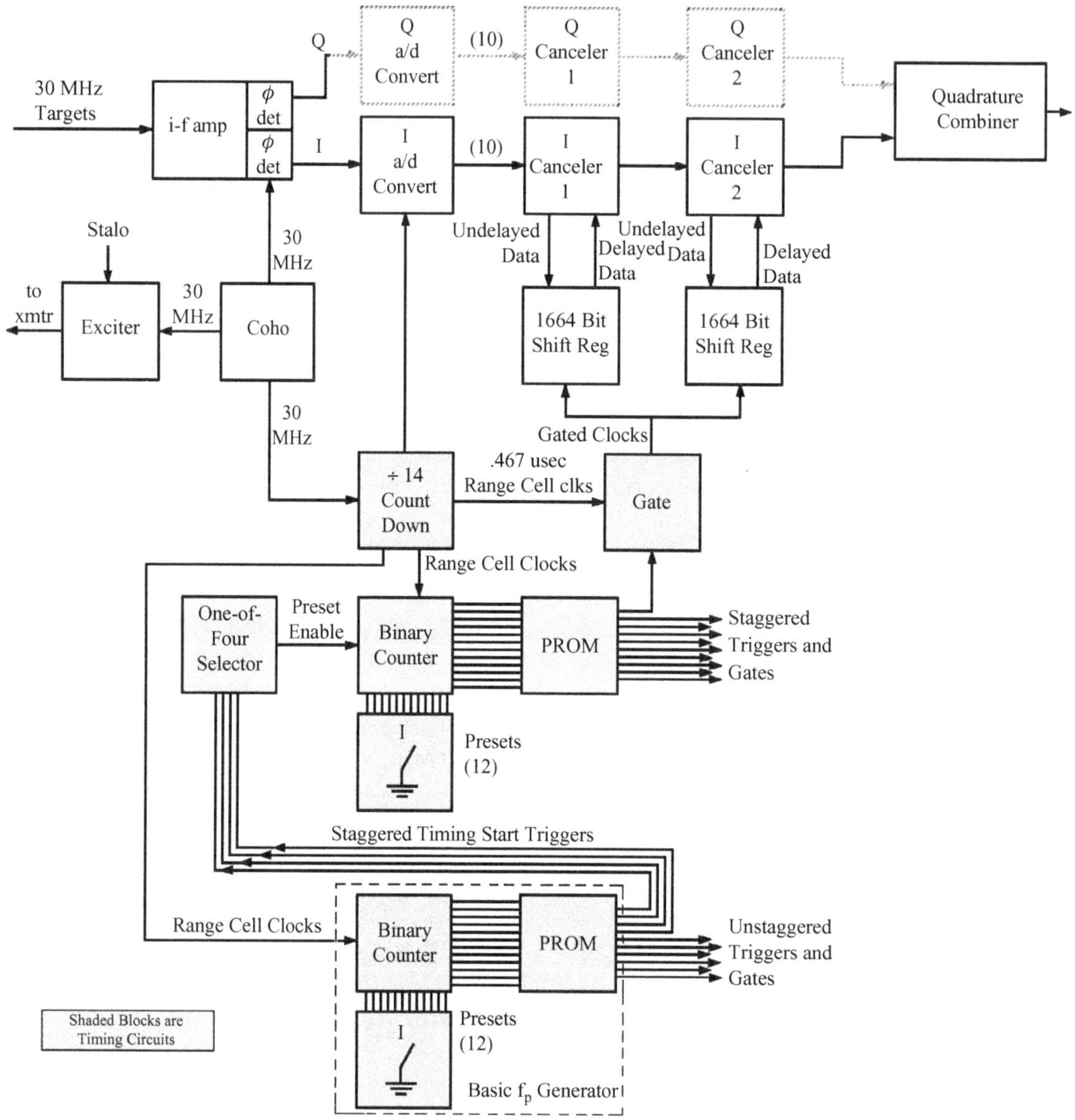

FIGURE 13–1

Synchronizer circuits.

detector, with a changing voltage, in which the rate of change will occur slowly enough to ensure that all possible binary number combinations will occur at the output of the quantizers in a single T_r; there is a 150% probability that each number will occur. Two types of ramp generation are available to the technician. In one mode, the ramp occurs once each T_r, and the digital resultant should be canceled in the cancelers since it simulates a maximum variety of fixed targets. This mode is, accordingly, called the ***fixed ramp.*** In another mode, the ramp is generated on alternate T_rs and should not cancel; this mode is called ***moving ramp.***

The Quantizers

Figure 13-3 is a simplified block diagram of the ***primary quantizer***. The intent of the A-D conversion is to create 10 parallel digital bits. Because of the range-cell rate, the conversion must be very fast, dictating the use of ***simultaneous*** conversion, also called ***flash*** conversion. In this type of conversion, all bits are developed at once, in contrast to another, slower, ***sequential*** type, in which the bits are developed in a one-at-a-time fashion.

At the time the ASR-8 was first manufactured, the hardware state of the art was such that simultaneous conversion did not offer the desired 10-bit voltage resolution. To achieve the resolution, a method called ***dual***

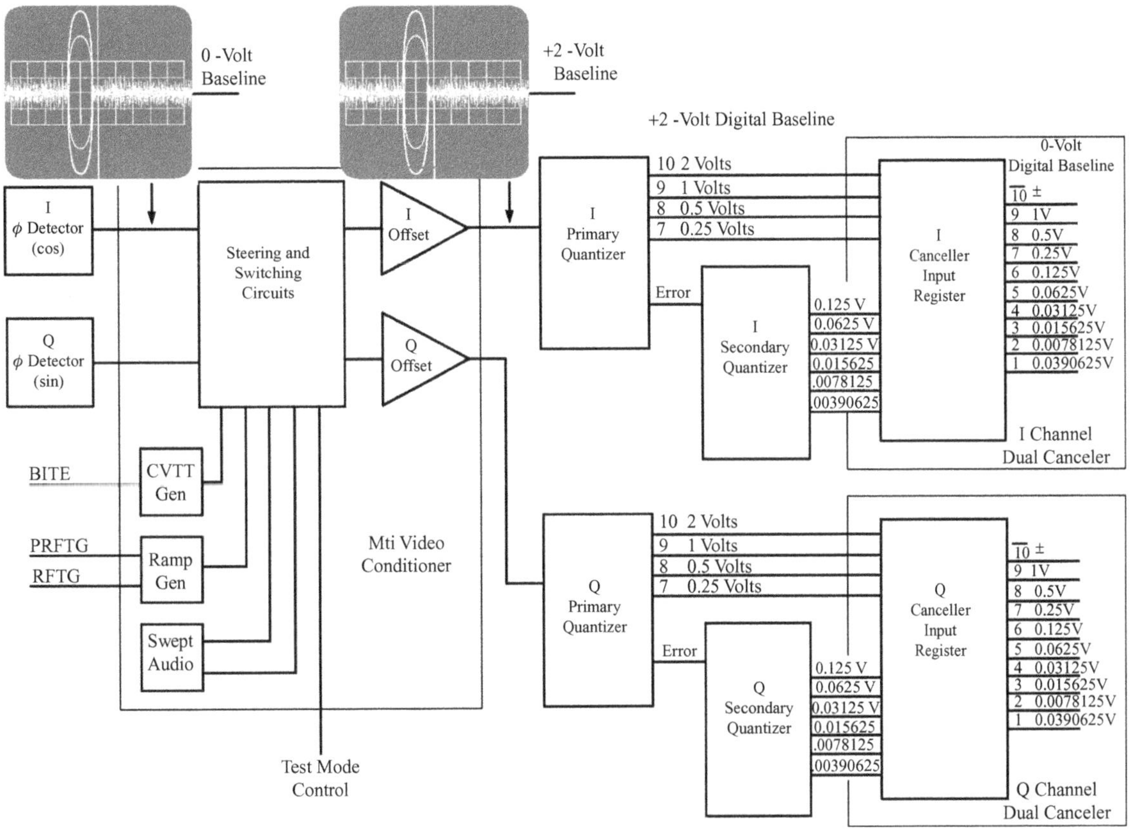

FIGURE 13–2

Digital "precanceler" circuits.

approximation, or ***flash conversion with subranging***, was employed. In this type of system, two simultaneous converters are used in a sequential fashion. In the primary quantizer illustrated in Figure 13-3, the four most-significant bits are generated first. Those bits are then converted back to analog voltage within the quantizer card and then compared to the input sample. Any difference between the two is an ***error voltage*** and represents a voltage to be more finely resolved by a ***secondary quantizer***, as shown in Figure 13-2.

The original primary quantizer input range is 4 V. Therefore, the msb weight is 2 V; the next msb, 1 V; the third msb, 0.5 V; and the lsb, 0.25 V. As the lsb weight is 0.25 V, the maximum error that can be sensed in the analog comparator is 0.24999999 V, essentially 0.25 V. By multiplying the error voltage by 16, the maximum error output to the secondary quantizer becomes –4 V, of opposite polarity, but otherwise, the same as the input to the primary quantizer. The secondary quantizer then develops an additional 6 bits from the error voltage input; in respect to the original primary quantizer input, the weight of those bits becomes 0.125 V, 0.0625 V, etc., down to an lsb, weighted at 0.00390625 V. Because of that lsb value, the quantizers are said to have a resolution of 4 mV.

In either quantizer, the actual voltage detection occurs in a several-step process, once each range cell. The following description uses the primary quantizer, illustrated in Figure 13-3, as an example. First, a 133-ns sample gate places a charge on a capacitor in a high-impedance circuit, which prevents rapid discharge after the gate has ended. The capacitor is, thus, quickly charged to the input voltage during the gate, and the charge then remains, until the next sample gate. Next, a 2-V threshold circuit will be energized if the input sample is 2 V or greater, and that output will become the most significant bit. The voltage sample is applied to an array of seven threshold circuits, each of which may be supplied with one of two precision voltages for comparison. The selection of a high or low comparison voltage is made by use of the 2-V comparator output. After sufficient time to set the comparator-pair output bits, an octal converter changes the 7 bits into a binary form.

The lower portion of Figure 13-3 illustrates the effects of an applied ramp to the primary quantizer. Each time the primary quantizer input ramp changes from a maximum of +4 V down to a minimum of 0 V, 16 error ramps

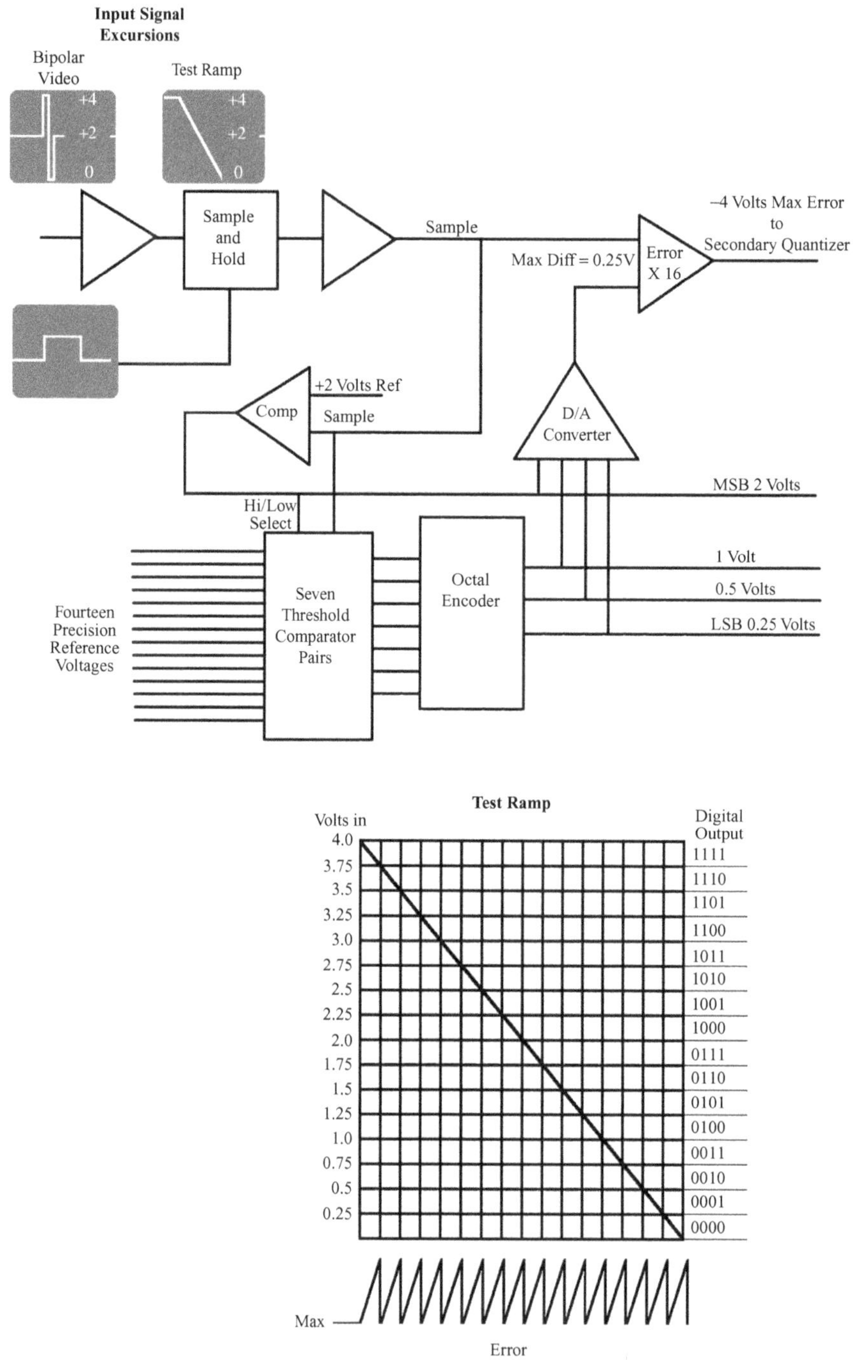

FIGURE 13–3

The primary quantizer.

are created for input to the secondary quantizer. Each of these error ramps represents a 0- to 0.25-V error in the 4-bit primary quantizer output. During the error ramp, the secondary quantizer produces all possible combinations of its 6-bit output.

Some clocked-circuit time is required to develop the 4-bit output from the primary quantizer, and then, still more time is required, to develop the 6-bit output from the secondary quantizer. The secondary quantizer output then, lags the primary quantizer by a fraction of a range cell. However, the time difference is inconsequential, because a clock to the canceler input registers occurs at a time which “captures” the entire 10-bit word at a singular time.

The Digital Canceler

Data Steering

A quadrature system contains two cascade-canceler systems, one each for the *I* and *Q* channels. Figure 13-4 is an initial, simplified, dual-canceler block diagram, intended to show the major functional areas and data steering in the canceler. There is a 10-flip-flop input register, to capture the quantizer 10-bit outputs. The outputs of the input register may go in two directions, either directly into the canceler-1 circuitry, or into a multiplexer circuit, used to select one of two data routes. In normal operation, the multiplexer circuit passes the canceler-1 output. Should the system be operating in a canceler-2-only mode, the multiplexer routes the input register data directly to the canceler-2 input circuitry. Similarly, a second multiplexer provides a means to bypass canceler 2. If the system is operating in either a canceler-1-only mode or in a canceler-off mode, the second multiplexer provides the necessary routing.

Feedback Scaling

The feedback-control circuits provide for velocity response shaping. The feedback is obtained from the canceler-2 delayed data. In any feedback mode, 100% of the data in an ***outer loop*** is added to the canceler-1 input. An ***inner loop*** provides for a selected fraction of the feedback to be added to the data, at the canceler-2 input. The inner-loop feedback may be 75%, 50%, 25%, or 0%. The feedback modes are labeled in terms of subclutter visibility. Velocity-response shaping and subclutter visibility were explained in Chapter 12 on Doppler and mti; remember that subclutter visibility decreases as the ratio between residue and a moving target over clutter decreases. As feedback increases, the velocity-response notch is made more narrow, and clutter residue increases. Therefore, as feedback increases, subclutter visibility decreases, and the shaping modes are called "25 dB SCV" for 75%, "30 dB SCV" for 50%, "35 dB SCV" for 25%, and "40 dB SCV" for 0%.

Hardwired Right and Left Shifting

Throughout the digital mti processor, and for a variety of reasons to be further addressed, there is a need to adjust the "scale" of the data. To achieve this, a technique best described as ***hardwired shift*** was employed. Just as in computer-data program techniques, a data right shift accomplishes division, and a data left shift accomplishes

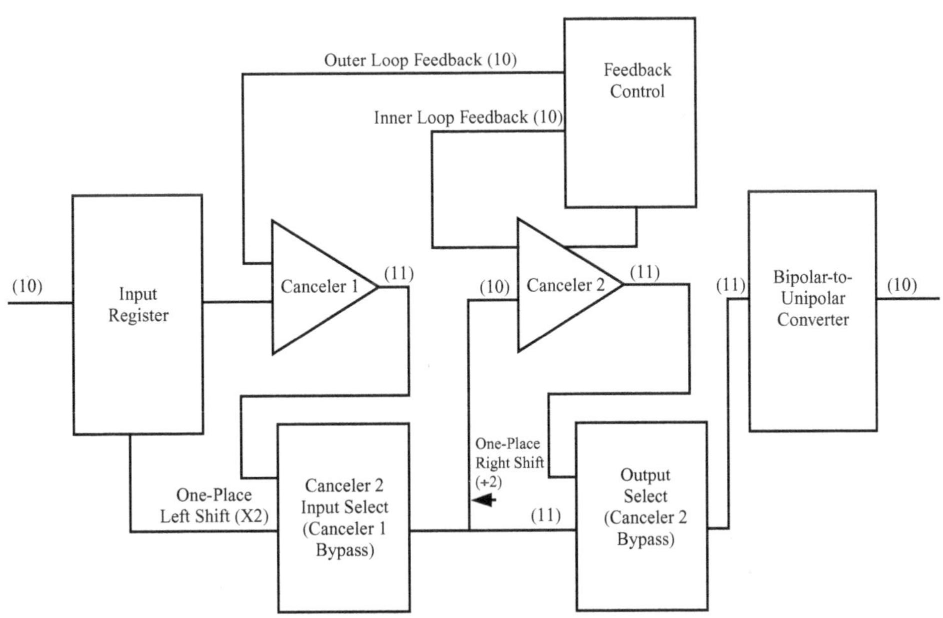

FIGURE 13-4

Canceler data paths.

multiplication. For instance, if a 10-bit data is transferred from one circuit to another in such a manner that the msb becomes the second msb, the msb is replaced with a "zero" via a hardwired ground, and the lsb is dropped, the data is divided by 2. In contrast, if the 10-bit data is transferred to an 11-bit input, with the data msb connected to the receiving-circuit msb, the data lsb routed to the 2nd lsb of the input, and the receiving-circuit lsb tied to ground, the data is left-shifted and multiplied by 2.

Figure 13-4 shows data paths of different numbers of bits in parallel words. The input to canceler 1 is 10 bits, but the comparator may produce twice as great an output as when an optimum-speed, optimum-phase target is present. The input to the canceler-2 input select multiplexer from the input register is therefore multiplied by 2, so that the multiplexer output will be an 11-bit data, no matter what the source be. The canceler-2 input must be on the same 10-bit scale as the canceler-1 input, so the multiplexer 11-bit data is right-shifted one place, as it is applied to the canceler-2 input. The canceler-2 11-bit output is not divided by 2 and is used in its entirety. If canceler 2 is bypassed, the 11-bit output from the canceler-2-input-select multiplexer becomes the output. In any case, an 11-bit data is applied to the bipolar-to-unipolar converter.

Greater Detail, Dual Canceler Operation

The canceler-bypass multiplexers have been omitted from the drawing shown in Figure 13-5 to avoid cluttering. The illustrated canceler would always be operating in cascade, and neither of the two cancelers could be bypassed.

The Input Register

The input register is simply 10, parallel flip-flops clocked at a time which captures the 10-bit data for a range cell; the data is provided by the primary and secondary quantizers. Remember that the data from the phase detectors was elevated to a +2-V baseline in the mti video conditioner. Before the quantizer data may be used in the cancelers, the 0-V baseline must be restored. This is accomplished by utilizing the "*Q*-not" register output for the msb and the "*Q*" outputs for all the others. The msb, previously the 2-V bit, then becomes a polarity bit. If the data input were in excess of 2 V, the polarity bit would be a zero, the conventional representation of a positive sign. Conversely, if the data input were less than 2 V, the polarity would be a 1, the conventional representation of a negative sign. The negative numbers are in the form of the "two's complement" of the equivalent and corresponding numbers in the positive direction (see Figures 13-5 and 13-6).

The Feedback Adder and Limiter

The output from the input register is applied to the feedback adder and limiter (Figure 13-7). Note that the msb is "double-carried" into the adder; since the msb is the polarity bit and is always the opposite state of the weighted bits, this has no effect on the value of the number. However, it has the effect of shifting the polarity bit one place to the left, and creating a new magnitude column into which carries may fall, without disturbing the polarity information.

Only the nine magnitude lsb's of the adder output are used, and the input to the canceler-1 subtractor is a 10-bit word. If a carry does fall into the new column, an "overflow" condition has occurred. Logic circuitry, to detect a difference between the tenth and eleventh bits, forces the nine magnitude bits to the state of the tenth bit. If the overflow occurred when the number was positive, the 10-bit number becomes 0111111111; if it occurred when the number was negative, the 10-bit number becomes 1000000000. These two numbers are the maximum positive and negative excursions of the number system and represent +2 V and –2 V, respectively.

The One-T_r Data-Delay Path

The data from the feedback adder and limiter takes two paths; one of these is to the delay shift register circuit cards, where the parallel bits will be delayed by 10 parallel shift registers, 1,664 range cells in length. The delayed data is captured on a 1,665th range cell in a register in the canceler. The "*Q*-not" outputs of the register are used for an input to the subtractor circuit, forming a "one's complement" of the data.

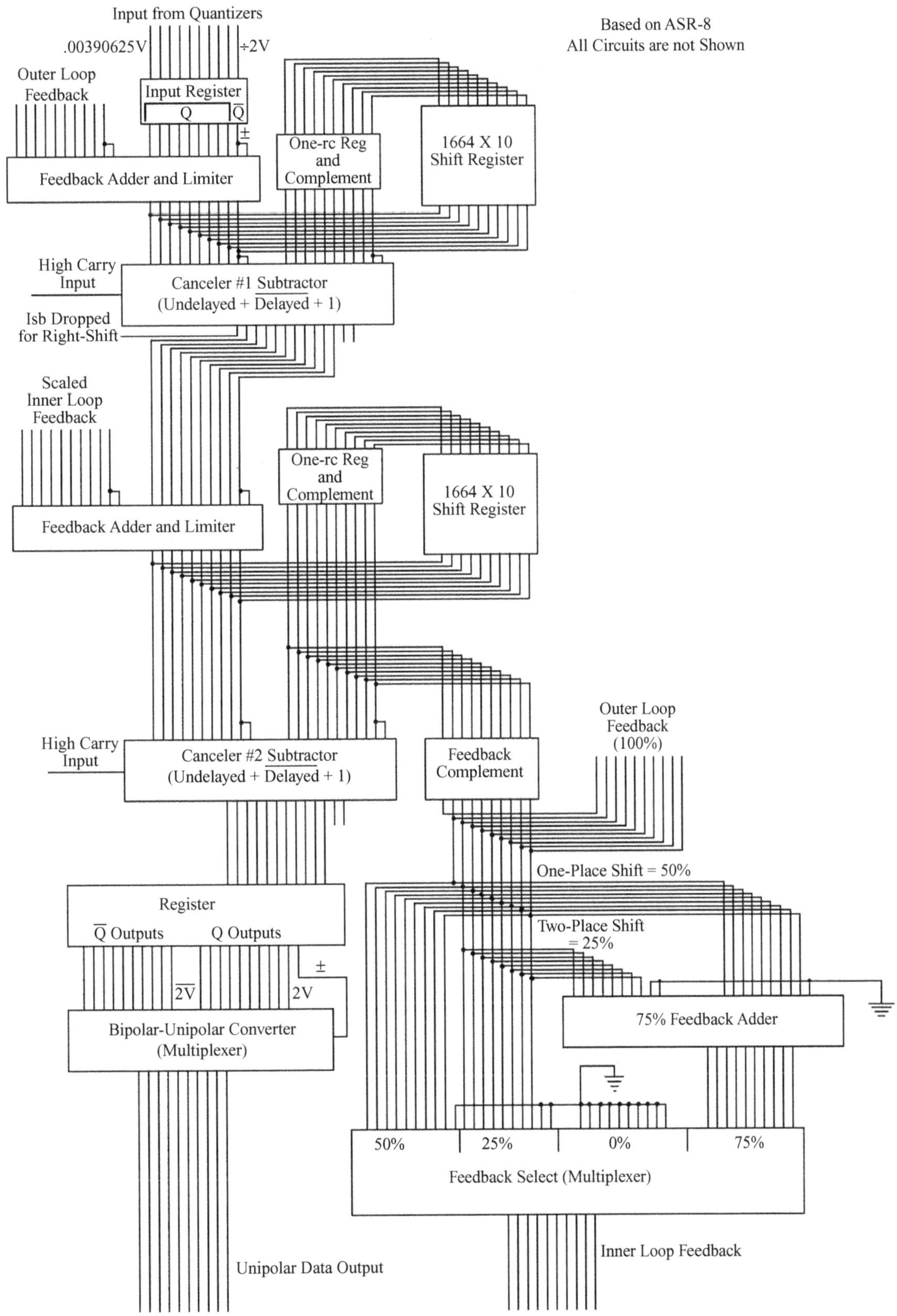

FIGURE 13–5

Details, a digital canceler.

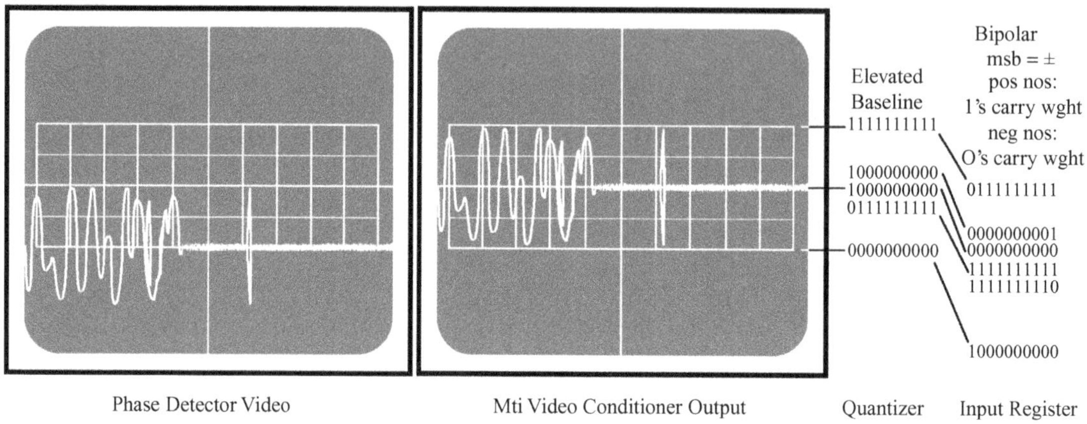

FIGURE 13-6

Baseline shifting, analog and digital.

Staggered f_p and Canceler Operation

Analog systems used a quartz delay line, and the delay time was fixed by the physical characteristics of the delay line. The canceler then required a fixed video T_r. When staggered f_p was employed, the video required "realignment," upstream of the canceler. Where shift registers are used for delay, the delay is variable, being 1,665 clocks, plus whatever time elapses, until the next T_0. The shift-register clocks always start at T_0, so the delayed data for a given range cell is always in temporal alignment with the data for the same range cell in the current T_r. The data in the canceler remains staggered, and destaggering occurs downstream. In the ASR-7, the data remains staggered, even as it is routed to the indicator site.

The Canceler "Subtractor" Circuit

The "subtractor" is an adder circuit, where the carry input, to the least-significant column, is a continual "1" (see Figure 13-8). Since the delayed input is a "one's complement," the high carry makes it a "two's complement." Binary subtraction requires that the subtrahend be converted to "two's complement," and then added to the number from which it is to be subtracted. Therefore, in the adder, the delayed data will be subtracted from the undelayed data.

The tables in Figure 13-8 illustrate several different possible conditions for the inputs. The two data-input msb's are "double-carried" to provide a new column for carries and to protect the polarity bit, just as described in the feedback adder. The double carry is depicted by the leftward arrows in the tables.

In the case of data of like polarity, it is possible for the subtractor to produce outputs greater than the inputs, so the outputs are 11-bit words. On becoming inputs to canceler 2, they will be right-shifted one place, keeping them on the same scale as the input to canceler 1.

The first table in Figure 13-8 illustrates the cancellation of a fixed target. Note that the input carry ripples through the entire sum to the S4 output of the most significant portion of the adder; it is not connected, and the 11-bit number is 00000000000. The second table in Figure 13-8 illustrates an optimum-speed, optimum-phase moving target. The result is twice as large as the inputs, but the positive polarity bit is retained. The other two tables illustrate inputs which differ by other values.

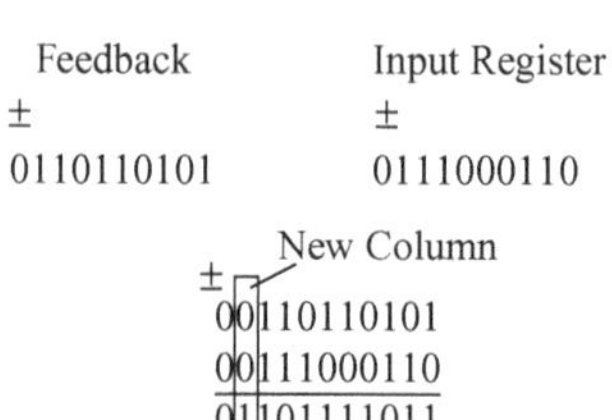

FIGURE 13-7

"Double-carry" to protect the polarity bits.

Canceler 1 to Canceler 2

Again, see Figure 13-5. The drawing shown does not provide for canceler bypass routings, and the data from the canceler-1 subtractor goes directly to the canceler-2 feedback adder and limiter. Note the hardwired right-shift to divide the canceler-1 output by 2. The feedback adder-and-limiter operation, the subtractor operation, and the delay routing are all identical to canceler 1.

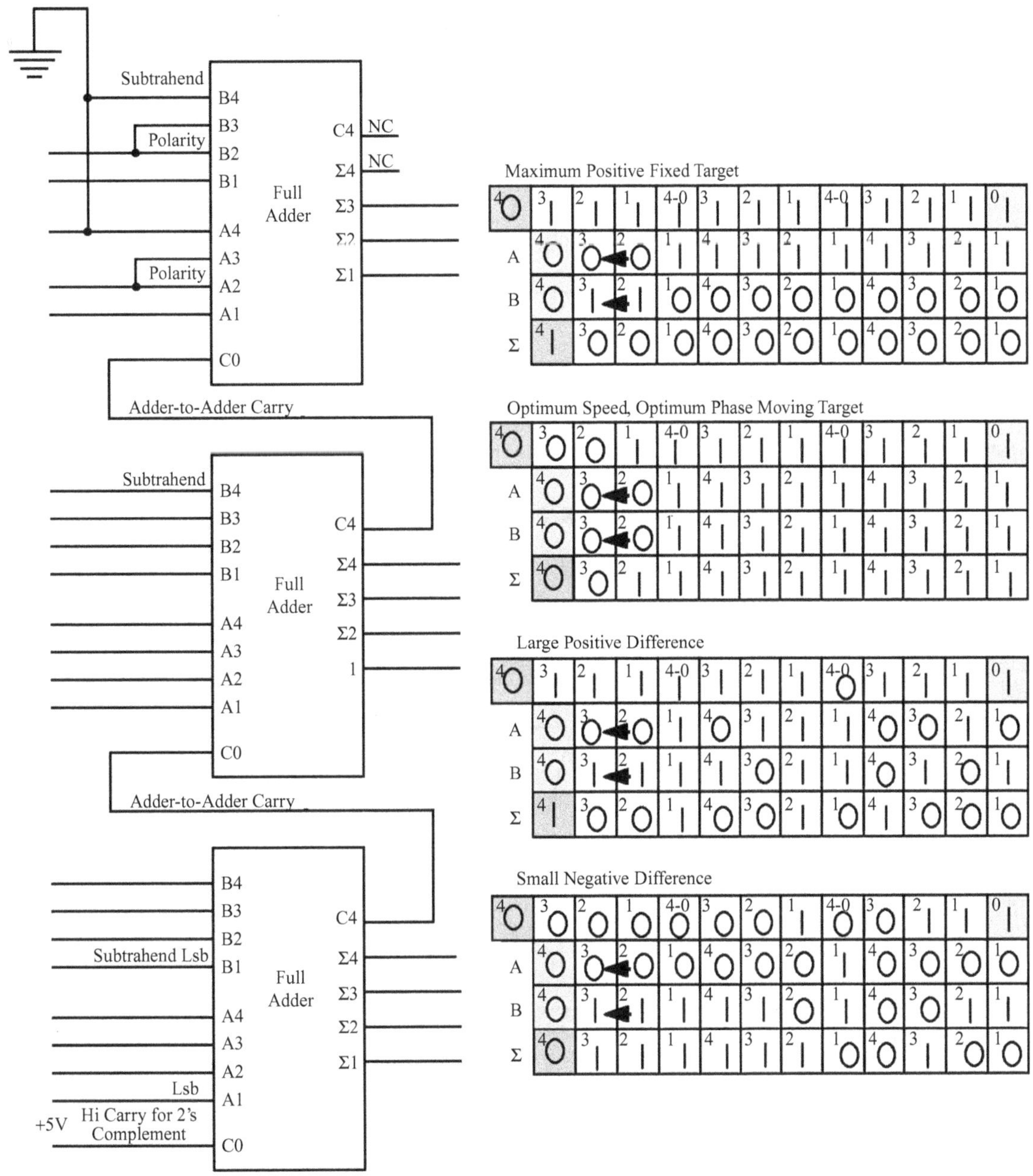

FIGURE 13–8

The canceler "subtractor" circuit.

Feedback Modes

Recall from Chapter 12 on Doppler and mti that velocity-response-shaping feedback within a canceler, because the feedback is bipolar, may be either degenerative or regenerative, depending upon the $\Delta\phi$. An additional circuit in the canceler-2 routing provides for the velocity-response-shaping feedback. The delayed data is again complemented, to give it the appropriate polarity relationship to the input data, and then routed directly to the canceler-1 outer-loop feedback adder. The entire feedback complement circuit is inhibited in nonfeedback operating modes. The feedback data is also applied to a four-input-to-one-output multiplexer circuit for inner-loop feedback scaling. The multiplexer selection is controlled by a 2-bit input from a cabinet front-panel switch. One set of inputs is grounded, for zero feedback, in the 40-dB SCV mode. Another set of inputs is hardwired right-shifted two

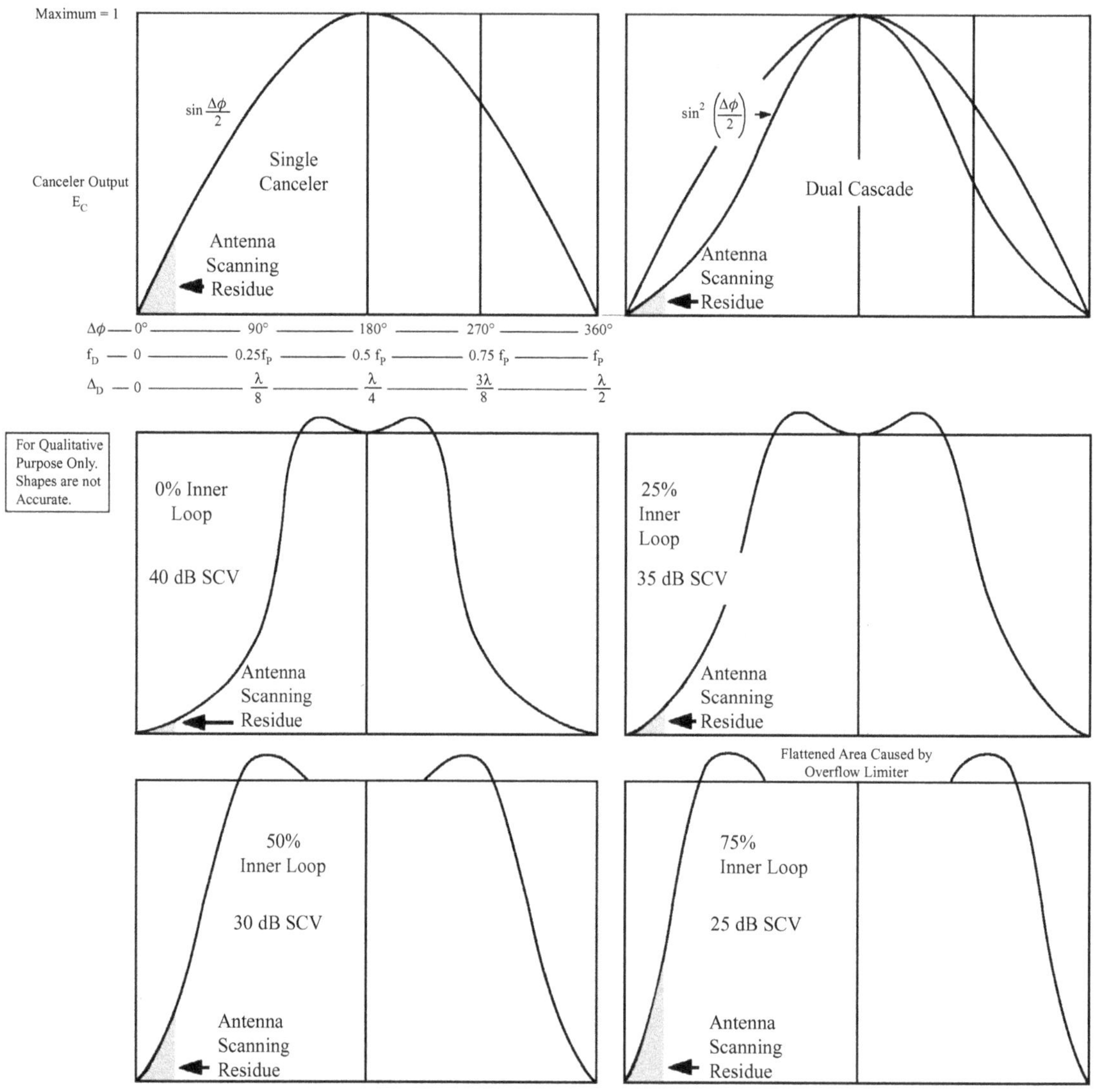

FIGURE 13–9

Velocity response shapes.

places, for 25% feedback, in the 35-dB SCV mode. Another set of inputs is right-shifted one place, for 50% feedback, in the 30-dB SCV mode. The fourth set of inputs incorporates an adder, which adds 25%-feedback data to 50%-feedback data for 75% feedback, used in the 25-dB SCV mode. Figure 13-9 is an artistic depiction of increasing width of the response and decreasing width of the notch, with increasing feedback.

The Register and Bipolar-to-Unipolar Converter

Again, see Figure 13-5. The eleven-bit output of canceler 2 is clocked into a register. The 10 magnitude bits, appearing on both the "*Q*" and "*Q*-not" outputs of the register, are routed to a two-input-to-single-output multiplexer, the "*Q*-not" bits to the "B" inputs and the "*Q*" bits to the "A" inputs. The polarity bit serves as the multiplexer switch signal, so that a high polarity bit, indicating a negative number, selects the "B" inputs, all of which are complemented because they are from the register's "*Q*-not" outputs. In short, if the magnitude data is negative, it will be complemented and if not, it will be unchanged. A minor correction circuit, not shown, called a "one's adder," adds 1 to numbers that had been negative since they had been "two's complement." Another minor circuit, also not shown, provides overflow protection for the one's adder.

The Canceler Output

The 10-bit, unipolar, magnitude-only data becomes the canceler output. It is significant that there is no output limiter, nor any provision to alter the scale of the signal for different velocity response shapes. Such limiting and unipolar gain controls are essential. In earlier systems, they were in the canceler; this system will incorporate them at a point downstream.

Canceler Testing and Troubleshooting

Clearly, there is potential for integrated-circuit failure in the digital canceler, and there must be a method to isolate such failures. Since the test ramp creates all possible bit combinations from the quantizers in each T_r, evaluation of the ramp at the canceler output provides conclusive information that the canceler is operating properly (see Figure 13-10).

Use of the ramp to test the canceler requires that the technician be able to view an analog representation of the canceler output. In the ASR-8, a 10-bit D/A converter is connected to the quadrature combiner output, making the analog monitoring possible.

The first illustration in Figure 13-10 is of a moving ramp at (1) the input to the quantizers and (2) the output of the combiner. Since the ramp is only present on alternate T_rs, it should not cancel. The ramp begins at +4 V, and then runs down to 0, during the T_r. Because of the baseline shift in the canceler input register, the +4-V ramp level becomes +2 V, the +2-V ramp level becomes 0 V, and the 0-V ramp level becomes –2 V. When the ramp passes through the center of its decline, the digital representation of the voltage becomes negative and the bipolar-to-unipolar converter complements the number, so the ramp again rises in the positive direction. The result is the display of a "vee."

When the fixed ramp is applied, it should cancel, because it is identical, each T_r. If there is a bit-failure problem in the canceler, residue will be created when the bit is incorrect, and cancellation will occur where the bit is correct. Inspection of the frequency of the residue will indicate the particular bit that has failed. For instance, as shown in the second illustration in Figure 13-10, if the msb is in error, the bit will be incorrect for either the first or last half of the T_r, and correct for the opposite half. In the third illustration in Figure 13-10, the second msb is in error, and the bit is in error twice during the ramp. If the third msb were in error, four pulses of lesser amplitude would appear, and if the fourth msb were in error, eight pulses would be displayed.

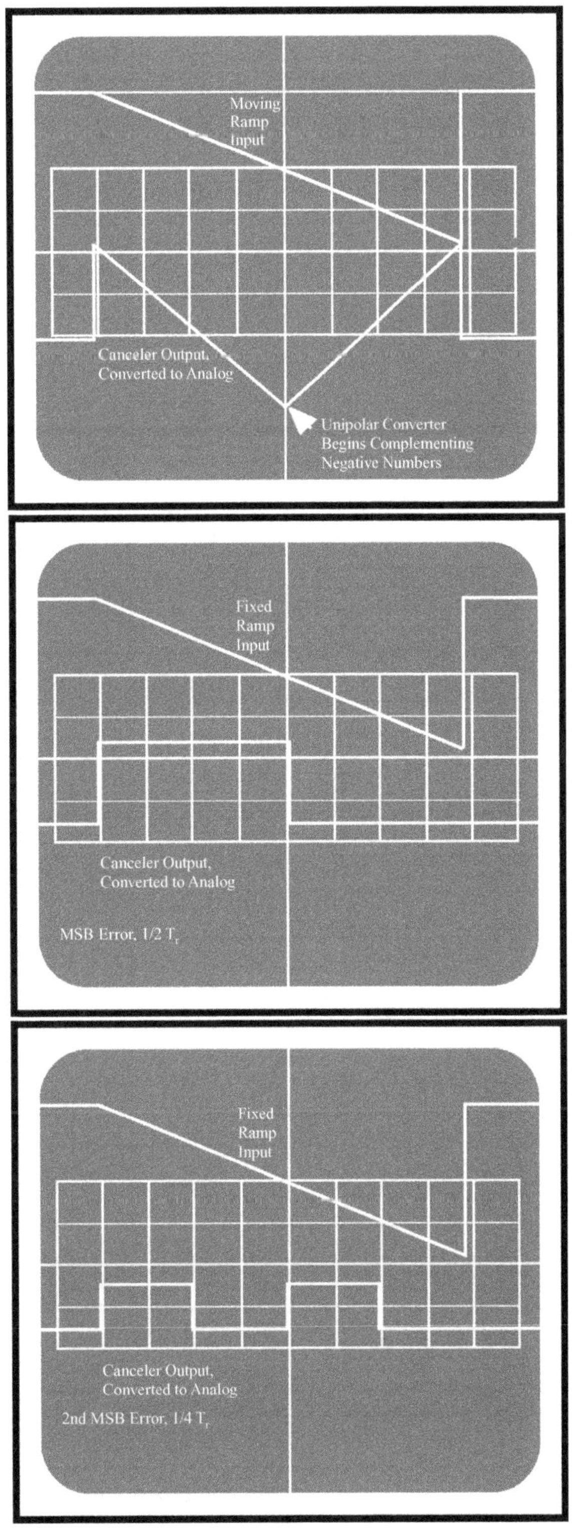

FIGURE 13-10

Ramp testing.

Post-Cancellation Circuits

The Quadrature Combine

Both the I and Q 10-bit unipolar magnitude data are applied to the quadrature combiner (see Figure 13-11). The purpose of the combiner is to simulate square root of $I^2 + Q^2$, in accordance with the Pythagorean theorem, or the trigonometric identity, $\sin^2\theta + \cos^2\theta = 1$; the two are closely related; the identity is the Pythagorean theorem, with sin and cos values of a given angle to describe the two sides of a right triangle. The objective is to overcome phase-detector ambiguities, called ***blind-phase effects***. This subject is explained in detail in the chapter on Doppler and mti. The combiner contains a magnitude-comparator integrated circuit, used to provide a switching signal to indicate which, I or Q, has the greater amplitude. By means of a multiplexer, adder, and hardwired right shifting, the combiner performs an addition of the greater, plus one-half the lesser, of I and Q. The operation is not precise, but is a very close approximation of the square root of $I^2 + Q^2$. This was an early attempt at quadrature combination, and was very effective for the purpose it served. However, latter-day states of the art have refined techniques to far greater accuracy, and employ PROMs, addressed by the I and Q data; the PROMs are loaded with tabular information to provide the square root of $I^2 + Q^2$ solution.

The "hole-filler" circuit is a modification, originated in the late 1970s by this author and two other FAA ASR-8 instructors, Lynn Garst and Harry Samford. An explanation of its function must follow a description of the "logarithmic scaler and ftc" circuit. Except during special-case instances in which the circuit becomes active, data from the combiner is unaffected by the hole filler.

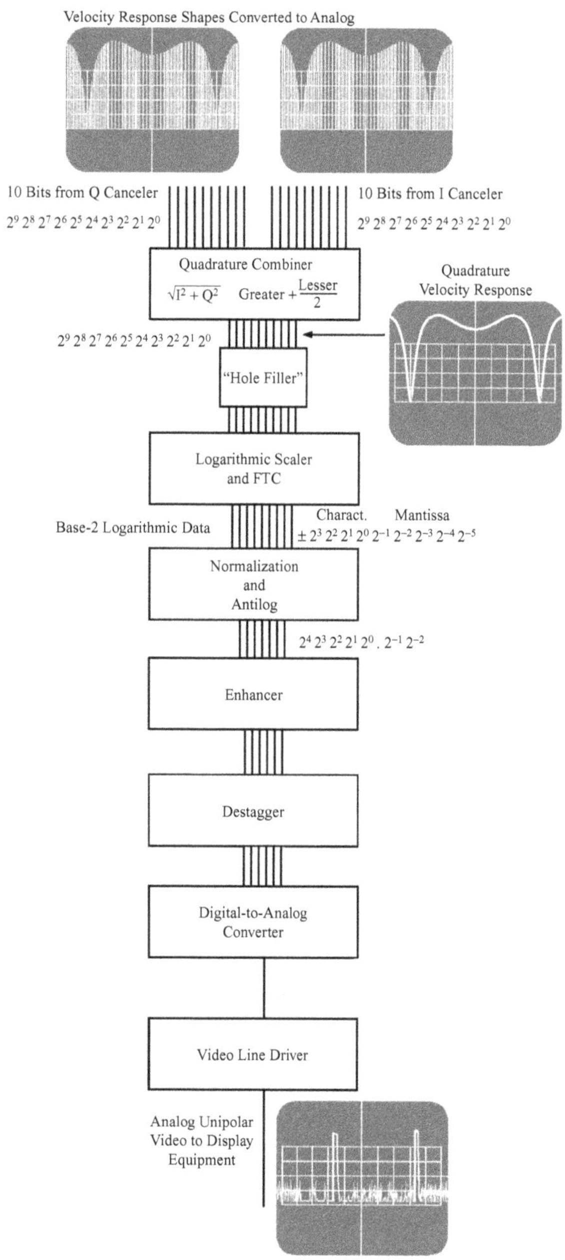

FIGURE 13-11

After-cancellation circuitry.

Logarithmic Scaler and ftc Circuit

Figure 13-12 illustrates the relationship between linear and logarithmic radar information. This illustration is of analog video, but it accurately depicts what will happen to the digital data from the log combiner. Recall, from earlier chapters, that log ftc is principally used as a weather-cancellation, ***constant-false-alarm-rate (cfar)*** device. Although a major purpose of this circuit is to provide a digital log ftc function, the circuit is absolutely essential, even when the ftc is not in use. The amplitude, or scale, of the signal must be made adjustable, so that there may be a unipolar gain before limiting. Since the signal is digital, conventional amplification is impossible, but is accomplished by digital means, in the normalization and antilog circuit. An encoder circuit in the log ftc circuitry converts the 10-bit magnitude data from the combiner to a 9-bit base-2 logarithmic data. Signal amplification or attenuation can then be achieved in the normalization and antilog circuitry, with the constant addition or subtraction of a fixed number. Recall that the addition or subtraction of logarithms affects the multiplication or division of the antilogs.

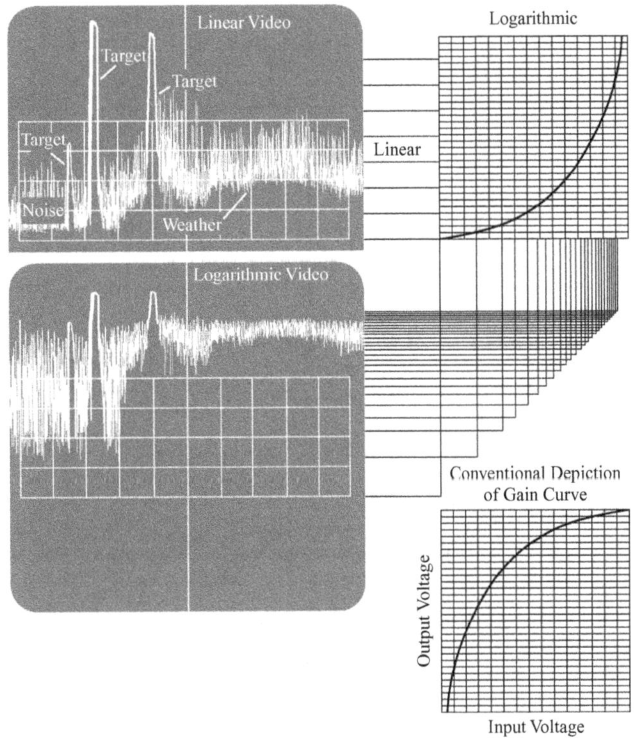

FIGURE 13–12

Logarithmic scaling, in analog for conceptual visualization.

Log Conversion

As in the quadrature combination, this operation is only a close approximation, but is perfectly adequate for the intended purpose (see Figure 13-13). Later states of the art use PROM tables for accurate logarithmic conversions. The characteristic encoder and controlled serial shifter perform the conversion to base-2 log approximation; all possible states are illustrated in the table in Figure 13-13, below the block diagram. In short, the power of 2 of the most significant high input bit becomes a 4-bit characteristic of the logarithm, and the next 5 bits to the right of that highest power-of-2 bit become the mantissa. The characteristic then generates a serial shift to the entire parallel word in the serial shift register; the number of shifts is determined by the value of the characteristic. Of the 9 bits into the shifter, only 5 bits are taken from the outputs, making up the 5-bit mantissa.

There is a five-range-cell delay to the logarithmic data before it becomes available at the output. If the ftc circuitry is not energized, the data at the output will be identical to the data from the logarithmic encoder. The subtractor circuit produces a polarity bit, which will always be zero, when the ftc circuit is de-energized.

ftc Operation

The intent of the ftc circuit is to approximate the operation of the log ftc circuitry used in conjunction with normal video. For mti, a logarithmic i-f is impossible, because of the hard limiting and phase detection; however, the need to remove weather data with some sort of an averaging technique is just as important. The data, of necessity to normalization, has already been placed in logarithmic scaling, and introducing the ftc process is convenient.

The outputs of the logarithmic encoder are captured in a "word 1 register." Several range cells of information will be used in the process, and they are assigned numbers according to age. Since word 1 is the last to arrive at the word 1 register, it is the newest, and the most distant in range. The output from the word 1 register is applied to both a 9-bit-wide, 8-bit-long shift register, and to a "word 1 adder," in an "accumulator" circuit.

To understand the accumulator operation, imagine that it begins with the entire accumulator empty. When the output of the word 1 register is applied to the word 1 adder, it is added to zero, and applied to the word 5 subtractor. Word 1 is also applied to the word 9 subtractor, but the 9 × 8 shift register is empty, so word 1 is available at the input to the one-range-cell-delay register, in the same form that it left the word 1 register.

On the next range cell, word 1 is clocked into the one-range-cell-delay register; because it is one cell older, it becomes "word 2." A new word 1 is available in the word 1 register, and the word 1 adder now adds word 1 to word 2. The output of the word 1 adder is the sum of words 1 and 2. The first stage of the 9 × 8 shift register contains word 2.

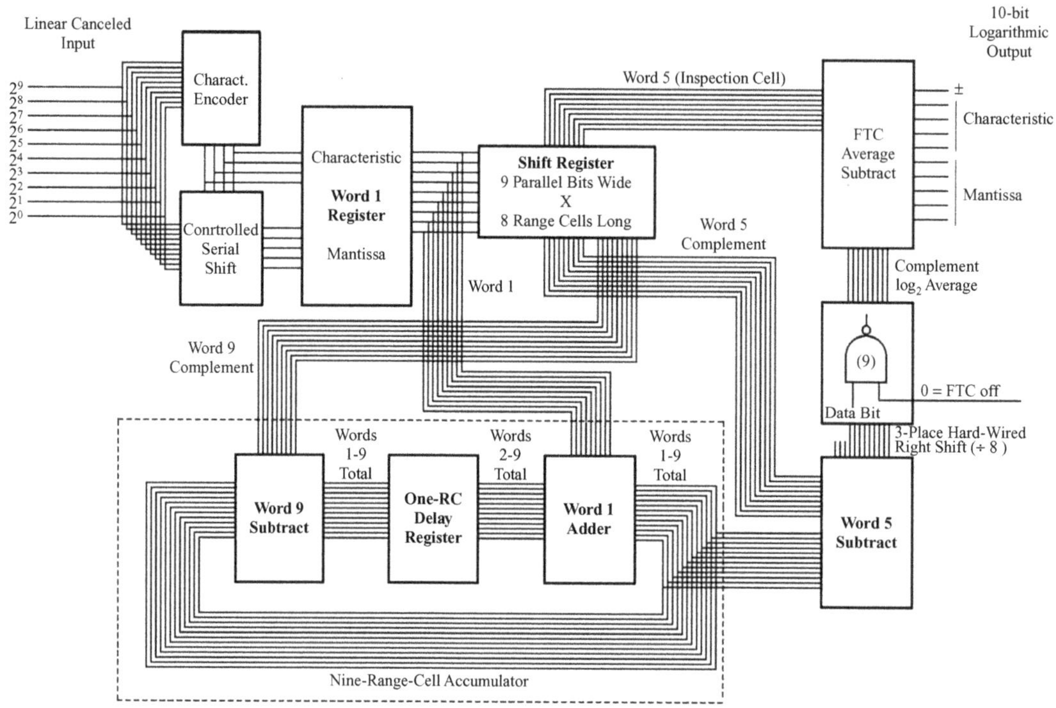

Input										Output								
2^8	2^8	2^7	2^6	2^5	2^4	2^3	2^2	2^1	2^0	Charact				Mantissa				
1	X	X	X	X	X	X	X	X	X	1	0	0	1	2^8	2^7	2^6	2^5	2^4
0	1	X	X	X	X	X	X	X	X	1	0	0	0	2^7	2^6	2^5	2^4	2^3
0	0	1	X	X	X	X	X	X	X	0	1	1	1	2^6	2^5	2^4	2^3	2^2
0	0	0	1	X	X	X	X	X	X	0	1	1	0	2^5	2^4	2^3	2^2	2^1
0	0	0	0	1	X	X	X	X	X	0	1	0	1	2^4	2^3	2^2	2^1	2^0
0	0	0	0	0	0	X	X	X	X	0	1	0	0	2^3	2^2	2^1	2^0	0
0	0	0	0	0	0	1	X	X	X	0	0	1	1	2^2	2^1	2^0	0	0
0	0	0	0	0	0	0	1	X	X	0	0	1	0	2^1	2^0	0	0	0
0	0	0	0	0	0	0	0	1	X	0	0	0	1	2^0	0	0	0	0
0	0	0	0	0	0	0	0	0	1	0	0	0	0	0	0	0	0	0

FIGURE 13–13

Logarithmic scaling and ftc circuitry.

One more clock makes the accumulation obvious. The previous sum of words 1 and 2 become the sum of words 2 and 3 in the one-range-cell-delay shift register. The new word 1 is added to the sum of words 2 and 3 in the word 1 adder to become the sum of words 1, 2, and 3. The 9 × 8 shift register contains the new word 2 in the first stage and word 3 in the second stage.

The accumulation process continues until the sum of words 1 through 9 is available at the word 1 adder output. At that time, word 9 is available at an output of the 9 × 8 shift register, and the word 9 subtractor removes word 9 from the total, so that the total of words 1–8 is available at the one-range-cell-delay register. On the next range cell, that total of words 1–8 becomes the total of words 2–9 at the output of the one-range-cell-delay register. From this point on, the word-1-adder output is the total of words 1 through 9.

This process has created a ***range sliding window***, which is nine range cells long (see Figure 13-14). It is called "sliding" because it "slides," one range cell at a time, from minimum to maximum range. The purpose of this will be to calculate an average of the data in eight range cells, two groups each of four cells, on each side of an ***area of inspection,*** the center range cell of the sliding window. Since the range cells are numbered, from maximum to minimum range, or in terms of age, from 1 to 9, the center range cell, and area of inspection, is word 5. Word 5 will be used as an output from the 9 × 8 shift register, both in true and complement forms. In the true form, it will be applied to the ftc average subtractor. In the complement form, it will be applied to the word 5 subtractor, so that the data in the area of inspection will be removed from the accumulated total.

The output of the word 5 subtractor is hardwired right-shifted three places for divide-by-8, enabled by the ftc NAND gates, and applied in complement to the ftc average subtractor, to be subtracted from the area of inspection, word 5.

Because word 5 was subtracted from the accumulated total, the accumulated total became eight words, words 1 through 4, and 6 through 9. With the hardwired three-place right shift, the eight-word total was divided by 8, providing an average of the data in the sliding window. Since this data is logarithmic, it could also be said that the process takes the eighth root of eight multiplications of the antilog. The result is the same; however it may be viewed. It is central to the concept that subtracting the logarithmic average from the logarithmic inspection area will have an end result of attenuating the antilog of the inspection area.

Ordinarily, the average is calculated predominately on noise, and the main effect is that noise is attenuated. In the presence of weather, the average increases, and the weather is attenuated as well. In those cases where the average is greater than the data in the area of inspection, the polarity bit is set, and will be used in the normalization and antilog circuitry to create a "bottom limit."

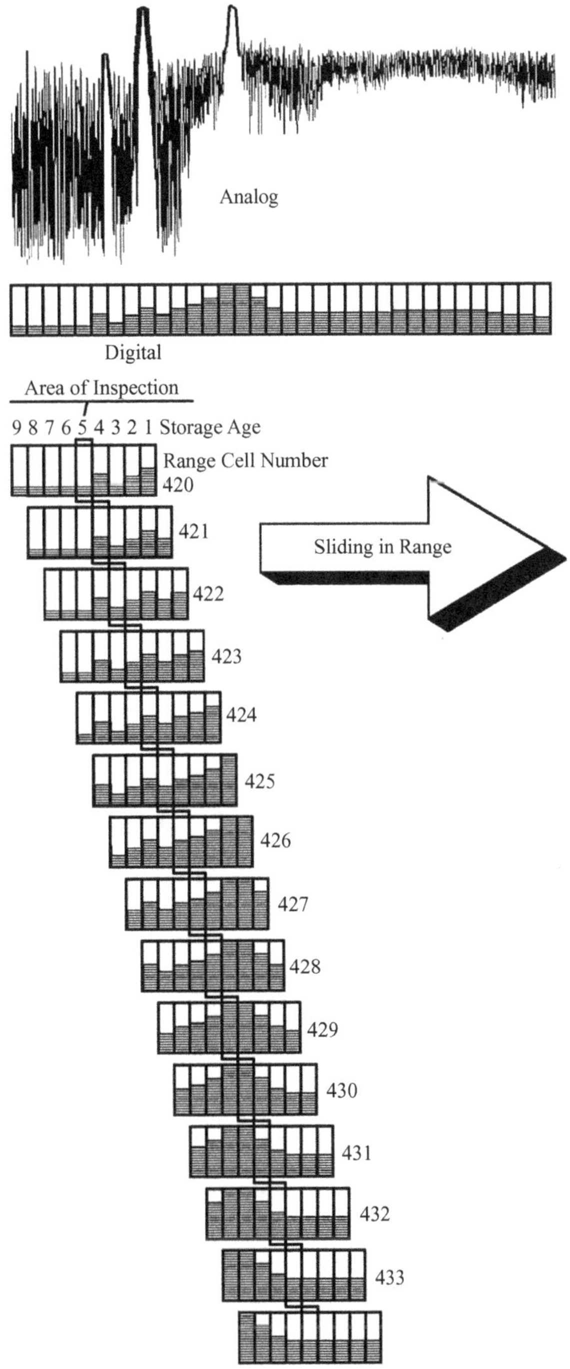

FIGURE 13–14

Range sliding window.

The "Hole Filler"

Early in the ASR-8 deployment, users began to complain that clutter residue increased, whenever the mti log ftc was enabled (see Figure 13-15). This author determined that this was a result of "holes," blank areas in the noise, which would cause the sliding-window average to drop. At the output of the ftc subtractor, clutter residue in those holes would then gain a higher amplitude, relative to the noise.

To overcome the problem, the "hole filler" became an employee suggestion, by a combined effort of three FAA Academy ASR-8 instructors, Bouwman, Garst, and Samford. Whenever the data from the combiner drops below a DIP-switch value, a magnitude comparator provides a switching signal for the multiplexer.

The input from the combiner is added to a DIP-switch value in a "baseline" adder. Whenever the magnitude comparator senses a hole, the multiplexer selects the output from the baseline adder. The final result is that whenever a "hole" is sensed, a DIP-switch value elevates the baseline to prevent the ftc average from "dipping," when the sliding window encounters the hole.

The Normalization and Antilog Circuitry

The final grass level for different operating modes could vary substantially, if there were not some means to compensate for mode changes (see Figure 13-16). For instance, when mti log ftc is enabled, the subtracting

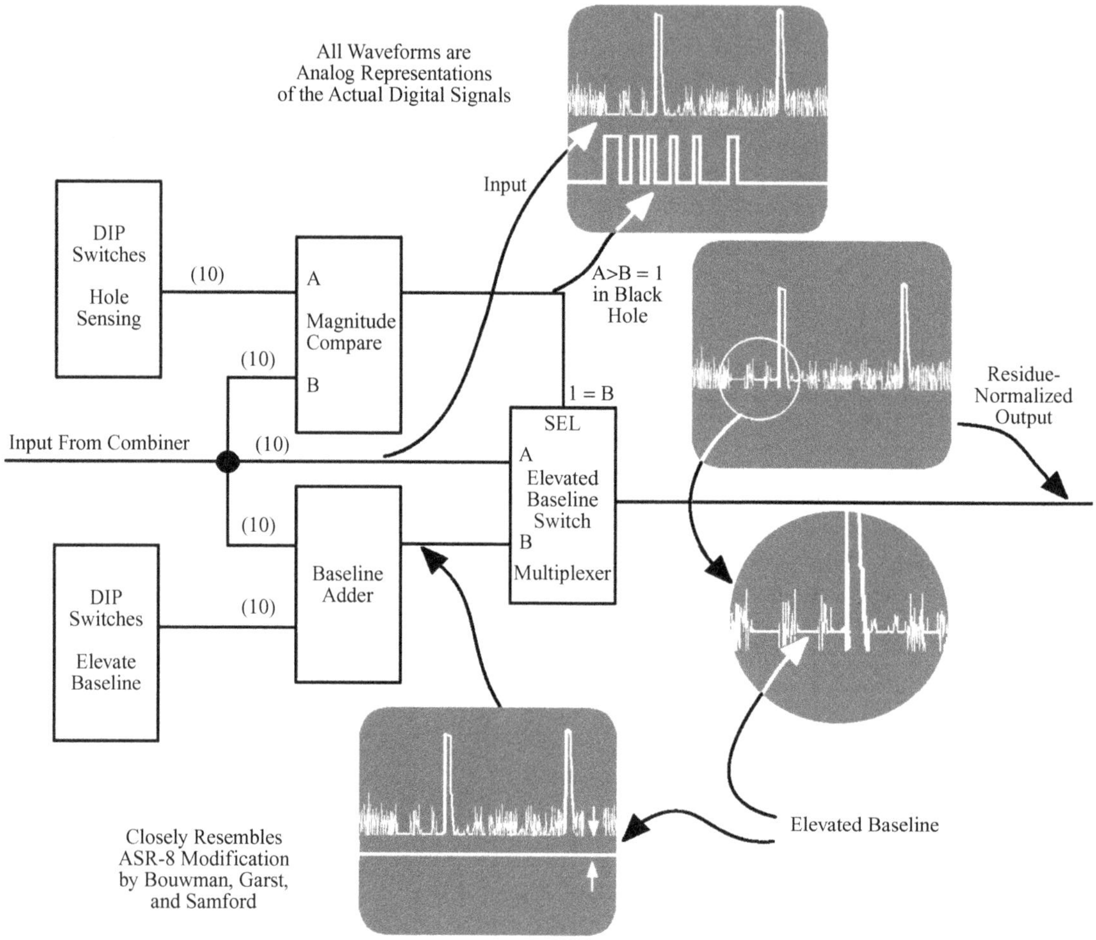

FIGURE 13–15

The "hole filler."

average reduces the grass level. When different canceler modes are selected, the grass level may vary; feedback modes increase grass. When the downstream enhancer is enabled, the final output grass level increases, and the texture becomes more coarse. There are many combinations of all these operational modes, all of which may yield different grass levels. Beyond the need to balance the noise output for the different operating modes, recall from Chapter 12 on Doppler and mti that there must be a post-cancellation unipolar gain-before-limiting adjustment. The normalization circuitry will serve both of these mode-balancing and unipolar-gain purposes.

Since the data from the log ftc circuit is in logarithmic form, the addition or subtraction of another logarithmic constant will have the effect of, respectively, amplifying or attenuating the antilog of the data. The log data is routed into the normalization plus data adder, where the scaling is modified by a number from the ftc/enhancer/canceler-mode normalization adder. The latter adds numbers from a variety of sources to produce a "normalization number," a logarithmic value, generally negative, to attenuate the log data.

One of the inputs to the ftc/enhancer/canceler-mode normalization adder is from a ROM, addressed by three bits, which are connected to the canceler-mode switch. As the canceler modes are changed, the ROM provides log numbers to keep output noise at a constant level. One address bit to the ROM is high in quadrature operation to decrease the noise level in that case.

The canceler-mode normalization is added to a number from one of four DIP switch packages. Each of these packages is adjusted to provide the appropriate noise level for a combination of enhancer, on or off, and ftc, on or off.

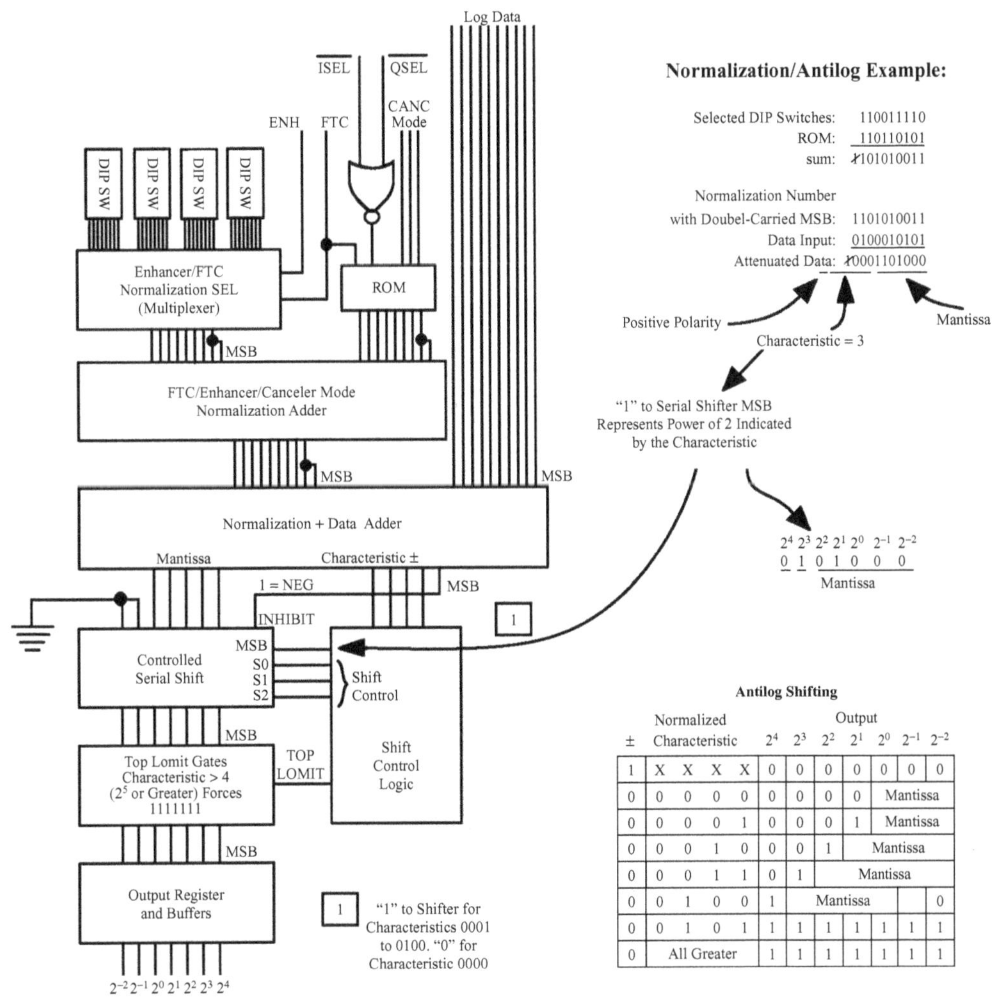

Antilog Shifting

±	Normalized Characteristic				Output 2^4	2^3	2^2	2^1	2^0	2^{-1}	2^{-2}
1	X	X	X	X	0	0	0	0	0	0	0
0	0	0	0	0	0	0	0	0	Mantissa		
0	0	0	0	1	0	0	0	1	Mantissa		
0	0	0	1	0	0	0	1	Mantissa			
0	0	0	1	1	0	1	Mantissa				
0	0	1	0	0	1	Mantissa					0
0	0	1	0	1	1	1	1	1	1	1	1
0	All Greater				1	1	1	1	1	1	1

FIGURE 13–16

Normalization/antilog circuitry.

The sum of the canceler-mode-ROM and selected DIP-switch package becomes the input to the normalization adder, which increases, or decreases, the logarithmic magnitude of the data. The data is then applied to two circuits, (1) a controlled serial shifter and (2) a shift control logic, which perform the reverse of the base-2 logarithm encoder in the log ftc circuitry.

The table at the lower right-hand corner of Figure 13-16 illustrates all conditions of operation. Should the data characteristic be negative, the polarity bit inhibits the controlled shifter to provide a final output of all "zeroes"; this is a "bottom limit" condition. Should the normalized data characteristic be 5 or greater, under a "top limit" condition, the final data is forced to "all ones"; this is the necessary unipolar data limiting, which is equivalent to the output limiting in analog cancelers.

For those conditions between bottom and top limits, the characteristic controls the serial shifter, to steer the mantissa bits to the appropriate output lines. In all cases, except for a characteristic of zero, the shift-control logic places a "1" on the msb parallel input to the shifter. The second msb of the shifter parallel input is the msb of the normalized mantissa, and the remainder of that mantissa is applied as the 3rd, 4th, 5th, and 6th msb's of the shifter parallel inputs. The characteristic will serially shift the inserted msb and mantissa by an amount determined by the numerical value of the characteristic.

In summary, for those conditions between limits, and except a 2^0 characteristic, a "1" is inserted in a bit position corresponding to the power of 2 indicated by the characteristic, and mantissa bits occur to the right of the msb. If the characteristic is 2^0, the operation is the same as for 2^1, except that a "0" is inserted as the msb. For example:

Normalized Data			Linear Output
±	Characteristic	Mantissa	$2^4 2^3 2^2 2^1 2^0 2^{-1} 2^{-2}$
1	0011	11010	0 0 0 0 0 0 0
0	0000	11011	0 0 0 0 1 1 0
0	0001	11011	0 0 0 1 1 1 0
0	0011	11011	0 1 1 1 0 1 1
0	0110	11011	1 1 1 1 0 1 1

The Enhancer

The output of the normalization/antilog circuitry is applied to an enhancer. The operation of this unit was explained in Chapter 6, but its effects on mti performance are important. Recall that the enhancer performs a T_r-to-T_r addition of synchronous pulses; targets which are weak, or "broken," will be boosted to strong, consistent, limit-level displays, very desirable to the user. One effect on mti performance is that lesser-case, blind-phase targets can be integrated into good, solid targets; weak targets over clutter are similarly improved. Unfortunately, clutter residue is synchronous, and it will also be boosted. The ultimate effect of the enhancer on mti performance then is that the overall velocity-response notch is narrowed. To reduce clutter residue then a lower-feedback canceler mode is required for enhancer operation. Figure 13-17 illustrates the appearance of the enhancer on–off velocity response shapes, viewed at the output of the D/A converter. The shapes are limited by the top limit in the normalization/antilog circuit.

The remainder of the circuitry shown in Figure 13-11 was explained in Chapter 6.

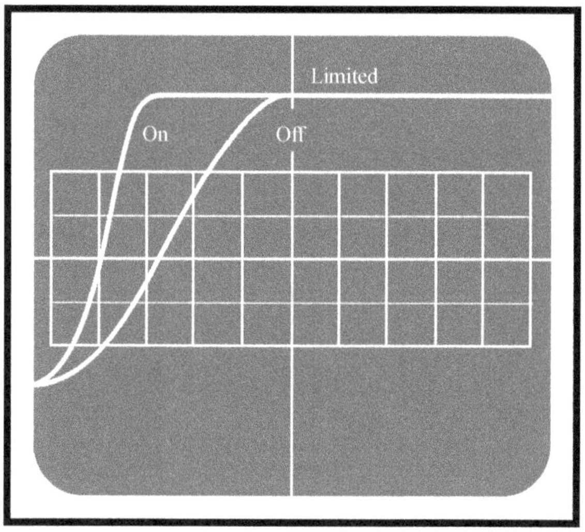

FIGURE 13–17

Leading edge, velocity response shape, enhancer on and off.

Review Questions

1. How many cancelers are in an ASR-8 radar system?
2. What determines the T_r of a digital synthesis system?
3. What determines the size of the range cell in a digital system?
4. Ignoring any test functions, what is the purpose of the video conditioner?
5. Name three test functions, and their purpose, in the video conditioner.
6. How much delay is applied to the data in the delayed channel of a digital canceler?
7. Why is a "sequential" quantizer not used in an ASR radar?
8. Name two major purposes of the canceler input register.
9. What is the requirement for digital subtraction?
10. The schematic of an ASR-8 canceler shows that the msb of an adder input is connected to two inputs. What is the purpose of this?
11. What technique is used to create the 25-dB SCV operation?
12. When clutter residue increases, SCV ____________________.
13. What are the payoff controls in the ASR-8 digital mti system?
14. An input to the log ftc card is 0001011011. The ftc is turned off. What will the output be?
15. A normalized input to the antilog circuit is 1001101011. What will the output be?
16. A normalized input to the antilog circuit is 0011101010. What will the output be?
17. A normalized input to the antilog circuit is 0001110101. What will the output be?
18. A fixed ramp is applied to an ASR-8 mti system. Four pulses appear on an analog representation of the output. What is indicated?
19. A user complains that residue increases when ftc is turned on. Comment.
20. A user complains that residue increases when the enhancer is turned on. Comment.

Answers to Review Questions

1. How many cancelers are in an ASR-8 radar system?

   Cancelers 1 and 2 = 2, I and Q cancelers in each = 2, four cancelers in two radar channels = 8.

2. What determines the T_r of a digital synthesis system?

   The DIP switch settings for the counter preset. Review Chapters 6 and 9, if necessary.

3. What determines the size of the range cell in a digital system?

   Optimum is 0.75 t_p. Smaller cells increase storage and frequency requirements. Larger cells could cause missed targets. Review Chapters 6 and 9, if necessary.

4. Ignoring any test functions, what is the purpose of the video conditioner?

   The bipolar video must be placed on an elevated baseline, to satisfy the input requirements of the quantizer.

5. Name three test functions, and their purpose, in the video conditioner.

   The composite video test target train is used for cancellation-ratio tests.

   The ramp is used for quantizer and canceler testing.

   The swept audio is used for velocity-response testing.

6. How much delay is applied to the data in the delayed channel of a digital canceler?

   The delay will be whatever the T_r may be; 1,665 range cells, plus the time until the next T_0.

7. Why is a "sequential" quantizer not used in an ASR radar?

   A sequential quantizer develops 1 bit at a time. It cannot be made to run at a high enough rate, given the current state of the art.

8. Name two major purposes of the canceler input register.

   To capture the quantizer output bits, and to shift the digital bipolar baseline back to zero.

9. What is the requirement for digital subtraction?

   Add the two's complement of the subtrahend to the number from which it is to be subtracted.

10. The schematic of an ASR-8 canceler shows that the msb of an adder input is connected to two inputs. What is the purpose of this?

    Addition carries can destroy polarity information. The "double-carry" creates a new column, into which carries may fall.

11. What technique is used to create the 25-dB SCV operation?

    Feedback is obtained from the canceler-2 delay path. 100% outer loop feedback is used at the canceler-1 input. For the inner-loop, canceler-2 input, 75% feedback is used; it is obtained by adding one-place right shift (50%) and two-place right shift (25%).

12. When clutter residue increases, SCV ***decreases.***

13. What are the payoff controls in the ASR-8 digital mti system?

    The i-f gain, and the normalization DIP switches.

14. An input to the log ftc card is 0001011011. The ftc is turned off. What will the output be?

2^9	2^8	2^7	2^6	2^5	2^4	2^3	2^2	2^1	2^0	*charact:*	$\pm$	2^3	2^2	2^1	2^0	*mant:*	2^{-1}	2^{-2}	2^{-3}	2^{-4}	2^{-5}
0	*0*	*0*	*1*	*0*	*1*	*1*	*0*	*1*	*1*		*0*	*0*	*1*	*1*	*0*		*0*	*1*	*1*	*0*	*1*

15. A normalized input to the antilog circuit is 1001101011. What will the output be?

    This is a bottom-limit condition: 0000000.

16. A normalized input to the antilog circuit is 0011101010. What will the output be?

    This is a top-limit condition: 1111111.

17. A normalized input to the antilog circuit is 0001110101. What will the output be?

charact:	$\pm$	2^3	2^2	2^1	2^0	*mant:*	2^{-1}	2^{-2}	2^{-3}	2^{-4}	2^{-5}	2^4	2^3	2^2	2^1	2^0	2^{-1}	2^{-2}
	0	*0*	*0*	*1*	*1*		*1*	*0*	*1*	*0*	*1*	*0*	*1*	*1*	*0*	*1*	*0*	*1*

18. A fixed ramp is applied to an ASR-8 mti system. Four pulses appear on an analog representation of the output. What is indicated?

    One pulse would be present if the msb were in error, two for the second msb, and four for the third msb.

19. A user complains that residue increases when ftc is turned on. Comment.

    The "hole filler" is not installed, or is improperly adjusted. Without the "hole filler" modification, the ftc average will drop when the sliding window runs across a canceled-clutter hole. Less average is subtracted from clutter residue than from noise, and the residue-to-noise ratio increases.

20. A user complains that residue increases when the enhancer is turned on. Comment.

    This is a normal phenomenon, and should be expected. The enhancer narrows the final, overall, velocity-response notch.

CHAPTER 14

The Moving Target Detector MTI System

MTD Definition

There have been other systems called "moving target detectors," and the term could lead to confusion to some experienced technicians unfamiliar with air traffic control radar. "MTI," discussed in previous chapters, is an abbreviation for "moving target indicator," a broad class of radar systems which "indicate" moving targets. An FAA MTD system, the system type hereafter simply called "MTD," is within the broader class of "MTI" radar systems, and may be called an "MTI system" in some literature.

Improvement over Canceler-Type MTI

Canceler-type MTI systems have had many recognized shortcomings since the earliest days. Among them have been subclutter visibility limitations, blind phase effects, blind velocity effects, tangential effects, weather effects, anomalous propagation, excessive target data from birds and insects, and antenna scan-motion residue. There are more. Solutions to these problems were envisioned long ago; however, they were not practical because of hardware technology limitations. As device speeds increased, and storage media became more compact and practical, MTD became possible. Several MTD systems have now been built, but the largest production effort was the FAA ASR-9, manufactured by the Westinghouse Corporation; it will be the basis for the discussion in this chapter.

Figure 14-1 depicts the contrast in displays for MTI and MTD systems. The MTD system, illustrated to the right, offers a very "clean" display, free of grass, residue, and other distractions. The display is not in real time, but in a special, "reconstituted," "synthetic" real time; the rotating sweep is considerably "behind" the radar antenna, perhaps as much as 120° or more. This is all necessary because the MTD system performs a time-consuming, complex analysis of target data, and computes the center-of-range and center-of-azimuth to produce a single digital target message, not compatible with real-time ppi displays. The data must then be "reconstituted" into analog video, so that it may be displayed in a form compatible with video mapper outputs, ARTS outputs, indicator-site timing, and more.

The MTD system offers many improvements over canceler-type MTI systems. Among these are a greatly improved detection of targets over clutter or in weather, a means to retrieve tangential targets, additional means to recover blind- and dim-speed targets, false target and interference reduction, reduction to the deleterious effects of weather, and more. A disadvantage is that all primary radar targets look alike, appearing in only two different forms; there is no grass on the displays, normal radar video is unavailable, and the controller is able to exercise only very limited judgment or discrimination in evaluating the likelihood that a target is real or false.

General Description

Although overly simplistic, it could be said for an opening discussion that the MTD system relies on two basic principles for its operation. First, target information is sorted, according to Doppler frequency, in a bank of narrow-band digital audio filters. Then, the data is processed by many steps of analysis, much of it comparing radar information over several scans. The final output is a digital message, describing the radar target ***range and azimuth center of target (centroid)*** along with other pertinent data.

Doppler filters are not new; the type used in MTD systems first appeared in about the late 1970s or early 1980s. Cancelers have even been called Doppler filters in textbooks. A canceler is a Doppler pass filter with a notch at zero Doppler and at all multiples of the f_p; they have also been called "comb" filters because of the

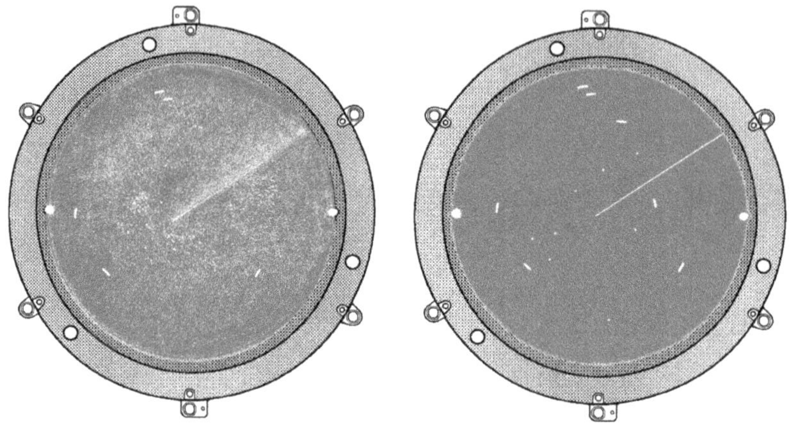

FIGURE 14–1

MTI versus MTD displays.

multiple notches. In the purest and simplest sense, a true filter would be an analog, reactive-component network, which passes a selected band of frequencies; earlier Doppler filters used in military equipment, as early as the 1960s, were of this type. A brief discussion of this earlier circuit will make the principle clear.

The Original Analog Range-Gated Doppler Filters

In Figure 14-2, a bank of range-gated Doppler filters, numbered 1 through n, is illustrated. The phase-detector output has been routed to range-gating circuits, in all these filters, in parallel. However, the range-gating circuits

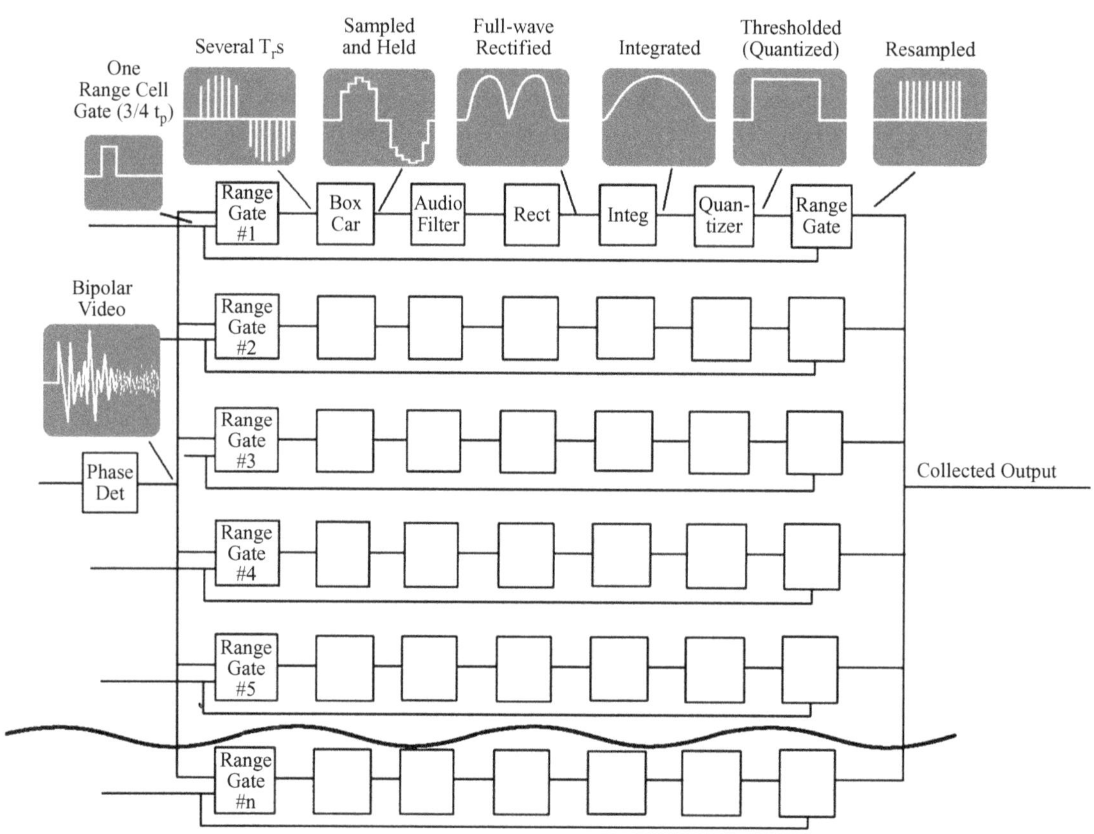

FIGURE 14–2

An analog doppler filter system.

are enabled sequentially, one range cell at a time. At the beginning of the T_r, range gate 1 is enabled for one range cell; on the next range cell, range gate 1 is disabled and range gate 2 is enabled; next, range gate 2 is disabled and 3 is enabled, etc. On the last range cell of the T_r, range gate $n - 1$ is disabled and range gate n is enabled; then, range gate n is disabled. In practice, there would be many Doppler filters; a system with which the author is familiar contained over 80.

Recall, from earlier chapters, that the objective in canceler-type MTI was to make pulse-to-pulse comparisons of bipolar video. For a specific target, the amount of voltage change from one T_r to the next was determined by the $\Delta\phi$ of the target, and the $\Delta\phi$ was determined by the T_r and Doppler frequency. By utilizing the canceler delay line or shift register, each bipolar echo could be compared with the echo at the same range from the previous T_r. In the Doppler filter bank shown in Figure 14-2, a delay line is not used, but the data in each range cell is applied to a separate filter circuit, and the boxcar detector will hold the voltage level that is present during the range gate.

The boxcar detector holds its set voltage until the next T_r, when it is reset to the new voltage at the designated range cell. Therefore, the output of the boxcar detector is an audio frequency, if there is a moving target in the range cell, and a dc level, if there is a fixed target. If there were no target at all, the boxcar detector output would be random, because it would be set by noise spikes.

After the boxcar detector, the information is applied to a bandpass filter. Fixed-target dc will be unable to pass, but Doppler audio, or noise, will pass. The rectifier and integrator then build the detected targets into a unipolar voltage level. The threshold circuit inhibits any data beneath the threshold level. The output range gates operate in sequence, in the same manner as the input gates, sequentially gating the output of each circuit onto a common bus. The information on the common bus is canceled, unipolar, MTI video, similar to the conventional, canceler-type MTI system, but quantized.

This type of analog filter bank was the beginning of Doppler filter technology, and was crude, in comparison with today's filters. Because of the rugged, discrete components that could be used in construction of the filters, this technique worked well where fragility was unacceptable, such as in portable radars used in military ground combat. Today's filters have been made possible by new storage devices, and by digital, integrated, arithmetic circuits. The system illustrated in Figure 14-2 required a dedicated circuit for every range cell; newer technology allows a few circuits to be used repeatedly.

An Overview of the MTD System

Timing Scheme

The MTD timing was discussed in Chapter 9 on synchronizers, but a brief review is appropriate here (see Figure 14-3). The MTD system requires a unique form of staggered f_p for the operation of the Doppler filters. The f_p operates at a high rate for 10 T_rs, and then at a lower rate for 8 T_rs. Each group of T_rs is called a ***coherent processing interval (CPI)***, and the two groups together are called ***a coherent processing interval pair (CPIP)***. There are 256 CPIPs in one revolution of the antenna, and each of these has a dedicated azimuth center.

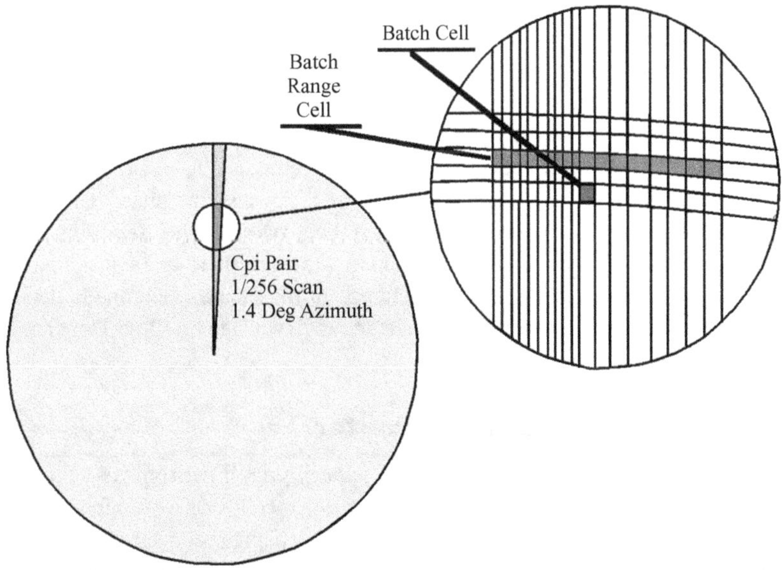

FIGURE 14-3

CPIP comparison.

When each CPI is complete, there must be a waiting period before the next one begins. To maintain the integrity of the transmitter spectrum and, subsequently, to insure reproduction of the first few echo pulses in the receiver, it is necessary to generate additional transmitter bursts at the low f_p during this time; they are called ***fill pulses***. Signal processing requires that each T_r be divided into 1/16-nmi range cells, and there are 973 of these per T_r (960 for 60 miles and 13 more to maintain mean-level-threshold operation after end of range). Each group of 18, azimuth-adjacent, 1/16-nmi range cells, for a specific range, comprises a ***batch range cell (BRC)***, and there are 973 BRCs in a CPIP.

Figure 14-4 illustrates the basic blocks of the MTD system. The manufacturer divided the hardware into major "functional" areas, called the ***rf/i-f*** function, ***filter/magnitude*** function, and the ***target-&-two-level detector*** function. The filter/magnitude function and target-&-two-level detector function comprise another, larger, functional area, called the ***digital signal processor (DSP)***.

The rf echoes are mixed with the stalo in the rf receiver to produce a 31.07 MHz $\pm f_D$ output to the i-f amplifier. The stalo must always be tuned below the transmitter frequency, unlike other synthesis systems, in which it may be either above or below.

The i-f Amplifier and Phase Detectors

The i-f amplifier is very different from those used in canceler-type MTI systems. The bandwidth is 923 kHz at the 6-dB points (Δf is usually measured at the 3-dB points) and the transmitter pulse width is 1 μs. The Δf is much less than $1/t_p$, and far below optimum. This reduces noise and signal harmonics, both of which would be objectionable in the circuitry to follow. A phenomenon, called ***coherent integration***, will recover most of the signal discernibility. The technician, familiar only with canceler-type MTI, would suspect something wrong when viewing the bipolar video of an MTD system; the signal-to-noise ratio would first appear astronomical (see Figure 14-5). The bipolar video maximum excursions are ±4 V; when the oscilloscope gain is set to display these, there does not appear to be any noise. Because the noise level is so low, the i-f amplifier contains, upstream of the phase detectors, a special logarithmic amplifier and amplitude-detection circuit, to provide a semblance of normal video for the maintenance monitor ppi display.

The i-f amplifier is also much different than one in a canceler-type MTI system, in another respect. A canceler-type system's MTI i-f amplifier contains a severe limiting circuit prior to the phase detectors. That severe limiting imposed a major limitation on subclutter visibility; it was necessary to ensure that the outputs for most targets would be representative only of phase, and not of signal strength. In the MTD system, signal strength information is essential, and phase angle can be determined from the cosine (called I) and sine (called Q) responses of the phase detectors. To ensure that the i-f amplifier does not easily limit, it has a wide dynamic amplitude range of 63 dB, which means that inputs can vary by that much from minimum discernibility to limit.

In canceler-type MTI systems, quadrature phase detectors were used to supply parallel I and Q canceler systems; the objective was the elimination of blind-phase effects (see Figure 14-6). The quadrature phase detectors, in an MTD system, serve an entirely different purpose. Because their outputs are electrically separated by 90°, they may be mathematically utilized as rectangular coordinates to describe vectors. And, because they have a sin–cos relationship, the magnitude of the vector they describe indicates the signal strength of a received echo (because $\sin^2\theta + \cos^2\theta = 1$). Because the phase detectors are representative of both phase and magnitude, some engineers have called them ***synchronous detectors*** to distinguish them from those phase detectors which utilize a hard-limited input. Although there is a considerable difference in the input from the i-f amplifiers, there is little or no difference in the phase/synchronous detector circuit operation. For further information on phase/synchronous detectors, review Chapter 12 on Doppler and MTI.

The Analog-to-Digital Converter

This is a high-speed ***flash converter with subranging***; it performs two operations per range cell, and produces 12 parallel bits of data, twice each 0.772-μs range cell, one word representing I and the other representing Q. This data is said to be ***range ordered***, because it is being produced at the A/D output in ascending range order as quickly as it is received, and is still in ***real time***. The data is ***rectangular coordinate data***, since each range cell contains two words, which describe phase angle and magnitude.

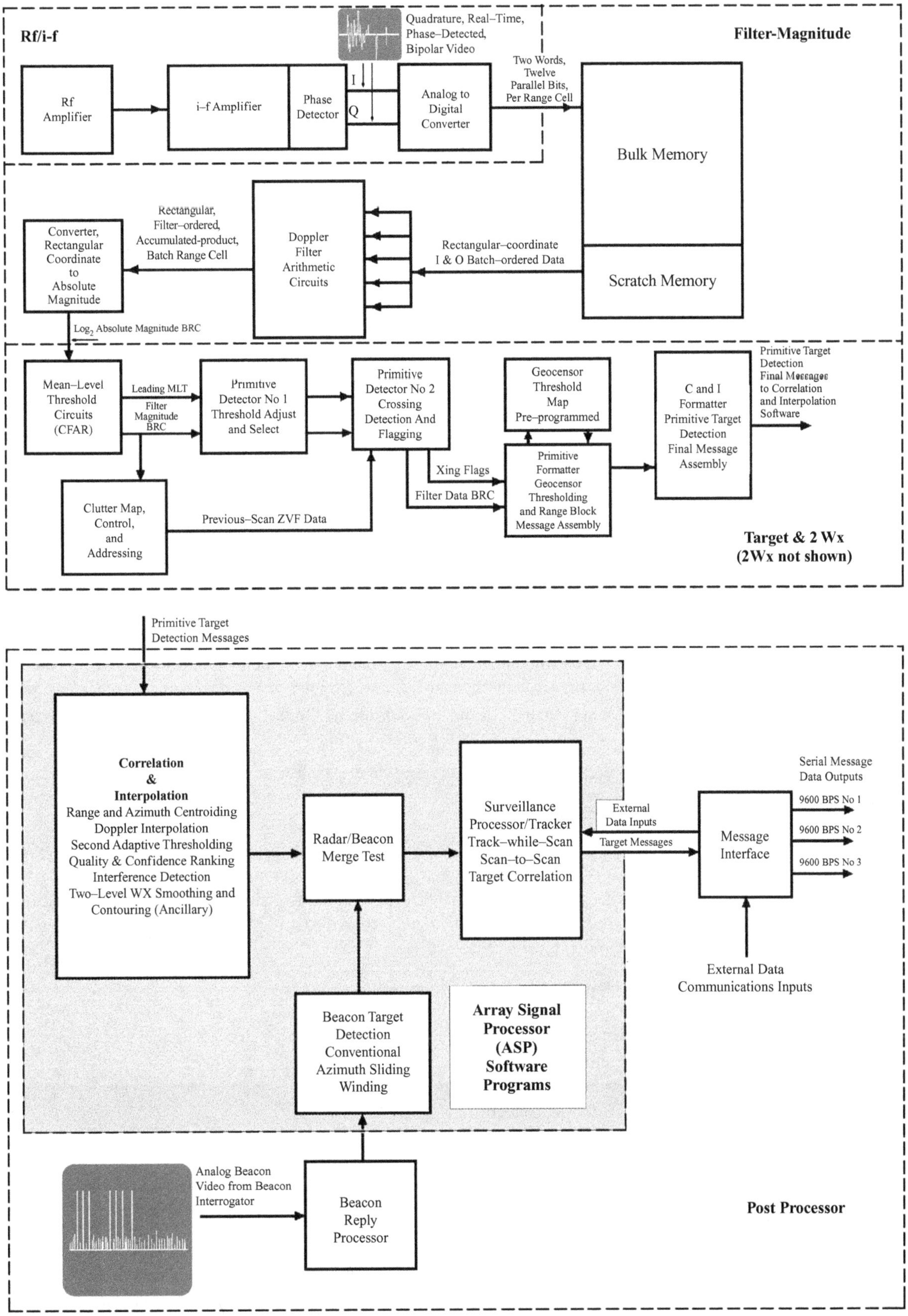

FIGURE 14–4

Simplified block diagram, ASR-9 MTD system.

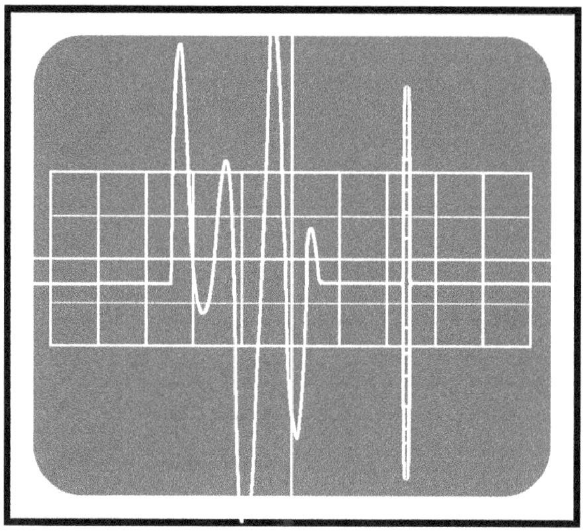

FIGURE 14–5

Synchronous detector output.

Memory and Filter Bank

The rectangular coordinate data is loaded into a memory for use on the next CPIP, as ***batch-ordered data***, which is, in turn, used to operate the Doppler filter bank. The Doppler filter bank provides 36 filter-data output words (18 rectangular-coordinate sets) for every 1/16 nautical mile × 1 CPIP; each of these 18-filter messages is 13.9 μs long, and that time is called a ***batch range cell***. There are 973 batch range cells in a CPIP for live radar data processing. The purpose of the filters is to sort the target data according to Doppler frequency. The Doppler frequency responses for the filters used for each batch range cell are illustrated in Figure 14-7.

Power Combiner (Rectangular-Coordinate Converter)

The data in the filter output message is still in rectangular coordinate form; all necessary vector mathematics are complete at this point, and the $\log_2(\sin^2 + \cos^2)$ is calculated in the power combiner, to solve for the absolute magnitude. The power combiner has additional, but auxiliary, functions, not shown in this simplified block. The data output from the power combiner is an 18-word, 13.9-μs, batch range cell, containing absolute magnitude (signal strength) information, in $\log_2$ form. Each filter data word describes the amplitude output of 1 of 18 Doppler filters in a 1 CPIP × 1/16 nmi batch range cell; there are 973 batch range cells in a CPIP. Four of the filter data words describe filters, the pass of which spans zero Doppler; they are called ***zero velocity filters (ZVF)***, and the clutter information is contained in them. Accordingly, the rest of the filters are called ***nonzero velocity filters (NZVF)***.

The Target and Two-Level Detector

This is a large functional area, in terms of circuitry; it consists of 11 major circuit cards. For the purpose of an introduction, it can be stated that it is a hardware processor. The name, "target and two-level detector," describes two functions which it performs: target detection and two-level weather detection. Actually, the target detector comprises nearly all the circuitry; only one card is totally dedicated to the two-level weather function. Among the things taking place in the target and two-level detector are:

Mean-Level Threshold Development for each Filter Data
Storage of ZVF Data in a Clutter Map
Threshold Development and Adjustment
Threshold Crossing Detection for NZVF data
Threshold Crossing Detection for ZVF data
"Geocensor" Threshold Crossing Detection
"Primitive" Target Message Formatting
A Heavy-Clutter Map to Automatically Alter the NZVF Doppler Response
High- and Low-Level Weather Detection on Alternating Scans

The ***geocensor*** is a preprogrammed map, contained in a memory, which can supply special thresholding, flagging, or both, in selected zones. It may inhibit, or flag, possible road traffic or clutter residue, and it may "flag" the MTI reflector, to ensure that it will not be inhibited by any additional processing to follow. The word ***primitive*** describes the target data messages which will be supplied to the postprocessor; they are called "primitive" because a great deal more processing must follow.

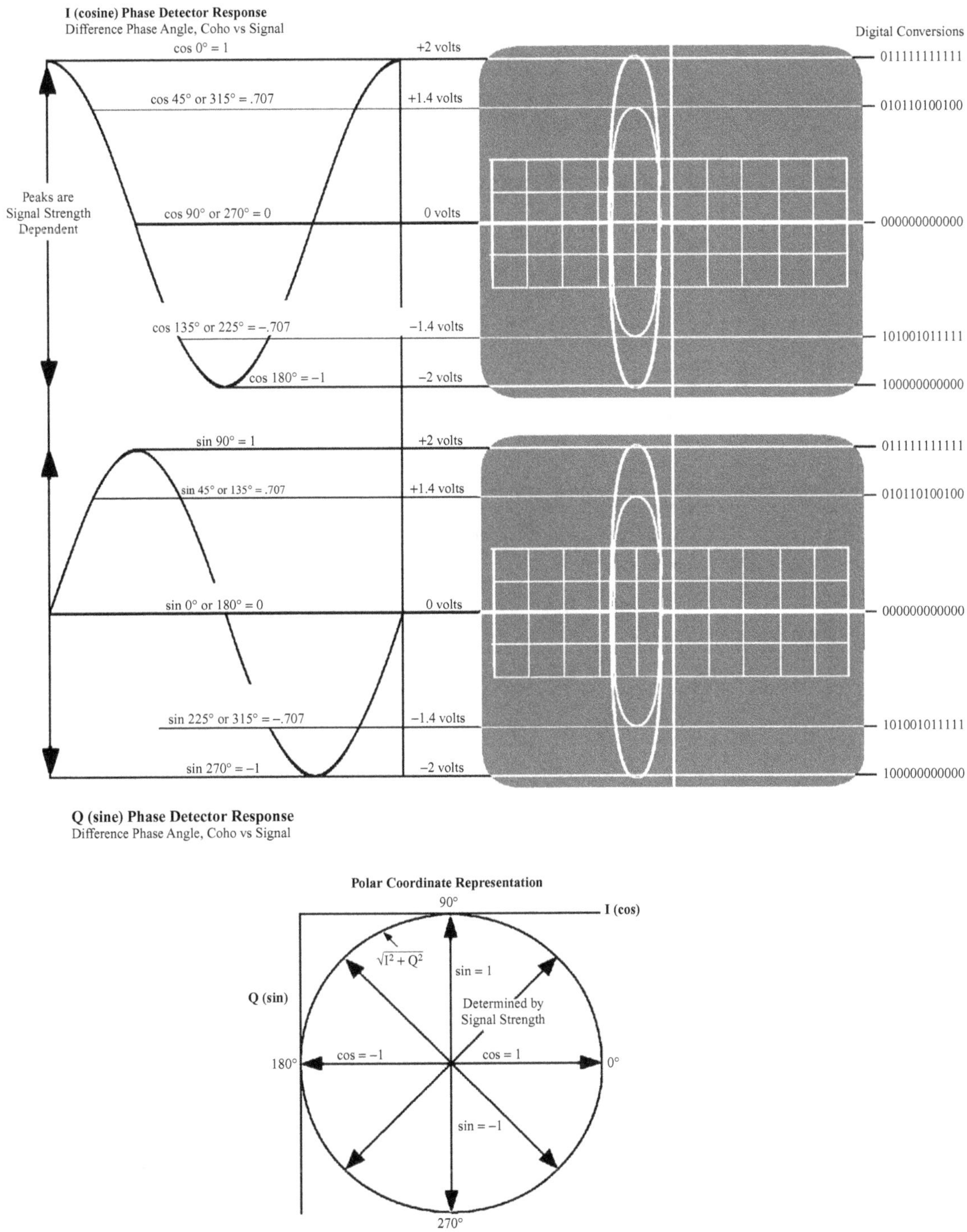

FIGURE 14-6

Synchronous phase-detector response.

The Postprocessor

There are three major hardware areas in the postprocessor. The largest, most complex, and most significant, to target processing, is called the ***array signal processor (ASP)***. This area is a computer, and the programs are more significant to target data, than is the hardware itself. Among the program modules are:

Correlation and Interpolation
Centroids Primitive Radar Targets
Performs Doppler Interpolation from Primitives
Ranks Targets according to Quality and Confidence
Performs Second Adaptive Thresholding
Performs Interference Elimination
Performs Two-Level Weather Smoothing & Contouring

Beacon Target Detector
Performs Conventional Azimuth Sliding Window
Validates Beacon Codes

Surveillance Processor
"Merge" Function Compares Beacon to Radar
"Track" Function Performs Scan-to-Scan Analysis

Also contained in the postprocessor is the ***beacon reply processor***, which receives raw beacon data from the ATCRBS and then converts it into a format for use by the beacon target detector module in the ASP. Still further, the postprocessor contains ***message interface*** circuitry, which formats target data messages into the CD/ASR format for transmission at 9,600 bps to the remote site.

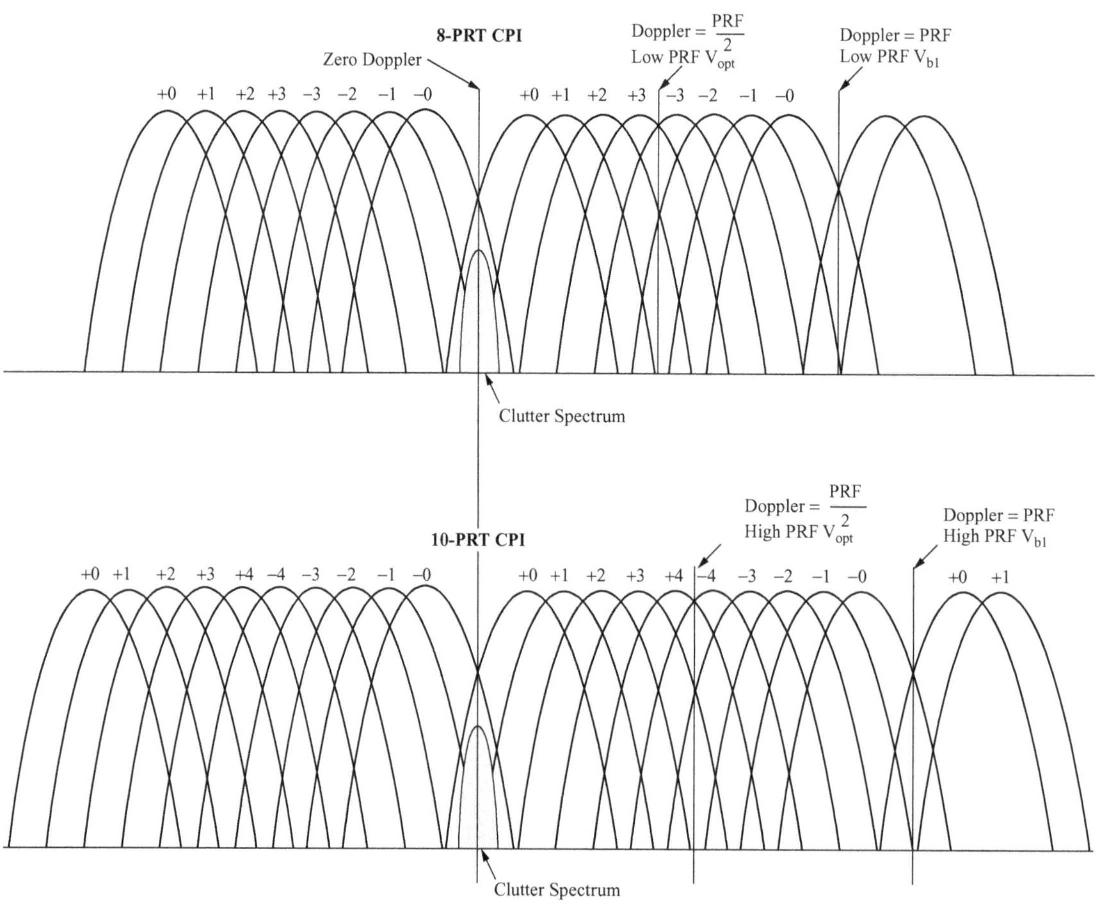

FIGURE 14–7

Filter responses for two CPIs.

Rectangular Coordinate Data Input

Recall that the two synchronous detectors operate in quadrature, that the *I* detector has a cosinusoidal response, and that the *Q* detector has a sinusoidal response. They produce, therefore, rectangular coordinates, describing i-f phase angle and magnitude. The detector output magnitudes, polarity, and proportional relationships describe the receiver output signal strength because $\sin^2\theta + \cos^2\theta = 1$, and because the hypotenuse of a right triangle = square root of $a^2 + b^2$. The amplitudes of the *I* and *Q* synchronous-detector outputs were converted to 12-bit parallel binary numbers in the A/D converter and then applied to the A/D interface card, which serves a number of utility purposes, such as restoring the zero baseline, adding weather test data, or accepting an input from the opposite channel's A/D interface card (in normal online/standby operation, both processors are operating, utilizing the data from the operational transmitter and receiver, and either processor may be placed on line, without changing transmitter channels).

The Input Data Buffer

The data from the A/D Interface card is in ***real time*** (see Figures 14-8 and 14-9). It is being converted into two, 12-bit, binary, words, one for *I*, and one for *Q*, every 0.772-μs range cell; this data is flowing into the Input Data Buffer in ascending ***range order***. The Input Data Buffer contains four memories, called "***bulk memory*** x", "bulk memory y", "***scratch memory*** x", and "scratch memory y". The ***range-ordered data*** is written into either

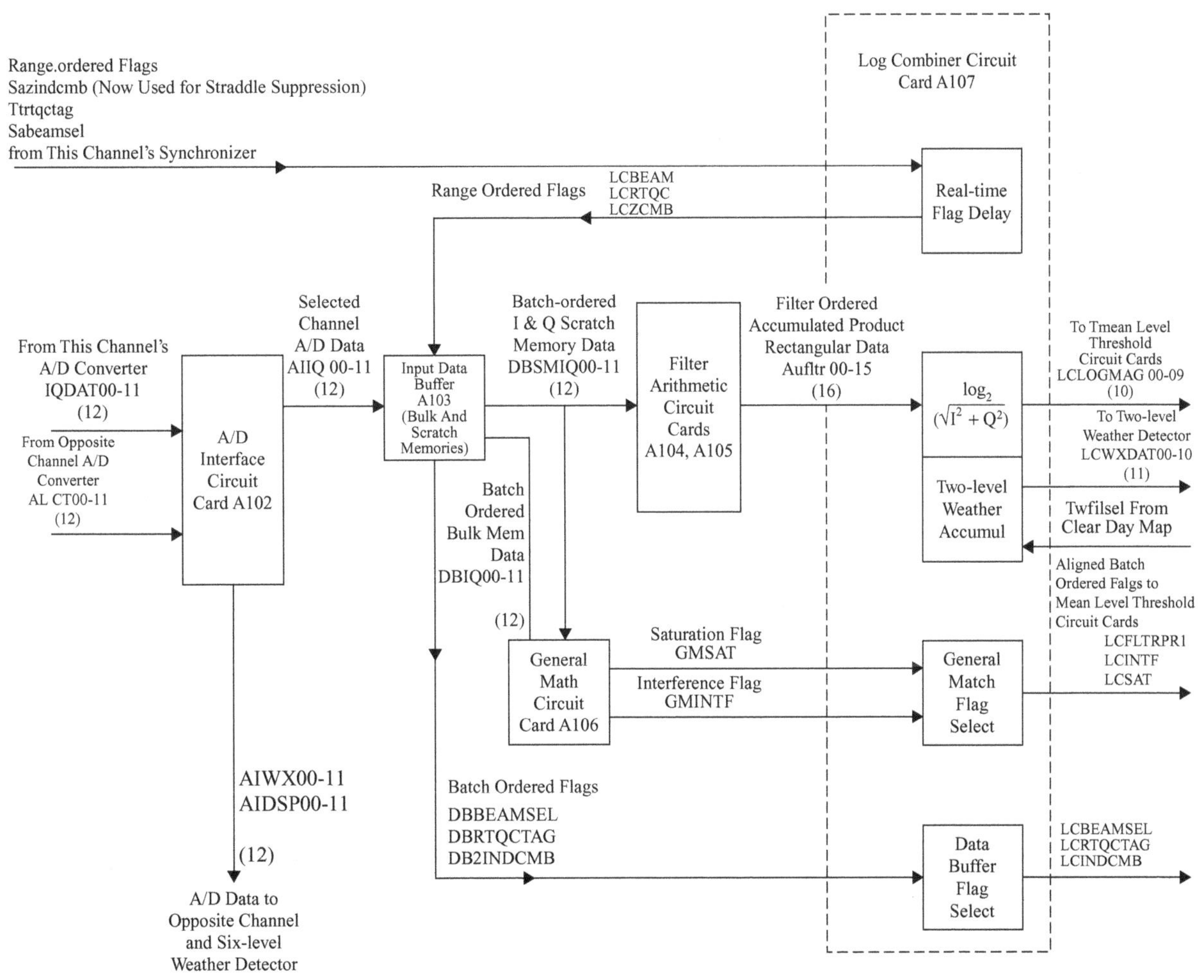

FIGURE 14–8

Filter-magnitude function.

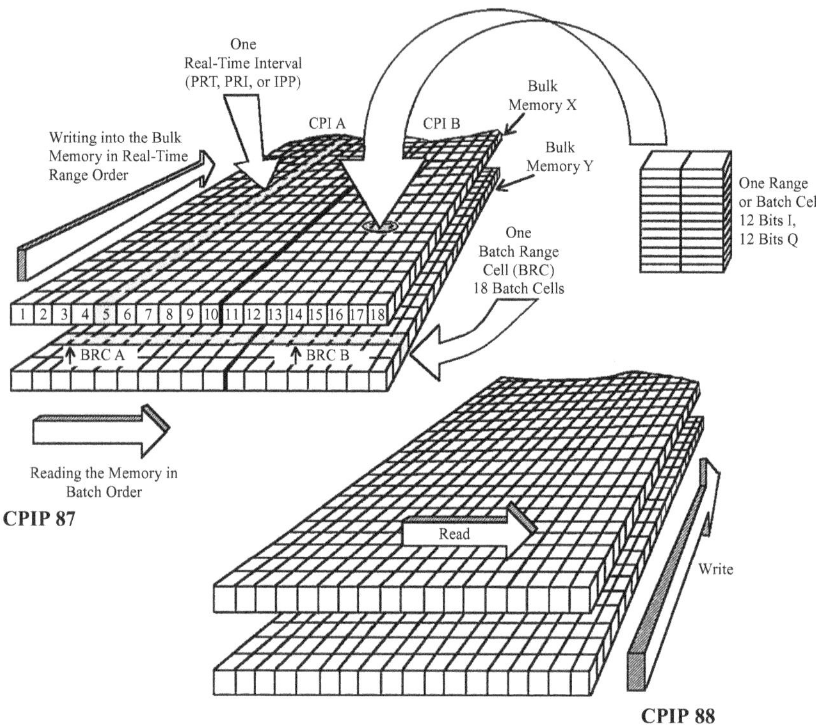

FIGURE 14–9

Range-order to batch-order conversion.

bulk memory, x or y, for an entire CPIP; the memory addressing selects the range-ordered locations. The two bulk memories are used in a CPIP flip-flop manner; during one CPIP, memory x is being written to and memory y is being read. On the opposite CPIP, memory y is being written to and memory x is being read.

On the following CPIP, the bulk memory is read sequentially, and at the same rate as it was written upon. However, the addressing scheme is changed, so that it is read "horizontally" or, what is called, ***batch order***. The effect of this is that 18 memory locations, containing all the data received in a 1/16 × 1 CPIP ***batch range cell***, is sequentially retrieved, over a period of 13.9 μs. For the purpose of maintaining definition, each memory location, called a "range cell" during the write operation, is now called a ***batch cell***. The batch range cell will hereafter be abbreviated to ***BRC***, the ten batch cells from range cells in CPIA will be called ***BRC A***, and the eight batch cells from range cells in CPIB will be called ***BRC B***.

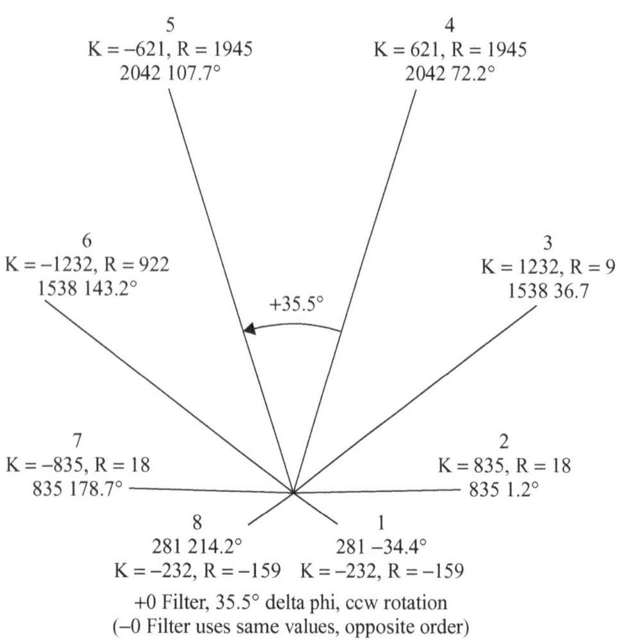

FIGURE 14–10

A rotating coefficient vector.

As the data for each BRC is read from the bulk memory, it is written sequentially into a scratch memory, again, either x or y. The scratch memories operate in a flip-flop write-and-read operation, also; however, the flip-flop occurs with each BRC, and each scratch memory contains one BRC of batch-cell data. Each batch cell contains 12 bits of *I* information and 12 bits of *Q* information.

The General Math Unit

This unit is used to detect saturation or interference conditions in a BRC. It contains absolute-magnitude conversion circuitry to enable signal strength comparisons, and utilizes both the bulk and scratch outputs, to accomplish this. The flags, GMSAT and GMINTF, indicate saturation or interference within a BRC, when low. GMSAT is set by excessive signal strength, and GMINTF is set when a single batch cell contains significantly greater signal strength than the others in the BRC.

The Filter Arithmetic Units

These cards are the Doppler filters. The filtering is accomplished by digital arithmetic in ***multiplier-accumulator (MAC)*** integrated circuits. There are two circuit cards, each containing five multiplier-accumulator circuits to total 10. Two MACs are required to facilitate a filter operation, so the two cards contain five filters. To obtain the 18 filters, the cards are used four times for each batch range cell; once, for five filters, for BRC A data; once, for five different filters, for BRC A data; once, for four filters, for BRC B data, and once more, for four different filters, for BRC B data. All 18 filter operations are completed in each 1/16 nmi, 13.9-μs batch range cell. The number of filters is equal to the number of batch cells in a BRC, so that the time required to output the filter data is the same as the time required to read one BRC from the bulk memory.

The rectangular coordinate multiplication equation is

$$(IR - QK) + j(QR + IK).$$

To analyze the filter operation, consider one filter only. Remember that there are five, operating simultaneously, but that the following analysis will consider only one of those five.

Figure 14-11 illustrates the multiplication of the I, Q data by the R, K data. In vector mathematics, conventional expression is with the sine of the angle on the Y-axis and the cosine as the "X"-axis; this places 0° in the "east" direction, and the vector progresses counterclockwise, with increasing angles. Having previously established that the Q synchronous detector is the sine of the phase angle, then it follows that Q must be on the "Y"-axis. R and K are intended as sine and cosine, respectively, so R is expressed as "Y," and K as "X" illustrated in figure 14-10.

An understanding of the process requires a step-by-step investigation of a few operations.

1. The bulk memory is read, one 0.772-μs batch cell at a time. 0.386 μs is required to read the I data, and 0.386 μs for the Q data.
2. The I and Q data for each batch cell is stored in the scratch memory. Ten I and Q batch cells will be written during BRC A and eight during BRC B.
3. On the next batch range cell, the scratch memory will be read, but at twice the frequency (0.386 μs per batch cell, 0.193 μs for I or Q).
   a. When the first batch cell is retrieved from the scratch memory location, the following occurs at a 10.35 MHz rate (0.0966 μs period):
      (1) An R and K rectangular coordinate set is retrieved from the PROM on the filter card.
      (2) I is multiplied by R in one MAC, and Q is multiplied by R in the other.
      (3) I is multiplied by K in one MAC, and Q is multiplied by K in the other.
      (4) The QK product is subtracted from the IR product in one MAC, and the QR and IK products are added in the other. The $IR - QK$ products are now contained in one MAC and the $QR + IK$ products in the other.
   b. When the second batch cell is retrieved from the scratch memory location
      (1) A new R and K rectangular coordinate set is retrieved from the PROM.
      (2) The $QR + IK$ and $IR - QK$ products are obtained as in step a.
      (3) The new $QR + IK$ and $IR - QK$ products are added to those from step a. The $IR - QK$ accumulated products are now contained in one MAC and the $QR + IK$ products in the other.
   c. When the third batch cell is retrieved from the scratch memory location
      (1) Another new R and K rectangular coordinate set is retrieved from the PROM.
      (2) The $QR + IK$ and $IR - QK$ products are obtained as in steps a and b.
      (3) The new $QR + IK$ and $IR - QK$ products are added to those from steps a and b. The $IR - QK$ accumulated products from three batch cells are now contained in one MAC and the $QR + IK$ products in the other.
   d. The process described in steps a through c continues for 10 batch cells, after which the accumulated $IR - QK$ and $QR + IK$ products are stored in registers as ***accumulated product data***. $IR - QK$ is one rectangular coordinate (called only Q on the output) and $QR + IK$ is another (called only I on the output).

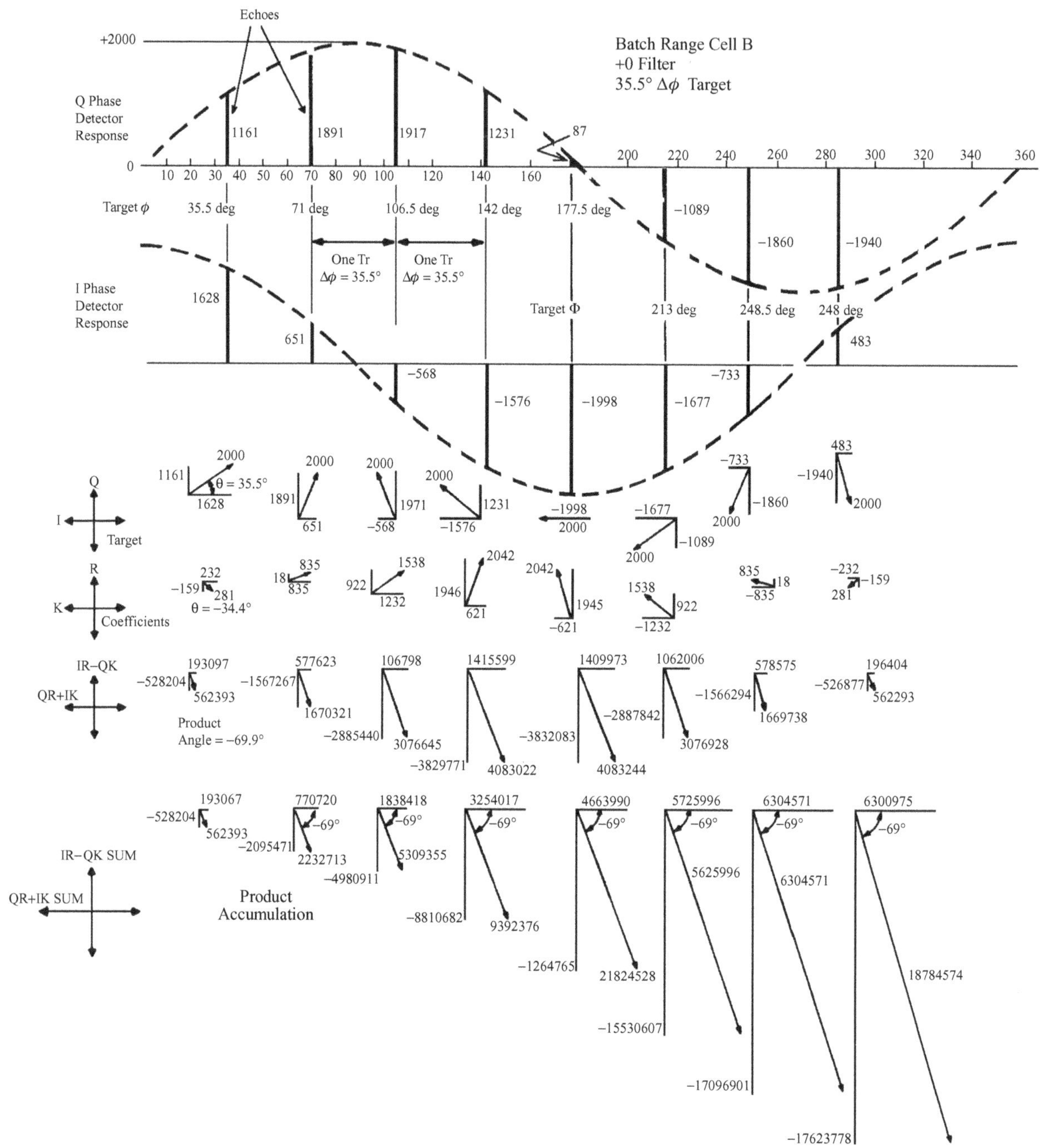

FIGURE 14–11

Filter math.

Together, they are rectangular coordinates describing the magnitude and angle of a final accumulated vector.

e. The same 10 scratch memory locations are read a second time, and the same multiply and accumulate process as steps a through d are applied, but different values of *R* and *K* are used, so different products are accumulated. The same hardware has, therefore, been used for two different computation sets for BRC A.

f. BRC B is now read from scratch, one batch cell at a time, and the cross-multiplication and accumulation occurs again. Different values of *R* and *K* are used, and there are only 8 batch cells instead of 10.

g. BRC B is read from scratch a second time, different values of R and K are used, and the same hardware has now been used for two computations for BRC A, two computations for BRC B, and four computations for the entire batch range cell.

4. Throughout the preceding operations, the accumulated products from all five hardware filters are being clocked from the output registers onto an output bus; this has to occur four times per BRC to ensure that each register's contents have been "captured" before a new accumulated product is placed in it.
5. During BRC B, only four of the five hardware filters are used. Therefore, the outputs for the entire BRC are in a sequence: 5 filters BRC A, 5 filters BRC A, 4 filters BRC B, 4 filters BRC B, for a total of 18 filters.

It takes 13.9 μs for all the accumulated product data, for one BRC, to occur, which is the same amount of time required to read the 18 batch cells from the bulk memory. All the filter data for one 1/16 nmi BRC is now contained in a 1/16 nmi BRC; the BRC time is unchanged, but its data contents are different.

The Accumulated Product

Figure 14-11 illustrates a case in which the vector described by the target rectangular coordinates was rotating at the same rate as the vector which is described by the coefficient rectangular coordinates. In that case, the angular difference between the I, Q and R, K vectors remained constant. The product rectangular coordinates describe a vector whose angle is the difference between the target and coefficient vectors. Even though the two input vectors (receiver data and coefficient) rotate, the product vector does not, because there is no change in the difference between target and coefficients.

As the product coordinates are accumulated, the vector they describe may "grow" at the same angle (where the target Doppler and coefficient change are equal), or it may change angles (where the target and coefficient rotations are mismatched) (see Figure 14-12). When a mismatch occurs, the final accumulated product will be less, providing a lower magnitude vector description. Although the angle of the product vector is described by the final accumulated rectangular coordinates, only the magnitude will, in the end, be important. The magnitude of the accumulated product vector will describe the signal strength of the filter output, and the angle represents only an angular difference between the coefficient and target vectors.

When a target is present, there will ordinarily be outputs from several of the filters, but one or two will contain peak magnitudes, indicating the center Doppler of the target. Because the BRC contains information from the same range cells in both CPIs, and because the CPIs are at different f_ps, additional information can be obtained, regarding the center Doppler, by the Doppler interpolation process in the C&I software, to the extent that the f_D can be resolved to 64ths of the f_p.

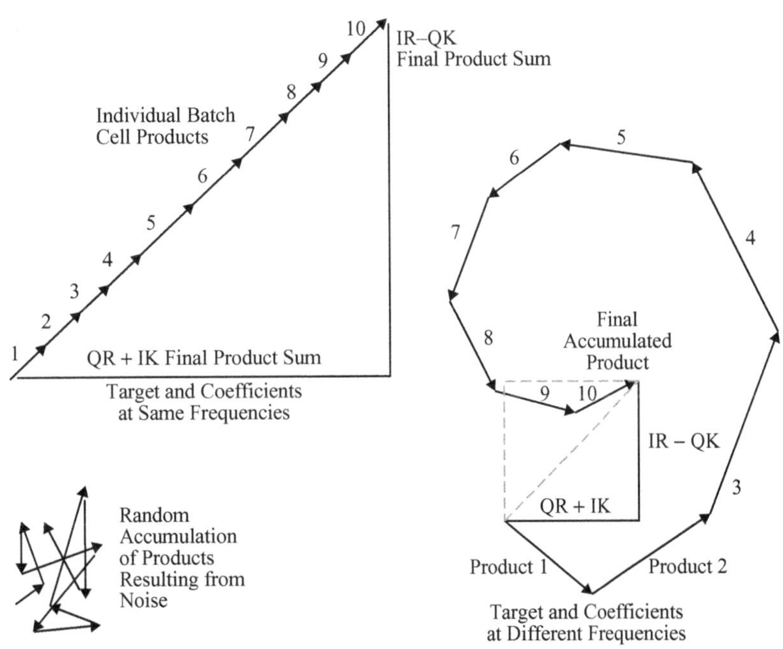

FIGURE 14–12

Product accumulation.

Figure 14-12 shows that noise in a filter adds in a random fashion, in comparison to targets, providing a low-level output. Because of this effect, the signal-to-noise ratio is enhanced, and an overall mds improvement of 8 dB is obtained. This overcomes the loss incurred by the very narrow-band i-f amplifier to produce a final sensitivity comparable to a conventional canceler-type MTI system. The signal-to-noise improvement in the filters is called ***coherent integration***.

The Filter Data Message

As the filters develop the final products, those final products must be promptly removed from the output registers to ready those registers for the next product. As the 18 ***implemented*** filter operations are completed in the five ***hardware*** filters, the data is removed from the output registers in a five-filter, five-filter, four-filter, four-filter sequence, as shown in Figure 14-13. There is a complete set of 18 filter rectangular-coordinate product data for each BRC, and this occurs every 13.9 μs. The filters are numbered in a meaningful manner, with the value of the number proportional to a Doppler value in terms of f_p (recall that $f_{DVbl} = f_p$). For instance, the −0 and +0 filters

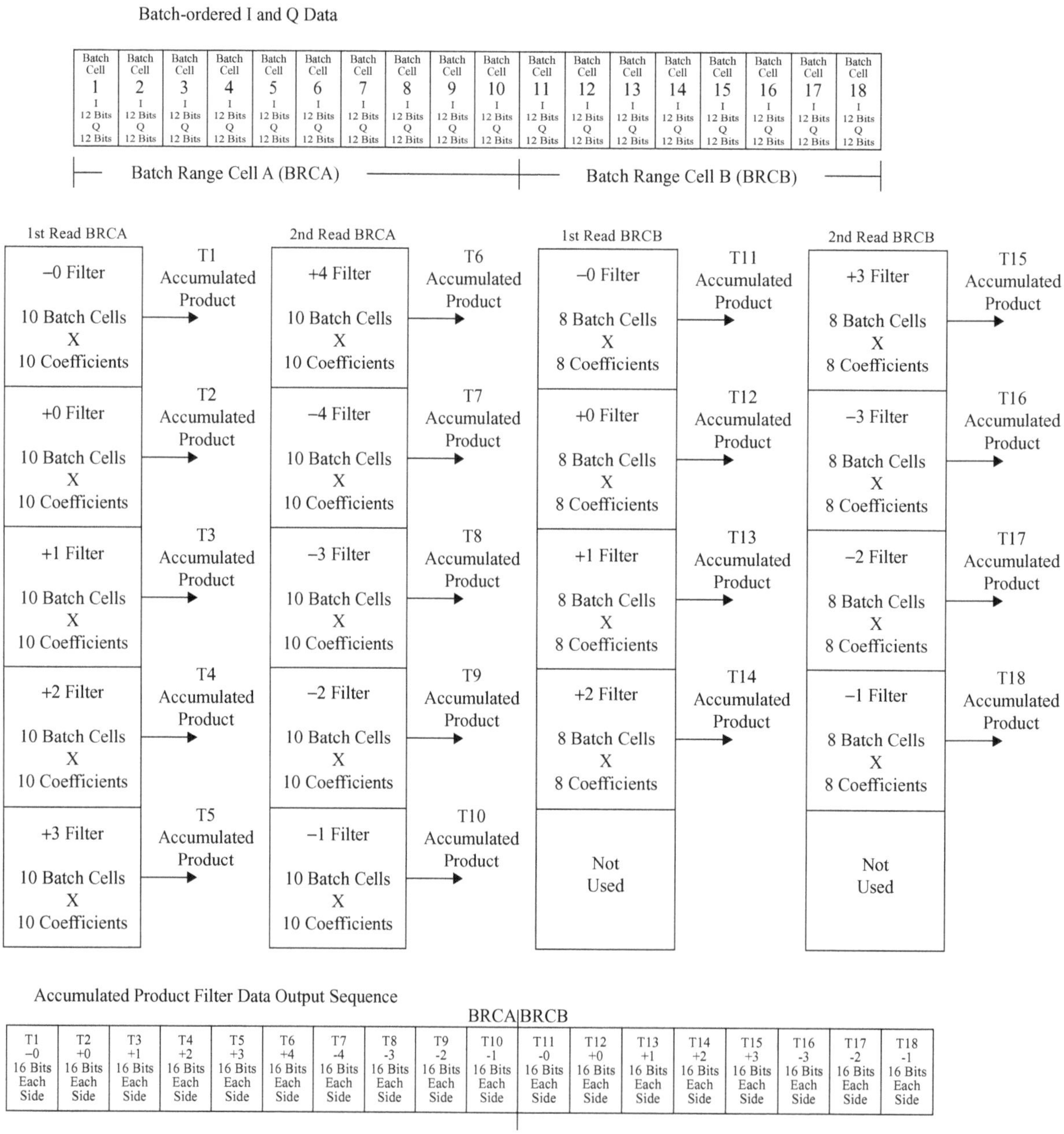

FIGURE 14–13

Filter data output message sequencing.

encompass zero Doppler (the region where clutter occurs), and also the V_b regions, where the Doppler equals the f_p or multiples. The two +0 and −0 filters for BRC A use identical coefficients, but in reverse order, so they are "tuned" for identical, but reverse, rotation. The two +0 and −0 filters for BRC B are also tuned for identical, but reverse, rotation. All other filters of the same number are also tuned for identical but reverse rotation, such as +1, −1, +2, −2, etc. At optimum velocity, the highest-numbered filters cross ±4 filters for BRC A, and ±3 filters for BRC B. Past 180° Δϕ, the Doppler becomes ambiguous; for instance, a 190°Δϕ target could appear to be a −170° Δϕ target.

The Power Combiner

The filter-product-data BRC message contains rectangular-coordinate data, which describes all the filter outputs in signal strength and angle. The angle is only a consequence of the difference angles between the input and coefficient data, and serves no further useful purpose after the filters. In the power combiner, the base-2 logarithm of the square root of the sum of the squares of the rectangular coordinates is found in a table. The data output of the power combiner is still in a 13.9-μs filter-ordered BRC, but it is in a final form, and is called the ***log magnitude BRC***, because each filter data now represents the signal strength in dB above noise, of each filter output.

Again, see Figure 14-8 and note the "flags" from "this channel's" synchronizer. These flags are signals which indicate real-time events, such as beam-switch gates, and the time of occurrence of an rf ***real-time quality control (RTQC)*** test target. These flags must occur simultaneously, with the data to which they apply. Since the data was placed into batch order, the flags must also be placed into batch order to maintain temporal alignment. This is accomplished by delaying them one real-time range cell, then writing them into the bulk memory in real-time range order, and reading them from the bulk memory in batch order. When the flags are applied to the log combiner output (DB flag select; "DB" stands for "data buffer"), they are in temporal alignment with the log magnitude BRCs, to which they apply.

The power combiner also contains a two-level weather (2Wx) accumulator circuit, which will accumulate log magnitude filter data throughout BRC A, and then again through BRC B, providing two single summed outputs per BRC. The 2Wx accumulations may be in one of two forms; they may be the sum of all the filter data, or the sum of only all the NZVF filter data. Since the NZVF filters do not include the ±0 filters, their sum is essentially lumped MTI information; since the sum of all the filters includes all Dopplers, their sum is essentially lumped log normal information. In short, the 2Wx accumulator produces batch-ordered log normal or MTI data. The accumulator operation is governed by a signal called TWFILSEL (not shown) from a ***clear-day map*** memory in the 2Wx detector in the target and 2Wx detector functional area to follow. The clear-day map has a granularity of 1 nmi × 1 CPIP. In those areas where clutter was detected when the map was made on a clear day, TWFILSEL causes the 2Wx accumulator to produce MTI data; otherwise, it produces log normal data. The purpose of this is to prevent clutter from being injected into the weather detector, where it would cause threshold breaks. This is a latter-day form of clutter-gated MTI. It differs from earlier types, in that the clutter gates are semipermanently stored in the clear-day-map memory, rather than being derived from quantized live normal video. Of course, it also differs, in that it does not operate in real time.

The Target and Two-Level Weather Detector

The Mean-Level Threshold (MLT) Cards

Figure 14-14 is a simplified block diagram of the ***target and two-level weather detector function***. The data input from the log combiner in the filter/magnitude function is first applied to the MLT cards in the target and two-level weather detector. The reader may find it helpful to review the digital log ftc circuit described in Chapter 13 on a sample MTI processor; the MLT cards operate on a similar principle, but are far more complex. In the digital log ftc circuitry, the thresholds were subtracted from the data in the "area of inspection." In these MLT cards, 13-BRC averages will be computed for each of the filter data, but will not be employed against the area-of-inspection data until further downstream, where adjustments have been made by the ***primitive detector number one (PD1)*** circuit card.

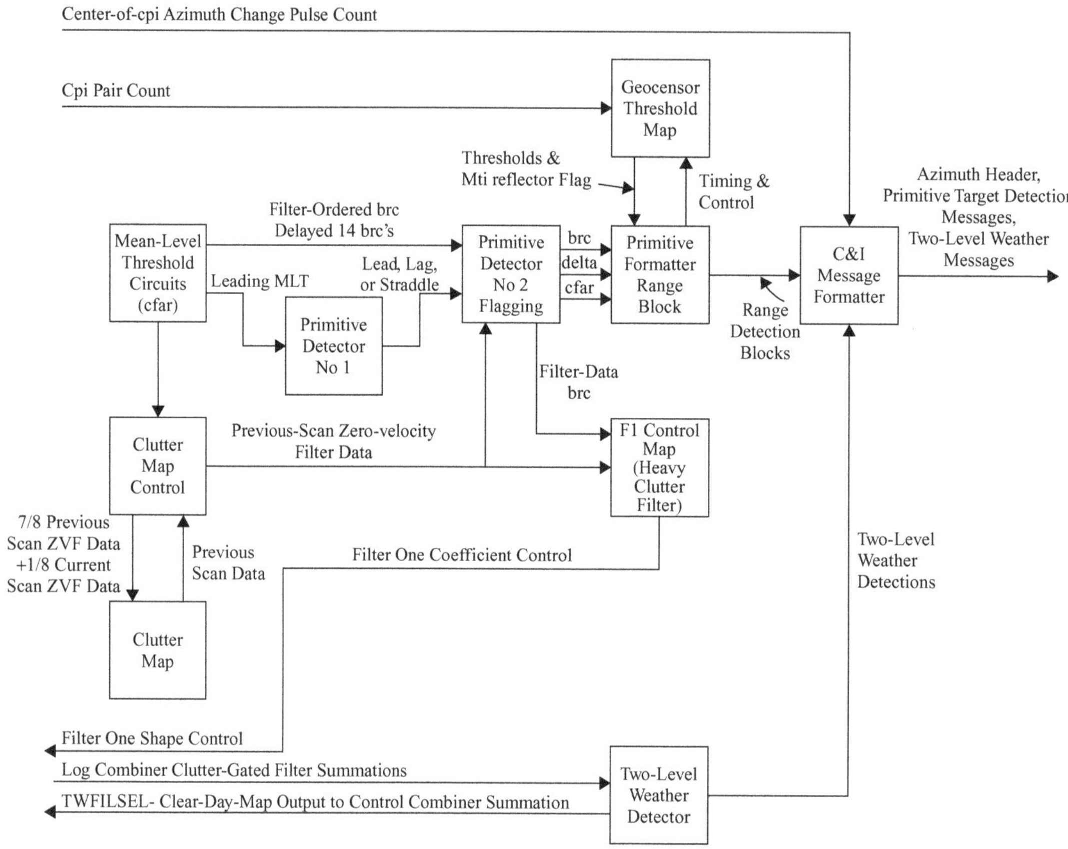

FIGURE 14–14

Target and two-level Wx, block diagram.

Improved Detection in Weather

An MTD system has a greater potential to detect targets in weather for two reasons. One is that the visibility of the target in strong weather is degraded by i-f limiting in an MTI system; however, the absence of that limiting, in the MTD system, allows for the target to be separated from the weather in the Doppler filters. Even further, a target at a different Doppler than the weather will be present in a different filter data word, and will be unaffected by the thresholds developed for the filters containing the weather data.

The two MLT cards operate in parallel for time conservation (see Figure 14-15); the 18 filter data in each BRC are alternately distributed to RAMs in the two cards, so that 1.54 μs are available to process each $\log_2$ data word, rather than 0.772 μs, the time duration of each filter data from the power combiner. Figure 14-15 illustrates the manner in which the filter data in the BRC is routed to the RAMs. The operation is accomplished with addressing techniques, but may be conceptually simplified to a row-and-column depiction. The filter data for each BRC are placed into a column, assigned by a 14-BRC counter, and all the data for each filter then become arranged in rows.

As shown in Figure 14-16, the cards contain 18 ***range sliding window*** circuits, which develop thresholds for each of the 18 log magnitude filter data, arranged in rows in the RAMs. The log-magnitude BRCs come out of the cards in the same form and order as they went in from the power combiner; however, they occur 14 BRCs later (13.9 μs × 14 = 194.6 μs) because they are held in the memory for that length of time. In time alignment with each output log magnitude filter data word in the BRC, is a mean-level threshold, applicable to that filter data; there are 18 thresholds for 18 log magnitude filter data. The threshold for each log magnitude filter data is obtained from an average of all the same filter data in the 13 BRCs that occurred after it. The data from that BRC immediately following the log-magnitude BRC is excluded from the average as a ***guard cell***, and the highest amplitude data and two cells adjacent, are excluded, as a ***peak cell*** and ***peak guard cells***. Because of the removal of the guard cell, peak cell, and peak guard cells, the average is of a maximum of nine BRCs, and may be reduced even further, if saturation or interference flags had been present.

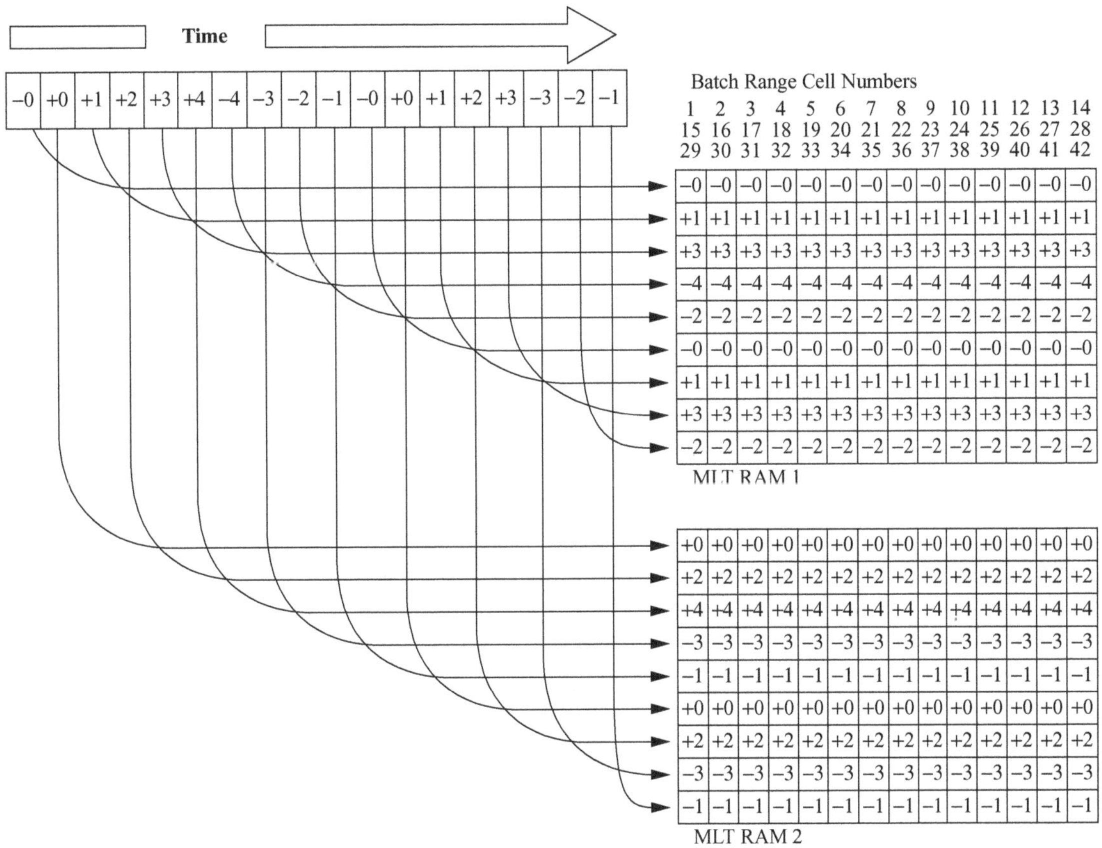

FIGURE 14–15

Distribution of filter data into the MLT RAMs.

Primitive Detector Number 1 (PD1)

The thresholds developed in the MLT cards are called ***leading mean level thresholds,*** which might seem erroneous, since they are developed from data which occurred after the area of inspection. As in log ftc or cfar circuits, the objective is to maintain an average of the data in the range sliding window, which ordinarily will be an average of noise. The average rises when weather is present, as is illustrated in Figure 14-17. An objective in the target detector is to remove as many low-level noise spikes as possible, as they cause false detections (false alarms). The mean-level threshold circuits cause the threshold to adjust with data, so as to cause a constant false detection rate called ***constant false alarm rate (cfar)***, but the MLTs are an average of the noise and below many of the spikes. In PD1, another threshold number is added to the MLT to raise the threshold above most of the spikes; it is adjustable by a ***variable site parameter (VSP)*** called ***PFA cfar*** (probability of false alarm for cfar data).

Again, see Figure 14-16. PD1 also contains a 14-BRC memory, but it is for the threshold data. Recall that the log magnitude data is aged by 14 BRCs, and was the first data in the window. If the threshold data is stored in a memory to be retrieved 14 BRCs later, it is aged more than the log magnitude data; therefore, it represents information that occurred before (at a lesser range than) the log magnitude data. This memory, then, provides thresholds which preceded the log magnitude data in time. They are called ***lagging mean level thresholds.***

A comparator circuit in PD1 compares the levels of the leading and lagging thresholds and chooses the greater of the two, to become the data threshold. The chosen threshold is called the ***greater mean level threshold (GMLT)***. Another threshold circuit obtains a threshold from both sides of the log magnitude BRC; it is called a ***straddle threshold*** and was added, as a "fix," to prevent false target detections on the leading edge of weather blocks. In normal operation, PD1 may choose between the leading, lagging, or straddle threshold; the mode is called ***three-sided threshold***. The straddle threshold may cause degraded target resolution, and can be inhibited

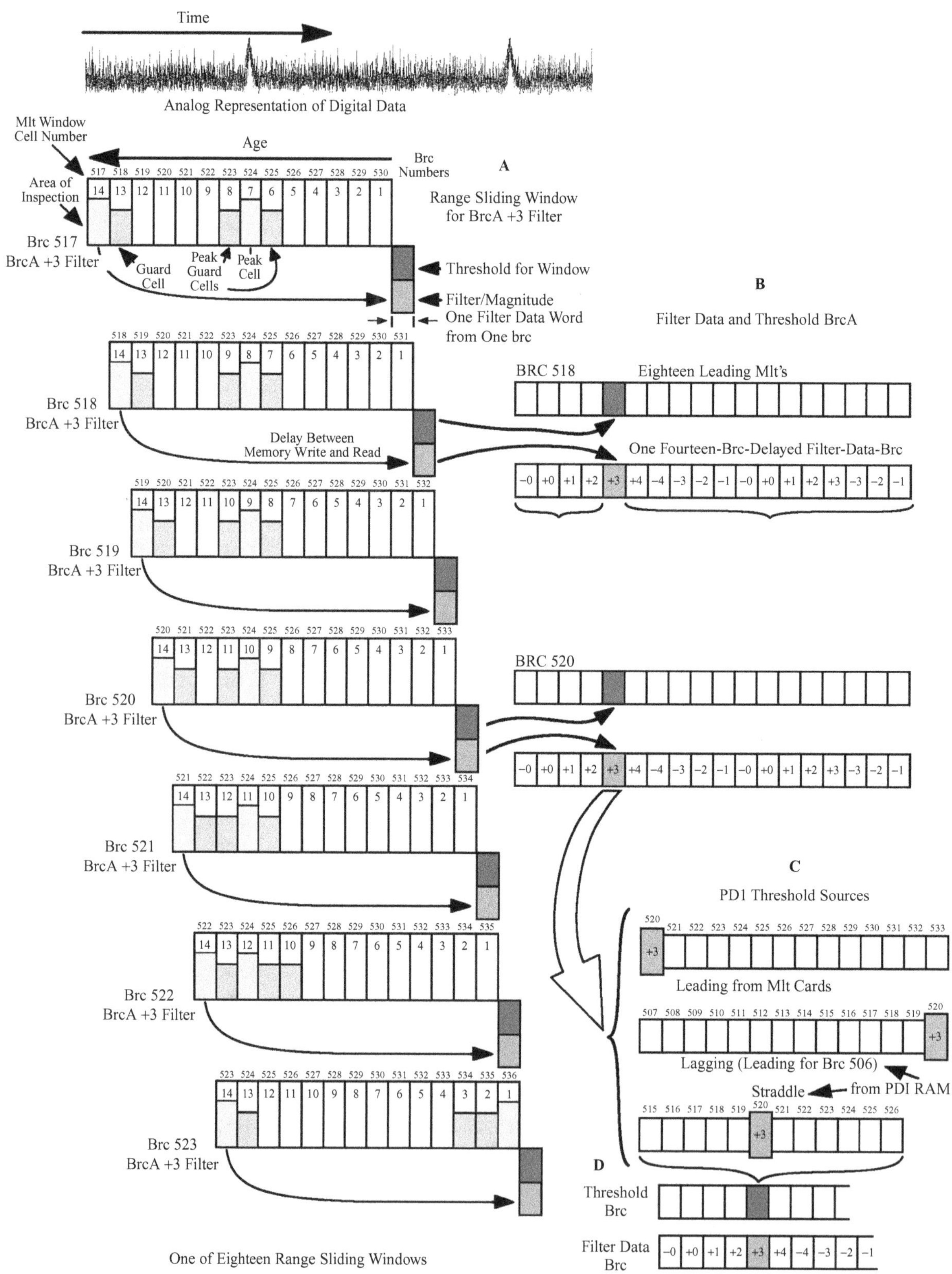

FIGURE 14–16

CFAR threshold development.

in critical areas, such as runway approaches, by a RAG program. This mode, in which straddle is inhibited, is called a ***two-sided threshold*** and only the GMLT is available. Another mode, called ***one-sided threshold***, is used during the first 13 BRCs of the CPIP; at those ranges, there is no data available for a lagging threshold, and only the leading threshold, or a memory ***residue map***, may be used.

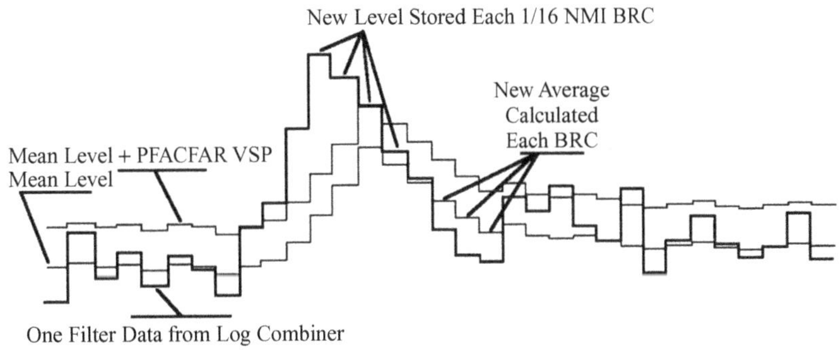

FIGURE 14–17

Mean level averaging.

Clutter Map and Clutter Map Control Card

Were it not for the need to recover tangential and blind velocity targets, these cards would not be necessary, and clutter could be removed by simply eliminating the ZVF data. The clutter map is a memory, with a 1/16 nmi × 1 CPI granularity, the maximum possible resolution. The main purpose is the scan-to-scan storage of ZVF data, achieved in a manner which smoothes the map to prevent moving ZVF targets from making severe changes to the map. That smoothing process is achieved by updating the map, each scan, with 1/8 of the current received ZVF data and 7/8 of the ZVF data recovered from previous scans. The map may be updated every *n* scans, where "*n*" is a VSP, which may be set from 0 to 4.

The radar transmitter must be in operation for several scans, before a smoothed clutter map can be achieved. Were no provision made to preclude the possibility, a transmitter-off condition, whether intentional, or because of failure, would cause grass-level ZVF data, and the clutter map would quickly be filled with a total absence of clutter. To preclude this, a signal from the transmitter will "freeze" the map, whenever the high voltage has been removed, and the map may not, then, be updated.

Primitive Detector No 2 (PD2)

This is the first point in signal flow where data is actually tested against thresholds. The data input, although delayed by the MLT RAM and clocked-circuit progression, is yet unmodified from its form, at the power combiner output in the filter-and-magnitude area, and is in the form of a filter-ordered BRC. In this card, the thresholds from both the clutter map control card and PD1, will be applied to the data; where a filter data exceeds a threshold, a flag is set. As illustrated in Figure 14-18, the actual operation is made possible, because each threshold is in time coincidence with the filter data to which it applies. If an NZVF filter data exceeds an adjusted cfar threshold, a ***CFAR flag*** is set; if a ZVF filter data exceeds an adjusted clutter map threshold, a ***Δflag*** is set. The flags and data are, then, routed to the ***primitive formatter*** card, where a final geocensor threshold test is made, and the preliminary assembly of an output data message begins.

Consider the significance of the CFAR and Δflags. If a Δflag has been set, it represents a ZVF threshold crossing; for that to occur, the current incoming ZVF data had to differ from that contained in the clutter map, from previous scans. Since the data contained in the clutter map had been smoothed over several scans, it predominately represents ground clutter, and any threshold crossing had most likely to be caused by something other than ground clutter, which may easily have been a ***tangential velocity target*** or a ***blind velocity target***.

The Geocensor Map

This circuit card contains two memories, one of which is range–azimuth addressed with a granularity (resolution) of 1 CPIP × 1/8 nmi. Each memory location in that map contains information regarding the type of threshold to be used, for the geographical location described by the range–azimuth address. There are two types of geographical

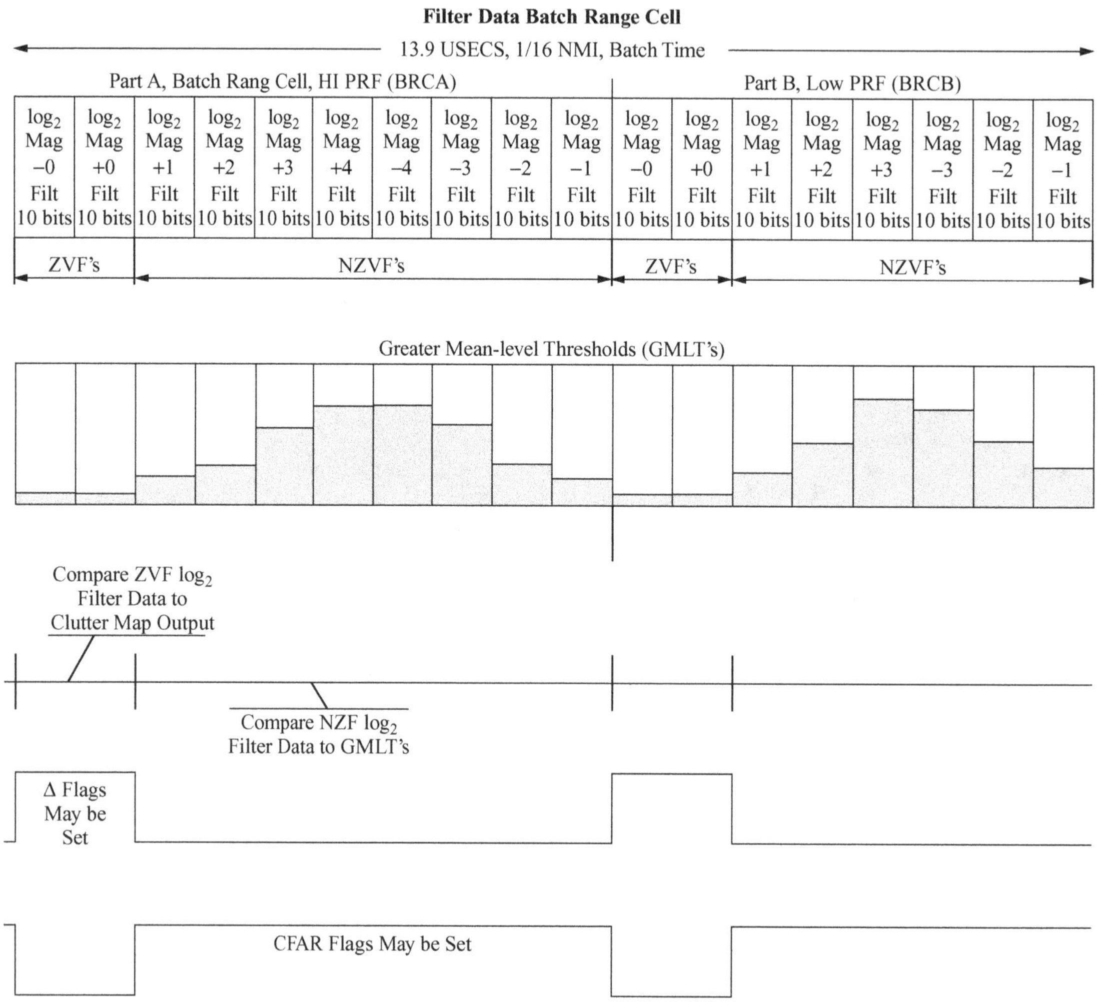

FIGURE 14–18

Filter data BRC versus thresholds versus flags.

thresholds, called ***flat*** and ***shaped***; the designations are misleading, since either may be shaped but in different manners, and for different purposes. The flat threshold is intended for use in discriminating against moving targets in predetermined areas; these would be caused by ***road traffic***. The shaped thresholds are intended for use in discriminating against ***special clutter problems***.

The range-addressed map does not contain threshold data, but only ***threshold type data***. When a flat threshold type is prescribed, an entire ***family of thresholds*** is retrieved from another ***threshold memory***; when a shaped threshold is prescribed, an entirely different family of thresholds is retrieved from the threshold memory (see Figure 14-19). The families of thresholds are sequentially retrieved from the threshold memory, in temporal coincidence with the filter data BRC, as it is applied to the ***primitive formatter card***, and the geocensor thresholding occurs in that primitive formatter card. A specific threshold is applied to each filter data, and the thresholds are adjusted by VSPs. There are only two families of thresholds; all flat-threshold families for all selected geographical locations are identical, and all shaped-threshold families for all selected locations are identical.

The address to the threshold memory is determined by the filter-data time within the BRC, and by information to indicate BRC A or BRC B. Since there are two more T_rs in CPIA than in CPIB, it is necessary that those thresholds for BRC A be slightly elevated to compensate for the additional accumulated signal power. The thresholds may or may not eliminate target detections, but have an effect on target processing, in either case. Targets which break the geocensor thresholds in the primitive formatter card will become messages, which contain flags to indicate that they were in threshold zones; this will influence downstream software processing. The geocensor map may, thus, influence data processing, even if all the thresholds are set to zero, and all data inputs to the primitive formatter card exceeds the thresholds. The geocensor range–azimuth, threshold-type memory locations may also contain an ***MTI reflector flag***. Because the MTD system does not provide normal radar information to the indicator facility, radar alignment depends entirely upon the MTI reflector targets, and the geographical areas containing those must be flagged, to prevent the signal processing, at any stage, from eliminating a target that has occurred in an MTI reflector flag zone. The MTI reflector flag will appear in the target detector output message, to be described in paragraphs to follow.

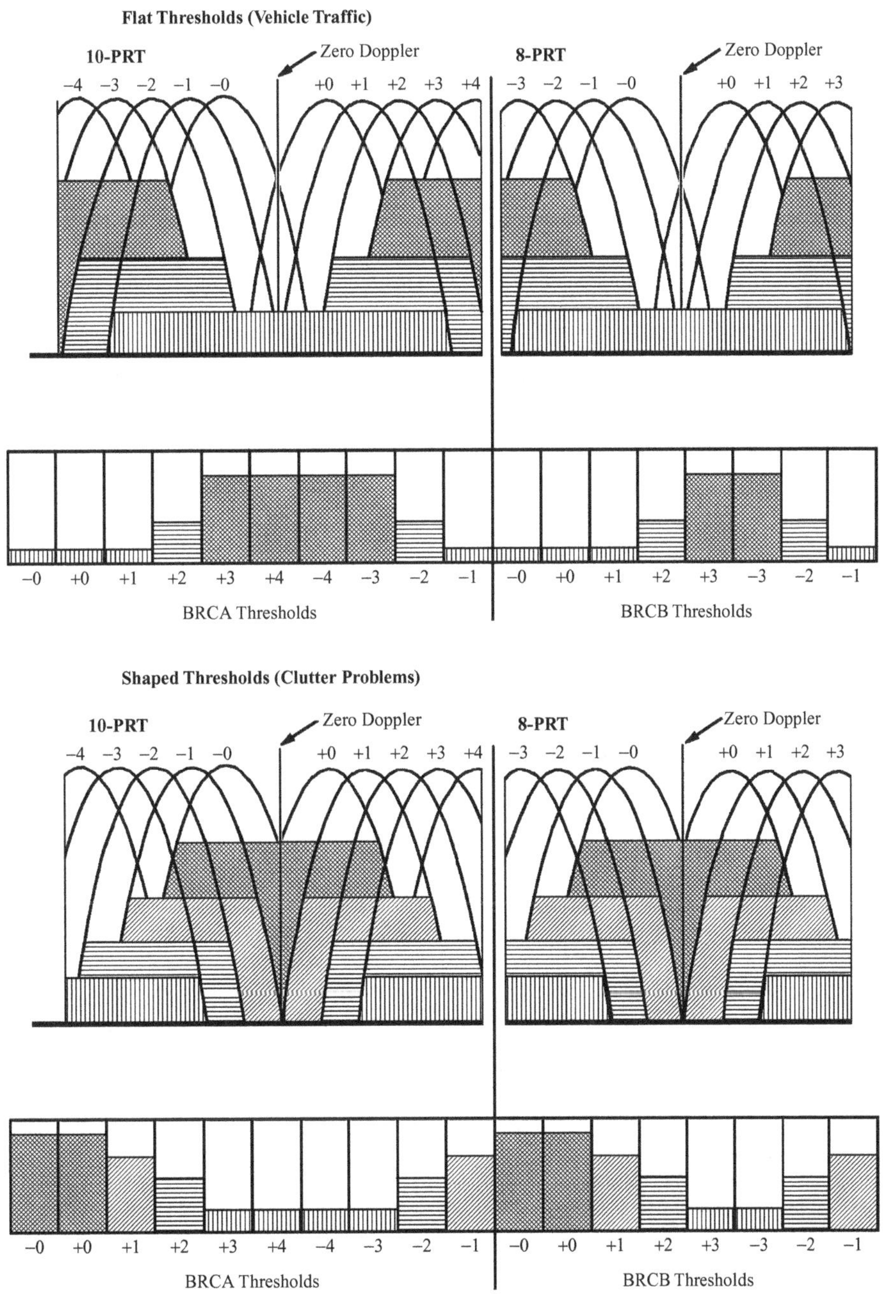

FIGURE 14–19

Geocensor threshold "amilies."

The Primitive Formatter Card

Here, the first formatting of the final ***primitive target detection message*** begins. Inputs to this card include the filter data BRC, CFAR and Δflags, geocensor threshold data, and a 2-bit geocensor threshold-type data. The card contains a timer/sequencer used for message assembly, a geocensor threshold comparator, a "range counter," which is actually a BRC counter counting a 1/16-nmi increment once every 13.9 μs, and other supporting circuitry.

If a filter data exceeds a threshold in the PD2 card, and then further exceeds a geocensor threshold in the primitive formatter card, a ***primitive target detection*** occurs. A single primitive target detection is a single filter-data threshold crossing, so there can be several primitives in a BRC since several filter data may break thresholds. Additionally, for a given target, there can be several range-adjacent BRCs with threshold crossings, and several azimuth-adjacent BRCs as well. For a given aircraft, a ***cluster*** of primitives will occur; the centroiding process in the C&I software module will find the center of density of this primitive cluster to declare a single azimuth and range.

The primitive formatter card assembles ***range block detection data*** (see Figure 14-20). For the threshold crossings in a given BRC, a message, consisting of header words, filter data, and filter identifiers, is assembled. Because the range from the BRC counter is contained in the header, timing relationships, between the detection data and CPIP start, are insignificant to radar range considerations past this point in data processing, and no further real-time temporal references are necessary.

The range header consists of the range count, a block count, and other data. The block count indicates the number of detection words that will follow; since that number is variable, the receiving buffer in the postprocessor ASP must "know" how many words are to be received, before it again "looks for" a new range header. Following the range header are 10-bit words, describing the log magnitude data that crossed the threshold, and additional bits to describe the filter number and to mark the ***peak filter*** (the strongest signal in the BRC). The range detection blocks are routed to the ***C&I formatter card***, so named because it formats all target and two-level weather detector information for use by the C&I software process in the postprocessor.

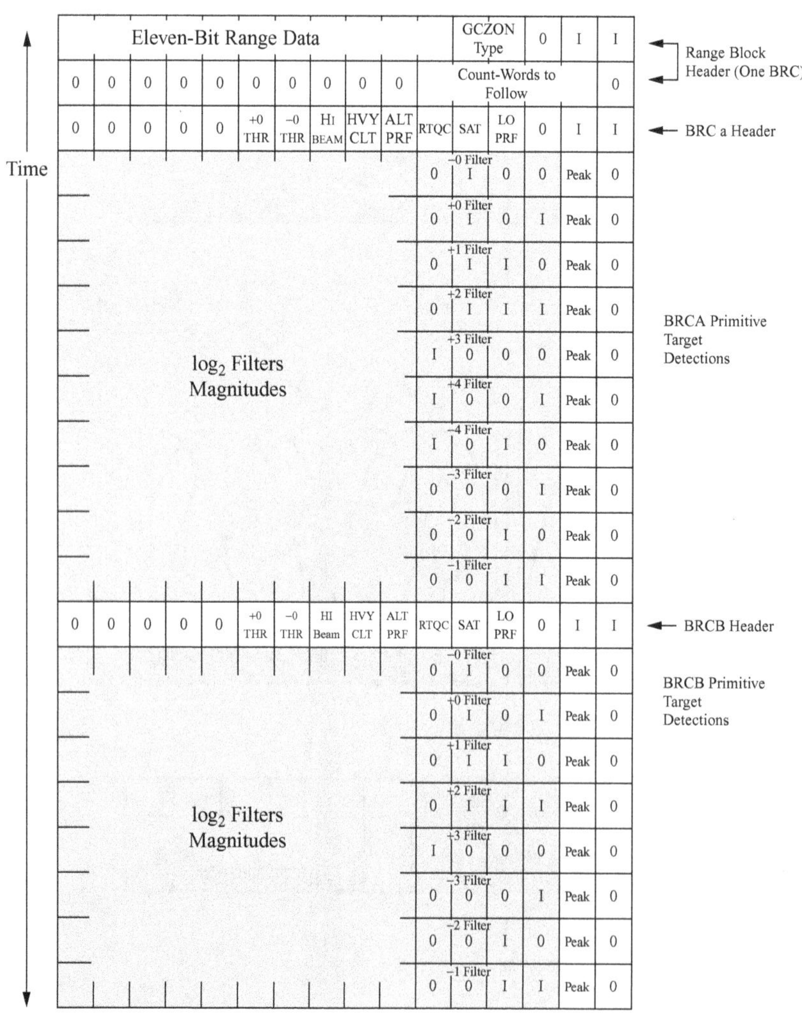

FIGUR]E 14–20

Range detection blocks.

The C&I Formatter Card

This card provides the final output messages to a high-speed interface buffer in the ***ASP***, for processing by the C&I computer program. The C&I formatter card operates at a high speed, transmitting a 16-bit word every 0.385 μs. It is capable of transmitting all 18 detections, in all 960 BRCs each CPIP, and cannot be overloaded. Of course, such a condition is not anticipated, and would be extremely unusual. Figure 14-21 illustrates the total message-content capability; a condition under which a single message would use all words illustrated would, again, be very unusual. The C&I formatter card contains a number of circuits, including a timer/sequencer, FIFOs, and others. Inputs include the range detection blocks, system status information, and two azimuth-counter inputs from the synchronizer; these azimuth-counter inputs describe the azimuth of the center of each CPI in the CPIP, and will be used downstream in target centroiding.

At the beginning of the CPIP, the C&I formatter produces an ***azimuth header***. That azimuth information is then retained in ASP memory until the next CPIP, and applies to all range blocks that follow. Range detection blocks are transmitted as detections occur, and are in BRC order.

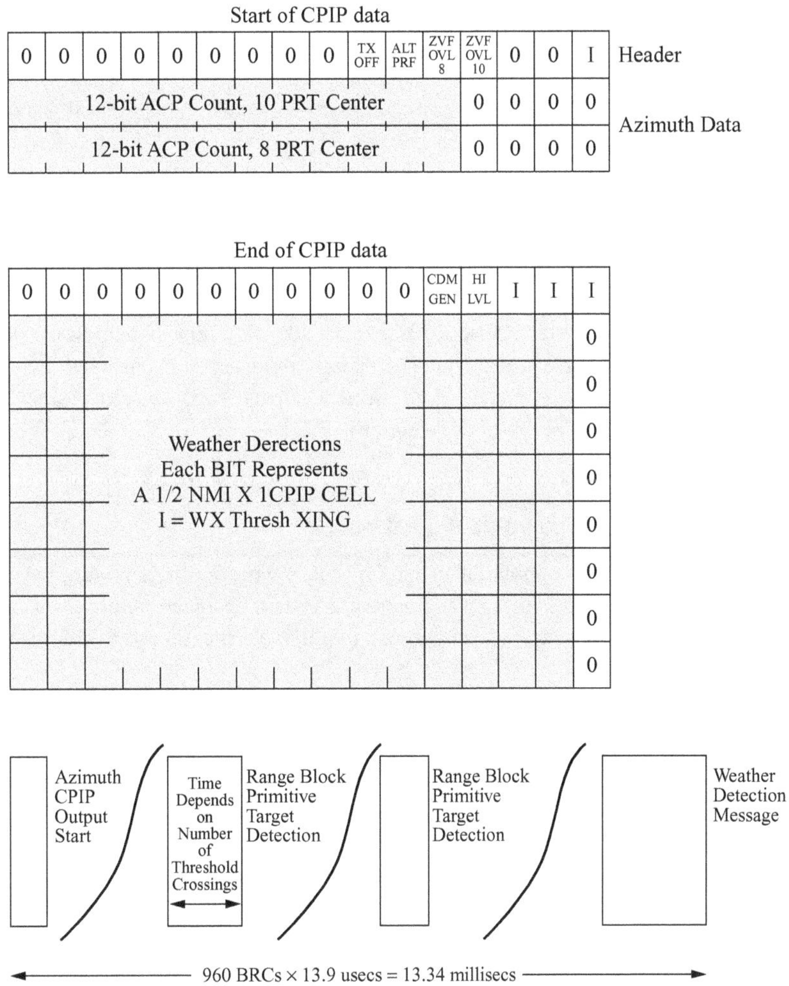

FIGURE 14–21

Output messages to the C&I process.

The Weather Data

At the end of the CPIP, the two-level weather (2Wx) message is transmitted from the C&I formatter. The two-level weather detector is a weather signal-strength threshold device. On alternating antenna scans, the detector threshold is at a low level; on opposite scans, it is set to a higher level. The thresholds switch each scan, and are two of the six national weather service recognized levels; they are chosen by VSP. A weather header identifies the high or low scan. Weather detections are made in 1/2-nmi increments, so the weather message contains 120 bits in several words; each bit, when set to a "1," indicates a weather detection.

The F1 Control Card

Recall that the ZVF filter responses are centered on clutter, and that the number 1 filter responses bracket the ZVF filter responses. The Doppler frequency separation between the number 1 filters is, then, somewhat equivalent to the velocity response notch in a canceler-type MTI system. When the clutter spectrum becomes very strong, the clutter begins to appear in the number 1 filter data, which could cause false moving-target detections. This is comparable to a canceler-type MTI velocity response shape, with too narrow a notch. The F1 control card provides a means to narrow the response of the number 1 filters, one scan later, in the presence of heavy clutter, creating an automatic velocity response change.

The F1 control card contains a memory map, with a 1/2 nmi × 1 CPIP granularity, and operates on smoothed inputs from the clutter map and all data from PD2. The inputs from PD2 are used to supply NZVF data. The NZVF data is used to detect the presence of weather; if there is excessive NZVF data, indicating weather, the

heavy clutter decision (logic circuit decision to change filter 1 coefficients) must be inhibited. The smoothed clutter-map information is used for ZVF data, to prevent heavy clutter decisions from being made on transient conditions.

When a threshold is broken by strong clutter inputs, and yet another threshold is not broken by strong NZVF inputs, a bit is set, and then written into the F1 memory map. Each location of the memory is read before being overwritten, and the output bit is used to select a different set of coefficients for the number 1 filters in the filter/magnitude circuitry. Another bit, delayed by two BRCs, is routed to the power combiner to flag the data, so that the heavy clutter information will appear in the range block headers of the C&I output message. Because the memory read operation occurs before the write, the filter 1 coefficients are changed on the scan following the one in which the decision was made.

Once the heavy clutter decision has been made, different switching thresholds for that zone are selected. Were this not done, the heavy clutter decision could become unstable, switching back and forth frequently. There are thus four VSP threshold adjustments to the card, two for clutter strength thresholds and two for ***Doppler decision*** (NZVF weather detection).

The Two-Level Weather Detector

The weather detector is an auxiliary circuit, but is worthy of discussion, because of the application of several principles, because it provides an example of a system designed purely for weather level detection, and because it contains a memory map, whose purpose requires clarification to avoid misunderstandings or confusion with other maps.

The Characteristics of Weather Echoes

Weather level detection depends upon signal strength, and an accepted national standard to describe that signal strength is recognized by the US Weather Bureau. In a conventional normal receiver, observation of receiver video would reveal that precipitation resembles an increased noise level. The measurements of weather strength, then, begin at the system noise level. As with other power measurements, deciBels are the preferred expression. The ***reflectivity*** of the weather, a measurement of its density, determines the level of the received signal.

The problem in detecting specific weather levels is that signal strength declines with range, and a simple constant threshold would cause close-range weather to appear more severe than distant weather. The expression of weather strength is then in dB_z, which is a measurement of the reflectivity related to the signal power at the receiver output, but not an expression of that power. A threshold designed to detect a specific level in dB_z must grow deeper with range, and zero dB_z is understandably at the system noise level, since system noise alone is evidence that no weather exists.

There are six national-standard weather levels. Any weather between 0 and 30 dB_z is described as ***level 1 weather***, the least significant, normally caused by light rainfall up to a rate of 0.2 inches/h. These levels are

level 1: 0 to 30 dB_z, 0 to 0.2 inches/h
level 2: 30 to 41 dB_z, 0.2 to 1.1 inches/h
level 3: 41 to 46 dB_z, 1.1 to 2.2 inches/h
level 4: 46 to 50 dB_z, 2.2 to 4.5 inches/h
level 5: 50 to 57 dB_z, 4.5 to 7.1 inches/h
level 6: above 57 dB_z, above 7.1 inches/h

Although weather return appears, on either normal or MTI video, to resemble noise, its characteristics are considerably more complex. Every raindrop provides a minuscule 1 t_p echo; as these echoes occur coincidentally, their power is additive, providing an accumulated strong echo. In fact, rain can cause a significantly stronger echo than an aircraft, since it exists over a wide span of elevations, rather than at only one. Additionally, most of the raindrops will have some radial velocity to create a Doppler shift. The Doppler shift will be related to the winds and turbulence within the rainfall, and because of the variations, will exist over a wide audio spectrum. Depending on the wind direction and velocity, the weather Doppler may encompass any frequencies, including a zero Doppler shift.

Weather Level Detection and MTI

Since rainfall may exhibit a zero Doppler shift, a clutter reduction system, such as MTI, can cancel some rainfall; such a condition is obviously undesirable in a weather-detection system. On the other hand, clutter must be removed from the radar information before it is subjected to a weather detection threshold; were this not done, false weather detections would occur over clutter. The requirement for a weather-level detector input is then for (1) information at all velocities to be applied except (2) MTI information, in those areas where clutter sufficient to break the threshold exists.

In the power combiner, the log magnitude filter data BRC contains signal-strength information, regarding all filters, and all Dopplers (see Figure 14-22). For the weather detector, these filter data words may simply be all summed together to provide two outputs for each BRC, one for BRC A and one for BRC B. Since all Dopplers, in all filters, have been added together, the addition creates a digital data equivalent to log normal. The requirement for MTI information may be satisfied by simply omitting the ZVF filters from the summation process.

Selecting Normal or MTI Inputs for the Two-Level Weather Detector

Dedicated circuitry in the power combiner provides the information for use by the weather detector. Since two summation modes are necessary, a switching signal called TWFILSEL is needed to cause the elimination of ZVF data in clutter areas. TWFILSEL is obtained from a range–azimuth addressed memory, called the ***clear-day map***. The clear-day map is a 1 CPIP × 1/16 nmi granularity, semipermanent map, deliberately made by technician operations on a day in which no precipitation is falling. Each memory location causes a TWFILSEL signal, which is either a "1" (where log normal data is required) or a "0" (where MTI data is required). The map is located on the weather detector card, since it is created by placing that card into a special, "reverse," mode of operation. During normal operation, the map outputs to the log combiner prevent clutter from breaking the threshold. During clear-day-map generation, clutter is allowed to break the threshold to create the MTI switching data.

Two-Level Weather Detector Operation

The weather detector receives two inputs per BRC from the log combiner. Since there is more signal power in BRC A, an adjustment to equalize the two is made in the detector input circuitry. 2Wx threshold "families" are contained in a threshold memory; one is used on one scan, the other on the opposite scan. The lower level threshold is at weather level 1 through 5, and the high-level threshold is at weather level 2 through 6. The thresholds are not readily changed, as the clear day map is made after they have been selected. The thresholds descend toward the grass level with range, and are altered by active–passive (low–high) beam shifting or linear–circular polarization modes. When weather data does not exist in both BRC A and BRC B, a range ambiguity is indicated, and no weather detection is generated. Weather detections caused by aircraft are omitted by eliminating detections less than 8 BRCs (1/2 nmi) in duration.

Weather detections occur with a duration of 1/2 nmi in BRC time (111.2 μs), and leave the detector in a serial stream, where a "1" indicates a weather detection, and a "0" indicates an absence of weather. Throughout the CPIP, this data is assembled into the weather message in the C&I formatter card. After the 960th BRC, the entire weather message is transferred from the C&I formatter to the postprocessor.

The Postprocessor and MTD Software

Primary Radar Data Inputs

Once the primitive target detections leave the C&I formatter card in the DSPs target and two-level detector, the messages become inputs to a data buffer in the ASP; at this point, there is no further reference, of any sort, to real time. The targets are referenced to range and azimuth because of the message headers. However, there may be many primitive target detections for a single target; there may be detections in several filters, in several CPIs,

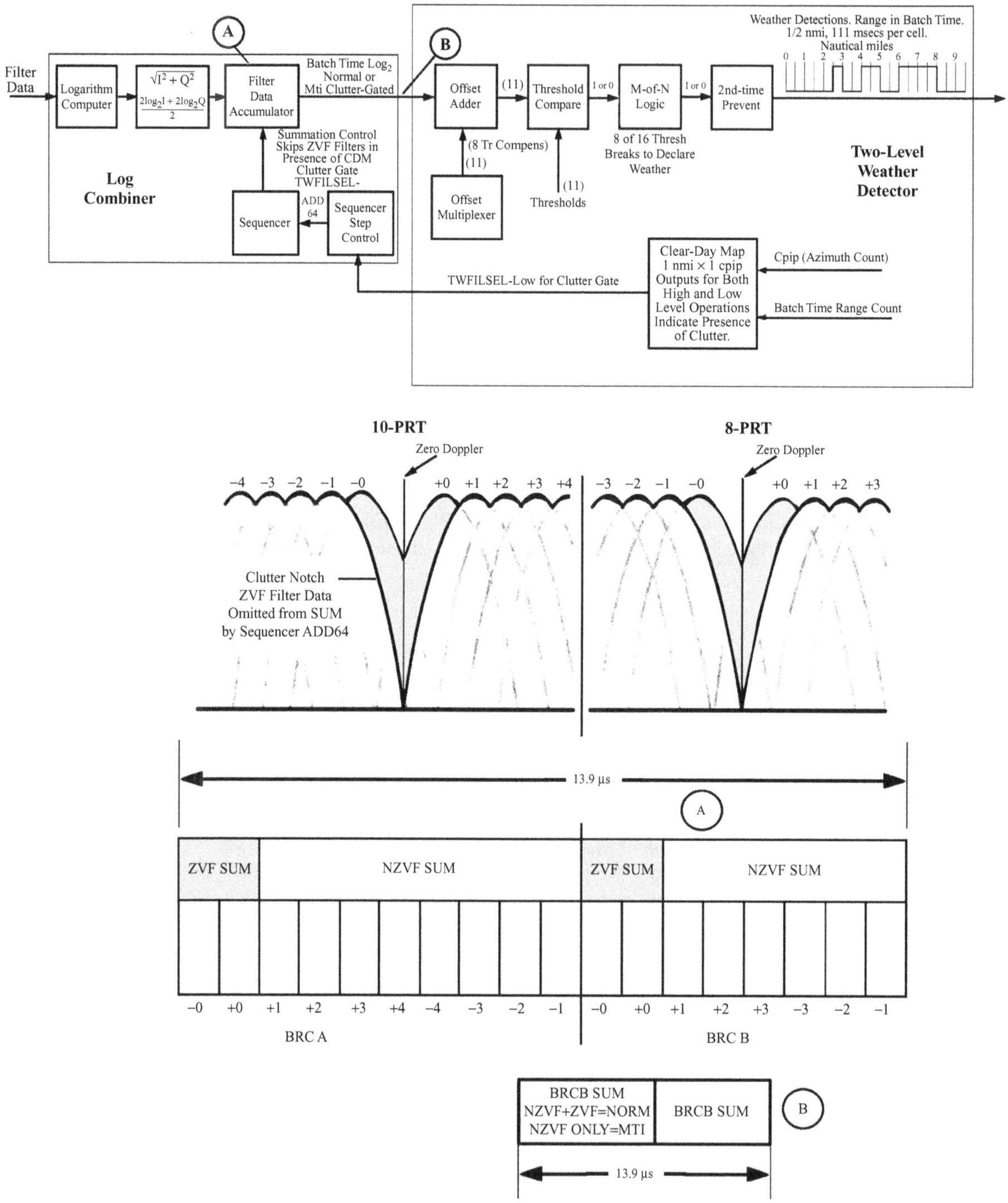

FIGURE 14–22

Two-level weather detector and clutter gating.

and in several range cells; some estimates in engineering documents have been as high as 35 possible primitives for a strong aircraft target. In the Doppler-filtering process, the target data lost precise azimuth reference. The eight or ten T_rs required for the filter operation resulted in a single filter-data message for BRC A or BRC B, and approximately 16 azimuth change pulses occurred during the CPIP. The azimuth data from the C&I formatter card represents only the center of the CPIs, only a gross approximation in regard to the center of target azimuth. It will be a major function of the C&I software to associate all the primitive-detection inputs into a target cluster, and then determine the centroided range and azimuth of the target, based upon signal strength and CPI azimuths.

Secondary (Beacon) Data Inputs

The MTD postprocessor contains two circuit areas other than the ASP. One of these is the ***message interface***, which provides formatted data outputs for use at the indicator site, and the other is a ***beacon reply processor***, which is supplied with raw beacon video from the ATCRBS interrogator. The beacon reply processor contains a code extractor, the purpose of which is to provide two 12-bit words for each beacon reply, in each beacon T_r; the beacon reply messages become inputs for another data buffer in the ASP. One of these words contains the range of the beacon reply, the other contains the reply code and garble status; the range is obtained from a range counter that starts with the P3 mode pulse, and the code extraction is obtained with four time-shared circuits (for range-overlapped beacon targets). An azimuth header, transmitted at the beginning of each beacon T_r, serves to identify the precise azimuth of each T_r and the beacon mode for that T_r. The azimuth information is obtained from the synchronizer, and the mode data is detected by P1–P3 pulse spacing; a counter is started on receipt of P1; the count present at the time of P3 receipt determines the mode to be declared. More information on the beacon reply processor is contained in Chapter 7 on secondary radar systems.

Recall that the primary radar data primitive target detections contain only coarse azimuth data, and special processes will have to take place to centroid the target to a precision of ±1 acp. Since a new beacon reply, azimuth, and mode data is available every T_r, no azimuth precision was lost in the beacon system, and target centroiding is achieved by a conventional ***azimuth sliding window***, similar in process to that used in many other radar data processors. The ASP software, which will accomplish beacon centroiding, code validation, and message assembly, is called the ***beacon target detector***.

The C&I Software Module

The C&I module performs several functions (see Figure 14-23):

Centroiding: Determining the center of mass of all associated primitives
Doppler Interpolation: Utilizing available primitives to resolve Doppler to 1/64 of f_p
Confidence and Quality Ranking
Second Adaptive Thresholding
Interference Processing and Elimination
2Wx Smoothing and Contouring

Centroiding

This process is unlike any other in earlier systems; it is not a conventional azimuth sliding window, as found in nearly all other FAA processors. The target data represents individual BRCs, with only coarse azimuth. The centroiding process, called ***correlation and interpolation***, utilizes signal strength of the peak Doppler filter data to calculate center of azimuth, and 27 algorithms are used to achieve this. To briefly describe azimuth centroiding in simplicity, the center of azimuth is determined from the rise and fall of echo power, as the beam moves across the target.

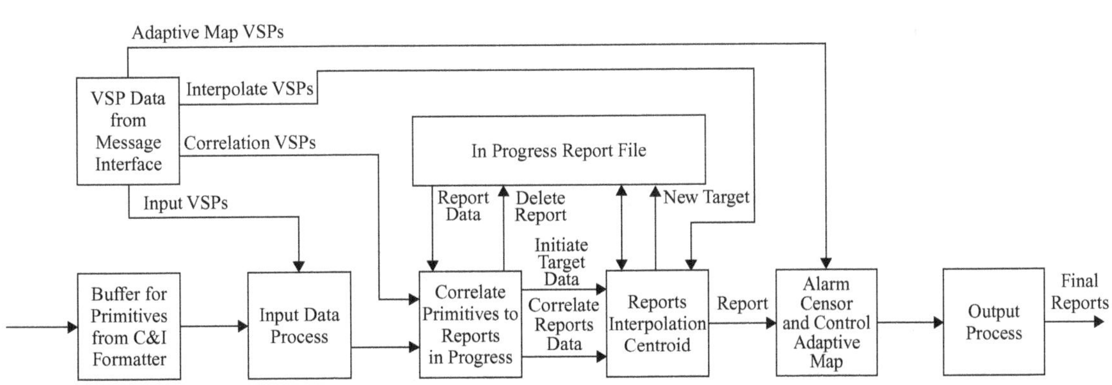

FIGURE 14–23

C&I program block diagram.

Data from the DSP is first received by a ring buffer. The input data process removes the data from the ring buffer for correlation with active in-process reports maintained in an "active target file." The active target file is a "linked list"; each target data message contains a pointer to the next message, in range order. Targets in process in the active target file are called ***target reports***. Each time the input data process provides an azimuth header to the correlation-of-primitives-to-reports function, that function scans the ring buffer from minimum to maximum range; any message in the ring buffer which can be associated with an existing report in the active target file is used to update the report; if no association may be made, a new target report is created.

Azimuth centroiding is achieved through one of three groups of algorithms. Where targets are overlapped in azimuth, a ***beamshape match*** is utilized; if more than one target exists, the algorithms find the two separate amplitude peaks, and then calculate the center of each. When there is available data in an inadequate number of azimuth-adjacent BRCs, an azimuth ***interpolation*** process is used. When the antenna pattern has been switched between high and low beam during a target report, a ***beam splitting*** algorithm may be used.

Range centroiding is also based on signal strength. When two hits in a report contains equal-amplitude target data at two adjacent ranges, the range centroid is adjusted by 1/32 nmi, improving the range accuracy to 1/32 nmi.

Quality and Confidence Ranking

Quality and confidence flags are created by the characteristics of the data contained in the target report. These flags are to be used in determining how the data is treated in further processing; they may cause a target report to be subjected to, or excluded from, the second adaptive thresholding process. They may also cause the target to become eligible, or ineligible, for track processing in the surveillance processor. Quality and confidence flags are

Quality
One CPI report: 00
Two CPIs, BRCs A and B: 01
Two or more CPIs, BRCs A only or B only: 10
Two or more CPIs, both BRCs

Confidence
Flat Geocensor flag set (roads): 0
Shaped Geocensor flag set (marked clutter): 1
Single CPI interference: 2
None of other four categories: 3
Maximum magnitude in ZVF data: 4

rf Interference (rfi) Testing

The presence of excessive single-CPI target reports is an indication that interference exists, most likely from another radar system. Such interference appears as multiple single-CPI targets, because the interference is non-synchronous, and will not dwell at the same range from one CPI to the next. Whenever the number of single-CPI reports in a 5.6° sector (four CPIPs) exceeds a VSP threshold, those single-CPI reports will be eliminated. There are two VSP thresholds, one for CPIA and one for CPIB.

The Second Adaptive Threshold Map

The second adaptive threshold process is among the very most complex in the entire MTD system (see Figure 14-24). The purpose of the map is to remove "poor" target reports that may have been generated by birds, insects, anomalous propagation, and other phenomena. The map provides for a scan-to-scan analysis of target reports in individual range-azimuth cells, increasing the threshold to data in each cell, as the average number of "poor" reports increase. It, further, performs analysis on individual filter data. The map is divided into 12, unevenly spaced range zones; as the range increases, the range dimension of the range zones decrease. The map is also divided into a maximum of 32 azimuth sectors. In range zone 1, there are no azimuth sectors; range zone 2 is divided into eight sectors, zones 3 and 4 into 16, and zones 5 through 12 into 32 sectors. The intent of the decreasing range dimensions, and the increasing azimuth-sector division, is to maintain a similar area in each cell.

It is not intended that the second adaptive map thresholds be applied to all target reports; targets certain to be aircraft must not be thresholded. An ***exclusion testing*** is performed on all reports; those targets possessing sufficient quality and confidence are excluded from the second adaptive map, neither affecting the thresholds nor being subjected to them.

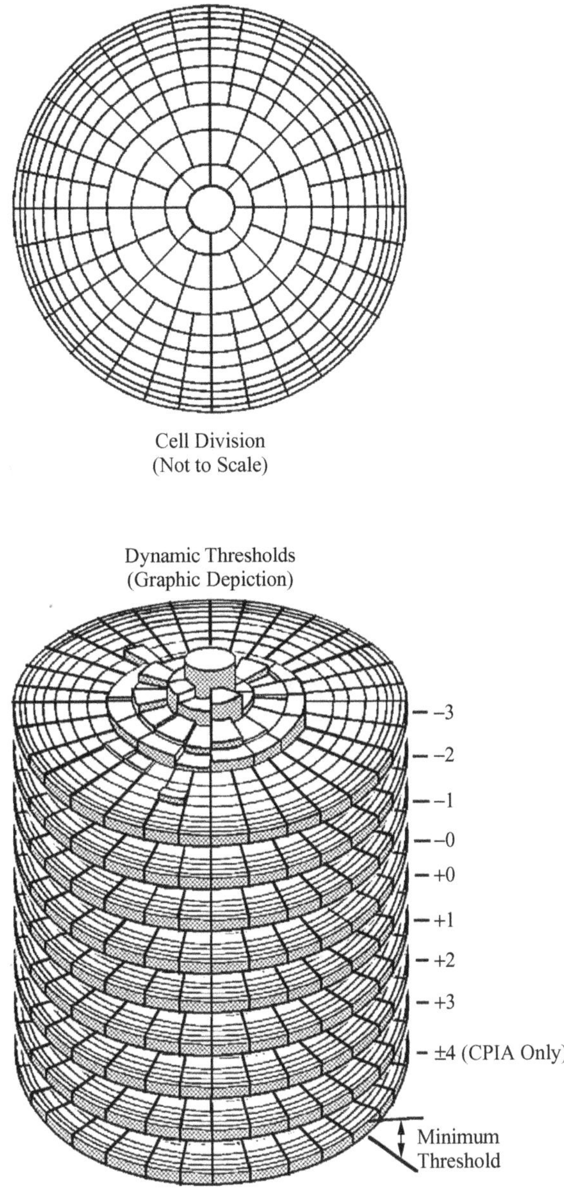

FIGURE 14–24

Graphic depiction, second adaptive map.

Track Eligibility

There are five quality/confidence bits in the target report, two for quality, two for confidence, and one to indicate that the rfi threshold had been crossed. Two more bits are added to indicate eligibility for tracking to the surveillance track process to follow. These bits will indicate (1) high probability aircraft; (2) low probability aircraft; or (3) probable vehicular traffic.

Performance Observation

The C&I module provides for a multitude of performance observations, by a crt monitor, via the ***remote maintenance subsystem (RMS)***. Target reports into, and out of, the second adaptive process are available. DSP performance is indicated by primitive detection input counts for each filter. The second adaptive map itself may be viewed, or the specific thresholds for any cell, in hundredths of a dB, may be displayed. A ***target performance window*** provides a means to center a range–azimuth window over a given target, such as an MTI reflector; the Doppler interpolation provides Doppler information to 1/64 of the f_p on targets within the window, and centroid information in the target reports provide range and azimuth to ±1 acp or ±1 range cell.

Second Adaptive Map Maintenance Display

The display of the second adaptive map itself is interesting because of the ingenuity that was used to create the display (see Figure 14-25). Because there are so many cells, because the threshold values contain up to five characters (two digits, a decimal point, and two digits), and because a computer monitor display is limited to 80 columns and 26 lines, a special technique was used. A single character is used to represent a single threshold. All the numbers, 0 through 9, and all the letters of the alphabet are used to represent 36 threshold steps. The scale of the map may be changed to increase or decrease the value of the display threshold steps. A common value to keep most of the data from "limiting" (exceeding the 36th level) is 3/4 dB per step.

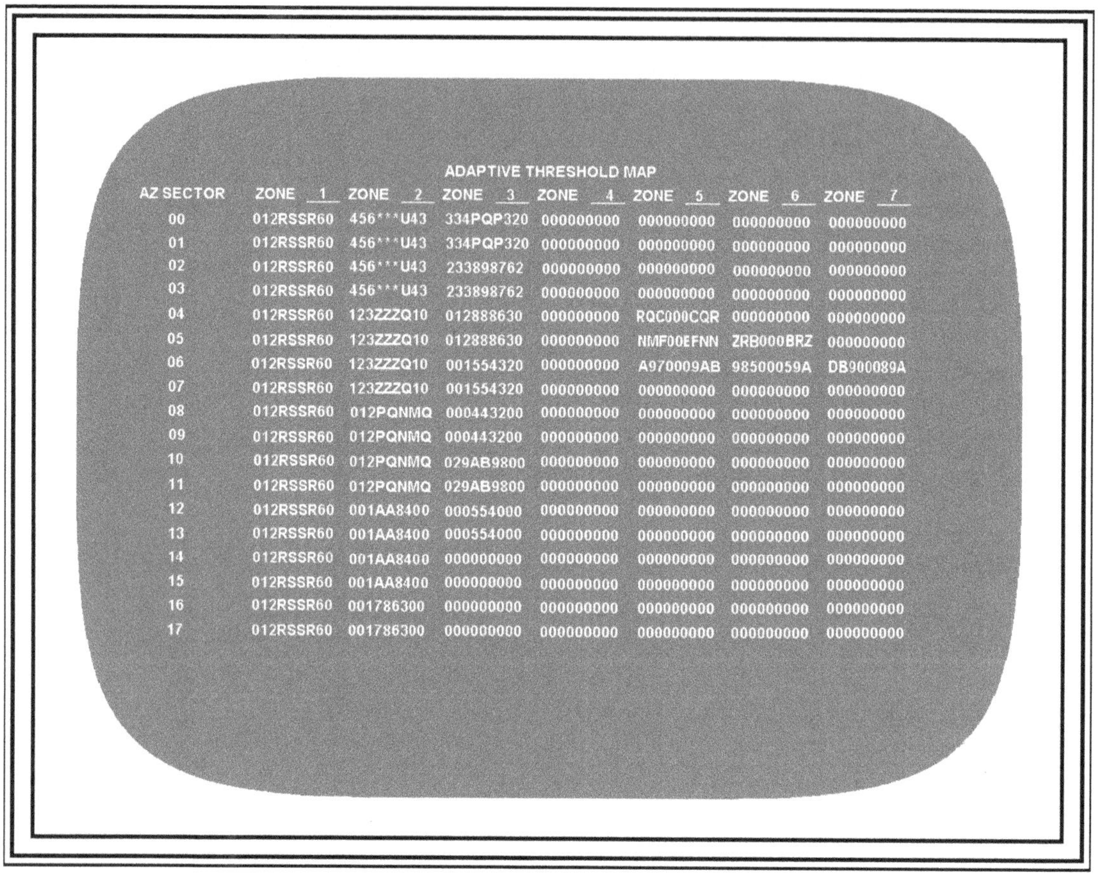

ADAPTIVE THRESHOLD MAP

AZ SECTOR	ZONE 1	ZONE 2	ZONE 3	ZONE 4	ZONE 5	ZONE 6	ZONE 7
00	012RSSR60	456***U43	334PQP320	000000000	000000000	000000000	000000000
01	012RSSR60	456***U43	334PQP320	000000000	000000000	000000000	000000000
02	012RSSR60	456***U43	233898762	000000000	000000000	000000000	000000000
03	012RSSR60	456***U43	233898762	000000000	000000000	000000000	000000000
04	012RSSR60	123ZZZQ10	012888630	000000000	RQC000CQR	000000000	000000000
05	012RSSR60	123ZZZQ10	012888630	000000000	NMF00EFNN	ZRB000BRZ	000000000
06	012RSSR60	123ZZZQ10	001554320	000000000	A970009AB	98500059A	DB900089A
07	012RSSR60	123ZZZQ10	001554320	000000000	000000000	000000000	000000000
08	012RSSR60	012PQNMQ	000443200	000000000	000000000	000000000	000000000
09	012RSSR60	012PQNMQ	000443200	000000000	000000000	000000000	000000000
10	012RSSR60	012PQNMQ	029AB9800	000000000	000000000	000000000	000000000
11	012RSSR60	012PQNMQ	029AB9800	000000000	000000000	000000000	000000000
12	012RSSR60	001AA8400	000554000	000000000	000000000	000000000	000000000
13	012RSSR60	001AA8400	000554000	000000000	000000000	000000000	000000000
14	012RSSR60	001AA8400	000000000	000000000	000000000	000000000	000000000
15	012RSSR60	001AA8400	000000000	000000000	000000000	000000000	000000000
16	012RSSR60	001786300	000000000	000000000	000000000	000000000	000000000
17	012RSSR60	001786300	000000000	000000000	000000000	000000000	000000000

FIGURE 14–25

Adaptive map maintenance display.

Interpretation of the map is as follows:

1. Each column represents a range ring and each row represents one of 32 azimuth sectors. Seven range rings and 18 sectors may be displayed on a single screen.
2. Each block of characters represents nine thresholds; the filter data from the high and low f_p CPIs are lumped together, and the ±4 filter data is lumped to place those ±4 thresholds on the same scale as the others (±4 filters only exist for one CPI). Each character represents a single threshold, in the order −3, −2, −1, −0, +0, +1, +2, +3, ±4.
3. A "0" represents a minimum threshold and a "Z" represents a maximum within the threshold display range. An asterisk indicates that the threshold is off scale, or past "Z."
4. The format requires that there must always be 32 sectors. In the first range ring, all sectors are the same, because there are no sector divisions in that range. In the second range ring, groups of four sectors are the same, since that ring is divided into eight azimuth sectors. In the third and fourth range rings, pairs of two sectors are the same, because that ring is divided into 16 azimuth sectors. In the fifth through twelfth range ring, all sectors are different, as the scan is divided into 32 parts.
5. High-level thresholds, near the center of the block, indicate that the threshold is acting on near-zero-Doppler data, such as clutter breakthrough. High-level thresholds, near the edges of the block, indicate that the thresholds are operating on moving data, such as weather.

Regular usage of the second adaptive map maintenance display will reveal that the map is quite sensitive, and will even show predictable daily variations; it can provide the technician with an indication of both system performance and atmospheric phenomena. Remember that the second adaptive map is a scan-to-scan operation. If the map is permitted to "die," the system may have to operate for several minutes, before it totally recovers. For this reason, a transmitter-off bit in the azimuth headers of the C&I input messages will "freeze" the map. In the event of a power failure, a battery backup for the cabinet will sustain power, long enough for an engine generator to start.

The Target Performance Window

From information obtained in the C&I process, the monitoring system software provides a screen for use by maintenance personnel to verify system alignment (see Figure 14-26). Alignment of radars from previous states-of-the-art was usually based on surveyed ***permanent echoes (PEs)***, such as water towers, mountain peaks, MTI reflectors, etc. Since there is no normal video available at the indicator site, there must be a means to align the system purely on MTI reflectors, or some known moving object, as a windmill, oil-well pump, rotating sign, or another radar antenna. The target performance window permits the technician to gather all target information within a window of adjustable dimensions.

Appearance of the MTI Reflector

In Figure 14-26, the first target is an MTI reflector, and shows a Doppler of "32," which is 32/64 × PRF, or $f_p/2$, an optimum-velocity target. That the target is actually the MTI reflector is made clear by the identical values of "32" for both the high and low f_ps. Only an artificial moving target, providing a $T_r - T_r \Delta D$ of $\lambda/4$, could appear to be at optimum velocity for two different T_rs.

The performance window provides signal-strength information in terms of dB above system noise, and this information is gained from the original filter-data magnitude information. This information provides a means for reflector alignment to be very precise.

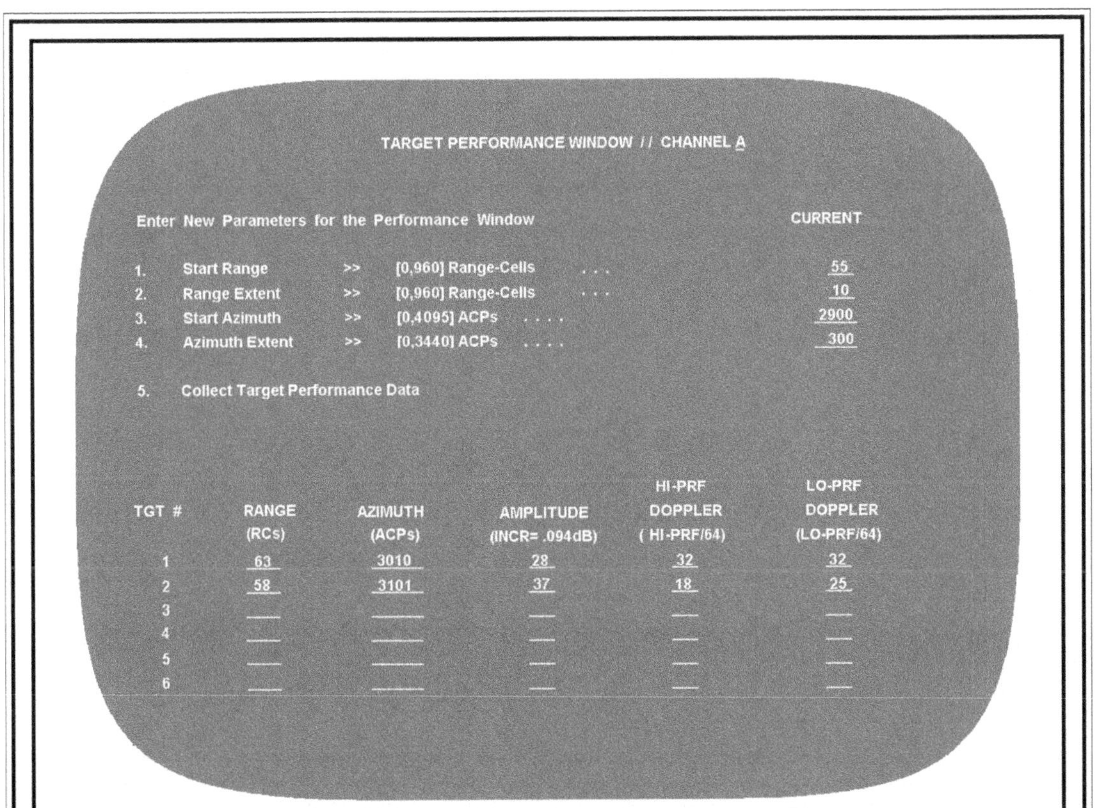

FIGURE 14–26

Target performance display.

Azimuth Alignment

Finding the MTI reflector of interest may prove difficult, should the azimuth be significantly in error. It is likely that the reflector will not be received at all, unless the antenna beam switching is programmed for low beam, over the reflector. And, since the beam-switch program depends upon the acp count, even a properly programmed gate will not occur over the target, unless the synchronizer azimuth preset is close to alignment. Still further, the reflector may not survive the processing in the C&I and SP programs, if it does not carry a geocensor MTI reflector flag, and the geocensor CPIP addressing is derived from acp counts. Although the synchronizer azimuth offset offers convenient final corrections, it will reduce the technician's efforts to ascertain the apg is near alignment, before he uses the performance window. A useful and convenient tool for this is a locally fabricated "pulse catcher," which lights a light-emitting diode, when the azimuth reference pulse occurs. If one apg is known to be correctly aligned, the maintenance ppi real-time videos can be observed, and the apg under alignment can be adjusted to closely match the one known to be aligned.

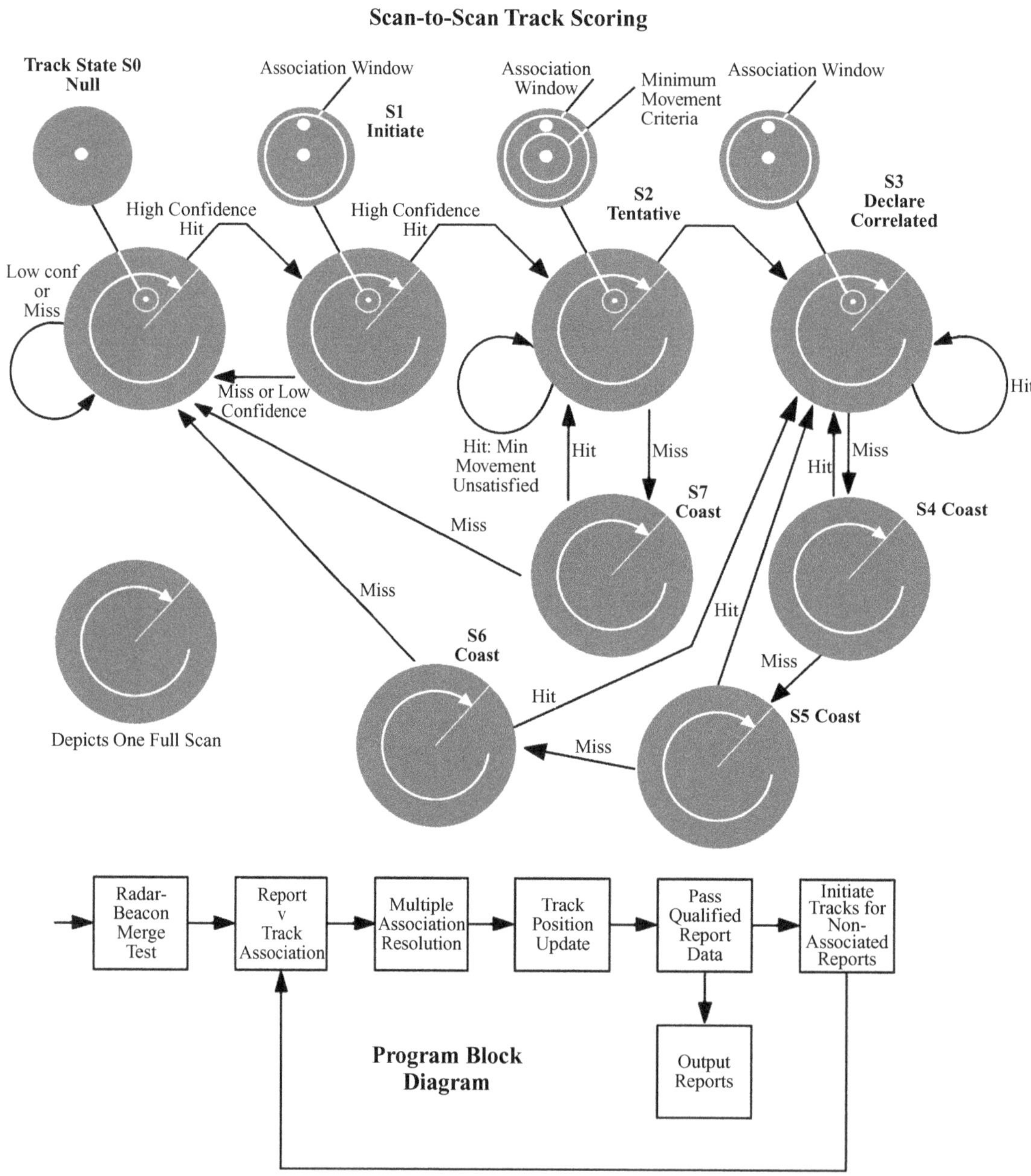

FIGURE 14–27

Surveillance processor track scoring.

The Surveillance Processor (SP)

This is a software ***track-while-scan*** program, so called to differentiate it from those radar tracking systems employing monopulse, or any other technique, in which the antenna "locks on" to a single target and follows it, continually positioning azimuth and/or elevation according to target data. Of course, track-while-scan techniques are necessary in air traffic control radars, because the surveillance scan cannot be interrupted.

The basic concept of the surveillance processor, also called the ***tracker***, is a scan-to-scan analysis of the target data from the C&I program module (see Figure 14-27). With each target report, a "window" is placed around the target; if another target report falls within that window on the next scan, the program updates a ***track state*** that accompanies the target data in a track list. If the track eligibility assigned the target in the C&I module is sufficient, the track state is updated. When the track is initially being established, the track state is upgraded from S0 until it reaches S3, which is called a ***firm track***. On the scan in which the target reaches track state S3, the target is said to have become a ***correlated target***. (There is a potential for misunderstanding of terms in this area. The centroiding process was also called "correlation," and there is no connection.)

In addition to the track window, another window surrounds targets, at ranges less than 20 miles. This ***minimum distance movement criteria*** window is smaller than the track window; tracks will not be established on targets which remain in this window from scan to scan. The minimum distance movement criteria does much to remove targets caused by the slow movements of any remaining false detections, helicopters, automobiles, trains, clutter residue, weather residue, etc.

Part of the SP program is a radar/beacon ***merge*** test. If the radar and beacon targets occur within reasonable, adjustable proximity, a ***radar-reinforced*** bit is set in the beacon output message. The source of range and azimuth in both output messages may then be from either radar or beacon, and is determined by a VSP setting. Generally, the preferable azimuth source is the radar, and the preferable range source is the beacon, because those sources provide the greatest resolutions.

Although used throughout the development and production of the ASR-9, the word "correlated" may have been an unwise choice. Other radar systems may use "correlated" to describe target data in which both primary and secondary radar targets have appeared at the same, or very near, range and azimuth. Still further, the word "correlation" is also used to describe both azimuth and Doppler operations in the C&I program. Primary radar target data produced by the MTD system finally is described by a single target message; the target message may be for either a "correlated" or "uncorrelated" target; only a single bit in the message differentiates between the two.

Although it is a user responsibility, the technician maintaining one of these systems should also be aware of the hazards in displaying only correlated radar targets (see Figure 14-28). This mode of operation provides a very clean display, free of distracting false targets, but many of those targets just acquired, or not meeting track criteria, may not be visible. A proper display is one in which both correlated and uncorrelated targets are visible, but the uncorrelated targets are obvious, because they have been programmed to appear to be smaller.

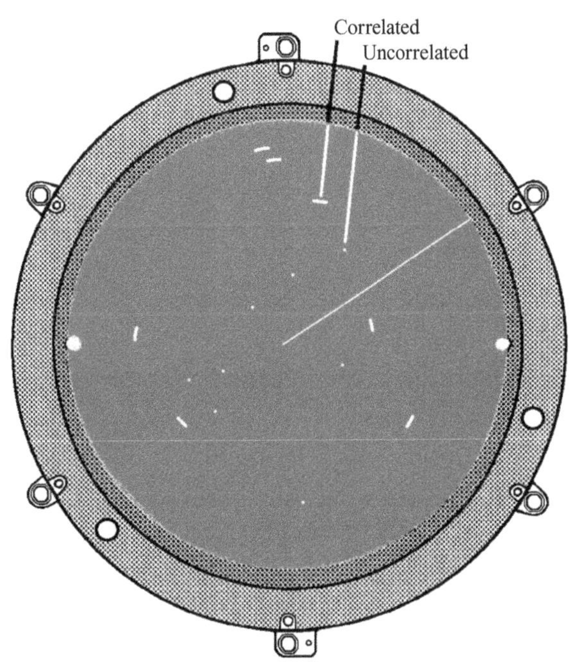

FIGURE 14–28

Correlated versus uncorrelated display targets.

Two-Level Weather Smoothing and Contouring

The purpose of the 2Wx detection system is to provide weather data for use as background video on air traffic control displays. Weather detections ordinarily have a very broken, "choppy," appearance, which would obscure aircraft targets. The smoothing and contouring process fills in "holes," and averages weather declarations into large, smooth blocks, which will not distract air traffic controllers.

The Message Interface

Target reports in the SP, whether correlated or uncorrelated, are available for output, after each scan. The ASP is in continual handshaking communication with the message interface, which transfers the parallel data from the ASP into a formatting and parallel-to-serial conversion for serial transmission to the indicator site. Because the data in the ASP is in parallel, but must be transmitted serially, there are two or three (usually three) modem channels for primary or secondary radar (surveillance) information, so that data transmission may be distributed over multiple lines. The output data is at a 9,600 bps rate.

Primary radar targets are contained in a 52-bit message, illustrated in Figure 14-29. The message contains a header, to identify it as a primary radar target, and a 51st bit to identify it as correlated or uncorrelated. Range, azimuth, quality, and confidence information are also contained in the message.

Msg Bit No	Word Bit No	Word 1	
1	1	1 = Wx from Unavailable Processor	
2	2	0	Search Message Identification Label
3	3	0	
4	4	1	
5	5	1	
6	6	0	
7	7	1	
8	8	1	
9	9	0	
10	10	0	
11	11	0	
12	12	0	
13	13	X = Parity	

Msg Bit No	Word Bit No	Word 2
14	1	1 = 32 nmi
15	2	1 = 16 nmi
16	3	1 = 8 nmi
17	4	1 = 4 nmi
18	5	1 = 2 nmi
19	6	1 = 1 nmi
20	7	1 = $\frac{1}{2}$ nmi
21	8	1 = $\frac{1}{4}$ nmi
22	9	1 = $\frac{1}{8}$ nmi
23	10	1 = $\frac{1}{16}$ nmi
24	11	1 = $\frac{1}{32}$ nmi
25	12	1 = $\frac{1}{64}$ nmi
26	13	X = Parity

Msg Bit No	Word Bit No	Word 3
27	1	1 = 2048 acp's
28	2	1 = 1024 acp's
29	3	1 = 512 acp's
30	4	1 = 256 acp's
31	5	1 = 128 acp's
32	6	1 = 64 acp's
33	7	1 = 32 acp's
34	8	1 = 16 acp's
35	9	1 = 8 acp's
36	10	1 = 4 acp's
37	11	1 = 2 acp's
38	12	1 = 1 acp
39	13	X = Parity

Msg Bit No	Word Bit No	Word 4	
40	1	C&I Quality	00 = 1 cpl 01 = 2 cpl's, Different f_p's 10 = 2 or More cpi's, Same f_p 11 = 3 or More cpl's, Different f_p's
41	2		
42	3	Confidence	000 = Geocensor Traffic Zone 001 = Heavy Clutter Zone 010 = Interference 011 = Angels and Aircraft 101 = Angels and Aircraft 100 = Max Doppler in Zero Filter
43	4		
44	5		
45	6	Track Eligibility	00 = Vehicles; Do not Track 01 = Track Correlation 10 = Track Initiation
46	7		
47	8	ARTISIIIA Quality (SRAP Emulation)	
48	9		
49	10		
50	11	0	
51	12	1 = Correlated 0 = Uncorrelated	
52	13	X = Parity	

FIGURE 14–29

Output primary radar message.

Review Questions

1. Why does an MTD system not hard-limit the i-f, upstream of the synchronous phase detectors?
2. For what purpose does the ASR-9 i-f amplifier contain a logarithmic amplifier/detector?
3. State the advantages of MTD over MTI.
4. State a disadvantage of MTD.
5. The ASR-9 employs an internal rf test-target generator to provide for an automated mds measurement. Why can a conventional rf test set not be used?
6. How does the appearance of MTD bipolar video differ from MTI bipolar video?
7. In the stream of signal flow from the phase-detector output to the postprocessor input, at what point does "real time" end?
8. Briefly summarize the principle of operation of the digital Doppler filters.
9. Name two ways in which blind-velocity targets may be recovered in an MTD system.
10. How does MTD improve upon subclutter visibility?
11. Why are there 18 filters, rather than 20?
12. What is "coherent integration"?
13. Is the threshold developed in the MLT cards obtained from data which occurs before, or after, the area of inspection?
14. How does the MTD system improve detection of targets in weather?
15. What is contained in the geocensor map?
16. If an MTI reflector flag were not properly programmed, what might be the result?
17. If an MTI reflector flag covered a large geographical area, what might be expected?
18. Where is data first tested against thresholds?
19. Where does the construction of an output detection message begin?
20. Name, and give the purpose of, the memory maps in the MTD system.
21. How does the MTD system differ from other processors in centroiding?
22. What is the software source for the information in the target performance window?
23. The technician has installed an apg, but cannot locate the MTI reflector on the target performance menu. What might he do?
24. The maintenance display of the second adaptive map contains two asterisks in the center of many of the blocks of characters, but other characters exhibit values in the order of "9" through "F." What is indicated?
25. The target performance window lists six targets, and the one of interest cannot be found. What might be done to correct this?
26. The target performance window shows a target with a Doppler of "32" for both f_p's. What is the most likely source of this target?
27. Name two major functions of the SP module.
28. Define "correlated, uncorrelated, and merge."
29. How often is the track scoring for a target updated?
30. Why is a correlated-only display hazardous?

Answers to Review Questions

1. Why does an MTD system not hard-limit the i-f, upstream of the synchronous phase detectors?
   It improves subclutter visibility and permits centroiding.
2. For what purpose does the ASR-9 i-f amplifier contain a logarithmic amplifier/detector?
   To decrease the signal-to-noise ratio, so that grass can be seen on the monitor ppi.
3. State the advantages of MTD over MTI.
   Better visibility over clutter, recovery of tangential and blind-speed targets, improved cfar, elimination of anomalous propagation, elimination of road traffic, reduction of targets resulting from birds and insects.
4. State a disadvantage of MTD.
   Reduces the ability of the user to use judgment in evaluating targets.
5. The ASR-9 employs an internal rf test-target generator to provide for an automated mds measurement. Why can a conventional rf test set not be used?
   A signal generator is incoherent, and produces a random-speed target. The Doppler filters would not pass the test signal, and would treat it in a similar manner as they treat noise.
6. How does the appearance of MTD bipolar video differ from MTI bipolar video?
   There is very little noise apparent on the baseline, and the "butterflies" vary in amplitude.
7. In the stream of signal flow from the phase-detector output to the postprocessor input, at what point does "real time" end?
   The last point that data timing resembles real time is when it is written into the bulk memory.
8. Briefly summarize the principle of operation of the digital Doppler filters.
   A maximum output is obtained when the rotation rate of the radar vector equals the rotation rate of the coefficient vector.
9. Name two ways in which blind-velocity targets may be recovered in an MTD system.
   Targets at blind speed in one CPI will not be at blind speed in the other. Blind speed targets will not be "canceled" by clutter-map outputs, because the targets will be in different locations, from one scan to the next.
10. How does MTD improve upon subclutter visibility?
    More target information is available when a target is over clutter, because the signal is not obliterated by limiting.
11. Why are there 18 filters, rather than 20?
    There are 18 batch cells read from the bulk memory in 13.9 μs. If there were more than 18 filter data, it would take more than 13.9 μs to get the information out of the filters, and the input would "overrun" the output.
12. What is "coherent integration?"
    Signals are integrated in the Doppler filters, but noise is attenuated. Therefore, the signal-to-noise ratio is increased.
13. Is the threshold developed in the MLT cards obtained from data which occurs before, or after, the area of inspection?
    After.
14. How does the MTD system improve detection of targets in weather?
    The unlimited i-f permits the target Doppler to "ride atop" weather Dopplers without distortion, and the target can then be detected in the filters. The mean-level-threshold circuits raise threshold levels for only those filters containing the weather; if a target is at a different Doppler than the weather, it will not be affected by the MLT thresholding.
15. What is contained in the geocensor map?
    Threshold type; flat, shaped, or MTI reflector.

16. If an MTI reflector flag were not properly programmed, what might be the result?

    It could be eliminated by the second adaptive map, or by the minimum distance movement criteria in the surveillance processor.

17. If an MTI reflector flag covered a large geographical area, what might be expected?

    Any small targets would be flagged and "protected," and a block of false targets might occur.

18. Where is data first tested against thresholds?

    In primitive detector number 2.

19. Where does the construction of an output detection message begin?

    In the primitive formatter.

20. Name, and give the purpose of, the memory maps in the MTD system.

    Clutter map: Provides data to cancel clutter, but pass tangential and blind-speed targets.

    F1 control map: Provides for automatic switching of ±1 filter shapes, in the presence of strong clutter, but not weather.

    Clear-day map: Provides MTI clutter gating for the 2Wx detector.

    Geocensor map: Indicates locations where geocensor threshold "families," or MTI reflector flags, are to be used.

    Second adaptive map: Reduces false targets caused by birds, insects, anomalous propagation, and low quality or confidence.

21. How does the MTD system differ from other processors in centroiding?

    It relies upon signal strength, and does not employ a conventional azimuth sliding window.

22. What is the software source for the information in the target performance window?

    The C&I program.

23. The technician has installed an apg, but cannot locate the MTI reflector on the target performance menu. What might he do?

    Ascertain that it is close to alignment. Temporarily increase the size of the target performance window, the low-beam gate, and the MTI reflector zones.

24. The maintenance display of the second adaptive map contains two asterisks in the center of many of the blocks of characters, but other characters exhibit values in the order of "9" through "F." What is indicated?

    The center of the character block is the zero-Doppler area. The presence of clutter from the DSP is indicated.

25. The target performance window lists six targets, and the one of interest cannot be found. What might be done to correct this?

    Decrease the window size.

26. The target performance window shows a target with a Doppler of "32" for both f_ps. What is the most likely source of this target?

    This is most likely an MTI reflector; only an artificial target is likely to produce the same Doppler for both f_ps.

27. Name two major functions of the SP module.

    Radar-beacon merge test and radar tracking.

28. Define: "correlated, uncorrelated, and merge."

    Correlated: A track has been established.

    Uncorrelated: A track has not been established.

    Merge: A beacon and radar target exist within close proximity of each other.

29. How often is the track scoring for a target updated?

    With each scan.

30. Why is a correlated-only display hazardous?

    All targets upon which tracks have not been established cannot be seen.

CHAPTER 15

Radar Displays

Preface

Radar display techniques have passed through even more stages of evolution than radar itself. Many of those stages will be addressed in this chapter, but some latter-day display techniques involving computer operating systems such as Lynx, Unix, and others, or which may encompass changing television methods are not included, and will require study of other material.

General

That equipment used to show radar information is called a ***display*** or an ***indicator.*** Figure 15-1 is a generic representation of an FAA ***planned-position-indicator (ppi)*** display of the type to be found at an ASR facility. Formal academic literature may refer to this as a ρθ ***(rho–theta)*** display, where ρ denotes the radius of a circle and θ the angle of the radius; these dimensions are, of course, the ***range*** and ***azimuth.*** Shown on the illustrated display are range marks, a video map, normal video, and ARTS alphanumerics. The basic principles of ppi displays were addressed in Chapter 4. This chapter will deal principally with the ppi display and associated equipment; however, some mention of other types is worthwhile, if only to provide the reader with some ability to participate in intelligent discussions with those in branches of the radar field other than air traffic control.

Many attribute microwave technology as the most important development in radar; others assign the importance to sensitive receivers. However, no part of radar technology could have been of any use without the ***cathode ray tube (crt),*** the means of displaying the received echoes. The history of this device began substantially earlier than one might expect; it is traceable to an English scientist, Sir William Crookes, in 1875. Crookes built a device in which a beam was made to pass through a space under the influence of high voltage, and he proved the particles in the beam to be negatively charged. In 1917, Nikolai Tesla unknowingly predicted radar and the crt when he said, ". . . and cause this intercepted ray *(we now call it the "echo")* to illuminate a fluorescent screen *(as in X-ray, already developed by Roentgen)*. . . then our problem of locating the hidden submarine will have been solved." *(excerpt from Tesla by Margaret Cheney copyright 1981).* Actual development of the phosphor-faced crt began in the 1920s.

Electromagnetic and Electrostatic Deflection

Crts may be divided into two major classes, depending upon the means used to deflect the electron beam, emitted from the ***electron gun*** at the "input" end of the tube (see Figure 15-2). The electrostatic-deflection tube contains internal deflection plates, and the beam is deflected purely by electric fields, created by the electric potential difference between those plates. Electrostatic deflection is superior in many respects, and electrostatic crts are widely used where precision deflection is essential, as in oscilloscopes. Electrostatic deflection also allows for a much wider variety of sweep speeds than electromagnetic deflection, since there are no limitations imposed by the inductive reactance of the deflection coils, which changes with sweep speed. Unfortunately, electrostatic deflection is not practical for large-diameter displays, such as those required for radar, and the tubes are considerably more expensive than the electromagnetic type. Electromagnetic deflection permits the use of larger and less expensive tubes, and in those applications where the sweep speeds are restricted within a known range,

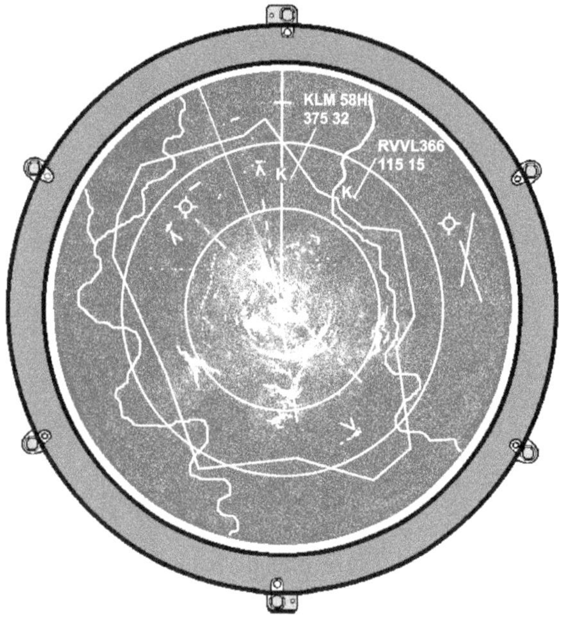

FIGURE 15–1

An FAA ppi/rappi display.

they are perfectly adequate. As an example, electromagnetic deflection is used in television receivers because the television sweeps are of a known and fixed value, and the impedance of the deflection coils can be designed into the system, even though the horizontal sweep occurs within about 63.5 μs.

The Three Dimensions

In describing the operations of the displays, three dimensions, axes, or planes are called the *X*, *Y*, and *Z* (see Figure 15-3). As in mathematics, *X* refers to the horizontal axis and *Y* refers to the vertical. The third, *Z*-axis, is perpendicular to both the *X* and *Y* dimensions, and describes a line from the "input" end of the tube to the face; it is the axis of the electron beam. The *Z*-axis then is normally used in reference to display intensity.

Display Types

Early in the development of radar, the many different means of displaying radar data were classified by letter designations. Among these were types "A," "B," "C," "E," "G," "J," and "P." The type "P" was the ppi, which also has been divided into two subdivisions, ***ppi*** and ***random-access ppi (rappi).*** Following is a brief description and application information regarding some of all these display types.

The A-Type Display

The simplest radar indicator is an "A"-type presentation, which operates in much the same manner as an oscilloscope (see Figure 15-4). Military, or ex-military, technicians, particularly those with US Navy backgrounds, often call oscilloscopes ***A 'scopes.*** On this type of radar display, a spot begins moving from left to right in the *X*-axis, across the face of the oscilloscope at, or shortly before, radar time zero. A single movement of the electron beam across the face of the crt creates a ***trace,*** usually called a ***sweep.*** A radar trigger initiates generation of a ***sweep sawtooth*** waveform, which is applied to the ***horizontal deflection*** circuitry. The deflection circuitry may differ considerably, depending upon whether the cathode ray tube is the electrostatic or electromagnetic deflection type. The "A" type presentation is sometimes said to be "deflection modulated," which means that the radar data deflects the beam in the *Y*-axis. In this type of display, there is no variation in the *Z*-axis; it is simply adjusted to a desired intensity level by the user.

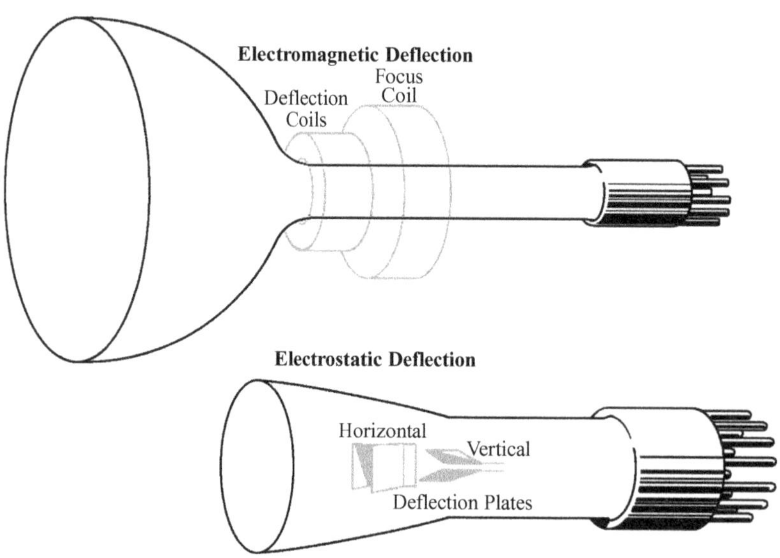

FIGURE 15–2

Electromagnetic and electrostatic crts.

The A-type display is of range versus signal strength and employs vertical deflection to indicate the presence of targets or noise. Figure 15-4 is a display of normal video. The time required for the sweep to move across the face of the tube is the display range of the indicator. The display range is the desired viewing range, not the maximum range of the radar. If the display range were set to 30 nmi, it would take 12.3552 × 30 = 370.656 μs for the sweep to cover the face of the crt, and, if a target occurred at 10 nmi, it would be displayed after 123.552 μs, one-third of the way from left to right. The main disadvantage of the A-type display is that it provides no ***bearing (azimuth)*** or ***altitude (elevation)*** information, which must be obtained from antenna position data. On some of the earliest radars, the antenna was manually moved to desired bearing, and the azimuth was known by virtue of the position of the stopped antenna. In tracking radars, bearing and elevation information was available by antenna position data, and only the range of the target under track was necessary on the display.

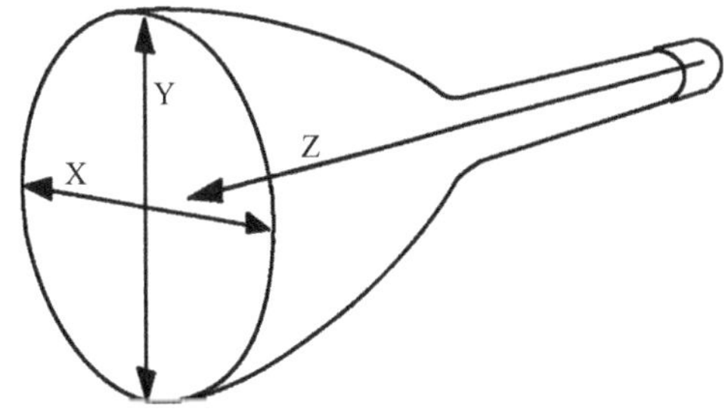

FIGURE 15-3

The three dimensions.

Real Time

An A-type display of this type, and any other display using a sweep synchronized to the radar f_p, and using radar video from the receiver output at the same rate it becomes available, is called a ***real-time*** display. This definition is assigned to differentiate it from digital displays of processed radar data, in which the data is displayed according to range and azimuth, contained in digital messages, which are not synchronous with the radar f_p.

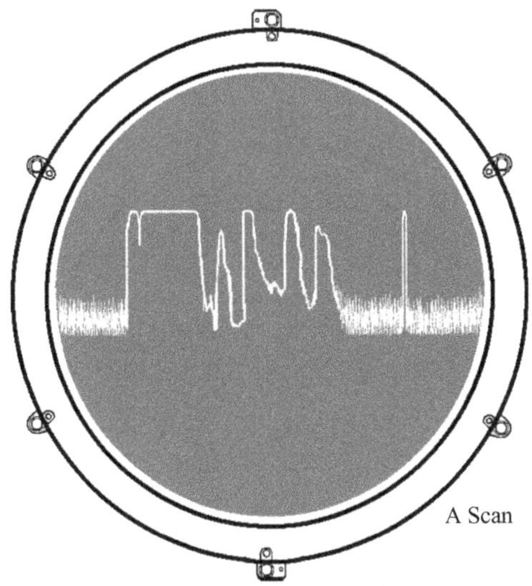

FIGURE 15-4

Type "A" display.

The J-Type Display

This display was introduced during World War II; it was the type used in the SCR-584 tracking radar, the main WWII system used for antiaircraft and antimissile fire control (see Figure 15-5). It was quite similar to the A-type, except that a circular sweep was created with a variation of a circular ***Lissajous pattern.*** When a sine and a cosine wave are, respectively, applied to the deflection in the *Y*- and *X*-axes, the deflection causes the electron beam to travel in a clockwise circle around the face of the crt. In the J-type display, the sweep would start at 0° (north, or "straight up"), and then travel in a circle; the radar video would act as a gain, proportionally increasing both the sine and cosine values to increase the right-triangle hypotenuse, so the video would "ride upon" the circle. Advantages of this were that (1) it allowed for the display of more range, and (2) range could be associated with the pointing angle of a target display, as though the display were in the form of a dial or gauge.

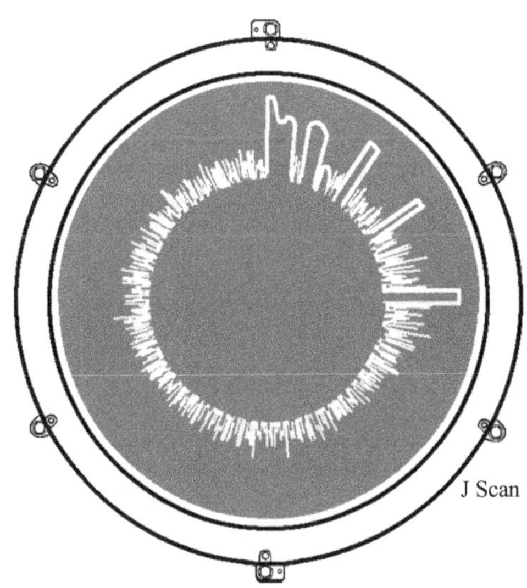

FIGURE 15-5

Type "J" display.

The B-Type Display

The B-type display is another real-time indicator, sometimes used with radars scanning a sector; a ***scan*** is a total antenna movement from one limit, or extreme, to the other

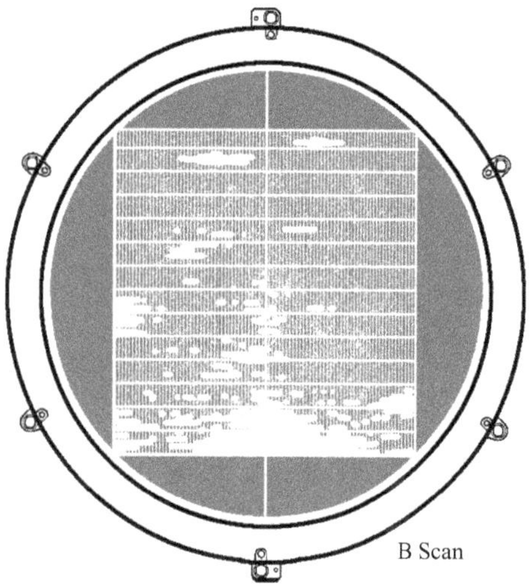

FIGURE 15-6

Type "B" display.

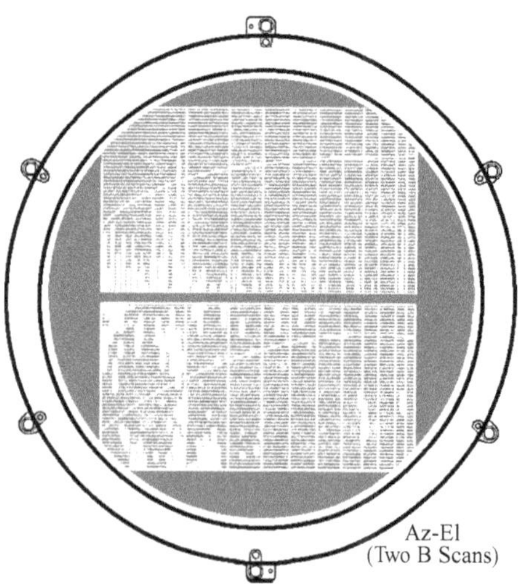

FIGURE 15-7

Az–El type "B" display.

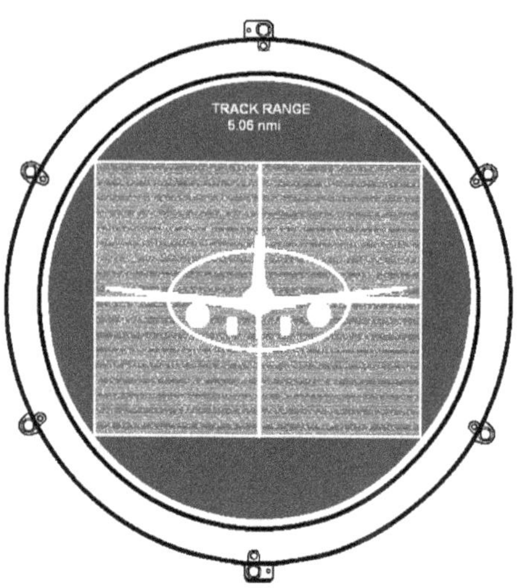

FIGURE 15-8

Type "C" display.

(see Figure 15-6). As shown, the range sweep moves from bottom to top in the *Y* dimension, and the antenna azimuth position varies the sweep position uniformly between left and right in the *X* dimension. Target data ***intensity-modulates*** the sweep, in the *Z*-axis, by bias on the crt electron gun, and the targets appear as brightened spots.

The B-scan may also be used for an elevation-versus-range-versus-intensity display, where the range is indicated from left to right, and the target height from bottom to top (see Figure 15-7). A quad-radar uses two B-scans on one display, a bottom one for azimuth versus range versus intensity, and a top one for elevation versus range versus intensity.

The C-Type Display

The type C display is a real-time display of targets only at a specific range; it is used in three-dimensional radars, where range, azimuth, and elevation information are simultaneously available (see Figure 15-8). The sweep is deflected in the *X* dimension by azimuth data, and in the *Y* dimension by elevation data. This type of indicator has also been called ***AHI,*** for azimuth versus height versus intensity. It has the advantage of displaying a cross-sectional view of a target, approximately as the silhouette it would present if seen by the eye; actual target definition depends upon the antenna beam dimensions. The AHI may be used with high-definition, pencil-beam radars, with the display range automatically tracking a target selected by a human radar controller. Some provision must be made to indicate the target range, such as a separate range count display. One version of this display incorporated a dual-mode display, in which the operator located a target on an AZ–EL-range B-scan, placed a ***cursor*** on the target of interest, and then switched to the AHI display. A ***cursor*** is an operator-positioned line or circle on a radar display. Cursors often operate in ***time-share,*** a method of switching the sweep generation from the radar scan to the positioned sweep. The words ***time-share*** are not limited to sweep generation, and may be used to apply to any circuitry used to alternate between two different information inputs; in digital logic, the use of the word ***multiplex*** is more common.

The E-Type Display

The type E display is another real-time display that bears considerable resemblance to the type B (see Figures 15-9 and 15-10). The main difference is that the sweep origin stays at a fixed point, and the sweep rotates about that point throughout an antenna scan. As with the type B, the type E can be either azimuth versus range versus intensity, or elevation versus range versus intensity. Type-E displays have been used for height-finder radars, as shown in Figure 15-9. A major significant use of the type E has been in azimuth-elevation ***precision***

approach radars (PAR) used in ***Ground Controlled Approach (GCA)*** radar systems. Figure 15-10 illustrates the two-E-scan PAR display in an AN/CPN-4 or its many descendants.

The G-Type Display

This is a real-time AHI display, somewhat similar to the type C, but there is no sweep or scan; the beam is continually positioned in the X- and Y-axes, and an electron beam is enabled in the Z-axis, each time the echo is received (see Figure 15-11). This type of display might have been found in monopulse radars, used for fire control. When the target is dead center of the antenna position, it will appear at the center of the display; if signal strength detected by multiple receive-beam patterns is greater in some beams than in others, azimuth or elevation errors cause the target to be deflected accordingly. Range data determines the target width; the "wings" spread out as range decreases.

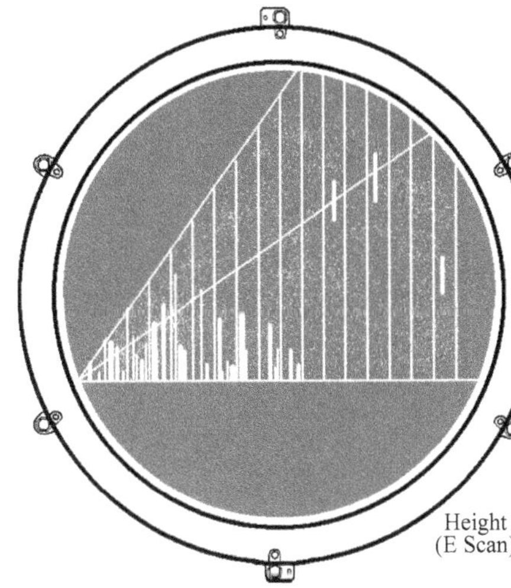

FIGURE 15-9

Type "E" display.

The P-Type Display

The type P display, commonly called a ***planned-position indicator (ppi),*** is probably the most common of all radar displays (see Figure 15-12). It is a range-versus-azimuth-versus-intensity display, and it is used predominately in surveillance radars with continuously rotating antennas. The sweep is intensity-modulated in the Z-axis by radar video, and the electron-beam movement begins the sweep trace at the center. The sweep "points" in a direction corresponding to the radar antenna position. For instance, when the antenna is pointed north, the sweep will be oriented at twelve o'clock. When used with shipboard or aircraft radar, the "twelve o'clock" position may represent "dead ahead," and the "six o'clock" position may represent "dead astern." In any case, the ppi display will resemble a map of radar information, or a picture, as viewed downward, from a very high point, directly above the radar.

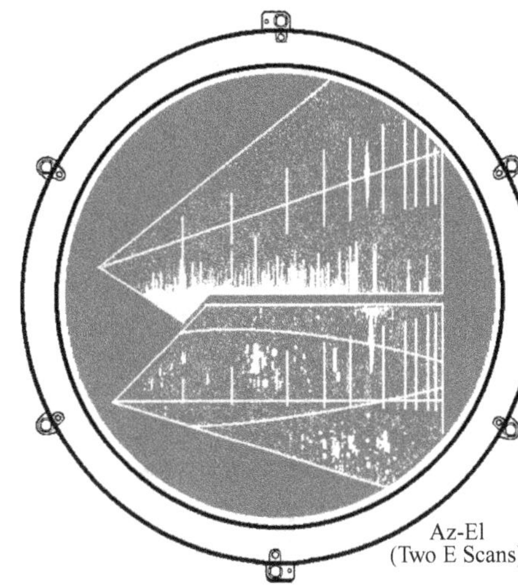

FIGURE 15-10

Type "E" Az–El display.

Variations of Type P Displays

The ppi illustrated in Figure 15-12 is a real-time display; the radar information may all be delayed from T_0 by some fixed amount of processing time, but it is placed on the face of the tube at the same rate that it is received. For instance, if the target range is 10 nm, the display range is 30 nm, and the diameter of the tube is 16 inches, the electron beam will move 2.67 inches from the center of the tube in 123.55 μs, and the target will then intensify the Z-axis electron beam to illuminate the phosphor material on the face of the crt. This rather laborious description of the seemingly obvious occurrence is intended to introduce the difference between the real-time display, and other types of displays in which radar echo time plays a much more indirect part in the location of the displayed target.

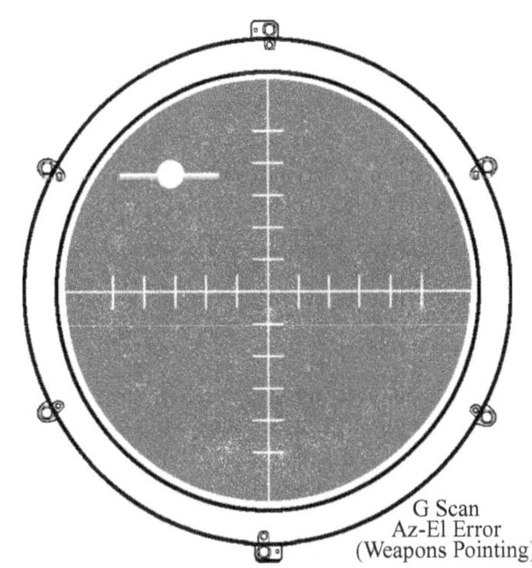

FIGURE 15-11

Type "G" display.

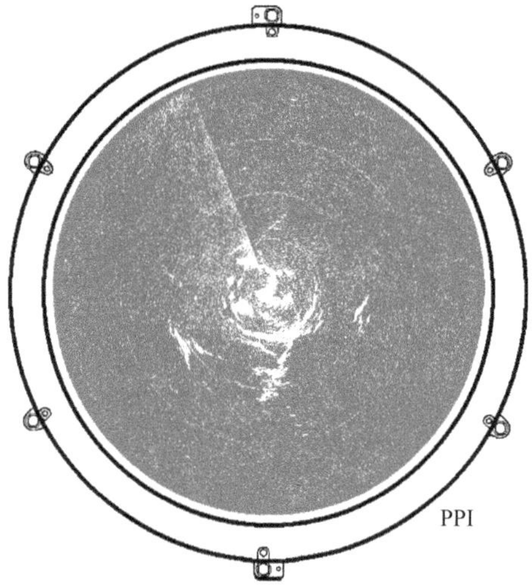

FIGURE 15-12

Type "P" display (aka "ppi").

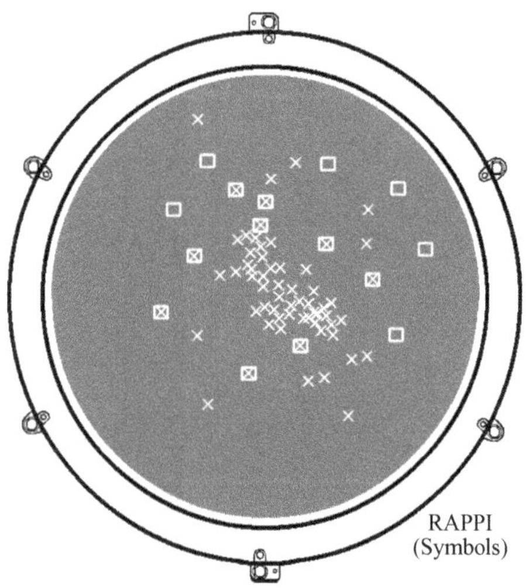

FIGURE 15-13

Random-access ppi (rappi) symbol display.

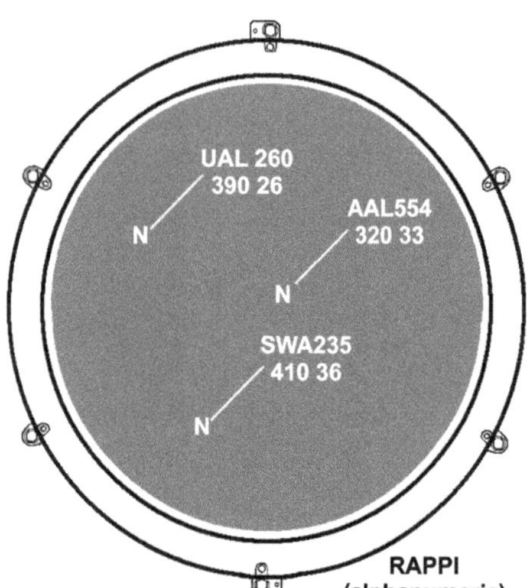

FIGURE 15-14

Air traffic control alphanumeric rappi.

The Random-Access ppi (rappi)

When azimuth sliding window or mtd-centroiding digitizers are employed for radar data, the real-time relationship with the transmitter "main bang" is lost, and the target may not be available for display until many degrees of antenna rotation after it has been received (see Figures 15-13–15-15). However, the necessary information for display is retained, in the form of digital range and azimuth data words in a target message. That information may then be used to instantaneously position ***("gross" or "major" positioning)*** the crt electron beam to the proper *X–Y* coordinates on the crt. Further programmed manipulation of the electron beam by small deflection "strokes," accompanied by programmed blanking, then creates symbols, as in Figure 15-13, or alphanumerics, as in Figure 15-14. Stroke deflections are illustrated in Figure 15-15, and described in more detail at a more advanced point in this chapter. One might visualize the process as one in which (1) the electron beam is moved to a location, much the same as if a hand holding a pen were placed at a location on a piece of paper, and then (2) the electron beam "writes" on the crt phosphor, as if by a pen, but much faster. The programmed blanking is equivalent to the lifting of the pen from the paper. The symbol display in Figure 15-13 resembles that used as a maintenance display for the common digitizer, and the alphanumerics display in Figure 15-14 resembles that used as an operational display for the ARTS. Terminal radar systems using the ARTS have employed real-time/rappi hybrid ppis, in which the alphanumeric data is written on the crt between real-time sweeps. Although many of these are still in use as this is being written, they are being replaced by purely digital television ppi displays, where the drawing of all characters and lines is accomplished with television pixel memory addressing.

Scan Conversion

Preface

Since World War II, there has been a continual effort directed toward improved radar displays. Television technology presented the best options, and an evolution of conversion of the rotating sweep scan to a television picture began in the 1960s. The use of the ppi and analog radar data is now in decline, and totally digital color displays such as the ARTS Color Display (ACD) and Remote ARTS Color Display (RACD) are replacing ppis and scan converters.

Radar Television Displays

On a real-time ppi for surveillance radars, such as ASRs and ARSRs, there is a major disadvantage to the user, in that he must wait for the target to be "refreshed," on the next revolution of the antenna, in order to determine the course (direction

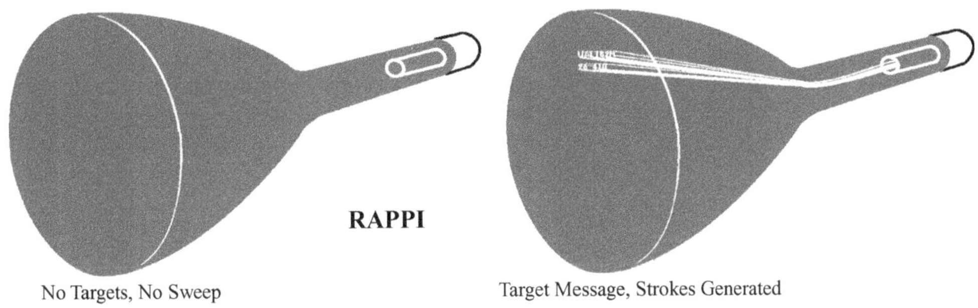

FIGURE 15–15

Drawing symbols or characters on a rappi with X and Y strokes.

of motion) of the target (see Figure 15-16). For an ASR with a 12.5 rpm antenna, the target only appears on the ppi once every 4.8 s, and quickly grows dim afterward, even though the crts with the greatest possible persistence are used in ppis. These shortcomings are tolerable in radar traffic control rooms, where the controllers' full attention may be devoted to watching the display. However, it is also necessary for radar displays to be available in air traffic control towers, and conventional ppis become unusable in that environment because (1) they do not offer enough brilliance to be visible in daylight, and (2) the controller cannot divert his attention to the display long enough to determine the aircraft course. Over the years, there have been several solutions to provide daylight-useful radar displays, but two elements are common to all. First, there was a method to "store" and hold the target, and second, the stored information would be "read" repeatedly with a television raster to create a television display. In the early 1950s, ***scan conversion*** technology was under development for the initial purpose of converting television pictures from one television scan rate to another. Since the necessity for such equipment was greatest in Europe, it was only natural that France pioneered the development. The French developed a special-purpose electron tube, in which a television picture could be "written" onto one side of a capacitive storage surface, and then "read" from the other side by ***secondary emission,*** as the surface was scanned by a television raster at the new rate.

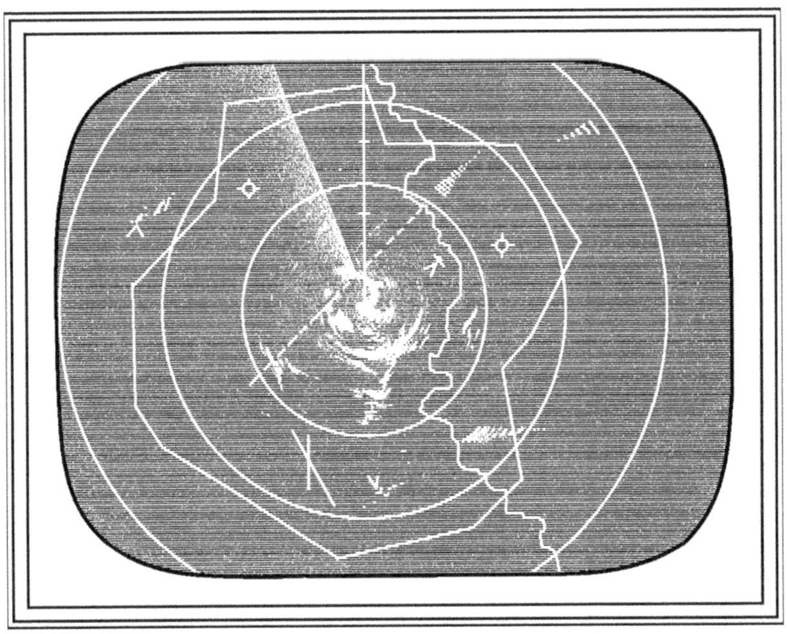

FIGURE 15–16

A radar television display.

Radar Bright Display Equipment (RBDE)

The value of the storage tube for use in radar was almost immediately apparent, and the "input" side of the tube was soon being written upon by $\rho\theta$ information (see Figure 15-17). One application was to broadcast the television information from a single radar in a harbor, so that all ships could view the harbor radar on television receivers. The FAA soon purchased equipment from France, and began installing it throughout the United States. The first major deployment was designated RBDE-2. There were several later versions, through RBDE-6, in use as recently as the early 1990s.

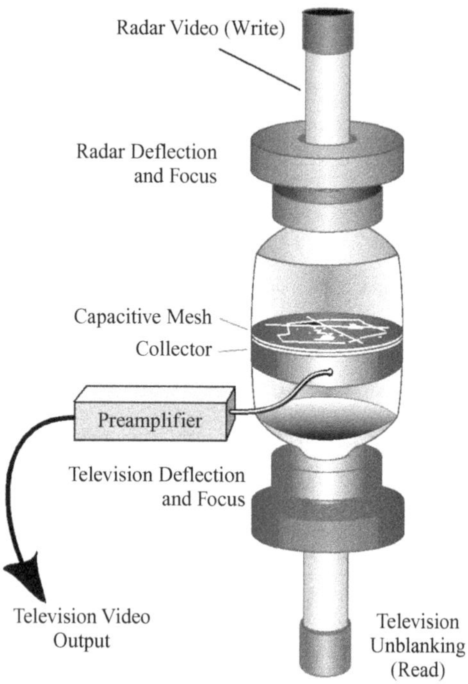

FIGURE 15–17

Scan conversion.

In the radar application of scan converters, the storage surface in the tube was capable of retaining a target "charge" for many read operations, and the target could remain visible for many antenna scans. This would cause the appearance of a "trail" on the display, and a user could immediately see the course of the target at a glance.

Storage Tube

Bright Radar Intensity Television Equipment (BRITE)

Eventually, RBDE equipment was replaced with a more economical type of scan conversion which eliminated the scan-conversion storage tube (see Figure 15-18). The BRITE equipment simply used a television camera, pointed through a dark tube at a small ppi display. The television camera employed a ***vidicon,*** but it differed from the conventional camera in that it exhibited a very high retention rate to provide the necessary storage. The ppi itself was of very low persistence, so that residual light from radar data would not add to the camera light input. To obtain better resolution, the RBDE and BRITE systems employed higher line rates than commercial television. The same frame and field rates, 30 Hz and 60 Hz, were employed, because of the compatibility with commercial power; however BRITE used a 945-line frame, rather than the commercial 545-line frame.

Synthetic Real-Time ppi Displays

At ASR Radar facilities all across the country, there are many millions of dollars invested in equipment associated with real-time radar displays. Among these are ppi displays, video mapping units, ARTS computers, and ***air traffic control radar beacon service (ATCRBS)*** decoders. This equipment has accumulated over a period of 40+ years, and all relies on common real-time radar inputs, such as system triggers, azimuth-change pulses, and mti and normal videos. When the ASR-9 was built, the self-contained digitizer in the mtd system eliminated the availability of analog real-time video. Radar data messages, much like those sent from the ARSR facilities to the ARTCCs, were the only ASR-9 output, and nothing else was available. No doubt, this was an improvement, and the new modems, requiring only 2,400-bps telephone lines, eliminated the need for expensive and troublesome video cables, line drivers, and compensators. Simply with telephone connections, the radar data could be sent any distance, opening a multitude of opportunities for future technology and expansions. However, the ASR-9 was years ahead of all the ancillary equipment it had to support, and a special equipment called ***Surveillance Communications Interface Processor (SCIP)*** was built to maintain compatibility.

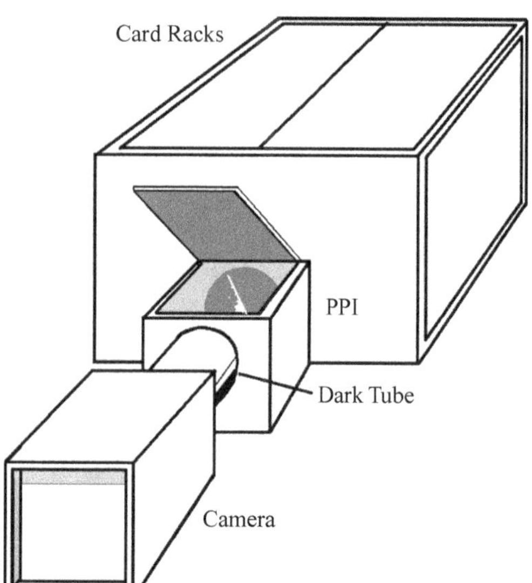

FIGURE 15–18

BRITE equipment.

Reconstituting Real Time

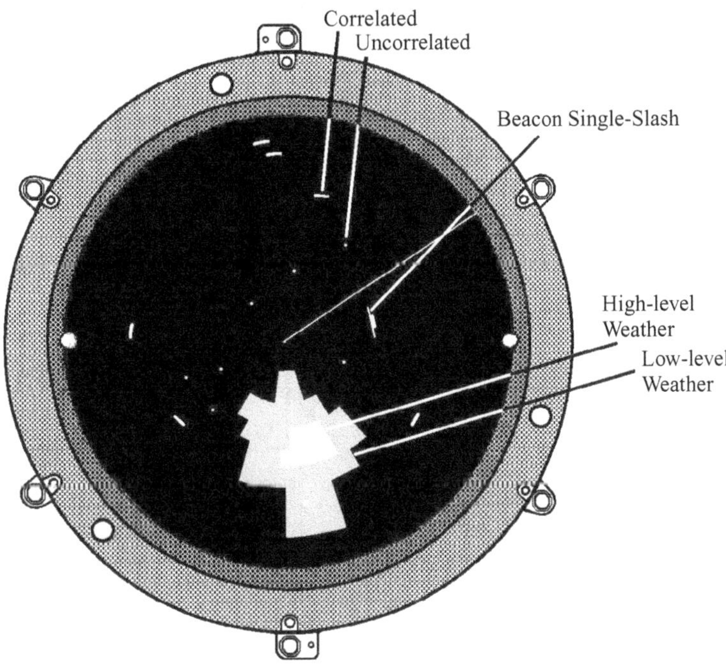

FIGURE 15-19

Reconstituted data, synthetic real-time ppi, ASR-9 SCIP.

Chapter 14 deals with the operational theory of the ASR-9 mtd system, and its data output (see Figure 15-19). At the user facility, the SCIP "reconstitutes" a "synthetic" real time, developing a synthetic display-site pretrigger; synthetic azimuth change and reference pulses; and synthetic real-time video, making all systems compatible, just as if they were connected to a real-time radar input. There is no normal video available, and the messages from the ASR-9 are converted to correlated, uncorrelated, beacon single-slash, and weather analog video. The azimuth and range width of each type of target is programmable, and is generally adjusted so the correlated targets are larger in both dimensions than the uncorrelated, and the beacon single-slash targets are wider in azimuth than either correlated or uncorrelated. The SCIP even reconstitutes beacon mode and code information, so it may be available for decoders.

Although the displayed SCIP video is pleasing to the user, there are some disadvantages. There is no grass apparent, and the controller cannot "spot" weak targets in the grass; this is more of a personnel confidence problem than a radar shortcoming; mtd coherent integration can detect such weak targets. If there are false targets, their appearance is identical to the real ones. Because no normal video is available, permanent echoes cannot be used for alignment checks; however, a ***target performance window*** offers the technician a means to verify alignment far more precisely than in previous systems, and programmable SCIP test targets provide a means to "mark" map alignment points. Because there is no normal video available, operational mti reflectors are absolutely essential. mti reflectors are discussed in Chapter 12, and the target performance window in Chapter 14.

Basic Operation of the ppi

General

These illustrations depict an early rotating-deflection-coil (yoke) system; that type is used for opening discussion of the operation, to provide the reader with the simplest possible foundation (see Figures 15-20 and 15-21). Rotating-yoke ppis are now very rare. Most of the circuitry, except the sweep drivers and deflection coil, is still very similar, but it has evolved from electron tubes to solid-state devices. The synchro system used to drive the rotating deflection coil is not shown in these illustrations, but that system, further addressed in the proceeding text, will cause the yoke to rotate in synchronization with the antenna.

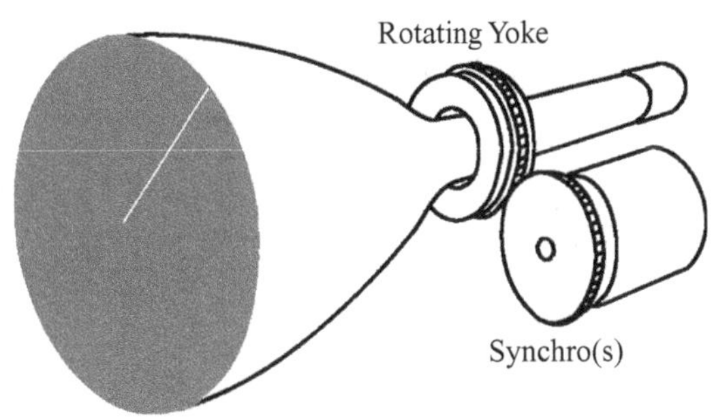

FIGURE 15-20

Mechanically rotated deflection coil.

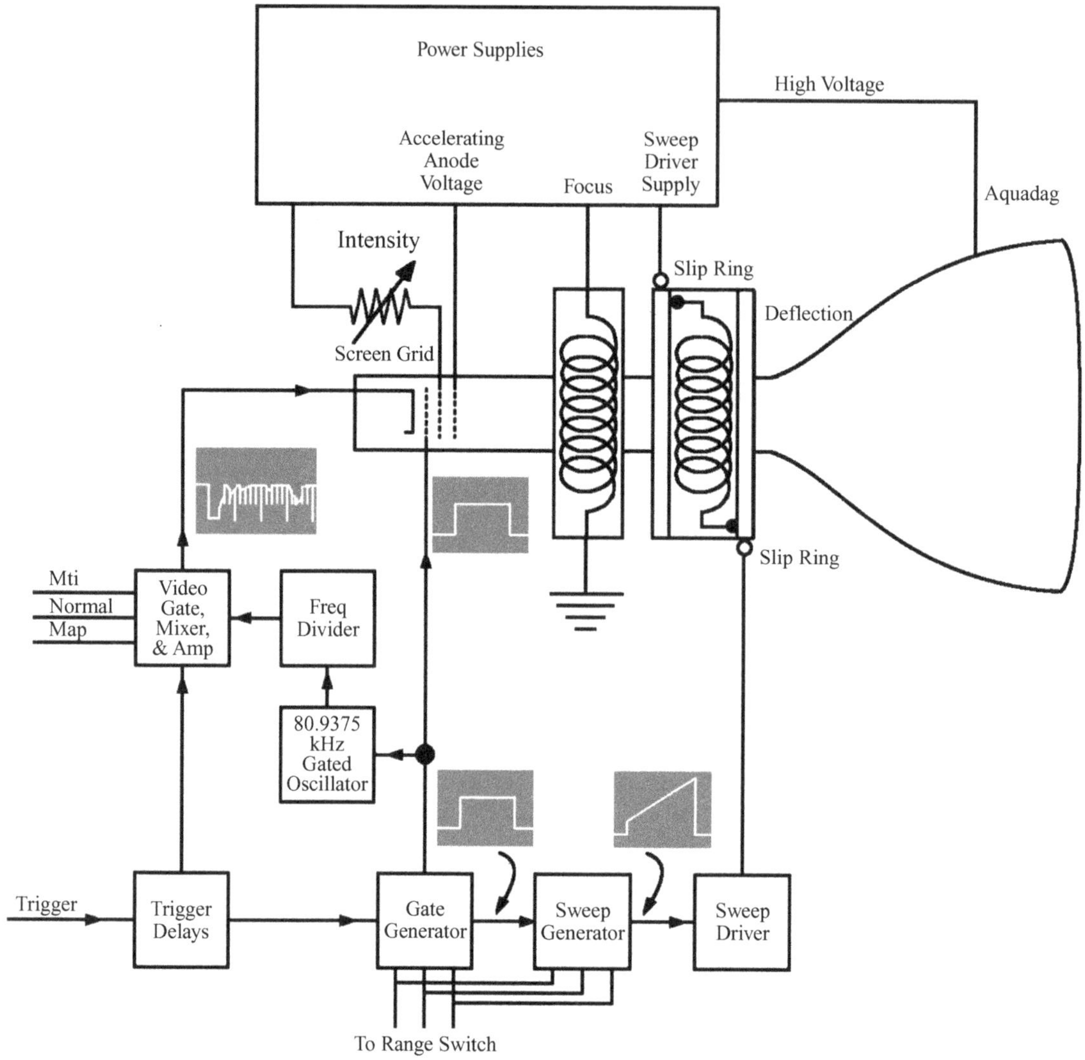

FIGURE 15–21

Simplified rotating-yoke ppi.

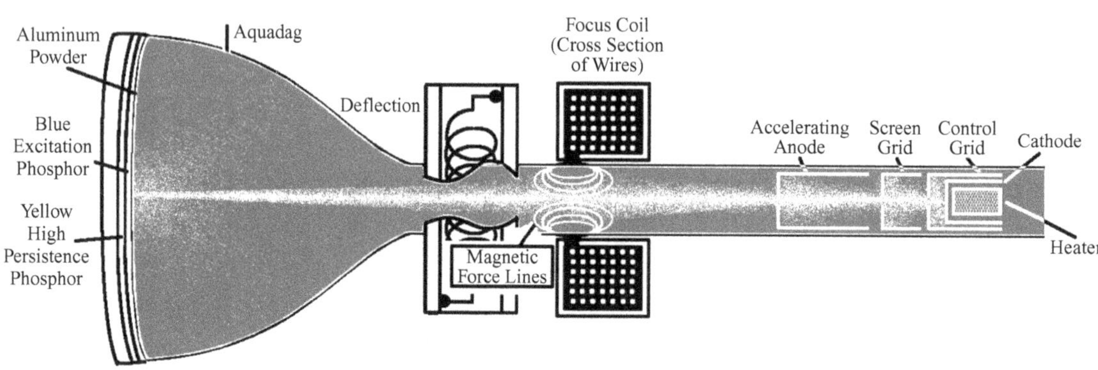

FIGURE 15–22

The electromagnetic crt.

The Electromagnetic Cathode Ray Tube (crt)

The major parts of the crt are (1) an electron gun, (2) the face, composed of very thick glass for safety, phosphor (s), and an aluminum-powder backing, on latter-day versions, and (3) an ***aquadag,*** a conductive coating over much of the forward part of the tube, intended to remove electrons from the phosphor, and the space behind it (see Figures 15-21 and 15-22). Two external parts are the deflection and focus coils, two electromagnetic devices of very different purpose, effect, and construction.

The Electron Gun

Earlier electromagnetic tubes employed a triode gun, where the principal elements were a cathode, control grid, and accelerating anode (see Figure 15-22). Of course, a filament was necessary to heat the cathode, so that it could emit electrons. Most latter-day tubes employ a tetrode gun. The fourth element of tetrode is a screen grid, which offers improved control of the electron beam.

The cathode is a circular element, surrounding and covering the heater; when heated, the increased molecular activity makes it possible for electrons to leave the cathode material under the influence of a repelling negative applied voltage. The electrons from the cathode are propelled toward the more-positive control grid, another circular element, which blocks many of those electrons not on a near-direct course down the center of the gun.

Although an electrical device, the control grid is analogous to a lens in an optical system, and begins to focus a concentrated beam. The control grid is so called because a voltage, more positive than the cathode, will control the electron concentration and acceleration in the beam. In many indicators, video is applied to the control grid; in others, the intensity control may be connected to the control grid.

The voltage on the screen grid is still more positive than that on the cathode, and more positive than the control grid. The screen grid further controls, focuses, and accelerates the beam; the additional influence on the beam dramatically improves the tube performance over the triode gun, in both intensity and focus. In some displays, the screen-grid control may be a technician INTENSITY LIMIT adjustment, serving as the principal means to regulate the amount of user-available control. The accelerating anode has a high positive voltage upon it, as much as 500 V, causing the electrons to be propelled toward the crt face at a high velocity.

When the cathode is heated, and a high negative voltage (in respect to the aquadag) applied, the objective is only to transfer electrons into the cathode material from a power source and then out of the cathode, and into the electron beam. Even under normal conditions, but particularly when the indicator is first energized, there is some probability that entire ions, charged atoms from the cathode material, or from gas molecules in the tube, will enter the electron beam. Compared to the electron, these ions are heavy, and can burn the phosphor face of the tube. Early-day tubes would require, as an exterior device, an "ion trap," a small magnet, affixed to the neck of the tube with a circular spring clip. The magnet would pull the heavier ions to the side of the tube. Improved tubes next used an "oblique" electron gun. All the elements, except the accelerating anode, were deliberately "aimed" slightly toward the tube neck. The course of the lightweight electrons would be "straightened" by the accelerating anode, but the heavier ions would go to the side of the tube. The oblique gun is obvious from the outside of the tube, and a technician, not aware of the intent, might mistake it for a manufacturing error. The latest tubes use "aluminized" screens, which further preclude ion burns.

The Phosphor(s)

For clarity, Figure 15-22 shows layers of material on the face of the tube. These layers actually exist, but are not nearly as thick as those shown. Phosphors are classified in terms of numerical persistence ratings, such as "P7," the greatest persistence, and the one most often used in radar. Additionally, illuminated phosphors may emit a color of light which may be related to the persistence. A phosphor may be illuminated either by electron bombardment, or by external light; it is this dual capability that permits the use of layered materials. The low-persistence blue phosphor, when struck by the electron beam, recovers quickly, but its light illuminates the more-fragile, but high-persistence, yellow phosphor, visible to the user. That the yellow phosphor can be partially illuminated by external light is yet another reason that the ppi is not desirable in other than dim or dark rooms. The contrast, the difference between light and dark areas on the face of the tube, is reduced by daylight.

A phosphor exhibits ***secondary emission.*** Any material, when struck by an electron, becomes negatively charged by that excess electron. In most cases, the most desirable condition for the phosphor is that each electron that strikes it causes another to be emitted; this is secondary emission (there are special-purpose phosphor meshes, used in storage tubes, where the secondary emission does not occur until the phosphor is "read" with a second, higher energy, electron beam).

The main objective in radar displays is a high-intensity target display, and the velocity and number of electrons in the beam, established largely by the accelerating anode voltage, are major factors in providing this. The secondary-emission capability of the phosphor, however, limits the intensity; when the phosphor cannot emit electrons as rapidly as it is struck, it becomes negatively charged and additional electrons are repelled, to contribute no more to the intensity. ***Aluminized phosphors*** raised the display intensities substantially. The use of powdered aluminum, sprayed on the "gun side" of the phosphors, improved the secondary-emission rate dramatically; being a conductor, the aluminum provided a path for electrons to flow from the phosphors to the high-voltage aquadag. The aluminum further serves as a mirror, sustaining illumination by light.

Cathode Ray Tube Hazards

Any device employing a high-velocity electron beam, created under the influences of high voltages, may be capable of ***X-rays.*** As a general rule, the aquadag potential is a good indicator of the hazard; as it exceeds 20 kV, the possibility of X-ray significantly increases. The greatest hazard is behind the front of the tube. Most ppi displays use voltages much below the levels that would create X-rays, but the only way to be certain is to check the manufacturer's instructions.

Because the crt is a vacuum tube, any shock that can create a crack will cause air to leak inward. Because of the high vacuum, that inward leak can cause such a violent rush of air that the remainder of the weakened tube envelope may collapse before the pressure has equalized; pieces of the tube strike each other and may even be propelled outward by the abrupt pressure change. Such an event is called an ***implosion*** and can cause serious injury to the handler. The very thick glass at the face of the crt is intended to reduce the risk of implosion to the user. Removal or replacement of the tube should always be done with protective clothing such as gloves, aprons, face masks, or whatever may be directed by current safety policy and instruction books.

The Focus and Deflection Coils

The effect of a magnetic field upon an electron in motion is addressed in Chapter 10; this effect is central to the construction and usage of the focus and deflection coils. Most fundamentally, if a magnetic field is at a right angle to an electron motion, it will change the course of the electron; if parallel to the course of the electron, it will have no effect.

The focus and deflection coils are radically different in one major aspect; the fields produced by them are at right angles to each other. The focus coil is wound so the wires encircle the neck of the tube. The coil is encased in metal, so that most of the magnetic field is contained, but a small opening permits the field to affect one small area in the neck of the tube. Electrons on a straight course down the center of the tube are unaffected by the focus-coil magnetic field. However, as the course of electrons deviates from that ideal course, the magnetic field begins to take effect, and the course is corrected. When the field is at precisely the correct strength, all electrons arrive at a single point on the phosphor.

Because the distance from the deflection coil to the phosphor is greater at the edges of the tube than at the center, some displays may incorporate a ***dynamic focus*** voltage, which adds to the focus-coil current as a function of the beam deflection. The curvature of dynamic focus voltage may be adjustable. Where this is not incorporated, best focus adjustment is attempted midway between the center and edge. Focus alignments will include instructions for initial physical positioning of the focus coil, and there may be circular ***precentering magnets*** on the backside of the focus-coil assembly, to ensure that the beam is centered in the neck of the tube before entering the focus-coil field.

The deflection coils are at right angles to the electrons in the beam, and the magnetic fields created by sweep currents directly pull or push the electrons.

The Sweep Generator Circuit

As with nearly all other radar design, the sweep generator has passed through a series of stages of evolution (see Figure 15-23). Until the advent of precision counters with A/D conversion, the usual practice was to, with a "sweep gate," enable an R–C network charging circuit a few microseconds before T_0. The RC product provided a time constant τ great enough that the capacitor would only partially charge before the end of the sweep gate. A feedback circuit would add a portion of the amplified charging waveform to the input to make the sweep linear, or to shape it as needed to meet the requirements of the crt and drivers. The shape could be adjusted with a resistive LINEARITY control in the R–C network, or with an amplifier GAIN control. The LINEARITY and GAIN controls were interactive. This method was used in a variety of circuits, including vacuum-tube "bootstrap generators," "phantastrons," transistor "Miller integrators," and integrated-circuit "operational amplifiers."

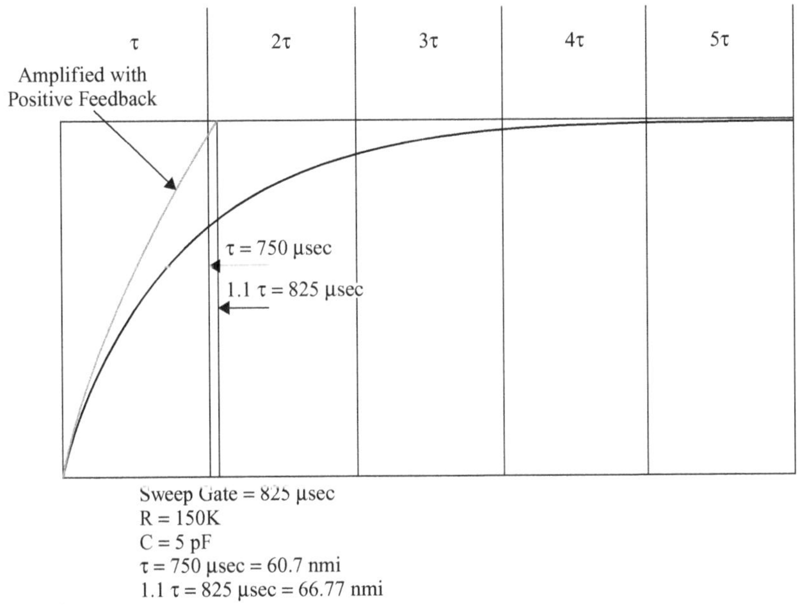

FIGURE 15-23

Using a portion of an R–C network charge for sweep generation.

The Phantastron

This paragraph should begin with a recognition that there have been several variations of the phantastron circuit over the 50+ years since its origin (see Figure 15-24). The circuit shown in this chapter is from USAF Manual 52-8, published in 1951. In the early days of radar circuit design with vacuum tubes, the exotic phantastron design was used to provide adjustable gates, sweeps, and/or sweep gates, with one special-purpose tube, a 6SA7. The 6SA7 contained seven elements: a plate, cathode, and five grids. The circuit contained a capacitance which served to make the circuit both a linear sweep generator and an adjustable gate generator. The overall circuit is roughly equivalent, but superior, to a single-shot multivibrator with both sides (the two triodes or transistors) contained in a single tube. The circuit is very stable because the length of the gate and sweep it produces depend on voltage divisions on the grids, and the ratios remain the same with power supply fluctuations.

In comparing the phantastron circuit to the single-shot multivibrator, it could be said that the cathode, G3, and plate are roughly equivalent to the first triode or npn transistor (normally drawn on the left), and is normally cut off. The cathode, G1, and G2 are roughly equivalent to the second triode or npn transistor (normally drawn on the right), and is normally conducting. The cathode-G3-plate circuit, cut off in the absence of a trigger input, causes most electron flow in that part of the circuit to be through R7, R6, CR1, and R4. C1 is charged (on the left) to the divider voltage at the wiper of R6 and (on the right) to the divider voltage at G1 in the R10, cathode, G1, R5 electron flow path.

To trigger the single-shot action, the cathode, G1, G2 circuit are cut off by a negative trigger input through C2, and the electron path is through an R8, G3, plate, R4 circuit. This new path, causing an immediate small plate current, creates the downward voltage "step" at the beginning of the plate waveform. The decreased voltage on the left side of C1 starts an electron flow through a path including R10, V1 cathode, G3, V1 plate, C1, and R5. The right side of C1 begins to go more positive.

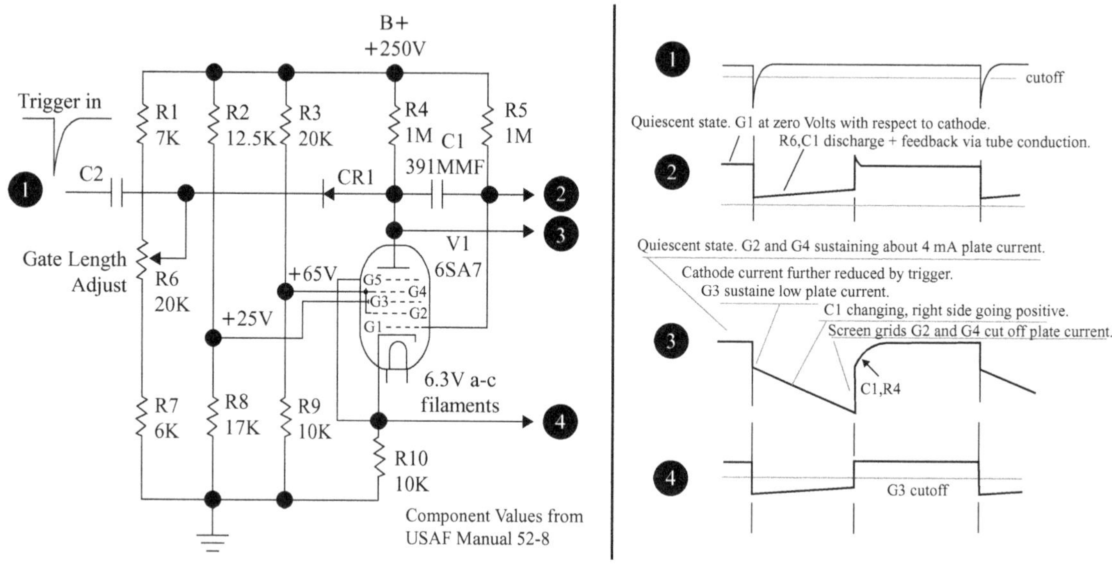

FIGURE 15-24

The phantastron circuit.

As the right side of C1 goes positive, the rising voltage on G1 increases the electron flow in the cathode-to-plate path, further decreasing the plate voltage, and further increasing the cathode-to-plate electron flow. However, the increasing flow raises the drop across R10. G3 becomes less positive in respect to the cathode, and more positive in respect to the plate. Those conditions tend to slow the electron flow through V1, and to slow the rate of change to the charge on C1. The result is a near-linear rise in current flow through V1.

The V1 conduction stops completely when the plate voltage drops below the point that there can be no more conduction through the G3 grid. At that time, the plate waveform returns first to the lower "step" value, and then back to the original quiescent state as C1 charges.

Ramp Generation by a Binary Counter

In latter-day design, a very precise sweep is created with a clocked binary register supplying a count to a D–A converter (see Figure 15-25). The degree of precision is then determined by the number of stages in the register, the clock frequency and stability, and the voltage resolution of the D/A converter.

Sweep and Unblanking Gates

Temporally adjusted triggers are applied to a gate generator, which will determine the sweep length. One gate-generator output is the ***sweep gate;*** during the time the gate is high, capacitance in the sweep generator charges, creating the voltage rise. Another gate-generator output is called the ***unblanking gate,*** or ***blanking gate*** (manufacturer's choice may differ); during the time the trace is to be "on," that gate enables the crt electron beam. Depending upon the manufacturer and type of crt, the unblanking gate may be applied to the cathode, control grid, or screen grid, of the crt, and it may, or may not, have video added to the active portion of the gate level. In Figure 15-26, the sweep gate and unblanking gate are of the same duration, and the video is applied to the cathode; this may also not be the case, and there may be separate and independent gate generators.

Timing

A principal objective in the display is to cause the video that occurs at T_0 to intensify the trace at precisely the same time the trace begins, normally called the ***sweep origin.*** The video could be "late" in relation to the

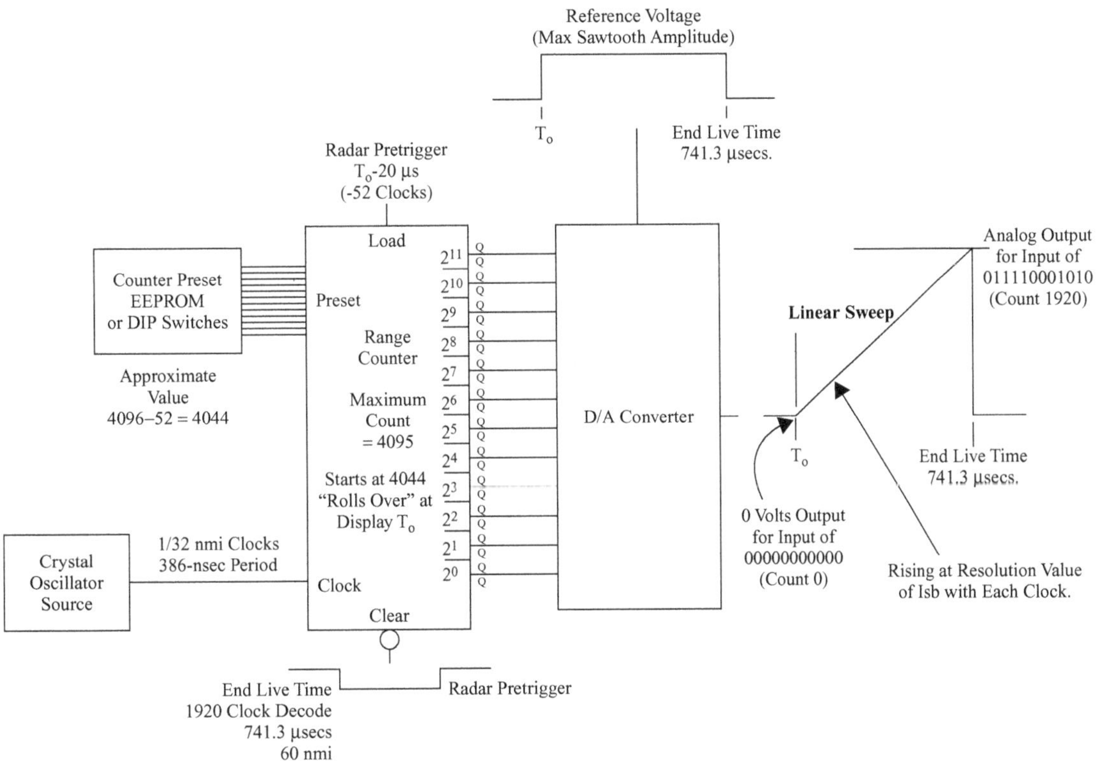

FIGURE 15-25

Generating a linear ramp with a clock counter.

beginning of the trace, or it could be "early." If it is early, some of the video may be lost, and, if it is late, a "hole" may appear in the center of the display. In either case, the range marks may be misaligned with the video, presenting a range error. To allow for these possibilities, the trigger input to the display may be considerably in advance of video T_0, and adjustments in the indicator provide the technician with a means to precisely match the video versus sweep-start timing.

Overscan Limit

The manner in which the sweep gate ends differs with systems. In older systems, the sweep gate time was established by a multivibrator, or phantastron, circuit; when the user changed the range with a panel switch, the gate-length determining components in the circuit were changed. In latter-day systems, an "overscan" limit circuit

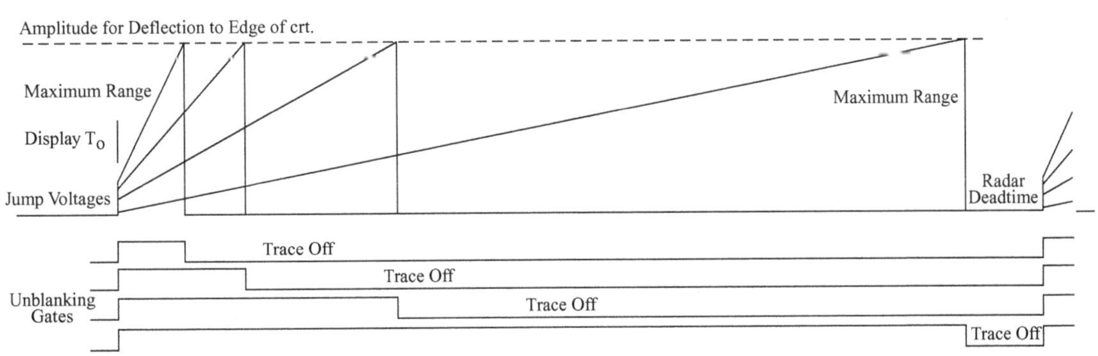

FIGURE 15-26

Sweep and unblanking gates.

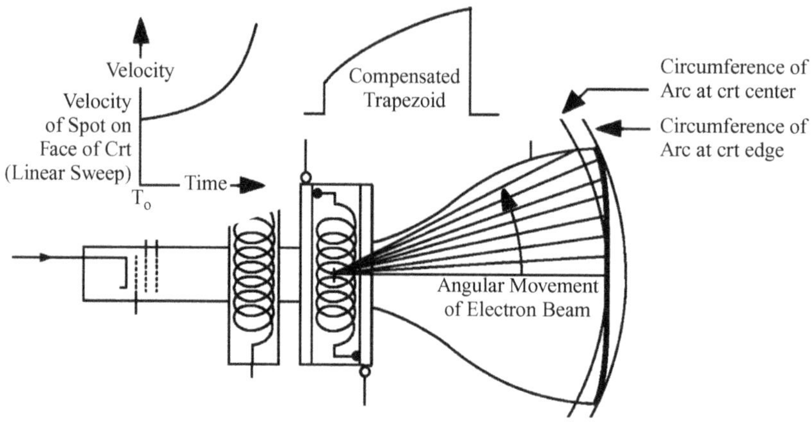

FIGURE 15-27

Sweep speed increases at outer edge.

senses the deflection coil currents, to generate a "gate-end" trigger when those currents reach a prescribed level. A common manner in accomplishing this is with a resistive summing circuit which algebraically adds samples of the n–s and e–w sweeps. When the sum exceeds prescribed threshold limits set by a potentiometer, a transistor circuit or operational amplifier produces a gate. The gate is of short duration, since the sweep ends immediately after the overscan is detected. The overscan limit circuit is particularly useful and necessary when the sweep origin is placed off center; then, the sweep must vary in length as it rotates.

The Sweep Generator and Driver

The sweep generator may produce a ***trapezoid*** waveform, which is actually the desired sweep, placed on a voltage pedestal (see Figure 15-26). The pedestal is called ***jump voltage.*** The deflection coil is in the current path from the power supply, through the sweep driver, to ground. The sweep driver is a current amplifier for the trapezoid, and causes a rising current through the deflection coil; the "jump pedestal" portion of the trapezoid does not appear in the sweep current; its purpose is to ensure an immediate, smooth, rise in current. The inductive reactance of the deflection coil opposes the initial sweep current, and the jump voltage creates an initial current flow, in spite of that initial opposition. The greater the range, the less the slope of the trapezoid, and the less the required jump voltage. An example of the need for jump voltage may become more obvious when the speed of deflection is considered. Consider, for instance, a 6-mile sweep, and a 16-inch diameter ppi. The spot creating the trace must move 8 inches, from the center of the crt to the edge, in just 74.1 μs, 9.267 μs/inch, 107,916.77 inches/s, 8,993.06 feet/s, 5,242 nmi/h. If a hypothetical value of 500 mA were required for full-scale deflection, and then the deflection-coil current would have to change at 62.5 mA/inch, and 6,744.8 A/s.

The final sweep voltage waveform may very well not be linear; the movement of the trace across the face of the crt would not be at a constant velocity, were that the case (see Figure 15-27). The electron beam is deflected by the deflection coil, so the point at which it passes through the coil becomes the pivot point of the angular movement of the beam. The face of the crt does not exhibit the same curvature as the arc of the angle from the deflection pivot point; purely because of this geometry, a linear sweep would cause the beam to move more rapidly at the edges of the crt, than it moves at the center. It is for this same reason that a dynamic focus voltage may be necessary.

Because of the high currents required through the deflection coils, the sweep drivers are among the units with the highest failure potentials in the entire display system; the technician can reduce his repair and troubleshooting efforts by developing a thorough understanding of the schematics and operation of those units, for the particular displays at his facility.

The Video Amplifier

This unit is controlled by panel adjustments on the display. Excessive video amplitudes will cause "blooming" on the face of the crt, increasing the target size, and worsening the range-azimuth resolution; therefore, the video amplifier normally contains technician-only limit and gain controls, in addition to the front panel controls. In addition to the amplifiers and mixer, for combining selected videos, the video amplifier may also contain a panel-controlled range gate, to provide for mti/normal gating. The preferable method of operation is for the mti/normal gate transition to normal to be set just beyond the clutter range, since the performance of the normal receiver, in clutter-free "clear" areas is superior to mti. In Figure 15-28, the video amplifier output also is applied, as negative-going video, to the cathode of the crt. The amplifiers must have sufficient bandpass to allow an adequate portion of the Fourier spectrum to provide adequate pulse reproduction. Recall that the first crossover of the Fourier spectrum is $1/t_p$. If the video bandpass is zero to $1/t_p$, there will be good pulse reproduction, but the t_p in that equation must be that of the most narrow inputs, which may include raw beacon video, with 450-ns pulses in the code train.

One input to the video amplifier is a range-mark video. An 80.9375 kHz oscillator provides 1-nmi range marks every 12.3552 μs. The frequency is divided by 2, 5, 10, 25, or other numbers, to provide marks of the selected spacing. Control of the divider or its outputs is by a front-panel switch. It is common practice to add range marks of two different frequency divisions, so that those which are coincident will be of greater intensity and brilliance. Examples are 2-mile marks intensified by 10-mile marks, 5-mile marks intensified by 25-mile marks, etc.

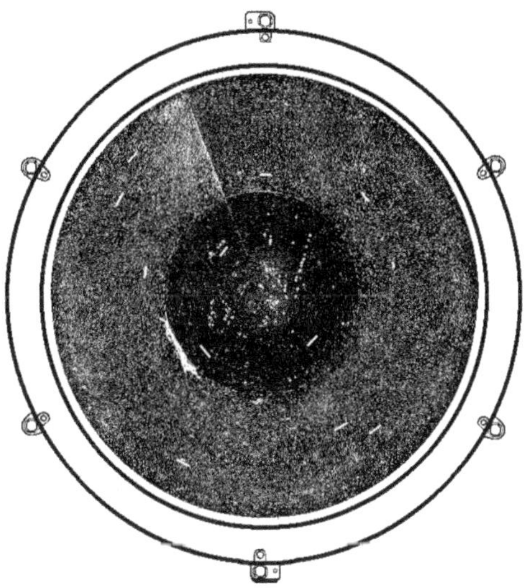

FIGURE 15-28

mti/normal range gate.

The Evolution of Azimuth Data

The first ppis with a rotating deflection coil obtained the necessary azimuth information from ***synchro systems,*** also called servomechanisms (see Figure 15-29). A servomechanism is defined broadly in dictionaries as any system containing devices or means to cause one mechanical or electrical device to change in correspondence with another. Even the electrical servomechanisms used in radar preceded its development, and were used on ships, to transmit the desired engine-speed information from the bridge to the engine room, and for other purposes. Used in radar, these systems were rugged and reliable, but were subject to the usual mechanical problems, such as noise, gear "play," and bearing failures and they could present alignment problems such as "hunting." Synchro systems still have useful applications, and will probably never become totally extinct, even though they are now rare in FAA surveillance radars. A synchro system has a distinct advantage, in that the azimuth remains accurate, no matter what the speed or direction of antenna rotation may be. This is not true of azimuth-pulse-generator systems, which must have continual, uninterrupted rotation in a single (usually clockwise) direction.

Basic Synchro System

The most basic synchro system consists of a synchro transmitter and receiver (see Figure 15-30). Synchros may also be called ***selsyns.*** "Selsyn" was originally a General Electric name for its own "self-synchronous" device, but has since become synonymous with "synchro." When a transmitter and a receiver are connected as shown in Figure 15-30, the mechanically driven rotor in the transmitter induces currents in the sta-

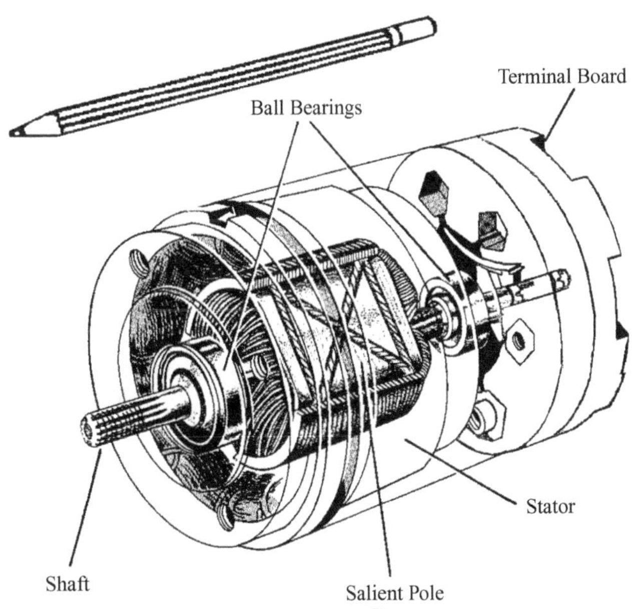

FIGURE 15-29

The synchro.

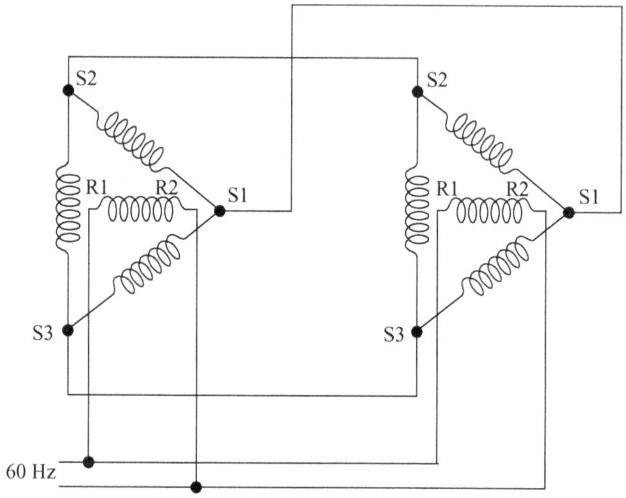

FIGURE 15-30

Simple synchro system.

tor windings; those currents are then induced in the receiver stator windings and cause the rotor in the receiver be positioned, by electromotive force, to the same orientation as the transmitter rotor.

Increasing Torque

Unfortunately, the system shown in Figure 15-30 does not provide great torque, and a means of ***torque amplification*** is necessary in operational radar systems (see Figure 15-31). To accomplish that a scheme was devised in which the torque is supplied by a ***two-phase motor,*** characterized by two sets of windings, separated by 90°. The synchro receiver was replaced by a ***control transformer***, and an a-c error voltage was obtained from the rotor. The rotor itself is not electrically connected to the a-c reference and is, therefore, not forced to alignment position; instead, it is mechanically driven by the two-phase motor.

The two-phase motor supplies the mechanical torque to re-position the rotor, but the real source of the torque amplification is the ***error amplifier,*** which gains additional electrical power from receiving-site power supplies. When the control-transformer rotor is 90° from the transmitter rotor, there is no error voltage output; as errors develop in either direction, the error voltage will increase, and its phase will depend upon the direction of error. The a-c reference for the two-phase motor is that for the rotor inputs of the synchro transmitter, and the a-c drive for the other phase input to the two-phase motor is the amplified error voltage from the rotor.

In the absence of error voltage, the two-phase motor would not run, and the result would be a "jerking" operation, in which the rotation would stop each time the error was lost, and then start again. To correct for this, a ***tachometer*** feeds an amount of 60 Hz back into the error amplifier, and that amount is related to the rotational speed, increasing proportionally. While the antenna is turning, the tachometer is maintaining the smooth rotation of the two-phase motor. When the antenna is stopped, the tachometer produces no output.

See the partial diagram in Figure 15-32. Even with multiplied torque, the system is subject to small errors, because the error voltage is not adequate for correction until the error reaches a few degrees. The incorporation of a second synchro, running at a greater speed, alleviates this. It is mechanically connected to the first synchro with gears, but the gear ratio causes it to turn several times faster, such as 10 times or 36 times. This second synchro is

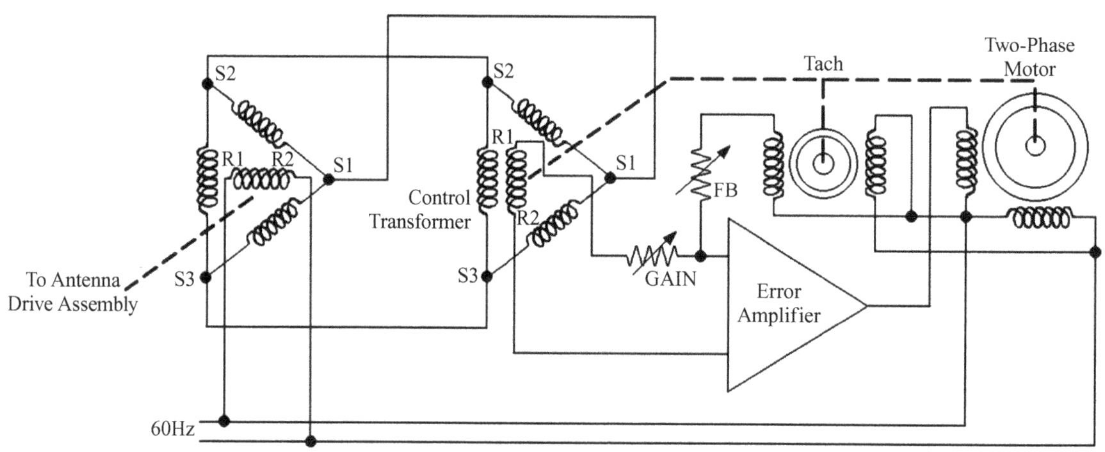

FIGURE 15-31

Control transformer, two-phase motor, and tachometer.

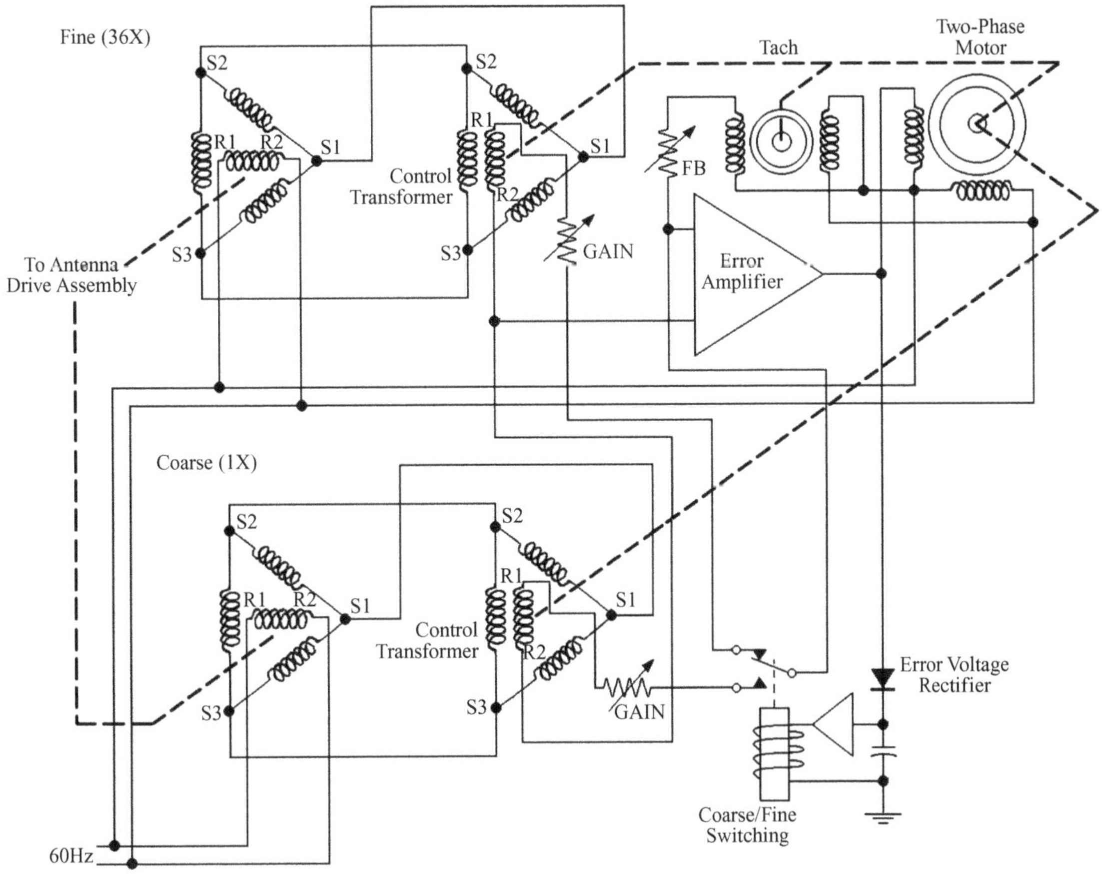

FIGURE 15-32

A two-speed synchro system, partial diagram.

called the ***fine*** synchro, and the original synchro is called the ***coarse.*** The coarse synchro is also called the **1** ×, as it makes one turn per antenna revolution. The fine synchros may be called **10** × or **36** ×**,** depending upon the gear ratio.

The error voltage from both coarse and fine control transformers cannot drive the two-phase motor simultaneously. If only the fine were used, the system could lock in at 10 or 36 different points. If only the coarse were used, the fine would be of no value. And if both were mixed, the algebraic combination of the two would provide the two-phase motor with useless, unintelligible, information. A switching system is therefore employed, so that, once the error voltage becomes low in coarse operation, a relay de-energizes to switch to fine. In fine operation, if the error becomes significant, the system switches back to coarse. Maladjustment of the GAIN or FEEDBACK controls can cause rapid switching between coarse and fine, making the system erratic.

Several refinements may be found in these systems. There may be a mechanical damper on the gear drive train. The system also exhibits an ambiguity; it is possible for it to "lock in" 180° from the correct point, as both error voltages would reach a null at that point. This possibility is precluded by mixing a portion of the 60-Hz reference with the 1 × control-transformer output. At 180°, the 1 × output is increased, but at 0°, the system operates normally. The added voltage is called "antistickoff."

Hybrid Systems

As technology evolved from synchro systems to azimuth-pulse systems, it was sometimes necessary to build special devices to convert a-c servo data to azimuth change pulses, or the reverse. Although now very rare, some of these may still be in existence.

The Trigonometric Identity, $\sin^2\theta + \cos^2\theta = 1$

The author's intent has been to deal as lightly as practical with the mathematics of radar, but the entire science is one of applied mathematics, and there is sometimes no way to logically present a theory of operation without the mathematical foundation. Such is the case in the relationship between latter-day radar azimuth systems and the trigonometric identity, $\sin^2\theta + \cos^2\theta = 1$. In fact, this equation is a cornerstone of many radar theories, including quadrature phase detection and circular polarization, to offer only two of many examples.

The ***Pythagorean Theorem*** states that the sum of the squares of the lengths of the two sides of a right triangle is equal to the square of the length of the hypotenuse (see Figure 15-34). The trigonometric identity, $\sin^2\theta + \cos^2\theta = 1$, is an application of the Pythagorean Theorem, with the two lesser sides of a right triangle set to $\sin\theta$ and $\cos\theta$ values. The sine or cosine of an angle θ can never exceed one, and is the value obtained by dividing either the side opposite or side adjacent θ by the larger hypotenuse. If the cosine is the X coordinate, and the sine the Y coordinate, a plot of the hypotenuse for all possible sin–cos values, always produces a hypotenuse length equaling unity, and the terminal end of the hypotenuse distant from the θ vertex will be a perfect circle for all values of θ. Sine and cosine values are obtained from azimuth θ, and range ρ is unity, and then multiplied by a scaling value in miles or meters.

The plot of sine and cosine in Figure 15-33 is the conventional academic representation in vector arithmetic (see Figure 15-34). In Cartesian coordinates, where $X = -1$ to $+1$, and $Y = -1$ to $+1$, the cosine $= +1$ and sine $= 0$ at $\theta = 0°$. As the angle is increased, the vector moves in the counterclockwise direction, toward a Y value of $+1$, and an X value of 0.

This counterclockwise movement of the radar sweep actually occurs in a ppi crt, but inside of it, on the phosphor. When viewed on the face of the tube, the trace appears to rotate clockwise, a mirror image of the phosphor side. To generate a radar presentation in conventional map form with north at the top, the deflection is rotated 90° to place the cosine function in the n–s dimension and the sine function in e–w. Because $\sin^2\theta + \cos^2\theta = 1$ the ppi crt electron beam is deflected in two planes, and the sawtooth voltages are modulated by sine and cosine azimuth data, the sweep will move in a circle. When the trigonometric functions are applied to the ppi, the n–s value is the side adjacent, and the e–w value is the side opposite.

Note that the tangent values can be obtained from rectangular coordinates, without angular information. This becomes useful in television display systems and data-processing software.

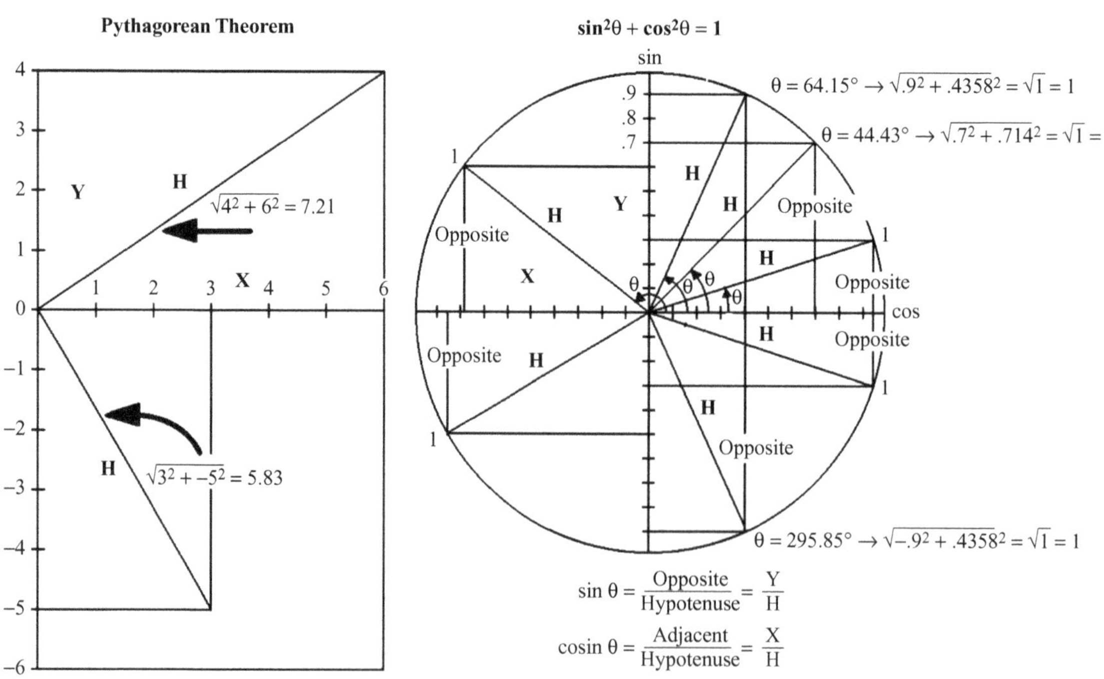

FIGURE 15-33

Pythagorean theorem and $\sin^2\theta + \cos^2\theta = 1$.

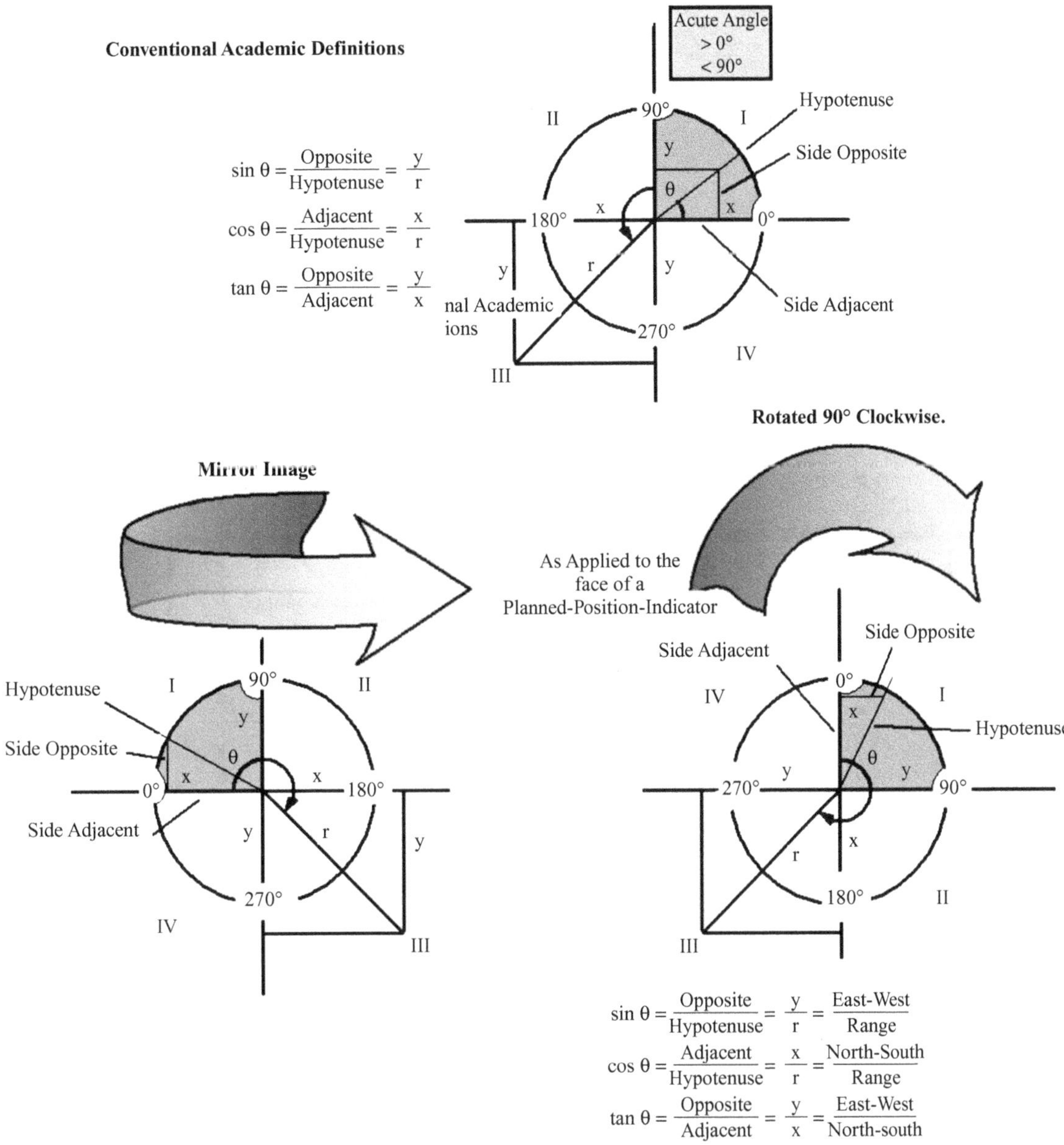

FIGURE 15–34

Visualizing the trigonometric functions applied to the azimuth of a ppi display.

Fixed-Coil Deflection

In the 1950s, the mechanically rotating coils began to be replaced with a more sophisticated deflection yoke (see Figure 15-35). The yoke did not move, but caused the sweep to rotate with magnetic fields in two planes, north–south (n–s) and east–west (e–w). The sweep amplitude and polarity in the n–s plane varied in amplitude and polarity as a function of the cosine of the radar antenna bearing, called azimuth. The sweep amplitude and polarity in the e–w plane similarly varied as a function of the sine of the azimuth. The resultant magnetic fields in the "neck" of the tube would therefore cause the beam to be deflected from the center to the perimeter, because $\sin^2\theta + \cos^2\theta = 1$, as shown in Figures 15-33 and 15-34.

Figure 15-35 is but just one example of a means to create sweep deflection. Four ***sweep driver*** circuits are shown. Two of these employ an npn transistor and positive collector voltage as the input stage, and the other two employ a pnp transistor and negative collector voltage as the input stage. One pair of npn–pnp input transistors

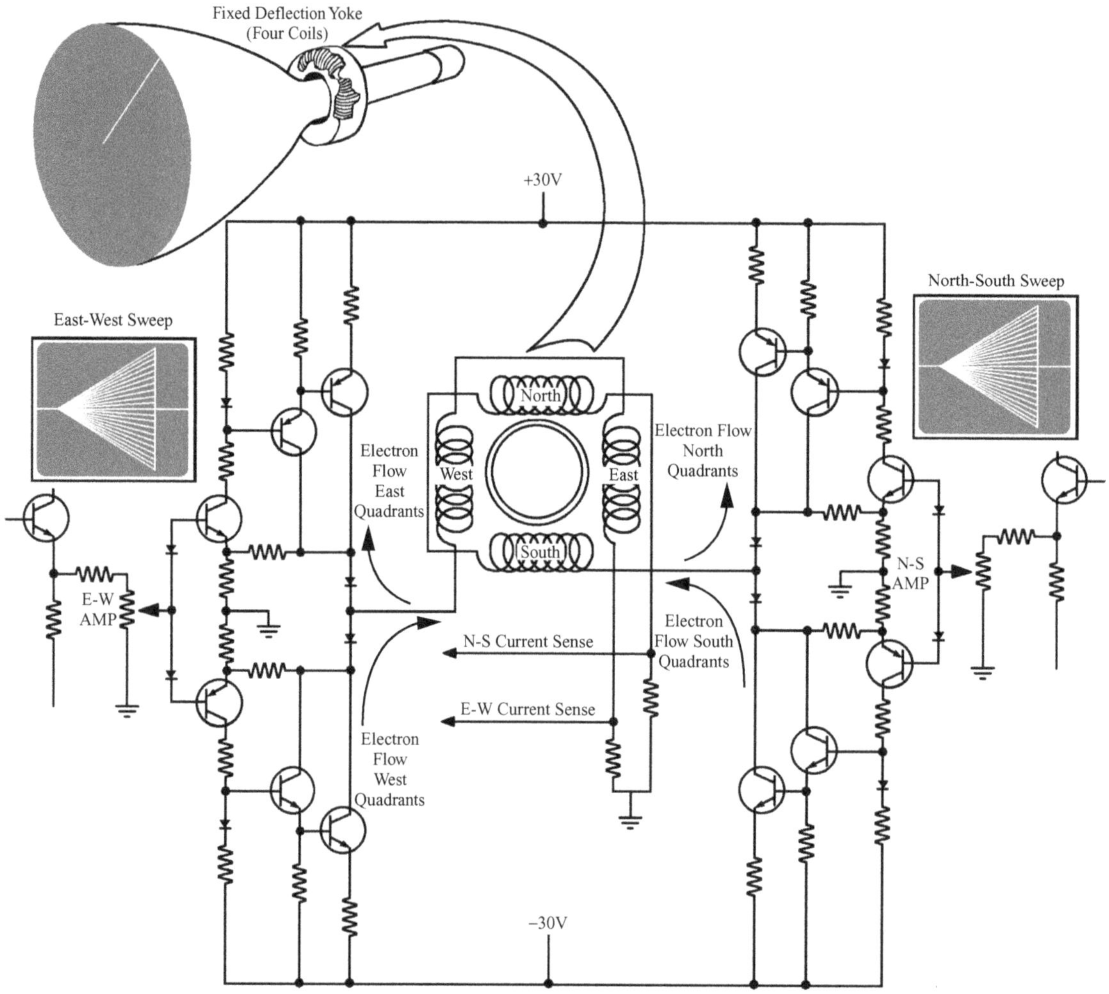

FIGURE 15-35

Fixed coil sweep deflection.

makes up the e–w driver input, the other, the n–s. The inputs to each vary in both the positive and negative directions, as shown in Figure 15-35, and in the waveforms shown in Figure 15-36.

The collector circuit of each input stage establishes the current in the Darlington pairs that follow. For each coil-driver set, east–west or north–south, the final current drive will be obtained from one Darlington pair, but not the other. The diodes in each of the final driver pairs cause this switching to take place, so that current may flow from either the positive supply, through two coils, to ground, or from the negative supply, through two coils, to ground.

The n–s coils are connected in series, as are the e–w coils; however, in either case, the two series coils are (1) physically oriented in reverse, and (2) wound in reverse, so that the magnetic lines of force in the neck of the tube will all be in the same direction, creating a mutually aiding deflection of the beam in one direction (see Figure 15-37). The polarization of the magnetic fields about the coils depends upon (1) the direction of electron flow through them, and (2) the direction in which the wires are wound. Chapter 10 contains a discussion of the effect of a magnetic field on an electron in motion, and basic courses on electricity and electronics address the magnetic fields created by current through a coil. Consider the manner in which the beam is deflected in the n–s dimension. If a magnetic field is established to make the lines of force in the neck of the tube move from right north pole to left south pole, the beam will be deflected upward.

Creating the sin–cos Modulation

This process has also gone through two major stages of evolution. See Figure 15-38. While synchro systems were still in use, an electromechanical device, called a ***resolver***, was used. The resolver could be functionally viewed as a transformer, with a mechanically rotating primary, and two stationary secondaries; the secondaries were physically oriented at 90° separations. The rotating primary was mechanically driven by the synchro system, and electrically supplied with the sweep sawtooth. As the rotor turned, the sweep induced into the two stators was "modulated" to provide the needed sine–cosine sweeps. Because the resolver was an a-c device, the baseline of the output sweeps would develop an a-c component, varying with the antenna rotation. This was corrected with gated clamping circuits, restoring the baseline to zero just before sweep start.

To replace the resolver, another system was employed for a period of time. The S1, S2, and S3 lines from the 1 × synchro were routed into a special ***Scott transformer*** circuit, which would convert the three 60-Hz voltages into two; those two 60-Hz voltages could then be rectified and filtered to provide a sine and cosine wave to be used for modulation of the sawtooth.

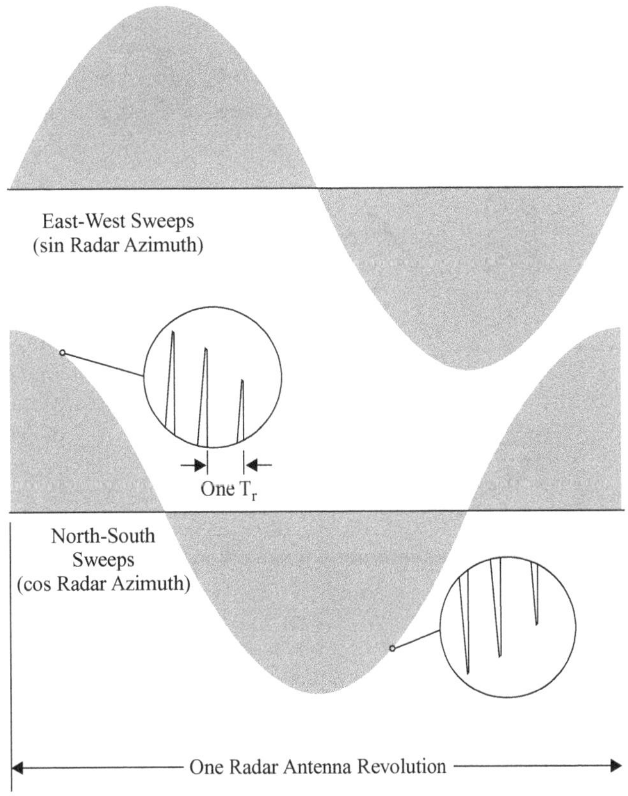

FIGURE 15–36

sin–cos sweep modulation with azimuth data.

Digital Azimuth Data

Synchro systems and resolvers were replaced by azimuth-pulse-generator (apg) systems, described in Chapter 6. One method of utilizing these was by use of a counter and PROM, as illustrated in the hypothetical, generic, system in Figure 15-39. As the acp counter counted the 4,096 azimuth change pulses, the counter outputs address the PROM, which provides digital sine and cosine data outputs. The PROM needs to contain only the data for the absolute numerical values of a single 90° curve. The same curve is used in all eight cases, four for sin, and four for cos. They differ only in that the single curve is used as stored, reversed and/or inverted for both the sin and cos. In Figure 15-39, the two most significant bits of the azimuth count, which represent the four quadrants, are used for selection of forward or reverse memory addressing, and positive or negative output polarity. The acp counter is reset to a preset value with each occurrence of the azimuth reference pulse (arp). Such an adjustment is often called the ***azimuth offset.*** With azimuth offset, acp counter zero is only related to, but is not necessarily coincident with, the arp. Such an adjustment gives the

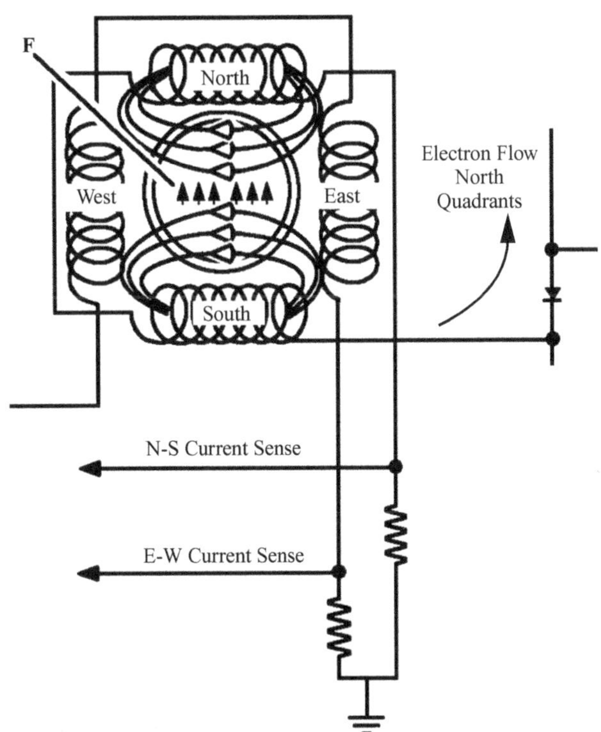

FIGURE 15–37

Electron beam forced upward.

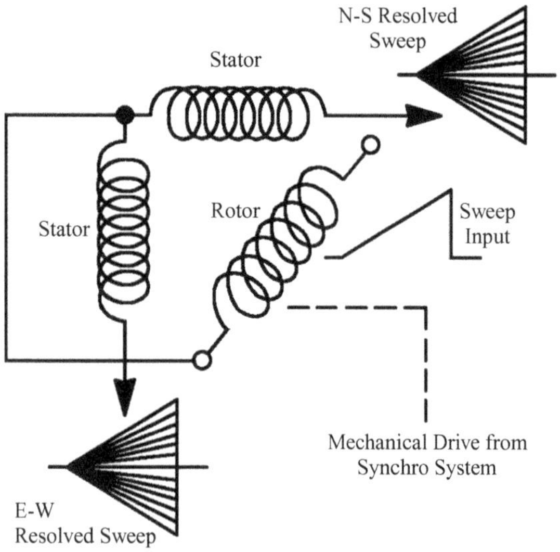

FIGURE 15-38

Sweep resolver.

technician the capability to make fine adjustments to the azimuth to both (1) correct for mechanical alignment error, and (2) to adjust the displays for variation in magnetic deviation. The antenna alignment ring on the pedestal is set to the constant true north reference on installation, but the displays must agree with aircraft compasses.

In Figure 15-39, the contents of the acp counter are "captured" and "converted" at radar pretrigger time for use during the entire T_r, by a rapid series of events, temporally separated only by the propagation delay of six buffers. Three clocks allow the development and capture of the sin value, and the next three allow the development and capture of the cos value. Once the six events have occurred, they are not repeated until the next radar pretrigger.

One Stored Curve can Produce All Sine and Cosine Data

To identify the quadrant for a proper rise, fall, true, or complement value, the logic at the top of Figure 15-39 provides four controlling outputs (see Figures 15-39 and 15-40). One output for sin and one for cos determine whether the curve PROM is to be read in forward or reverse direction. If the required curve is one which begins at zero and rises to unity, then the "forward" or "true" address is used. If the required curve is one which begins at unity and falls to zero, then the "reverse" or "complement" address is used. The same acp counter provides both, using the "Q" outputs for "true," and the "Q-not" outputs for "complement." The complemented addresses serve as a "downcount," and the "true" addresses serve as an "up count." The address "direction" is determined when the D-type flip-flops are set or cleared at clock 1 for sin and clock 4 for cos. By the time when clock 4 occurs, the sin value has been captured in the sin output register by clock 3.

The outputs from the PROM are captured in a single register at clock 2 or clock 5 time. That register provides both "Q" and "Q-not" outputs to a sin and cos multiplexer, where the positive "Q" or negative "Q-not" outputs may be selected by outputs from the quadrant control logic. During the cos clocks 4 through 6, all data from the sin operation is already held in the sin output register, and nothing is lost by the cos operation.

Alphanumeric Display

Conventional rotating-sweep ppis operate in real time during radar range, but are switched into a rappi operation during dead time (see Figure 15-41). For each data message, the electron beam in the crt is deflected to a ***gross position*** by the coordinates of the data to be displayed. A "stroke generator" then produces a stream of ***minor deflections*** for each character in the message. The stroke information is obtained from a memory, and the memory is addressed by each character. In addition to aircraft data, the messages may contain several types of lists, system status, and more.

"Accelerating" Real Time

The ARTSIIE was designed for lower-density airports and has a lower traffic capacity than ARTSIII systems, but increasing traffic placed new demands on both systems (see Figures 15-42 and 15-43). To increase the data handling capability, more time for alphanumerics was successfully achieved in the ARTSIIE by ***video time compression (VTC)***, a method of storing video in memory, and then (effectively) reading the memory at a higher rate than it was written. This may not be as simple as it might first appear. The gross positioning for alphanumeric data must change to correspond with the selected range and subsequent (accelerated) real-time beam movement across the face of the crt. The real-time video must be converted from analog to digital for storage, and then back

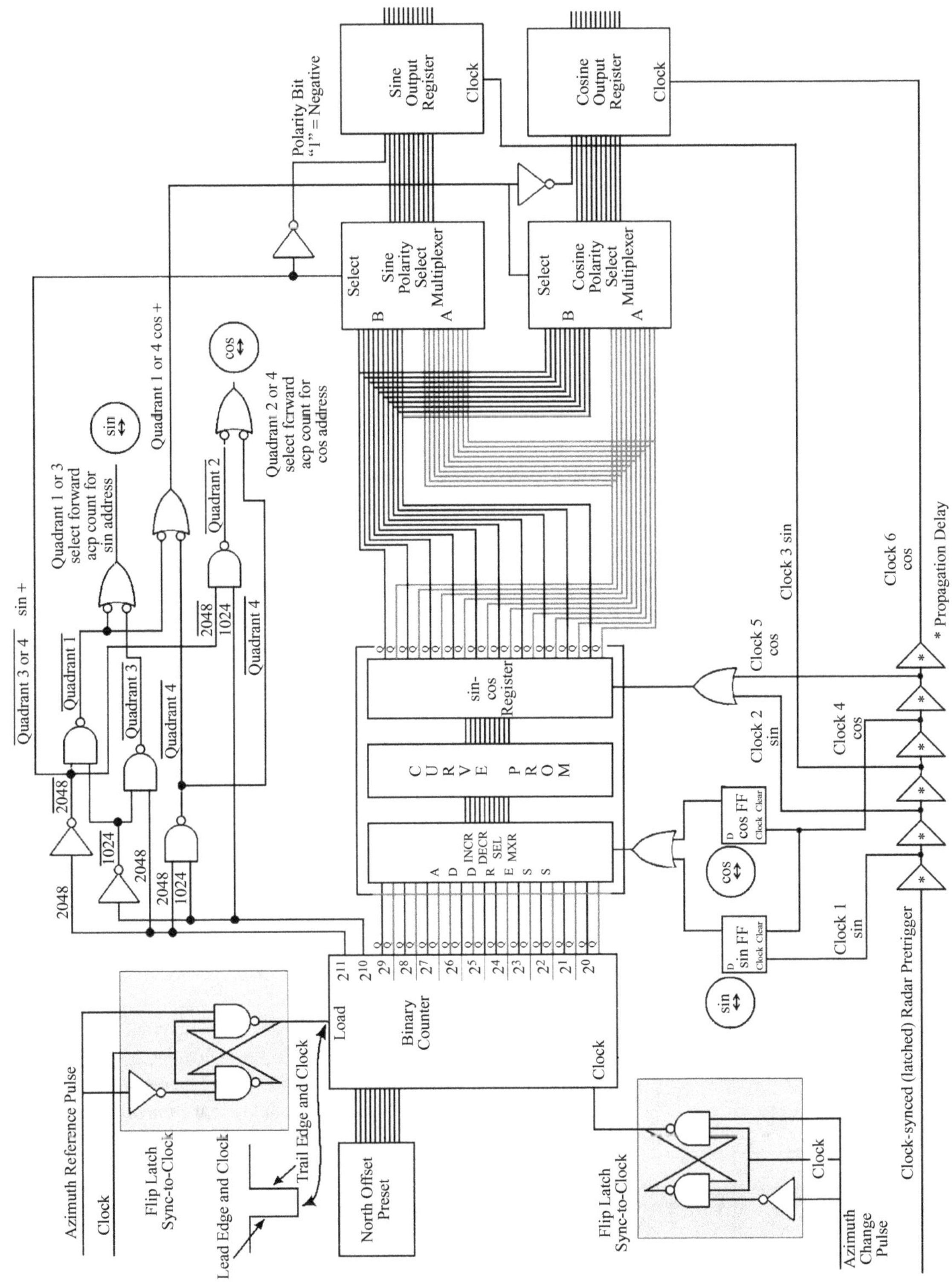

FIGURE 15–39

Generic sin–cos azimuth data converter.

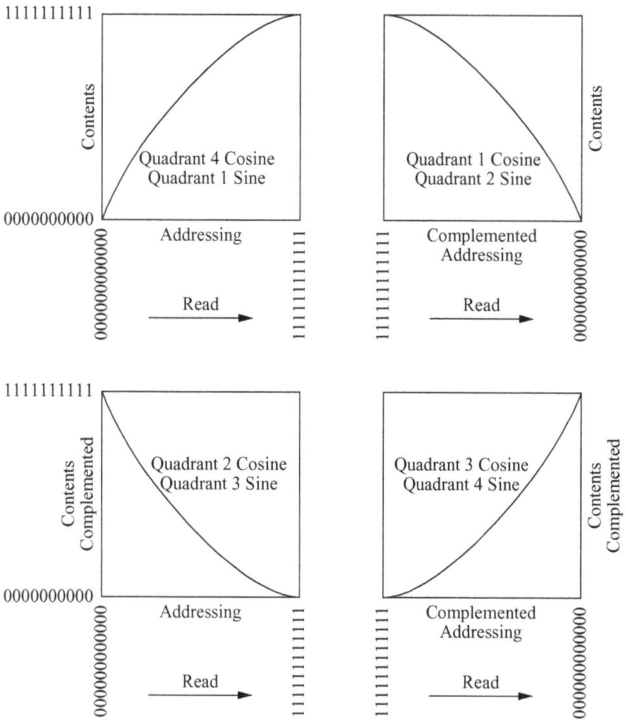

FIGURE 15–40

Complementing address and/or output produces all four quadrants of sine–cosine data.

to analog after reading. Variable range control of the ppi must be incorporated into the "window size" and memory addressing. It is noteworthy that VTC was never successful in the ***ARTSIII Data Entry and Display System (DEDS)*** equipment.

Actually, the read speed is no faster than the write, but a more coarse addressing in the read operation makes it seem to be (see Figure 15-43). A "window size" derived from the control PROM and range switch allows data "stretching" as the memory is written, so that reading will not miss any data.

Two A/D converters place input videos into 6-bit parallel word format for eventual storage in two dual-port memories. One A/D handles mixed (combined) normal, mti, or beacon data. The beacon data may be analog raw or decoded video, seldom used except in the absence of ARTS data. The other input is for video maps or other real-time video.

Noise filters at the A/D outputs remove transition spikes from the data, and the peak selectors set all the data in a window to the level of the highest peak in the window, thus "stretching" all data to the width of the window. Were this not done, targets displayed at the accelerated 60-mile range for instance would be too narrow to see, even if they could pass through a video amplifier, which is also unlikely.

Range marks are generated at the accelerated rate, controlled by the range switch and translation PROM. They are applied to a "nonadditive" mixer so as not to intensify map lines.

Many innovations through temporal acceleration or deceleration have appeared in recent years. In the 1970s, the sampling oscilloscope allowed clear display of high frequencies and very short pulses. A training radar at the FAA Academy uses a 1-ns transmitter burst and an interval less than a city block in radar range, but storage and deceleration provides a video resembling that from an ARSR, with clutter from furniture and other objects in the laboratory.

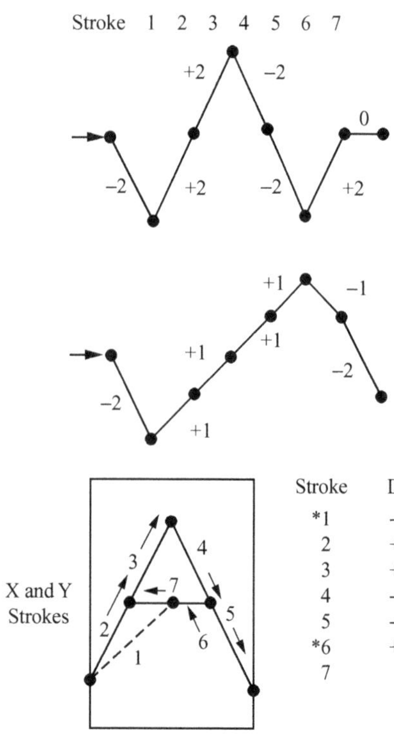

Stroke	Deflection
*1	−2Y, −2X
2	+2Y, +1X
3	+2Y, +1X
4	−2Y, +1X
5	−2Y, +1X
*6	+2Y, −1X
7	0Y, −2X

*Blanking

FIGURE 15–41

Deflection strokes for the character "A."

Other Uses of the Azimuth Counter

Whether or not the intent will be to develop sine–cosine information, any system, containing an azimuth pulse generator, must have an azimuth counter; in digitizer systems, the azimuth-counter output must be utilized in the sliding window or centroiding process. In the ASR-9, the azimuth offset is entered via a computer terminal, and is a ***variable site parameter (vsp)*** for the synchronizer. The azimuth ring on radar antennas is normally oriented to true north, since

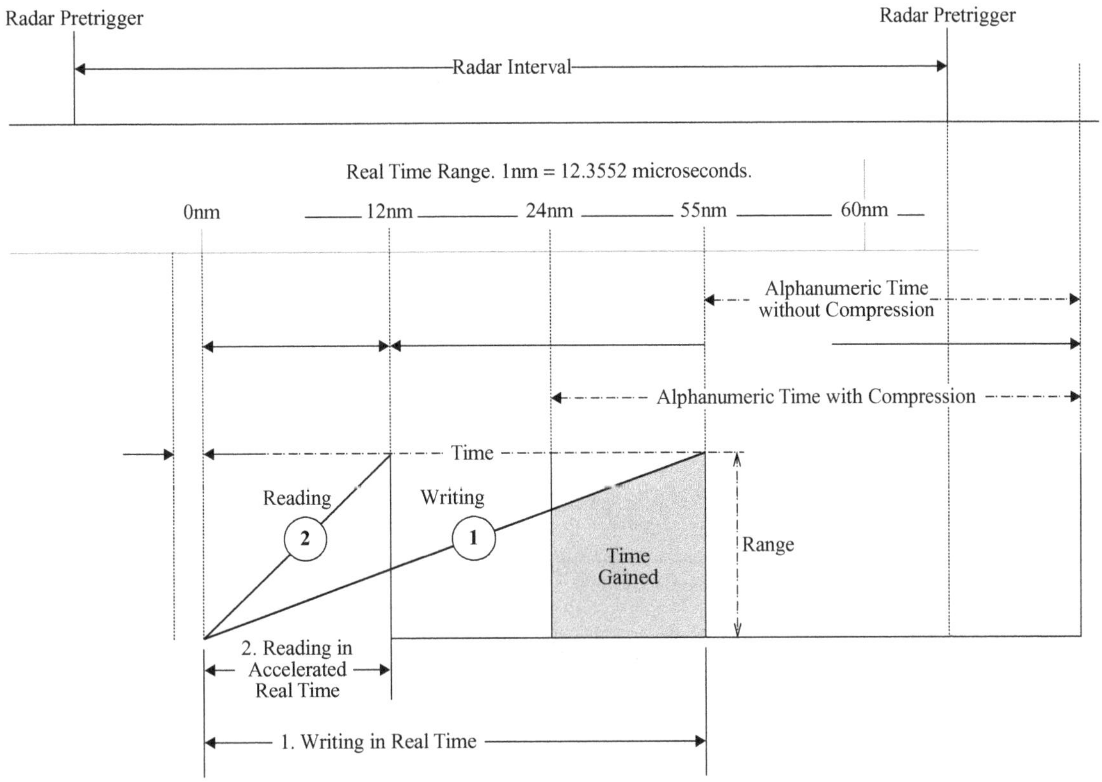

FIGURE 15–42

"Accelerating" real time to display 55 nm of data in 1/5 the time.

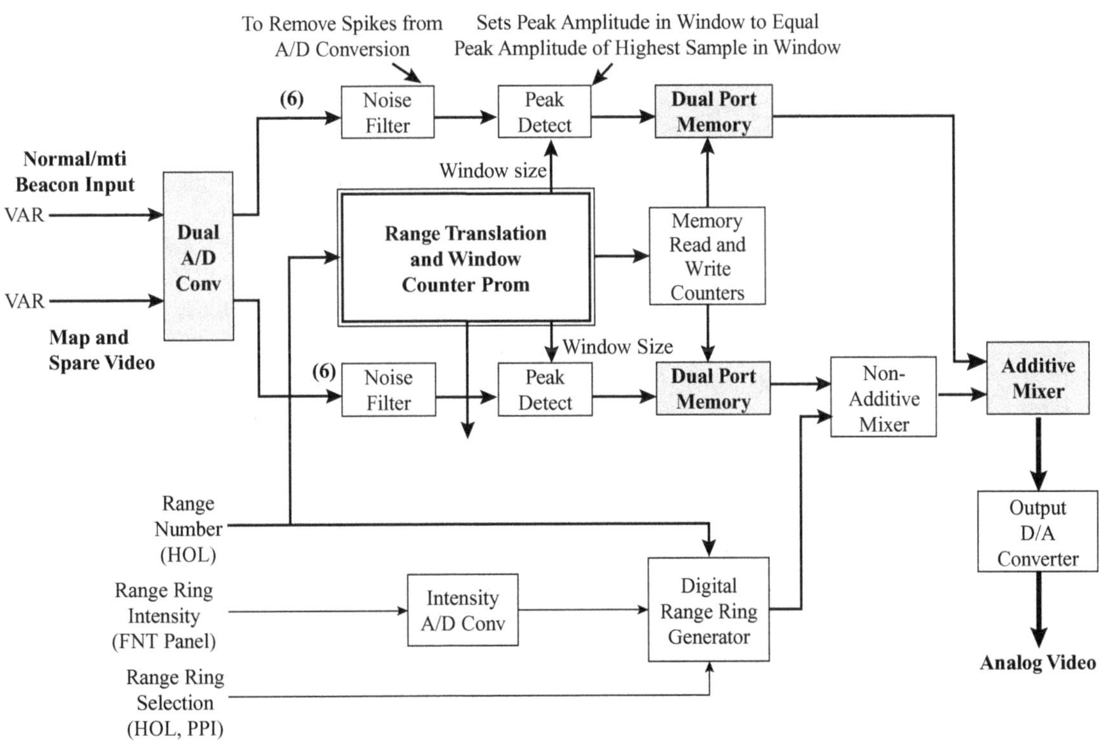

FIGURE 15–43

ARTSIIE video time compression.

that does not change, as does magnetic north. However, air traffic controllers communicate with aircraft which can only rely upon magnetic headings, and user-facility displays must, then, be referenced to magnetic north. In the ASR-9, the magnetic deviation from true north must be entered at the radar site, so that target messages will contain the correct azimuth data; the azimuth preset by means of a computer–terminal interface, offers a convenient means to accomplish that. See the information in Chapter 14 regarding the ASR-9 target performance window. By viewing the azimuth of the mti reflector in the target performance window, the azimuth preset can be adjusted to within ±1 acp of accuracy.

Arithmetic Calculations to Convert Azimuth Data

It will often be necessary for the technician to make conversions between acp counts and actual radar bearing in degrees. It may further be necessary for him to make conversions between CPIP's, acps, 16-sector systems, 12-sector systems, bearing in degrees, etc. Rather than trying to remember all the conversion numbers, it is convenient to simply construct a ratio-and-proportion equation, as shown:

$$\frac{\text{ACP Count}}{4{,}096} = \frac{\text{Bearing}}{360^\circ} = \frac{\text{CPIP}}{256} = \frac{\text{Sector No}}{\text{No of Sectors}}$$

$$\text{ACP Count} = \frac{\text{Bearing} \times 4{,}096}{360^\circ}$$

$$\text{Bearing} = \frac{\text{ACP} \times 360}{4{,}096}.$$

Azimuth Alignment

Actual procedure for individual facilities will vary according to policies and equipment types; the technician should abide by those (see Figure 15-44). Precise azimuth alignment begins on installation of the antenna, when the azimuth ring is positioned for alignment with true north. This may require a professional survey of a distant known point, in respect to the North Star. The antenna may be pointed at the surveyed point with a "boresight" telescope affixed to the reflector, if the point is within sight. When the surveyed point is quite distant, the antenna may be positioned by finding the peak of the beam with electronic means. The azimuth pulse generators are then adjusted for correct occurrence of the azimuth reference pulse; that part of the alignment may be very difficult. Use of a fabricated "pulse catcher," a hand-held box, containing a flip-flop triggered by the arp, and a light-emitting diode, is recommended. In the case of the ASR-9, great precision is not essential, since errors can be corrected by the vsp offset. In other systems, it may be necessary to install the magnetic offset at either the radar or indicator sites, depending upon equipment type. Once the azimuth data produced at the radar site is verified correct, any offsets at the indicator site must be installed. At the indicator site, each ppi yoke alignment must be verified. The video map azimuth must also be properly aligned. The final check of the alignment is done at the ppi, where (1) known permanent echoes and mti reflectors are at the proper azimuth in accordance with a cursor or

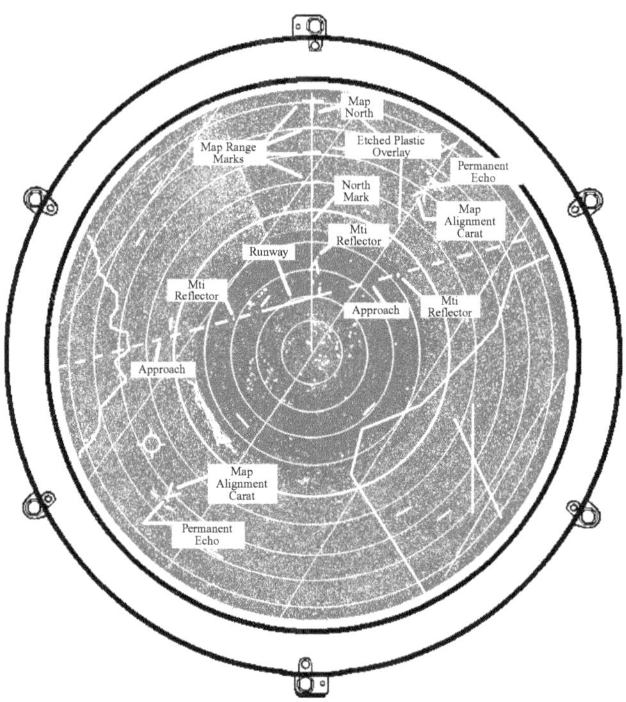

FIGURE 15-44

Alignment verifications.

map overlay, (2) the video map car-ats point to those permanent echoes, and (3) the runway approaches are in alignment with any mti reflectors used to mark them. The permanent echoes and mti reflectors may also be used as a range-mark alignment verification, and the map range marks may be used to verify the map linearity.

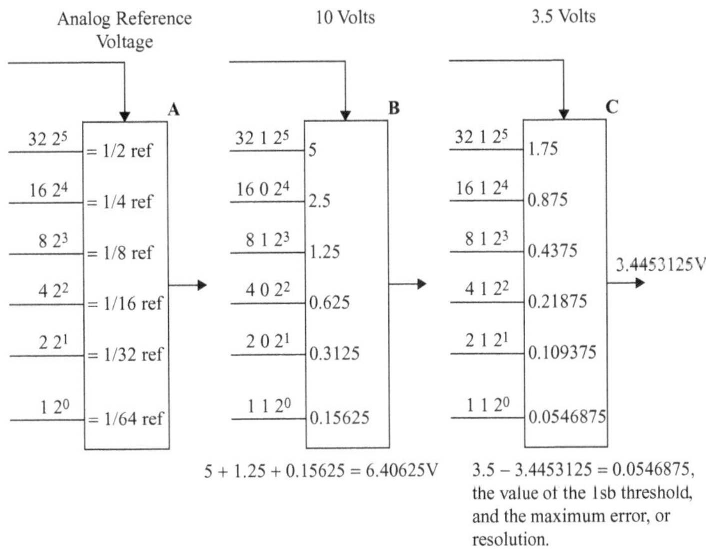

FIGURE 15-45

Multiplying D/A converters.

Sweep Modulator Circuitry Operated by sin–cos Information

There have been a wide variety of methods to accomplish the sin–cos variation of e–w and n–s sweeps. The sin–cos data has been used in both digital and analog forms, sometimes as the analog reference for a digital-to-analog converter for digital sweep amplitude data. In some cases, the analog data was used as a voltage supply for an amplifier stage, and in other cases, it was used as an input to an operational amplifier. Whatever be the circuit type, the result will be that each peak sweep amplitude will be a function of the sine and cosine voltages.

Figure 15-45 illustrates the application of a multiplying D/A converter, used in several display circuits. The "A" part of the illustration shows that the D/A output is dependent on the set bits at the input. The bits are assigned weights, with the lsb being equal to, in binary, 2°, in decimal, 1, or in voltage, 1/64 the reference voltage.

If only the msb is set, the output will be one-half the reference voltage. If the lsb is set, the output will be 1/64 the reference voltage. The "B" part of the illustration shows that a 10-V reference with a digital input of 101001 will provide an output totaling 1/2 the reference, 1/8 the reference, and 1/64 the input reference.

The "C" part of the illustration shows that if all the input bits are set, the voltage will be the reference minus the value of the lsb.

If both the reference voltage and the binary input is varied, the circuit becomes a type of modulator. If the binary input is the digital sine data word, and the reference voltage is an analog sweep sawtooth, the output as the antenna rotates is a sine-modulated sweep.

ppi Display Phenomena

Range Resolution

The minimum separation in range at which two targets may be identified as separate is the range resolution (see Figure 15-46). For a single-frequency pulse, the range resolution is simply the radar range of the transmitter pulse width. Pulse width is normally measured at the 3-dB power, 0.707-voltage points, and the separation of two pulses which merge at the 3-dB points is, by definition, the range resolution. This is a figure of merit; video limiting,

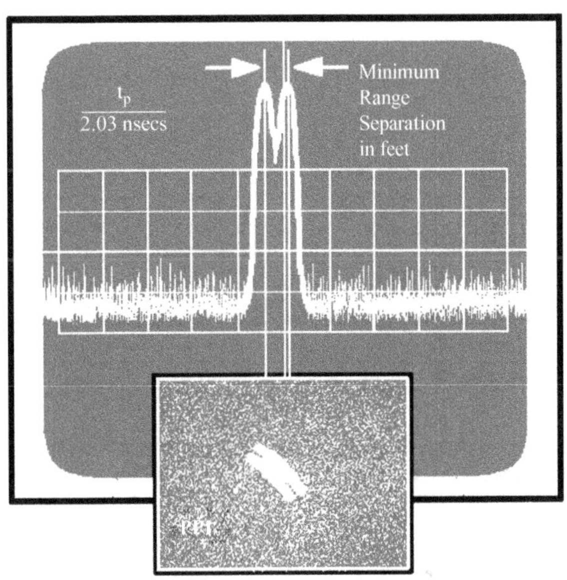

FIGURE 15-46

Range resolution.

"defocus," or crt blooming may worsen the resolution. In calculating the radar range resolution, it is usually more convenient to use the radar foot, 2.03 ns, rather than the radar nautical mile, 12.3552 μs:

$$R_{\text{resolution}} = \left(\frac{t_{\text{p}}}{2.03 \times 10^{-9}}\right) \times 1 \text{ foot}.$$

Azimuth Resolution

Azimuth resolution is also a figure of merit, and is, by definition, the azimuth in degrees between the 3-dB power points at the sides of the beam. This also may be worsened by video limiting, defocus, or crt blooming, or by antenna side lobes. Obviously, the necessary azimuth separation in terms of distance is a function of range. Radar range is normally expressed in nautical miles; to convert to feet, multiply by 6,076.11549:

$$D_{\text{Az}_{\text{sep}}} = 2\,\text{Range}\left(\sin\left(\frac{\theta}{2}\right)\right).$$

Summary of Digital Deflection

Major portions of this generic block diagram were addressed in preceding paragraphs (see Figure 15-47). Figure 15-47 combines those into a composite, along with additional information. Some areas are abbreviated or omitted simply for the sake of simplicity.

Selected Azimuth Data

In real time, this is the sine–cosine data derived from the arp and acp inputs. In rappi (a.k.a. "synthetic" or "alphanumeric") time, it is the gross position (a.k.a. "major deflection"), the n–s and e–w rectangular range coordinates for the beginning of an alphanumeric data display. In real time, the multiplying D/A "sweep modulators" employ the linear sweep as the analog reference, and the sine–cosine digital inputs as the modulation. In alphanumeric time, the selected gross position data provides the position deflection to start the display of alphanumeric data.

Range Expansion or Minor Deflection

In real time, this multiplying D/A utilizes the range switch data to amplify, or expand, the sweep so that it extends to the edges of the crt. In off-centered operation, the sweep always continues to the edge, because it is terminated by an overscan detection, addressed in a preceding part of this chapter. In alphanumeric operation, the minor deflection causes the "drawing" of alphanumeric characters.

Baseline Control

Because the A/D converters have obtained real-time sweep information from four versions of sine–cosine outputs from the sine–cosine azimuth data registers, the extremes of the negative curves (sine or cosine 0 to −1) will be the same as the extremes of the positive (sine or cosine 0 to +1). The polarity bits will serve as the means to correct this, commanding a reduction of the sweep baseline for the negative numbers. Another way to accomplish this in an earlier state of the art was to first utilize the intended negative shapes in positive noncomplemented form until the final analog stages, and then invert them, on command by the sign bits, to provide negative-going sweeps.

Display Circularity

Circularity may be a periodic test to be performed on ppis by technicians. Since all video is similarly distorted by poor circularity, it might superficially appear unimportant, but a noncircular sweep can create an azimuth error. Given the appropriate set of circumstances, this could cause an air traffic controller to provide potentially hazardous information to an aircraft pilot. Figure 15-48 shows an exaggerated noncircular condition to emphasize the effect.

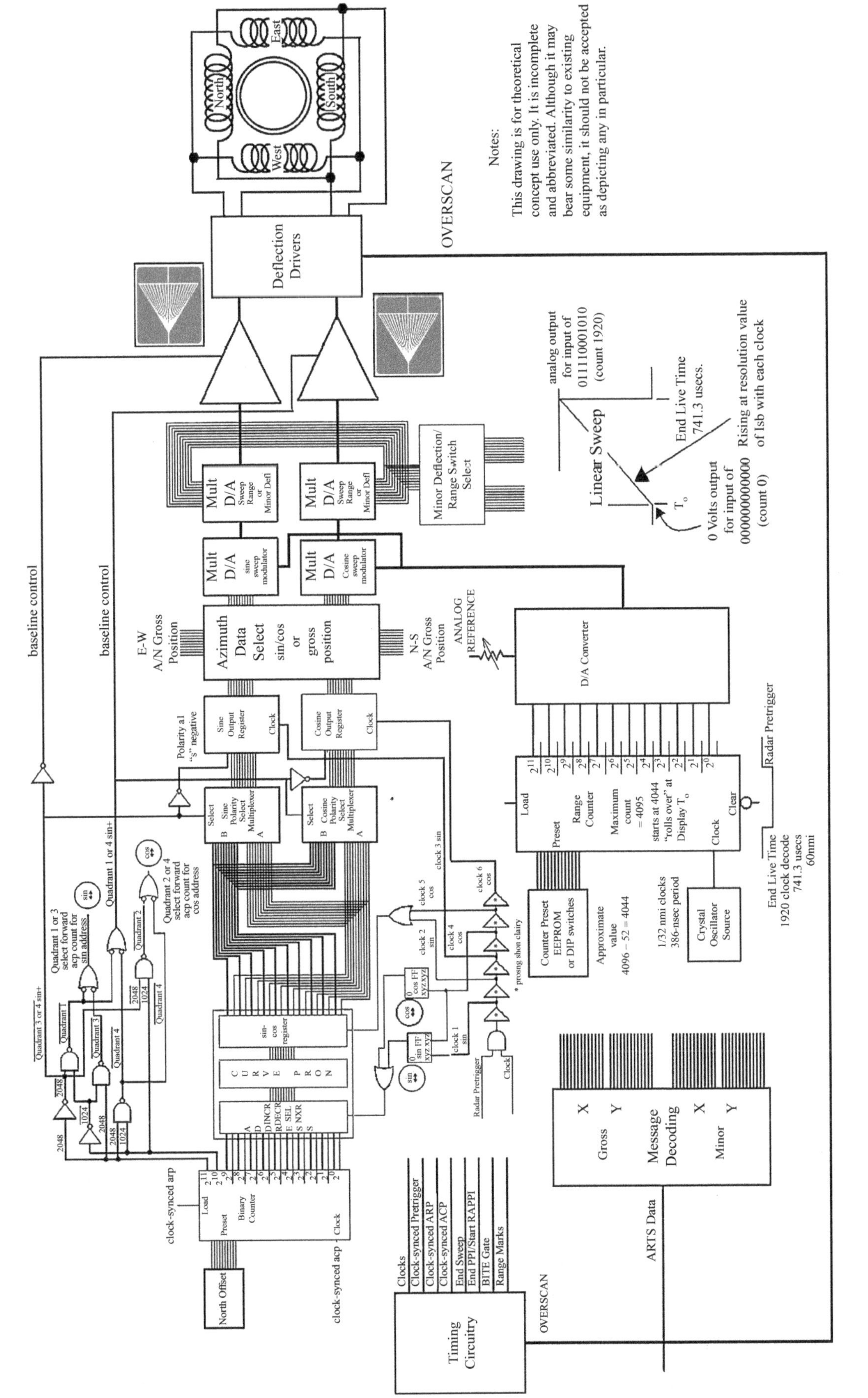

FIGURE 15–47

Generic digital deflection block diagram, ppi/rappi display.

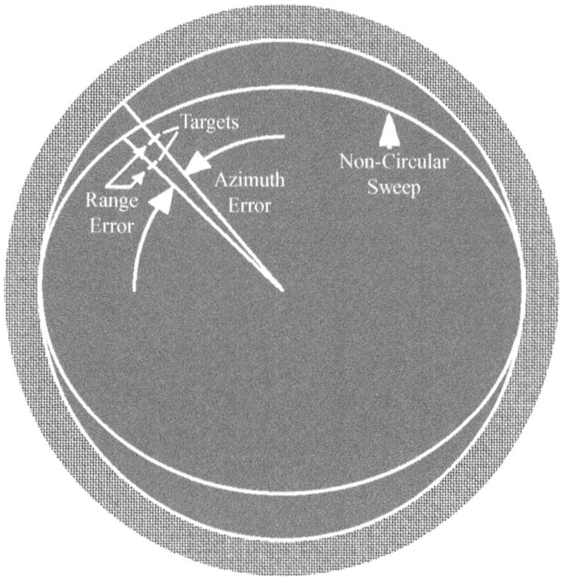

FIGURE 15–48

Noncircular display errors.

The "Slant Range" Effect

Particularly at great ranges, the distance from the radar antenna to the target may be significantly more for aircraft at high altitude than for those on a line from the antenna and tangent to the earth's curvature. Thus, two targets directly above a single geographic point, but at greatly different altitudes, may be displayed at different ranges.

Range-Ambiguous (Second-Time) Targets

If the time required for an echo to return to the radar exceeds the T_r, the transmitter will have "fired" a second time when the echo is received, and the target will occur at a range which is the actual range, minus the range of the T_r. In earlier radars, this was only significant in the case of large echoes from mountains and very strong weather. As ASR systems began employing higher power klystron transmitters, second-time echoes from high-altitude aircraft began to appear. Where these second-time aircraft appeared in the terminal traffic pattern, a serious confusion factor was introduced. The effect of second-time echoes can be greatly worsened by staggered f_p; each different T_r will cause a second-time target to appear at a different range, so multiple f_p's cause multiple targets.

Second-Time Weather

The effects of second-time weather may present a curious effect, in which storms are displayed in a series of "wedges" at the same azimuth. First of all, the weather is visible because it contains a variety of Doppler frequencies and would survive mti cancellation. The number of wedges will be dependant on the number of intervals in the staggered f_p pattern. The wedge-shaped appearance occurs because the azimuth width of the blocks of weather is "compressed" when erroneously displayed at closer ranges. For instance, a storm that is really 40 miles wide might appear as a second-time block that is only 10 miles wide.

The Video "Mapper"

A beginner in the field of air traffic control radar might dismiss the small video map unit as an accessory. He would discover the importance in short order; when the video map unit fails in an air traffic control facility, it is so serious a matter that the radar will be called "out of service." The video map provides the controller with extensive information, marking the location of airspace type and control boundaries, navigational fixes, permanent echoes, runways, approaches, VORs, remote airports, geographical features such as rivers, coastlines, hazards, and much more. The equipment is absolutely essential to safe air traffic control.

Figure 15-49 is a simplified diagram of what the author will call a "second generation" of three video map methods. The first generation incorporated a large horizontal turntable (perhaps 12" or more in diameter) driven by a synchro system. The second generation operated on a similar principle, but the rotating turntable was replaced by a smaller fixed map in front of a rotating ppi sweep. The central element of both the first and second generations was the map itself, a photographic negative. To ensure accuracy and fine, clear, lines, the maps were first drawn on large sheets by cartographers, and then photographed, for reduction to the appropriate size. The third generation maps are purely digital, drawn by software, contained in memory, and displayed in pixel television format.

Mapping units have also gone through an evolution since the 1950s. In the earlier, first-generation, units, the map negative was placed on a rotating turntable, driven by the indicator-site servo data. Beneath the turntable was a crt, employing a single left-to-right sweep, synchronized to the radar f_p; the sweep was of constant

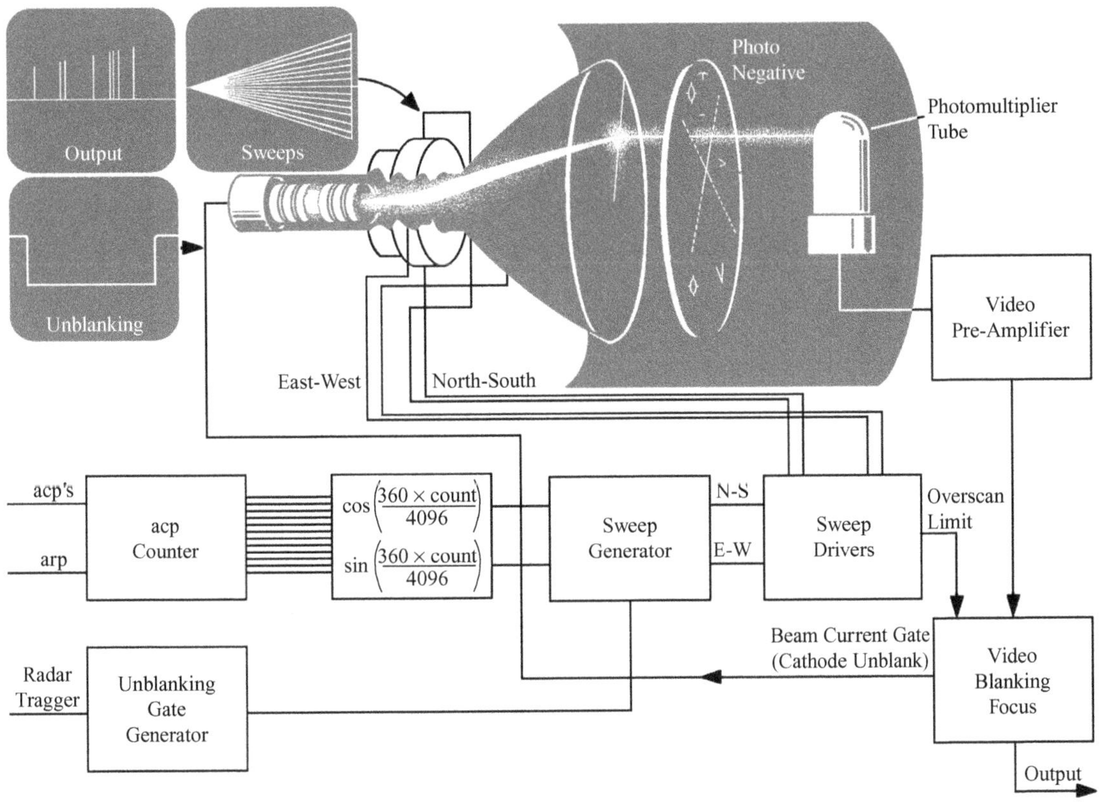

FIGURE 15-49

Second-generation video map unit.

intensity and positioned to start at the center of the rotating plate. As the electron beam moved across the low-persistence crt, the light would pass through any clear spaces in the negative, striking a ***photomultiplier tube,*** to ultimately provide map video. Because the crt sweep is synchronized to the f_p, the elapsed time from T_0 until the crt trace passes under a clear space on the map is predictable, and related to radar range. Because the video is developed from the crt "spot," passing beneath a clear space on the map, the map has often been called a ***flying-spot scanner.***

The second-generation system illustrated in Figure 15-49 operates in the same manner as described in the preceding paragraph, except that the map negative is stationary and much smaller, and the crt sweep rotates electrically, with a fixed deflection coil, similar to a ppi. There is an acp counter, and a sin–cos converter, to generate the modulation for the n–s and e–w sweeps.

The output of the photomultiplier tube is noisy, and the map video pulses are of unequal amplitudes. The video must, therefore be baseline-clipped and quantized. The video finally used for display is increased in amplitude with range to ensure that distant intensities equal those of the closer ranges. At greater ranges, sweeps are further apart, and video appears to have less intensity.

Alignment

Manufacturers provide test-pattern maps for initial alignment, but the final test is the proper matching of the map video to the real-time radar data. Map range marks must align with radar range marks, and carats must align with permanent echoes. In both the ppi and map units, sweep linearity adjustments must be performed, but the result is opposite. For instance, if a ppi had no sweep-speed correction, the electron beam would sweep through the outer parts of the crt at a faster rate. Since the ppi radar range marks occur at a fixed rate, and the sweep would move further between marks, the marks would be physically further apart at the distant ranges on the ppi. If the same condition existed in a video map unit, and, as the spot moved faster at the crt extremities, it would pass under the map etchings in shorter time, and the map video extremities would appear to be compressed on an oscilloscope or ppi display.

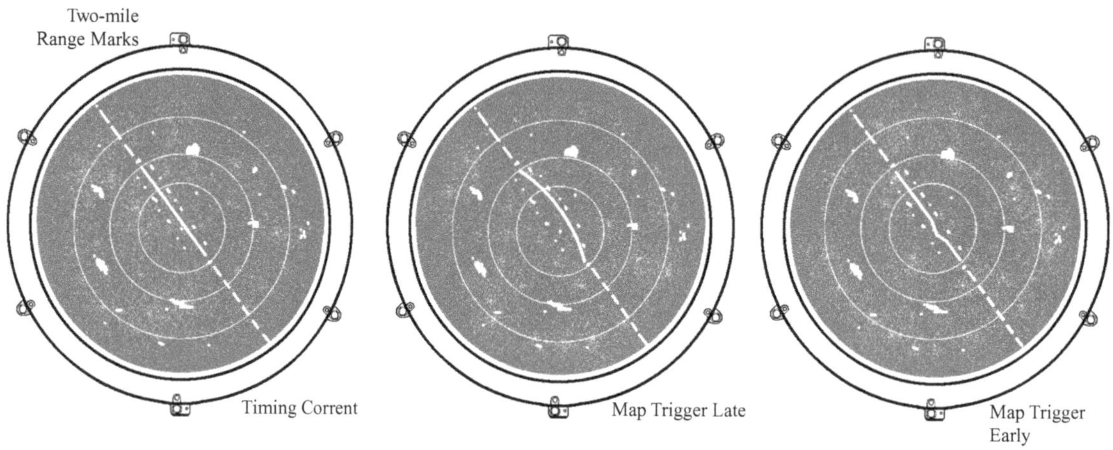

FIGURE 15-50

Map timing errors.

Timing

The timing of the sweep start in the map is very critical; map time zero must match radar video time zero, and even a half-microsecond error is sufficient to misalign such an important feature as a line marking a runway (see Figure 15-50). Where a runway line passes near the center of the map, the effect of time zero misalignment may be dramatic. If the map sweep starts earlier than the radar sweeps, the line may appear to be "pulled," or "bowed," inward, and if the sweep start is later, the line may appear to be bowed outward.

Orthogonality

In manufacturing the fixed deflection yokes for the small mapper crt's, it is impossible to ensure that the n–s and e–w coils are precisely separated by 90°, and even if they were, residual magnetic fields could introduce an error after use. To ensure azimuth accuracy, an ORTHOGONALITY adjustment was provided; a small, adjustable, sample of the vertical sweep was added to the horizontal (or vice versa) for correction.

Focus

To be useful, the video map must produce fine, sharp, lines, on the displays. Since it is an optical device, the physical distances between the crt, map plate, and photomultiplier tube are very important to good focus. Further, there are focus-coil voltages, both static and dynamic.

Digital Maps

As radar displays evolved into television systems with pixelmemory storage, the usefulness of the optical maps heretofore described began to decline (see Figure 15-51). Digital maps could be written to memory with greater precision and relieved technicians of the requirement for time-consuming alignments.

Digital Scan Conversion

The BRITE equipment has now been logically replaced by a digital scan-converter form called ***DBRITE*** (see Figures 15-51 and 15-52). In the DBRITE, radar data is written to a memory by a digital addressing technique that calculates television pixel locations.

Conventional Sine and Cosine Azimuth Data

The digital scan conversion requires calculation of X–Y pixel coordinates, and use of the familiar identity, $\sin^2\theta + \cos^2\theta = 1$, does not offer adequate precision. In conventional ppi use of that identity, the acp count is captured in a

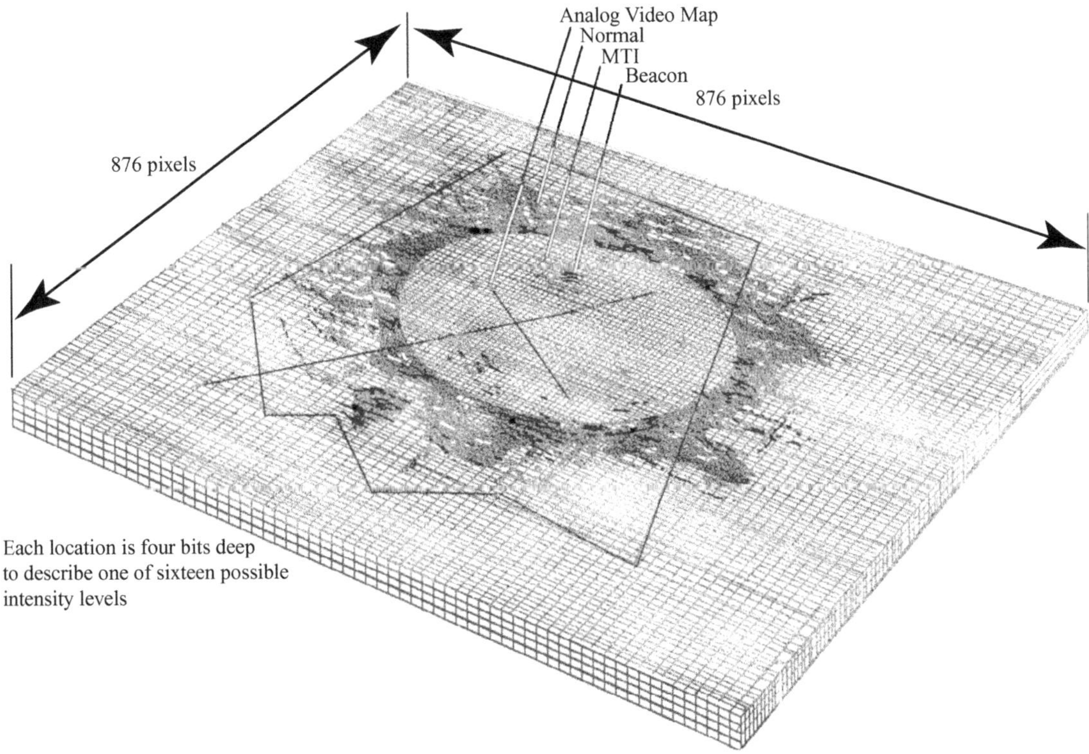

FIGURE 15–51

Visual depiction of a digital scan converter memory.

register at radar pretrigger time. A 90° sinθ curve is stored in a memory and addressed by the captured acp count for the remainder of the T_r. If the values in the curve rise from zero to unity as the memory is addressed from 0 to 1,024 acps, the first quadrant of the sine shape is produced. If the address bits are complemented as the acp count runs from 1,025 to 2,048 acps, the memory is addressed in reverse order and the curve returns to zero. From 2,049 to 3,072, the memory output is inverted and the address is not complemented, and from 3,073 to 4,096, the address is complemented. The cos quadrants operate similarly: reverse address, true output, forward address, complement output, reverse address, complement output, forward address, and true output. The sin and cos data change polarity and/or rise/fall shape for each quadrant, and the quadrant selections are made by the two msbs of the acp count. The captured sine and cosine values for each T_r become multipliers to the ascending radar sweep sawtooth to determine the sweep angle.

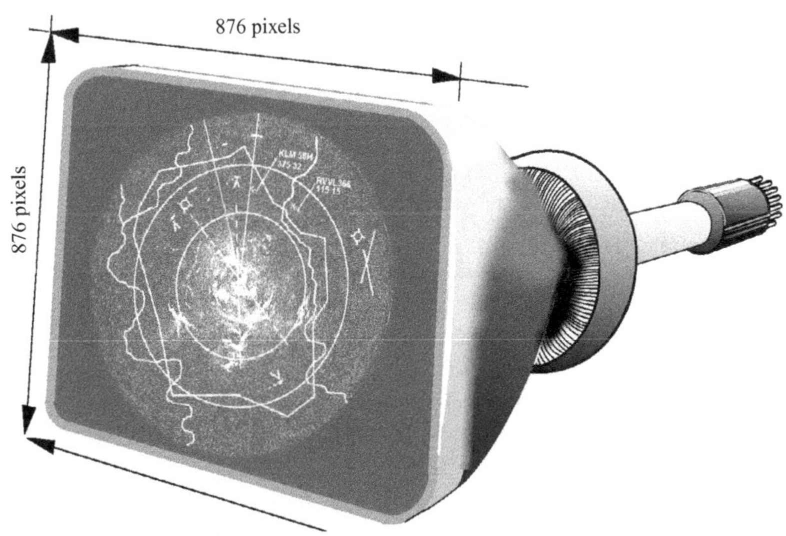

FIGURE 15–52

Square, flat-faced crt.

Octants, Tangents, and Cotangents

Figure 15-53 compares the first, and a portion of the second quadrant, of a cosine curve to a third, and a portion of a fourth, octant of tangent/ cotangent curves. Those with

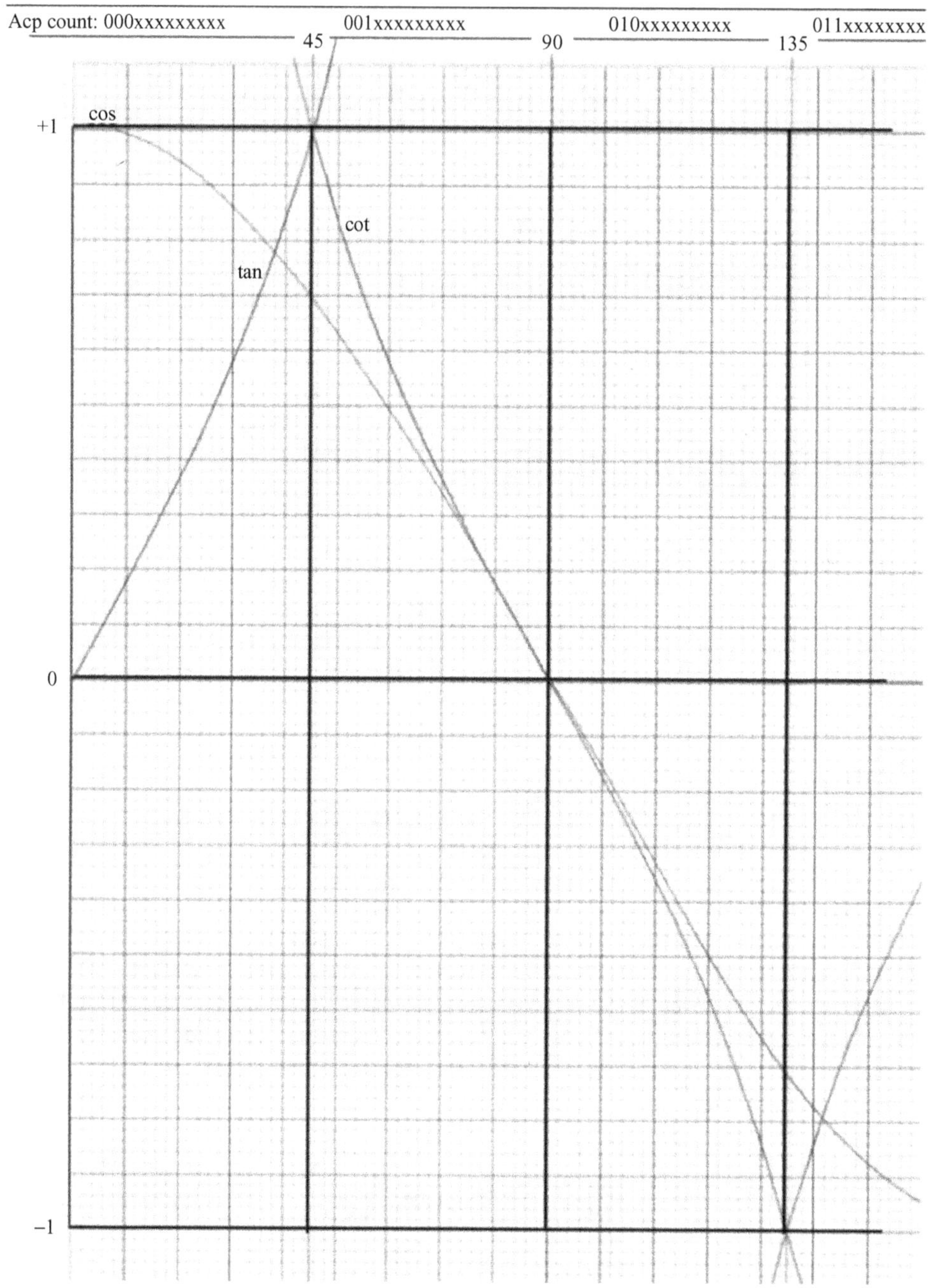

FIGURE 15–53

Using the tangent or cotangent for maximum excursion.

memories of trigonometry might recall that tangent and cotangent plots exhibit a significant "bend." However, most of that bend is beyond the values of +1 and −1, and tangent–cotangent values within those limits are nearly linear, making them useful in scan conversion. The tangent can be obtained from the acp count, representative of θ in 0.088° increments. The cotangent is the reciprocal of the tangent, and when one value exceeds +1 to −1, the other is within that range. The tangent or cotangent makes a complete, nearly linear, transition between +1 and −1 in just 90°, in contrast to the 180° for sine or cosine. Hence, a selection between tangent and cotangent offers greater-precision linear positioning for data storage in memory. Still further, the use of the tangent–cotangent functions can be useful in digital-message storage because a division of the rectangular coordinates yields the values.

One Tangent Curve in Storage

When sinusoidal curves near their peaks, the small differences in values do not provide adequate information for reliable television pixel selection, and the more precise tangent and cotangent are utilized. As in the $\sin^2\theta + \cos^2\theta$ method, a new acp count is captured in a register at radar pretrigger time. To incorporate tangent or cotangent data, the ***three*** msbs of the acp count select the true or complement form of the address or output data. As in sine and cosine data, only one curve must necessarily be stored. The memory may be addressed in either a forward (true) or reverse (complement) manner, and its output may be either positive (true) or negative (complement). In the first octant, the tangent rises from zero to unity. In the second octant, the tangent is past unity, so the cotangent is retrieved from the same memory by reverse addressing. The third octant is also a cotangent curve, derived from forward addressing and complemented output.

Selecting the Pixels to be Written

Through the duration of each T_r, an accumulator circuit continuously adds a representation of the ascending radar range count "sweep" to the captured tangent or cotangent value. Each time the accumulated sum provides a carry output, an X or Y pixel write address is incremented or decremented, and the accumulator is reset. Whether the calculation of the greater change to the pixel write address is in the vertical or horizontal depends upon the octant. For instance, in the eighth, first, fourth, and fifth octants, there is more change in the vertical pixel location than to the horizontal, so the Y vertical pixel write addresses are derived directly from the range count values, but the X vertical addresses are incremented or decremented only by the accumulator carry output. For instance, at due north, the pixel addresses decrement at a maximum vertical rate with the radar range, but there is no change to the X pixel address, which remains near the horizontal pixel center. At due east, there is no change to the Y pixel address, but the X address increments according to the radar range count. At odd multiples of 45°, there is a one-for-one pixel change ratio as the range count ascends. At 22.5°, there is a 2:1 vertical-to-horizontal pixel change ratio with the ascending radar range count.

Timing, Real-Time Storage, and Television Reading

At 60 nmi in range, the real time period is 741.3 μs. The pixel memory and square crt provide for 876 pixels per line, and 876 lines per frame, even though there are 961 lines in a frame in the BRITE-4 and DBRITE equipments. Given 30 frames per second, 33.33 ms are required per frame. There are two 480.5-line fields interlaced into 961 lines per frame. The fields either begin or end with a half-line, accounting for the odd number of lines in the frame. Each line then takes 34.68 μs.

38 μs is required per line, and (ignoring radar deadtime), at the very least, 19.5 lines occur during one T_r. Because the constant stream of real-time radar information must be immediately accepted and written without interruption to the stream or the television operation, a time-sharing and memory-access scheme is required. The television raster accesses the available radar pixel memory for transfer during television blanking periods.

Trails

In the RBDE equipment, the ppi data was stored on the capacitive mesh, and the mesh would be repeatedly read by the television raster, scanning the other side of the mesh to cause secondary emission. The mesh would be only partly discharged by the television read operation, and the time required to remove target data depended on voltages and signal levels. When target data was permitted to remain on the mesh for several antenna scans, a slowly decaying aircraft target would appear to have a trail becoming dimmer and smaller as it aged. In the BRITE equipment, a similar effect was achieved with the combined operation of a small ppi and a special-purpose vidicon tube in the television camera. In the DBRITE equipment, trails are achieved in memory operations with (1) a "read" of the pixel location, (2) an addition of the new radar information to a digital fraction of the data having been read, and (3) a "write" operation to rewrite the pixel. A front-panel TRAILS control permitted the air traffic controller to adjust the addition process.

Beyond the real-time to television-raster conversion, DBRITE pixel memories are also used for ARTS alphanumeric data, digital maps, range marks, and more. Among all the other advantages gained, the range marks can be off-centered from the radar sweep origin, providing air traffic controllers with an option to center the range marks on remote airports or navigational fixes, rather than being limited to the location of the radar antenna.

Alphanumeric Operation

ARTS alphanumerics data is not included in the real-time radar scan-conversion process because (1) its display is in rappi form, and (2) subjecting it to the trails process would blur and smear the alphanumeric information. With the BRITE-4 equipment, alphanumerics was added to the scan-converted radar data with the introduction of a ***Bright Alphanumeric System (BANS)***. Although it did satisfy the requirement to place alphanumeric information on the BRITE displays in the towers, it required meticulous alignment to match the alphanumeric data to the real-time radar presentation. Even before introduction of the BANS, alphanumeric data had been stored on separate but matched BRITE systems, to be mixed with real-time radar data. This equipment was principally used by the USAF, and by civilian airports supporting Air National Guard operations. The military provided AN/TPX-42 beacon digitizer equipment to create a limited alphanumeric beacon display. The TPX-42 was not comparable to ARTS computer equipment, and only centroided beacon data and created alphanumeric display of code and altitude.

ARTS data messages contain, expressed in radar miles, the rectangular coordinates of rappi target data. Adding display range and off-centering data to those coordinates, the pixel location for the data block is computed.

The ASR-9 Surveillance Communications Interface Processor (SCIP)

Synthetic real time was briefly described in the more general parts of this chapter. Because there are so many ASR-9 installations, more than any other ASRs, it is appropriate to provide more information on the SCIP. Chapter 14 addressed the ASR-9 mtd system, and the primary radar output message format, illustrated in Figure 15-54. There are also ASR-9 output messages for beacon targets and for six-level weather detections. There is one special message, called the ***azimuth sector mark message,*** which transmits the real-time radar azimuth 32 times per scan.

Message Data Transfer

To provide for data transfer from the radar facility to the user facility, the parallel data is converted to serial form in the ASR-9 Message Interface Processor (MIP). The serial data is then transmitted over modems or fiber-optic devices.

Since the ASR-9 development in the middle 1980s, modem technology has advanced considerably, and many improvements have appeared. As the ASR-9 was in development, those modems recognized as most reliable operated at 9,600 bps. Today, personal computers are employing 56 kbps modems.

Before the introduction of digital broadband availability, telephone lines were not reliable above 2,400 bps. Technology advanced to a method of representing several bits at once with phase and amplitude information, so that data can be transmitted rapidly over the 2,400 bps lines. Because of the time required in a radar processor to convert target data to serial form, it was necessary to use more than one modem, so that a second, or third, may begin message assembly while one or two are already engaged.

SCIP Block Diagram

Figure 15-55 is an abbreviated diagram of one channel of the ASR-9 SCIP. The modem data is all routed into an interface processor, which converts serial data back to parallel, to be placed on the multibus. The interface processor actually consists of two ***Programmable Interface*** circuit cards, each containing three data channels. The serial-to-parallel conversion in the cards operates on a priority basis, with the azimuth sector mark message carrying the highest priority, which means it will be processed by the first available circuit.

The SCIP contains two self-contained computers, called ***single-board computers (SBC).*** The original design incorporated three of these, but design progress eliminated the need for one, leaving SBC 1 and 3. SBC 1 works in communication with the interface processor to place parallel data on the multibus, and store it, according to

Msg Bit No	Word Bit No	Word 1	
1	1	1 = Wx from Unavailable Processor	
2	2	0	
3	3	0	
4	4	1	
5	5	1	
6	6	0	Search Message Identification Label
7	7	1	
8	8	1	
9	9	0	
10	10	0	
11	11	0	
12	12	0	
13	13	X = Parity	

Msg Bit No	Word Bit No	Word 3
27	1	1 = 2048 acp's
28	2	1 = 1024 acp's
29	3	1 = 512 acp's
30	4	1 = 256 acp's
31	5	1 = 128 acp's
32	6	1 = 64 acp's
33	7	1 = 32 acp's
34	8	1 = 16 acp's
35	9	1 = 8 acp's
36	10	1 = 4 acp's
37	11	1 = 2 acp's
38	12	1 = 1 acp
39	13	X = Parity

Msg Bit No	Word Bit No	Word 2
14	1	1 = 32 nmi
15	2	1 = 16 nmi
16	3	1 = 8 nmi
17	4	1 = 4 nmi
18	5	1 = 2 nmi
19	6	1 = 1 nmi
20	7	$1 = \frac{1}{2}$ nmi
21	8	$1 = \frac{1}{4}$ nmi
22	9	$1 = \frac{1}{8}$ nmi
23	10	$1 = \frac{1}{16}$ nmi
24	11	$1 = \frac{1}{32}$ nmi
25	12	$1 = \frac{1}{64}$ nmi
26	13	X = Parity

Msg Bit No	Word Bit No	Word 4	
40	1	C&I Quality	00 = 1 cpi 01 = 2 cpi's, different f_p's 10 = 2 or more cpi's, same f_p 11 = 3 or more cpi's, different f_p's
41	2		
42	3	Confidence	000 = Geocensor Traffic Zone 001 = Heavy Clutter Zone 010 = Interference 011 = Angels and Aircraft 101 = Angels and Aircraft 100 = Max Doppler in Zero Filter
43	4		
44	5		
45	6	Track Eligibility	00 = Vehicles; Do not Track 01 = Track Correlation 10 = Track Initiation
46	7		
47	8	ARTISIIIA Quality (SRAP Emulaton)	
48	9		
49	10		
50	11	0	
51	12	1 = Correlated 1 = Uncorrelated	
52	13	X = Parity	

FIGURE 15–54

ASR-9 message formats.

the azimuth contained in the message and modified by parameter data, in queue, in the dual-port memory. That the target data goes to memory locations associated with azimuth is central to the process of synthetic real-time reconstruction.

The ***General Interface Card*** generates the synthetic real-time azimuth change and reference pulses, and f_p, creating 18 display-site pretriggers for every 16 azimuth change pulses, as if to approximately divide the scan into 256 CPI pairs. The synthetic azimuth is not in agreement with the radar sector marks, and must necessarily be several degrees behind to allow for mtd processing and message transmission. Further, the synthetic azimuth will not always bear the same relationship to the radar antenna azimuth; however, within reasonable limits, that is unimportant, because the target azimuth is contained in the messages. It is, however, essential that the synthetic azimuth rotate at the same rate as the radar antenna, and SBC1 continually compares the synthetic azimuth to the sector mark messages to accomplish that. In the general interface card, the azimuth change pulses are developed by a counter, which produces a pulse for each complete count. SBC1 produces a preset for that counter; the larger the preset, the faster the counter reaches maximum, and the faster the acp rate.

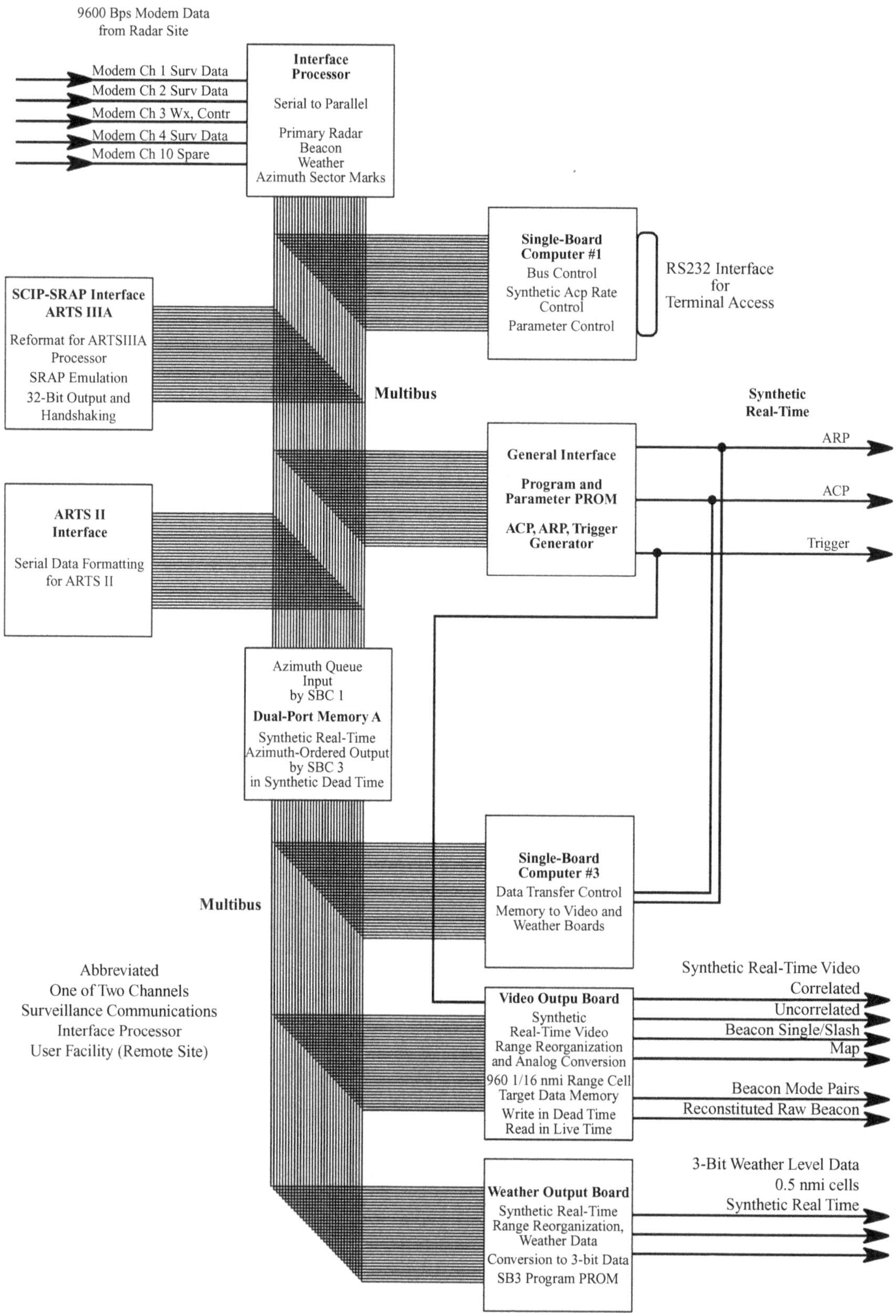

FIGURE 15–55

Abbreviated diagram, ASR-9 SCIP.

The general interface card also contains a PROM, which contains the SBC1 program, and an EEPROM, which contains the SCIP parameters. The EEPROM parameters may be rewritten from a computer terminal via an RS232 interface connector. Although there are many parameters, among them are the size, in azimuth and range, of the different target types.

The ***SCIP/SRAP*** and ***ARTSII Interface*** cards provide a means to reformat data for those processors. In the case of the ARTSIIIA, the installation of an ASR-9 obviates the need for the ***Sensor Receiver and Processor (SRAP),*** and the interface is necessary only to place the data in the same 32-bit form as the SRAP had previously done.

SBC3 runs on a program stored in the weather output card. It receives acp and arp information from the general interface card, providing the program with continual information regarding the synthetic azimuth. Targets are stored in the dual-port memory, according to the message azimuth, but in advance of that azimuth, by an amount determined by the SCIP parameters. When the memory address equivalent to the azimuth of the target, minus half its programmed width, is reached by the synthetic azimuth count, the target will be placed on the memory output multibus. This will occur at each synthetic T_r, during synthetic deadtime, a period of about 300 μs. The memory-retrieval operation is initiated by circuitry in the ***Video Output Card.***

The video output card contains 1/16-nmi-resolution range-ordered RAMs, and targets retrieved from the dual-port memory are placed in range-ordered RAM at locations corresponding to the range in the target message. By the time a primary radar target is placed in RAM, the message has become much smaller, as azimuth was no longer present after azimuth "queuing," and range was no longer present after range "queuing." The messages in the primary-target RAM memory contain only target-type and parameter information. In synthetic live time, the RAM is read, in ascending real-time range, one 1/16-nmi location at a time from range cell 0 to range cell 960. Each time the read operation encounters a target, a video output is produced, and the range width will be determined by the target type and parameter. Accompanying the target data in RAM is a count to indicate the number of times the target has been read; when the number of read operations match the number directed by the azimuth width parameter, the target is ended. Video outputs are correlated, uncorrelated, and beacon single slash. There is a map video output, but this has no connection with the previously discussed video map used for air traffic control geographical relationships; it is a SCIP-parameter range-azimuth-gated correlated/uncorrelated video.

Synthetic Real-Time Beacon Mode Pair and Code Trains

The video output card also contains a beacon real-time reconstitution circuit. In a manner similar to the primary, but more complex, beacon code trains and mode pairs may be reconstructed from a beacon RAM. This circuitry is necessary to provide for the operation of beacon decoders, used as a backup system for the ARTS.

Weather Data

Weather detections are processed by the ***Weather Output Card,*** which places the weather data into synthetic real time. A weather detection requires an entire CPI pair, so each weather detection is repeated 18 times to represent a CPI pair. Weather data is loaded in FIFOs, with one always being loaded, while the opposite one is being unloaded. Before display, further processing is required of the weather data; the 3 bits are used to represent weather detection levels zero through six.

Review Questions

1. The ppi is sometimes called a "ρθ" display in formal literature. The "ρ" refers to ______, and the "θ" refers to ______________.
2. Radar displays are likely to incorporate ______________ crts, and oscilloscopes are likely to incorporate _____________ crts, because_______________________________.
3. An oscilloscope is an _____ type display.
4. Identify the inputs to the *X*-, *Y*-, and *Z*-axes on the following radar displays:

   A: *X* ______________________ *Y* ____________________________ *Z* ____________

   B: *X* ______________________ *Y* ____________________________ *Z* ____________

   C: *X* ______________________ *Y* ____________________________ *Z* ____________

   E: *X* ______________________ *Y* ____________________________ *Z* ____________

   G: *X* ______________________ *Y* ____________________________ *Z* ____________

   J: *X* ______________________ *Y* ____________________________ *Z* ____________

   P: *X* ______________________ *Y* ____________________________ *Z* ____________
5. A rappi display differs from a ppi in what regard?
6. Synthetic real-time equipment is necessary to?
7. Give two reasons that BRITE equipment is used in air traffic control towers.
8. Name the two requirements for any type of radar television display equipment.
9. Define "secondary emission."
10. Name the three major steps in the evolution of radar television technology.
11. How does the BRITE display differ from conventional television?
12. How does the BRITE camera differ from conventional television?
13. A crt electron gun appears to be "crooked." Explain.
14. If a sweep current waveform were perfectly linear, what would be the effect upon (1) a ppi display, or (2) a video map?
15. At short ranges, a ppi display exhibits a bowed map runway line. Name a possible cause.
16. The output from the sweep generator is a trapezoid, and the sloped portion is not a perfectly traight line. Explain.
17. Any experienced radar technician should recognize the frequency, 80.9375 kHz. Why?
18. In preference to a synchro receiver, a synchro control transformer is usually used at the receiving end of a synchro system. Why?
19. To provide sin–cos sweep data, the synchro system may drive another electromechanical device. What is it called?
20. Briefly explain the difference between the Pythagorean Theorem and the trigonometric identity, $\sin^2\theta + \cos^2\theta = 1$.
21. What is the most significant difference in the manner in which the focus and deflection coils are wound?
22. "Overscan" limiting is _________________.
23. Define "orthogonality."
24. Briefly explain "synthetic real time."
25. It is necessary that the synthetic real-time azimuth significantly lag the radar antenna azimuth, and the amount of lag is adjustable by a SCIP parameter. Why?

Answers to Review Questions

1. The ppi is sometimes called a "ρθ" display in formal literature. The "ρ" refers to ***range,*** and the "θ" refers to ***azimuth.***
2. Radar displays are likely to incorporate ***electromagnetic*** crts, and oscilloscopes are likely to incorporate ***electrostatic crts***, because ***the electromagnetic crts offer larger displays, and the electrostatic crts offer more precise deflection, and a wider variety of deflection frequencies.***
3. An oscilloscope is an ***A*** type display.
4. Identify the inputs to the *X*-, *Y*-, and *Z*-axes on the following radar displays:
   A: *X*__***Radar Sweep***__________*Y*__***Radar Video***______________*Z*__***Intensity Adjustment***
   B: *X*__***Radar Sweep***_________*Y*__***Azimuth or Elevation***______*Z*__***Radar Video***
   The X and Y inputs may be reversed in some applications.
   C: *X*__***Azimuth***____________*Y*__***Elevation***_______________*Z*__***Radar Video***
   E: *X*__***Radar Sweep***_________*Y*__***Azimuth or Elevation***______*Z*__***Radar Video***
   G: *X*__***Azimuth***_______________*Y*__***Elevation*** ______________*Z*__***Radar Video***
   J: *X*__***sine + Video***________*Y*__***cosine + Video***____________*Z*__***Intensity Adjustment***
   P: *X*_***e–w sin-mod sweep***____*Y*__***n–s cos-mod sweep***_________*Z*__***Radar Video***
   Assumes fixed-coil deflection.
5. A rappi display differs from a ppi in what regard?
   There is no real-time sweep. The electron beam is placed to the appropriate X–Y coordinates according to message data, and then the screen is written upon with "strokes."
6. Synthetic real-time equipment is necessary to?
   Make digital message data compatible with real-time equipment.
7. Give two reasons that BRITE equipment is used in air traffic control towers.
   The television refresh rate of the target data makes daylight display possible, and the target storage leaves trails, to make the aircraft course immediately obvious.
8. Name the two requirements for any type of radar television display equipment.
   There must be a method to store and hold target data, and there must be a means to retain that data through repetitious reading by the television raster.
9. Define "secondary emission."
   The emission of electrons from a material struck by an electron beam.
10. Name the three major steps in the evolution of radar television technology.
    Scan-conversion RBDE equipment, BRITE equipment, and DBRITE equipment.
11. How does the BRITE display differ from conventional television?
    It employs 945 lines, instead of 545 lines.
12. How does the BRITE camera differs from conventional television?
    The vidicon recovers at a much slower rate, to provide storage.
13. A crt electron gun appears to be "crooked." Explain.
    It is probably an oblique gun, to prevent ion bombardment of the screen.
14. If a sweep current waveform were perfectly linear, what would be the effect upon (1) a ppi display, or (2) a video map?
    On the ppi, more distant range marks would be spread apart because of the faster movement of the beam. On the map, more distant range marks would be too closely spaced, because the beam passes through the map clear spaces more quickly.
15. At short ranges, a ppi display exhibits a bowed map runway line. Name a possible cause.
    The map trigger timing may be incorrect.

16. The output from the sweep generator is a trapezoid, and the sloped portion is not a perfectly straight line. Explain.

    The "step" portion of the trapezoid gives a smooth start to the current flow in the deflection coil, and the variation in sweep sawtooth slope is intended to keep the trace velocity constant.

17. Any experienced radar technician should recognize the frequency, 80.9375 kHz. Why?

    It is the reciprocal of 12.3552 μs, the radar mile, and is used for 1-mile range marks.

18. In preference to a synchro receiver, a synchro control transformer is usually used at the receiving end of a synchro system. Why?

    The synchro receiver cannot provide enough torque, and the synchro control transformer is part of a torque-multiplication system.

19. To provide sin–cos sweep data, the synchro system may drive another electromechanical device. What is it called?

    Resolver.

20. Briefly explain the difference between the Pythagorean Theorem and the trigonometric identity, $\sin^2\theta + \cos^2\theta = 1$.

    The identity is a restricted version of the Pythagorean theorem; because the sides are restricted to the sines and cosines of given angles, the hypotenuse will always equal 1.

21. What is the most significant difference in the manner in which the focus and deflection coils are wound?

    The focus coil lines of magnetic force are parallel to the electron beam, but the deflection coil force lines are at right angles to the electron beam.

22. "Overscan" limiting is ***a method of sensing the amplitudes of the north–south and east–west sweep currents to end the sweep gates, when the beam reaches the edge of the cathode ray tube.***

23. Define "orthogonality."

    "Orthogonal" refers to a 90° orientation of two entities. In fixed-coil deflection, the ORTHOGONALITY adjustment provides a means to correct for slight errors in the coils' north–south and east–west alignment.

24. Briefly explain "synthetic real time."

    Where a real-time range and azimuth timing environment has been re-structured to make a real-time user facility compatible with digital message data.

25. It is necessary that the synthetic real-time azimuth significantly lag the radar antenna azimuth, and the amount of lag is adjustable by a SCIP parameter. Why?

    Any digitizing processor requires time to provide target outputs. If a message of, for instance, 50° azimuth, were transmitted when the SCIP scan was already at 60°, the target would be missed.

APPENDIX A

Annotated Glossary

A

"A" (Type Display): Military origin, now mostly US Navy. A cathode ray tube display, showing range, or time, versus amplitude, where range is the horizontal and amplitude is the vertical (see Figure A-1). When used in fire-control radar, normally, the range is only a small portion of the overall range of the radar. The display is similar to an oscilloscope, and an oscilloscope display may also be called an "A" scope. More on pp 386, 387.

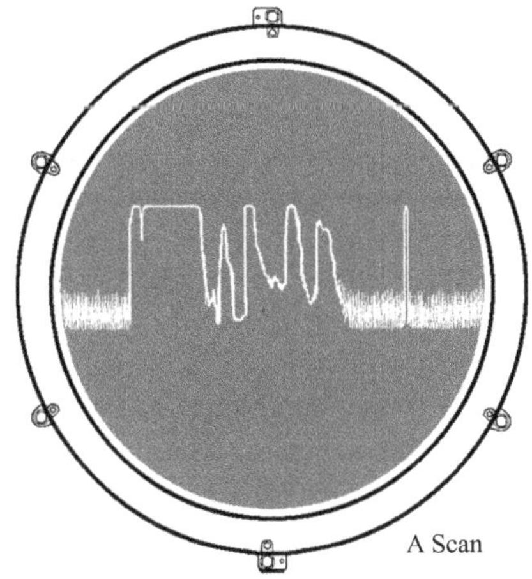

FIGURE A–1

Type "A" display.

AAW: US Navy. Anti-Aircraft Warfare. Used to describe that combat that takes place above the surface of the ocean. Various surface ship systems that support this type of warfare area include Fire Control Radar, Three Dimensional Radar, Tactical Air Navigation and Phased Array Radar (Multi-Function Array) systems.

ACD: ARTS Color Display (ARTSIIIE). Later generation displays are scan-converted, and the television video is colored to better alert controllers to conditions. For instance, aircraft at excessively low altitudes, or on potential collision courses, cause aircraft data blocks to turn red. White, yellow, green, and blue also are used for a variety of purposes. More on pp 135, 390.

Acquisition Gate: US Navy. In tracking radar, a gate defined in range and bearing, derived fromtwo circuits, one, the azimuth change pulse, and the other, the range counter. The gate defines the coarse, two-dimensional, position of a target of interest. This gate is normally used to position a smaller, and more accurate, gate called the ***track gate.***

AC& W: USAF. Aircraft Control and Warning. Formerly used to describe installations containing ARSR-class long-range radars, height-finder radars, computer systems, and intercept control. More on pp 10.

ACP: See "Azimuth Change Pulse (acp)."

Active: Universal. Broadly, "active" and "passive" describe an electronic component that operates, respectively, with or without external voltage, in addition to the signal passing through it. In air traffic control surveillance radar, "active" or "passive" may refer to the receiving system waveguide path. Often, two antenna feedhorns are used. One is the usual bidirectional path with a duplexer, and is called the "active" system. The other is a receive-only path, called "passive," and the feedhorn is positioned to receive only higher elevation targets at near ranges. The principal reason for this arrangement is to improve mti performance by reducing close-in clutter strength. "Active" and "passive" may also be called "low beam" and "high beam," respectively. More on pp 75.

ADAC (module): A unit containing both analog-to-digital (A/D) and digital-to-analog (D/A) conversion circuits.

Adaptive Gain: US Navy. See also "adaptive thresholding," FAA, USAF. Rapid modification to the thresholding values on a range-cell-by-range-cell basis, vice the typical PRI to PRI.

Adaptive Thresholding: FAA, USAF. In radar, a "threshold" may be a voltage level or digital number. A signal or number which exceeds a threshold and permitted to pass through the circuitry is said to "break" the threshold. An adaptive threshold automatically adjusts in accordance with the data environment, increasing or decreasing as necessary. An adaptive thresholding may utilize surrounding ranges, surrounding azimuths, or both. Also, see "second adaptive," "CFAR," "MLT," "Log FTC," and "sliding window." More on pp 351, 354, 373, 374.

AEGIS: US Navy. Combat system installed on Ticonderoga Class Cruisers, and Arleigh Burke Class Destroyers. Capable of tracking over 100 targets simultaneously, directing and guiding missiles to intercept those targets, and performing evaluation on the outcome of those engagements. More on pp 130, 131.

AEW: US Air Force. Airborne Early Warning, a type of aircraft normally fitted with a two-dimensional long-range search radar utilized for detecting, tracking, and identifying aircraft in area defense or air superiority roles. Also see "Airborne Warning and Control System (AWACS)."

AES: Azimuth Encoder Synchronizer. Used with latter-day Inductosyn (trademark) 16,384-acpazimuth pulse generators. More on pp 106, 110, 132, 133.

AGC: Automatic Gain Control. More on pp 86, 260, 266.

AFC: Universal. Automatic Frequency Control. In magnetron radar systems, an automatic frequency control is necessary to mechanically drive either the stalo or transmitter tuning, to maintain the appropriate difference between the local oscillator and transmitter frequencies, ordinarily 30 MHz. More on pp 57.

AFC Lock Pulse: FAA, USAF, USN, others. May also be called "coho lock pulse." See "lock pulse." More on pp 58.

Agile-Beam: A radar system in which the radiated beam is quickly directed without physical movement. Also see "phased array" and "sinuous-fed." More on pp 35, 38, 53, 223.

AHI: Military. Azimuth v height v intensity. "See 'C' Type Display." More on pp 388.

Airborne mti: USAF and US Navy. Aircraft moving target indicator. See "TACCAR."

AIMS: A military-use beacon interrogator–transponder system. Air Traffic Control Radar Beacon System Identification Friend or Foe Mark XII System. More on pp 12.

Airborne Warning & Control System (AWACS): A USAF radar long-range early-warning radar system for intercept control and other related purposes. Contained in an E3A Boeing four-engine jet, resembling early 707's.

Air Route Surveillance Radar (ARSR): FAA. A radar used for the monitor and control of high-altitude "enroute" cross-country air traffic. Usually 200-mile range, L band. More on pp 10.

Airport Surveillance Radar (ASR): A radar used for the control of airport arrivals, departures, and local traffic. Usually 60-mile range, S band. See "terminal (radar)." More on pp 8.

AMASS: Airport Movement Area Surveillance System. Also Terminal Automation Interface Unit (TIAU). Together, the two systems provide an interface between digital radar outputs (ASR), automated radar tracking systems (ARTS) outputs, and Airport Surface Detection Equipment (ASDE). The overall system provides potential incursion warnings to air traffic controllers. More on pp 12.

Ambiguity: General. A single representation of more than one possible condition, the second, or additional, being undesirable. More on pp 81,126, 253, 291, 299, 300.

Amplitron: Universal. A final power amplifier in radar transmitters. Also called "crossed-field amplifier (CFA)." More on pp 11, 53, 195, 223.

AN/SPY-1 Radar: US Navy. Multifunction, computer-controlled phased-array radar, used as the primary sensor, fire control radar, and search radar, on board AEGIS Class and Arleigh Burke Class ships. More on pp 130.

Analog-to-Digital (A/D): Conversion process, voltage levels to digital "words." In radar, care must be exercised not to confuse "analog" with "antilog." An A/D converter may be called a "quantizer." More on pp 65, 83, 325, 350, 351, 397, 410, 411, 414.

Antilog: FAA, military, and others. Same meaning as in mathematics, but specifically applicable to radar data that has been converted to base-2 logarithmic form and then restored to a linear form. The linear form is called "antilog." Do not confuse with "analog." More on pp 25, 41, 81, 84, 205, 250, 336, 339, 341.

Aperture: Scientific and engineering, FAA, military, and others. The electrical, physical, or both size of a receiving antenna. See also "synthetic aperture." More on pp 39.

Applications: Computer programs written for use with a prescribed operating system. More on pp 185.

APG: See "Azimuth Pulse Generator (apg)."

APM: Antenna Pattern Measurement. In FAA monopulse beacon systems, the received 1,090-MHz replies are initiated by a PARROT transponder when interrogated by mode B. The PARROT is located at a ground facility near the radar site.

APP: US Navy. Antenna Position Programmer (also referred to as a Beam Steering Controller). Specific piece of equipment used in a phased-array radar to deliver phase taper commands to the phase shifter drivers located on the array.

Area of Inspection: FAA, military. That cell of data upon which a surrounding-cell average is calculated for subtraction or thresholding. See "adaptive thresholding." More on pp 338, 362.

ARP: See "Azimuth Reference Pulse (arp)."
ASR: See Airport Surveillance Radar (ASR).
ARSR: See "Air Route Surveillance Radar (ARSR)."
ARTS: FAA. Automated Radar Tracking System or Automated Radar Terminal System. Operates on radar data message inputs from a "digitizer," which determines azimuth data either from a "sliding window" process, or a monopulse determination process. Radar data is displayed in alphanumeric form. Accepts entries from local or remote air traffic controller keyboards, digital altimeter setting indicators, flight plan sources, air route traffic control centers (ARTCC), and more. Maintains and starts tracks on aircraft, detects, and warns of minimum safe altitude (MSAW) or conflicting aircraft courses (conflict alert) and altitudes that present a danger of collision. More on pp 10, 12, 97, 100.
ASDE: FAA. Airport Surface Detection Equipment. A short-range radar for observation of airport surface traffic. More on pp 12.
ATC: FAA, military, others. Air Traffic Control.
ATCRBS: FAA. Air Traffic Control Radar Beacon System. See "beacon," "IFF," and "secondary radar." More on pp 97 through 139.
AT-R: Anti-transmit-receive gas tube once used in duplexers to isolate, or "close the path to," the transmitter from the receiver during "listening" time. Now replaced by four-port circulators and ferrite load isolators. Latter-day use of this term by a manufacturer of US Navy equipment is conflicting; the device that serves as a "T-R" is called the "ATR." More on pp 70.
AUI to RJ245: Ethernet to LAN.
Azimuth: The direction of the pointing angle of the radar antenna, often called "bearing." More on pp 35.
Azimuth Change Pulse (acp): USAF, FAA, other military. Created by an azimuth pulse generator, mechanically driven by the radar antenna. In earlier systems, each antenna revolution generated 4,096 azimuth change pulses, and one azimuth reference pulse. Newer systems now use Inductosyn (trademark) azimuth encoders to generate 16,384 azimuth change pulses. More on pp 64, 73, 74, 90, 101, 102, 106 through 110, 124, 125, 128, 132 through 135, 369, 373, 375, 378, 380, 407, 408, 409, 412 through 421, 423, 424, 425.
Azimuth Gate: US Navy and others. In tracking radars, a gate measured in bearing, normally mils or degrees and minutes, and defined by the start and stop pulses derived from an acquisition gate. May be used in surveillance systems as a stagger, or stc, gate, and may be in terms of acp count. See "Range-Azimuth Gate (RAG)."
Azimuth Pulse Generator (apg): FAA, USAF, other military. Azimuth Pulse Generator. Surveillance-antenna assembly unit to create azimuth change and reference pulses. See "acp" and "arp." More on pp 35, 73, 132, 378, 407.
Azimuth Reference Pulse (arp): FAA, USAF, other military. The single pulse from a surveillance-antenna azimuth pulse generator; occurs once during each revolution of a surveillance radar antenna. More on pp 64, 73, 101, 124, 132, 407, 412, 414, 415, 417, 424, 425.
ASTERIX: All purpose STructual Eurocontrol Radar Information eXchange. A data communication protocol.
Azimuth Resolution: All radar. The minimum bearing separation, between two targets of equal range, at which the targets may be identified as two. Usually given as the 3 dB points of the antenna radiation pattern, but this is a figure of merit. Because targets may be detected and displayed at lesser ranges in the pattern, where side lobes exist, the azimuth resolution may be a greater angle than the 3 dB points. See pp 15-23, 400, 414.
Az-El: The radar display in precision-approach (PAR) systems AN/CPN-4 and many descendants (see Figure A-2). The display is of range, azimuth, and elevation. See "'E' (Type Display)" and "'B' (Type Display)." More on pp 388, 389.

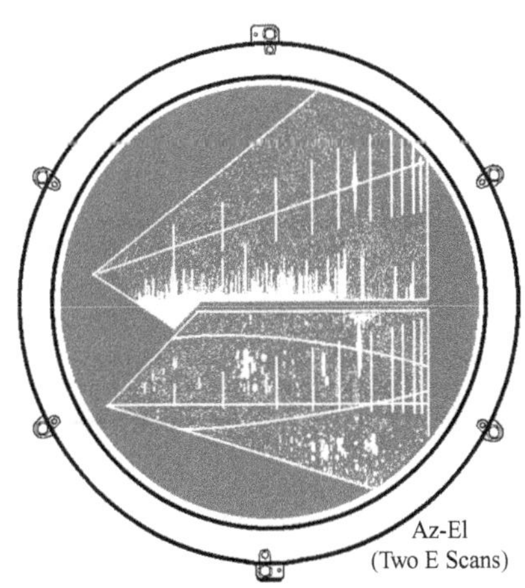

FIGURE A–2
Az-El display.

B

"B" (Type Display): Military, and air traffic control entities other than FAA. A CRT that displays a small area, or sector (see Figure A-3). Several applications, US Navy, US Army, US Air Force, and US Marine Corps. Azimuth or elevation data may be in either the *X* or *Y* dimension, range is in the opposite dimension, and radar video controls intensity in the Z dimension. More on pp 387, 388.

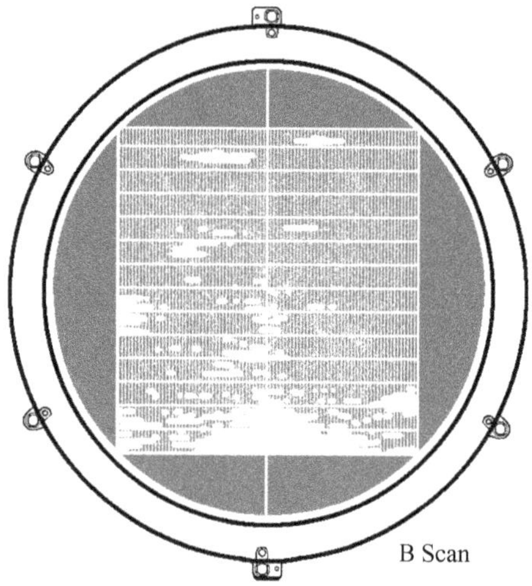

FIGURE A-3

Type B display.

Back Porch: A television term. The portion of the horizontal blanking pulse following the horizontal sync pulse (see Figure A-4). See also "front porch," "serrations," "Electronic Industries Associates (EIA)," and "National Television Systems Committee (NTSC)."

Backward Wave Oscillator (BWO): Microwave industry, military. See "Traveling Wave Tube (TWT)."

BAM: US Navy. Binary Angle Measurement; The breakdown of a circle into a number of radii that can be defined by a binary digital word. For instance, 8 bits would provide up to 128 different angles; that would equal 2.8125 degrees of azimuth resolution.

Bandpass: Universal. The band of frequencies that may be passed by a circuit or device. Ordinarily measured between those frequencies, of a leveled, swept, input band, which produces a half-power decline, in power level, at the output. Not to be confused with "spectrum," which is the simultaneous existence of signals at a variety of frequencies. Simplistically, envision "bandpass" as a "door," and "spectrum" as that which must pass through that door. May also be called "passband" or "bandwidth." More on 40, 58, 59, 71, 77, 79, 81, 203, 205, 206, 245, 246, 251, 256, 257, 259, 260 through 269, 292, 294, 296, 305, 313, 314, 349, 401. Also, see "passband".

Barker Code: A phase-coding technique used in radar transmitter burst expansion and compression, called "compressed high-resolution pulse (CHIRP)." Echoes not matching the transmitted code are attenuated by the receiver processor, reducing signal processing of weather clutter, interference, and electronic countermeasures radiation. More on pp 48, 195.

Batch Cell: FAA mtd. One cell of an mtd batch range cell See Batch Range Cell (BRC).

Batch Ordered: The mtd data which represents the information in a batch range cell. Batch-ordered data may be a group of batch cells, filter accumulated product data, or $\log_2$ filter magnitude data. Not related to "batch files" in personal computers See Batch Range Cell (BRC).

Batch Range Cell (BRC): FAA mtd. A message block of data, one range cell X 1 CPIP. Differs from a CPIP, which describes all the batch range cells in a CPIP. The BRC contains two smaller messages, BRCA and BRCB, derived from the information at a single range for CPIA and CPIB, respectively. More on pp 187, 350, 351, 356 through 66, 368, 369, 371, 372.

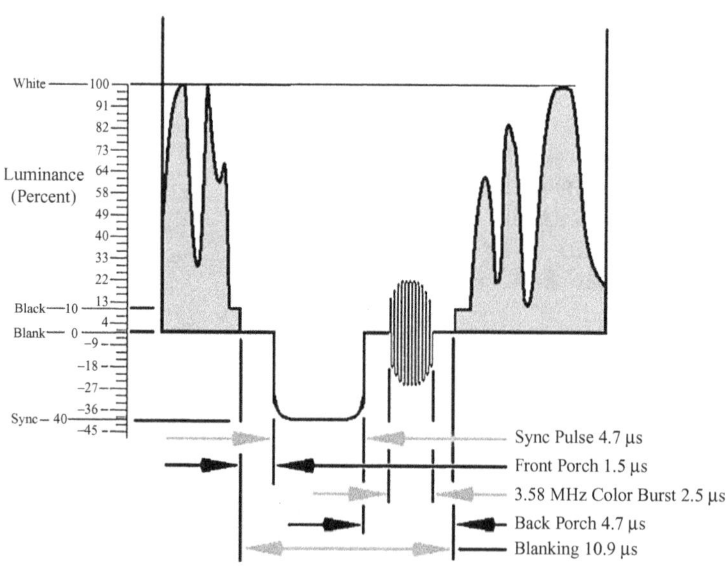

FIGURE A-4

NTSC sync signal.

Battery Alignment: US Navy. The process of aligning all the search, fire control, electronic support systems, and weapons, to a common reference, to increase weapons delivery accuracy.

Beacon: FAA, military, and aircraft industry. See ATCRBS, IFF, SIF, and secondary radar. More on pp 97 through 138. See also, "Secondary radar," and "ATCRBS."

Beacon Code Extractor: FAA, USAF. Digitizing circuitry to convert beacon reply code trains into messages describing the code. The code extractor is necessary to convert data to messages compatible with sliding-window digitizers. More on pp 123.

Beacon Code Validation: FAA, USAF. A digital process to confirm that the reported beacon code is repetitive and consistent. More on pp 102.

Beacon Data Acquisition System (BDAS): In FAA, this refers to that portion of the ARTS sensor receiver and processor dedicated to digitizing secondary radar transponder data. See also "sensor receiver and processor (SRAP)."

Beacon Test Set: Special-purpose test equipment providing 1,090-MHz rf signal generation, operator-selectable codes, and many other features. More on pp 127.

Bearing Resolution: A figure of merit to quantify the minimum distinguishable azimuth separation between two adjacent targets. The antenna beam half-power points are used for the figure. Same as "azimuth resolution." More on pp 4, 37.

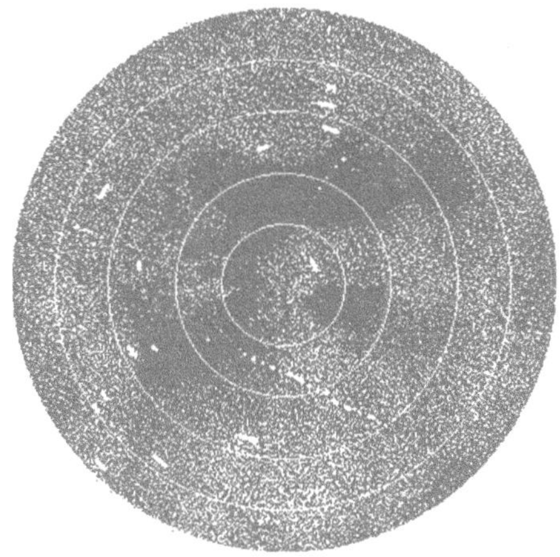

FIGURE A-5

Black-hole video.

Bipolar: Analog or digital data with excursions in both the positive and negative directions from a static level, or baseline. In digital form, negative-going data is the two's complement of an equal positive-going value. In mti systems, the output of a phase detector is called "bipolar video." See "butterfly." More on pp 56.

BIT: Built-in, automated performance testing, usually performed on online systems (see also FIT). More on pp 127.

BITE: Built-in test equipment. See BIT.

Black-Hole Video: Where "black holes" appear on the ppi display in ground-clutter areas (see Figure A-5). Caused by excessive mti i-f gain; worsens subclutter visibility. More on pp 308.

Blind Phase: An ambiguity, where two consecutive phase-detector outputs of different i-f/coho phase comparisons are the same, in both amplitude and polarity. Alleviated with quadrature phase detectors. More on pp 291.

Blind Speed: An mti ambiguity, where the target Doppler is equal to the f_p, or any multiple, and appears to have no Doppler. More on pp 290.

Blip: Used more by the news media than by professionals; this term refers to a target "popping on" to radar display, usually with a ppi in mind.

Blocked Oscillator: Now very rare. Bears some resemblance to the blocking oscillator, except that it can only operate when triggered, and contains a capacitance in the cathode circuit to prohibit dc current through an electron tube. See "blocking oscillator".

Blocking Oscillator (see Figure A-6): A vacuum-tube circuit used to produce high-amplitude triggers for high current-demand distribution. Characterized by a transformer in the plate circuit, a feedback from the plate to the grid, and a capacitance in the grid circuit to inhibit retriggering until the charge on the capacitor decays to an amount enough positive The transformer primary drives a secondary and tertiary winding, one for the high-amplitude output, the other for feedback. Can be made to work in solid-state circuits, but those more often use power transistors in "Darlington pairs." The blocking oscillator was once used as the main trigger generator for many radar systems, and was called the "master trigger blocking oscillator (MTBO)." The blocking oscillator will "free-run" at its natural period determined by the circuit reactance, but is often synchronized by a shorter period trigger input. In mti systems, the MTBO was synchronized by a "circulating trigger," the original MTBO trigger passed through the canceler quartz delay line to make the T_r equal to the delay.

Bracket Pulses: FAA, military. Also called "framing pulses." Beacon code train pulses spaced by 20.3 μs. See "reply code." More on pp 110.

Breakthrough: Slang term for "mti clutter residue," often used by air traffic controllers. More on pp 376.

Brightness (BRITE Equipment): Universal. The control in television-type equipment, such as BRITE, which determines the cathode-ray-tube control grid bias, and therefore, the light intensity of the displayed presentation. Not to be confused with ***contrast***, which adjusts the television video pedestal, to establish the dark-to-light video ratio.

BRITE: FAA, USAF. Bright Radar Indicator Tower, or Television, Equipment. A television camera is pointed directly at a radar ppi, to create a television radar display, for use in air traffic control towers. Newest version, DBRITE, uses digital storage, rather than a television camera. More on pp 391.

BRITE Alphanumeric System (BANS): FAA. Scan-conversion equipment to place RAPPI ARTS alphanumeric data into television form compatible with BRITE radar television video. More on pp 422.

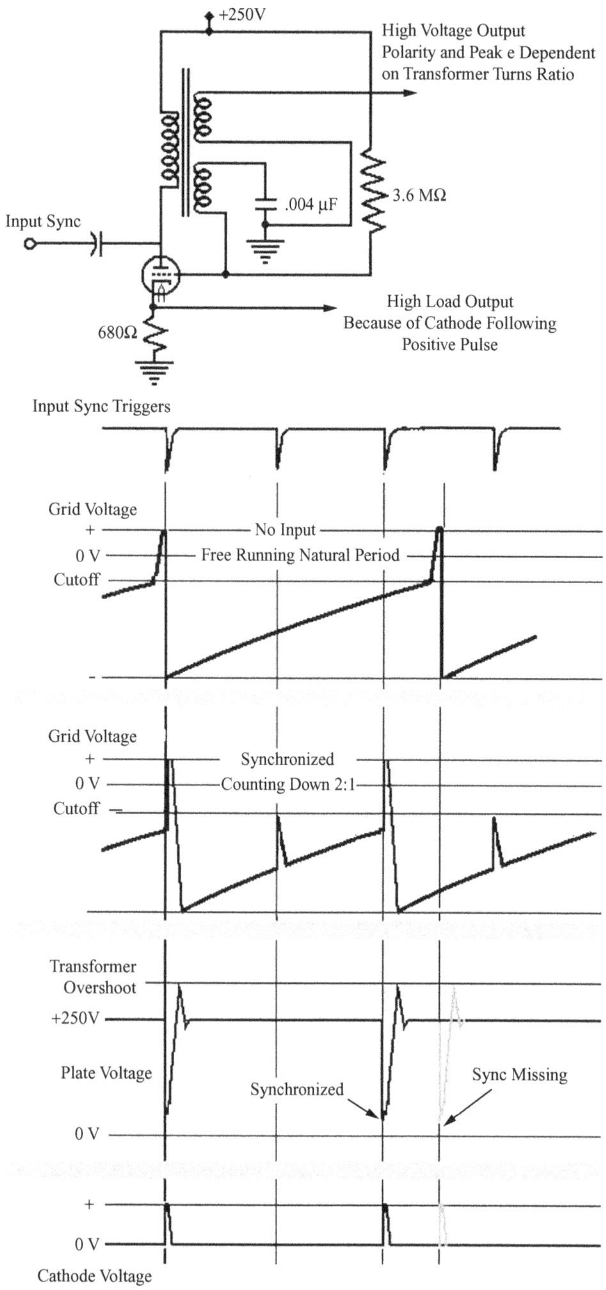

FIGURE A–6

Blocking oscillator.

Broad Band: Data transfer, for signals containing frequencies above audio, requires broad-banded high-frequency responses. Analog radar video is, therefore, "broad-band" information. In contrast, radar data messages are transferred by "narrow-band" audio-frequency modems. More on pp 63, 157.

Bucket: Norden, manufacturer of FAA ASDE-3. A slang name for a card rack. May also be expanded to "minibucket," "maxibucket," and "megabucket." Card racks may also be called "crates," "cages," and others.

Bull Gear: General mechanical term. The large, driven, gear in a surveillance radar antenna pedestal. More on pp 72.

Burst: Universal. An emission of rf energy for a short time duration, such as, for instance, 1, 5, or 10 µs. Often erroneously called a "pulse," which, in radar, is the modulating signal to create the burst.

Butterfly (see Figure A-7): US Army Air Corps, USAF. Slang from the earliest mti radar history. The rhythmic, fluctuating, bipolar, moving-target phase-detector output resembles the wings of a butterfly. See also "bipolar video." More on pp 56.

C

"C" (Type Display) (see Figure A-8): Military. Also called AHI, for azimuth v height v intensity. Provides a head-on view of an aircraft on GCA final approach. More on pp 388.

Canceler: FAA, miltary. The functional part of the mti system that performs T_r-to-T_r comparisons of phase-detector bipolar video to attenuate ground clutter. May be spelled "canceller," "cancellor", or "canceler;" both "canceller" and "canceler" supported by Webster's. More on pp 294, 296, 329, 334.

Cancellation Ratio: FAA, military. The ratio between canceled and uncanceled fixed targets in an mti system. Different methods may be employed to use an echo box, lock test pulses, or video test targets. In any case, the measurement will be expressed as 20 times the log of the voltage ratio between the same signal, canceled and uncanceled, viewed on an oscilloscope at the canceler output. More on pp 303, 304, 312, 313, 314.

Cascaded Cancelers: FAA, miltary. mti cancelers connected in series. More on pp 83, 84, 291, 294, 295, 296.

Cavity: Universal. A metal container used as a tuned microwave circuit. May have several shapes. More on pp 5, 53, 56, 58, 69, 79, 143, 162 through 171, 207 through 216, 220, 221, 249, 256, 314.

CCA: Universal and US Navy. May stand for "circuit card assembly," or "carrier-controlled approach." Carrier-controlled approach radars are air traffic control radars used on aircraft carriers, and may resemble some GCA systems.

CD, CD-1, CD-2: Common digitizers, so named because they are common to FAA and military users, and may be used by a variety of surveillance radars. See also, "Common Digitizer," and "Digitizer."

Center of Gravity: US Navy. The center of a radar track in range, bearing, and elevation, should not be construed as the true COG, as utilized in aviation.

Centroiding: FAA. A computer program or hardware data process to calculate a single range and azimuth of a radar target from all the echoes received. Performed by a conventional azimuth sliding window, where real-time radar data is available. Performed by a "correlation and interpolation" process in mtd systems, where filter data and signal strengths must be utilized. See "sliding window." More on pp 63, 102, 109, 115, 134, 264, 351, 368, 369, 373, 374, 379, 389, 410.

CFA. Cross-field amplifier, also called "amplitron." See amplitron.

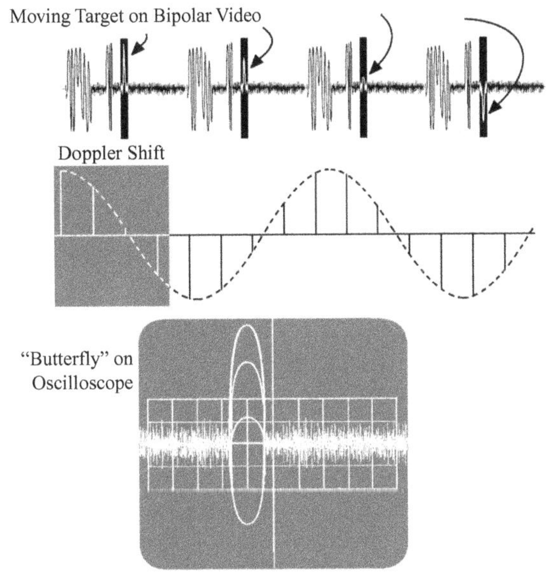

FIGURE A–7

The "Butterfly."

CFAR: Military, FAA. Constant False Alarm Rate. A circuit used to keep the rate of false radar threshold breaks constant. See also "Log FTC," "False Alarm Rate," "sliding window," "mean level threshold," and "adaptive threshold." More on pp 80, 86, 89, 280, 336, 351, and 362 through 368.

Challenge: Military. Once described the transmission of a beacon mode pair, and synonymous with "interrogate." Evolving to mean "mode 4 challenge." More on pp 11, 98, 101, 104, 105, 109, 113, 116, 124, 126, 131.

CHIRP: Military. Compressed high-resolution pulse. A means of expanding a transmitted pulse with linear frequency modulation, stepped frequency modulation, or Barker coding, then compressing the pulse in the receiver. See "pulse compression." More on pp 44, 45, 47, 48. See also, "Barker code."

Circular Polarization: Military, FAA. An antenna feedhorn arrangement (or radiator configuration) which causes radiation in both the horizontal and vertical polarizations; the rf energy cycles bear a $\sin^2\theta + \cos^2\theta$ shared-power relationship; the rf phase then determines the pointing angle of a vector, described by the sin and cos rectangular coordinates. Circular targets, most importantly raindrops, will be severely attenuated by the polarizer. Also see "polarization." More on pp 74, 130, 131, 371, 404.

Clear-Day Map: FAA. In mtd systems, a semipermanent, programmable, map memory used in the ASR-9 to prevent weather declarations from ground clutter. Programmed on a clear day. The memory locations contain switching information comparable to mti/normal clutter gates. Used to inhibit clutter from weather detector inputs. Also see "clutter-gated." Not the same as "clutter map." More on pp 80, 355, 361, 362, 371, 372.

Clutter: FAA and Military. Radar returns which "clutter" a display. May apply to ground clutter, sea clutter, or weather clutter, but usually accepted as meaning "ground clutter" unless otherwise specified (see Figure A-9). More on pp 1, 6, 34, 35, 43, 44, 55, 70, 72, 76 through 89, 177, 181, 206, 258 through 264, 277 through 329, 339, 342, 347, 351, 352, 354, 361 through 380, 400, 401.

Clutter Gated: Military. A circuit operation, in some systems, to provide normal radar data in all areas where ground clutter is not present; in those areas, mti data will be provided. The normal receiver is not vulnerable to blind speeds, blind phases, and tangential effects; has a greater sensitivity; and offers superior target detection. More on pp 80, 361, 362, 372.

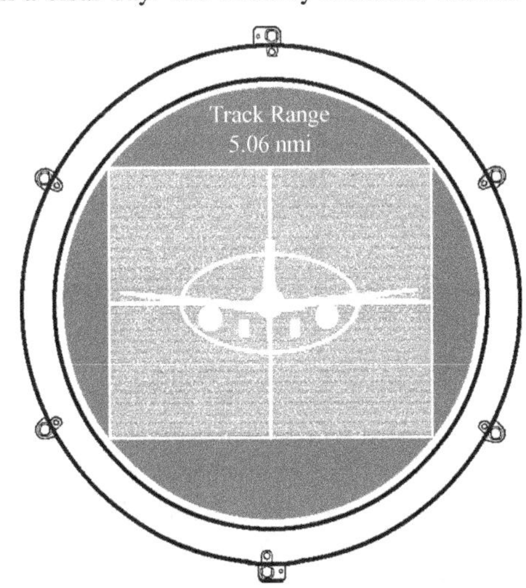

FIGURE A–8

Type C display.

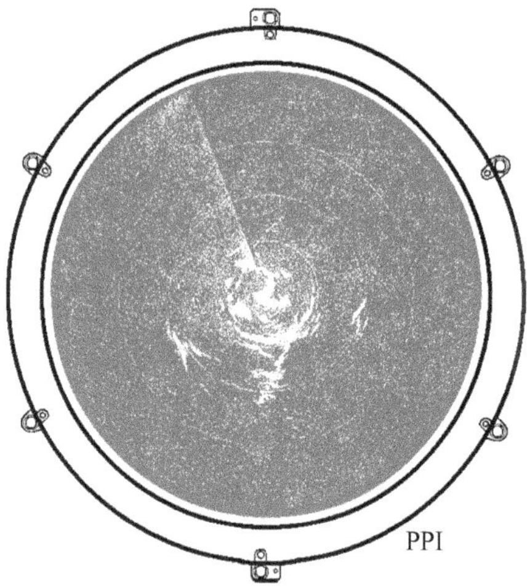

FIGURE A-9

Ground clutter.

Clutter Map: FAA and military. This term has been used to describe several different things. In an FAA mtd system, the clutter map is a memory for storage of zero-Doppler filter data. The main purpose in the mtd case is to provide a means to recover tangential and blind-velocity targets via scan-to-scan comparisons. Also used in military systems as a storage device where the detection threshold of the radar has been exceeded for greater than a pulse length, usually stored in a computer memory, and used for adaptive thresholding magnitudes. More on pp 291,351, 352, 362, 365, 366, 369, 370.

Clutter Residue: FAA and military. The remaining ground clutter, not canceled or attenuated, by the mti system. More on pp 70, 72, 81, 85, 206, 258, 262, 294, 296, 303, 306 through 311, 314, 315, 329, 339, 342, 352.

CMOS: Complementary Metal Oxide Semiconductor.

Code Extraction: FAA. A process to prepare beacon data for use by sliding-window digitizers. The raw beacon code trains must be converted to message form. More on pp 373.

Coherence: FAA, military, scientific and engineering. Where all the frequencies in the radar system bear intelligible relationships, either inherently, as in a synthesis system, or by phase-locking circuitry, as in a magnetron system. More on pp 53 through 58, 174, 223, 255, 256, 257, 280, 286, 287, 298 through 303, 312, 313, 314.

Coherent Gain: US Navy. Gain control that is effective only in a very narrow bandwidth, in an effort to reject similar frequency interference. See Coherent Integration.

Coherent Integration: FAA, scientific and engineering. An increase in signal-to-noise ratios, inherent to the Doppler filter bank, in an mtd system. The filter arithmetic process causes radar targets to accumulate and integrate, but the random phase of noise lessens integration. More on pp 44, 264, 350, 360, 393.

Coherent Oscillator (coho): US Army Air Corps, USAF, CAA, FAA, others. The reference oscillatorin an mti system. The coherent oscillator operates at i-f and is compared to the i-f echoes in the phase detector. Called "coherent" because it bears a coherent relationship with both the transmitter and stalo frequencies. More on pp 53, 58, 64, 256, 281, 301.

Coherent Processing Interval: FAA and USAF. See "CPI/CPIP." More on pp 184, 188, 349.

Coherent Video: FAA, military. The output of an mti phase detector. The video is said to be "coherent" because the i-f echoes and coho oscillator compared in the phase detector bear an intelligible relationship. The video is bipolar because the phase-detector response is the cosine of the angular difference between the two inputs. See "butterfly."

Coho: US Army Air Corps, CAA, USAF, FAA, others. See "coherent oscillator." Very old original acronym for "Coherent Oscillator" from WWII work at MIT Radiation Laboratories.

Coho Gate: US Army Air Corps, USAF, CAA, FAA. A signal used in a magnetron mti system, to disable the coho oscillator, immediately before the lock pulse burst arrives, to establish the coho phase. See "lock pulse." More on pp 58, 177, 179, 301, 312.

Coho Lock Pulse: In magnetron systems, the coho lock pulse is an i-f representation of both the transmitter and stalo frequencies; it is used to phase-lock the coho oscillator to achieve coherence in the phase detector. Also called "afc lock pulse" and "afc/coho lock pulse" See also "lock pulse" and "coho gate."

Collimation: Multiple use and varied meaning. Dictionary definitions give little hint of the radar application of this term. The word is used predominately in connection with ARSRs supplying data to ARTCCs, and is connected with the proper range and azimuth alignment of primary radar targets and beacon information from those targets. Utilized in fire control radar to align its track beam to the illumination beam. Also see "registration." "Collimation" may have other meanings, as in the mesh of a vidicon tube.

Common Digitizer: FAA, USAF. Called "common" because the output data may be jointly supplied to different users, or with any ARSR. Latter-day radar systems have self-contained digitizers. Supplies narrow-band radar data messages to users. Also see "digitizer." More on pp 63, 89, 126, 390.

Compensation Amplifier: FAA, military. Compensates for high-frequency line losses with highfrequency emphasis. See "Line Driver and/or Compensator."

Composite Video Test Target Train (see Figure A-10): FAA. Used in mti radar systems to monitor cancellation performance. A train of fixed targets and optimum-speed moving targets, occurring in deadtime, and generated by built-in test equipment. More on pp 178, 303, 325.

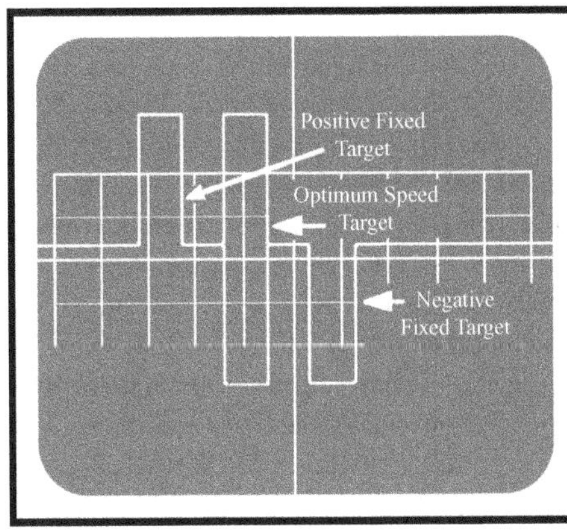

FIGURE A–10

Composite video test target train.

Compression Counts: US Navy. Used as a coarse determination of target amplitude.
Cone of Silence: The radiation pattern of a surveillance radar exhibits a gradual rise in elevation with range, creating a "cone of silence" directly above the radar site.
Contrast (BRITE): Universal. In television equipment, the control which effects the adjustment of the dark-to-light ratio. See also "brightness."
Control Transformer: FAA. The receiving device in a synchro system employing torque multiplication with an error amplifier and two-phase motor. More on pp 403.
Cookie Jar: General slang. See "wavetrap."
Correlation and Interpolation (C&I): FAA. One of the three major programs in mtd. The "correlation" refers to the comparison and connection of all "primitive" targets to form a single, centroided report. "Interpolation" is a Doppler interpolation process, to calculate the center target Doppler. In the ASR-9, the Doppler interpolation program divides the Doppler spread, between zero frequency and the f_p, into 64ths of the f_p. A Doppler value of 32 represents an optimum speed target. More on pp 359, 362, 368 through 373, 375, 377 through 380, 423.
COTS: Commercial Off The Shelf (Equipment).
CPI/CPIP: USAF and FAA. mtd terms, acronyms for "coherent processing interval" and "coherent processing interval pair." A CPI is a group of T_rs, and a CPIP contains two CPIs at two different f_p's. The first and second CPIs in a CPIP are frequently called "CPIA and CPIB." More on pp 184 through 188, 349, 350, 354, 356, 362, 365, 369, 372, 374, 376, 423, 425.
Cross Gating: US Navy, and other military. In tracking radar, a method of determining a target's range to within 1/2 of a pulse width by placing the bipolar return so that the "early" and "late" gates divide the return equally between the gates. May also be given other meanings, in other applications, by manufacturers.
Crossed-Field Amplifier (CFA): Military. A transmitter power output tube. See also "amplitron."
CRT: Universal. Cathode Ray Tube. Developed in the 1920s; origin dates back to Sir William Crookes, 1875. More on pp 3, 35, 203, 204, 375, 385, 386 through 390, 393 through 401, 404, 408, 414, 416 through 419, 421.
CTS: Clear To Send. A signal used in data-transmission devices.
C-W Doppler: Universal. A nearly continuous-wave radar, used principally for velocity measurement, and without range capability.

D

Darlington (Pair): A high-current transistor circuit used in output drivers. In an NPN type, one transistor's emitter is connected to a power transistor's base, so that a small current in the driver becomes a large current in the power transistor.
Deadtime: FAA and military. That time in a radar interval that is beyond the intended range, and before the next transmit time T_0. The time from T_0 to range maximum is, similarly, called "live time." For instance, the live time for a 60-mile ASR radar is 741.3 μs, and the deadtime is the T_r minus 741.3 μs. Live time for a 200-mile ARSR is 2,471 μs. More on pp 34, 36, 124, 125, 174, 177, 178, 264, 399, 408, 421, 424.
Defruiter: FAA and military. Beacon auxiliary equipment to remove "fruit." Principle is somewhat similar to video enhancers or integrators; pulse-to-pulse comparisons allow only repetitive code trains. There were three stages of evolution: tracking delay line (1950s), storage-tube (1960s and 1970s), and digital (1970s to present). Use is being gradually discontinued; latter-day digitizer equipment, employing azimuth sliding windows, makes the defruiter unnecessary. So also does new monopulse equipment. More on pp 126, 128.
Delta Phi **(Δϕ):** FAA, USAF, USAF, engineering and scientific. Pulsed coherent mti systems sample the Doppler shift at the f_p. The portion of a Doppler cycle between successive samples, when expressed in degrees or radians, is the Δ (delta for "change in") ϕ (phi for "phase angle"). More on pp 288, 356. The symbolic representation is used extensively throughout the book.

De Moivre's Theorem. A trigonometry equation used in multiplication of rho–theta coordinates. Academic Trigonometry textbooks.

Dicke-Fix Receiver: Military hard-limited, wide-band receiver, followed by narrow-band amplification. Used to recover low-level echoes from noise or nonsynchronous interference.

Digital mti: FAA and military. In the FAA, this term refers to those canceler-type mti systems which employ range-cell processing, A/D conversion, and shift registers, for T_r-to-T_r delay. Latterday systems in FAA ARSR 1/2 and FPS-20/66/67 employ a system in which radar data is stored in a memory with an interval identifier, and three past intervals are compared with the current live time interval in an arithmetic equation. More on pp 325 through 342.

Digitizer: FAA and military. Equipment which converts real-time radar data into digital messages for narrow-band modem transmission to a user facility over telephone lines. Messages may describe the range, azimuth, and beacon mode and code of targets. Far more complex than "quantizers" or "analog-to-digital (A/D) converter" circuits. Also see "common digitizer." More on pp 12, 17, 19, 41, 63, 85, 86, 89, 126, 132, 138, 264, 390, 392, 410, 422.

Dim Speed: FAA, USAF, and others. Definition may vary. In staggered-f_p mti systems, the dim speeds occur at those radial velocities which cause a lesser canceler output; normally those points at which blind speeds occur, in unstaggered operation. Sometimes used interchangeably with "blind speed," and erroneously assumed to have an identical meaning. More on pp 260, 317, 347.

Diplexer: FAA, US Navy, USAF, and others. A frequency-steering device, tuned to two different frequencies, to provide two different paths. Beacon interrogators employ diplexers for steering the 1,030-MHz mode pair to the antenna, and the 1,090-MHz reply codes to the receiver. Diplexers in "diversity radars (two online channels at different frequencies)" allow two radars to use the same waveguide paths to the antenna. More on pp 98, 103, 118, 120, 190.

Directional Coupler: Universal. A device providing a "straight-through" path, and an attenuated path to a test port. Used principally for connecting test equipment to the radar waveguide assembly. More on pp 26, 27, 29, 34, 36, 41, 58, 70, 71, 75, 76, 99, 100, 118, 171, 172, 173, 202, 206, 210, 212, 246, 249, 250, 268, 301, 302, 310, 312.

Display: Universal. In radar, equipment that provides a radar display; also called "indicator." More on pp 3, 8, 9, 17, 33 through 36, 59, 64, 65, 73, 81, 84, 85, 89, 90, 97, 100, 101, 105, 111, 113, 114, 121, 122, 125, 127, 137, 138, 177, 197, 202 through 206, 264, 266, 267, 289, 291, 294, 295, 298, 303, 304, 308, 314, 315, 316, 317, 335, 336, 347, 350, 375, 376, 377, 379, 385 through 425.

Discrete Code: FAA. Beacon code trains incorporating the "C" and "D" pulses. More on pp 98, 99, 102, 110, 111.

Doppler: In 1842, Christian Johann Doppler offered a mathematical formula to quantify the frequency shift caused by motion. Because it is a proper noun, the term should always be capitalized in use. See "Doppler Frequency." mti theory is all based on Doppler shift (see Figure A-11). More on pp 1, 8, 55, 56, 59, 60, 63, 65, 84, 85, 134, 135, 174, 177, 179, 184, 187, 206, 252, 263, 264, 266, 277 through 318.

FIGURE A–11

Doppler shift.

Doppler Filters: Universal. Bandpass filters which pass radar targets with audio Doppler frequencies. mti cancelers may be Doppler "comb" filters because of the repetitive response, but the term is generally reserved for range-gated Doppler filters, which filter targets for one selected range cell at a time. Latter-day Doppler filters are built around digital multiplier–accumulator circuits. More on pp 348, 351.

Doppler Filter Bank: FAA, USAF, US Navy. A group of Doppler filters tuned to different audiofrequencies, usually for the purposes of separating data according to the Doppler spectrum. A group of Doppler filters tuned to different audiofrequencies.

Doppler Frequency: Universal; physical science. Often used to describe "Doppler shift," the difference in transmitted and received frequencies caused by motion. Widely used and accepted, a purist might insist that "Doppler frequency" is incorrect, and may only be called the "apparent Doppler frequency." The argument is based on the premise that only transmitted and received frequencies exist in the propagation medium. Doppler frequencies are in the audio range, and are expressed as f_D. More on pp 188, 278, 279, 280, 285, 288, 347, 349, 352, 369.

Down Conversion: Universal. In superheterodyne receivers, the process of reducing a received frequency "down" to a lower intermediate frequency. May also apply to frequency synthesizers, where the a transmitter frequency is reduced to be lower than the receiver local oscillator frequency.

Ducting: Universal. The effect of temperature differences between the ambient surface and air, where the duct causes "bending" or "trapping" of the radar beam. More on pp 47.

Duplexer: Universal. A waveguide device to protect the radar receiver from the high-power transmitter energy, and to isolate the received energy from the transmitter. Not to be confused with "diplexer," a tuned circuit used to steer energy according to frequency. See "AT-R" and "T-R." More on pp 33 through 36, 54, 70, 71, 74, 118, 161, 162, 169, 209, 225, 281, 286.

Duty Cycle: Universal. In radar, the transmitted burst duration, divided by the pulse interval, T_r. In deciBels, it is 10 times the log of the division result. Some textbooks may define duty cycle and duty ratio differently, claiming one is the reciprocal of the other. More on pp 28, 29, 106, 107, 118, 120, 201.

Dwell: US Navy. The time from the stimulus message transfer from the computer to the response message transfer to the computer. This time would include the computer output, data distribution, transmission, reception, data convergence, and computer input.

E

"E" (Field): Universal, microwave technology. The electric field in a microwave waveguide, cavity, or other device. The "H" field is the magnetic field. More on pp 149, 151, 152, 153, 155.

E (Type Display)(see Figure A-12): Military. A radar display in which the sweep pivots on a center point and moves between limits in a sector. More on pp 388, 389, 380.

ECL: Emitter-coupled logic. Sometimes used in data-synchronizing circuitry.

ECM: Military. Electronic Countermeasures. Generally, equipment to interfere with radar operation. More on pp 6, 11.

ECCM: Military. Electronic Counter Counter Measures. Generally, equipment to defeat ECM. More on pp 11, 47, 95.

Echo: Universal. One reflected transmitter rf burst. A "target" is either a group of echoes, or the object that causes a group of echoes. More on pp 1, 2, 3, 12, 19, 29, 33, 35, 39, 40 through 47.

Echo Box: FAA and military. A precisely-calibrated, resonant, cavity, connected it the INCIDENT port on the directional coupler, and which rings for an extended period of time after excitation by the transmitter burst. A very useful tool, one of the oldest test equipment pieces in radar. Now in declining use. More on pp 71, 168, 169, 170, 173, 174, 258, 301, 310, 311, 312, 313.

EIA: An abbreviation for "Electronics Industries Association." EIA adopted standard methods, signals, and terms to be used throughout the industry. For instance, the serrated vertical sync pulse in television is often called "EIA sync." See also "serrations," "front porch," "back porch," and "National Television Systems Committee (NTSC)."

Elevation: The angle of a line, in respect to level ground, from the radar to a target See "GCA".

Enhancer: An equipment which performs T_r-to-T_r integrations of synchronous and repetitive echoes, to improve the video-to-noise ratio, and remove nonsynchronous interference. Also called "video integrator." See "video integrator." More on pp 84 through 88, 126, 127, 336, 340, 341, 342.

Enroute (Radar): FAA. Enroute air traffic is that which is traveling cross-country, from one airport control area to another, usually at high altitude. Enroute traffic is controlled by Air Route Traffic Control Centers (ARTCC), and monitored by ARSRs. See "Air Route Surveillance Radar (ARSR)".

Ethernet: A bidirectional data transmission system and protocol.

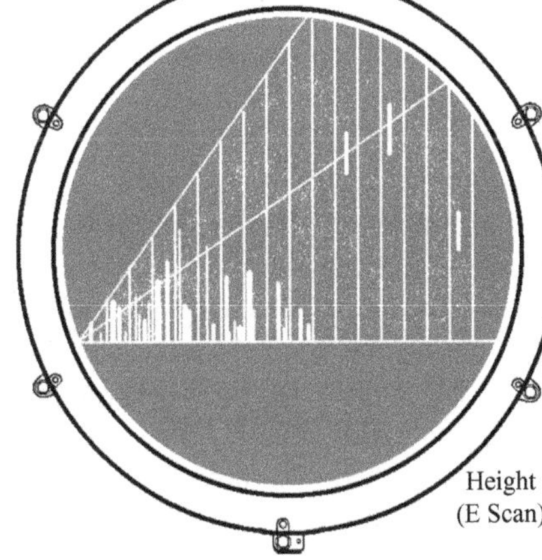

FIGURE A–12

Type E display.

F

"F" (Type Display): US Navy. A cathode-ray-tube display showing the elevation versus azimuth tracking errors, where the horizontal is the azimuth and the vertical is the elevation. Used to determine the tracking accuracy, in elevation and azimuth, of three-dimensional tracking radars.

False Alarm Rate: FAA and military. Originated in military radar systems, employing digitizers, which use thresholding, to inhibit weaker targets or noise. Noise or weather which crosses thresholds are said to cause "false alarms." Range sliding windows to calculate average levels are used to create an automatically variable threshold to keep the rate of false alarm down and constant; such circuitry is therefore called "constant false alarm rate" (CFAR). Also see "FTC," "Log FTC," and "Mean Level Threshold." See "CFAR".

Fill-Pulse mti: US Navy. Utilized to defeat the effects of second-time-around clutter.

Filter-Ordered Data: FAA. In mtd, all the accumulated filter product data for each batch range cell is sequentially taken from the filter circuitry, in a specific filter-number order.

FIFO: First-In-First-Out ordered digital memory. Not to be confused with Random Access Memory (RAM), though the same chip can perform the function, depending on configuration. Differs from shift registers in additional controls, and ability to handle parallel words. More on pp 123, 124.

FIT: Automated fault isolation testing, usually performed on off-line systems, as contrasted to built-in testing (BIT), usually performed on operating systems. More on pp 127.

Flag: A signal accompanying another, or others, to provide some sort of identification. Often a single bit in a radar target data message, such as a "radar-reinforced" flag in a beacon target message to indicate that a primary radar detection had also been made.

Flip-Flop: Also called "bistable multivibrator." A basic part of all modern digital equipment. A two-state tube or transistor circuit in which either one side conducts while the opposite side is cut off. The plate or collector of the cutoff side produces a binary "one," which changes to a "zero" when "flipped" or triggered. Different types include different additional features, such as preset, reset, load, and clear. Used in registers and counters. Basic electronics textbooks

FM Radar: Frequency modulated radar, as that used in radar altimeters.

Four-Port Circulator: A bidirectional waveguide steering device; one use was in upgrading the radar duplexer to eliminate the need for an AT-R tube. More on pp 58, 70, 75, 107, 134, 249, 252, 301.

Fourier: Jean Baptiste Joseph Fourier was a French mathematician and physicist who developed the Fourier trigonometric series of harmonic analysis. This has enormous implication in pulse radar, since the harmonics of a pulse determine the transmitter frequency spectrum, and receiver design must be with attention to the echo spectrum. More on pp 40, 197, 199, 201, 203, 206, 211, 230, 245, 401.

Four-Horn Comparator: US Army, Navy and Marine Corps. Used in older monopulse tracking radars, to determine the offset from beam center in elevation and bearing, normally through algebraic addition of resultant rf.

fp: Scientific and engineering, FAA. Expresses "frequency of pulse." Preferred expression for pulse repetition frequency, because it cannot be confused with any other, and is an acceptable mathematical representation. Same as PRF, PRR, f_r, etc. More on p 37, table 4-1.

FPA: Final power amplifier, such as a klystron or amplitron (aka CFA cross-field amplifier).

fr: Scientific and engineering, FAA. A common textbook term for repetition frequency. Some care must be exercised to note the manner of use, since the term may be used in basic electronic courses to describe resonant frequency. Same as PRF, PRR, f_p, etc. More on pp 37, table 4-1.

Framing Pulses: FAA and military. Also called "bracket pulses." The beacon reply code pulses, spaced by 20.3 μs, which "frame" the code data pulses. See "reply code." More on pp 110, 121, 125. See also, "bracket pulses," "reply codes."

Frame/Field: Television terms. A "frame" is a complete picture, containing two interleaved fields. Frames occur at 30-cycle rates, and fields occur at 60-cycle rates. The 60-Hz field rate is necessary to avoid flicker, and the 30-Hz frame rate is necessary to avoid excessive video bandwidth requirements. See "NTSC".

Frequency Agility: Military. The ability of a radar system to quickly change frequencies during operation, and as a necessary function of the operation. See "Agile Beam." More on pp 11, 130.

Frequency Diversity: FAA, USAF May have different meanings. The 1960's "Frequency Diversity" ARSR/GCI joint-use radar program deployed systems operating in different bands, VHF, UHF, L, and S. In other radars, "frequency diversity" is a mode of operation in which two radar channels transmit at slightly different times, and at different frequencies; the object is to combine the videos from both, to gain the advantages of each frequency and effectively double the received signal power in ideal cases. More on pp 11, 146.

Frequency-Scanned Array: An antenna whose radiated beam is formed by an array of single radiating elements, spaced in such a manner as to point the beam in accordance with the rf energy phasing of the elements. Changes of frequency change the element phasing, therefore the pointing angle of the beam. The US Navy describes this type of antenna as "sinuous-fed." See "agile beam".

Front Porch: A television term to describe the portion of the horizontal blanking signal occurring in advance of the horizontal synchronization pulse (see Figure A-13). See also "serrations," "back porch," "National Television Systems Committee (NTSC)," and "Electronic Industries Associates (EIA)."

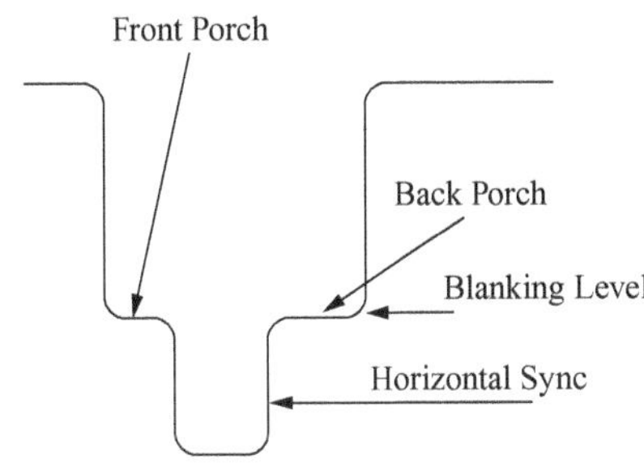

FIGURE A–13

EIA television sync and blanking.

FRUIT: FAA, USAF, US Navy, other air traffic control entities. Acronym for "false replies unsynchronized in time." Secondary radar code trains received at one facility as a result of interrogations by another, or by others. More on pp 104, 106, 117, 126, 128, 133.

FTC: Fast Time Constant. A means of reducing weather clutter on radar displays. The earliest circuits were capacitive–resistive differentiating networks, the basis for the name. Later improvements led to the use of logarithmic i-f amplifiers, and average-subtraction circuits. Still later, digital circuits employing range-sliding-window circuitry made it possible to apply the technique to mti data. See "CFAR".

G

"G" (Type Display)(see Figure A-14): US Army, US Army Air Corps, USAF. Old radar display used for weapons-aiming. More on pp 122 through 126, 128, 132, 373, 389.

Garble: FAA, military. Used in regard to secondaryradar (beacon) systems, to characterize a condition, in which the code pulse positions of two reply code trains are in alignment and can represent a third code. Also see "phantom" and "interleave." More on pp 122, 123, 124, 128, 132, 373.

Gate: Universal. A radar-industry reference to a rectangular-wave signal, of any time duration, intended to enable or inhibit another signal, as if to open or close a gate (see Figure A-15). Differs from "trigger," which is used to initiate, synchronize, or "trip" some electronic action. A trigger is usually used to begin a gate, and a device may be "triggered" by the leading or trailing edge of a gate. Also see "trigger" and "time-share." More on pp 33, 43, 45.

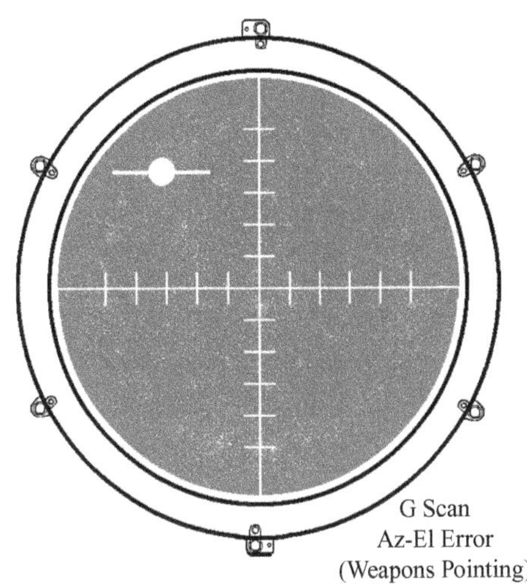

FIGURE A–14

Type "G" display.

Gain Time Control (GTC): Military. Used by manufacturer of the UPX-14 beacon interrogator; same as stc. See "stc." More on pp 104, 119.

Gauss: Scientific and Engineering. One line of magnetic force per centimeter. Named for Karl Friedrich Gauss (1777–1855). More on pp 217, 218, 219, 221, 222.

Geocensor Map: FAA. An mtd system memory, containing threshold and flag data, identifying range-azimuth locations, where detected targets might be ground traffic, clutter residue, or mti reflectors. More on pp 291, 351, 352, 362, 365 through 368, 374, 378, 380, 423.

Granularity: FAA, military. The minimum range-azimuth dimensions of a radar data storage area. For example, if the granularity of a memory is 1 nmi × 2°, the same memory location would be in use from 32 nmi to 32.999 nmi and from 126° to 127.999°, each scan. More on pp 361, 365, 369, 371 aka "resolution."

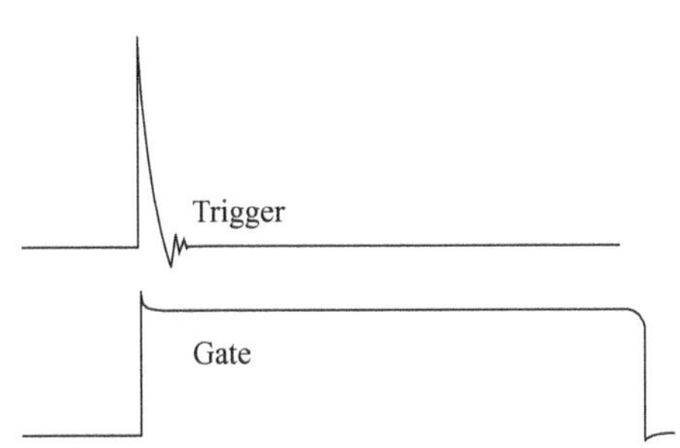

FIGURE A–15

Trigger and gate.

Grass: US Army Air Corps, US Navy origin; now universal. Slang, to characterize the received noise, that appears on the radar video displayed on an oscilloscope. The noise closely resembles

uncut lawn grass, moving under a breeze. More on pp 34, 35, 79, 80, 81, 86, 173, 245, 255, 260, 261, 308, 309, 312, 339, 340, 347, 365, 371, 382, 393.
Gray Code: FAA, military, aircraft industry. Named for its inventor. The Gray code was devised to provide a means for electromechanical encoding, without encountering large errors at the transition points. Used for beacon altitude encoding. See page 112, 113.
Ground Controlled Approach (GCA): Originally US Army Air Corps, then USAF. The original air traffic control radar system, first used in WWII, to land aircraft in poor visibility; comprises two radar systems, called "search and precision." Also see "ASR" and "PAR." More on pp 7 through 10, 18, 86, 97, 129, 130, 388.
Ground Controlled Intercept (GCI): USAF. Long-range, early-warning, radar systems for intruder detection and intercept control. Also see "AC&W." More on pp 10.
GUI: Graphic User Interface. Now becoming the primary means of controlling, adjusting, and analyzing a radar system with a computer terminal. GUI operating systems vary from DOS to Windows XP, UNIX, and others. The radar control and monitoring is accomplished with application software written expressly for that purpose.
Gunn Diode: Universal. A solid-state device, capable of generating microwave frequencies, well into K_a band. More on pp 249.

H

"H" (Field): Universal. The magnetic field in a waveguide, cavity, or other microwave device. See also "E" field, the electric field. See Chapter 8. More on pp 69, 74, 149, 150, 151, 152, 155, 157, 158, 161 through 166, 254, 255.
Hard-Limit: Universal. A condition in which much of an i-f or video signal is severely limited, so as to substantially reduce the signal-to-noise ratio. In contrast, the term "soft limit" has been used to describe minor current limiting in a receiver, intended only to prevent signals from exceeding levels, which they only rarely reach. More on pp 86, 262, 263, 282, 306, 337.
Hertz: Universal. Replaced "cycles-per-second" frequency-value expression in honor of Heinrich Rudolph Hertz, who first proved and demonstrated, in 1888, that electromagnetic energy traveled through space at the same velocity as light. More on pp 1, 2, 13, 59, 146, 277.
Hits per Scan: An expression of value, of the number of radar echoes, that will be received from a single-point target, between the half-power points, of the rotating beam, of a surveillance radar. Only a figure of merit; the number of echoes is not limited by the beam half-power points. More on pp 12, 72, 126, 315.
HPA: High power amplifier. Also see "LDPA".
Hub: A distribution device to provide multiple USB ports.

I

IAGC: A fast-acting, automatic, i-f gain control, to attenuate echoes of excessive length. Superior to ftc because effect occurs upstream of i-f limiting. More on pp 86, 260.
ICAO: International Civil Aviation Organization. An organization for standardizing and coordinating air traffic control throughout the world. Headquartered in Montreal, Canada.
Ident: All air traffic control and aircraft entities. Can refer to the appearance of a beacon target on a ppi when the aircraft pilot engages the transponder IDENT switch, or may refer to the special-position-identifier (SPI) pulse which occurs 4.35 µs after the beacon code train, also by engagement of the aircraft IDENT switch. More on pp 97, 101, 110 through 113, 121, 122, 123, 125. Synonomous with "special position identifier" and "SPI".
IFF: Identification Friend or Foe, the original military designation for secondary radar equipment, now called "beacon," and "Air Traffic Control Radar Beacon System." See "secondary radar" and "beacon." See Chapter 7. More on pp 6, 11, 47, 98, 105, 121.
Indicator: Universal. Radar display equipment, such as a "planned position indicator (ppi)." See Chapter 15. More on pp 35, 56, 63, 65, 88, 89, 114, 125, 332, 347, 367, 373, 377, 380, 385 through 389, 394, 396, 399, 405, 412, 416.
Inductosyn: Manufacturer of a new 16,384-acp generator used in several types a primary and secondary FAA radar systems. See "apg".
Interlace (Mode): FAA, military, aircraft industry, international. The process of "time sharing," or alternating, between different mode pair interrogations. See "Mode Pair".

Interlace (Television): Universal. The occurrence of two interleaved fields. A commercial television picture is called a "frame," and that "frame" is made of two "fields," which occur alternately. The lines of one field occur between the lines of the other. See "Frame/Field." See "National Television Systems Code (NTSC)".
IPA: Intermediate Power Amplifier. Usually a high-power driver for a final power amplifier.
IPP: FAA. Interpulse period, a Westinghouse Corporation name for the system interval. More common usage is "PRI," "PRT," or "T_r." Table 4-1, p 37.
I, Q, or I&Q: Universal. The quadrature phase or synchronous detectors in mti receivers, color television receivers, and other equipment. One phase detector, called "I," exhibits a cosine response\ to the phase difference between the received signal and a reference oscillator. The other detector, called "Q," exhibits a sinusoidal response. Also see "quadrature," "rectangular coordinate data," and "orthogonal." More on pp 83, 291, 292, 295, 325, 329, 336, 350, 355, 357, 360.
Interleave: FAA, military. In secondary radar reply code data, the condition where two reply code trains occur within the times of one another, but the code pulse positions are not aligned to cause a "garble." Also see "phantom" and "garble." More on pp 122, 124, 125, 126.
Interrogator: FAA, military. The ATCRBS ground-based equipment that transmits a 1,030-MHz "mode pair" interrogation to initiate 1,090-MHz aircraft transponder "reply code" responses. Although uncommon, the interrogator may be in an aircraft, and the transponder on the ground; this may be encountered outside the continental United States. See Chapter 7. More on pp 59, 97 through 108, 114 through 120, 126 through 129, 134, 136, 351, 373.
IP: Usually for "interface processor." May be hardware, software, or both.
I\O: Input/output.
Isotropic Source: Scientific and engineering. An imaginary, single-point electromagnetic radiation source, with a perfectly spherical pattern. The basis for gain and directivity calculations. More on pp 37, 38, 39.
IU: Interface Unit. Usually preceded by another acronym/abbreviation.

J

"J" (Type Display) (see Figure A-16): US Army Air Corps and US Navy origin. An early-day type of radar display, used particularly in tracking radars. Displays range and echo amplitude, similar to an oscilloscope display, but with a circular video baseline. More on pp 387.

K

Klystron: Universal. Two very different electron tubes, one used in transmitters and the other used in receivers. See "klystron drift tube" and "reflex klystron." More on pp 5, 10, 16, 53, 66, 68 through 71, 79, 154, 167, 173, 195, 202, 207 through 214, 223, 224, 236, 237, 249, 256, 314, 416.
Klystron Drift Tube: Often misused to describe the entire klystron power amplifier, the "drift tube" is really only a part of a larger, entire electron-tube assembly. The entire tube is a high-power, final power amplifier, employed in synthesis-class transmitters. Not an oscillator, as is a magnetron, the klystron amplifies a burst of rf drive. An electron beam "drifting" toward a collector is velocity-modulated by the drive; the electrons in the beam are then further "bunched," as tuned cavities are excited by the electron movement toward the collector. See "klystron".
KVM: Keyboard-Video-Mouse. Usually associated with interface to a graphic user interface (GUI) or remote monitoring (RMS) computer.

L

LA/CA/MCI: An ARTS term to describe safety systems, Low Altitude (LA), Conflict Alert (CA) and Mode C Intruder (MCI). LA is also called Minimum Safe Altitude Warning (MSAW).
LAN: Local Area Network.
LDPA: Low Duty Power Amplifier. Used in ATCBI-6 to describe a transmitter amplifier used for the P2 and P5 1,030-MHz bursts in side-lobe suppression. Also see "HPA".

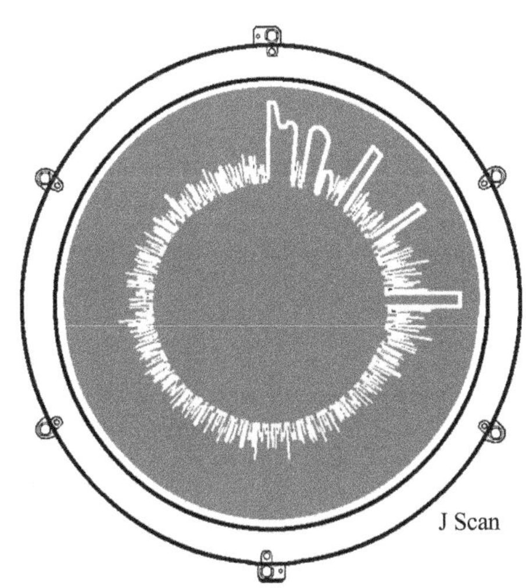

FIGURE A–16

Type "J" display.

Lead Edge (Azimuth Sliding Window): FAA, military. The signal that is generated when the window has filled with sufficient echo "hits" to equal the established criteria. Also see "trail edge" and "run length." See "sliding window".

Line Driver and/or Line Compensator: FAA, USAF, others. The line driver and compensator are transmitting and receiving equipment units for analog radar information, such as normal, mti, and beacon video; and for triggers and azimuth data. Most FAA terminal facilities, prior to the installation of the ASR-9's, relied on long coaxial cables to route information from the radar site to the user facility. The cable length, the high-frequency content of the video pulses, and the potential for noise addition to the data, resulted in attenuation and degradation of the signals. The line driver and compensator operate interdependently to overcome these degrading effects. More on pp 64, 89, 230, 336.

Line-Of-Sight (LOS): US Navy and others. The direction that a radar beam travels in unless effected by subrefraction, super-refraction, or ducting. Potential of confusion with "LOS" for "loss-of-signal" in microwave link or other equipment.

Live Time: Immediately after the radar transmitter has fired, the system must enter a sustained "listening time," until sufficient time has elapsed for receipt of all received echoes from within the intended range. This listening time is called "live time," and the time from the end of live time until the next transmitter burst is called "dead time." If, for instance, the system interval, T_r, were 1,000 μs, and the radar maximum intended range were 50 nautical miles, the live time would be 617.76 μs, and the dead time would be 382.24 μs. More on pp 33, 34, 36, 90, 177, 178, 181, 325, 399, 415, 424, 425.

Lock Pulse: USAF, USN, FAA, others. In mti radar systems, a burst of i-f energy representing the relative phases of a magnetron transmitter and stalo. Used to lock the phase of the coherent oscillator at T_0 to establish mti coherence. Also used to operate afc circuitry. More on pp 56, 57, 58, 64, 256, 257, 258, 300, 301, 302, 312, 313, 314.

Lock Test Pulses: USAF, USN, FAA, others. A long train of regularly spaced i-f bursts, injected into an mti i-f amplifier, for the purpose of testing mti performance, in circuitry that precedes the canceler. More on pp 303, 313, 314.

Log I-f Amplifier: Universal. An i-f amplifier with a logarithmic gain characteristic; the gain decreases as the signal input increases. Used in spectrum analyzers, for weather-level detection, and in conjunction with log ftc circuits. Useful for those and other purposes wherever necessary to represent signal strength without limiting. More on pp 64, 80, 81, 86, 87, 260, 261.

Log ftc: USAF, USN, FAA, others. ftc is an acronym for "fast time constant." The original ftc circuits were r-c differentiating networks; ftc circuits are intended to reduce long-duration weather or chaff echo blocks. A log ftc circuit exhibits a similar effect as the differentiating network, but has a better potential to recover targets in that weather. A log ftc circuit employs logarithmic information, since that may represent signal strength in deciBel-type scaling, and is less susceptible to limiting. The logarithmic data is averaged over a range "window," and the average is subtracted from a central "area of inspection." Log ftc is accomplished with both analog and digital techniques. Also see "CFAR," "mean level threshold," and "sliding window."

LRU: Line Replaceable Unit, Lowest Replaceable Unit, Lowest Reference (designator) Unit.

LVA: Large Vertical Aperture. A 7-foot tall, 32-column dipole array used for monopulse beacon systems. More on pp 116, 128, 129, 133.

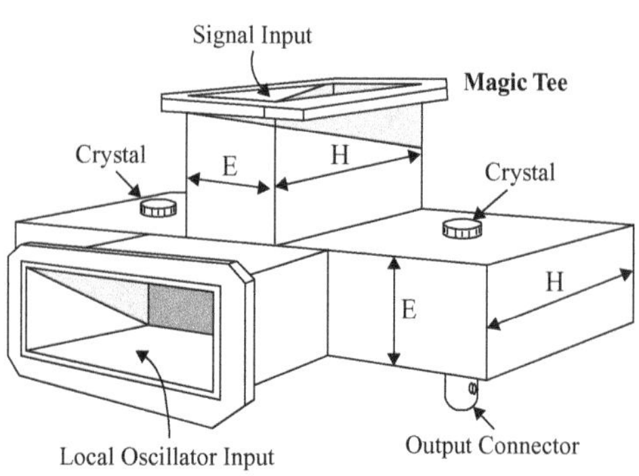

FIGURE A–17

Magic tee.

M

Machine Language: The actual codes used for direct instruction to a computer processor, and the end objective of any software of a higher, more programmer-friendly level. More on pp 186.

Magic Tee (see Figure A-17): Universal microwave technology use. A special waveguide t-joint in both the e-field and h-field planes. Used in receiver double-balanced mixers as part of a local oscillator noise-cancellation scheme. More on pp 78,161, 254, 255.

Magnetic Modulator: A radar pulsed modulator, such as the type in the FAA ASR-8, using magnetic saturable-transformer and choke design for current steering. More on pp 69, 236.

Magnetron: A high-power microwave transmitter tube invented by British Birmingham University radar team scientists, Randel and Boot, when Great Britain was at war with nazi Germany, and before the United States had entered the war. The tube was secretly taken to the United States by a British radar team in 1940, and precipitated the development of microwave radar in the United States. More on pp 1, 4, 6, 10, 11, 34, 36, 53 through 64, 76, 79, 83, 85, 143, 169, 169, 170, 173, 177, 180, 183, 195, 207, 214 through 227, 230, 256, 257, 280 through 303, 312 through 316.

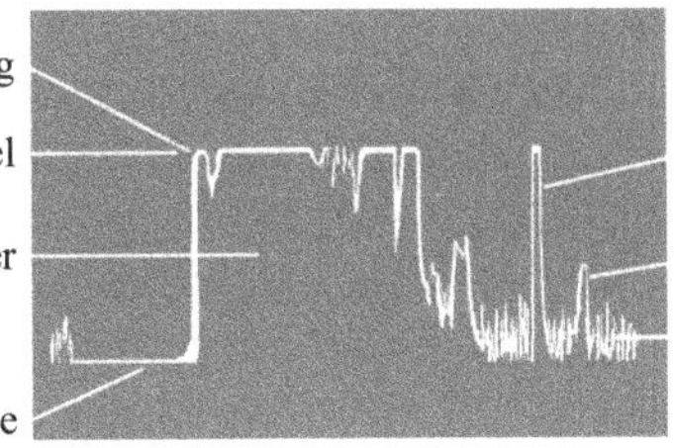

FIGURE A–18

Normal video and terms.

Main Bang (see Figure A-18): US Army Air Corps and US Navy. Early-day radar slang, depicting the transmitter pulse displayed on radar video. The pulse may be present because of a high-power echo from the antenna reflector, T-R tube leakage, or both. More on pp 34, 35, 76, 114, 173, 190, 224, 389.

Manchester Encoding: Phase-shift encoding of a carrier to describe serial digital data bits. Each time the carrier phase is shifted, a digital "one" is represented. For instance, shift-shift-no shift-shift describes 1101. Used in the P6 interrogation burst in mode S. More on pp 108.

Map: General. May have several meanings. A "video map" for a ppi display may be created by a "flying-spot scanner," employing a crt, photographic negative, and photosensitive device, such as a photomultiplier tube. A video map for display may also be contained in a memory. Other maps may also be contained in a memory, where each location contains data regarding a specific range-azimuth cell; the size of the cell determines the "granularity" of the map. In computer discussion, a "memory map" may also refer to a drawing of memory use allocations. For the video mapper, see pages 15-23 through 15-25. Several memory maps are described in Chapter 14. More on pp 80, 137, 291, 351, 352, 355, 361, 362, 365, 366, 367, 369 through 377, 416, 417 through 419, 424, 425.

Master Trigger Blocking Oscillator (MTBO): See "blocking oscillator."

MAU: Multiple meanings. Media adapter unit, maintenance adapter unit, megabucket assembly unit.

MBTS: Monopulse Beacon Test Set. See Chapter 7. See "Beacon test set" and "Monopulse"

mds: Universal. Minimum discernible signal; the lowest level of the test signal visible on a ppi or an oscilloscope, conventionally measured in dBm. A test rf burst of a duration equal to the transmitter burst is injected into the incident port of the directional coupler intended for test purposes. The result for a tangential mds (twice noise level) should nearly equal -174 dBm + $10(\log \Delta f) + F_{dB}$. ($\Delta f$ is overall receiver bandwidth, and F_{dB} is the overall receiver noise figure in decibels.) For an mds where the signal is attenuated into the grass until barely visible, use -171 dBm. See pages 4-9 through 4-12, and page 11-2. More on 41, 72, 75, 77, 80, 119, 174, 246, 247, 248, 250, 260, 264, 305, 313, 360.

MDS Test Set (see Figure A-19): FAA and military. Slang for the pulsed rf signal generator used at the incident port of the directional coupler, to inject a receiver test signal. See "signal generator."

Microwave: Universal. A broad term originated during World War II and describes all those higher frequencies which require waveguide-and-cavity technology. Not clearly or officially defined, microwave frequencies begin at approximately 1 GHz, the lower end of the L band; however, waveguides and cavities have been used at frequencies as low as 400 MHz. When this was first published in 1996, the author had knowledge of commercial work in the 55-GHz region. More on pp 1, 4, 5, 6, 11, 25, 28, 34, 36, 37, 53, 56, 60, 63, 69, 73, 75, 77, 78, 79, 89, 118, 138, 143, 144, 161, 163, 166 through 171, 207, 213, 214, 216, 248, 249, 251, 254, 255, 256, 278, 279, 298, 385.

Mean Level Threshold: FAA and military. A range-sliding-window circuit used in mtd systems to develop an automatic adaptive threshold for each filter data; resembles digital log ftc. Thresholds vary with noise or weather. Also see "FTC," "CFAR", "Log FTC," and "sliding window."

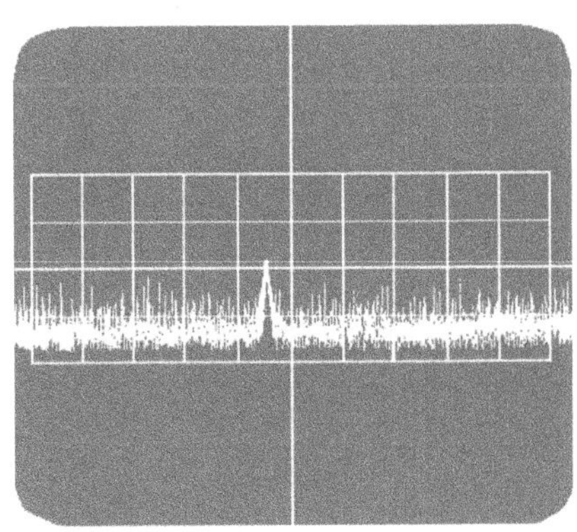

FIGURE A–19

Test signal slightly greater than mds.

Message: Universal. A group of digital words that could represent any number of things, but most frequently understood in radar data communications to define a target, complete with beacon mode and code, center of range and azimuth, primary radar reinforcement, and more. Also see "broad band" and "narrow band." More on pp 80, 89, 90, 101, 108, 109, 113, 114, 116, 123, 124, 125, 132, 135, 347, 351 through 354, 360, 361, 362, 365 through 374, 379, 380, 390, 391, 408, 415, 420, 422, 423, 425.

Minimum Range: Multiple usage, but with variances. In FAA systems, the minimum range is the t_p plus the T-R recovery time. Some confusion may exist in certain US Navy equipments, as "ATR" may be used in place of "T-R." The US Navy may define minimum range as equal to the pulse width PW, plus the ATR recovery time, plus the time it make take to reset or apply phase commands (agile-beam systems) to the antenna, if applicable (STP). Finally, the formula would be the distance traveled by the speed of light in the sum time of PW + ATR + STP. Closely related to range resolution. More on pp 59, 113,115, 338, 413.

MIP: More than one meaning. Message Interface Processor, Mode Interlace Pattern, Mode Interlace Program. More on pp 109, 422.

Mode 2: Military beacon interrogate mode. Three-microsecond P1–P3 spacing. See "Mode Pair".

Mode 3/A: Military mode 3/Civilian mode A beacon interrogate mode. Eight-microsecond P1–P3 spacing. See "Mode Pair".

Mode 4: Military challenge for potential enemy or other unidentified aircraft. Uses special codes which are changed frequently. See "Mode Pair".

Mode B: Used to interrogate a PARROT for APM (Monopulse Antenna Pattern Measurement). Seventeen-microsecond P1–P3 spacing. See "Mode Pair".

Mode C: Interrogates the altitude-reporting aircraft transponder. Nineteen-microsecond P1–P3 spacing. See "Mode Pair".

Mode D: Currently not used in the United States. Twenty-five-microsecond P1-to-P3 spacing. See "Mode Pair".

Mode Pair (see Figure A-20): Military, FAA, aircraft industry. Describes the Air Traffic Control Radar Beacon System (ATCRBS) interrogation modes 1, 2, 3/A, B, C, and D. The two 1,030-MHz bursts radiated by a secondary radar directional antenna. The first of the two bursts is called "P1," and the second is called "P3." The P3 burst is referenced to the secondary radar range zero, and the position of the P1 burst in advance of the P3 determines the "interrogation mode." A third, "P2," burst is radiated from an omnidirectional antenna, 2 μs after the P1 directional burst, for side-lobe suppression by the transponder. Other modes 4 and S are not defined as ATCRBS modes, and differ significantly. More on pp 75, 98 through 105, 113 through 120, 125, 127, 134, 425.

Modulator: Universal. As in any other radio-frequency transmitter, the modulator provides an intelligence signal to the final power amplifier. In a magnetron transmitter, the magnetron oscillates when excited by a high-voltage modulating pulse. In a synthesis transmitter, both the high-voltage pulse and an rf drive pulse must be applied to the final power amplifier. Radar transmitters are either pulse amplitude, pulse coded, or frequency modulated, and emit a spectrum of frequencies commensurate with the Fourier content of the repetitive pulse modulation. The spectrum becomes more complex in agile-beam, pulse-code-modulated (PCM) and CHIRPed radars. More on pp 33 through 36, 54, 58, 65 through 70, 79, 87, 103, 107, 179, 195, 196, 208, 212, 223 through 236, 255, 256, 257, 281, 286, 301, 316, 413, 415.

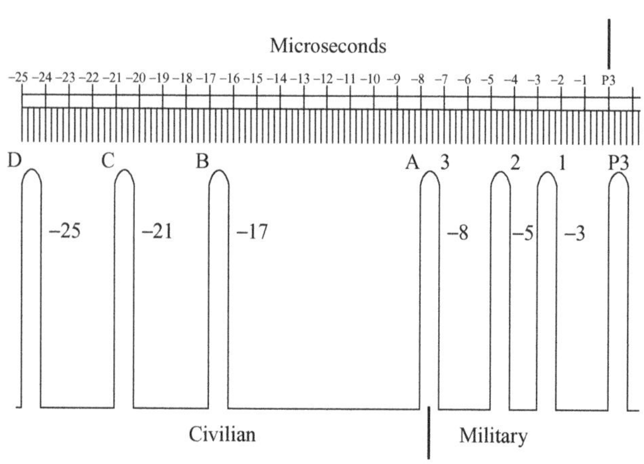

FIGURE A–20

Mode pair spacings. (Modes S and 4 not shown.)

Monopulse: Universal. A radar that can offer accurate azimuth and/or elevation determination from a single echo, in contrast to those which require a scan across the target, to be followed by centroiding. Monopulse techniques have previously been primarily used in fire-control systems, where the antenna must track the target. In FAA beacon Mode S systems, monopulse is now used in preference to azimuth sliding-window techniques. Target positional information may be obtained from rf signal strength in multiple antenna elements, or from Doppler differences in multiple elements. Antennas may be phased arrays, or multiple-feedhorn reflector types. More on pp 64, 73, 89, 90, 98, 102, 105, 106, 109, 110, 113, 114, 126 through 135, 379.

Monostable Multivibrator: See "single shot."

Moving Target Detector (mtd): FAA. A complex mti system which employs a bank of Doppler filters, and both firmware and software processing. This is in contrast to the older, conventional, delay-line, or shift-register, mti cancelers. More on pp 44, 63, 83, 174, 177, 179, 183, 184, 185, 187, 206, 211, 259, 263, 264, 282, 347 through 380.

Moving Target Indicator (mti): US Army Air Corps and MIT Radiation Laboratories origin, first developed during World War II. A radar receiving system and processor which attenuates ground clutter of zero, or near-zero, Doppler. Most moving aircraft, above or within the clutter, remain visible. Whether or not aircraft may be visible is a very complex subject. More on pp 1, 8, 9, 10, 19, 34, 40, 43, 44, 53 through 58, 63, 64, 65, 70, 72, 76 through 90, 173, 174, 177 through 188, 206, 252 through 264, 277 through 319, 325 through 342.

MTBF: Mean Time Between Failures. A reliability figure expressed in averaged time.

mti Memory: US Navy and FAA. Originally a First-In, First-Out (FIFO) memory device, now a RAM in latter-day systems. Stores one, two, three, or more PRIs of bipolar data for use in a summing comparator to determine Doppler shift and attenuate low-velocity clutter data. Using varied PRI lengths reduces the probability of target loss due to blind velocities. May have other, but similar, meanings depending on the equipment manufacturer. More on pp 325 through 342.

Mti Reflector: FAA, USAF, other air traffic control entities. A stationary radar reflector, containing a multivibrator-operated feedhorn assembly, which changes the electrical distance to the radar by $\lambda/4$ at each T_r, to simulate optimum-velocity motion. More on pp 300, 362, 412.

Multiple Conversion: Universal. Any superheterodyne receiving system which utilizes more than one "down conversion" or "signal mixing" and more than one intermediate frequency. More on pp 34.

MSSR: Monopulse Secondary Surveillance Radar. May be ATCRBS, Mode S, or Mode 4. More on pp 64, 102, 105, 106, 107, 113, 116, 117, 127, 128, 134.

N

National Television Systems Committee (NTSC). A standardized method of synchronizing 525-line broadcast color television frames, each made of two 262.5-line interlaced fields, with each line taking 63.5 μs from one horizontal sync pulse to the same point on the next horizontal sync pulse. Replaced Electronics Industries Associates (EIA) sync used for black-and white television by the introduction of the 3.58-MHz, 2.5-μs color burst on the 4.7-μs horizontal "back porch" of the horizontal sync-and-blanking signal. The starting phase of the color burst reverses with horizontal lines, and with frames, to provide acceptable gray-scale presentations on black-and-white televisions. There are two interlaced fields per frame, but four fields are unique because of the color burst phase-reversal pattern. In field 1, the burst phase is reversed; field 2, not reversed; field 3, not reversed; and field 4, reversed (see Figure A-21).

Narrow Band: Universal. Audio-frequency, modem, digital, radar data transfer. See "broad band" for additional information. More on pp 63, 86, 203, 347, 360.

Nautical Mile: Used in navigation; now by international agreement, 6076.11549 feet; 1/60 (minute), of one degree, of the circumference of the earth. Believed to be traceable to The Netherlands sixteenth-century "Golden Age." The speed of a ship was measured by throwing a log off the stern of a ship. The log was tied to a long rope with carefully spaced knots, and the rate at which the knots passed over the stern indicated speed. May be abbreviated in many ways, among them nmi, NMI, nm, NM, naut. m, naut. M, $\text{mile}_{\text{naut}}$, and more. In latter days, NMI or nmi can be confused with "no middle initial." "Mile" and "feet" are traceable to the Roman Empire, but were standardized to 5,280′ per mile and 12″ per foot by the Queen of England. More on pp 114, 123, 180, 352, 414.

NEXRAD: Next Generation Weather Radar. Pulsed Doppler, 200-mile radar for weather velocity measurement.

Noise Figure: A figure to indicate the amount of internal noise that will be contributed to a receiver or receiving system device. The noise figure is an input signal-to-noise (S/N) ratio, divided by the resulting output S/N ratio. Since the input S/N ratio must always be the greater number of the two, the noise figure F must always be greater than 1. Always expressed as "F," so as not to be confused with "*f*," for "frequency." (There is still a minor conflict with the "F" for "force.") The noise figure is directly related to the minimum discernible signal; see "minimum discernible signal" for a mathematical relationship. More on pp 246 through 250.

Nonzero Velocity Filter (NZVF): FAA. In mtd, the range-gated Doppler filters not encompassing zero Doppler in an mtd Doppler filter bank. More on pp 352, 361, 365, 369, 370, 372.

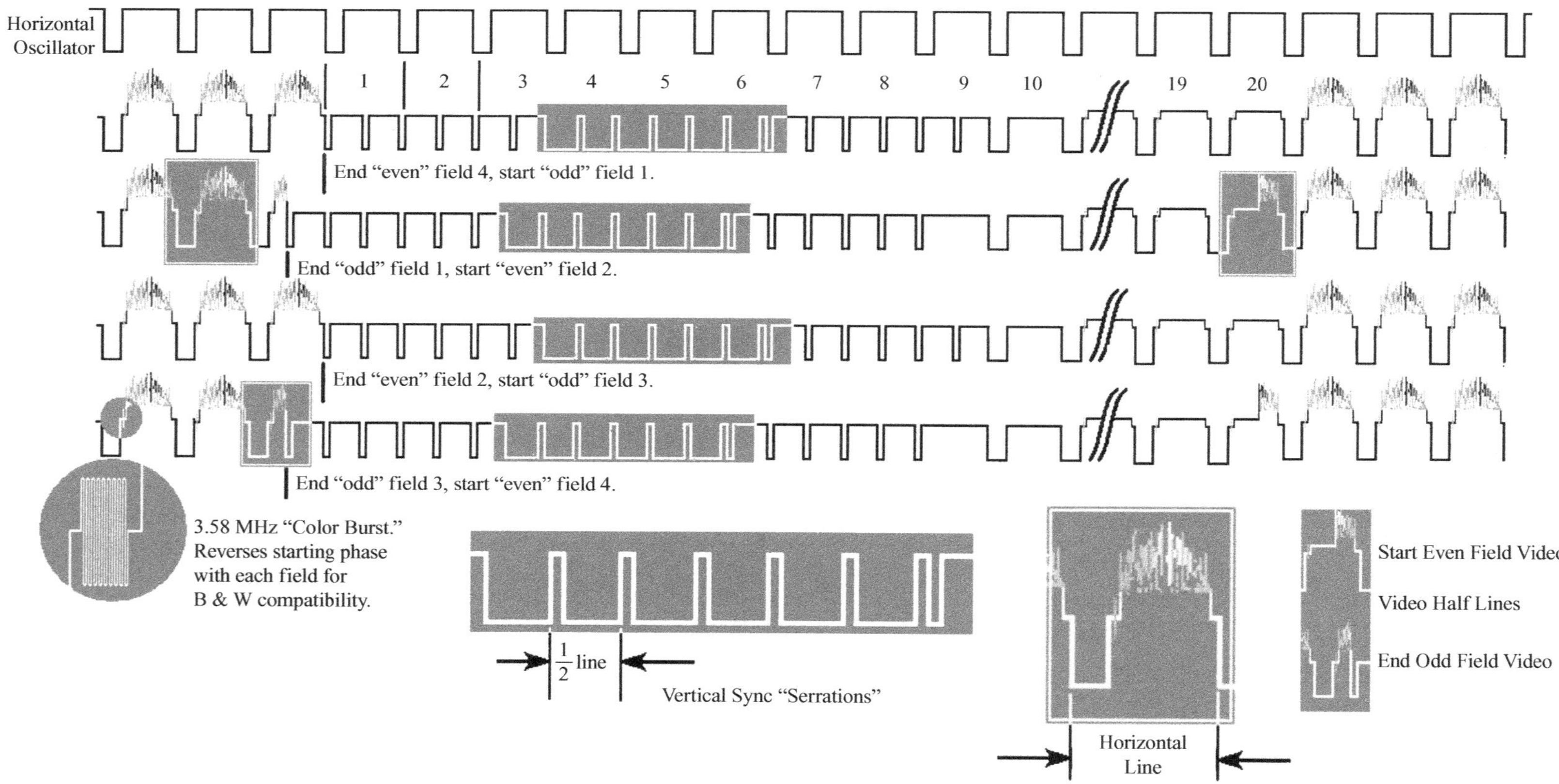

FIGURE A–21

NTSC commercial television standard.

Normal or Normal Video: FAA and Military. Amplitude-detection radar video of the earliest, simplest form (see Figure A-22). Assigned the name "normal" when mti systems were first built, so as to distinguish it from the mti receiver video. More on pp 8, 34, 35, 40, 56, 58, 64, 75 through 90, 119, 168, 173, 177, 259 through 264, 268, 281, 298, 301, 305, 310 through 313, 337, 347, 350, 361, 370, 371, 372, 377, 385, 387, 392 through 395, 400, 401, 410, 411, 419.

North Mark or Crossing: FAA. Often incorrectly used to describe the azimuth reference pulse (arp), the north mark was used with radar systems remoted by a radar microwave link (RML) to cause a line to appear at north on the user facility ppi's; the north mark was generated at the radar site, and mixed with mti video. The purpose was to provide user facilities with a constant assurance that the azimuth synchronization had not been lost. See "arp" and "apg." More on pp 73, 412.

NUNIO: (Network Universal Input/Output) Communication Link

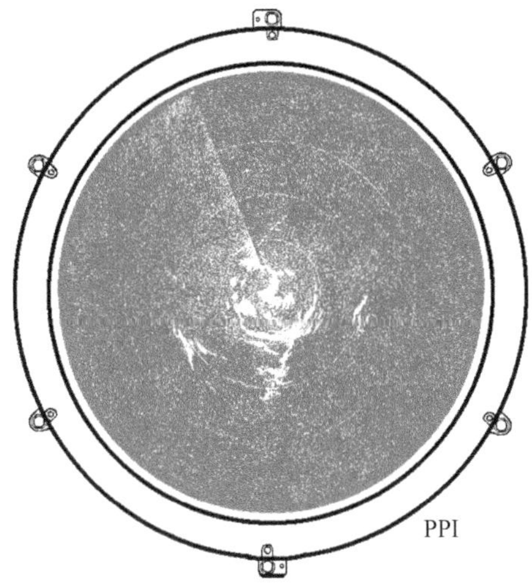

FIGURE A–22

Normal video on a ppi.

O

OBA: Off boresight angle. "Boresight" is the center azimuth of the directional radiation pattern of a monopulse antenna. The OBA provides a correction to beacon reply information received from aircraft not in the center of the pattern. Although the antenna moves during the mode interlace pattern, all replies from a single aircraft show the same azimuth because of an OBA correction to the azimuth change pulse count. More on pp 106, 107, 133, 135.

ODU: Operational Display Unit. Used in the FAA ASDE equipment, but likely not limited to that equipment. More on pp 12.

Operating System: Universal. The "platform" language used computer systems to provide for interface with the application software, input/output devices, and machine code. MSDOS, Microsoft Windows, OS2WARP, LYNX, UNIX, and several others are operating systems. Now a part of the radar vocabulary because of graphic user interfaces (GUI). More on pp 186.

Optimum Bandwidth: Scientific and engineering. A radar receiver bandwidth to offer optimum compromise between minimum noise and acceptable reproduction of the Fourier harmonic content of the received echo. Equal to $1.2/t_p$. Radar receivers are not necessarily tuned to the optimum bandwidth, as there are often other considerations. More on pp 40, 41, 59, 119.

Optimum Phase: FAA and military. In mti discussion, the pulse-to-pulse sampling of Doppler shift may be considered in terms of the angular change of a Doppler cycle in degrees ($\Delta\phi$). Optimum phase occurs when both the $\Delta\phi$ and single echo phases of a target produce a maximum possible T_r-toT_r difference at the canceler comparator. More on pp 289, 291, 303, 311, 330, 332, 333.

Optimum Velocity: FAA and military. A radial velocity that produces a Doppler shift equal to one-half the f_p, consequently producing the maximum possible output from a single mti canceler. Optimum velocity for most ASRs and ARSRs is roughly 50 nmi/h. More on pp 289, 290, 291, 298, 361, 377.

Orthogonal: Universal. Physical or electrical right-angle orientation of two components, elements or signals. Orthogonal orientations create x and y rectangular coordinates with cosine and sine relationships, so the trigonometric identity $\sin^2\theta + \cos^2\theta = 1$. Applied, for instance, in creating N-S and E-W fixed-coil deflection ppi sweep rotation, in the separation of radiated electromagnetic energy into horizontal and vertical polarizations in circular polarizers, and in the 90° phase-shifting applied to the reference oscillator input of the "Q" detector in an I-Q quadrature phase-detector circuit. Also see "quadrature."

Orthogonality: FAA and military. Predominately used in video mapping equipment, military and FAA. An adjust ment to the 90° sweep relationships in a ppi or video map unit. See "orthogonal."

P

Passband: Scientific and engineering, British, formal. Synonymous with "bandpass" and "bandwidth." See "bandpass".

Passive: Any electronic device must be either "active" or "passive." An "active" circuit component requires external power, and a "passive" device passes a signal without application of external power. In radar, somewhat in contradiction to the general rule, a "passive" receiver is one which receives inputs from an antenna element or feedhorn, other than that used by the transmitter. Also see "active." More on pp 43, 45, 75, 77, 297, 298, 312, 371.

PA: Power Amplifier. Usually a transmitter final stage. Also called "FPA" for "final power amplifier." Also see, FPA, LDPA, HPA.

PAL: Programmable Array Logic

PARROT: Position Adjustable Range and Reference Orientation Transponder. The acronym may have preceded its definition. A remote stationary beacon transponder used for confidence and performance testing. Also called "Site Monitor." More on pp 127. Also see, "Permanent echo."

Payoff Controls: US Army Air Corps, CAA, USAF, FAA. Early slang from the first use of mti, the payoff controls are an mti i-f gain adjustment, and a gain adjustment to the canceler output video; the name stems from the absolute importance of their final adjustment. More on pp 81, 262, 305, 308, 309, 310, 312.

Permanent Echo: FAA and military. A stationary radar target, such as a building, water tower, or mti reflector, used to verify radar alignment. More on pp 412.

PC: Printed Circuit or Personal Computer. Printed circuit cards may also be called "PCCA" for "printed circuit card assembly," or CCA for "circuit card assembly."

PEC: Printed Electronic Circuit (Raytheon).

PFN: See "Pulse Forming Network (pfn)." More on pp 64, 66, 68, 224, 226, 227, 233, 235, 236.

Phantastron: A tube-type circuit used to produce both long, stable gates and a trapezoidal waveform. The linear trapezoid is sometimes used in sweep generation. Original phantastron circuits used a five-grid, seven-element electron tube. The interaction between the grids with different voltages, circuit capacitance, and feedback create a series of events that begin only with an external trigger. The phantastron has been replaced by digital counters that address a D/A converter. More on pp 178, 397, 398, 399.

Phantom: FAA and military. A secondary radar reply-code condition, where two codes at close azimuths are at such range proximity that there is a 20.3-μs separation between a pulse in one code train, and a pulse in the other, causing a false "bracket detection." Also see "garble" and "interleave." More on pp 122, 124, 125, 126, 128.

Phased Array: Formerly military, and generally associated with three-dimensional radars. The technology has evolved to a great state of sophistication from the earliest pre-WWII fixed-phase dipole arrays. One of the earliest, if not the first, variable phased arrays was the "squeezing waveguide" developed for precision "landing radars" early in WWII. Now, a radar antenna system in which the beam is directed by electronically adjusting the phase of the rf applied to the individual elements, normally accomplished through phase delay. The types of phased arrays are active, in which the individual radiating elements are controlled independently, and passive, where the delay is applied to the rf, prior to application to the array. The definition originally included frequency-scanned antennas, but usage has now evolved to apply predominately to those antennas using digital phase-shifting devices. Passive, fixed nonvariable phased arrays are used in many applications. See "Three Dimensional (3D) Radar." More on pp 7, 10, 11, 38, 47, 48, 53, 129, 130, 131, 153, 195, 223, 252.

Phase Detector: Universal. A circuit which compares two input signals to produce an output which is, instantaneously, a voltage representative of the cosine of the phase difference. Over time, the output is a frequency, representative of the difference frequency, between the two inputs. Used in color television to determine color and hue (depth of color). Used as the second detector in a phase-coherent system, as in mti and mtd. Sometimes called "synchronous detector" because of the coherent relationship of the two inputs. More on pp 43, 55, 56, 58, 65, 76, 78, 81, 82, 83, 106, 133, 134, 135, 174, 179, 180, 181, 207, 252, 253, 258, 262 through 265, 281 through 289, 291 through 295, 299 through 315, 325, 332, 336, 348, 351, 353, 358.

Phase Detector Ambiguity: Scientific and engineering. The cosinusoidal response of the phase detector yields two equal voltages for two angles of equal cosine values. In mti radar, where a single phase detector is used, this causes a "blind phase" condition. More on pp 81, 253, 291, 299.

Phase Lock Loop (PLL) (see Figure A-23): Universal. Long used in many forms as a means to synchronize oscillators to a reference signal. The horizontal sweep circuitry in a television set may use a phase lock loop to synchronize the horizontal lines to those in the transmitted signal. The horizontal sync signal may be used as the sample gate, the horizontal sweep the ramp, and the horizontal oscillator the voltage-controlled oscillator. Digitally controlled phase lock loops have gained popularity with advent of integrated circuits. Where the reference oscillator is a high-frequency crystal, the division of its frequency provides excellent stability.

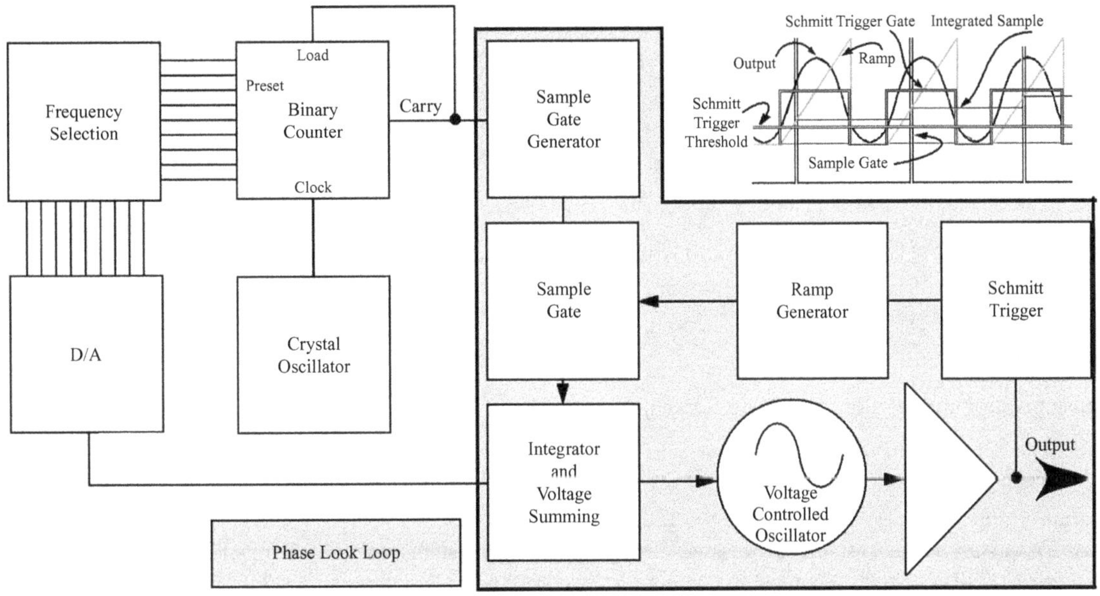

FIGURE A-23

Generic phase lock loop in a frequency generator.

Pixel: The smallest element of a television picture. In color screens, a pixel is made of three dots: blue, green, and red. Very slight positioning of the electron beam causes a mixture of the three colors to display the appropriate one. An equal mixture of all produces white. The minor changes in *X* and *Y* beam deflections are derived from two quadrature phase detectors, each comparing the phase-modulated 3.58-MHz received color signal to a 3.58-MHz color oscillator. The 3.58-MHz color oscillator is locked to the correct phase at the beginning of each horizontal line, by a 3.58-MHz "color burst" on the "back porch" of the horizontal sync-blanking signal.

Pincushion: Television industry. Probably from color television test equipment used to generate a display for the deflection alignment of the three colors. Linear time relationships must appear as linear optical relationships on the face of a cathode ray tube.

Planned Position Indicator (ppi) (see figure A-24): Universal. Originally, a military ***type "P"*** display. Sometimes formally called a ***rho–theta*** ($\rho\theta$) display, where ρ is the radius of a circle, and the range of the radar; θ is the bearing angle. Has also been called a ***radial*** display, but that usage is rare. A circular radar display of range, from the center of the display to the outer edge, and of bearing, measured from the 12 o'clock position, clockwise through 360°, back to the 12 o'clock position. Displays the echoes or compressed video (CHIRP systems) received from the area surrounding a radar site, which is located at the center. The 12 o'clock position is generally north for ground radars; it may be "dead ahead" for ships and aircraft. May be a real-time display, where a sweep, with an origin at the center of the display, maintains an angle corresponding to the pointing direction of the antenna and rotates in synchronization with the antenna, as a hand of a clock. May also be a target-message-display ***random-access ppi (rappi),*** or a synthetic real-time display. See Chapter 15. More on pp 8, 35, 36, 45,72, 79, 81, 84, 85, 86, 97, 98, 100, 120, 125, 184, 190, 246, 250, 251, 255, 258, 264, 291, 303, 304, 308, 309, 310, 347, 350, 378, 385, 386, 389, 390, through 396, 400, 405, 410 through 413, 415, 416, 417, 421.

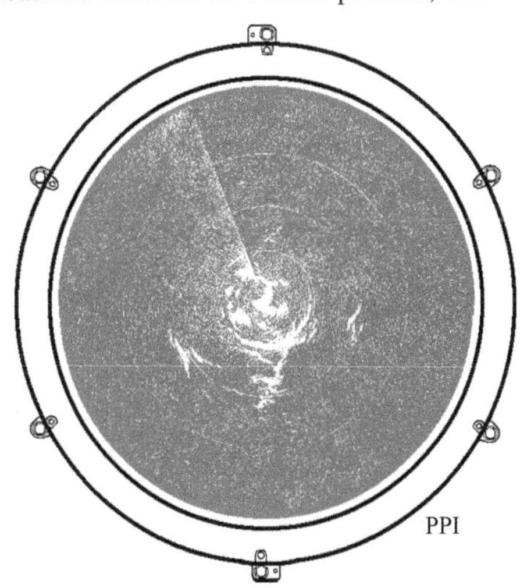

FIGURE A-24

ppi display.

Polar Coordinates: Expression of the length (rho) and angle (theta) of a vector. Also see "rectangular coordinates."

PPM: Pulse Position Modulation (see Figure A-25). Subject to confusion with pulses per minute. The Mode S transponder reply

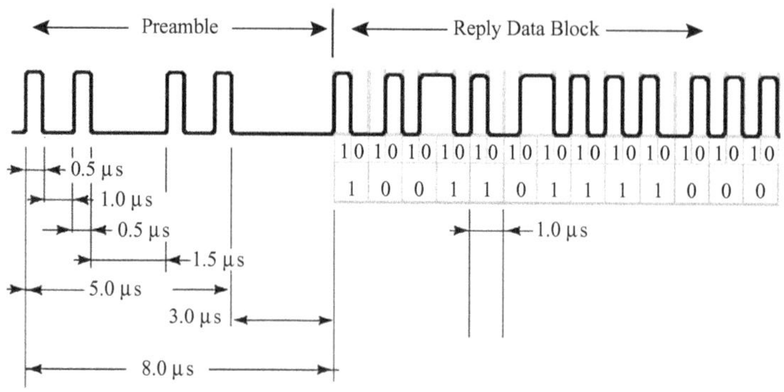

FIGURE A–25

Pulse position modulation in a Mode S transponder reply.

uses PPM. Data pulses are assigned to a specific code "window" in the data train. The pulse position in the window describes a "0" or a "1." See Chapter 7. More on pp 112.

Polarization: Universal. Refers to the orientation of the "E" lines in the transmitted rf energy, with respect to the earth's surface; may only be horizontal or vertical. "Circular" polarization utilizes both orientations, and the energy is power-shared between the two orientations in a sine–cosine quadrature manner. Returns from circular targets are of equal amplitudes in both orientations, and that characteristic permits rf cancellation in the polarizer assembly. See "Circular Polarization." More on pp 74, 75, 130, 131, 208, 220, 371, 404, 406.

Post Processor: FAA. The final processing element in an mtd system, receiving inputs from the digital signal processor (DSP). Includes a computer, containing the correlation-and-interpolation (C&I), surveillance processor (SP), and beacon target detector (BTD) programs. The ASR-9 postprocessor also contains a beacon data acquisition system called "beacon reply processor (BRP)," and message-formatting and data-exchange hardware. See Chapter 14. More on pp 351.

Precision Approach Radar (PAR): US Army Air Corps, CAA, USAF, FAA, US Navy, other military. A short-pulse, short-range, X-band, radar system, scanning a small segment of azimuth and elevation, to provide for verbal air traffic controller radio guidance to pilots on the final approach to the runway. More on pp 7, 8, 9, 60, 63, 97, 129, 130, 131, 135, 388, 389.

Primary Radar: FAA. A true pulsed radar which relies upon echoes. In contrast, a "secondary" radar is an "answer-back" system; the transmitter is an "interrogator," and a "transponder" generates "replies" for the receiver. More on pp 47, 64, 97 through 101, 103 through 107, 109, 112 through 120, 123, 125 through 128, 131, 132, 134, 135, 347, 371, 373, 379, 380, 422, 424, 425.

Primitive (Targets): Scientific and engineering, FAA. (mtd) Those target detections from the Digital Signal Processor (DSP). A single target detection from any Doppler filter in a batch range cell at any range. There may be many primitive detections from a single target providing echoes. See Chapter 14. More on pp 101, 351, 352, 354, 362, 368, 369, 371, 372, 375.

PSR: Primary Surveillance Radar. Also called "search," "surveillance," "primary," and simply "radar." More on pp 98.

Pulse: Universal, scientific, physical science. A rapid change in voltage, or physical force, during a much longer static period, for example, a 1-μs voltage of different value during a 1,000-μs static period. Often incorrectly used to describe a "burst," which is a brief occurrence or emission of rf energy during a longer static period; the misuse may create confusion.

Pulse Forming Network (pfn): Universal. The high-voltage, energy-storage component in a radar modulator. The pfn is charged by the high-voltage power supply during most of the system interval. It is then rapidly discharged through the pulse transformer and thyratron, SCR, or other type of switch, at transmit time, to provide the high-voltage pulse for the magnetron or final power amplifier. See Chapter 10. More on pp 64, 66, 68, 224, 226, 227, 233, 235, 236.

Pulse Compression: Military. A method of distributing lower peak transmitter power over a longer pulse duration and then restoring the received echo to a narrow, high-range-resolution detected pulse by use of the frequency/phase/coding contents of the transmitter pulse and rf echo. Permits use of lesser transmitter voltages, and more economical transmitting devices, while still maintaining the peak received pulse power and range resolution of a higher voltage, narrow-pulse transmitter. Often called "CHIRP," an acronym for "compressed high-resolution pulse." Other uses and advantages are obtained by the pulse coding. Also see "compressed video" and "compression." More on pp 11, 45, 46, 47, 48, 43. Also see, "Video time compression."

Pulsed Doppler: General. A radar system designed to provide range, azimuth, and velocity measurements, as for weather detection or monopulse tracking purposes. Differs from mti radar in major classification, because the repetition frequency is necessarily higher to minimize Doppler ambiguity; a velocity which causes a Doppler shift greater than half the repetition frequency is ambiguous. See also "Range Ambiguity." More on pp 63, 286, 287, 315.

Pulse Recurrence Time (PRT): Military. See "T_r." More on p 37, table 4-1.

Pulse Repetition Frequency (PRF): Military. See "f_r" and "*f*p." Reciprocal of PRT.

Pulse Repetition Interval (PRI): US Navy and other military. The time measured from the leading edge of the transmit pulse to the leading edge of the next transmit pulse. A "dwell" (US Navy) can encompass several PRIs. See "dwell." More on p 37, table 4-1.

Pulse Transformer: Universal. Usually assumed to refer to that transformer in a radar modulator, which utilizes the pulse-forming network discharge current, to produce the high-voltage pulse for the magnetron, klystron, amplitron, or other final-power transmitting tube. See Chapter 10. More on pp 66 through 70, 208, 224, 226, 227, 231, 233, 234, 235, 236.

Q

Quadradar: FAA, military, and worldwide air traffic control entities. An air traffic control radar, widely used throughout the world, because of its versatility and low cost. Characterized by two reflector-type antennas, which scan in the azimuth and elevation planes. The name is derived from its multiple capability as an ASDE, ASR, height finder, or PAR. Designed and manufactured by ITT Gilfillan, once among the most popular overseas civilian air traffic control radars. Wide usage by rapid-deployment military entities. Variations in the general principle have been used as carrier approach control and military tactical radar. Still in use. More on pp 8, 9, 10.

Quadrature: Universal. A 90° phase relationship between two equal frequencies. Quadrature relationships are frequently created and employed for the purpose of utilizing the trigonometric identity, $\sin^2\theta + \cos^2\theta = 1$. The use of quadrature phase detectors, or quadrature synchronous detectors, provides rectangular coordinate data, representing magnitude and/or phase angle in color television, mti, and mtd systems. In radar circular polarization, the transmitter rf is divided into horizontal and vertical polarizations in sin–cos power relationships. Also see "orthogonal." More on pp 81 through 84, 253, 262, 263, 264, 291 through 295, 325, 326, 329, 335, 336, 337, 340, 350, 351, 355, 404.

Quantize: Universal. Has two somewhat different definitions, depending mostly upon the age of the equipment or the literature in which the word appears. First referred to an amplitude standardization of radar or beacon-code pulses, where a Schmitt-trigger circuit or overdriven amplifier forced all pulses above a threshold level to a maximum amplitude. Later used in reference to analog-to-digital converters, which generate several parallel binary bits, to comprise a "word," occurring once each range cell. More on pp 83, 86, 89, 119, 120, 121, 124, 181, 246, 325 through 330, 332.

Quieting (3 dB Quieting): A receiver sensitivity measurement for frequency-modulated systems. The receiver noise output is first measured in the absence of any input; then, an injected continuous-wave test signal is increased, to decrease the noise power by 3 dB. The test signal level at that point is the 3 dB quieting sensitivity. More on pp 109.

R

RACD: Remote ARTS Color Display (ARTSIIIE DBRITE replacement). Television display equipment used in air traffic control towers (ATCT). Provides ATCT controllers with interfaced use of ARTS equipment. More on pp 135, 390.

Radar Approach Control (RAPCON): A 1950's USAF/CAA/FAA designation for a joint-use radar air traffic control facility where both military and civil ASR controls were accomplished by civilian controllers, but precision approaches to the military bases were performed by military personnel. Most RAPCON ASR radars were AN/CPN-18 located on air bases. More on pp 9.

Radar Bright Display Equipment (RBDE): FAA. Radar equipment which utilized a storage tube, to convert a ppi display into a television presentation (see Figure A-26). The storage tube operated by capitalizing upon a secondary-emission principle. A phosphor-mesh surface was written upon by the ppi sweep and then read when scanned by a television raster. Secondary emissions created television video pulses, where the mesh had been charged by the ppi scan. Often called "scan converters." More on pp 391, 392, 421.

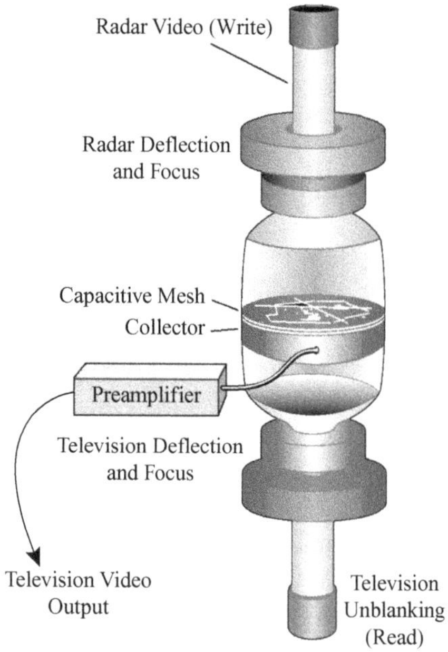

FIGURE A-26

Scan conversion tube.

Radar Cross Section (RCS or A_o): Universal, but acronyms differ. "RCS" is used by the US Navy and some manufacturers. The FAA Academy and other formal academics use "A_o" in received-echo-strength equations. RCS or A_o represents the surface area of a given object reflecting rf energy back to the source. The value is often expressed in square meters, but must be in the same terms as the other entities in the equation in which it is used; one square meter equals 10.7584 square feet, or 2.91×10^{-7} square nautical miles. A one-square-meter target is a scientific standard in defining expected target detection capability. More on pp 41 through 44, 47, 104, 245.

Radar Data Acquisition System (RDAS): FAA and military. In the FAA, this refers specifically to the digitizer in an ARTS system sensor receiver and processor (SRAP). Also see "sensor receiver and processor (SRAP)". The SRAP also contains a beacon data acquisition system (BDAS); see "Beacon Data Acquisition System (BDAS)."

Radar Meter: Military and foreign. Echo time for 1 m; 6.67 ns. See "radar mile." More on pp 33.

Radar Microwave Link (RML): FAA. Microwave receiver–transmitter repeaters to transfer radar information over great distances to a user facility. More on pp 89.

Radar Mile: FAA and military. Semantically incorrect; truly a time, rather than a distance, but, nevertheless, commonplace. The time required for electromagnetic energy to travel one nautical mile, be reflected by an object, and then return to the radar receiving antenna. First rounded to 12.4 μs in slide-rule days, now commonly defined as 12.36 μs, a radar mile is more precisely 12.3552 μs, when based upon the most precise definitions of the speed of light and nautical mile. More common in air traffic control radar applications than others, which use the metric system. A radar meter is 6.67 ns. More on pp 33, 54, 90, 113, 280.

Radial Velocity: Scientific. A resultant vector velocity toward or away from any point, such as a radar antenna. More on pp 59, 278, 279, 280, 285, 286, 288, 297, 320, 370.

Ramp: Universal. A more formal term for "sawtooth" waveform, a signal which slowly rises, or falls, at a linear rate, to a maximum or minimum level, and then abruptly returns to the starting value. Author's personal experiences with British engineers in the 1960s suggest this term is of English origin. More on pp 48, 268, 325 through 328, 335, 398, 399.

Random-Access ppi (RAPPI): FAA and USAF; first known use was in the first common digitizer. A radar display similar to a ppi, in which the radar is located at the center, but which does not employ a rotating sweep, and which does not display analog video. Radar data appears on the display in the form of symbols or alphanumerics; *x–y* positioning is obtained from range-azimuth data in digital data messages, and the crt beam is instantaneously deflected upon receipt and decoding of the message. Because the messages have been produced by digitizers, the display is not in "real time." There are hybrid ppi–rappi displays, in which radar live time is dedicated to real-time ppi display, and dead time is dedicated to RAPPI display (see Figure A-27). More on pp 386, 389 through 391, 408, 414, 415, 422.

Range: Universal. The physical distance from a radar antenna to an object producing an echo. Although semantically incorrect, radar range is synonymous with time-for-echo-for-distance, as 12.3552 μs/nmi, and 6.67 ns/m. See "radar meter" and "radar mile."

Range Ambiguity: US Navy and formal scientific literature. Although applicable to FAA surveillance radars, this term is generally used in reference to pulse-Doppler weather and tracking radar. An "ambiguity," by dictionary definitions, is an unintended misrepresentation due to a third factor. In range ambiguity, that third factor is a T_r that is less than the time-for-echo-for-distance. In FAA surveillance radars, range ambiguity is simply the range error introduced by second-time echoes, and is undesirable. In pulse-Doppler radars, different f_p's must be used, with high rates, to preclude Doppler ambiguities, and which naturally introduce range ambiguities. In these Doppler radars, the system must perform a "range ambiguity check," varying the f_p on a pulse-to-pulse basis, to determine the actual range of a target. More on pp 126, 315, 371.

Range-Azimuth Gate (RAG): FAA and military. A signal used to change some mode of operation of the radar system during a defined "window" in the coverage pattern. Used to switch between beam elevations, mti and normal video, different receiver gain settings, and more. More on pp 77, 316, 365.

Range Cell: FAA, military, scientific, and engineering. In digital radars, or radar data digitizers, the received data is sampled and converted to digital words at a regular rate. The reciprocal of the rate, the period, is called the range cell. When used in the radar itself, the usual range-cell time duration is approximately $3t_p/4$. More on pp 65, 83, 85, 87, 179 through 188, 304, 319, 325 through 328, 330, 337, 338, 339, 348 through 52, 355, through 361, 363, 366, 375, 424, 425.

Range-Ordered: FAA, especially in reference to ASR-9, but universal. The ascending distance order of receivedt real-time echo and noise information. Applies to all real-time radar data. More on pp 355, 356, 425.

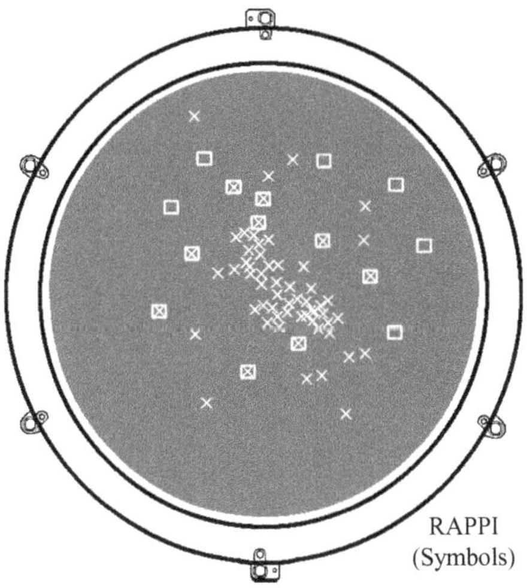

FIGURE A–27

RAPPI symbol display.

Range-Gated Doppler Filters (RGDF): Military, scientific and engineering. See "Doppler filters." See Chapter 14. More on pp 348.

Range Resolution (see Figure A-28): Universal. A figure of merit, placing a value on the minimum range separation of two adjacent targets. Formal literature expresses it as $ct_p/2$. More practically, resolution is equal to the transmitter burst width in radar distance. For example, if $t_p = 1$ µs, then the minimum range resolution $= 1 \times 10^{-6}/2.03 \times 10^{-9}$ feet. Of course, other time-for-echo-for-distance numbers may be used to express results in nautical miles, meters, etc. In CHIRP systems, the compressed pulse width, rather than the transmitter burst width, determines the range resolution. More on pp 5, 46, 59, 195, 413, 414.

Raster: Universal. Applicable to some 3D radar scans, but usually associated with television. A scanning technique in which the scanner moves both horizontally and vertically to completely cover a square, or rectangular, area. A television picture is accomplished by a raster scan with intensity variations. A precision radar may scan a raster-type pattern to detect targets in a rectangular window. Defined in Webster's as "the pattern of illuminated lines formed on a television picture tube when no signal is being received." More on pp 130, 131, 390, 391, 421.

Reactance-Tube Modulator: FAA. A tube which contributes a controllable amount of capacitive reactance to a tuned circuit. Used in "swept-bandpass" afc circuits. More on pp 58, 256, 257.

RDC: Multiple meanings. Radar Data Converter, Radar Display Console, Receiver-Decoder-Correlator, and more.

Real Time: FAA and military. The condition in which radar echoes are displayed or used, without appreciable delay, during the same interval in which they are received. Used in contrast to digital systems, in which azimuth sliding windows are used, to calculate azimuth centroids, for radar data messages, which may occur after considerable delay (as 120° of antenna scan), and which bear no temporal relationship with radar range or transmitter pulse time. More on pp 35, 47, 90, 101, 113, 114, 123, 124, 135, 184, 187, 188, 260, 264, 266, 347, 350, 351, 355, 356, 361, 368, 371, 378, 387 through 394. 408 through 411, 414, 417, 421 through 425.

Real Time Quality Control (RTQC): FAA and USAF. Synthetic target data injected into a radar system, or digitizer, to verify the digital and/or signal-processing performance. More on pp 361, 368.

Recovery Time: Universal. The measured time duration between the beginning (leading edge) of the transmitter burst and the point at which the receiver sensitivity is degraded precisely 3 dB from maximum. The degradation is caused by a delay in deionization of the T-R device. Some manufacturers of US Navy equipment may call the device an "ATR," which may lead to confusion with the AT-R tube of the 1940s and 1950s. More on pp 75, 76, 170, 174.

Rectangular Coordinates: Academic. Expression of the length of the *X* and *Y* sides of a right triangle of a vector. Also see "polar coordinates."

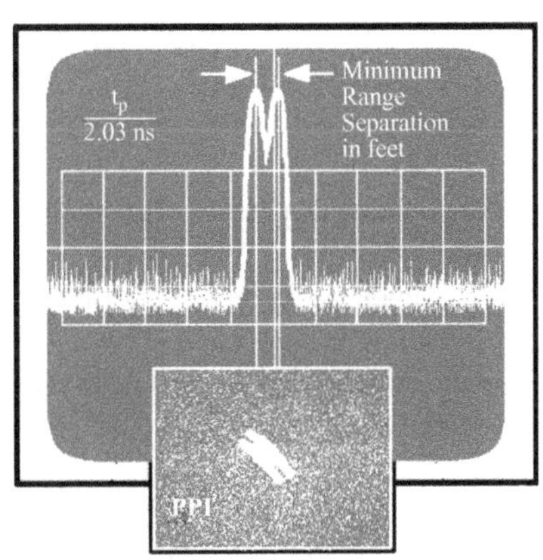

FIGURE A–28

Range resolution.

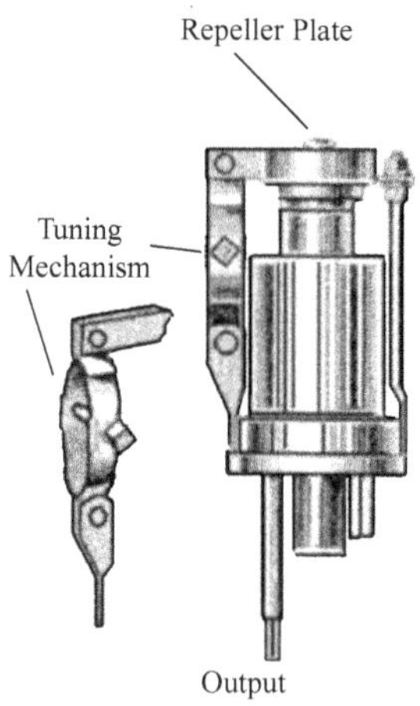

FIGURE A–29

Reflex klystron. Redrawn from USAF Manual 101.8

Rectangular Coordinate Data: Scientific and engineering. Any information based on the sine and cosine values of an angle, often in reference to the mti phase. The synchronous detectors in an mtd system produce rectangular coordinate data, which describes the relative phase angle and magnitude of the i-f output. The synchronous detectors in a color television receiver produce rectangular coordinate data, which describes the color (phase) and hue (magnitude) of the color signal. More on pp 108, 134, 145, 264, 265, 292, 350, 351, 352, 355 through 361, 404, 414, 420, 422. Radar display equipment uses rectangular coordinate data in messages to determine the data display location on a crt or liquid crystal display. See Chapter 15.

Reflector: Universal. The part of a radar antenna assembly which is illuminated by a feedhorn or array. Called, in slang, the "sail." More on pp 38, 72, 169. See also, "mti reflector."

Reflex Klystron (see Figure A-29): Universal. A microwave oscillator once used as a local oscillator in many radar systems. Developed nearly in parallel with the magnetron. Used both as a transmitting tube and a receiver local oscillator in microwave link systems. More on pp 79.

Registration: Webster's dictionary definition is simply "a registering or being registered," followed by others which imply that there is a connection to the register entries and a numerical listing record. In FAA radar, "registration" describes the accuracy of positional agreement between the data from two different radars. See also collimation, the alignment of primary and secondary radar data from an aircraft.

Remote/Local: Universal. Used to define system locations, control functions, or system user/provider sites. May be used differently, depending upon intent; to a radar manufacturer, a radar site is "local" and the indicator site is "remote." To a manufacturer of indicator-site equipment, the radar site may be called "remote." For outage reporting purposes, the radar site may be called "remote" because it is remote to the users. Controls not on the equipment they affect may be called "remote." More on pp 72, 89.

Remote Monitoring System/Subsystem: FAA. A system to provide radar (or other type) operating performance and status information to a facility geographically located elsewhere.

Reply Code: FAA, military, and aircraft industry. The original aircraft transponder codes, still in use, and called "ATCRBS," are a train of 450-ns, 1,090-MHz bursts, transmitted on receipt of a 1,030-μs "mode pair." Two "bracket" or "framing" pulses are spaced 20.3 μs between lead edges and surround up to 13 code data pulses. Adjacent data pulses are spaced by 1 μs from the trail edge of the first to the lead edge of the second. A 450-ns identification pulse may follow the second framing pulse by 4.35 μs (measured between lead edges). Latter-day Mode S and Mode 4 transponders use different schemes, and the reply codes may be lengthy messages. See Chapter 7. More on pp 98, 101, 103, 105, 113, 116, 119, 122, 124, 126, 128, 136, 373.

Resolver (Sweep): FAA and military. Now obsolete. An electromechanical device, rotated by a servomechanism, "modulating" sweep sawtooth voltages in a sin–cos fashion. More on pp 406, 407, 408.

RFCO: RF Changeover (coaxial switch for ATCBI-6 performs a function similar to a radar channel waveguide switch).

rf Signal Generator: A signal generator producing a test signal to, for one example, measure the minimum discernible signal of a radar receiver. See "signal generator".

Rho–theta *(ρθ)****:*** Scientific and engineering. Range and azimuth dimensions. A ppi is a ρθ display. More on pp 11, 35, 385.

RMM: Radar Maintenance Monitor, or Remote Maintenance Monitor. See "Remote Maintenance Monitor System/Subsystem".

RMS: Remote Monitoring System (or subsystem). See "Remote Maintenance Monitor System/Subsystem".

Rotary Joint: Universal. The part a rotating antenna which couples rf energy between the stationary waveguide from the radar equipment, and the rotating section. More on pp 19, 29, 71 through 75, 99, 100, 103, 107, 132, 144, 160, 170, 174.

RTADS: Multiple meanings. Remote Tower Alphanumeric Display System/Service. Real Time Aircraft Display System. Royal Thai Air Defense System.

Run Length (Azimuth Sliding Window): FAA, military. The azimuth span between lead and trail edges. See also "lead edge" and "trail edge." More on pp 85, 90, 125. See also, "sliding window," "centroid," lead edge," and "trail edge."

Running Rabbits (see Figure A-30): FAA and military. Slang. Spiral interference on a ppi, usually from the opposite radar channel, when it is not synchronized to the operational channel. More on pp 190, 310.

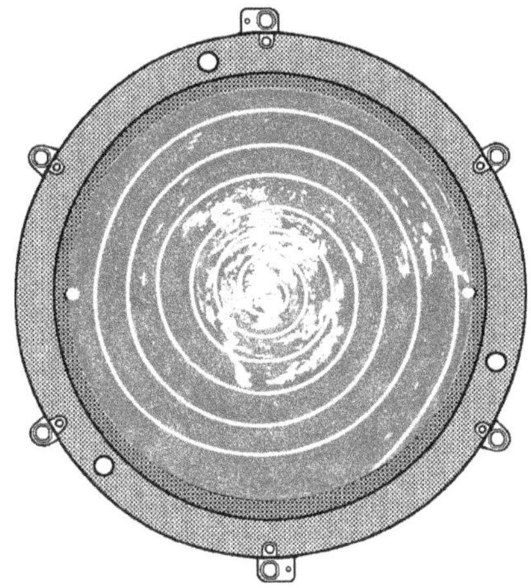

FIGURE A–30

Running rabbits.

S

Sail: Slang for "antenna reflector." More on p 72.

Sawtooth: FAA, military, and others. A waveform which when displayed on an oscilloscope resembles the tooth of a saw blade. Also called "ramp." A sawtooth rises or falls slowly and then abruptly changes to its starting value. Among its most frequent applications are in sweep generators for crt displays and as the sweeping signal for swept-frequency generators. More on pp 48, 203, 204, 268, 325, 386, 399, 404, 407, 413, 419.

Scan: Universal. One revolution, start-to-stop or limit-to-limit, movement of a radar antenna. More on pp 35, 387.

Scan Conversion: FAA and television. See "Radar Bright Display Equipment (RBDE)." See Page 15-5 and 15-8. More on pp 10, 390, 391, 392, 418, 420, 422.

SDR: Sum Difference Ratio. Used in monopulse systems. See Chapter 7 (Difference Amplitude versus. Sum Amplitude). See "monopulse".

Schmitt Trigger (see Figure A-31): Made obsolete in new design work by the operational amplifier, the Schmitt Trigger circuit was a bistable circuit that would change states when the input exceeded a threshold level. Some included a "hysteresis" adjustment to slow the rate of change, so that unwanted spikes would not "flip" the circuit. More on pp 258.

Search: FAA and military. A radar with an antenna that rotates in a full circle, either mechanically or electronically. May be one mode of a multimode system, as in Quadradar or Aegis. Also called "surveillance." More on pp 7, 8, 9, 89, 90, 91, 102, 114, 256 through 259, 380, 423 See also, "afc" and "surveillance".

Second Adaptive Threshold (Map): FAA. A major mtd computer program which performs a scan-to-scan analysis of range-azimuth cells to regulate threshold levels for each cell. Reduces false target detections resulting from birds, insects, and anomalous propagation. More on pp 351, 354, 373 through 376.

Second-Time-Around (Targets): FAA and military. Slang, range-ambiguous target. An echo from a transmitter burst arrives at the receiver input after a second system interval has begun. See "range ambiguity."

Secondary Radar: FAA. A system which relies on the bidirectional speed of light to determine radar range, but does not depend on echoes at the same frequency, as does a "primary" radar. A secondary radar employs an "interrogator" and a "transponder;" the transponder transmits a "reply" at a frequency different from that of the interrogator. Chapter 7 of this book deals entirely with secondary radar.

Selective Identification Feature (SIF): USAF, other military. From military applications of secondary radar. See "secondary radar." See Chapter 7. More on pp 11.

Selsyn: See "synchro." More on pp 401.

Sensitivity Time Control (stc): A receiver circuit to apply an attenuation to echoes; the attenuation decays with radar range. One of the earliest radar "fixes," intended to prevent receiver "blocking" by strong echoes. More on pp 42, 43, 44, 47, 64, 71, 75, 76, 77, 104, 119, 120, 177, 179, 188, 260 See also, "GTC".

Sensor: FAA and military. Sometimes used by data processing equipment manufacturers in reference to a radar system. More on pp 425.

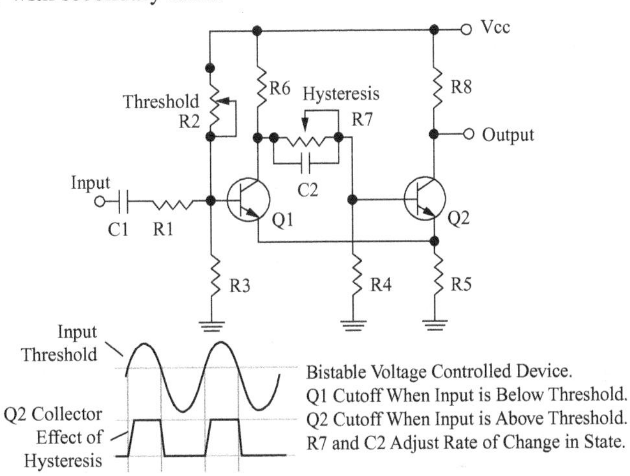

FIGURE A–31

Schmitt trigger.

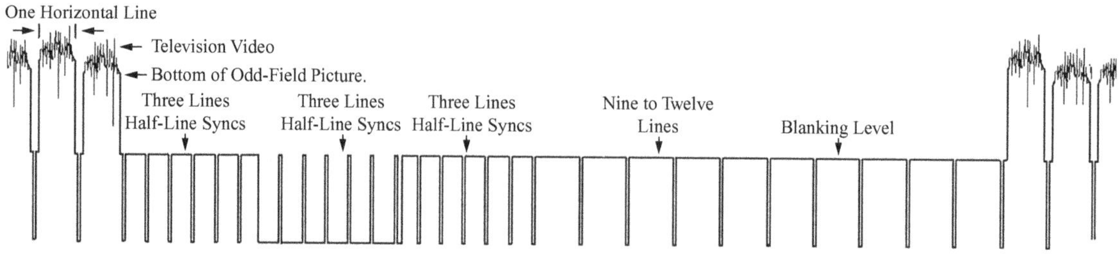

FIGURE A-32

Vertical serrations.

Sensor Receiver and Processor (SRAP): FAA. In data processing equipment applications, the radar system is sometimes called a "sensor." A radar digitizer used in conjunction with ARTS equipment. Uses an azimuth sliding window and data thresholding techniques to develop both primary and secondary radar data messages for the ARTS. The primary digitizer is called a "radar data acquisition system (RDAS)," and the secondary digitizer is called a "beacon data acquisition system (BDAS). See "digitizer," "sensor," "BDAS," and "RDAS." More on pp 380, 423, 424, 425.

Serpentine (Waveguide): Military, particularly US Navy. A folded waveguide used for phased rf energy distribution in frequency-scanned arrays; folding the waveguide offers considerable electrical length in short physical distances. More on pp 130, 131.

Serrated (Vertical Serrations) (see Figure A-32): Television industry; applicable to radar television displays. The combination of synchronizing pulses used in Electronics Industries Associates (EIA) standard television signals to achieve interlaced vertical synchronization. The signal appears "serrated" as the edge of a steak knife. EIA synchronization was upgraded to National Television Systems Committee (NTSC) standards for broadcast color television and compatibility with black-and-white television. Many new television and computer-monitor display techniques are being introduced, and this should no longer be assumed to be the only method. See also National Television Systems Committee (NTSC).

Servo/Servomechanism: Universal. Webster's dictionary defines a servomechanism as "an automatic control system in which the output is constantly or intermittently compared with the input through feedback so that the error or difference between the two quantities can be used to bring about the desired amount of control." Although usually describing electrical-mechanical devices, the word "servo" has also been used to describe purely electronic feedback-loop circuitry; see pages 11-21 through 11-23. A common use of the term is in reference to synchro systems for transmitting angular position information. More on pp 264, 403, 416. See also, "synchro," and "selsyn".

Signal Generator: Universal. Vague when used without other adjectives. At a radar facility, technicians assume the term to refer to the rf signal generator. See "rf signal generator." More on pp 19, 26, 27, 41, 74, 75, 76, 173, 250, 260, 261, 267, 268, 269, 310, 311.

Signal-to-Noise Ratio (see Figure A-33): Universal. The ratio of rf, if, or video signal level to the level of noise, contributed by the atmosphere, equipment, or other sources. May be expressed in voltage or power. More on pp 40, 41, 76, 80, 81, 246, 247, 261, 264, 304, 308, 309, 350, 360.

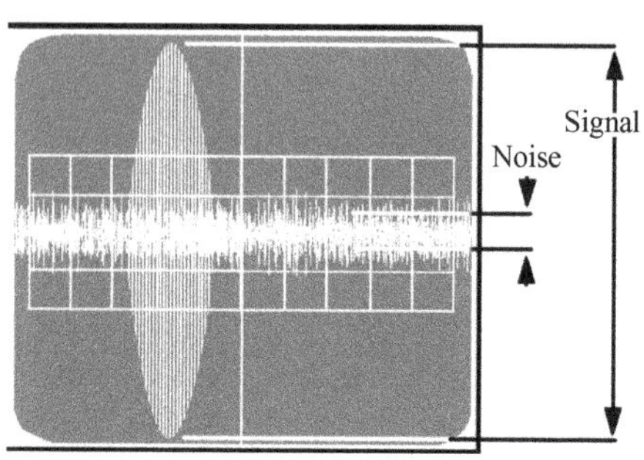

FIGURE A-33

Signal-to-noise ratio.

Single-Shot: Universal. A multivibrator that produces a single gate when triggered. The gate length is determined by a resistor–capacitor network, and the resistance is often variable. The trailing edge often triggers a second single-shot multivibrator, so that the first provides a delay. See also "phantastron." Basic electronics textbooks.

Sinous-Fed: US Navy. Distribution of energy to a frequency-scanned array antenna. Normally, the frequency change will allow a beam to form, some distance from the antenna, at an angle that is inversely proportional to wavelength. Hence, an increase in frequency would, if the antenna waveguide slots are so displaced, decrease the angle at which the beam leaves the antenna. More on pp 38 See also, "phased array."

Sliding Window: FAA, USAF, and others. An area of radar data under examination by a hardware or software process, and which moves at a regular rate in either range or azimuth. An "azimuth sliding window" provides for processing of surveillance azimuth-adjacent radar echoes over multiple T_rs (also called PRTs) to determine the azimuth centroid. A "range sliding window" provides for averaging of radar echoes, over several adjacent range cells, for use in constant-false-alarm-rate (CFAR) circuitry. More on pp 63, 89, 90, 101, 102, 106, 109, 115, 124, 126, 127, 128, 134, 338, 339, 354, 362, 363, 364, 373, 390, 410. See also, "centroiding."

Slip Rings: An assembly in the antenna rotary joint to connect electrical signals from the fixed to the rotating antenna assemblies. More on pp 74.

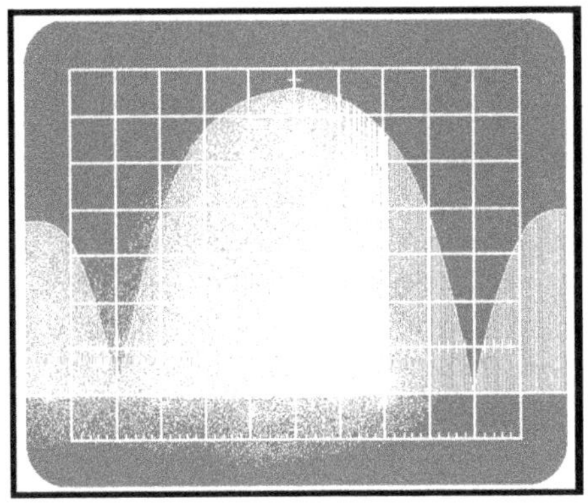

FIGURE A-34

Pulsed radar spectrum.

Soft Limit: Limiting, as in an i-f amplifier, intended to attenuate signal excursions past an established level. Differs from the hard limit in the severity and circuit method. Soft limiters may retain some amplitude variations, while hard limiters clip extreme signal excursions. See "hard limit." More on pp 86, 168, 264, 266, 317.

Special Position Identifier (SPI): Aviation industry, military. The "ident" pulse following a beacon code train. See "ident." More on pp 110, 111, 112, 116, 121,122,123. Synonomous with "ident".

Spectrum: Universal. The band of distributed frequencies emitted by any radio-frequency or microwave transmitter, or any other device. The spectrum of a pulsed radar contains the center frequency and the Fourier harmonics of the modulating pulse (see Figure A-34). Differs from "bandpass," the frequencies that a device may pass. Simplistically, the bandpass may be envisioned as a "door," and the spectrum as that entity which must pass through the "door." More on pp 19, 29, 36, 40, 47, 55, 59, 63, 70, 71, 157, 170, 173, 185, 195, 106, 197, 199, 200 through 207, 211, 213, 222, 230, 234.

Spectrum Analyzer: Universal. A swept receiver, with a crt display, in which the "*x*" dimension is a trace created by the same sweep applied to the local oscillator. The "*y*" dimension is deflected in proportion to the received signal power. Spectrum analyzers were first built for radar use, but now have wide application. More on pp 70, 173, 196, 197, 202 through 206, 251.

Squeezing Waveguide: US Army Air Corps (USAAC), USAF, FAA. Slang for "variabledimension" waveguide. Probably the first phased-array sinuous-fed antenna, built initially for precision approach radar in the AN/MPN-1 GCA system in World War II, and still in use today. Scanning was achieved via a variable phasing of dipole radiators on a waveguide, in which one wall was placed in motion by a motor drive. See Page 4-6 and 4-7. More on pp 7, 38, 129, 130, 131.

SSR: Secondary Surveillance Radar, usually ARCRBS only, no mode S.

Stabilized Local Oscillator (Stalo): US Army Air Corps origin, now FAA and military. The local oscillator in an mti system. The word "stabilized" is added to "local oscillator" because of the essential characteristics in early magnetron mti systems; the stalo served as the single source to provide phase and frequency coherence between the magnetron and receiver. Sometimes called the "coherence link." More on pp 53 through 58, 64, 70, 71, 78, 79, 173, 174, 182, 207, 251, 253, 255 through 258, 281, 288 through 299, 313, 314, 326, 350.

Stagger (Prf): Military and FAA. A repetitive selection of two or more radar T_rs to reduce the effects of mti blind velocities. Also see "blind speeds" and "dim speeds." More on pp 65, 68, 87, 88, 127, 181, 182, 183, 188, 226, 236, 259, 290, 291, 314, 316, 317, 326, 332, 349, 416.

Stagger (Tuning): Universal. A series of cascaded amplifiers, tuned to different frequencies, to achieve a broader passband. Contrasts with "synchronous tuning," where all stages are tuned to the same frequency (see Figure A-35). More on pp 119, 211. See also, "synchronous."

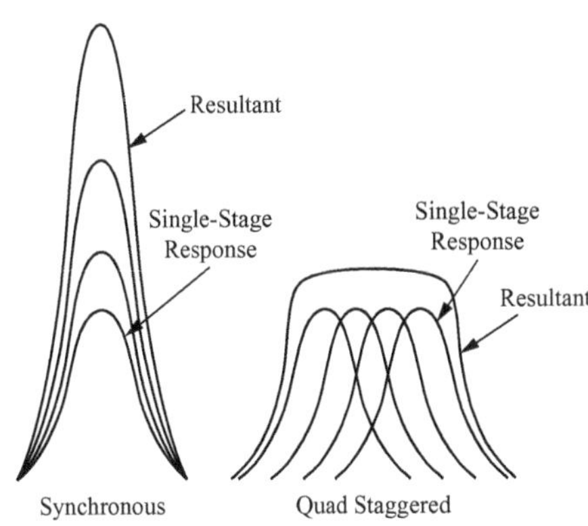

FIGURE A-35

Synchronous versus stagger tuning.

stc: See "Sensitivity Time Control (stc)." More on pp 42, 43, 44, 47, 64 71, 75, 76, 77, 104, 119, 120, 177, 179, 188, 260.

Stealth: Universal. An aircraft design technique based on angular, oblique surfaces, which reflect electromagnetic radiation in directions other than its origin. The principle is somewhat analogous to the manner in which a mirror may be used to direct a spot of light at an angle.

Subclutter Visibility: FAA, military, scientific and engineering. The measurement of the ability of an mti system to detect a moving target over, or in, ground clutter. When echoes of different phases occur at the same radar range, they compete for a resultant phase-detector output; the stronger input has the greater influence. As clutter becomes stronger, moving targets become less detectable, because the phase detector is "pulled" by decreasing degrees. More on pp 19, 44, 77, 86, 260, 262, 264, 305, 310, 311, 312, 316, 329, 347, 350.

Surveillance: Universal. A radar which continuously scans 360° of azimuth for the purpose of detecting any targets within the beam elevation spread. May be one of several modes, or the only manner in which the radar may operate. See also "search." More on pp 6 through 10, 12, 48, 59, 60, 63, 64, 72, 78, 90, 98, 101, 102, 108, 125, 127, 128, 130, 351, 354, 374, 375, 378, 379, 380, 389, 390, 392, 401, 422, 424.

Surveillance Processor or Surveillance Tracker: FAA, USAF. An mtd postprocessor program module to perform scan-to-scan target tracking. More on pp 351, 354, 374, 378, 379.

SVA: Sum Video Amplitude. Used in beacon monopulse systems.

Sweep (Display Equipment): Universal. The continual, steady deflection and movement of the crt electron beam by a ramp (also called "sawtooth") from an origin to a fixed end point, usually at the edge of the crt. In a ppi, the origin is at the center, and in an oscilloscope, it is on the left-hand side. More on pp 385 through 381, 393, 394, 396 through 400, 404 through 408, 413, 414, 416 through 419, 421, 422.

Sweep (Frequency): Universal. A continual, steady increase or decrease of an oscillator frequency created by a ramp, triangle, or even a sine wave, from an origin frequency to an end frequency, and back. More on pp 267.

Sweep Generator: FAA, military, others. The circuit in a ppi display, video mapper, or oscilloscope, which is used to generate the ramp (also called "sawtooth") voltage waveform ultimately used to deflect the electron-beam trace on the crt. Sometimes improperly used in reference to a "swept-frequency generator," a test-signal generator, used to create a spectrum of test frequencies, for examination of the passband, of an rf or i-f circuit. More on pp 203, 204, 205.

Swept-Audio Generator (mti System): FAA, military. Beyond the obvious, a swept-frequency audio generator is included in many radar systems as built-in test equipment to create a swept spectrum of simulated Dopplers, for the cancelers. The Doppler passband is typically called the "velocity response shape." More on pp 325.

Swept-Bandpass afc: FAA, military. An automatic frequency control system employed by older magnetron radars and by other microwave systems, to maintain the transmitter–receiver frequency difference at precisely the intended intermediate frequency (i-f). The automatic adjustment may be to either the transmitter or receiver local oscillator; generally, the local oscillator will be the receiving element of a servo loop, but modified systems, not previously incorporating afc, will adjust the magnetron frequency. The "swept-bandpass" term arises from the use of a tuned circuit, the center frequency of which is determined by the reactance of an electron tube, which in turn is swept with 60 Hz. Also see "reactance tube modulator."

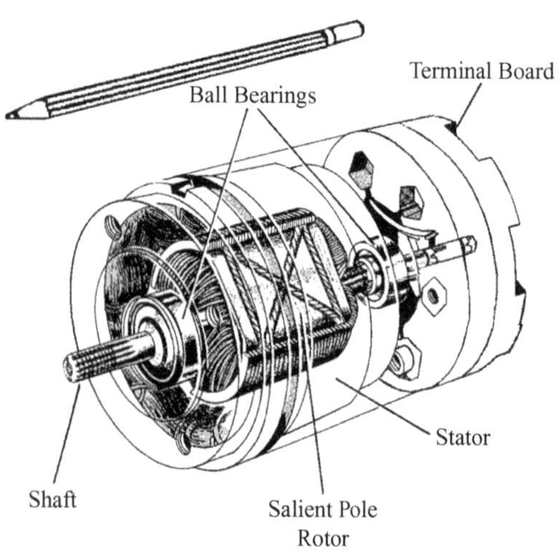

FIGURE A–36

Synchro. From US Navy Basic Electricity Rote Training Manual NAVPERS 10086-B 1969 Edition.

Swept-Frequency Generator: Universal. A test-signal generator of a spectrum of frequencies for response examination of rf, i-f, video, or audio circuits. Sometimes incorrectly called "sweep generator," which may be confused with crt deflection circuitry. More on pp 81, 213, 267 See also, "sweep frequency".

Synchro (see Figure A-36): Universal. Also called "selsyn," once a General Electric name for "self-synchronous" device. An ac servomechanism used for electrical transmission of angular position data, such as the azimuth of a radar antenna. A synchro system comprises a transmitter and receiver, each containing a fixed body, with three stator windings, and a rotating shaft, with a single winding. Currents are induced into the transmitter stator windings by a field around the rotor; the currents into the receiving stator windings create fields which force the receiving rotor to the same position as the transmitting rotor.

In surveillance radar systems, synchros have been largely replaced by azimuth pulse generators, but synchros are still necessary in those systems in which the antenna does not continuously scan in the same direction, usually clockwise. More on pp 35, 393, 394, 401 through 408, 416.

Synchronizer: Universal. The element of a radar system that creates a master trigger, at the basic repetition frequency, and all triggers and gates, referenced to, and synchronized by, the master trigger. In latter-day systems, the synchronizer also provides all range-cell and analog-to-digital triggers and gates. See Chapter 9. More on pp 33, through 36, 58, 65, 88, 124, 177 through 190, 264, 301, 310, 312, 313, 326, 355, 361, 378, 410.

Synchronous (Detection): Phase detection, called "synchronous" because the two inputs to the phase detector circuit are "coherent" and bear an intelligible phase relationship. More on pp 266, 282, 307, 350, 352, 355, 357.

Synchronous (Harmonics): Called "synchronous" because all harmonic frequencies of a pulse bear an intelligible temporal and spectral relationship to the fundamental period and frequency of the pulse repetition rate. More on pp 197, 198, 213, 219.

Synchronous (Timing): Called "synchronous" because radar triggers, gates, pulses, etc. all bear a designed and predictable relationship to a single event, often the pretrigger or transmitter burst. More on pp 85, 126, 342, 387, 401.

Synchronous (Tuning): Where the stages of an amplifier are all tuned for a peak response at the same frequency. In contrast is "stagger tuning," used to obtain a wider bandwidth for improved reproduction of the transmitted or received intelligence. More on pp 213, 259.

Synchroscope: US Army Air Corps, USAF. Obsolete term, once used to differentiate triggered, delayed-sweep oscilloscopes from those other early oscilloscopes which could not be synchronized by an external trigger. Most quality, latter-day "oscilloscopes" could qualify as "synchroscopes."

Synthesis: FAA. One Webster's definition is "a whole made up of parts or elements put together." In radar, the "whole" is the transmitter frequency and the "elements or parts" are the stalo, coho, and any intermediate local oscillator, combined to create that transmitter frequency. The earliest known FAA/USAF radar use originated with the AN/FPS-35 frequency "synthesizer," used to produce basic frequencies for that double-conversion-receiver system. Briefly, synthesis is the process of mixing frequencies to achieve a resultant frequency. Generally, in radar, a "synthesis system" is one in which the transmitter frequency is created with two or more oscillators, and then amplified. The term is used to differentiate from a magnetron system in which the transmitter tube is also the rf oscillator. See Chapter 5. More on pp 10, 53 through 58, 60, 63, 64, 70, 79, 83, 85, 119, 180, 181, 183, 207, 222, 223, 230, 236, 256, 260, 281, 285, 286, 287, 298 through 303, 312 through 316, 325, 350.

Synthetic. One Webster's definition is "artificial; not real or genuine." In radar, several uses corresponding to that definition may appear. "Synthetic real time" is a radar display timing technique in which digital target data is reconstructed into a 12.3552 μs/nmi time base for use in analog ppi displays. Alphanumeric data generated for display is sometimes called "synthetic," since it is not in real time.

Synthetic Aperture: Military, scientific, and engineering. An antenna with a variable aperture, so as to decrease beam dimensions with increasing range to maintain the target definition. Used for radar terrain mapping and other purposes. Not used in air *traffic control radars* at the time this book was written; for further information, See Merrill Skolnik's ***Radar Handbook.***

T

Time Averaged Clutter-Coherent Airborne Radar (TACCAR): Scientific and engineering; USAF and US Navy. An mti system in which the coho frequency is regulated by the averaged echo Doppler to compensate for the aircraft velocity, so that the mti system perceives the aircraft and ground clutter to be stationary. For further information, See Merrill Skolnik's ***Radar Handbook.***

Tangential (Target): USAF, FAA. The condition in which the course of a moving target is at a tangent to a circle around the radar, and is perpendicular to a line from the radar. In mti, a tangential target exhibits zero Doppler shift and will be canceled as if it were a fixed target. In mtd, scanto-scan analysis of zero-Doppler data provides for recovery of tangential targets. More on pp 19, 260, 280, 306, 312, 347, 365.

Target: US Army, US Navy origin; now universal. One of the earliest uses of radar was for "gunlaying," which evolved to "fire control." Because of this purpose, echoes received on this type of radar were called "targets." Use of the term expanded into other radar systems, so today a "target" is any object which causes radar echoes, or any detectable radar data caused by echoes from a single object or source. In surveillance radar, a "target" differs

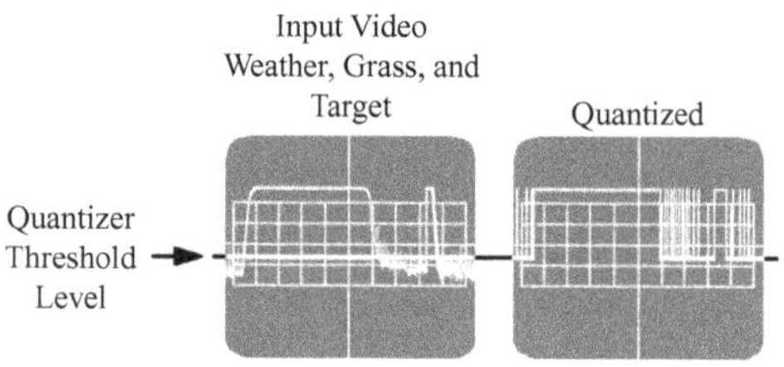

FIGURE A–37

Thresholding.

from an "echo" in that a target offers, or comprises, a number of single echoes.

Television Microwave Link (TML): FAA. Microwave link equipment to transmit television scanconverted radar data to air traffic control user facilities, such as control towers.

Terminal Doppler Weather Radar (TDWR): FAA. A C-band, pulsed-Doppler radar used for weather detection and velocity measurements at different elevations.

Temporal: Universal. A universal English word of Latin origin, meaning "of or limited by time" (Webster's). In radar, a temporal relationship is one expressed in time, as by microseconds. Sometimes appears as the name of a time-delay adjustment. More on pp 35, 55, 65, 69, 97, 104, 109, 113, 114, 119, 125, 126, 177, 179, 180, 183, 213, 229, 277, 314, 316, 319, 332, 361, 366, 368, 410.

Terminal (Radar): FAA. A radar system used at an airport, also defined as a "terminal," as that is the point of flight origins or terminations. Usually a 60-mile, S-band system. See "airport surveillance radar (ASR)" and "ASR."

Three-Dimensional (3D) Radar: A radar system which simultaneously provides radar data in azimuth, elevation, and range. See "phased array," "frequency-scanned array," and "sinuous-fed." More on pp 11.

Threshold: Universal. A voltage or a digital number to establish a point beneath which data may be inhibited or allowed. Thresholds may be fixed, or automatically variable, in accordance with data. A threshold which varies in accordance with data environment is said to be "adaptive." Thresholding devices are frequently called "quantizers" (see Figure A-37). More on pp (electronic switch point) 19, 58, 86, 116, 119, 120, 246, 258, 264, 305, 308, 327, 328, 349, 350, 351, 352, 355, 361 through 376, 400, 413.

Time-Shared: Universal. A means of using one circuit to provide different functions at different times. For instance, a radar processor may be "time shared" to process radar information during live time, and then test signals, during dead time. A time-shared circuit is ordinarily operated by a gate or gated logic. In digital logic circuitry, a multiplexer integrated circuit may be used to "timeshare" digital words. More on pp 373, 388.

Track-While-Scan: Usage is rare, largely academic, but nevertheless, very definitive. The tracking process that is used in digital equipment operating on surveillance radar data inputs. FAA ARTS and mtd Surveillance Processors are Track-While-Scan systems. See Chapter 14. More on pp 351, 379.

Transmit-Receive (T-R) Tube or Device: Universal. A waveguide gas device ionized by the high-power transmitter burst to create an rf short across the waveguide, to protect the delicate receiver "front end." Also see "ATR," "duplexer," and "four-port circulator." Research for this book has revealed a potential conflict in this definition. One manufacturer of equipment for the US Navy has named a device equivalent to the T-R, an "ATR." See "AT-R." More on pp 58, 64, 70, 75, 76, 77, 249, 252, 301, 302.

T-R Recovery Time: Universal. The measured time from the leading edge of the transmitter burst to the point where a test signal is attenuated 3 dB from maximum sensitivity. Some US Navy entities may call this "ATR recovery time." More on pp 75, 76, 170, 174.

t_p: FAA. The duration of the transmitted rf burst, measured at the half-power lead- and trail-edge points, or at the reciprocal of one-half the frequency span, between the nulls on either side of the main spectral lobe. Also called PW, T, t, T_p, and others. This book has used only t_p. More on p 37, table 4-1.

T_r: FAA. A radar system interval; the time between transmitter pulses, the reciprocal of the pulse repetition frequency. Called "PRT" by the military. Use of "PRT" implies the multiplication of three quantities, and is poor scientific annotation, but is nonetheless commonplace. There is some potential for confusion between the intended "time of repetition" and "time of one cycle of a resonant frequency." More on p 37, table 4-1.

TRACON: FAA. Terminal Radar Control. An airport radar arrival, departure, and local traffic air traffic control facility. See "airport surveillance radar (ASR)" and "terminal (radar)." More on pp 100, 102, 135, 136, 137, 138 See also, "Radar Approach Control (RAPCON).

Trail Edge (Azimuth Sliding Window): FAA, USAF. That signal generated when the azimuth sliding window has "emptied" down to the established criteria. More on pp 90, 123. See also, "sliding window."

Transparent Mode: Sometimes associated with graphic user interface (GUI) computers, where the operator may exit the GUI application software to work in the operating-system language, such as DOS, Unix, etc.

Transponder: FAA, military, aircraft industry, all air traffic control entities. The "answer-back" equipment in a secondary radar system; located in the aircraft in standard air traffic control beacon equipment. An air traffic control transponder responds to 1,030-MHz interrogations, from radar sites, with 1,090-MHz beacon code trains. Also see "beacon," "interrogator," "mode pair," "reply code," and "secondary radar." See Chapter 7. more on 42, 47, 75, 76, 90, 97, 99 through 119, 124, 127, 128, 133 through 138.

Traveling Wave Tube (TWT): Universal. An electron tube used for a microwave rf amplifier, relying upon the transit time of rf, for amplification. More on pp 12, 195.

Triangular (Phase Detector Response): FAA. The phase-detector response in nonquadrature mti systems; the shape ensures that echo targets of any $\Delta\phi$ will produce the same amplitude differences near the response peaks, as on the slopes. See pages 6-20, 6-21, 11-20, 12-18. More on pp 81, 263, 268, 292, 294.

Trigger: Universal. A very short-duration signal with an extremely fast risetime; the lead edge is intended to cause, "trip," or start, an event or signal, in a radar system. Also see "gate," which differs from "trigger" because of purpose and time duration. More on pp 33.

TWT: See "Traveling Wave Tube (TWT)." More on pp 12, 195.

U

Unblanking Gate: FAA and military. The gate that is applied to either the cathode or the control grid of a ppi cathode ray tube, during the "live" time when video may be displayed. For instance, an indicator set to a 30-mile range will require an unblanking gate of about 371 μs. Some manufacturers may use "blanking gate," which simply refers to the display "off time," instead of the "on" time. See page 15-13. More on pp 398, 417.

Unipolar (mti Systems): FAA, USAF, and others. Video or data with excursions in only one direction from a static baseline. Used to distinguish from the "bipolar" information from a phase detector; bipolar information exhibits excursions in both the positive and negative directions from a static baseline. See pages 6-23, 12-21, 13-10. More on pp 54, 56, 58, 84, 85, 87, 180, 253, 281, 286, 294, through 298, 301, 304, 305, 306, 308, 309, 311, 312, 329, through 336, 340, 341, 349.

V

Valve: British word for "electron tube."

Varactor (Diode): Universal. A microwave solid-state diode. The junction capacitance changes with applied voltage, permitting the potential drop to be "pumped." The varactor diode is the main working component in low-noise parametric amplifiers. See page 11-5. more on pp 77, 249.

Velocity: Scientific. The rate at which a body in motion traverses a distance. In radar, "velocity" is often assumed to be the radial velocity V_r, a cosine function of the aircraft course in relation to the radar bearing. See Chapter 12. More on pp 277, 278, 277 through 280, 285 through 299, 304, 305, 315, 316, 319, 325, 329, 333 through 336, 342, 347, 352, 361, 362, 365, 369, 370, 377, 395, 396, 400.

Velocity Ambiguity: FAA, military, scientific, and engineering. The "Doppler ambiguity" condition in which more than one velocity may be represented by the canceler voltage output E_c. The phenomenon occurs when the target Doppler exceeds f_p's. It is of little consequence in mti systems, but of great consequence in pulsed Doppler, where velocity measurement is the intent. Pulsed Doppler systems, therefore, employ high repetition rates to prohibit velocity ambiguities. See "range ambiguity." More on pp 315, Doppler Ambiguity.

Velocity Response (mti Systems): FAA, USAF, other military. The audio Doppler response of the canceler(s). Displayed by use of a swept audio oscillator. See Chapter 12. More on pp 19, 85, 292, 293, 294, 295, 297, 299, 304, 305, 315, 316, 319, 325, 329, 333, 334, 335, 336, 342, 369.

Velocity Response Shaping: FAA, USAF, other military. Cascaded-canceler mti systems employing feedback to alter the canceler Doppler bandpass. See pages 12-18 through 12-22, and 13-9. More on pp 319, 329, 333.

Video Integrator: FAA, USAF. A unit to perform pulse-to-pulse additions of radar video; synchronous signals accumulate to high levels, random noise accumulates to lesser levels, and nonsynchronous pulses do not accumulate. The end result is an enhancement of radar targets and an elimination of nonsynchronous interference. Used with both normal and mti radar information. Also called an "enhancer." See pages 6-23 through 6-26. More on pp 19, 84, 85, 88, 126. See also, "enhancer".

Video Mapper: FAA, USAF. Radar user-facility equipment to display runways, air traffic routes, terrain features, and other related information. Named "flying spot scanner," arising from the basic concept of permitting the light, from a crt sweep, to pass through the clear areas of a photographic negative, to a photomultiplier tube; the photomultiplier output becomes "map video." See pages 15-23 through 15-25. more on pp 137, 385, 412, 413, 416 through 419, 425.
Versa Module Europe (VME). A standardized bus structure used in many systems. Off-the-shelf circuit cards that may be plugged into VME card "crates" offer economical and versatile combinations for many purposes. Many VME crates contain a computer module on the left side, to control and communicate with the other modules in the crate. The backplane provides for connections to P1 and P2 card-edge connectors. The P1 connectors are generally for communications between modules within the crate, and the P2 connectors are for local and specific-purpose communications.
Video Time Compression. A method of storing real-time radar video in a memory on one interval and then reading the memory at an accelerated rate on the next interval. The purpose is to shorten display "live time" to allow more time for an ARTS alphanumeric display in "dead time." More on pp 408, 411.

W

Watchdog Timer: A computer timer that starts running when the computer ceases to execute instructions.
Waveguide: Universal. Hollow microwave transmission lines, rectangular or circular, to propagate energy at the speed of light, and with minimal loss. Originally developed by Massachussets Institute of Technology Radiation Laboratories, early in World War II. See Chapter 8. More on pp 99, 100, 113, 129 through 131, 143 through 174.
Wavetrap: Universal. A sharply tuned, precisely calibrated cavity used in conjunction with a power meter for frequency measurements. A little less precise than a counter, but far less expensive and less delicate. Often called "cookie jar" in slang. See Chapter 8.
Weber: Scientific. A measurement of magnetic flux, named for Wilhelm Eduard Weber, a colleague of Karl Friedrich Gauss. One Weber per second produces 1 Volt. A Weber is a Volt-second. See page 10-27. More on pp 217, 218, 237.
Word: Universal, originating in data-processing and computer industry. A group of binary bits, usually representing a single value, and usually occurring in time coincidence on several lines, but can also be in serial form. In radar, a "word" differs from a "message," which contains many words. More on pp (data processing application) 123,124,125, 133, 179, 181, 319, 328, 330, 337, 338, 339, 350, 352, 362, 364, 368, 380, 388, 410, 413, 423.
Wow: General electronics term. A repetitive variation in a signal or meter indication, usually associated with a mechanical rotation. Origin is in audio work and stems from the sound of fluctuating audio, as in a tape recorder or record player. Used in radar to describe, for one, the regular variation in reflected power from a rotary joint, which normally occurs with antenna rotation. More on pp 174.
WRTADS: Windows Real-Time Aircraft Display System. For displaying real-time radar data on a computer with a Microsoft Windows operating system.

Z

Zero-Velocity Filter (ZVF): FAA. Those Doppler filters in an mtd system which pass zero and near-zero frequencies. Those other filters, passing other Dopplers, are called ***nonzero velocity filters (NZVF)***. See Chapter 14. More on pp 351, 352, 362, 365, 366, 369 through 372, 374.

APPENDIX B

Major Equations

Antenna Gain

$$G_t = \frac{4\pi\ A_e\ k}{\lambda^2}$$

A_e is the effective aperture of the antenna, the surface area exposed to the echo.
k is the antenna efficiency.
The squared wavelength λ^2 and aperture A_e must be in the same quantities, as meters or feet.

Azimuth Change Pulse to Bearing Conversion

$$\angle_{\text{bering}} = \frac{360 \times N_{\text{acp}}}{4{,}096} \quad \text{or} \quad \angle_{\text{bering}} = \frac{360 \times N_{\text{acp}}}{16{,}384}$$

$\angle_{\text{bearing}}$ = radar azimuth in degrees
N_{acp} = acp count
Choice of 4,096 or 16,384 dependent on the azimuth pulse generator type.

Beacon Received Power Calculation

$$P_{r_{\text{sec}}} = \frac{P_{\text{tt}}\ G_{\text{tt}}\ A_e}{4\pi R^2}$$

$P_{r_{\text{sec}}}$ = interrogater received signal power from transponder
P_{tt} = peak transmitter power of a radar beacon transponder
G_{tt} = beacon transponder antenna gain
A_e = interrogator antenna effective aperture
R = target range.

DeciBels

$$\text{dB} = 10\ \log_{10}\left(\frac{x}{y}\right)$$

To express gain or loss in dB, x and y are power input and output of a device.
To express dB as a voltage gain or loss in dBV, use 20 log.
If $y = 1 \times 10^{-3}$ Watts, then dB is dBm.

Doppler Shift of an Aircraft at a Radial Velocity

$$f_{\text{D}} = \frac{V_{\text{acft}}\ [\cos\ (\angle_{\text{beaing}} - \angle_{\text{course}} + 180^\circ)]\ f_{\text{xmtr}}}{291.375 \times 10^6}$$

V_{acft} = the actual aircraft groundspeed
$\angle_{\text{bearing}}$ = the magnetic radar azimuth of the aircraft
$\angle_{\text{course}}$ = the aircraft course by compass heading

f_{xmtr} = the radar transmitter microwave radio frequency in Hz
f_D = the Doppler shift, or apparent Doppler frequency
The 180° reverses the angle from target-to-radar to radar-to-target.
The cosine function reduces the resultant velocity from a maximum where the target course is directly toward or away from the radar.

Echo Power

$$P_r = \frac{P_t G_t A_o A_e}{16 \pi^2 R^4}$$

P_r = received echo power in Watts
P_t = peak transmitted power in Watts
G_t = antenna gain
A_o = target cross-sectional area in same units as A_e and R
A_e = antenna effective aperture (area exposed to the echo) in same units as A_o and R.

Fourier Harmonic Series

$$E_{ac} = 2E \left(\frac{t_p}{t_r}\right)\left(\frac{\frac{\sin \pi t_p}{T_r}}{\frac{\pi t_p}{T_r}}\right) \cos 1\,(2\pi f_p) t_x$$

$$+ 2E \left(\frac{t_p}{T_r}\right)\left(\frac{\frac{\sin 2\pi t_p}{T_r}}{\frac{2\pi T_p}{T_r}}\right) \cos 2\,(2\pi f_p) t_x$$

$$+ 2E \left(\frac{t_p}{T_r}\right)\left(\frac{\frac{\sin 3\pi t_p}{T_r}}{\frac{3\pi t_p}{T_r}}\right) \cos 3\,(2\pi f_p) t_x + \cdots$$

E = pulse amplitude
E_{ac} = sum of all the harmonic amplitudes
t_p = width of a repetitive pulse
T_r = pulse repetition period
t_x = time after pulse center
π, 2π, 3π, etc. represent a radian multiplied by a harmonic number.

Hits per Scan

$$N_s = \frac{T_{scan} \beta}{360\, T_r}$$

N_s = number of hits per scan, a figure of merit defining the number of radar echoes or beacon reply codes from a surveillance radar antenna scan between the antenna radiation pattern 3 dB points. The number of echoes or replies is not limited to N_s.
T_{scan} = time required for an entire 360° rotation of the antenna
T_r = pulse repetition period
β = antenna beamwidth at the 3 dB horizontal dimensions in degrees.

Maximum Expected Radar Range

$$R_{max} = 0.282\sqrt[4]{\frac{P_t\, G_t\, A_o\, A_e}{P_{r_{min}}}}$$

P_t = transmitter peak power in Watts
G_t = antenna gain
A_o = target cross-sectional area reflecting toward the radar in the same units as A_e
A_e = antenna effective aperture in the same units as A_o
$P_{r_{min}}$ = minimum discernible signal in Watts.

Use logarithms to solve.
See extensive information and derivations on Range Equation in Chapter 4.

mds Prediction

$$P_{r_{min}} = k\, T_k \Delta f\, F$$
$$= -174\text{dBm} + 10\log\Delta f + F_{dB}$$

$P_{r_{min}}$ = minimum discernible signal in Watts
Δf = receiver overall passband in Hz. This equation is invalid if Δf is less than $1.2/t_p$.
F = receiver overall noise figure
F_{dB} = receiver overall noise figure in dB
T_k = standard temperature of 290° Kelvin
k = Boltzman's constant: 1.38×10^{-23} Joules per degree Kelvin.

Overall Noise Figure

$$F_t = F_1 + \frac{F_2 - 1}{G_1} + \frac{F_3 - 1}{G_1 G_2} + \frac{F_4 - 1}{G_1 G_2 G_3} + \cdots$$

F_t = overall noise figure
F_1, F_2, F_3, etc. = noise figure of individual stages
G_1, G_2, G_3, etc. = gain of individual stages.

Wavelength

$$d = rt \qquad \lambda = c\left(\frac{1}{f}\right) = \left(\frac{c}{f}\right)$$

$$\lambda_{(cm)} = \frac{3 \times 10^{10}}{f} \quad \Rightarrow \quad \lambda_{(inches)} \quad \frac{1.18 \times 10^{10}}{f}$$

c = velocity of light
d = distance
f = frequency
r = rate
t = time
λ = wavelength.

VSWR (Voltage Standing Wave Ratio)

$$\mathrm{VSWR} = \frac{1+\sqrt{\dfrac{P_r}{P_i}}}{1-\sqrt{\dfrac{P_r}{P_i}}}$$

P_i = power incident (or forward)
P_r = power reflected (or reverse).

APPENDIX C

Conversions and Constants

Work and Force

1 dyne = 1g/s/s
1 erg = 1 dyne-cm
1 J = 10^7 ergs
1 J = 10^7 dyne-cm
1 J = (1 N) (1 m)
1 W = 1 J/s
1 horsepower = 746 W = 550 ft-lbs/s

Electrons and Charges

Charge on an electron, e, = 1.6×10^{-19} coulombs
1 coulomb = 1 A for 1 s
1 coulomb = 6.2422×10^{18} electrons

Distances

1 meter = 3.28 feet
1 foot = 30.8 cm
1 inch = 2.54 cm
1 cm = 0.3937 inches
1 nautical mile (nmi) = 6076.11549 feet = 1.852000 km = 1.15077 US statute miles
1 US statute mile = 5280 feet = 1.609 km = 0.8689 nmi
1 statute mile per hour = 1.150779449 nmi/h
1 foot/s = 0.59248 nmi/h = 0.3929448 statute miles per hour

Speed of Light (c)

582.75×10^6 nmi/h
670.6 US statute miles/h
161,875 nmi/s
186,282 US statute miles/s
983.571×10^6 feet/s
$11{,}802.8 \times 10^6$ in/s
2.9979×10^{10} cm/s
2.9979×10^8 m/s

Radar Echo Time for Distance

1 US statute mile (5,280 feet) = 10.736 μs
1 international nmi (6,076.11549 US feet) = 12.3552 μs
1 US foot = 2.03 ns

1 US inch = 169.45 ps
1 cm = 66.7 ps
1 m = 6.67 ns
1 km = 6.67 μs

Common Radar Times and Frequencies

58 nmi : 716.6 μs
60 nmi : 741.3 μs
200 nmi : 2471 μs
250 nmi : 3138 μs
frequency of a 1-mile range-mark oscillator: 80.9375 kHz
intermediate receiver frequencies: 30 MHz, 31.07 MHz, 60 MHz, 1.25 GHz

Frequency Bands and Wavelengths

L : 25 cm or 9.8 inches @ 1.2 GHz
S : 10 cm or 3.9 inches @ 3 GHz
C : 6 cm or 2.4 inches @ 5 GHz
X : 3 cm or 1.2 inches @ 10 GHz
K_u : 2 cm or 0.8 inches @ 15 GHz
K_a : 1 cm or 0.4 inches @ 30 GHz

Temperature Conversions

Temperature Kelvin = Centigrade or Celsius plus 273
Temperature Fahrenheit = 1.8 Celsius + 32 = 1.8 Kelvin – 459.4
Temperature Celsius = 0.555 (Fahrenheit – 32)
Freezing and boiling points of water at sea level:
Celsius: 0°, 100°
Fahrenheit: 32°, 212°
Kelvin: 273°, 373°

Boltzman's Constant and Power

Boltzman's constant: 1.38×10^{-23} J/°K temperature
1 W = 1 J for 1 s

Magnetic Force

Weber = Field produced by 1 V for 1 s
1 Weber per square meter = 1×10^4 Gauss
1 Tesla = 1 Weber per square meter

APPENDIX D

Government Nomenclatures

FAA and ICAO

ARSR: Air Route Surveillance Radar, 200-mile range, 5 rpm antenna rotation
ASDE: Airport Surface Detection Equipment, 4-mile range, 60 rpm antenna rotation
ASR: Airport Surveillance Radar, 60-mile range, 12.5 rpm antenna rotation
ATCRBS: Air Traffic Control Radar Beacon System without Mode S
MSSR: Monopulse Mode S Secondary Surveillance Radar. Range and rotation dependent on associated primary radar
PAR: Precision Approach Radar, 20-mile range
SSR: Secondary Surveillance Radar Modes 2, 3/A, and C. Range and rotation dependent on associated primary radar
TDWR: Terminal Doppler Weather Radar

Military:

(Those radars designated "SCR," for "Signal Corps Radio," preceded the current nomenclature system.)
Designation is three characters plus system number assignment, as follows:

The first character is the radar site type:

A: Airborne
C: Cargo Transportable
F: Fixed
G: Ground Stationary
M: Mobile
S: Shipboard

The second character "P" indicates "pulsed radar"

The third character shows the purpose:

N: Navigational
Q: Fire control, Mortar Locating, Bomb-Sighting, etc.
S: Surveillance
Y: Special purpose

Example: The SPY-1C is a shipboard, pulsed radar, special-purpose, third major modification.

APPENDIX E

Radar Frequencies

Note:
The frequency band designators in this list are principally based on those first used in WWII. Newer designators in alphabetical order are now being used by some entities.

IEE Band Designator	Frequency	Usage
HF	22–28 MHz	Great Britain's Chain Home circa 1938.
VHF	30–300 MHz	AN/FPS-24 ARSR. 1960s. US Army SCR-268, SCR-270. WWII. USN CXAM WWII.
UHF	300–1,000 MHz	AN/FPS-35 ARSR. 1960s.
L	1,000–2,000 MHz	ATCRBS Interrogators. WWII to now.
	1,030 MHz	ATCRBS Transponders. WWII to now.
	1,090 MHz	Mode S Interrogators. 1980s to now.
	1,030 MHz	Mode S Transponders. 1980s to now.
	1,090 MHz	ARSR-1, -2, -3, -4. 1950s to now.
	1,215–1,450 MHz	AN/FPS-20, -66, -67, etc . 1940s to now.
	1,215–1,450MHz	
S	2,000–4,000 MHz	
	2,700–2,900	ASR-1, 2, 3, 4, 5, 6, 7, 8, 9, 11. 1940s to now. AN/FPS-27. 1960s. AN/MPN-1. WWII to 1960s. AN/CPN-4. 1948 to 1960s. AN/MPN-11,13,14,15. 1960s to now.
C	4,000–8,000 MHz	
	5.6–5.65 GHz	Terminal Doppler Weather Radar. 1980s.
X	8,000 MHz–12 GHz	
	9–9.6 GHz	WWII to present. Microwave Data Links
K_u (under H_2O vapor resonance)	12–18 GHz	
	14–17 GHz	ASDE-3. 1970s to now. Microwave Data Links
K	18–27 GHz	
	23.8–24.27 GHz	ASDE-1,2. 1950s, 1960s.
K_a (above H_2O vapor resonance)	27–40 GHz	

APPENDIX F

Greek Alphabet in Radar

A,α "Alpha," α, angles, component characteristics.
B, β "Beta," β, angles, component characteristics.
Γ, γ "Gamma," Γ, reflection coefficient.
Δ, δ "Delta," Δ, "change in."
E, ε "Epsilon"

ε = 2.718281828, used as a power in raising or decaying numbers as in nature, called "natural" logarithms. Often replaced with "e."

$$\log_{\varepsilon} N = 2.30259 \log_{10} N$$
$$\log_{10} N = 0.43429 \log_{\varepsilon} N$$

Z, ζ "Ζετα," Z, impedance.
H,η "Eta."
Θ,θ "Theta," θ, the central angle upon which calculations are based. May reference azimuth when used in navigational electronics. A ρ–θ (rho–theta) system is a range–azimuth system.
I, ι "Iota."
K, κ "Kappa," κ, "constant." Often used in equations to represent Boltzman's constant, 1.38×10^{-23}.
Π, λ "Lamda," Π, wavelength.
M, μ "Mu," μ,"micro," 1×10^{-6}, when preceding a number. μ may indicate gain, as the gain of an amplifier.
N, ν "Nu."
Ξ, ξ "Xi."
O, o "Omicron."
Π, π "Pi," π = 3.141592654, the result, when the circumference of a circle is divided by its radius.
P, ρ "Rho ρ." Often used in regard to radius and range, as the radius and range of a circle around the radar. The pointing angle, or azimuth, of a radar may be called θ and range–azimuth may be called ρ–θ.
Σ, σ "Sigma," Σ, used for sum, or summation.
T, τ "Tau," τ, time constant, as in R-C circuits.
Y, υ "Upsilon."
ϕ, φ "Phi," φ, phase angle.
X, χ "Chi."
Ψ, ψ "Psi," ψ, phase difference, as between two signals.
Ω, ω "Omega," Ω, ohm. ω is an abbreviation for 2π*f*/ angular velocity.

For example, $X_L = 2\pi/L = \omega L$.

APPENDIX G

Technical Expression

Means of expression in technical literature for government use is determined by military or other specifications, and government contractors must abide by these. Nongovernmental corporate entities or publishers may adopt their own writing standards. Some practices are common, and some are not, but there is usually a carefully considered reason behind whatever choice the technical documentation authority has made.

The first objective in technical writing is clarity, and some grammatical rules are discarded to achieve that. Unlike conventional expression, the names of controls, indicators, test points, and mnemonics are often capitalized, while common abbreviations are not. The practice assists the reader to spot the significant items in the text, without distractions. In the late 1960s, IEEE recognized an importance in capitalizing the first letter of names, so "V" in "Volt," "W" in "Watt," "B" in "dB," were capitalized, and "cycles per second (cps)" became "Hz" for "Hertz."

In procedural instructions, wording of the steps begin with "On the (unit or panel)" followed by "connect (meter, oscilloscope, etc.) to (test point)," and "adjust (control)" for (results).

Alternating current: "ac."
Amperes: "A."
Assembly designators: Begin with largest assembly on left. Example: 2A4A5A6A9R22. Channel 2, 4th Assembly, 5th subassembly, 6th subsubassembly, circuit card 9, resistor R22.
Automatic frequency control: "afc."
Automatic gain control: "agc."
Bipolar video: do not hyphenate between "bi" and "polar."
Clockwise: "cw."
Continuous wave: "c-w."
Coherent oscillator: "coho."
Component reference designators: Capitalize, as "R49, C2, L12, CR4, etc."
Control designators: All capitals, as "FREQ ADJUST."
Counterclockwise: "ccw."
Cycles per second (formerly cps): "Hz," for Heinrich Rudolph Hertz. Also "kHz," for kiloHertz, and "MHz," for megaHertz.
DeciBels: Capitalize "B" for "Bell" in Alexander Graham Bell; abbreviate as "dB."
Direct current: "dc."
Doppler: Capitalize "D" for "Doppler" in "Christian Johann Doppler," as "f_D."
Indicator lamp reference designators: All capitals, as "POWER."
Intermediate Frequency: "i-f." The hyphen prevents confusion with "if."
Major equipment designators: All capitals, as "ASR-9" or "AN/FPS-67."
Moving target detector: "mtd."
Moving target indicator: "mti."
Mnemonics: All capitals, as "TWFILSEL-."
Ohms: Greek omega, "Ω," or "ohms."
Planned-position indicator: "ppi."
Pulse width: "t_p" (FAA Academy).
Radio frequency: "rf."
Receiver: "rcvr."
Regularly used abbreviations: do not capitalize (stc, ftc, i-f, rf, aka, mti, etc.).

Stabilized local oscillator: "stalo."
Switch reference designators: All capitals, as "POWER ON."
Testpoint designators: All capitals, as "TP2 Q VIDEO."
Transmitter: "xmtr."
Unipolar video: Do not hyphenate between "uni" and "polar."
Volts: Capitalize first letter.

APPENDIX H

Trigonometry

Some Trigonometry Equations And Information

Proofs not offered. Consult a textbook.

$$\sin\theta = \frac{y}{r} \qquad \sin\theta = \frac{1}{\csc\theta}$$

$$\cos\theta = \frac{x}{r} \qquad \cos\theta = \frac{1}{\sec\theta}$$

$$\tan\theta = \frac{y}{x} \qquad \tan\theta = \frac{1}{\cot\theta}$$

$$\cot\theta = \frac{x}{y} \qquad \tan\theta = \frac{\sin\theta}{\cos\theta}$$

$$\sec\theta = \frac{r}{x}$$

$$\csc\theta = \frac{r}{y} \qquad \cot\theta = \frac{\cos\theta}{\sin\theta}$$

$$\sin^2\theta + \cos^2\theta = 1$$

$$\sec^2\theta = 1 + \tan^2\theta$$

$$\csc^2\theta = 1 + \cot^2\theta$$

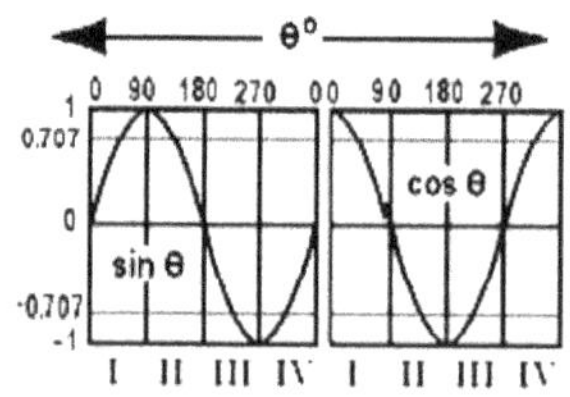

Θ	0°	30°	45°	60°	90°
sin	$\frac{1}{2}\sqrt{0}$	$\frac{1}{2}\sqrt{1}$	$\frac{1}{2}\sqrt{2}$	$\frac{1}{2}\sqrt{3}$	$\frac{1}{2}\sqrt{4}$
cos	$\frac{1}{2}\sqrt{4}$	$\frac{1}{2}\sqrt{3}$	$\frac{1}{2}\sqrt{2}$	$\frac{1}{2}\sqrt{1}$	$\frac{1}{2}\sqrt{0}$

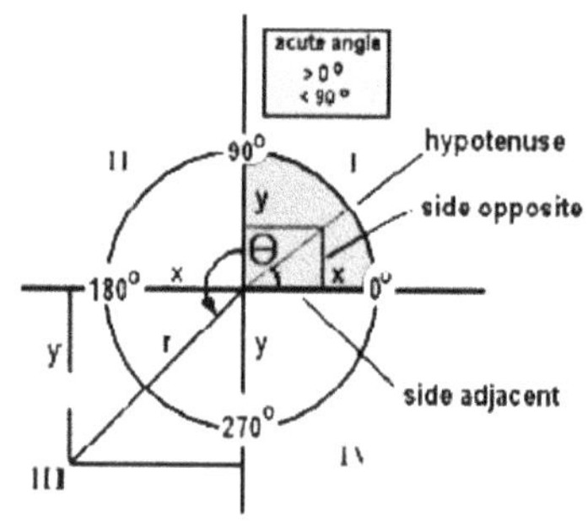

$$\sin\theta = \frac{\text{opposite}}{\text{hypotenuse}}$$

$$\cos\theta = \frac{\text{adjacent}}{\text{hypotenuse}}$$

$$\tan\theta = \frac{\text{opposite}}{\text{adjacent}}$$

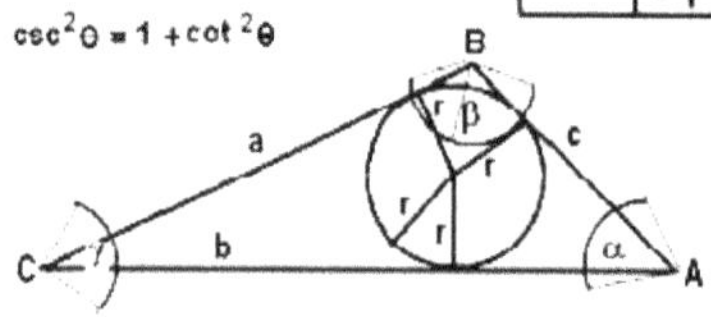

$$s = \frac{a+b+c}{2}$$

$$r = \sqrt{\frac{(s-a)(s-b)(s-c)}{s}}$$

$$\tan\frac{\alpha}{2} = \frac{r}{s-a}$$

$$\frac{\tan\frac{\alpha-\beta}{2}}{a-b} = \frac{\tan\frac{\alpha+\beta}{2}}{a+b}$$

$$rs = \sqrt{s(s-a)(s-b)(s-c)}$$

law of cosines

$$a^2 = b^2 + c^2 - 2bc\cos\alpha$$

$$b^2 = c^2 + a^2 - 2ca\cos\beta$$

$$c^2 = a^2 + b^2 - 2ab\cos\gamma$$

law of sines

$$\frac{a}{\sin\alpha} = \frac{b}{\sin\beta} = \frac{c}{\sin\gamma}$$

law of tangents

$$\frac{a-b}{a+b} = \frac{\tan\frac{1}{2}(\alpha-\beta)}{\tan\frac{1}{2}(\alpha+\beta)}$$

$$\frac{b-c}{b+c} = \frac{\tan\frac{1}{2}(\beta-\gamma)}{\tan\frac{1}{2}(\beta+\gamma)}$$

$$\frac{c-a}{c+a} = \frac{\tan\frac{1}{2}(\alpha-\gamma)}{\tan\frac{1}{2}(\alpha+\gamma)}$$

Pythagorean Theorem

$$x^2 + y^2 = r^2$$

Applied to Sine and Cosine

$$\sin^2\theta + \cos^2\theta = 1$$

$$\sin\theta = \frac{\text{opposite}}{\text{hypotenuse}} = \frac{y}{r}$$

$$\cos\theta = \frac{\text{adjacent}}{\text{hypotenuse}} = \frac{x}{r}$$

$$\tan\theta = \frac{\text{opposite}}{\text{adjacent}} = \frac{y}{x}$$

$$\sin(\alpha \pm \beta) = \sin\alpha\cos\beta \pm \cos\alpha\sin\beta$$

$$\cos(\alpha \pm \beta) = \cos\alpha\cos\beta \mp \sin\alpha\sin\beta$$

$$\tan(\alpha \pm \beta) = \frac{\tan\alpha \pm \tan\beta}{1 \mp \tan\alpha\tan\beta}$$

$$\sin 2\alpha = 2\sin\alpha\cos\alpha$$

$$\cos 2\alpha = \cos^2\alpha - \sin^2\alpha = 2\cos^2\alpha - 1 = 1 - 2\sin^2\alpha$$

$$\tan 2\alpha = \frac{2\tan\alpha}{1-\tan^2\alpha}$$

$$\sin\frac{\alpha}{2} = \pm\sqrt{\frac{1-\cos\alpha}{2}}$$

$$\cos\frac{\alpha}{2} = \pm\sqrt{\frac{1+\cos\alpha}{2}}$$

$$\tan\frac{\alpha}{2} = \pm\sqrt{\frac{1-\cos\alpha}{1+\cos\alpha}}$$

Radian Measurement

2π radians = 360°

one radian = 57.29577951°

$$\sin x + \sin y = 2\sin\left(\frac{x+y}{2}\right)\cos\left(\frac{x-y}{2}\right)$$

$$\sin x - \sin y = 2\cos\left(\frac{x+y}{2}\right)\sin\left(\frac{x-y}{2}\right)$$

$$\cos x + \cos y = 2\sin\left(\frac{x+y}{2}\right)\cos\left(\frac{x-y}{2}\right)$$

$$\cos x - \cos y = -2\sin\left(\frac{x+y}{2}\right)\sin\left(\frac{x-y}{2}\right)$$

DeMoivre's Theorem
(Polar Coordinate Multiplications)

z: single ρθ product. $z_1 = r_1(\cos\theta_1) + i(\sin\theta_1)$

$z_2 = r_2(\cos\theta_2) + i(\sin\theta_2)$

multiplication: $z_1 z_2 = (r_1 r_2)[\cos(\theta_1+\theta_2) + i\sin(\theta_1+\theta_2)]$

$$[r(\cos\theta + i\sin\theta)]^n = r^n(\cos n\theta + i\sin n\theta)$$

www.ingramcontent.com/pod-product-compliance
Lightning Source LLC
LaVergne TN
LVHW080056220926
848484LV00007BA/1773